AF335217

Physical Metallurgy
of Cast Iron V

Physical Metallurgy of Cast Iron V

Proceedings of the 5th International Symposium
on the Physical Metallurgy of Cast Iron,
Nancy, France, October 3-5, 1994

Editors:

G. Lesoult

Institut National Polytechnique de Lorraine, Nancy, France

and

J. Lacaze

Institut National Polytechnique de Toulouse, Toulouse, France

Scitec Publications

Member of the Trans Tech Group of Publishers

ISBN 3-908450-28-4

Volumes 4-5 of
Advanced Materials Research
ISSN 1022-6680

Distributed in the Americas by

Trans Tech Publications
PO Box 699, May Street
Enfield, New Hampshire 03748
USA

Phone (603) 632-7377
Fax   (603) 632-5611

and worldwide by

Scitec Publications
Brandrain 6
CH-8707 Uetikon-Zuerich
Switzerland

Fax +41 (1) 922 10 33
e-mail: ttp@ttp.ch
http://www.ttp.ch

Committees and sponsors

Scientific advisory committee :

Fredriksson H.	(Royal Institute of Technology, Sweden)
Jones H.	(The University of Sheffield, U.K.)
Kozlov L.	(Moscow Institute of Steel and Alloys, Russia)
Kurz W.	(Polytechnic Institute of Lausanne - EPFL, Switzerland)
Liu B.	(Tsinghua University, China)
Loper C.R.	(The University of Wisconsin-Madison, USA)
Lux B.	(Technische Universität, Wien, Austria)
Ogi K.	(Kyushu University, Japan)
Sahm P.	(Giesserei Institut - RWTH, Germany)
Stefanescu D.	(The University of Alabama, USA)

Local organizing committee :

Mme. S. Parent-Simonin	(French Technical Foundry Institute - CTIF)
D. François	(Ecole Centrale de Paris)
J. Lacaze	(Ecole des Mines de Nancy, Secretary)
G. Lesoult	(Ecole des Mines de Nancy, Chairman)
A. Pineau	(Ecole des Mines de Paris)
J.M. Schissler	(Université Nancy-I)

Sponsors :

Association Université Entreprise pour la Formation - AUEF (Programme COMETT)

District de l'Agglomération Nancéienne

Institut National Polytechnique de Lorraine - INPL

Laboratoire de Science et Génie des Matériaux Métalliques - LSG2M

Foreword

SCI5 was the fifth symposium of a series which began in Detroit in 1964. The aims of the pioneers who organized the first seminar entitled "Recent research on Cast iron" were to stimulate further research, to attract capable scientists to the field of cast iron, and to encourage foundrymen to "pause and sort out knowledge from practice".

Each of the four previous symposia acted as an efficient catalyser for bringing into contact different communities, both technological and scientific, to compare and up date their knowledge on various aspects of cast iron. More than half of the papers presented in Detroit were related to solidification, nucleation and growth of graphite from the melt. In fact, these questions are among those most frequently and constantly discussed in all of these symposia. In 1974, the second symposium resulted in definite refinements of previously developed ideas and in the enrichment of the experimental information on the metallurgy of cast iron. More attention was paid to engineering properties of cast iron, such as mechanical properties and the quality of as-cast materials. Among new items discussed in 1984 in Stockholm during the third symposium, the development of new alloys such as bainitic cast irons and vermicular cast iron must be mentioned. Moreover for the first time in this series of symposia, a few contributions were devoted to the numerical simulation of the formation of the microstructure during the solidification process. The fourth symposium which was held in 1989 in Tokyo revealed the rapidly growing importance of the activity of modelling based upon basic knowledge of the physical metallurgy of cast iron. Rapid solidification and detection of casting defects by using new tools were two other new items developed at the last symposium.

To keep to the spirit of this series of symposia, SCI5 was aimed at reflecting and stimulating the development of knowledge and changes of scientific interest in the field of cast iron.

This volume contains the complete proceedings of the oral contributions and posters which have been presented during the symposium.

A key factor for the success of the symposium has been the quality of the invited lectures the texts of which are collected in the first part of the volume. Some invited speakers offered synthetic surveys of the state of the knowledge on important properties or processes which are specific to cast irons :
  • Current status of austempered cast irons,
  • Fracture characteristics and fracture toughness of spheroidal graphite cast iron,

• Behavior of oxygen and nucleation of graphite in production of spheroidal graphite cast iron,

      • Methodologies for and performance of macro transport-transformation kinetics modelling of cast iron.

Other ones discussed very basic concepts which could be used more systematically than done till now in the world of cast irons :

• The mechanisms of spherulitic growth in polymer and iron melts,

• The cast iron as a composite.

Finally, a few speakers from industry gave indications of what engineers are expecting from cast iron in the future :

• The status in cast iron microstructure engineering,

• The trends and metallurgical factors involved in automative cast iron parts.

The four other parts of these proceedings present the more focused scientific contributions. The choice of the subdivision corresponds to a reasonable balance between the items in terms of number of contributions :

• mechanical behaviour of cast iron,
• melt, solidification and product processing,
• structural analysis,
• structure and process modelling.

It is worth to notice that the first concern in this symposium, about one third of the contributions, was related to the mechanical behaviour of the cast iron. This corresponds to a steady growth of interest since 1964 along the series of symposia : spheroidal graphite cast iron is now recognized as a potentially good structural material.

One fourth of the papers was devoted to the modelling of microstructures or of processes. The rapidly growing activity in the numerical modelling of the formation of microstructures from the liquid or in the solid state is the sign that the metallurgy of as-cast alloys, including cast irons, is no more just a matter of experience, empirism, and hypothetical thoughts but becomes a computer-assisted scientific domain. The obvious interest of several foundries for softwares dedicated to the simulation of filling, solidification, and cooling of cast alloys proves that foundry is no more pure art but can and must be a scientifically quantified processing technology.

The attraction which is exerted at present on searchers by the activities of numerically modelling the microstructures may cause temporarily a slight decrease in the related experimental studies although they should be even more necessary. This trend results in a constant decrease since 1974 of the number of papers devoted to experimental studies upon the development of microstructures in cast iron.

The few previous comments suggest probable directions for the future, either simple extrapolations or possible bifurcations. At first, since the mechanical properties of SG cast iron can be as good and as various as those of steels, more fundamental work on the control of the microstructure of the iron-rich matrix would be useful. In fact, the good mechanical properties of cast irons are mainly due to the properties of the matrix the microstructure of which is obtained by controlling the austenite decomposition after solidification. The optimization of the final microstructure involves optimization of the austenite structure which involves the microsegregation originated from the solidification and its local consequences on the solid state transformations. In the second place, confidence in complex models predicting microstructural development requires a better understanding of graphite growth from the liquid phase, with and without austenite shell, and in the solid state, in contact with either austenite or ferrite.

The growing use of predictive numerical models implies the rapid development of basic data in thermochemistry of multicomponent systems, including interfacial tensioactive properties, and in kinetic and mechanical properties of transforming solid phases.

These expected progresses in the fundamental understanding of the structure and the properties of cast irons should help to create new materials based on the concept of cast iron and to improve processes to produce such materials.

Thanks are due to the members of the scientific advisory committee, and of the organizing committee. Special thanks are given to the members of these committees who accepted the difficult task of reviewing the manuscripts.

      G.Lesoult   J.Lacaze

Table of Contents

Impact Properties

Fatigue and Fracture Behaviour

Wear Behaviour

MELT, SOLIDIFICATION AND PRODUCT PROCESSING

Melt Processing and Related Properties of Cast Iron

Solidification and Product Processing

STRUCTURAL ANALYSIS

Solidification Microstructures

Solid State Transformations

Quantitative Analysis

STRUCTURE AND PROCESS MODELLING

Microstructure Formation

Modelling Related to Processes

Advanced Materials Research Vols. 4-5 (1997) pp. 1-16
© *1997 Scitec Publications, Switzerland*

Current Status of Austempered Cast Irons

R. Elliott

Manchester Materials Science Centre, University of Manchester,
Grosvenor St., Manchester M1 7HS, UK

Keywords: Austempered Cast Irons, Properties, Uses, Austempering Process, Heat Treatment Processing Window, Standards, Structure-Property Relationships, Unalloyed Austempered Cast Iron, Alloyed Austempered Cast Iron

ABSTRACT

Mechanical and physical properties obtained by austempering different types of cast iron are reviewed and present and potential uses indicated. Reasons for the slow development of ADI are identified and ways forward suggested. Our current understanding of the austempering reaction in unalloyed irons is described and the heat treatment processing window for optimising mechanical properties is defined. The influence of casting and heat treatment parameters on austempered properties and the ability to satisfy ADI standards is discussed. Finally, the difficulties encountered in austempering alloyed cast irons are examined.

INTRODUCTION

Three significant developments in cast iron technology are the practice of inoculation to ensure grey iron solidification, the use of Mg and Ce to control graphite morphology and the heat treatment of the matrix to give a wide range of mechanical properties. Each of these developments has taken time to come to fruition. The latest heat treatment process, austempering is no exception. This paper discusses the current position and considers the way forward.

TYPES OF AUSTEMPERED CAST IRON

Refer to austempering and the subsequent discussion will probably be centred on austempered ductile iron, ADI. However, austempering is a matrix heat treatment and may be applied to grey flake irons, AGI, compacted graphite flake irons, ACI or malleable irons, AMI with some interesting results.

The austempering cycle in a grey iron is likely to differ from a ductile iron because of the lower Si content, a higher S content and a greater graphite -matrix interface area. Si has an important influence on austempering because it decreases C solubility in austenite, decreases the rate of bainitic ferrite formation and suppresses and prolongs carbide formation. The lower Si content of AGI reduces austempering time and narrows the processing window. S segregates to the graphite-matrix interface where it creates a strong C diffusion barrier. As a result the as-cast matrix is usually pearlitic and it takes a longer time to saturate the matrix with C during austenitisation. Some recent exploratory measurements of the properties of AGI [1] are shown in Fig. 1. The mechanical properties show behaviour typical of an austempered matrix but clearly graphite shape is the prime factor determining and limiting elongation. A unique relationship of increased damping with increased strength was observed. An inverse, linear relationship was observed between resonant frequency and tensile strength such that resonant frequency could possibly be used for quality and process control. AGI has been used in heavy duty cylinder liners, gears and

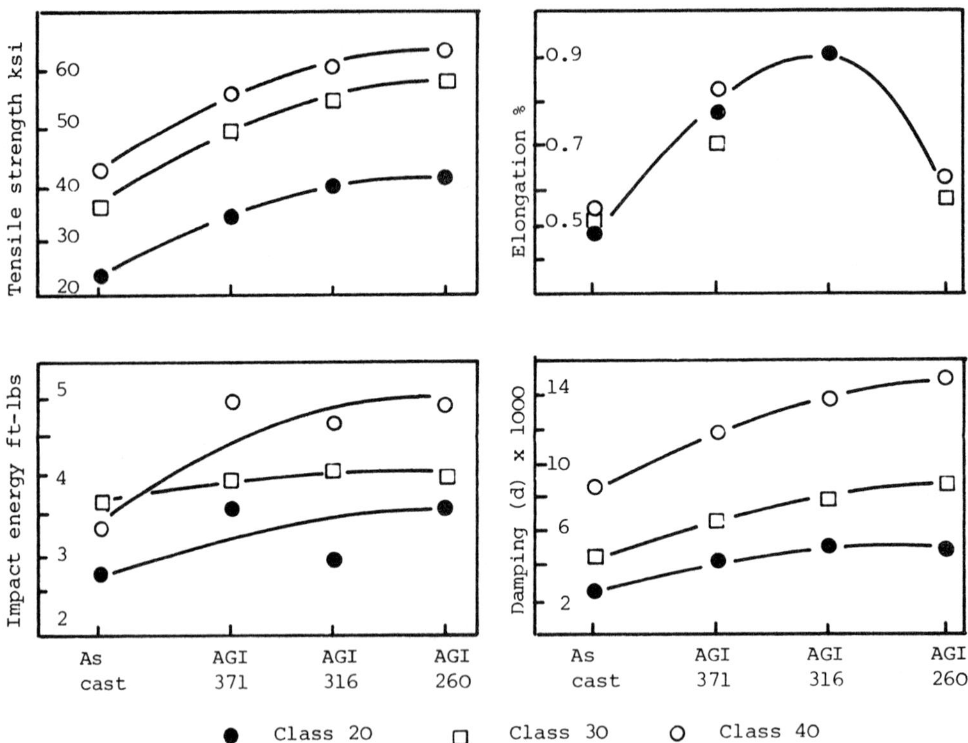

Figure 1.   Properties of grey flake irons austempered at 371, 316 and 260°C [1]

high load bearing mounts. Austempering offers an alternative method to
quenching and tempering for producing high strength grey irons. The better
wear resistance of AGI compared with grey iron together with increased strength
and good thermal conductivity suggests potential applications in brake
components. Coupled with good damping it may allow designers to use more
aggressive brake pad material without sacrificing drum life at a lower
vibration and noise level. An important consideration in automobile
applications is weight. Recently, it has been argued that the value of a
pound of weight saved in 1995 would be worth two dollars and in 2000, three
dollars. To retain the excellent casting properties of grey irons perhaps
design changes based on AGI properties should be considered. Higher strength
AGI could mean thinner sections in applications such as turbine casings,
engine blocks, intake manifolds and brake drums.
      Compacted graphite irons already offer a challenge to grey irons in
several of these applications because of their increased strength due to their
modified graphite morphology. This challenge has increased [2] now that
reliable methods such as the Sintercast Process can guarantee a fully compacted
graphite structure. Figure 2 shows some recent measurements [3] of the
processing window for a compacted iron. The kinetics show the dependence on
heat treatment parameters shown by ductile irons with times modified compared
to ductile irons by the lower Si content and compared to grey flake irons by
the much lower S content. Mechanical properties of the ACI are shown in
Fig. 3. Tensile strength, proof strength, elongation and impact energy are all

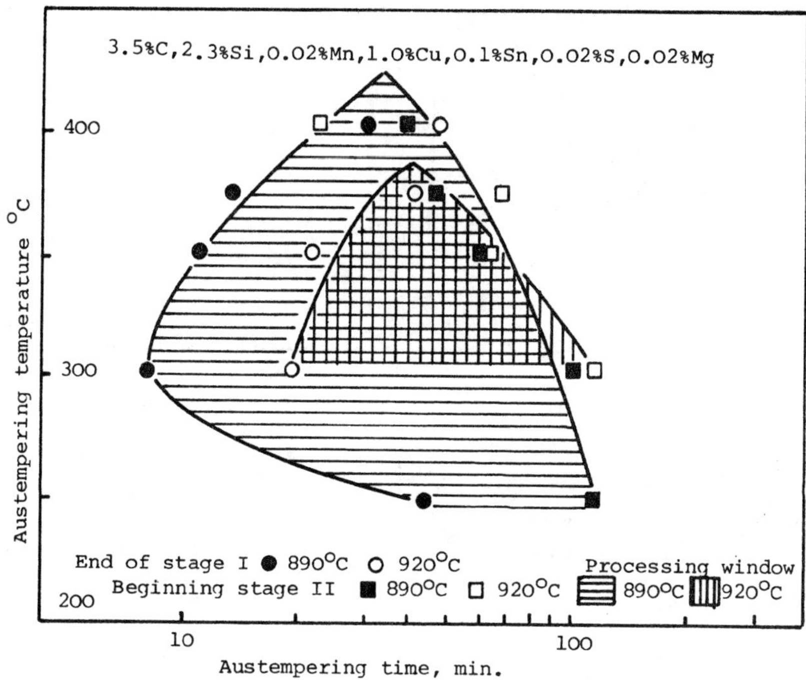

Figure 2.   The variation of time of the end of the stage I reaction and
the beginning of the stage II reaction with austempering
temperature for austenitising temperatures of 890 and 920°C.
The shaded areas show the corresponding processing windows.

increased compared to the as-cast condition and compared to the optimum
properties displayed in Fig. 1 for the austempered grey iron. For example,
the tensile strength of the ACI is more than doubled without loss in elongation
compared to the AGI austempered at 260°C and compared to the as-cast state.
It is envisaged that ACI could challenge AGI for the applications suggested
above with extra weight saving.

Malleable iron has a much lower C and Si content than ductile iron. It
offers some advantages compared to ductile iron due to its inherent properties
of higher modulus, high toughness at low temperatures, more matrix per unit
volume and cost. In addition, the nodule distribution across various sections
of the casting and the matrix structure are more uniform. Preliminary
studies [5] of the use of AMI for gear applications have shown encouraging
results as indicated by the comparison of contact fatigue measurements of AMI
with ADI and carburised steel shown in Fig. 4.

Undoubtedly the most well known of the austempered products is ADI.
Figure 5 shows the minimum tensile strength as a function of minimum elongation
for forged C, C-Mn and low alloy steels specified to BS 970:1983. Also shown
are the minimum properties of ductile iron according to BS 2789:1985 and the
most stringent of the national and proprietary ADI standards ASTM A897M:1990,
the grades of which are shown in Table 1. Figure 5 shows that the properties of
ferritic ductile iron are comparable to those of low strength forged steels
and the properties of the more ductile grades of ADI approach those of the
medium strength low alloy steels. However, the high strength grades of ADI
have a much lower ductility than wrought steels of similar strength and are not

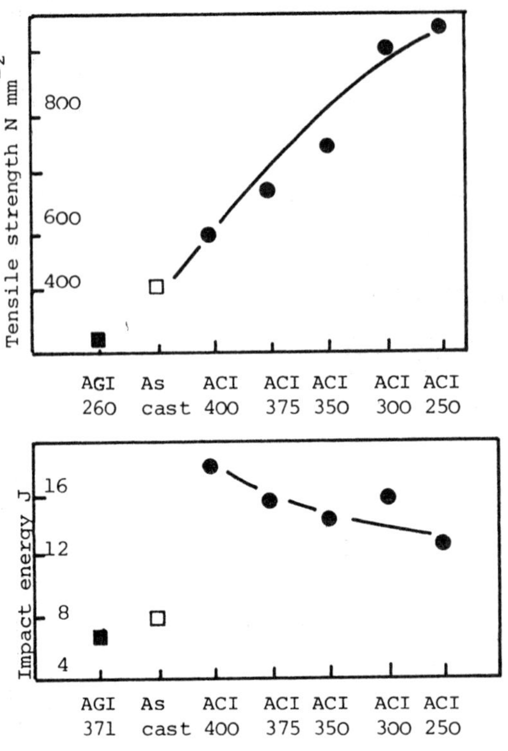

Figure. 3. The variation of the mechanical properties of ACI with austempering temperature after austenitising at 890°C. (ref [4]).

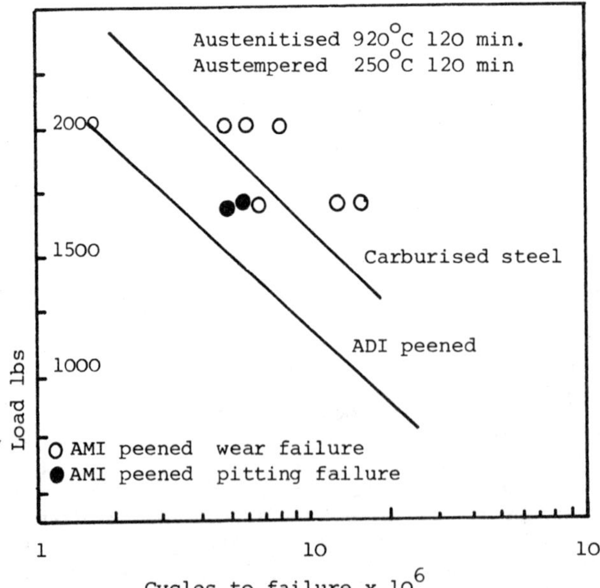

Figure 4.

Contact fatigue data for an AMI compared to data for an ADI and a carburised steel. [5]

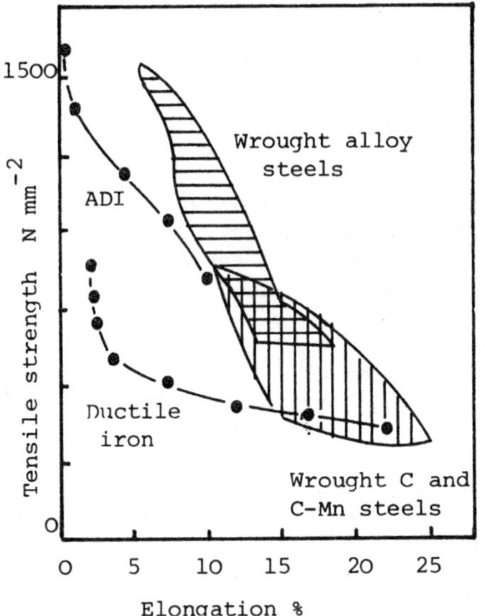

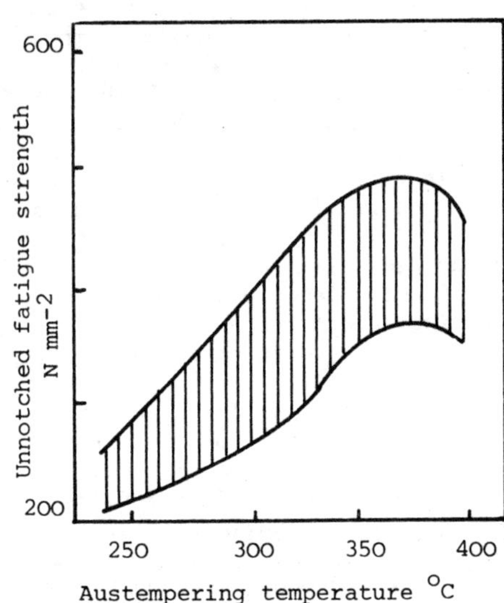

Figure 5. Comparison of the minimum tensile properties of wrought steel (BS970:1983), ductile iron (BS2789:1985) and ADI (ASTM A897M:1990).

Figure 6. The variation of unnotched rotating bending fatigue strength with austempering temperature. (ref [6]).

Grade	Minimum yield strength N mm^{-2}	Minimum tensile strength N mm^{-2}	Minimum elongation %	Minimum unnotched Charpy value J	Typical Brinell hardness
1.	550	850	10	100	269-321
2.	700	1050	7	80	302-363
3.	850	1200	4	60	341-444
4.	1100	1400	1	35	388-477
5.	1300	1600	–	--	444-555

Table 1.   ADI grades to ASTM Standard A897M:1990 (metric).

likely to compete with forged steels for components requiring high levels of tensile strength and ductility. Many applications in which ADI is used to replace high strength cast and forged steels are likely to be subjected to fatigue loading. Figure 6 shows rotating bending fatigue data compiled by BCIRA [6]. The scatter band reflects differences in alloy content, test bar diameter, austempering treatment and definition of fatigue limit. However, the results show that austempering increases fatigue strength and that it decreases as the tensile strength increases. The endurance ratio of ADI is in the range 0.3-0.4 comparable to alloyed steel castings of comparable tensile strength and

overlaps the bottom of the range for wrought steels. Figure 6 also suggests
that a maximum fatigue strength occurs in the range 350-400°C. The high
strength grades of ADI may be competitive when wear resistance is a
consideration. The wear resistance of high strength ADI is superior to that of
many steels and good wear properties can be achieved with lower strength ADI if
the austenite undergoes strain hardening or transforms to martensite when cold
worked locally.

Since the late 70's the properties outlined above have led to ADI being
used successfully in several hundred different applications world-wide.
Details of many of these applications have not been disclosed because foundries
and end-users are reluctant to divulge development details. However, a
comprehensive description of these uses which include gears, crankshafts,
transmission and suspension components, earthmoving equipment and railway
engineering has been given recently[7]. One of the most reported applications
is the use of ADI to replace carburised steel forgings for hypoid ring and
pinion gears on rear drive GM cars. Castings were annealed, machined,
austempered at 235°C to about grade 5, shot peened and assembled onto the
axles. This conversion provided a total energy saving of about 50%, allowed
greater machining speeds and feeds, reduced gear weight by 10% and operating
noise by 14%. Crankshafts, tripot housing (coupling devices in the steering
system of four wheel drive light trucks) and several parts of pneumatic tools
are examples of components currently made to grade 1 or 2. The completely new
designed Ingersoll-Rand paving breaker has a durable one piece ADI housing which
eliminates the need for side rods, head bolts, anvil block and other high
maintenance parts giving greatly improved reliability as well as lower
production costs.

CURRENT STATUS OF ADI

When the outstanding mechanical and wear properties of ADI were first
revealed, confident predictions were made for its rapid growth. Although there
is always a lag time between research findings and field applications, in the
case of ADI it has been longer than usual, particularly in the U.K. Reasons
have been given for the overdue growth in ADI applications at several recent
conferences and may be summarised as follows
1. Many lower strength steel castings and forgings have been converted to
ductile iron leaving ADI to compete against heat treated alloy steels for high
duty applications where there is an uncompromising requirement for product
reliability.
2. A lack of national standards which has been rectified recently.
3. The mechanical property data bank is not as extensive for ADI as for steel.
Further data are required for impact, fracture toughness, rotating-bending and
axial fatigue, thermal conductivity and wear properties. There is a need to
convey this data to designers in a coherent form.
4. Many foundry sale teams have an inadequate understanding of the material,
austempering process and the requirements of the end-user. In addition, ADI
is often competing in markets which are not traditional ones for ductile iron.
Foundry salesmen are often inadequately trained or given an unbalanced view that
can lead to the promotion of ADI in areas where its use is not justified.
5. Laboratory trials only give an indication of service performance and field
trials are expensive and time consuming.
6. ADI is sold on properties obtained from test bars. Actual components may
display inferior properties due to section size and surface effects. Further
information on this aspect is required.
7. Many contract heat treaters are reluctant to process ADI because it will
undermine profitable steel heat treatment business. Foundries are reluctant to
invest in their own facilities until they are certain that a demand exists for
their products.

8. There is a need for foundrymen and heat treaters to be more knowledgeable
of each other's role in producing a sound, correctly heat treated ADI component.
9. It is not clear whether machining should be performed before and/or after
austempering.
10. There is a lack of convenient inspection techniques. Hardness does not
indicate a correctly austempered iron. Usually test bars are austempered and
tested with each batch of castings.
11. Financial pressures on engineers are resulting in their retaining traditional
materials.
12. Few foundries are equipped to machine and austemper their own castings and
many are reluctant to co-ordinate these activities on a subcontract basis. At
the same time there is an increasing resistance from customers to do so as they
increasingly are looking to be supplied with finished, ready to use components.

SOME SUGGESTED SOLUTIONS.

    Fulfilment of the early predictions for ADI will be achieved only when the
barriers described above are overcome. This will involve continued
1. Research and development
      to create a greater understanding of the austempering process, the influence
      of casting size and the effect of alloying additions
      to establish a more extensive data base
      to co-ordinate component and field trials
      to develop suitable non destructive testing techniques.
2. Training
      with the purpose of transfering ADI technology to production and sales teams.
3. Marketing
      focussed particularly on design engineers and material specifiers.
4. Heat treatment
      installation of high volume heat treatment plant to overcome the current
      vicious circle of low production volumes and high heat treatment costs.

SOME RECENT DEVELOPMENTS.

<u>Unalloyed ADI</u>: role of the heat treater.

    Our understanding of the heat treatment of unalloyed ADI to produce the
different grades of the ASTM standard has been well established over the past
few years. The iron casting is austenitised to convert the as-cast, ferrite-
pearlite matrix structure into austenite. It is then cooled sufficiently
quickly to the chosen austempering temperature (higher austempering temperatures
to promote the lower grades of the ASTM standard and lower austempering
temperatures for the higher grades of the standard) to avoid the formation of
probainite products such as ferrite and pearlite which reduce mechanical
properties [8]. The iron is then held at the austempering temperature for
a specified time followed by air cooling to room temperature.
    The microstructural changes during austempering occur in two stages.
In the stage I reaction low C austenite transforms into ferrite and high C
austenite or ferrite/carbide and high C austenite at lower austempering
temperatures. The stage I product has been termed ausferrite. In the early
part of the stage I reaction ferrite platelets form close to the graphite
nodules or on austenite grain boundaries and C diffusion enriches the
surrounding austenite. In unalloyed irons the stage I reaction occurs
continuously through the eutectic cell and into the intercellular regions with
a continuously decreasing driving force. The stage I reaction is considered
completed at time, t_1, when the austenite in the ausferrite structure has been
sufficiently enriched in C for it to be retained on cooling to room temperature.
The microstructural changes can be observed with the optical microscope and

quantitative metallography can be used to determine the time, t_1, defined as the austempering time when the unreacted austenite volume fraction, UAV, has fallen to a level of 1%. This is considered to be the level at which the martensite (formed during cooling from any unreacted, low C austenite) is not continuous in the structure and harmful to mechanical properties. The metastable, ausferrite structure, which is that required in a correctly austempered iron, remains the austempered structure until an austempering time, t_2, when the stage II reaction commences.

The stage II reaction, delayed by the high Si content of the iron, is the more familiar bainite reaction in which the high C austenite breaks down into ferrite and carbide. The carbide product embrittles the structure and reduces mechanical properties particularly ductility. Unfortunately, the onset of carbide formation cannot be observed with the optical microscope; only the later stages can be identified when the carbide precipitation destroys the clarity of the ausferrite structure. Only a few TEM studies [9-11] of carbide formation have been reported. Further studies are required to identify the carbides formed, define the stage II reaction and the role of carbides in embrittling the iron. In the absence of detailed kinetic measurements the onset of the stage II reaction has been related to the fall in the volume fraction of high C austenite as measured by x-ray diffraction or from suitably etched micrographs. A time, t_2, corresponding to the austempering time beyond which the mechanical properties of the austempered iron are impaired can be defined. The time, t_2-t_1, is the processing window and austempering times within the window must be used by the heat treater to ensure optimum mechanical properties.

Figure 7 shows recent measurements of the variation of times, t_1 and t_2, with austempering temperature for an iron containing 0.9%Cu and 1%Ni which may be considered as "unalloyed" for the present discussion.

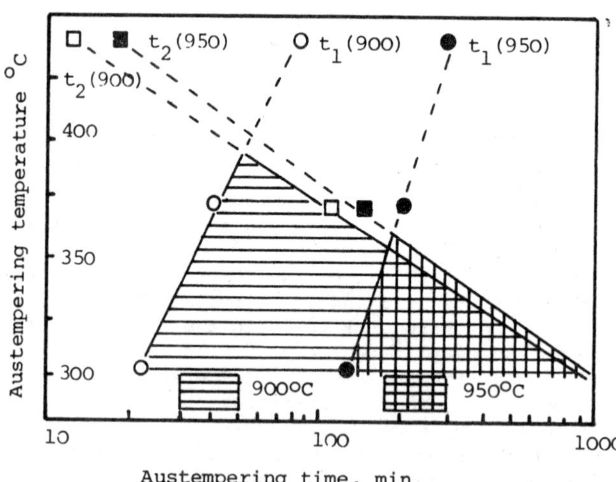

Figure 7.

The processing windows as a function of austempering temperature for austenitising temperatures of 900 and 950°C. ref [12-14].

The shaded areas show the processing windows as a function of austempering temperature for austenitising temperatures of 900 and 950°C. This figure shows clearly that the width of the processing window decreases as the austempering temperature increases and that the window closes at a certain temperature making it impossible to obtain optimum mechanical properties above this temperature. It also shows that the choice of austempering time is more critical for the more ductile grades of the standard (higher austempering temperatures). Austenitising temperature is another parameter at the disposal of the heat treater. Decreasing the austenitising temperature reduces the parent austenite C content and increases the driving force for the stage I reaction. This is reflected in the displacement of the t_1 line in Fig. 7 to the left as the austenitising temperature decreases. Decreasing the austenitising temperature also raises the window closure temperature and may, therefore, be of benefit in obtaining better mechanical properties if the appropriate adjustment in austempering time is made. However, it must be remembered that decreasing the austenitising temperature reduces the hardenability (austemperability). Figure 8 shows the results of mechanical property measurements made in the study.

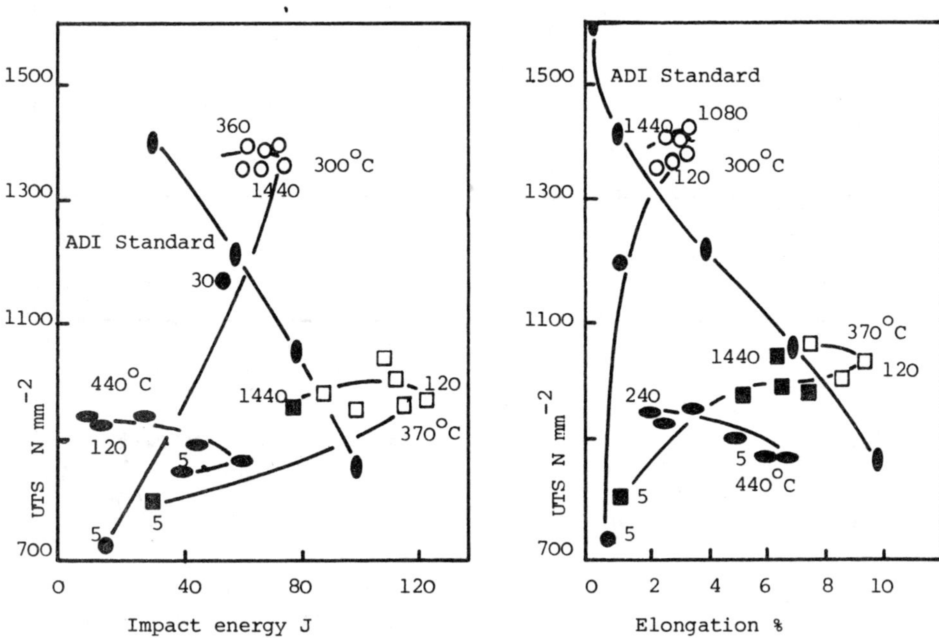

Figure 8.   Comparison of mechanical properties with ASTM Standard. Austempering temperatures of 300, 370 and 440°C after austenitising at 900°C for 120 minutes. Open symbols are for austempering times within the processing window. Numbers next to symbols indicate austempering times.

It can be seen that the high strength grade of the standard can be satisfied at 300°C with a wide processing window. The low strength grade can be satisfied by austempering at 370°C but the processing window is narrower. If an attempt is made to achieve the lowest strength grade by austempering at 440°C, the mechanical properties fail to satisfy the standard at all austempering times. This is because the processing window is closed. A heat treatment procedure can be defined using the information given above. However, can the laboratory results be used in the field?. Will all areas of a casting with an average section thickness of 2.5 cm respond to heat treatment as described by the results in Figs 7 and 8 which were obtained on specimens taken from the bottom section of a 2.5 cm Y block?.

<u>Unalloyed ADI: role of the foundryman</u>

Rundman and co-workers have addressed this question [15]. Mechanical

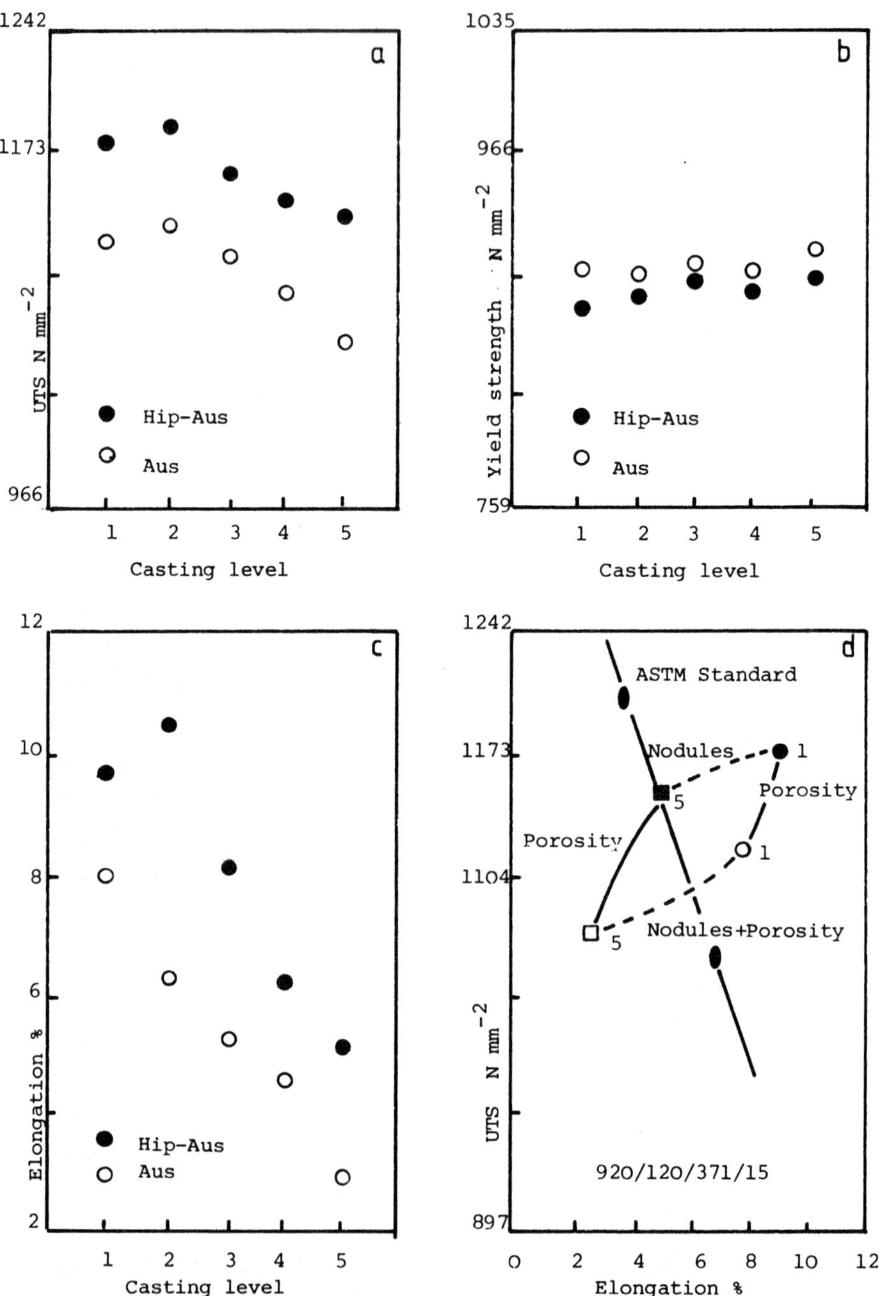

Figure 9. The dependence of (a) UTS, (b) yield strength and (c)
elongation on position in the Y block of austempered and
hipped and austempered unalloyed ductile iron.
(d) The relation of tensile properties at levels 1 and 5
to the ASTM Standard. (ref[15]).

properties of an unalloyed, austempered iron are shown in Fig. 9.

The iron composition was 3.46%C,2.37%Si,0.045%Mn,0.004%Mo,0.033%Cu,0.041%Ni.
Specimens were taken from 5 different levels in a 2.5 cm Y block. Level 1 was
close to the bottom of the block and level 5 close to the riser. Half the
specimens were hipped before austempering at 371°C for 15 minutes. Figure 9
clearly shows that the tensile strength and elongation but not yield strength
reduce significantly both with and without hipping as the specimen location
moves closer to the riser. As the iron was unalloyed and all specimens
received identical austempering treatment the difference in mechanical
properties with position can be attributed to differences in the solidified
structure. The solidification time increases with change in position from
level 1 to level 5 and it is well documented that properties decrease with
increasing solidification time. The decrease is attributed to
1. low nodule count [16-18]
2. low nodularity [19]
3. porosity.
The importance of Fig. 9 is that it quantifies these effects and allows a
measure of their significance. The differences in properties between levels 1
and 5 in austempered irons is shown by the dashed line joining the unfilled
symbols in Fig. 9(d). The fall in properties is due to contributions from the
3 effects listed above and shows that the properties in areas close to risers
fail to satisfy the standard. The effect of porosity can be estimated at each
level by comparing measurements on hipped and austempered and austempered
irons and is shown by the full lines joining circles at level 1 and squares at
level 5. This shows, not unexpectedly, that the decrease in properties due to
porosity at level 5 is greater than that at level 1. More importantly, it
shows that these effects are significant, emphasises the important role of the
foundryman in producing a desired structure for austempering and reinforces
the statement that heat treatment cannot be used to rectify casting defects.
The foundryman must exercise control over the solidification process to ensure
high nodule count, high nodularity and low porosity. He must be careful with
riser placement to ensure that inferior properties do not occur in any
critically stressed areas. Selective use of chills can be helpful.

Alloyed ADI: role of the foundryman
==

If larger section castings are to be austempered successfully alloying
additions must be made to maintain austemperability. Mo, Mn, Ni and Cu are
elements commonly used in decreasing order of potency. As casting size and
solidification time increase so do the effects described in the previous
section. In addition, Mo and Mn segregate strongly during solidification to
intercellular areas where carbide formation with associated porosity is often
seen. Similar measurements to those shown in Fig. 9 have been reported for
large Y blocks [20]. Some of these results are shown in Fig. 10. The filled
circles show the properties at different levels in a 10 cm Y block for an iron
of composition 3.61%C,1.94%Si,0.19%Mn,1.41%Ni,0.17%Mo,0.95%Cu. As for the
unalloyed iron the properties decrease as the riser is approached in level 4.
It is interesting to note that only areas close to the casting surface with
the shortest solidification times satisfy the standard after austempering at
352°C. The data points represented by the filled squares show the maximum
properties obtained at level 1 with irons containing different Mn contents.
The numbers close to the points 1,2,3,4 correspond to Mn contents of 0.19%,
0.29%,0.38% and 0.54%, respectively. The open squares show corresponding
property measurements at level 4 approaching the riser. These two sets of
results illustrate the decrease in properties with position and also the
decrease in properties with increasing Mn content.

It is important that this problem is treated at source because although
austenitising may reduce the carbide content it does not eliminate the
carbide from the intercellular areas. Degree of segregation (influences the
amount of carbide) depends on solidification time and nodule spacing

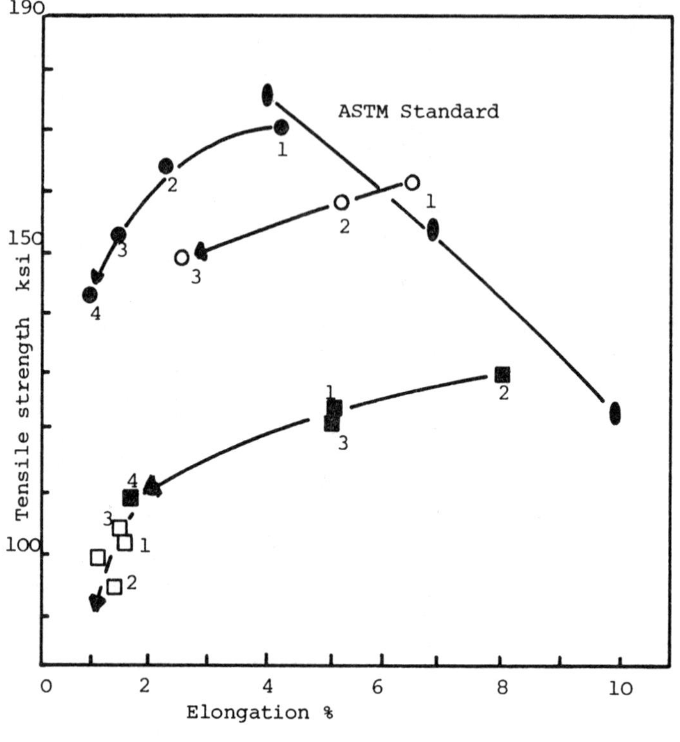

Figure. 10.

A summary of
mechanical property
data obtained from
10 cm Y blocks of
austempered,alloyed
ductile irons. [20]

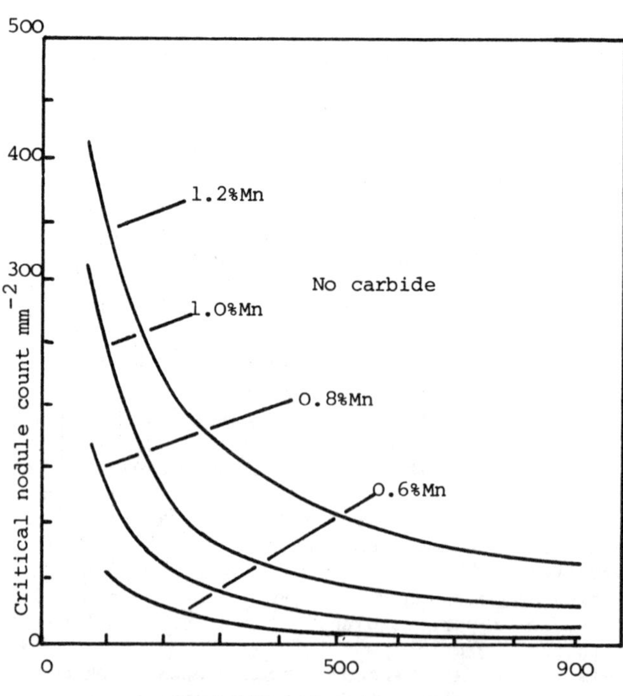

Figure. 11.

Critical nodule count
as a function of the
solidification time
for avoiding carbide
forming in the as-cast
structure .[21]

(controlled by inoculation). Recent increased interest in solidification and microsegregation modelling is allowing quantitative evaluation of these dependencies. For example, it has been shown[21] that the amount of carbide formation is a function of nodule count and solidification time in a 1%Mn iron. Modelling allows the definition of a critical nodule count to avoid carbide formation in irons of different Mn contents as a function of solidification time as shown in Fig. 11.

Thus we see in Fig. 12 that whilst the heat treater can select austenitising and austempering temperature and austempering time to give a range of properties generated by the varying proportion of and scale of the ausferrite structure, the quality of the casting moves the property point in a direction at right angles. This emphasises that the production of good quality ADI components is achieved by the combined efforts of the foundryman and the heat treater.

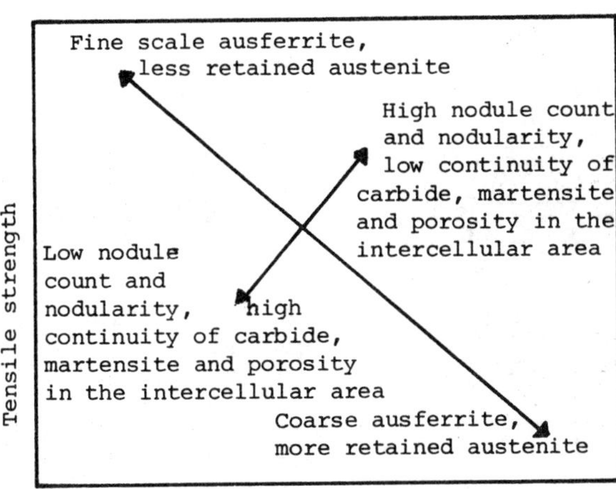

Figure. 12.

Structural features influencing the mechanical properties of an austempered alloyed iron.

Alloyed ADI: role of the heat treater

Alloying elements have a further effect on the austempering process. They, in particular Mn, slow the transformation process. Mn affects the stage I reaction significantly at higher austempering temperatures. Aided by solidification segregation, the delay of the stage I reaction in the intercellular areas can be considerable. This is evident in Fig. 13 which shows t_1 and t_2 times measured for an alloyed iron containing 0.67%Mn, 0.25%Mo and 0.25%Cu[22]. This figure shows that the processing window is closed above 370°C even for carefully selected austempering conditions. At temperatures just below 370°C the processing window is narrow requiring careful control over the austempering treatment. The mechanical properties obtained in specimens machined from close to the bottom of a 2.5 cm Y block are shown in Fig. 14. It can be seen that the properties measured on specimens taken from areas of a thin Y block where casting defects discussed above would be expected to be absent fails to meet the standard after austempering at 400°C with an austenitising temperature of 920°C and satisfies it over a narrow range of austempering times at an austempering temperature of 375°C after austenitising at 870°C.

Thus obtaining the higher ductile grades of the standard can be a serious problem in thick section castings. The normal solution is to limit the amount of Mo and Mn added with the possibility of compromising austemperability.

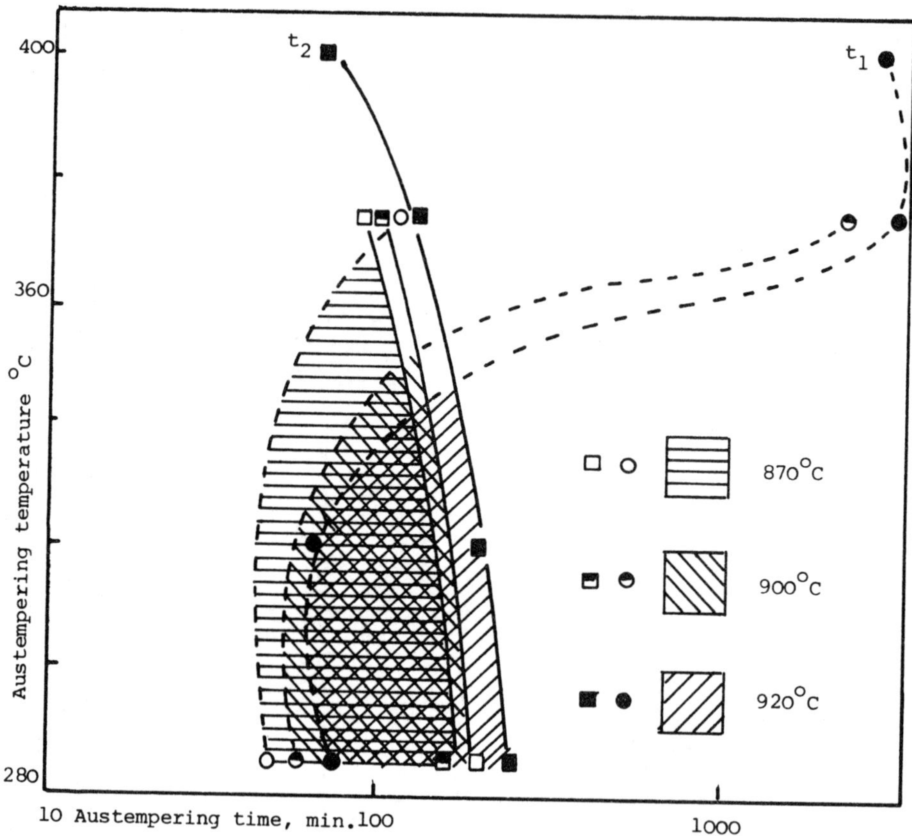

Figure 13. The variation of times, t_1 and t_2 and the width of the processing window with austempering temperature.

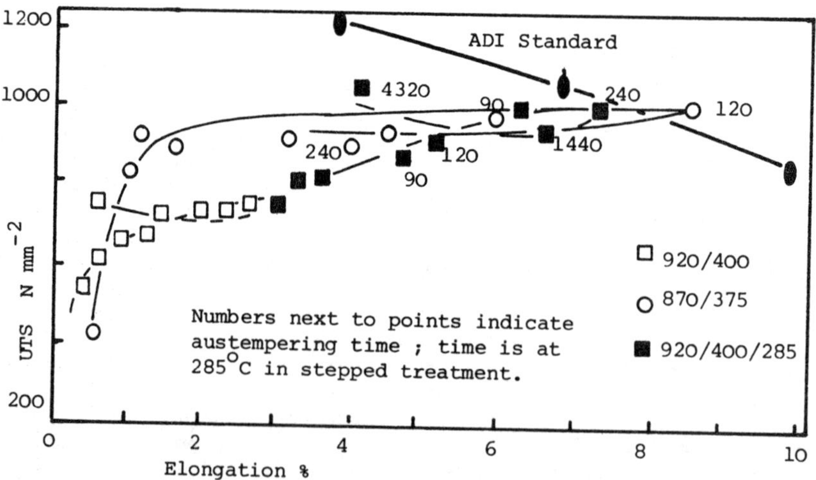

Figure 14. Mechanical properties for single and stepped heat treatment.

For example, Kovacs [23] has suggested that Mn should be limited to 0.5% in castings with a section size less than 5 cm and to 0.35% in larger castings; Mo should be limited to 0.3%. The difficulty in austempering alloyed irons arises because of the delay in the stage I reaction in the intercellular area. In many cases the stage II reaction is well under way in the eutectic cell before the stage I reaction is completed in the intercellular region. A modified, stepped heat treatment has been suggested [24] in which austempering is continued in the 1st step until the stage I reaction is completed in the eutectic cell. The treatment temperature is then dropped in the 2nd step into the lower bainite range. This increases the driving force for the stage I reaction in the intercellular region and delays the stage II reaction in the eutectic cell. It is possible to complete the stage I reaction before the stage II reaction occurs in the eutectic cell. The results of applying this stepped heat treatment with a 1st step of 400°C and a 2nd step of 285°C are shown in Fig. 14. A considerable improvement in mechanical properties is obtained compared to the single treatment at 400°C. Further studies to define the effect of 1st and 2nd step austempering temperatures and times are in progress. These studies suggest that the heat treater may have to adopt a different approach to the austempering of thick section castings.

CONCLUSIONS.

Examples are given which show that interesting and potentially useful properties can be generated in different types of cast iron by austempering. Reasons are given for the overdue, predicted rapid growth of ADI applications. These range from an incomplete mechanical property data base and poor communication of measure properties to designers and material specifiers to a complex heat treatment process that is not attractive to heat treaters unless there is a guaranteed market for products. Some recent research and development studies are described which increase our understanding of the austempering process in unalloyed and alloyed ductile irons and define the respective roles of the foundryman and heat treater in the production of an austempered component.

REFERENCES.

1. B.V. Kovacs and J.R. Keough, AFS Trans., 101, 283 (1993)
2. M. Lampic, Giesserei, 79, 871 (1992)
3. Hamid Bayati, A.L. Rimmer and R. Elliott, Cast Metals, 7, 11 (1994)
4. A.L. Rimmer and R. Elliott, Cast Metals, in press
5. P.H. Mani, AFS Trans, 101, 377 (1993)
6. Y.J. Park et al, AFS Trans., 92, 395 (1984)
7. R.A. Harding, BCIRA report, (1991)
8. A.S. Hamid Ali, K.I. Uzlov, N. Darwish and R. Elliott, Mater. Sci. Technol., 10, 35 (1994)
9. Z.K. Fan and R.E. Smallman, private communication.
10. Z.K. Fan and R.E. Smallman, Scripta Met, 31, 137 (1994)
11. K.B. Rundman et al., Proc. 2nd International ADI Conference, Michigan, USA, p. 157, (1986)
12. N. Darwish and R. Elliott, Mater. Sci. Technol., 9, 572 (1993)
13. N. Darwish and R. Elliott, Mater. Sci. Technol., 9, 586 (1993)
14. N. Darwish and R. Elliott, Mater. Sci. Technol., 9, 882 (1993)
15. K.L. Hayrynen, D.J. Moore and K.B. Rundman, AFS Trans., 101, 119 (1993)
16. T. Ohide and S. Izui, Trans JFS, 10, 55 (1991)
17. R.W. Reesman and C.R. Loper, AFS Trans., 75, 109 (1968)
18. Y. Tanaka and H. Kaze, Materials Trans. JIM, 33, 543 (1992)
19. H.W. Hoover, Jr., AFS Trans., 94, 601 (1986)
20. G.P. Faubert, D. Moore and K.B. Rundman, AFS Trans., 99, 551 (1991)

21. M. Nili Ahmadabadi, E. Niyana and T. Ohide, Cast Metals, 6, 182 (1994)
22. Hamid Bayati and R. Elliott, Mater. Sci. Technol., in press
23. B. Kovacs, AFS Trans., 99, 281 (1991)
24. Hamid Bayati and R. Elliott, Mater. Sci. Technol., in press.

Advanced Materials Research Vols. 4-5 (1997) pp. 17-30
© *1997 Scitec Publications, Switzerland*

On the Mechanisms of Spherulitic Growth in Polymer and Iron Melts

G. Faivre*

Groupe de Physique des Solides, CNRS URA 17, Universités Paris 6 et 7,
Tour 23, 2 Place Jussieu, F-75251 Paris Cedex 05, France

Keywords: Spherulitic Growth

1. INTRODUCTION

Spherulitic crystallisation (SC) is a mode of growth of crystals from the melt encountered in many different types of liquids. Because some of these (high-polymer and cast-iron melts) have a great industrial importance, considerable attention has been given to SC during the last decades. Careful studies using modern means of investigation have provided us with a wealth of detailed information about the conditions of appearance, growth rate, internal microstructure of spherulites in several high polymers and in cast irons [1,2]. Nevertheless, the nature of the mechanisms responsible for this peculiar mode of crystallisation of liquids still is an open question.

With the aim of gaining some insight into the basic mechanisms of SC, we carried out a detailed experimental study of the crystallisation of liquid selenium (ℓ-Se) as a function of the undercooling ΔT of the liquid. The present article is essentially a brief survey of the results and conclusions of this study (detailed reports can be found elsewhere [3,4]). Among spherulite-forming liquids, ℓ-Se is a somewhat exotic case (it is one of the rare liquids belonging to the class of "equilibrium polymers"[1] [5,6]), but it turned out that, because, in particular, of the high purity at which this elementary liquid is available (about 10 ppm total impurity content), this choice allowed us to clarify some pending questions concerning all spherulite-forming melts.

2. EXPERIMENTAL CONSIDERATIONS

Most spherulite-forming systems are thermodynamically complicated (e.g., cast irons, in which spherulites only appear near the austenite-graphite eutectic point) and/or ill-defined. Polymer melts, in particular, are never chemically very pure, and, more important, never free from a variety of specific "defects", (polydispersity, stereo-irregularities, chemical branchings of the chain molecules). Furthermore, in high molecular-weight polymers, the phenomenon called "entanglement" prevents crystallisation from being complete (high-polymer spherulites always contain a noticeable fraction of amorphous material). By choosing ℓ-Se for our investigation, and by showing that Se spherulites are quite similar to those of, e.g., polyethylene, we got rid of these complications, and furthermore proved that they are unessential to SC. Indeed, the above-mentioned equilibrium-polymer nature of ℓ-Se could introduce other complications, but we showed that, as far as crystallisation is concerned, ℓ-Se essentially behaves like a low molecular-weight polymer. In particular, because, ℓ-Se chain molecules have a finite lifetime, entanglements do not come into play, and crystals (which, roughly speaking, consist of an hexagonal close-packing of helical Se_n chains) are not prevented from growing in the direction parallel to the chains, entailing that there is no non-crystalline material left in the spherulites.

There exists several different modifications of crystalline Se, but only one of them (trigonal selenium) grows from the melt. The symmetry elements of trigonal selenium are a ternary **c** axis (parallel to the helical chains, it is the optical axis of the crystal) and three polar binary **a** axes perpendicular to **c**. The growth habit, known by direct observation of isolated melt-grown single

---

* The experimental study on which this article is based was performed in collaboration with G. Ryschenkow and J. Bisault at the LPMTM, CNRS UPR 9001, Villetaneuse, France, and completed in 1991. I am grateful to Gerard Lesoult, who suggested me to present this talk at the SCI-5 Conference, and provided me with a bibliography about spheroids in graphitic cast irons. Figures 2 5, 6, 7, 9, 10 are reproduced from the Journal of Crystal Growth **110** 889 (1991) with the permission of Elsevier Science , Amsterdam, Netherlands.
[1] Roughly speaking, an equilibrium polymer is a liquid consisting of chain molecules (here, Se_n), which are continually reacting according to reversible reactions of the type $Se_{n+p} \rightarrow Se_n + Se_p$.

crystal (see fig. 8 below), is sketched in figure 1. Selenium, either in the crystal, liquid, or glassy state, is a semi-conducting (thus non-transparent) material.

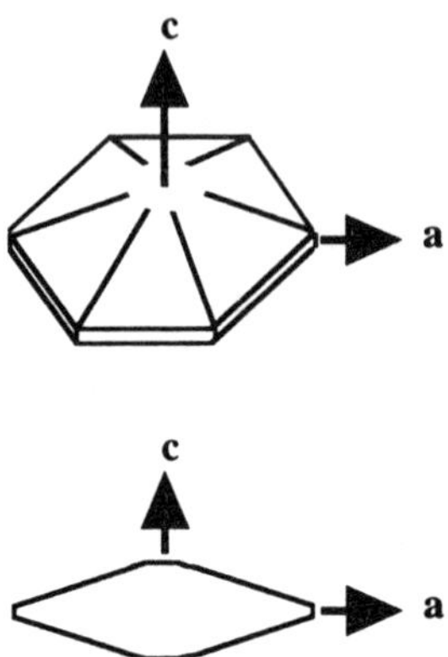

Figure 1. Growth habit of selenium crystals at medium- and low-T_c: (a) perspective view; (b) "prismatic" cross-section (the elongation of this shuttle-like shape increases as T_c decreases).

The experimental study of SC in ℓ-Se is facilitated by the fact that ℓ-Se is a glass-forming liquid (like most spherulite-forming liquids), whose glass-transition temperature is above room-temperature ($T_g = 30\ °C$). This implies that (i) the number of active heterogeneous-nucleation centres is small (it is temperature-dependent, and, at medium-T_c, of the order of 1 mm^{-3}), and (ii) the crystal growth rate is very low at any undercooling ($< 0.35\ \mu m\ s^{-1}$). Spherulites (and all other forms of crystallisation) are obtained as follows. An isothermal treatment is applied at a temperature T_c below the melting point ($T_m = 221\ °C$), and abruptly interrupted by quenching the sample to room temperature. In the quenched samples, crystals are embedded in the glassy matrix. The samples can be metallographically cross-sectioned and observed by reflection polarised-light microscopy. The contrast exhibited by crystal aggregates then reveals the local orientation of the **c** axis. Electron microscopy (TEM and SEM) and X-ray diffraction on crystals extracted from their glassy gauge can also used. The temperature of the annealing bath was controlled to within a few 10^{-2} K, allowing us to scan values of ΔT ranging from 0.05 K to 200 K.

3. DEFINING FEATURES OF SPHERULITIC CRYSTALLISATION

It will be seen below that there are several known "variants" of spherulite and a controversy about the basic mechanisms of SC. It is therefore not easy to formulate a satisfactory definition of SC. To introduce what we consider to be such a definition, let us begin by describing the variant called "ringed spherulite". For the moment, we restrict ourselves to spherulites growing in an isothermal, one-component liquid.

Figure 2 shows an equatorial cross-section of a low-T_c ringed spherulite of selenium observed between nearly-crossed polars. Three main characteristic features are observable under such conditions:

1- the internal microstructure of the spherulite consists of rough radial "spherulitic fibres" of constant average diameter ϕ (hence, the traditional –but unsatisfactory– definition of spherulites as "spherical arrays of fibres radiating from a common centre");

2- each spherulitic fibre grows with a given type of crystal axis (an **a** axis) parallel to the radial direction of the spherulite, while another crystal axis (the **c** axis) regularly rotates during the course

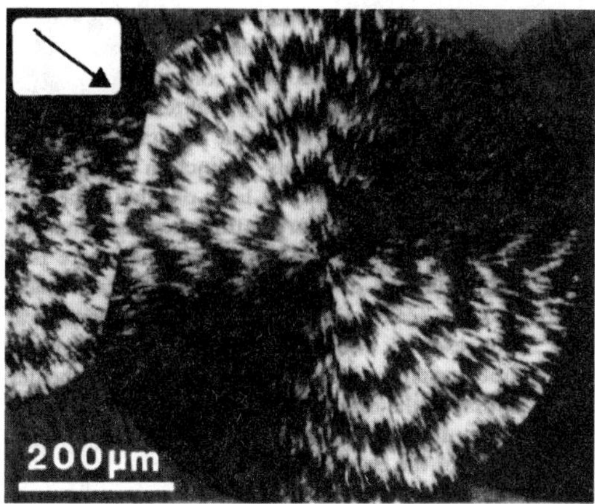

<u>Figure 2</u>. Spherulite of selenium grown at 119 °C (ΔT = 102 K). Polished equatorial cross-section observed in reflection polarised-light microscopy. Arrow: surface of impingement of two spherulites.

of growth. This striking phenomenon, called the fibre "radial twist", explains the periodic alternation of black and white contrast observed along each fibre. The global contrast in dark concentric rings —to which ringed spherulites owe their name— indicates that, roughly speaking, the radial twist is "coherent" in the whole spherulite (note, however, that it is the slight difference of twist phase between near-by fibres that reveals the fibres). The twist pitch λ_{twist} is constant at given T_C, and a well-defined function of T_C (fig. 3). Note that the microstructure presents several characteristic lengths (ϕ, λ_{twist}, and others to be defined below) which are of different orders of magnitude.

3- the outer envelope of the spherulite (the "spherulitic growth front") appears smooth and circular. The "radial growth rate" dR/dt (R: radius of the spherulite; t: time) is constant at given T_C, and a well-defined function of T_C (fig. 4). Note that dR/dt remains unaffected when near-by spherulites impinge on each other (upper left corner in fig. 2).

Generalising the above properties, we can formulate the following empirical characterisation, valid for all spherulite variants: considered on a sufficiently large scale, spherulitic crystallisation is a steady (or, at least, permanent) mode of growth characterised, at fixed T_C, by the fact that the growth front advances at a constant velocity, and leaves behind a fine microstructure exhibiting a hierarchy of constant characteristic lengths and a regular crystallographic arrangement of the microcrystals. It can be almost immediately inferred from this that spherulites are generated by the regular recurrence, at the interface of a growing crystal, of a branching event creating a new crystal, crystallographically misoriented with respect to its parent crystal. This "spherulitic branching"[2] operates on a very fine scale, so that the entities observable with an optical microscope (e.g., twisted fibres of ringed spherulites) are the result of a large number of elementary branchings.

On this basis, many early authors (in particular, Keller and co-workers [7]) have developed the so-called "regular-branching (R-B) model". We show below that this model satisfactorily accounts for many features of SC. Formally, it is a purely geometrical model, which leaves the question of the physical origin of the elementary branching events entirely open. However, the number of reasonable conjectures on this subject is much reduced if three additional experimental facts, common to all known spherulite-forming systems, are introduced in the list of the characteristic

---

2) We think preferable to avoid the term "non-crystallographic branching" sometimes used in the literature, since, in the case of growth twins (which, as will be seen below, is one of the mechanisms at play in SC), a definite crystallographic relation exits between the branch and the parent crystal.

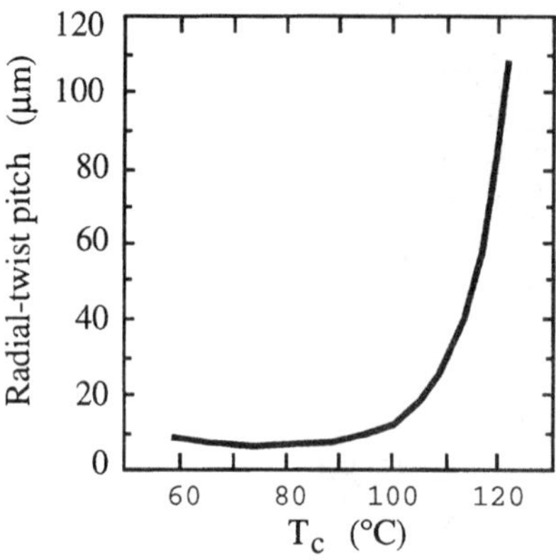

Figure 3. Pitch of the fibre radial twist of Se ringed spherulites (Mode A), as a function of crystallisation temperature.

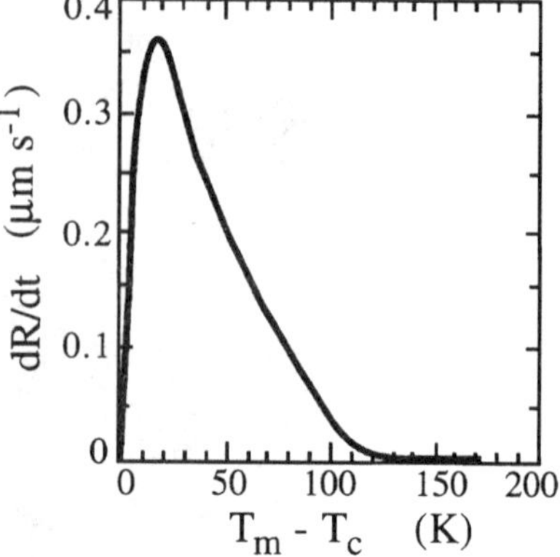

Figure 4. Radial growth rate of Se spherulites (Modes A and B), as a function of undercooling.

features of SC, namely,

4- on a sufficiently fine scale, the spherulitic growth front is not smooth, but composed of crystal spikes protruding into the liquid (fig. 5). These spikes present essentially the same anisotropic growth habit as isolated single crystal growing at the same temperature;

5- the initial nucleus of spherulites is either a single crystal, or an aggregate of a few crystals, e.g., a multiple twin;

6- all the characteristic distances of the spherulitic microstructure increase very rapidly as T_c approaches T_m (see, e.g., fig. 3).

Points 4 and 5 were established for various systems either by electron microscopy (at low T_c), or by optical microscopy (at high T_c) in taking advantage of point 6. In ℓ-Se, the simultaneous

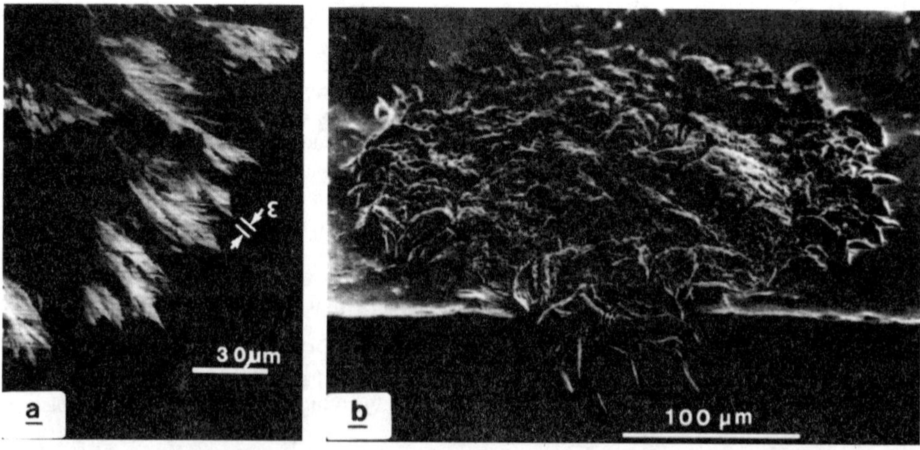

Figure 5. Medium-T_c Mode A spherulites ($\Delta T \approx 50$ °C): (a) equatorial cross-section, observed in optical microscopy (ε: see fig. 6); (b) sample first cross-sectioned and then etched with a selective solvent of the glassy phase, observed in SEM (the picture shows an edge of the sample).

growth of spherulites and single crystals allows one to directly check the validity of assertion 4.

Stating the existence of a definite (T_C-dependent) growth habit (property 4) is almost equivalent to saying that crystal growth is essentially kinetic-controlled (see Appendix) in all spherulite-forming systems. So, ultimately, two properties are characteristic of SC: (i) a kinetic-controlled, anisotropic crystal growth habit, and (ii) a (quasi-)periodic appearance of crystallographically misoriented branches on the growing crystals (spherulitic branching).

It must be finally mentioned here that this "classical" conclusion (it dates back to the beginning of this century [8]) was challenged by Keith-Padden [9] in 1964, who put forward the idea that spherulitic branching is nothing but a dendritic branching —thus driven by solute diffusion in the liquid— coupled with some unknown dislocation-generating mechanism in the solid. We shall not consider the question of SC in multi-component melts until Section 7, but we can already state that the Keith-Padden conjecture, although very attractive, seems in fact impossible to reconcile with the observation of typical spherulites in an ultra-pure melt.

3. THE REGULAR-BRANCHING MODEL

The basic assumptions of the R-B model are simple. It is assumed that, (i) at fixed T_C, all the crystals grow with the same, well-defined growth habit. They do not "feel" the near-by crystals until they impinge upon them; (ii) new crystal periodically appear at definite sites of the interfaces of the growing crystals; (iii) these "branches" have a definite misorientation with respect to their parent crystal. The parameters of the model are thus the location of the branching site, the rotation between branches and parent crystals, and the period separating two successive branching events on a given crystal. Note that assuming the branching events to be separated from one another in space and time does not necessarily imply that they are physically independent of each other (they may correspond, for instance, to successive turns of a growth super-spiral; see below)

Figure 6 shows the crystal aggregate formed, according to the model, after a time of growth corresponding to a few branching periods, assuming the initial nucleus to be a single crystal and the growth habit to be the hexagonal plate sketched in figure 1. Such "incipient" spherulites do not

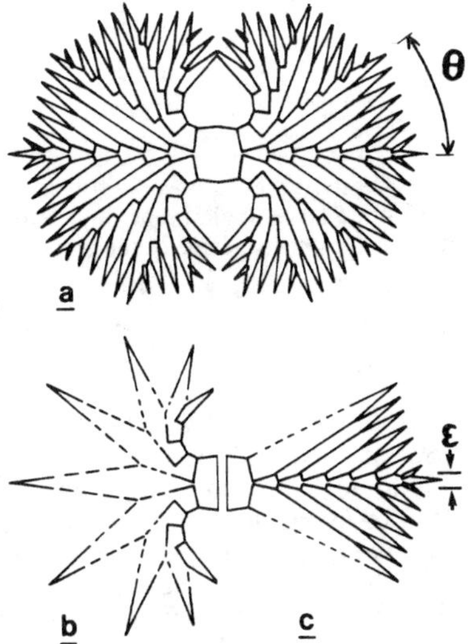

Figure 6. Regular-branching model of an incipient spherulite., An **a** axis in normal to the plane of the figure. Note the sheaf-like (a), circular (b), and feather-like (c) motives. Viewed from above, this aggregate would exhibit an hexagonal outer shape.

exhibit the features of well-developed spherulites (in particular, their outer envelope is polyhedral, and not spherical, in shape; hence the name "hedrites"[3] sometimes given to them) only because an insufficiently large number of branchings has taken place. In many systems, including ℓ-Se, incipient spherulites presenting the characteristic sheaf-like aspect illustrated in figure 6 were observed (statement 5 above follows from these observations).

In spite of the simplicity of its ingredients, it is a very difficult task to develop the R-B model beyond the early stage of growth represented in figure 6, especially if, for the sake of realism, one introduces some irregularity in the recurrence of the branching. To our knowledge, no one has been yet able to model a well-developed 3-dimensional spherulite, by hand or by computer. Nevertheless, it is clear that regular branching gives rise to a hierarchy of specific local geometrical motives, composed of a (possibly large) number of basic entities. In figure 6, the basic motives are a feather-like motif and a circular one . Their combination gives rise to a sheaf-like motif. It can be shown that, on a larger scale, and in 3 dimensions, motives composed of portions of sheaves can be built[4]. Such a partial "3D sheaf" is sketched in figure 7. In ℓ-Se, clear examples of partial 3D sheaves emerging at the growth front can be seen in medium- and high-T_c Mode-A spherulites. This shows that the regular branching model encompasses most of the phenomena involved in SC, even far beyond the early stages of growth. Further pieces of evidence will be given later on.

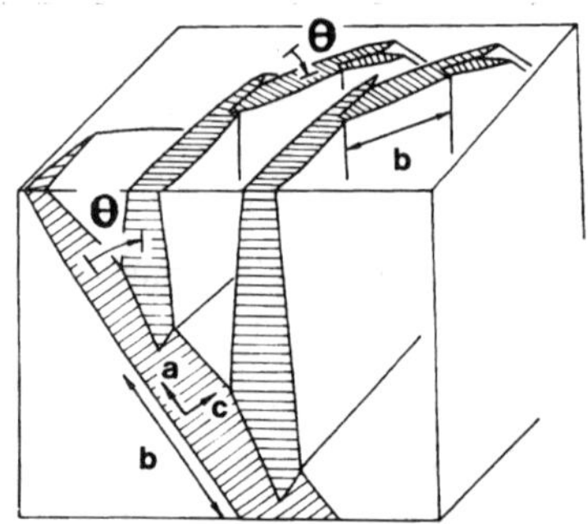

Figure 7. Model for the partial 3-dimensional sheaves observed in medium-T_c Mode A spherulites (see fig. 5b). These motives are generated by the large-angle branching (branching period b, rotation angle θ). The small-angle branching responsible for the curved shape of the crystal is not represented in the sketch.

---

[3] At sufficiently high T_c, the characteristic distances of SC are generally larger than the average distance between nucleation centres, so that only hedrites can be observed. This could wrongly induce one to think that a morphological transition from spherulites to hedrites occurs as T_c increases. It is therefore preferable to avoid the term "hedrite", and replace it by, e.g., (early) incipient spherulite. The same is probably true for other terms appearing in the literature. In fact, the term "spherulite" itself is somewhat unfortunate, since the spherical outer shape of the aggregates generated by SC is, in some sense, an artefact stemming from the dispersion of the nucleating agents in the liquid. Experiments show that, when nucleation takes place on surfaces (walls of the container, interface of a pre-existing single crystal), the "spherulitic" crystallisation is replaced by a "columnar" crystallisation, which presents all the characteristics of SC, expect that the growth front is macroscopically flat.

[4] The 3D generalisation of the "circular motif" is a conical helix (a continuous version of it is represented the study on graphite spheroids by Double and Hellawell [10]).

4. COEXISTENCE OF SEVERAL VARIANTS OF SPHERULITES

One of the striking characteristics of SC (which should in fact be added to the above list of 6 properties) is the (co)existence, in a given system, of several variants of spherulites. In ℓ-Se (but also, for instance, in polypropylene [11]), two microstructurally different types of spherulites (called A and B) can grow side by side in the same sample. Since, as mentioned, single crystals can also grow in parallel with spherulites, three "modes of growth" coexist in ℓ-Se at medium T_C. The T_C-ranges over which the various modes of growth were observed when single-step thermal treatments were applied to the samples are indicated in figure 8. They more or less coincide with

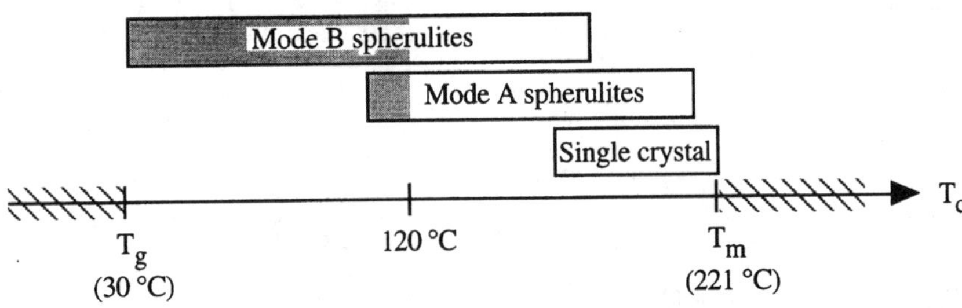

Figure 8. Table of coexistence of the modes of crystallisation of ℓ-Se as a function of T_C. Both A and B spherulites are ringed below about 120°C (shaded region).

the stability ranges of the modes, since (i) no transformation from a mode to another ever occurred during single-step thermal treatments, and (ii) a mode spontaneously transforms to another as the result of a T_C-jump out of the indicated range. Figure 9 shows a low-T_C "mixed" ringed spherulite, composed of an alternation of Mode-A and Mode-B sectors. At low-T_C, the radial-twist pitches of the two modes differ by a large factor. At medium T_C, the two coexisting SC modes apparently loose their ringed microstructure (see fig. 8). The origin of mixed spherulites and that of the ringed → non-ringed transition are explained below.

We cannot enter here into a detailed analysis of these phenomena, which, in fact, we do not yet entirely understand. We were however able to establish that [4]:
1- Mode A results from the combination of two types of branching mechanisms, namely, (i) a

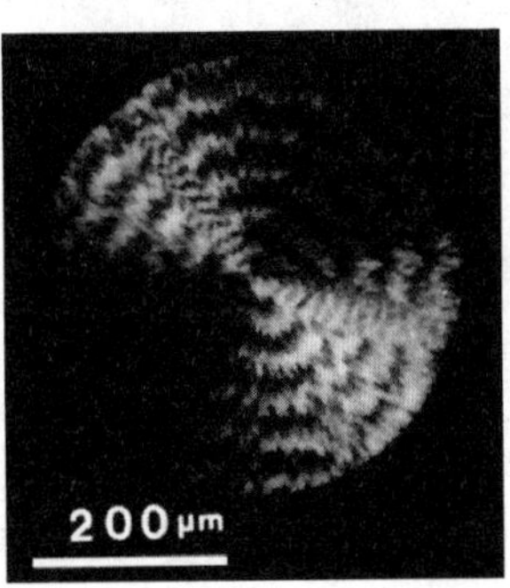

Figure 9. Mixed spherulite of selenium. $T_c = 114$ °C.

small-angle branching with a period of the order of 10 nm (responsible for the crystal curvature; see below), and a large-angle branching with a larger period (a "homoepitaxy" or growth twin; in ℓ-Se, the existence of growth twins is directly proved by the fact that high-T_C crystals are sometimes twinned);

2- Mode B only includes a small-angle branching, different from Mode-A small-angle branching (i.e., giving rise to a different arrangement of the microcrystals);

3- the disappearance of the ringed contrast above about 120 °C is initial-condition dependent. Ringed spherulites are obtainable even at medium T_C with the help of *ad hoc* multi-step heat-treatments. A closer look at the microstructure of non-ringed medium-T_C spherulites shows that their fibres are still exhibiting a radial twist (with very large pitches, so that millimetric spherulites had to be grown to bring this fact to light), but the "coherency" between near-by fibres is lost.

Points 1 and 2 (together with the "similarity law" explained in Section 6 below) allows us to identify the searched spherulitic branching with a small-angle branching operating on the nanometric scale. They further show that, in a given system, there can exist several, geometrically different, types of spherulitic branching. They finally indicate that spherulitic branching can trigger a large-angle, homoepitaxial branching. Whether it does so, or not, depends on its geometrical characteristics. Thus, in ℓ-Se, Mode-A spherulitic branching most probably allows a favourable site for the homoepitaxy to appear at the solid-liquid interface, while Mode-B spherulitic branching does not. Point 3 shows that the radial twist is a local effect, implying no coupling between neighbouring fibres. It is also an example of sensitivity to initial conditions.

5- SENSITIVITY TO INITIAL CONDITIONS

What are the factors responsible for the initial choice (selection) of the mode of growth, at temperatures at which there exists several stable modes? Figure 10 shows the result of a two-step

Figure 10. Result of a two-step treatment (see text). The **c** axis of the initial crystal is normal to the figure.

thermal treatment, in which the sample was first maintained at a low undercooling ($\Delta T \sim 0.5$ K), and then quenched at 115 °C. The single crystal formed during the first stage of the treatment serves as initial condition for the SC taking place during the second stage. It can be seen that the selected SC mode depends on the site considered on the interface of the initial single crystal. More precisely, the edges of the (pseudo-)hexagonal prism give rise to Mode B, while Mode A grows on the faces of the prism. From this, we can infer that the three Mode-A sectors appearing in mixed spherulites (see fig. 9) originate from three equivalent sites of a microscopic single crystal.

It is thus clear that the selection of the mode of growth at given T_C only depends on features of the initial nucleus such as its shape and lattice-defect content. These, in turn, depend on (i) the chemical nature and shape of the nucleating foreign particles, and (ii) the thermal history of the sample. So, the selected mode of growth (and, for a given mode, some aspects of the macroscopic edifice, e.g., the coherency of the radial twist) are in fact determined by factors (chemical nature, shape and dispersion of nucleating impurities) which are usually difficult to control in practice. Low-concentration impurities have a striking influence on the growth morphology, but this influence is not related to any diffusive instability.

6. NATURE OF THE SPHERULITIC SMALL-ANGLE BRANCHING

At medium T_C, a single-crystal mode of growth coexists with the spherulitic modes of growth (fig. 8). This indicates that the features of the initial nucleus necessary for the selection of SC can be altogether lacking. It is therefore very probable that they are linked with some type of lattice defect (e.g., twin boundaries). The fact that, below 160 °C, the single-crystal mode is unstable (i.e., spontaneously transforms to a spherulitic mode) suggests that this defect can be created by a thermally-activated process, whose activation energy decreases as ΔT increases. Nucleation of a misoriented islet at some favorable site (possibly linked to a lattice defect) of the interface is a plausible example of such a process.

The stability of the spherulitic modes of growth further indicates that, once the defect necessary for the initiation of these modes of growth has been created, it is periodically re-created. In agreement with this, we observed that, contrary to single crystals, the crystals composing spherulites ("spherulitic crystals") are heavily plastically distorted, in a way which is by no means random. Optically, we showed that each spherulitic crystal is curved, the radius of curvature ρ_{curv} being a well-defined function of T_C. By TEM and X-ray diffraction, we established that spherulitic crystals contain a relatively uniform density of subgrain boundaries. We measured the mesh λ_{mesh} of the polygonisation network (i.e., the size of the mosaic blocks) as a function of T_C. λ_{mesh} is of the order of 10 nm, and increases as T_C increases. The ratio of λ_{mesh} with the other characteristic distances of the spherulitic microstructure (ϕ, ρ_{curv}, λ_{twist}) is roughly independent of T_C. In other words, there exist a (rough) "similarity law" relating spherulites of the same mode grown at different T_C. Knowing the length-range covered by the spherulitic characteristic distances ($\lambda_{mesh}/\lambda_{twist} < 10^{-3}$), this is indeed a striking result, which confirms both the validity of the R-B model and the conclusion that small-angle branching is the mechanism responsible for SC.

The details of the small-angle branching mechanism are not known. At each occurrence of this mechanism, a subgrain boundary is created in the growing crystal and a branch on its interface, but it is not clear which of these phenomena is the leading one. We are for the moment reduced to conjectures. Let us give an example of the kind of mùechanisms that may be though of. In solution-grown polymer crystals, stacks of slightly misoriented plate-like crystals are sometimes observed, evoking a Frank growth spiral in which the rotating step would be a facet (instead of having an atomic height) [12]. Lotz *et all* moreover showed that a twinned crystal is probably at the origin of these "super growth spirals" [13]. It is therefore tempting to think that the spherulitic small-angle branching consists of the nucleation of a growth twin rapidly transformed to a super growth spiral (fig. 11). However, a direct proof of this conjecture is lacking. Moreover, the transformation

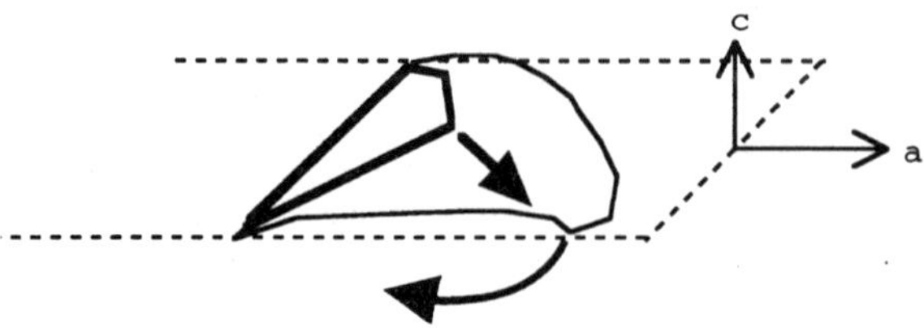

Figure 11. Sketch of the hypothetical transformation of a growth twin to a "super" growth spiral (dashed line: parent crystal; continuous thick line: undeformed branch-crystal; continuous thin line: deformed branch-crystal beginning a spiral growth.

growth-twin → super-spiral, if it exists, remains to be explained. Several investigators imagined that some forces are acting on the growing crystals, for instance, the viscous forces accompanying the advection flow of the liquid towards the growing crystal due to the different densities of the two phases. Caroli *et al* showed that such forces indeed exist, and are non-negligible in very viscous melts (as are glass-forming liquids) [14], but, for the moment, we have no direct experimental proof that they play a part in SC.

7- SPHERULITIC CRYSTALLISATION IN MULTI-COMPONENT SYSTEMS

The following speculative considerations are largely disconnected from the above results. They do not concern all the possible effects of impurities on SC, but only those effects which have been thought to support the Keith-Padden conjecture [9] according to which spherulitic branching essentially is a diffusion-controlled phenomenon.

To our best knowledge, there is no exception to the rule that, in spherulite-forming liquids, crystals grow with a very anisotropic habit. The relevant basic question is thus that of the morphological instabilities of crystals with slow, anisotropic interface kinetics growing from a solution. Theoretically, this problem is far from being well understood. Let us, for specificity, consider the case of a plate-like growth habit, i.e., the case in which there exists one very unfavourable direction of growth. In many systems (among which graphite-flakes growing from cast-iron melts), it was observed that, above a certain size, such crystals exhibit a dendritic pattern confined in the favourable plane of growth. The edifice as a whole then remains a quasi-2-D single crystal. Thus, even taking for granted that, in SC from impure melts, spherulitic crystals undergo a dendritric instability, we still do not explain spherulitic branching. As recognised by Keith and Padden themselves, an additional ingredient is needed, namely, that, for some unknown reason, the appearance of dendritic sidebranches is accompanied by the formation of dislocations (which moreover have to be arranged into subgrain boundaries) giving a (well-defined) misorientation to the branch —which is precisely the problem that we did consider in the preceding section. So, unless supported by direct experimental evidence, the Keith-Padden conjecture seems a rather arbitrary one. Let us now turn to experimental observations.

First of all, let us again stress that the primary result of our study on ultra-pure ℓ-Se is to show that, contrary to what was thought by some ancient authors, the presence of impurities is not necessary for the occurrence of SC. Moreover, in many cases, solutes in moderate concentration have only a weak influence on the characteristics of SC (this may simply be due to the fact that, at the high undercoolings characteristic of SC, solute trapping takes place) [15]. Nevertheless, the question of the fate met by solute molecules when SC takes place in multi-component systems is not entirely clarified. In some polydisperse polymer melts, a fine-scaled segregation is observed within the spherulitic microstructure between molecules of different molecular weight [1]. On the other hand, the microstructure of, say, ringed spherulites of polydisperse polyethylene, observed with an optical microscope, is completely similar to those of ultra-pure ℓ-Se. It thus seems that the basic microscopic entity which is to undergo the spherulitic branching is a composite structure of average concentration equal to that of the liquid. In other words, the ultra-microscopic segregation existing in polymer spherulites can be considered, on a larger scale, as a solute-trapping mechanism (this remark also concerns the incorporation of amorphous layers in polymer spherulites).

On the other hand, several authors, in particular Keith and Padden [16], showed examples of "anomalous" spherulites growing in concentrated solutions of polymers. The radial growth rate of these anomalous spherulites decreases with time. When two such spherulites approach each other, their advance is stopped before they touch. Moreover, these spherulites do not remain spherical. They are divided in broad fingers separated by grooves in which the rejected species segregate. All these features indubitably pertain to a mode of growth controlled by diffusion of rejected species. But the question is whether the scale on which the solute segregation takes place is connected, or not, with the characteristic scales of the spherulitic microstructure defined in the preceding sections. As far as we can see, this question must be answered in the negative. The distance at which near-by anomalous spherulites "feel" each other is, in order of magnitude, the diffusion length $\ell_d \equiv D / V$, (D: solute diffusion coefficient in the liquid; $V \equiv dR/dT$: growth rate), which gives the range over which the solute concentration gradient extents in the liquid ahead of the advancing front. In the experiments by Keith and Padden , ℓ_d is several order of magnitudes above the nanometric scale which we believe to be the characteristic scale of spherulitic small-angle branching. We therefore think that their observation can neither support nor contradict their conjecture.

In summary, the idea that we are putting forward is that the ratio between the diffusion length ℓ_d of the controlling solute and the period β of the branching at play in SC is the important parameter to consider. More specifically, the ultra-microscopic segregation in high-polymer spherulites would correspond to solutes for which $\ell_d/\beta < 1$, while the anomalous spherulites would be due to the presence of a solute such that $\ell_d/\beta \gg 1$. In the latter case, from the point of view of the rejected solute, the spherulitic growth front is smooth. The spherulite then behaves like an isotropic kinetic-controlled "average" crystal, and undergoes a morphological instability when its radius reaches some critical value. This suggestion can be stated more precisely as follows.

Let us recall that the isothermal (radial) growth rate V_{dif} of a spherical crystal whose growth is diffusion-controlled varies with time as $t^{-1/2}$. The thickness of the solute layer ahead of the interface thus increases as $t^{1/2}$, i.e., in the same way as R. However, as shown by Mullins and Sekerka a long time ago [17], such a crystal cannot remain indefinitely spherical. It must undergo a morphological instability (namely, the cellular/dendritic instability, discovered by Rutter and Chalmers in 1954 [18]). For systems with a very fast interface kinetics, this instability occurs for radii only a few times larger than the critical nucleation radius. Let us now turn to the case of systems with slow, isotropic interface kinetics, which was studied theoretically by Coriell and Parker [18]. In this case, the spherical shape remains stable, and the radial growth rate practically constant ($dR/dt = V_{kin}$) up to a much larger radius R_{inst}. In order of magnitude, $R \sim R_{inst}$ at the time t_{inst} at which $V_{kin} = V_{dif}$, where V_{dif} is the growth law that would be followed if the interface kinetics was infinitely fast. The transposition of this analysis to SC, supposing $\ell_d = D/V_{kin}$, to be much larger than the branching period, is sketched in figure 12. Presumably, beyond the instability threshold, the spherulitic growth front is divided in broad fingers, whose characteristic sizes are of order ℓ_d .

8- REMARKS ABOUT GRAPHITE SPHEROIDS

There seems to be a general agreement in the literature for considering graphite spheroids in cast irons melts as basically similar to, e.g., polymer spherulites. Indeed, not only spheroids, but also the other forms of graphitic growth in cast irons can be explained in the same terms as in other spherulite-forming systems. In particular, "graphite flakes" in fact are graphite single crystals, which exhibit a very anisotropic, hence kinetic-controlled, growth habit. A specific difficulty arises from the fact that graphite spheroids (contrary to all the other known spherulites) have their "fibres" growing with the most unfavourable direction of growth (which, like in ℓ-Se, is the ternary c axis) parallel to the radius of the spherulite. However, it was shown by several authors (see, e.g., ref. [8])

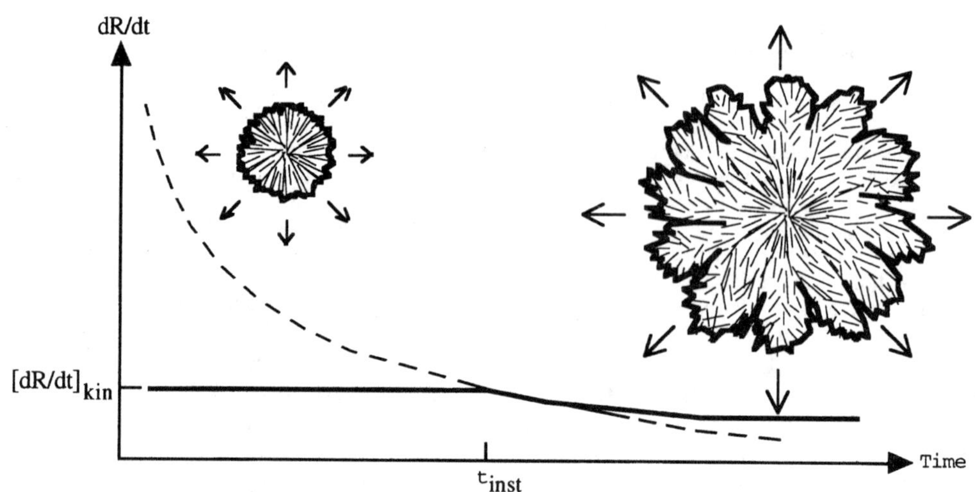

Figure 12. Hypothetical cellular instability of spherulites growing in multicomponent melts (sketch).

that this difficulty can be solved, at least hypothetically, within the frame of the regular-branching model.

Another specific difficulty concerning graphite spheroids is that their instantaneous growth rate is not known with accuracy. In fact, since graphite spheroids generally grow in an heterogeneous environment (namely, concentration and temperature gradients due to the continuous cooling of the ingots and the simultaneous growth of austenite), their growth rate is not easy to measure. This leads us to our last remark. In most cases, when spherulites grow in a macroscopically heterogeneous medium, because of their basically kinetic-controlled, slow growth rate, they "adiabatically" adapt their characteristics to the instantaneous concentration and temperature prevailing at the spherulitic growth front. For instance, if the temperature of the melt changes with time, the characteristic distances of the structure "follow" the variations of T_c. This property is sometimes used by geologists, who examine the variations of the radial-twist pitch of some (ringed) spherulite-forming minerals in order to gain information about the thermal history of rocks. Similar methods could perhaps be of interest in the case of cast irons.

APPENDIX

Our purpose in this Appendix is to provide the unfamiliar reader with a rough idea of what controls the anisotropic growth habit at the high undercoolings characteristic of SC. Generally speaking, two phenomena limit the rate of advance of a crystallisation front, namely, (i) diffusion in the liquid of latent heat and solutes rejected (or absorbed) by the crystal; (ii) phenomena delaying incorporation into the solid of molecules adsorbed at the solid-liquid interface (interface kinetics). In most practical cases, growth is entirely controlled by only one of these phenomena (the one imposing the slowest growth rate), the other playing a negligible role.

Let us focus on kinetic-controlled growth. In this case, the growth rate of a crystal in a given crystallographic direction can be viewed as the product of several different terms, namely, (i) the frequency at which a molecule situated on the liquid side of the interface jumps onto the interface. This term is essentially orientation-independent, being more or less equal to the diffusive jump frequency in the liquid, which is proportional to η^{-1} (η: bulk viscosity of the liquid);

(ii) a geometrical (steric) factor representing the probability for the jumping molecule of presenting itself with such a conformation and an orientation that it can be adsorbed on the interface; (iii) a term governed by the physics of the interface itself, strongly dependent on the interface structure (rough or faceted, perfect or intersected by dislocations, etc.). It is common practice to eliminate the jump-frequency term by considering the reduced velocity ηV, instead of the growth velocity V. Experimentally, it is found that ηV is a monotonously increasing function of ΔT, contrary to V itself (see fig. 4).

The physics of the solid-liquid interface at high undercoolings ($\Delta T > 10$ K) is far from being well understood. It is very likely that, at high undercooling, some type of dynamical roughening takes place, so that the reduced growth velocity is essentially controlled by geometric factors. For instance, in ℓ-Se, the fact that the preferential direction of growth is the **a** axis (perpendicular to the chain molecules) primarily means that the attachment of a molecular segment onto a **a** facet occurs more frequently than the attachment of a chain end onto a **c** facet, because of the low density of reactive chain ends in the liquid (in a more detailed discussion, some assumption about the unknown atomic structure of the **c** facets would have to be made).

REFERENCES

[1]   For a review, see D.C. Bassett: Principles of Polymer Morphology (Cambrigde University Press, Cambridge 1981).
[2]   For a review, see B. Lux: Giessereiforschung **22** 65 and 158 (1970)..
[3]   G. Ryschenkow and G. Faivre: J. Cryst. Growth **87** 221 (1988). See also G. Ryschenkow, thèse, Univ. Paris 13, 1985.
[4]   J. Bisault, G. Ryschenkow and G. Faivre: J. Cryst. Growth **110** 889 (1991). See also J. Bisault, thèse, Univ. Paris 13, 1988.
[5]   A. Eisenberg and A.V. Tobolsky. J. Polymer Sci. **46** 19 (1960).
[6]   G. Faivre and J.L. Gardissat,: Macromolecules **19** 1988 (1986).
[7]   A. Keller and J.R.S. Waring: J. Polymer Sci. **17** 447 (1955).
[8]   G. Friedel, Leçons de Cristallographie, (1926), 2^{dn} ed. (Blanchard, Paris 1964), p. 478.
[9]   H.D. Keith and F.J. Padden: J. Appl. Phys. **34** 2409 (1964).
[10] D.D. Double and A. Hellawell: in The Metallurgy of Cast Iron (Georgi, St-Saphorin, 1975) p. 509.
[11] D.R. Norton and A. Keller: Polymer **26** 704 (1985).
[12] F. Khoury and E. Passaglia: in Treatise on Solid State Chemistry, Vol. 3, Ed. N.B. Hannay (Plenum, New York, 1976) ch. 6.
[13] B. Lotz, A.J. Kovacs and J.C. Wittmann: J. Polymer Sci. **13** 909 (1975)
[14] B. Caroli, C. Caroli, B. Roulet and G. Faivre: J. Cryst. Growth **94** 253 (1989).
[15] In ℓ-Se, the addition of relatively large amounts (a few %) of Tellurium or Sulfur has no marked effect on SC (unpublished work).
[16] H.D. Keith and F.J. Padden: J. Appl. Phys. **35** 1270 (1964).
[17] W.W. Mullins and R.F. Sekerka, J. Appl. Phys. **34** 323 (1963).
[18] J.W. Rutter and B. Chalmers, Can. J. Phys, **31** 51 (1953).
[19] S.R. Coriell and R.L. Parker: in Crystal growth, ed. H.S. Peiser (Pergamon Press, New York, 1967) p. 703.

Advanced Materials Research Vols. 4-5 (1997) pp. 31-46
© 1997 Scitec Publications, Switzerland

Cast Iron as a Composite:
Conductivities and Elastic Properties

G. Grimvall

Theoretical Physics, Royal Institute of Technology,
S-100 44 Stockholm, Sweden

Keywords: Thermal Conductivity, Elastic Properties, Thermal Expansion, Composite, Mathematical Modelling, Graphite Morphology

ABSTRACT

The paper reviews mathematical models that yield estimates of the effective thermal conductivity and the effective elastic properties of statistically isotropic two- or multiphase materials, exemplified by cast iron. Such theories may express the effective property in terms of upper and lower bounds or in the form an estimate based on a particular approximation scheme. Input information is the volume fraction and morphology of the phases present, and the conductivity (or elastic properties) of each of these phases. Here particular attention is paid to the dilute-limit of such approaches, e.g., when the volume fraction of graphite is small. In this way, the difference between the properties of grey and nodular cast iron is quantitatively accounted for. Brief remarks are given on the thermal expansion coefficient.

INTRODUCTION

It is the purpose of this paper to give a brief introduction to theories for the overall conduction and elastic properties of composite materials. The emphasis is on applications to cast iron, considered as a two- or three-phase material, with an iron alloy matrix and inclusions of graphite and in some cases also cementite. We shall refer to previous theoretical work on the thermal conductivity of cast iron by Helsing and Helte [1] and Helsing and Grimvall [2] and theoretical or experimental work on elastic properties by Speich, Schwoeble and Kapadia [3], Anand [4], Fuller, Emerson and Sergeant [5], Sergeant and Fuller [6], Papakakis et al. [7], Löhe, Völhringer and Macherauch [8], Dryden, Deakin and Purdy [9], Dryden and Purdy [10], Bussière, Piché and Leclerc [11] and Ledbetter and Datta [12]. Many of the basic formulae are given in a monograph by Grimvall [13], or they can be found in the works cited above. For brevity, we therefore do not give complete references to the original theoretical work on general aspects of effective properties of composites.

Cast iron shows three main morphologies for the graphite. In grey cast iron, the graphite forms flakes (discs) in which the graphite crystals are layered parallel to the disc. (Their crystallographic c-axis hence is perpendicular to the plane of the disc.) In nodular cast iron the graphite forms spherical inclusions in which graphite layers are roughly arranged in concentric shells, with the c-axis pointing radially. Finally, in compacted cast iron the graphite inclusions have some resemblance to a rod-like shape. The crystallographic orientation of the graphite grains is of crucial importance since the physical properties of single-crystal graphite are highly anisotropic. The room-temperature thermal conductivity along the c-axis typically is 10 W/(m·K) and the in-plane conductivity is about 500 W/(m·K). Under hydrostatic pressure, a single crystal of graphite is compressed 56 times more along the c-axis than in the perpendicular direction (Appendix 1).

The effective conductivity and elastic properties of a composite material depend on the properties of the constituent phases, and their geometry, as expressed by their relative amount and morphology. The crystallographic orientation can be important in phases that have anisotropic single-crystal properties. The more that is known about the morphology, the better is of course the estimate of the effective value of a property. Such an estimate may be expressed through upper and lower bounds, or as a single value whose accuracy is more or less uncertain but may still be of great value. We first exemplify this by considering the thermal conductivity λ in a case simpler than that of cast iron, *viz.* a composite made up of two phases, 1 and 2, having isotropic conductivities λ_1 and λ_2. For instance, phase 1 could be ferrite and phase 2 be cementite, in a carbon steel. If we do not know anything about the relative amounts of the phases in the composite, expressed as the volume fractions f_1 and f_2 $(= 1 - f_1)$, all we can say is that the effective conductivity λ_{eff} lies between λ_1 and λ_2. Next, assume that we know the volume fractions, but nothing about how the phases are arranged. One can then show that λ_{eff} must lie between bounds λ_{series} and $\lambda_{parallel}$ given by the series and parallel coupling, respectively, and sometimes referred to as the Wiener bounds;

$$[f_1/\lambda_1 + f_2/\lambda_2]^{-1} \le \lambda_{eff} \le f_1\lambda_1 + f_2\lambda_2 \tag{1}$$

These bounds can be attained and then correspond to conduction perpendicular or parallel to the planes in a layered composite, Fig. 1. (Conduction along fibers in an aligned fiber composite also yields the case of parallel coupling.)

Many composite engineering materials have their phases more or less randomly arranged so that the composite has overall isotropic properties. Then there exist more narrow bounds to the conductivity, referred to as upper and lower Hashin-Shtrikman bounds [14], λ_{HS+} and λ_{HS-}, with

$$\lambda_{HS-} = \lambda_1 + \frac{f_2}{1/(\lambda_2-\lambda_1) + f_1/(3\lambda_1)} \tag{2}$$

Here labels 1 and 2 are chosen so that $\lambda_2 > \lambda_1$, and both phases are assumed to have isotropic λ. The upper bound λ_{HS+} is obtained by interchanging the labels 1 and 2 in Eq. 2. These bounds allow any geometrical distribution that is consistent with an isotropic overall behavior. With additional geometrical information, one may construct a hierarchy of more narrow bounds, and much current research is devoted to that task. If the properties of the constituent phases are very different, the bounds will be far apart and of limited value in practical work. One may then try to get a direct estimate λ^* of λ_{eff}. There are several schemes to accomplish this. We shall exemplify it with a so-called effective-medium theory. It expresses λ^* in the volume fractions and conductivities of the constituent phases and assumes that each individual grain of a phase does not deviate systematically from a spherical shape. Then λ^* is the solution to the equation

$$\frac{\lambda_1 - \lambda^*}{\lambda_1 + 2\lambda^*} f_1 = \frac{\lambda^* - \lambda_2}{\lambda_2 + 2\lambda^*} f_2 \tag{3}$$

Finally, one can focus on a direct calculation of λ^* in the "dilute limit", e.g., when the volume fraction of the inclusions 2 in the matrix 1 is so low that we can neglect any interaction of two neighboring inclusions on the overall properties. Such dilute-limit results can be derived for inclusions of ellipsoidal shape, which contains as special cases suspensions of spheres, discs (oblate ellipsoids) or bars (prolate ellipsoids) in a matrix. The cases just discussed are summarized in Table 1. Our example dealt with conduction, but similar results hold for elastic properties, with an extension also to thermal expansion. They will be considered in some detail in this review.

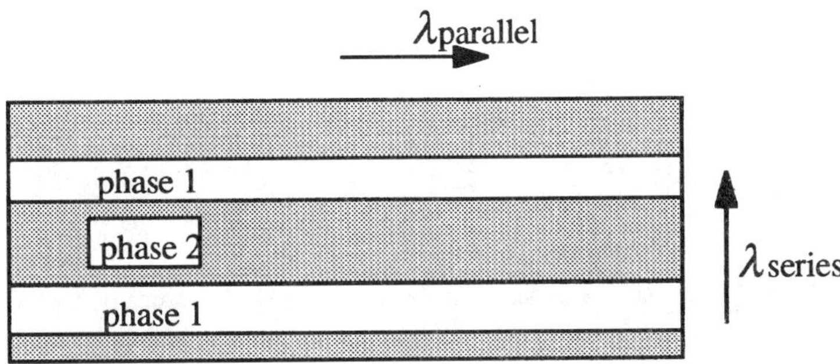

Figure 1. Conductivities λ corresponding to series and parallel coupling of phases in a layered or fibrous composite. This geometry may describe pearlite, as layers of ferrite and cementite.

Table 1. The effective conductivity λ_{eff}, or its estimate λ^*, in a two-phase composite of phases 1 and 2, with volume fractions f_1 and f_2, and conductivities λ_1 and λ_2.

Only λ of phases known	$\lambda_1 \leq \lambda_{eff} \leq \lambda_2$	
Amount of phases, f_i, known	$\lambda_{series} \leq \lambda_{eff} \leq \lambda_{parallel}$	(Eq. 1)
Overall statistical isotropy	$\lambda_{HS-} \leq \lambda_{eff} \leq \lambda_{HS+}$	(Eq. 2)
Additional geometrical information	higher-order bounds	(e.g., [15])
Overall isotropy, estimate	λ^*	(Eq. 3)
Dilute limit, spheres	$\lambda_{eff} \approx \lambda_1 \left[\, 1 - f_2 \dfrac{3(\lambda_1 - \lambda_2)}{2\lambda_1 + \lambda_2} \right]$	(Eq. 14)
discs	$\lambda_{eff} \approx \lambda_1 - f_2 \dfrac{(\lambda_1 - \lambda_2)(\lambda_1 + 2\lambda_2)}{3\lambda_2}$	(Eq. 15)
rods	$\lambda_{eff} \approx \lambda_1 - f_2 \dfrac{(\lambda_1 - \lambda_2)(5\lambda_1 + \lambda_2)}{3(\lambda_1 + \lambda_2)}$	(Eq. 16)

EFFECTIVE PROPERTIES OF A ONE-PHASE POLYCRYSTAL

Most of the simple estimation procedures for the effective properties of a composite assume that the relevant properties of the constituent phases can be described as isotropic. An iron alloy single-crystal of cubic lattice symmetry has an isotropic thermal conductivity and bulk modulus, while the shear modulus and Young´s modulus are slightly anisotropic. A single crystal of graphite, on the other hand, is *highly* anisotropic. One may therefore ask if it is meaningful first to average a property over all crystallographic directions for iron and graphite considered separately, and then use these values in a consideration of the cast iron composite. (Such an averaging is used, e.g., in Refs. [3, 4] in a treatment of elastic properties.) This section deals with the averaging of an anisotropic property in a statistically isotropic polycrystal.

Conduction properties

In a solid of cubic lattice symmetry, the conductivity is isotropic already in the individual single crystal. In an anisotropic material with a hexagonal lattice structure, the conductivity λ_c along the c-axis is different from the (isotropic) conductivity λ_a in a plane perpendicular to the c-axis. The effective conductivity λ_{eff} is bounded by ($\lambda_a > \lambda_c$) [1] (cf. [16] for improved bounds)

$$\lambda_c \frac{7\lambda_a + 2\lambda_c}{\lambda_a + 8\lambda_c} < \lambda_{eff} < \lambda_a \frac{4\lambda_a + 5\lambda_c}{7\lambda_a + 2\lambda_c} \tag{4}$$

An effective-medium type result [1] gives λ_{eff} as the solution to

$$\lambda^* = (1/4) [\lambda a + (\lambda_a^2 + 8\lambda_a\lambda_c)^{1/2}] \tag{5}$$

The application of Eq. 4 to graphite gives upper and lower bounds λ_u and λ_l that are so widely apart (see Table 2) that they are of little practical use, and also make the estimate λ^* from Eq. 5 uncertain.

Pearlite, in its ideal morphology, has alternating parallel layers of iron and cementite. The conductivities perpendicular and parallel to the lamellae are given by λ_{series} and $\lambda_{parallel}$ (Eq. 1) i.e., with a conductivity symmetry equivalent to that of a single crystal of hexagonal lattice symmetry. We may then consider pearlite as *one* phase, and apply the results of this section to a larger grain of pearlite, in which separate regions of parallel lamellae are more or less randomly oriented in all directions, Fig. 2. In this case, Eqs. 4 and 5 result in a very accurate prediction for the averaged conductivity of pearlite (Table 2), provided that $\lambda_{Fe\ alloy}$ and $\lambda_{cementite}$ are accurately known. The entries for pearlite in Table 2 are based on $\lambda_{Fe\ alloy} = 30$ W/(m·K) and $\lambda_{cementite} = 8$ W/(m·K).

Figure 2. Schematic possible morphology in a polycrystal where a physical property is anisotropic in each single crystal (each grain), but isotropic in a statistical sense in the entire polycrystal.

Table 2. The thermal conductivity λ, at 300 K, in W/(m·K), of graphite, pearlite, pure ferrite, a typical ferrite alloy with 1.5 wt.% Si and 2 wt.% Ni, and cementite. Ferrite and cementite have cubic lattice symmetry and hence an isotropic λ. Data from ref. [2].

	λ_c	λ_a	λ_u	λ_l	$\lambda*$
graphite	10	500	61	291	260
pearlite[†]	22.5	27.3	25.618	25.627	25.624
ferrite					78.5
Fe alloy					30
cementite					8

† The bounds and the estimate are given with an unphysical accuracy, only to show the numerical result based on the assumed λ_c and λ_a.

<u>Elastic properties</u>

In a polycrystalline (one-phase or multiphase) material where the individual grains are randomly oriented so that the material is overall isotropic, there are four commonly used engineering elastic constants; the bulk modulus K, the shear modulus G, the Young's modulus E and the Poisson number v. However they are interrelated, so that when knowing any two of them, the other two are completely determined. Assume that K and G are known. Then

$$E = \frac{9KG}{3K + G} \tag{6}$$

$$v = \frac{3K - 2G}{2(3K + G)} \tag{7}$$

Consider a one-phase polycrystal with randomly oriented grains that have hexagonal lattice symmetry. The individual grains are described by *five* independent elastic coefficients c_{ij}, but the polycrystal has only *two* independent elastic constants, e.g., K and G. There are many microstructures that would yield statistically isotropic K and G. If we only require overall elastic isotropy for the polycrystal, there is no unique mathematical expression for how the c_{ij} are weighted to form K and G.

If one makes the assumption that either the strain or the stress is uniform throughout the polycrystal, one obtains mathematically simple results for K and G expressed in c_{ij}. These two extreme cases are referred to as the Voigt and Reuss assumptions and would not be realized in a polycrystal with randomly oriented grains. However, they represent upper and lower bounds,

$$K_R < K_{eff} < K_V \tag{8}$$

$$G_R < G_{eff} < G_V \tag{9}$$

K_R, K_V, G_R and G_V all have simple forms when expressed in c_{ij}. We exemplify with K_V, and give the other expressions in Appendix 2;

$$K_V = \frac{1}{9}[\, 2(c_{11} + c_{12}) + c_{33} + 4c_{13}\,] \tag{10}$$

Hill [17] suggested that one estimates the overall elastic moduli as the average of the Voigt and Reuss results. The corresponding values, usually referred to as the Voigt-Reuss-Hill (VRH) estimates, thus are

$$K_{VRH} = (K_V + K_R)/2 \tag{11}$$

$$G_{VRH} = (G_V + G_R)/2 \tag{12}$$

In a single crystal of cubic lattice symmetry, such as fcc or bcc iron, there are only three independent elastic coefficients; c_{11}, c_{12} and c_{44}. Then $c_{13} = c_{21}$ and $c_{33} = c_{11}$, which yields

$$K_V = K_R = (c_{11} + 2c_{12})/3 \tag{13}$$

Hence the bulk modulus K of a polycrystal is in this case uniquely determined by the single-crystal elastic coefficients, irrespective of any texture. However, G is not uniquely determined even in a cubic lattice symmetry, since $G_V > G_R$, cf. relations in Appendix 2.

There are Hashin-Shtrikman type upper (subscript +) and lower (-) bounds to the elastic moduli of statistically isotropic polycrystals of various lattice symmetries [18]. They are mathematically complex, except for the bulk modulus in the case of cubic lattices that is given exactly, by Eq. 13.

It is obvious that the averaging methods just described do not give a useful estimate of the elastic properties of polycrystalline graphite. (Dryden and Perdue [10] arrive at the same conclusion.) However, the simplest theories for the elastic properties of cast iron, considered as a composite of ferrite and graphite, require that graphite is assigned isotropic elastic moduli. Speich et al. [3] suggest $K_{graphite} = 6.7$ GPa and $G_{graphite} = 3.3$ GPa. These values are also used by Anand [4] and Dryden and Perdue [10].

Table 3. Elastic constants for graphite. Entries are in GPa, from Ref. [18]. The Voigt-Reuss-Hill (VRH) entry is defined as $(K_V + K_R)/2 \pm (K_V - K_R)/2$ and the Hashin-Shtrikman (HS) bound entry as $(K_+ + K_-)/2 \pm (K_+ - K_-)/2$, with analogous definitions for the shear modulus.

c_{11}	c_{44}	c_{12}	c_{13}	c_{33}
1060	0.18[†]	180	15	36.5

K_{VRH}	K_{HS}	G_{VRH}	G_{HS}
161.1±125.3	120.2±84.0	109.2±108.7	73.5±72.7

[†] Uncertain value. Note that here $G_R \approx (5/2)c_{44}$.

DILUTE SUSPENSION OF INCLUSIONS

Consider a composite in which the inclusions in a matrix take up such a small volume fraction, and are so arranged, that the overall properties can be obtained from a superposition of results for a single inclusion. The influence of an ellipsoidal inclusion on the conductivity and the elastic properties is well known [13,19,20]. There are three interesting special cases of an ellipsoidal form; the sphere, the very oblate ellipsoid that well approximates a circular disc-like inclusion and the very prolate ellipsoid that well approximates a rod or fiber-like inclusion.

Conduction properties

Let index m denote the matrix and index i inclusions, with both $\lambda_m > \lambda_i$ and $\lambda_i > \lambda_m$ allowed. Then, for $f_i \ll 1$ and spherical inclusions,

$$\lambda_{eff} \approx \lambda_m \left[1 - f_i \frac{3(\lambda_m - \lambda_i)}{2\lambda_m + \lambda_i} \right] \tag{14}$$

For thin, and randomly oriented, discs

$$\lambda_{eff} \approx \lambda_m - f_i \frac{(\lambda_m - \lambda_i)(\lambda_m + 2\lambda_i)}{3\lambda_i} \tag{15}$$

and for randomly oriented rods

$$\lambda_{\text{eff}} = \lambda_m - f_i \frac{(\lambda_m - \lambda_i)(5\lambda_m + \lambda_i)}{3(\lambda_m + \lambda_i)} \qquad (16)$$

If the discs conduct much better than the matrix, the conductivity of the composite is mainly affected by conduction in the plane of the disc. (If the disc conducts much less than the matrix, it is the resistance in the direction normal to the disc that is significant.) In grey cast iron, the graphite planes are parallel to the graphite flakes and have a high thermal conductivity in that plane. Hence we replace λ_i by λ_a, take $\lambda_i \gg \lambda_m$ in Eq. 15, and get

$$\lambda_{\text{eff}} \approx \lambda_m + (2/3) f_{\text{graphite}} \lambda_a \qquad (17)$$

Dilute rod-like inclusions have a strong effect on λ_{eff} only when they conduct much better than the matrix. Then it is conduction along the rods that is significant. Hence, as an example suppose that $\lambda_m, \lambda_c \ll \lambda_a \approx \lambda_i$. Then, from Eq. 16,

$$\lambda_{\text{eff}} \approx \lambda_m + (1/3) f_i \lambda_a \qquad (18)$$

This result represents an (approximate) upper bound to the effect of rod-like inclusions of an anisotropic material i, Fig. 3.

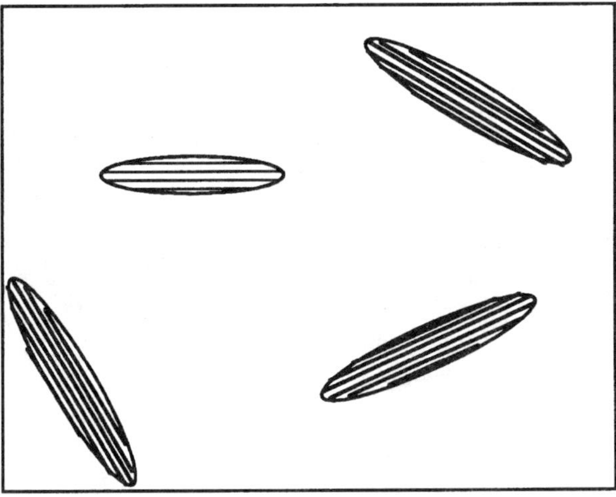

Figure 3. Schematic morphology of a dilute composite with inclusions shaped like prolate ellipsoids and having a definite crystallographic orientation relative to the ellipsoid axes. A significant effect on the effective conductivity results only if the conductivity along the ellipsoid axis is very high.

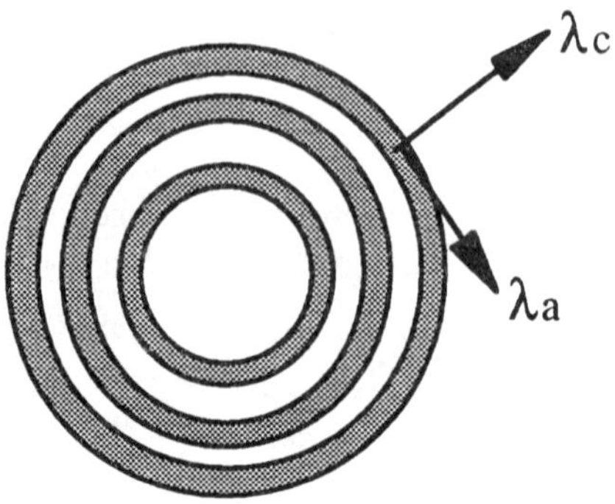

Figure 4. A schematic structure of the phase geometry in nodular graphite, with concentric graphite layers.

In nodular cast iron, with spherical graphite inclusions forming concentric shells and the crystallographic c-axis pointing radially (Fig. 4), we can apply an effective medium-type approach [1]. The result depends on a combination of λ_a, λ_c and λ_m. In the dilute limit, and with conductivity ratios λ_a/λ_c and $\lambda_c/\lambda_{\text{Fe alloy}}$ as in Table 2 we get

$$\lambda_{\text{eff}} \approx \lambda_{\text{Fe alloy}} + 0.07 f_{\text{graphite}} \lambda_a \tag{19}$$

<u>Elastic properties</u>

The dilute-limit results for spherical inclusions and isotropic elastic properties in each separate phase are

$$K_{\text{eff}} \approx K_m - f_i (K_m - K_i) \frac{3K_m + 4G_m}{3K_i + 4G_m} \tag{20}$$

$$G_{\text{eff}} \approx G_m - f_i (G_m - G_i) \frac{5(3K_m + 4G_m)}{9K_m + 8G_m + 6(K_m + 2G_m)(G_i/G_m)} \tag{21}$$

For randomly oriented discs one has

$$K_{\text{eff}} \approx K_m - f_i (K_m - K_i) \frac{3K_m + 4G_i}{3K_i + 4G_i} \tag{22}$$

$$G_{\text{eff}} \approx G_m - f_i (G_m - G_i) \frac{G_m + F}{G_i + F} \tag{23}$$

where

$$F = (G_i/6)\frac{9K_i+8G_i}{K_i+2G_i} \tag{24}$$

For randomly oriented rods, the dilute-limit result is

$$K_{\text{eff}} \approx K_m - f_i(K_m - K_i)\frac{K_m + G_m + G_i/3}{K_i + G_m + G_i/3} \tag{25}$$

The corresponding expression for G_{eff} is algebraically complicated.

Although graphite does not have isotropic elastic moduli, it can still be of some interest to express the dilute-limit results using the matrix values $K_m = 162$ GPa and $G_m = 81$ GPa, and the values $K_{\text{graphite}} = 6.7$ GPa and $G_{\text{graphite}} = 3.3$ GPa that have been used by, *inter alia*, Speich, Schwoeble and Kapadia [3] and Anand [4]. Then one obtains for spheres

$$K_{\text{eff}} \approx K_m(1 - 2.3f_i) \tag{26}$$

$$G_{\text{eff}} \approx G_m(1 - 1.8f_i) \tag{27}$$

It is interesting to compare with the inclusions being voids, with $K_i = G_i = 0$, and $K_m/G_m = 2$, as assumed above for the iron matrix. Then $K_{\text{eff}} = K_m(1 - 2.5f_i)$ and $G_{\text{eff}} = G_m(1 - 1.92f_i)$. For discs, the relations analogous to Eqs. 26 and 27 are

$$K_{\text{eff}} \approx K_m(1 - 14f_i) \tag{28}$$

$$G_{\text{eff}} \approx G_m(1 - 9f_i) \tag{29}$$

For rods, finally,

$$K_{\text{eff}} \approx K_m(1 - 2.6f_i) \tag{30}$$

We can summarize these expressions as

$$K_{\text{eff}} \approx K_{\text{ferrite}}(1 - af_{\text{graphite}}) \tag{31}$$

$$G_{\text{eff}} \approx G_{\text{ferrite}}(1 - bf_{\text{graphite}}) \tag{32}$$

From Eq. 6 we get, with K_m and G_m given the values above for alloyed ferrite,

$$E_{\text{eff}} \approx E_{\text{ferrite}}(1 - 0,143af_i - 0,857bf_i) \tag{33}$$

The elastic properties of the matrix (ferrite) do not vary much with small amounts of alloying elements. The elastic properties of pearlite are close to those of ferrite. Hence we can concentrate on the effect of the graphite inclusions. Dryden and Purdy [10], and also Ledbetter and Datta [12], considered not only the extreme limits of spherical, disc- or bar-shaped inclusions but studied ellipsoids with varying aspect ratios. Dryden and Purdy [10] suggest that in Eqs. 31 and 32 $a \approx 0.43$ and $b \approx 0.83$ for nodular cast iron, i.e., considerably lower values than in Eqs. 26 and 27. This is discussed below.

THERMAL EXPANSION

There is an *exact* relation for the effective thermal expansion coefficient α_{eff} of statistically isotropic composites, expressed in the expansion coefficients of the constituent phases, their bulk moduli and the effective bulk modulus of the entire composite [21-23];

$$\alpha_{\text{eff}} = \alpha_{\text{m}} - f_{\text{i}}\,(\alpha_{\text{m}} - \alpha_{\text{i}}) + \frac{\alpha_{\text{m}} - \alpha_{\text{i}}}{1/K_{\text{m}} - 1/K_{\text{i}}}\, [1/K_{\text{eff}} - f_{\text{m}}/K_{\text{m}} - f_{\text{i}}/K_{\text{i}}] \tag{34}$$

Since K_{eff} can be estimated with the methods described above, we get simple estimates also for α_{eff}. Often the upper and lower bounds to α, obtained by a Hashin-Shtrikman approach to K_{eff} lie *very* close, e.g., in cast Al-Si alloys [13]. However, these relations require that the constituent phases have isotropic properties. Then one has for spherical inclusions in the dilute limit, using our previous estimate (20) of K_{eff},

$$\alpha_{\text{eff}} \approx \alpha_{\text{m}} - f_{\text{i}}\,(\alpha_{\text{m}} - \alpha_{\text{i}})\, \frac{K_{\text{i}}(3K_{\text{m}} + 4G_{\text{m}})}{K_{\text{m}}(3K_{\text{i}} + 4G_{\text{m}})} \tag{35}$$

Very little seems to have been done on the theory of the effective thermal expansion in systems with highly anisotropic inclusions, such as cast iron, cf. refs. [9,24]. We also remark that Eq. 35 is not very good for nodular cast iron, since it was noted above that K_{eff} is not well described by Eq. 20.

DISCUSSION

The preceding treatment in this review has mainly focused on a dilute suspension of inclusions. The graphite volume fraction in cast irons typically is 0.10. At such concentrations there are deviations from the dilute-limit results. That effect will be more pronounced if the inclusions have a shape and arrangement so that they form more or less connected regions through the matrix, in spite of their moderate volume fraction. It is obvious that discs and rods more easily can form such a structure. As mentioned in the introduction, one can treat any volume fraction by theories that yield bounds, or by some modelling that yields an estimate which of course has to fall within the bounds. The mathematical forms of the bounds or estimates that have been developed are usually quite simple. For the benefit of the reader, the most important relations in simple and quick applications to cast irons will now be pointed out.

The termal conductivity λ_{eff} of gray cast iron may be roughly estimated from eq. (17), where λ_m is the conductivity of the metallic matrix and λ_a is the conductivity of graphite along the crystallographic a-axes where conductvity is very high. The corresponding relation in nodular cast iron is given in eq. (19). The thermal conductivity was considered more accurately by Helsing and Helte [1], who relied on mean-field approximation to derive an expression for the effective conductivity λ^* in a multi-phase composite where phases may have an anisotropic conductivity. That result was used by Helsing and Grimvall [2] in a consideration of various cast irons and steels. The theoretical predictions typically differ from the experimental values by ± 5 %.

The bulk modulus K, the shear modulus G, and the Young's modulus E may be easily, but somewhat crudely, estimated by eqs. (31), (32) and (33), respectively. The parameters a and b in these equations are not accurately known, but may typically be given as $a \approx 14$, $b \approx 9$ for gray cast iron and $0.4 < a = 2.3$ and $0.8 < b = 1.8$ for nodular cast iron. However, it must be stressed that an attempt at a more accurate modelling of the effective properties of cast irons, considered as composites, should be based on the experience from previous work where models and experimental results are compared. Here follows a brief review of such work.

The elastic properties have been considered by several authors. The Young's modulus E and shear modulus G of nodular and grey cast irons were modeled by Speich, Schwoeble and Kapadia [3], and Anand [4], using spheres or discs as inclusions, and with assumed isotropic elastic properties of the graphite. They can well account for E and G at volume fractions of about 0.10. The results for spheres and discs come close to the upper and lower Hashin-Shtrikman bounds, respectively. Dryden, Deakin and Purdy [9] considered nodular graphite with a more realistic treatment of the graphite properties, assuming that nodules are layered as in Fig. 4. When a theory for an estimate is based on ellipsoidal inclusions, one can vary the aspect ratios of the ellipsoid and cover inclusion shapes intermediate between those of spheres, thin discs and long rods. Such estimates of elastic moduli of cast iron have been performed by Löhe, Vöhringer and Macherauch [8], Dryden and Purdy [10], Bussière, Piché and Leclerc [11] and Ledbetter and Datta [12]. One may also note experimental work in Refs. [5-7, 25].

A major uncertainty in the treatment of the elastic properties is caused by the possibility of decohesion between the matrix and graphite inclusions. All the discussions of elastic properties and the thermal expansion in this review assume that there is solid bonding between the matrix and the inclusions. This may be a less good approximation in the case of, e.g., graphite flakes. Concerning nodular graphite, it was noted from a comparison with Ref. [10] that Eq. 20, with some average K_{eff} and G_{eff} of graphite, was a poor approximation. K_{eff} and G_{eff} do not decrease as rapidly with the graphite volume fraction as suggested by Eqs. 26 and 27. This is easy to understand from the fact the graphite forms more or less concentric shells, with very high moduli in the planes of the shells, i.e. perpendicular to the crystallographic c-axis. The properties of compacted graphite are still difficult to analyse, because of a complicated growth pattern with poorly known crystallographic orientations of the graphite. The pearlite is very well described by the models of this review.

We conclude that considering cast iron as a composite provides as framework that qualitatively, and often also quantitatively, accounts for transport and elastic properties of the material.

APPENDIX 1

A single crystal of hexagonal lattice symmetry, under a hydrostatic pressure p, is subject to strains [13]

$$\varepsilon_a = \varepsilon_b = -p \left[\frac{c_{33} - c_{13}}{(c_{11} + c_{12})c_{33} - 2c_{13}^2} \right] \tag{A1}$$

$$\varepsilon_c = -p \left[\frac{c_{11} + c_{12} - 2c_{13}}{(c_{11} + c_{12})c_{33} - 2c_{13}^2} \right] \tag{A2}$$

i.e., a strain ratio $\varepsilon_c/\varepsilon_a = (c_{11} + c_{12} - 2c_{13})/(c_{33} - c_{13})$. For graphite, this ratio becomes 56. The bulk modulus of *a single-crystal* is [13]

$$K = \frac{(c_{11} + c_{12})c_{33} - 2c_{13}^2}{c_{11} + c_{12} + 2c_{33} - 4c_{13}} \tag{A3}$$

With c_{ij} for graphite as in Table 3, we get $K = 36$ GPa. (Do not confuse this with the bulk modulus of a one-phase aggregate of randomly oriented single crystals, see Eqs. 10 and A4.)

APPENDIX 2

The Voigt and Reuss expressions for the bulk and shear modulus of a polycrystal are

$$K_R = C^2/M \tag{A4}$$

$$G_V = \frac{1}{30} \left[12c_{66} + 12c_{44} + M \right] \tag{A5}$$

$$G_R = \frac{5}{2} \left[\frac{c_{44} c_{66} C^2}{(c_{44} + c_{66})C^2 + 3K_V c_{44} c_{66}} \right] \tag{A6}$$

Here we have introduced three auxiliary quantities expressed in the elastic coefficients;

$$c_{66} = (c_{11} - c_{12})/2 \qquad\qquad (A7)$$

$$M = c_{11} + c_{12} + 2c_{33} - 4c_{13} \qquad\qquad (A8)$$

$$C^2 = (c_{11} + c_{12})c_{33} - 2(c_{13})^2 \qquad\qquad (A9)$$

ACKNOWLEDGMENT

This work is supported by the Swedish research councils NFR and NUTEK.

REFERENCES

[1] J. Helsing and A. Helte, J. Appl. Phys. **69**, 3583 (1991).

[2] J. Helsing and G. Grimvall, J. Appl. Phys. **70**, 1198 (1991).

[3] G. R. Speich, A. J. Schwoeble and B. M. Kapadia, J. Appl. Mech. **47**, 821 (1980).

[4] L. Anand, Scripta Metall. **16**, 173 (1982).

[5] A. G. Fuller, P. J. Emerson and G. F. Sergeant, Trans. American Foundrymen's Soc. **88**, 21 (1980).

[6] G. F. Sergeant and A. G. Fuller, Trans. American Foundrymen's Soc. **88**, 545 (1980).

[7] E. P. Papakakis, L. Bartosiewics, J. D. Altstetter and G. B. Chapman, II, Trans. American Foundrymen's Soc. **91**, 721 (1983).

[8] D. Löhe, O. Vöhringer, E. Macherauch, Z. Metallkde. **74**, 265 (1983).

[9] J. R. Dryden, A. S. Deakin and G. S. Purdy, Acta Metall. **35**, 681 (1987).

[10] J. R. Dryden and G. R. Purdy, Acta Metall. **37**, 1999 (1989).

[11] J. F. Bussière, L. Piché and G. Leclerc, in *Nondestructive Characterization of Materials,* Proceedings of the 3rd International Symposium, Saarbrücken, 1988. Eds. P. Höller, V. Hauk, G. Dobman, C. O. Ruud and R. E. Green. p. 353.

[12] H. Ledbetter and S. Datta, Z. Metallkde. **83**, 3 (1992).

[13] G. Grimvall, *Thermophysical properties of materials,* North-Holland, Amsterdam (1986).

[14] Z. Hashin and S. Shtrikman, J. Appl. Phys. **33**, 3125 (1962).

[15] S. Torquato, Appl. Mech. Rev. **44**, 37 (1991).

[16] J. Helsing, J. Appl. Phys. **74**, 5061 (1993).

[17] R. Hill, Proc. Phys. Soc. A **65**, 349 (1952).

[18] J. P. Watt and L. Peselnick, J. Appl. Phys. **51**, 1525 (1980).

[19] E. C. Stoner, Phil. Mag. **36**, 803 (1945).

[20] J. G. Berryman, J. Acoust. Soc. Am. **68**, 1820 (1980).

[21] B. M. Levin, Mech. Tverd. Tela **1**, 88 (1967).

[22] R. A. Schapery, J. Composite Materials **2**, 380 (1968).

[23] B. W. Rosen and Z. Hashin, Int. J. Eng. Sci. **8**, 157 (1970).

[24] H. Fredriksson, P.-A. Sunnerkrantz and P. Ljubinkovic´, Mater. Sci. Tech. **4**, 222 (1988).

[25] T. Okamoto, A. Kagawa, K. Kamei and H. Matsumoto, Imono, **55**, no 2 (1983).

Advanced Materials Research Vols. 4-5 (1997) pp. 47-60
© 1997 Scitec Publications, Switzerland

Fracture Characteristics and Fracture Toughness of Spheroidal Graphite Cast Iron

T. Kobayashi

Department of Production Systems Engineering, Toyohashi University of Technology,
1-1, Tempaku, Toyohashi, 441 Japan

Keywords: Spheroidal Graphite Cast Iron, Elastic-Plastic Fracture Mechanics, Fracture Toughness, Transition Temperature, Austempered Ductile Iron

ABSTRACT

Spheroidal graphite cast iron is used for various structural components which need high strength and toughness. Evaluation of fracture toughness and understanding of fracture mechanism are necessary for the design and use. Since spheroidal graphite cast iron behaves generally in a non-linear manner, elastic-plastic fracture mechanics should be applied to this iron. Especially, reasonable evaluation of dynamic fracture toughness is needed strongly. Ductile fracture changes to brittle fracture with lowering testing temperature in this iron and transition temperature is affected by the loading rate, notch shape and specimen geometry. The average value of $J(\Delta C)$ and $J(R)$, that is $J(mid)$, is proposed to define the fracture toughness of this iron. $J(mid)$ is considered to be a reasonable measure for the fracture toughness of this iron in the upper shelf temperature range, irrespective of loading condition and the pearlite content in the matrix. The austempered ductile cast iron with greater toughness and good fatigue properties has been successfully developed utilizing microsegregation effect of alloying elements and applying special austempering treatment.

1. INTRODUCTION

Spheroidal graphite (SG) cast irons have good strength and ductility compared with the other cast irons. Consequently, they are currently used widely as structural components in place of gray cast iron or cast steel. Casting is a single manufacturing process that produces a complicated shape while requiring less energy than many alternative production routes. Furthermore, the matrix of SG iron can be varied from that with high ductility such as ferrite to that with high strength such as pearlite or ausferrite by control of casting process, heat treatment and chemical compositions. Thus, the spheroidal graphite cast irons satisfy a wide range of properties required for engineering materials, and should become increasingly essential even in the modern manufacturing industry. The evaluation of fracture toughness and understanding of fracture mechanics which are different from the situation in steel are, therefore, necessary for the design and use of this iron. For instance, ferritic spheroidal graphite cast iron has been explored as a cask (container for spent nuclear fuel) material because of its low cost and good formability. The cask which is a huge casting with 400 mm thick and 100 Mg weight envelops the nuclear material. Therefore, the valid fracture toughness of cask must be evaluated not only under the static loading condition, but also under the dynamic loading condition to ensure its safety against an accident during the transport[1]. In this paper, fracture mechanics and fracture characteristics of spheroidal graphite cast iron including ADI will be also reviewed and summarised.

2. Fracture Mechanism and Characteristics

Fundamental fracture mechanism of spheroidal graphite cast iron is introduced as follows. Figures 1 and 2 show three kinds of typical fracture surface and the schematic illustration of ductile and cleavage fracture processes in the ferritic SG iron, respectively. In the ductile fracture obtained at room temperature, since the crack propagates through the growth and coalescence of voids nucleated at the interface between graphite nodule and matrix, many graphite nodules are observed on the fracture surface. If inclusions exist in the matrix, voids are connected through micro-voids. While, the cleavage fracture surface obtained at lower temperature and fatigue fracture surface are flat compared with the ductile one, and less graphite nodules are observed. The cleavage fracture tends to initiate at the inclusion around the eutectic cell boundary rather than at the graphite-matrix interface.

Ductile fracture changes to brittle fracture with lowering testing temperature in this iron and transition temperature is affected by the loading rate, notch shape and specimen geometry.

Figure 3 shows energy transition curves of unnotched, V and U notched Charpy specimens in comparison with the one of large-scale dynamic tear (DT) test ; energy scale in the case of DT test is shown on the right of the figure. In the U and V notched specimens, energy absorption is lower in V notched specimen because of its large stress concentration effect. However, the transition behavior of both the specimens is identical and the transition temperature is 10°C. While, the transition temperature locates at -40°C in the unnotched specimen. The author has already reported on the effect of notch shapes in SG iron[2,3].

On the other hand, the transition curve of DT specimen rises largely to higher temperature side and it is 40°C. This may be due to the high plastic constraining effect in the large size DT specimen

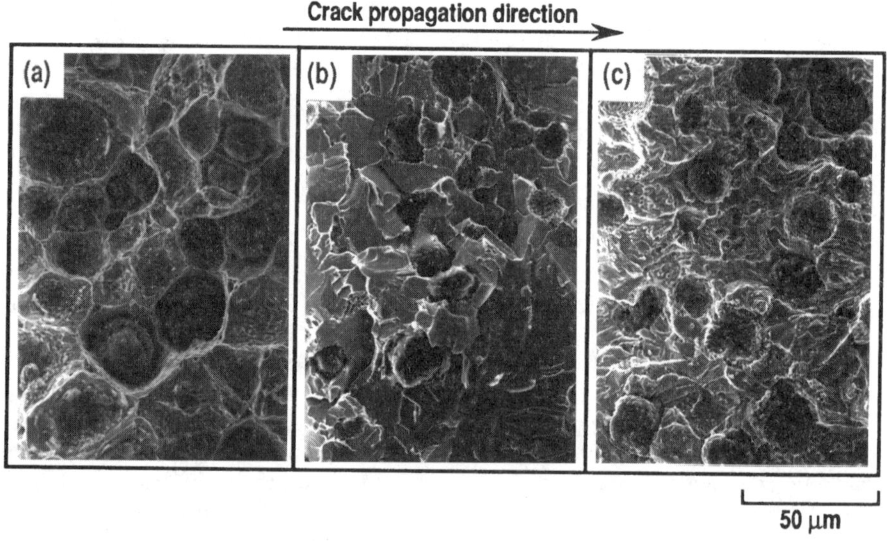

Fig. 1   Typical fracture surfaces of ferritic spheroidal graphite cast iron.
(a) Ductile fracture surface
(b) Cleavage fracture surface
(c) Fatigue fracture surface

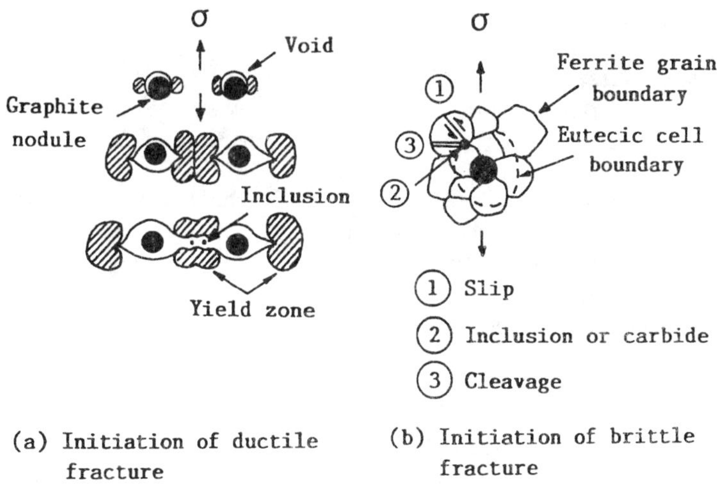

Fig. 2　Schematic illustration of fracture process in ferritic spheroidal graphite cast iron.

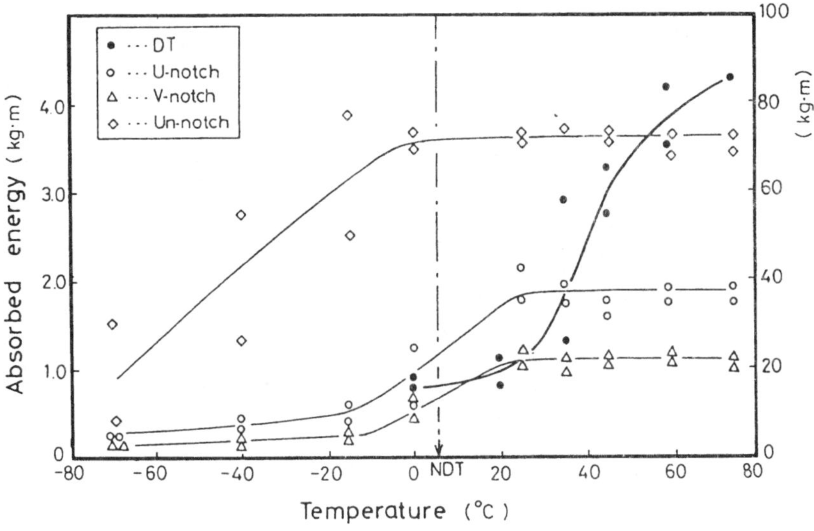

Fig. 3　Transition curves of the absorbed energy in various specimens (ferritic SG iron).

(25×76×305mm). It is a very important point that the transition temperature changes with the specimen geometry, and this fact must be taken into consideration when a design of a large structure is made.

According to the classical great work of Pellini[4], it will be necessary to determine nil ductility transition temperature (NDT) in the DT test, where both the yield strength and fracture will coincide at first in the lower temperature test. The nil ductility transition temperature of a steel has been well correlated with the absorbed energy of Charpy V notch test by many investigators; however, on SG iron, it has been only reported by Pellini *et al*[4] that the temperature, where 5 ft-lb (0.7 kg-m) energy absorption is obtained, may correspond to NDT. It was decided as 5℃ in my study from a fracture appearance transition curve, taking also the above suggestion into consideration.

Figure 4 shows the fracture analysis diagram obtained in this way[5,6]. Crack arresting temperature (CAT) curve in the figure shows the critical temperature where brittle unstable crack propagation is arrested at each working stress level, even if a large flaw is existed or initiated in a structure. If structure is operated under right side of this CAT curve, it will be absolutely safe against catastrophic brittle failure. In this context, we must know the fact that K_{IC} value gives only a safety criterion for brittle fracture initiation.

In the next, an empirical method to know about such measures on the design condition from the result of the standard Charpy V notch test is shown in Figure 5[5]. If the design stress is assumed as $\sigma_y/2$, the CAT of the material is 20℃, this temperature corresponds to the lower limit of upper shelf temperature range in the Charpy V transition curve shown in Figure 3. Such correlation and temperature criterion, the absolute fracture safe range is above 20℃, are similar to the result of mild steel[7].

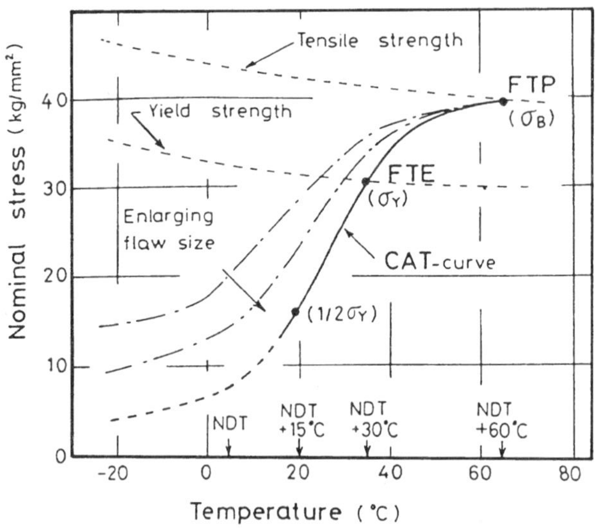

Fig. 4   Fracture analysis diagram of spheroidal graphite cast iron.

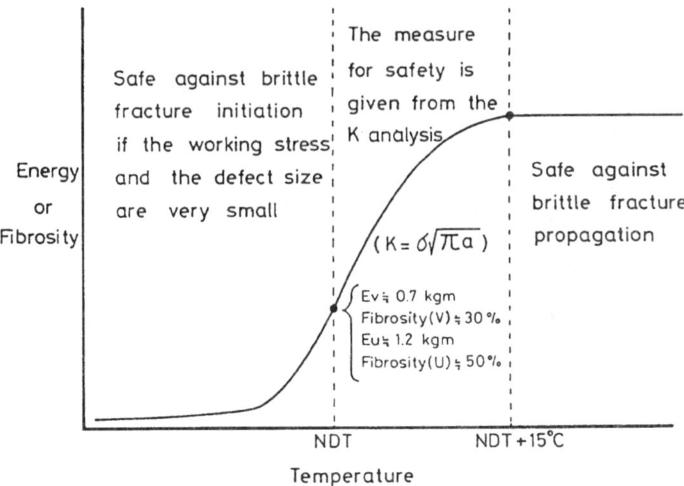

Fig. 5   Schematic illustration of Charpy V transition curve from the view-point of safety against brittle fracture.

3. Evaluation of Static and Dynamic Fracture Toughness

Fracture toughness is a resistance for crack initiation. However, crack initiation is a problem of the definition, and it has not been clarified yet in this iron. Moreover, elastic-plastic fracture mechanics must be applied to this iron expecting for the case at very low temperature.

The loading rate influences not only the mechanical properties and transition behavior but also the crack growth behavior. A ductile crack initiates at the middle of specimen thickness, and then grows under both static and dynamic loading conditions. Figure 6 shows the relationship between crack extension, Δa, and the extension ratio of ductile fracture surface width, η, in the static and dynamic fracture toughness tests for heavy wall ferritic SG iron as cask material[8]. η (%) is defined as a ratio of ductile fracture surface at the precrack-tip to the specimen thickness. The figure indicates relationships between crack growth in the direction of specimen width (Δa ; 9 points averaged value) and that in the direction of specimen thickness (η). The crack tends to grow first laterally under the static loading condition. While under the dynamic loading condition, since fracture process is sensitive to uniformity of strength in the material, the crack growth occurs in both direction.

J-Δa curves in static and dynamic fracture toughness tests for heavy wall SG iron are shown in Figure 7[8]. The acoustic emission (AE) method , electrical potential (EP) method and compliance changing rate (ΔC) method[9,10], where crack initiation is defined as a knick point in changing rate of assumed elastic compliance at any load-deflection curve point against initial elastic one, were used for detecting the crack initiation. J_{in} and J_d in the figure indicate the J values obtained for both tests, respectively. All methods were effective for detecting the strict crack initiation at the middle of specimen thickness under both static and dynamic loading conditions. $J_{in}(\Delta C)$ is nearly equal to $J_{in}(AE)$, $J_{in}(EP)$ or $J_{in}(\eta=20)$ under the static loading condition, while $J_d(\eta=20)$ is nearly equal to $J_d(P_{max})$ under the dynamic loading condition. Under static loading condition, J(R) is nearly identical

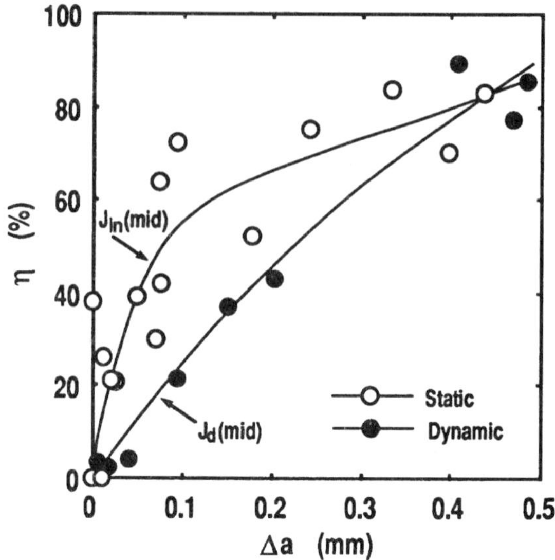

Fig. 6   Relationships between extension ratio of ductile fracture surface width, η, and crack extension, Δa, in static three-point bend test and instrumented impact test of heavy wall ferritic spheroidal graphite cast iron[8].

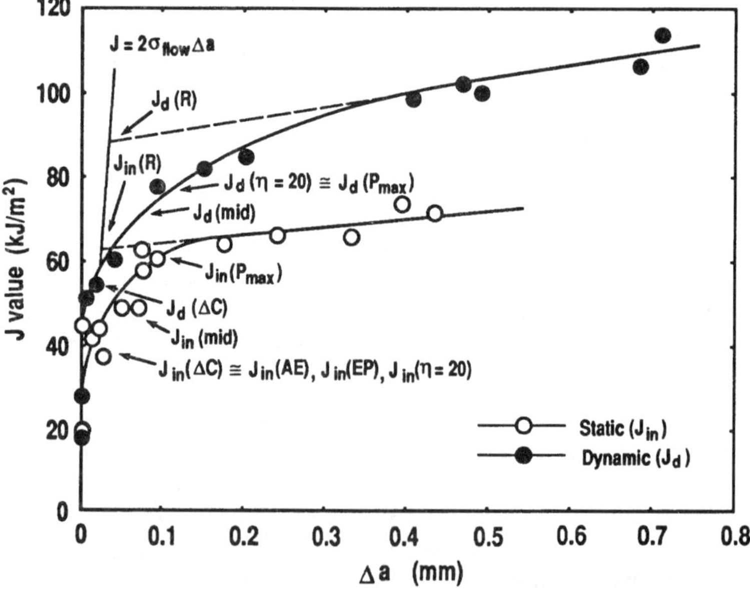

Fig. 7   J-Δa curves in static three-point bend test and instrumented impact test of heavy wall ferritic spheroidal graphite cast iron[8].

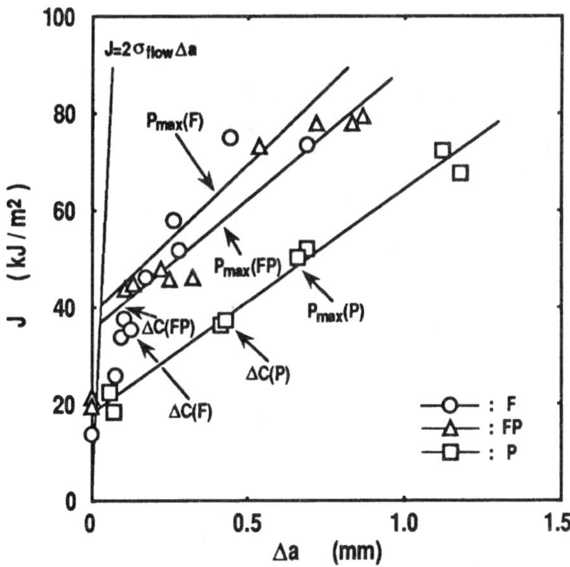

Fig. 8   J-Δa curves under the dynamic loading condition
in materials F, FP and P at 423K[8].

with the J value obtained at maximum load point, $J(P_{max})$, while under the dynamic loading condition $J(R)$ is larger than $J(P_{max})$. Moreover, both the 2% crack growth point from the initial crack length defined by K_{IC} test[11] and intersection of the 0.15mm exclusion line and the smoothed crack extension resistance curve through the data proposed by Salzbrenner *et al*.[12] indicate larger values than $J(P_{max})$ under both static and dynamic loading conditions. For data behaving in this manner, an alternative definition of J_{IC} is required. In this study, $J(\Delta C)$ obtained from the compliance changing rate method corresponds to $0.02\sim0.03$mm of Δa under the both loading conditions. Although $J(\Delta C)$ might be a lower-bound of J value concerning the crack initiation, corresponding values of Δa are too small as required for the crack initiation point. Therefore, $J(\Delta C)$ is believed to underestimate the fracture toughness, i.e., $J(\Delta C)$ involves the some amount of resistive ability of materials against the applied load. On the other hand, since the maximum loading point, P_{max}, is considered to be a critical point under the load controlling condition, $J(R)$ overestimates the fracture toughness, although $J(R)$ involves some amount of crack extension resistance of materials. Therefore, the average value of $J(\Delta C)$ and $J(R)$, that is J(mid), is proposed[8], since one reasonable J value must be selected in the practical design. J(mid) is considered as the reasonable fracture toughness of this iron.

The fracture toughness and crack growth are influenced by the pearlite content in the matrix. Figure 8 shows J-Δa curves in the dynamic fracture toughness test at 423K for the ferritic material (F), ferritic-pearlitic material (FP) and pearlitic material (P)[8]. For the ferritic and ferritic-pearlitic materials, $J(\Delta C)$ is nearly equal to $J(R)$ and these points locate near Δa=0.1mm point. For the pearlitic material P, however, $J(\Delta C)$ locates at Δa=0.4mm point and overestimates the fracture toughness, while $J(R)$ corresponds to about 65% loading point of P_{max} and is considered to underestimate it from the view point of applied load.

The relationship between J(ΔC) and J(R) varies with pearlite content. J(ΔC) is smaller than J(R) for fully ferritic material, while J(ΔC) is larger than J(R) when a pearlite content is larger than approximately 20%. The difference between J(ΔC) and J(R) is enhanced with increasing pearlite content in the matrix. Thus, it is not reasonable that only one criterion is used to evaluate the fracture toughness of SG irons with various matrix microstructures. J(mid) proposed above is, however, applicable irrespective to the matrix structure. J(mid) is considered to be effective especially for the pearlitic material P which has a great difference between overestimating J(ΔC) and underestimating J(R). However, valid condition on the specimen size requirement must be studied more hereafter[1,3].

4. ADI toughened by special heat treatment

It is wellknown that austempered ductile cast iron, so-called ADI , shows good toughness and high strength[13]. Nowadays, the usage of ADI for structural applications is increasing. Therefore, it is necessary to strengthen and toughen ADI furthermore. The author has already succeeded in developing ADI having greater toughness, that is TADI (Toughened ADI), utilizing microsegregation of alloying elements and applying special heat treatment, QB'[13,14].

The detail of the idea for toughening using microsegregation of alloying elements is schematically shown in Figure 9[14]. On the right side of the figure, silicon concentrates near a graphite nodule while manganese concentrates into an eutectic cell boundary. As a result, the austenitising temperature (A_s, austenitising starting temperature ; A_f, austenitising finishing temperature) near the

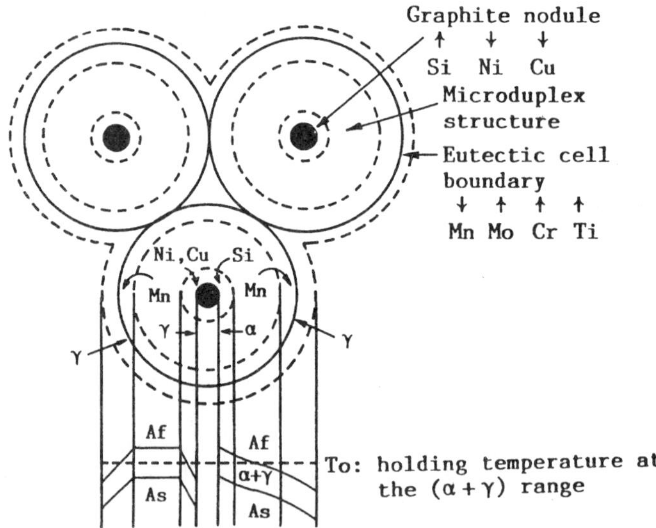

Fig. 9   Schematic illustration of the microsegregation of alloying elements and phase distributions induced by austempering from (α+γ) temperature range[14].
↑ indicates elements which increase A₁ or A₃ point.
↓ indicates elements which decrease A₁ or A₃ point.

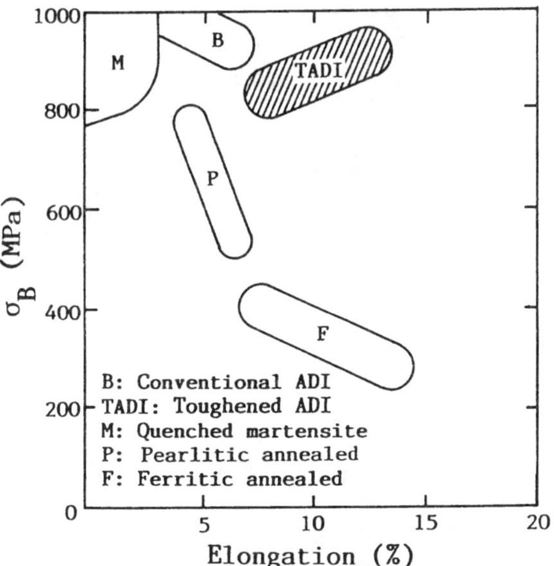

Fig. 10　Summarised comparison of tensile properties among various
spheroidal graphite irons at room temperature[14].

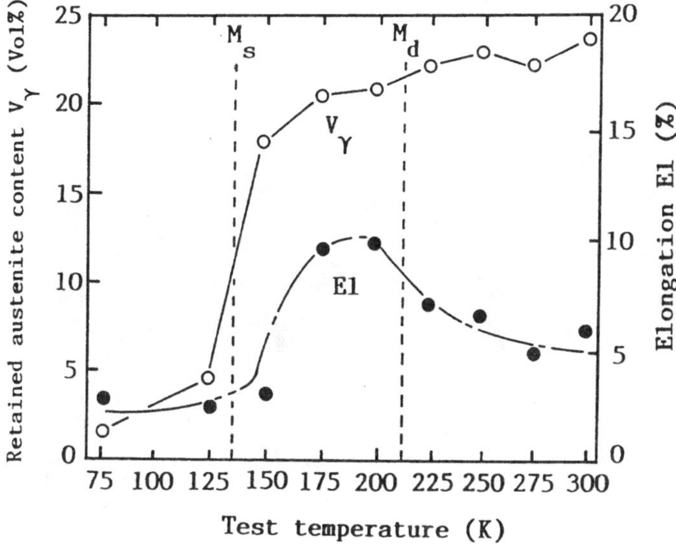

Fig. 11　Effect of temperature on retained austenite content
and total elongation in TADI[14].

graphite will increase, while that close to the eutectic cell boundary will decrease. However, like the left of the figure, by adding Ni, Cu etc., which have a trend to segregate around graphite nodules followed by lowering the austenitising temperature and then heating up into ($\alpha+\gamma$) temperature range, the fracture initiation sites such as a graphite-matrix interface and an eutectic cell boundary as shown in Figure 2 should be preferentially austenitised and then will become stable against external stress. Therefore, this method is expected to be effective for improving the toughness of ADI.

Figure 10 shows that tensile properties of ductile cast iron with various matrices[14]. TADI shows good strength which is equal to ordinary bainitic material and its ductility matches for ferritic material.

Figure 11 shows the effects of testing temperature on the total elongation and the amount of retained austenite on the fracture surface of the TADI in the static tensile test[14]. An abnormal elongation was observed between M_s and M_d temperatures. This phenomenon is considered to be a feature of TRIP. However, the change in retained austenite content is small at the temperature where the peak of elongation was observed. From this result, the phenomenon is different from the TRIP caused by the deformation induced martensite. The typical TEM micrographs of thin foils prepared near the fracture surface of the tensile tested specimen at the temperature where the peak of elongation appeared are shown Figure 12[14]. The deformation induced martensite was not observed in the retained austenite. However, the deformation twin is observed in the retained austenite.

Figure 13 shows the fatigue crack propagation behavior of TADI with that of the ordinary austempered ductile iron, SADI[15]. The threshold value of stress intensity range of TADI is higher than that of ordinary one. This result shows the excellent crack initiation resistance of this iron. Moreover, the rate of fatigue crack propagation in Paris region is slower compared with the ordinary one. Stress induced transformation from the retained austenite to martensite was observed in the microstructure around fatigue crack path. Stable retained austenite in TADI is considered to be effective for reducing the fatigue crack propagation rate.

By the way, in general, the heat treatment variables such as temperature and time should be optimized for getting the material with desired properties. In QB' treatment which is applied to TADI

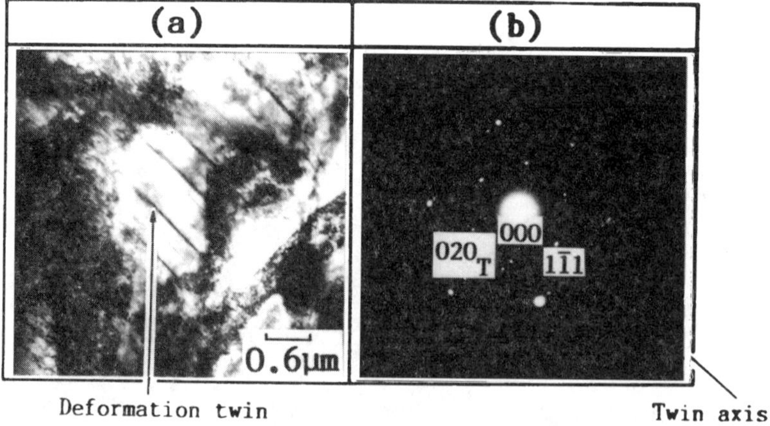

Fig. 12   TEM micrograph showing the deformation twin formed
in the retained austenite (tested at 198K) [14].

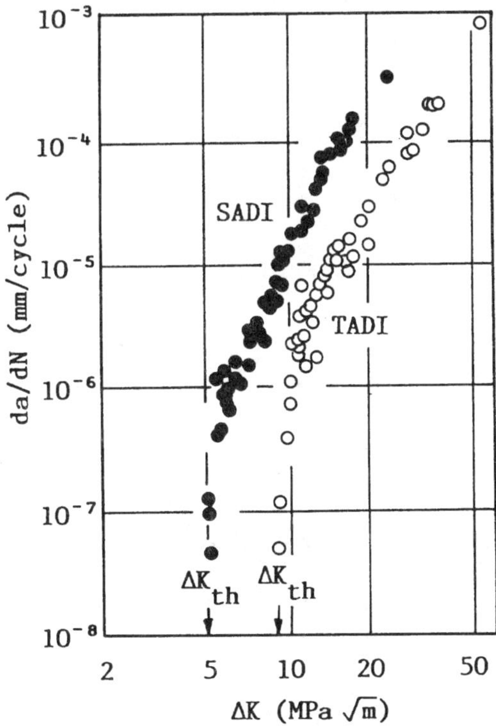

Fig. 13   da/dN-ΔK curve of TADI and SADI.

and Q means pre-quenching for refining the subsequent austenite phase (holding time at γ range in Q treatment must be short enough to maintain the microsegregation effect) and B' means reaustenitising at ($\alpha+\gamma$) range followed by austempering[14], it has been reported that the toughness can be varied with the holding time at ($\alpha+\gamma$) temperature range[16]. The above stated case was only 0.5hr. holding at ($\alpha+\gamma$) range. However, the trend and reason of the change in the toughness has not been reported. Recently, I have studied on such subjects. Load-displacement curves obtained by the instrumented Charpy impact tests at room temperature are shown in Figure 14[17]. Little change in the strength level with holding time was observed, while deflection to failure increased remarkably with holding time. Thus, the increase in toughness depends mainly on the increase in ductility. It was observed that the materials conducted with the QB' treatment included carbide particles and the amount of carbide decreased with holding time. In general, the carbide acts as initiation site.

It has been reported that carbides can be formed during the holding at the austempering temperature[15]. However, since the carbides were observed also in the material quenched from ($\alpha+\gamma$) temperature range, they had already existed before the material was quenched to the austempering temperature. TEM micrographs of the material with the holding time of 0.5hr. are shown in Figure 15. (a) and (b) were obtained from thin films and (d) was obtained from extraction replica. The carbides were identified as Fe_3C and their diameter was $0.1 \sim 0.5\mu m$. In the material before reaustenitising, as pre-quenched, fully martensitic microstructure without carbide was

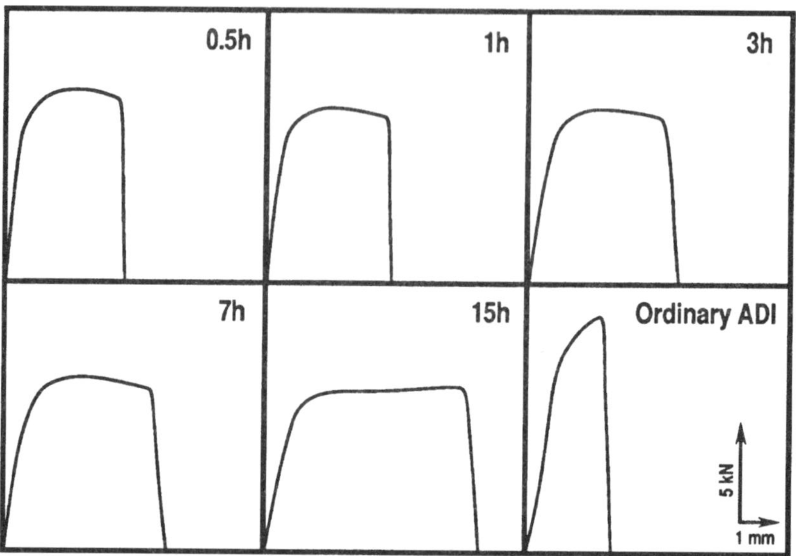

Fig. 14 Load - load-point displacement curve of the materials with various holding times at $(\alpha+\gamma)$ range and ordinary ADI tested at room temperature.

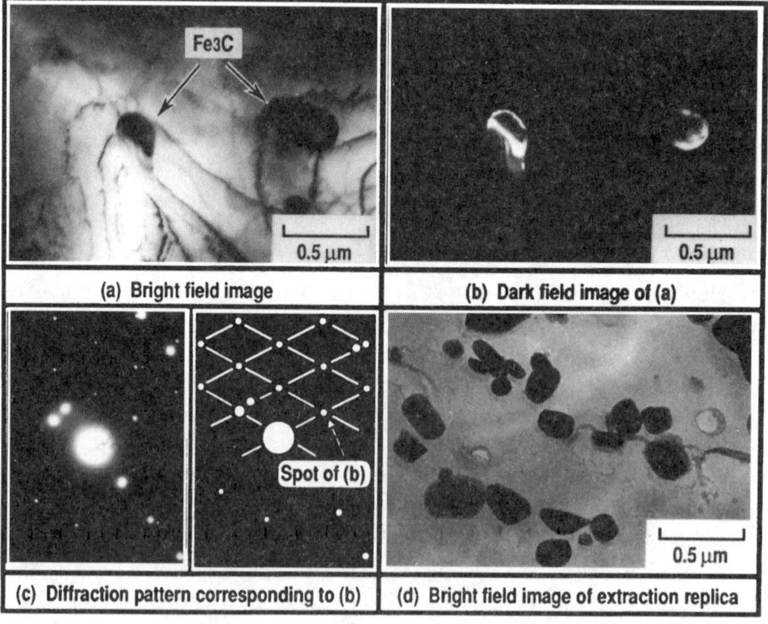

Fig. 15 TEM micrograph of thin foils and extraction replica obtained from the specimen reaustenitised for 0.5 hr.

observed. While, in the material reheated to ($\alpha+\gamma$) temperature range, carbide particles were observed. Thus, it is clear that the carbide forms during the reheating process of the martensite. Prolonged heating at ($\alpha+\gamma$) range dissolves the carbide and stabilizes the austenite phase and then it brings about very high toughness.

CONCLUDING REMARKS

1. Absolute safety criterion on brittle fracture can be given by crack arresting temperature. K_{IC} gives a safety criterion on brittle fracture initiation.
2. Fracture toughness depends on the definition of crack initiation. J(mid) value which is an average of J(R) and J(ΔC) was proposed as a meaningful engineering fracture toughness value in spheroidal graphite cast iron.
3. It is possible to toughen ADI more by applying special heat treatment. Refining and increasing the stability of retained austenite phase and to avoid the precipitation of carbide are the key factors.

REFERENCES

1. T. Kobayashi, H. Yamamoto and K. Matsuo, *Eng. Frac. Mech.*, 30, 397 (1988).
2. T. Kobayashi, *Trans. ISIJ*, 19, 676 (1979).
3. T. Kobayashi and H. Yamamoto, *Adv. in Frac. Res.*, 837 (1989), Pergamon.
4. W. S. Pellini, G. A. Sandoz and H. F. Bishop, *Trans. ASM*, 46, 418 (1954).
5. T. Kobayashi and S. Nishi, *J. Jpn. Foundrymen's Soc. (Imono; in English)*, 52, 76 (1980).
6. T. Kobayashi and S. Yamada, *Mech. Behav. Met. VI*, 6-57 (1991), Pergamon.
7. T. Kobayashi, K. Takai and H. Maniwa, *Trans. ISIJ*, 7, 115 (1967).
8. T. Kobayashi and S. Yamada, *Met. Mat. Trans.*, Now Publishing.
9. M. K. Tseng and H. L. Marcus, *Eng. Frac. Mech.*, 16, 895 (1982).
10. T. Kobayashi, I. Yamamoto and M. Niinomi, *J. Test. Eval.*, 21, 145 (1993).
11. ASTM E399-83,"Standard Test Method for Plain-Strain Fracture Toughness of Metallic Materials", (1983).
12. R. J. Salzbrenner, J. A. Van Den Avyle, T. J. Lutz and W. L. Bradley, *ASTM STP868*, 328 (1985).
13. H. Yamamoto and T. Kobayashi, *Conf. Proc. Cast Iron IV*, 243 (1990, MRS).
14. T. Kobayashi and H. Yamamoto, *Met. Trans. A*, 19A, 319 (1988).
15. S. Yamada, T. Kobayashi and K. Matsuo, *Proc. 8th Int. Cong. Heat Treatment of Materials*, 203 (1992).
16. M. Aoyama, K. Matsuo and T. Kobayashi, *J. Jpn. Foundrymen's Soc. (Imono)*, 62, 101 (1990).
17. S. Yamada, T. Kobayashi and K. Matsuo, *J. Jpn. Foundrymen's Soc. (Imono)*, 66, 477 (1994).

Advanced Materials Research Vols. 4-5 (1997) pp. 61-72
© 1997 Scitec Publications, Switzerland

Behavior of Oxygen and Nucleation of Graphite in Production of Spheroidal Graphite Cast Iron

T. Kusakawa

Emeritus Professor, Waseda University, Tokyo, Japan
Home address: 19-9 Saginomiya, 1 chome, Nakano-ku, Tokyo, 165 Japan

Keywords: Low Frequency Induction Furnace, Spheroidal Graphite Cast Iron, Total Oxygen, Carbon Deoxidizing, Soluble Oxygen, Spheroidal Graphite Formation, Graphite Nucleus

Abstract

Melting processes in low frequency induction furnaces are advantageous for the production of spheroidal graphite cast iron in both technological and economical viewpoints. The technological merits are the production of melts with extremely low contents of total oxygen due to refining by carbon deoxidizing at high temperatures. The economical merits are the stability of spheroidal graphite cast iron production, low consumption of graphite spheroidizing agents, and excellent products quality. During melting, measurements were made on changes in the chemical composition of melts in furnace, structure of chills, number of eutectic cells, total and soluble oxygen concentration with the temperature of melts. In particular since the presence of nuclei is necessary for graphite spheroidization in the core of graphite, it is noticed that the soluble oxygen content decreases with drop of the temperature of molten metal after the post inoculation following the graphite spheroidization. This is contrary to the theory and may be explained by the formation of oxides, presumably silicate, in reaction of soluble oxygen with Mg and Si in molten metal resulting in formation of the nuclei of spheroidal graphite. Furthermore a method is developed to reveal the presence of nuclei in the core of graphite by ion milling etching. The nuclei are analyzed to consist of the graphitization elements (Mg, Ca and Ce), Si, O and S with EPMA. Analysis of the crystal structure of the nuclei has not been done but it may be a type of silicate. A proposal is made on the formation of spheroidal graphite that, in the state of total oxygen, especially of low soluble oxygen, graphite nuclei occur easily around the oxides and grow rapidly to make a radial structure.

1. Introduction

In production of spheroidal graphite cast iron, the use of low frequency induction furnaces is very advantageous in both technological and economical aspects. In this paper I will discuss the advantages in terms of the chill structure, number of eutectic cells, chemical composition, and oxygen content and the spheroidizing mechanism of graphite on the basis of my experimental results on the processing conditions of graphite spheroidization, the inoculation effects and the formation of graphite nuclei.

2. Changes in melting temperature, chill structure, number of eutectic cells, chemical composition and oxygen content during melting

It is said that higher quality spheroidal graphite cast irons are produced by melting in low frequency induction furnaces than in cupola. To clarify this reason I examined the changes in the chill structure by plate chill test, the number of eutectic cells, the chemical composition and the total oxygen and oxygen activity with the temperature change during melting. The effects of Si addition and Mg spheroidizing treatment were paid particular attention in consideration of the merits of low frequency induction furnaces. On the basis of these results I will discuss the problems associated with graphite spheroidizing.

2-1. Experimental methods

(a) Specifications of the low frequency induction furnace

Table 1 shows the specifications of the low frequency induction furnace and Fig. 1 the dimension of the furnace with acid lining of silica. This furnace was used for commercial production of spheroidal graphite cast iron. The raw materials were mainly of steel scraps and pig irons for spheroidal graphite cast iron and melted for obtaining predetermined compositions.

Table 1 Capacity, electric power, frequency and lining material of low frequency induction furnace.

Capacity	5000 Kg
Electric power	1500 KW
Frequency	50 Hz
Lining	Silica

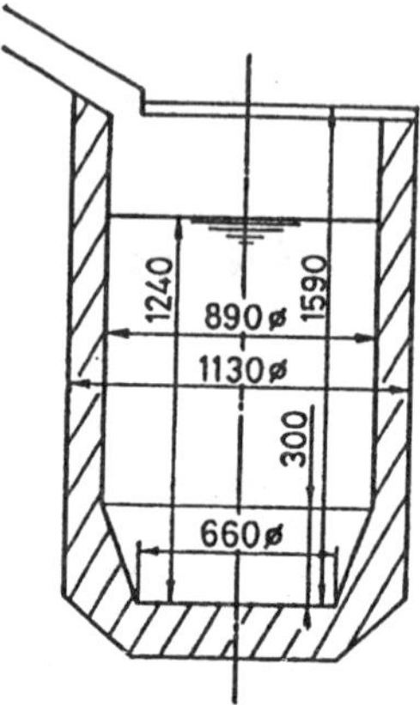

Fig. 1 Shape and size of low frequency induction furnace used in this experiment

(b) Shape and dimension of chill test plate

Figure 2 shows the shape and dimension of the chill test piece. CO_2 gas silica molds were used. The chill plate was a water-cooled copper plate of 30 mm thickness. The test specimens were fractured at the center for observation of the chill structure on the fractured surface.

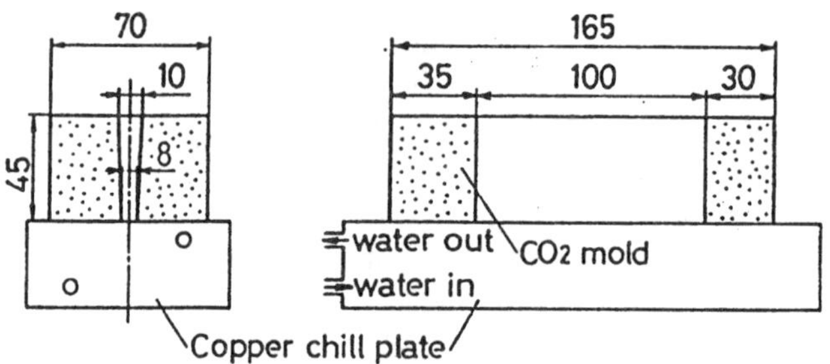

Fig. 2 Shape and size of mold for chill test piece.

(c) Methods and specimens for optical microstructure observation and counting of eutectic cells

Specimens, 20mm in diameter and 50mm in length, for observation both of the optical microstructure and the eutectic cell number were prepared by casting in a CO_2 gas mold. The specimen was cut at the height of 20mm from the bottom and polished for microstructural observation. The etchant was 5% picral. The eutectic cell structure was revealed by immersing in the Stead agent ($CuCl_2$: 1g, $MgCl_2$: 4g, HCl: 2ml and ethanol: 100ml) for 30 sec and by cleansing in a special rinsing solution ($CuSO_4$: 40g, HCl: 200ml and H_2O: 200ml) to remove the attached Cu. The number of eutectic cells was counted by using a universal projector.

(d) Measurement of total oxygen

The total oxygen in molten cast iron is the sum of the soluble oxygen representing oxygen activity and the oxygen as oxides. The specimens were sucked into a quartz tube (5mm in inner diameter), kept in a vacuum desiccator, cut into a block of about 0.5g, polished with a hand grinder, rinsed in acetone with ultrasonics. The total oxygen was determined by argon gas carrier ICP analysis.

(e) Analysis of soluble oxygen (oxygen activity)

Until now the oxygen activity was determined as an infinitely dilute solution in the iron-carbon binary system. However the oxygen activity has not been determined for the cast iron which is a condensed system. In addition the coexistence of silicon increases the complication in scientific treatment. Considering the factors influencing the nucleation and crystal growth of graphite in cast iron, the presence of a small amount of oxygen will be necessary for formation of the graphite nuclei as heterogeneous nucleation. With progress in the accuracy of oxygen sensors the amount of soluble oxygen less than 1ppm can be detected. Figure 3 shows the principle of oxygen sensor used in the present study with stabilized zirconia for the solid electrolyte and Cr and Cr_2O_3 powder for the normal electrode. The electromotive force, E, generated by the difference in oxygen partial pressure between the normal electrode and the Mo molten steel electrode was measured together with the temperature of molten iron. By using these values the oxygen activity, a_O, was calculated from Eq.1.

$$a_O = K\{(P_e'^{1/4} + P_{O2}^{1/4})\exp(-EF/RT) - P_e'^{1/4}\}^2 \quad \dots\dots\dots\dots(1)$$

where K is the equilibrium constant $(1/2 O_2 (g) = \underline{O})$, P_e' the electron conduction parameter, P_{O2} the oxygen partial pressure of the standard material (Cr_2O_3), and F the Faraday constant. The values of K, P_e' and P_{O2} were those derived by Janke and Fisher[2]. The value of a_O represented by concentration in wt% is the oxygen activity based on the infinitely dilute solution and is equal to the concentration in wt% in the cast iron. In contrast to the conventional method, this equation takes into account the correction term, P_e', for the error due to the electron conduction on solid electrolyte when the value of $\underline{O}$ is small. Although the values of a_O calculated from Eq. 1

(f) Analysis of chemical composition

The chemical compositions were analyzed in accordance with the Japanese Industrial Standards.

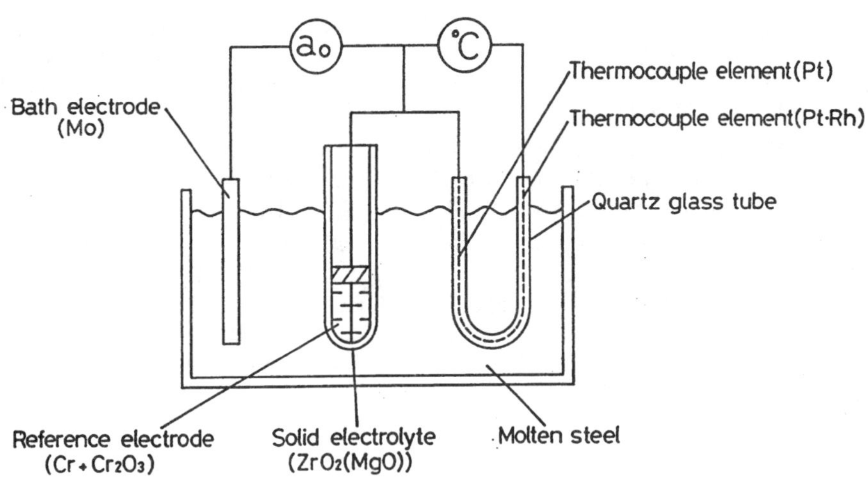

Fig. 3 Principle of oxygen sensor

2-2 Study of Experimental Results

   Figure 4 shows the changes in temperature, chill structure, eutectic cell number, soluble oxygen, total oxygen and chemical composition during the melting in the low frequency induction furnace. After the first sampling from the melt the procedure followed; stop of melting at about 3min, addition of SiC for controlling of the Si content, supply of 1.2% of Fe-Si-6%Mg alloy at 29min, inoculation of 4% ferrosilicon (75%Si) at 32min and pouring. The rise of temperature of molten iron during the procedure was from 1367°C at the start to in the range of 1508 and 1510°C after 23 to 29min.

Time (min)	0	6	11	15	19	23	26	29	32	38
The treatment of liquid metal							SiC	spheroidization	inoculation	
Temperature(°C)	1367	1390	1434	1460	1484	1510	1508	1510	1435	1350
The structure of chill 1cm 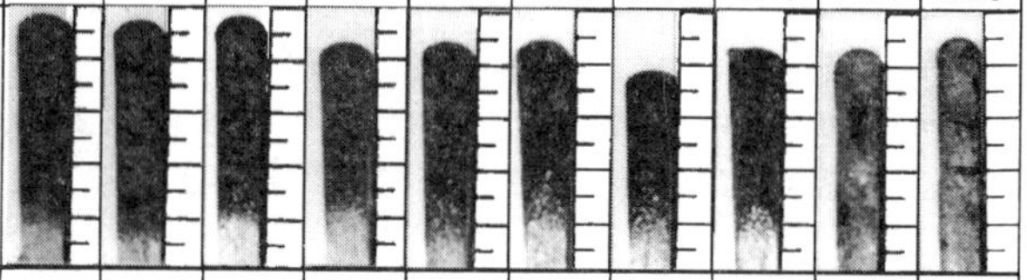										
Eutectic cell numbers(/cm^2)	692	317	250	149	122	105	94	72	23537	33889
Soluble O (ppm)	0.94	0.53	0.78	0.89	1.22	1.70	1.87	1.22	0.32	0.15
Total O (ppm)	18	14	—	15	6	5	4	4	4	5
Chemical composition (wt%) — C	3.70	3.73	3.71	3.70	3.70	3.71	3.71	3.72	3.56	3.56
Si	1.41	1.41	1.43	1.43	1.43	1.46	1.50	1.52	2.41	2.75
S	0.016	0.017	0.017	0.017	0.017	0.017	0.017	0.017	0.016	0.016
Mg	—	—	—	—	—	—	—	—	0.061	0.056
Mn: 0.27 P: 0.012~0.013 Cr: 0.07 Sn: 0.005~0.006										

Fig. 4   Relation among time, temperature, chill shape and length, eutectic cell number, total and soluble oxygen and chemical composition.

   With the temperature rise the chill grows in size exhibiting an inverse triangle shape, a type of the inverse chill phenomenon. After the graphite spheroidization treatment the triangular chill remain on the fractured surface besides the steel like fractured surface. After inoculation the chills disappear to be a totally steel like fractured surface.

   The number of cells decreases abruptly with elevation of the melting temperature; from 692/cm^2 at the start to 72/cm^2 after 29min. This means the decrease in graphite nuclei.

   The total oxygen decreases with the temperature rise from 18ppm at the start to 4ppm at 1510°C after 23min and does not change on Mg addition for spheroidization at 29min. On the other hand, the soluble oxygen increases with the temperature rise from 0.94ppm at the start to a maximum of 1.87ppm at 26min, but it decreases abruptly to 0.32ppm on spheroidization and to 0.15ppm on inoculation. This suggests that the abrupt decrease of soluble oxygen is related with the mechanism of graphite spheroidization. No appreciable change is detected for the other chemical composition except for added C and Si.

2-3. Discussion

The present experiment performed in a foundry provides interesting information on the problem of graphitization.

(a) Relation among chill testing results, melting temperature, eutectic cell number, oxygen content and chemical composition.

The chill test was generally used as a rapid test for examining the change in graphitization with the contents of C and Si. However in the present experiment a remarkable decrease of total oxygen is found with temperature rise without significant changes in the content of C and Si. Also the morphology of chills on the chill test plate varies from straight line to triangle with increase in the length. I will discuss how to understand these phenomena.

At first I consider as to the decrease of total oxygen. Most of the total oxygen in cast iron is the oxygen as oxides; the remaining part is the soluble oxygen. The oxides are considered to be silicon oxides from the chemical composition. Heine[3] considered that the nucleus material for graphite is silica. The silica is reduced by C at high temperatures and the number of graphite nuclei decreases resulting in the suppression of graphitization. In the molten metal the following reaction proceeds.

$$SiO_2(s)+2C=Si+2CO(g) \quad \dotsfill (2)$$

$$\log K = \log\{a_{Si}\cdot P_{CO}{}^2/a_{SiO_2}\cdot a_C{}^2\}= \log\{a_{Si}\cdot P_{CO_2}/a_{SiO_2}\cdot a_C{}^2\}= 27792.5/T = 15.40 \dotsfill (3)$$

where a_{Si} =1 and P_{CO} =1atm. Substitution of the concentrations, [%C]=3.70% and [%Si]=1.41%, of the present sample into Eq.3 gives the equilibrium temperature of 1423°C. By taking into account the interaction coefficient, the lower equilibrium temperature will be derived.

In practice the molten metal in a low frequency induction furnace is subjected to a large stirring force. As evident from Fig. 5 showing the state of stirring on a section, in the central part of the furnace the flows of stirred liquid metal collide and the reaction of Eq. (2) will proceed extensively. Thus the number of SiO_2 particles necessary for the graphite nuclei decreases. This corresponds with a decrease in the number of eutectic cells with temperature rise; the number of eutectic cells can be considered to coincide with the number of nuclei for graphitization.

The fractured surface after the spheroidization appears totally steel-like, but some triangular cell can be found by close observation in particular clearly by etching after polishing.

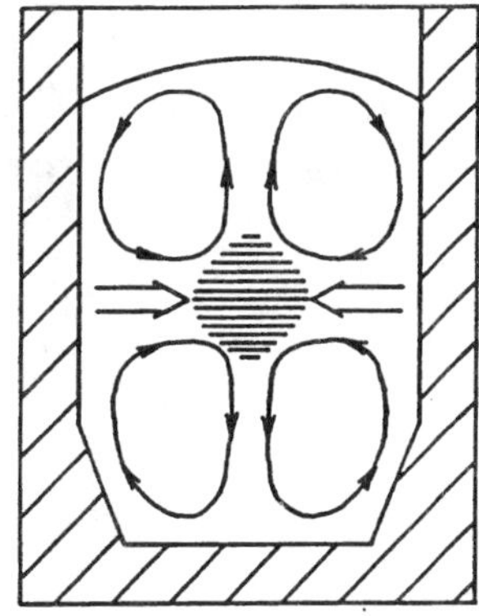

Fig. 5   Stirring state of molten iron in low frequency induction furnace

(b)   Graphite spheroidization and eutectic cells

In this section I will discuss the number of eutectic cells after graphite spheroidization following the discussion on the change in the number of eutectic cells in (a).   In the present experiment the number of eutectic cells increased enormously on spheroidization.   One spheroidal graphite particle is counted as a cell.   For graphite flakes in cast iron, one flaky crystal originates from a heterogeneous nucleus and grows and spreads into branches to be a shape of double flower.   When graphite forms a compact crystal around a heterogeneous nucleus in supercooling, a spheroidal graphite crystal is formed; thus a spheroidal graphite crystal corresponds with an eutectic cell.   The remarkable increase in the number of graphite crystals or eutectic cells on graphitization implies an increase in the number of graphite nuclei; this is considered to be equivalent to the inoculation.

(c)   Formation of triangular chills

The conventional chill testing is used for measurement of the chill length and the graphitization rate in terms of C and Si contents.   The shape of specimen is usually in wedge rather than plate.   In the present study attention was paid on the shape of chills as well as the length; the chills get in length and take a triangular shape with the temperature rise of molten metal without a significant change in chemical composition (in particular C and Si).   This is a type of inverse chill phenomena.   With progress of refining of molten metal in this way, the number of SiO_2 as inclusion decreases by carbon reduction and consequently the number of nuclei for graphitization decreases; in compensation the nuclei were supplied from the silica in the mold.   Thus the graphitization starts from the both edges of a specimen and forms a triangular chill that represents an inverse chill.   This phenomenon involves a type of refining effect leading to the decrease in total oxygen.

(d)   Behavior of oxygen

The behavior and reaction of oxygen in molten iron discussed in the section (a) was concerned with C deoxidation in the reaction in steelmaking.   Until now the oxygen in cast iron has been discussed in a number of reports but only a few papers gives clear explanation.   In the results reported in (a) on the temperature rise, chill structure and eutectic cell number, the abrupt change in soluble oxygen concentration is noticed on Mg addition by the following reaction.

$$Mg + O = MgO \dots\dots(4)$$

The added Mg in part dissolves in molten iron but mostly moves as vapor to form solid MgO by reaction with soluble oxygen.   At high temperatures (1400 to 1600K) the value of $-\Delta G$ for C and O reactions is fairly large.   The presence of soluble oxygen at the temperature of graphite crystallization in molten iron suppresses the growth of graphite by CO reaction; the soluble oxygen is considered to suppress the graphitization.   Therefore the smaller amount of soluble oxygen is favorable.   In the present experiment it is less than 0.3ppm, which seems to reduce the probability of the reaction between O and C and to promote the growth rate of graphite crystals.

3.   Correlation between post inoculation and soluble oxygen

As shown in Fig. 4, the post inoculation to add the inoculant like ferrosilicon is employed after spheroidization in the production of spheroidal graphite cast iron.   This process is intended to improve the mechanical property by eliminate cementite formed in the presence of Mg.   As evident from Fig. 4, inoculation decreases the soluble oxygen down to 0.15ppm and remarkably increases the number of eutectic cells that is the number of graphite grains.   On the basis of this phenomenon I made an experiment to clarify the relationship between inoculation and soluble oxygen .

3-1   Experimental methods

A base metal, of which chemical composition is shown in Table 2, was prepared by melting the raw materials of pig iron for spheroidal cast iron, steel scrap and metallic silicon in a 100KW high frequency induction furnace.   The molten metal was kept more than 10min above 1500°C, desulfurized by a desulfurizer and cast in a metal mold.   The base metal, 2.5kg, was remelted in a graphite crucible in a Kryptol furnace, kept at 1500°C for 10min, and added 1wt% of Fe-Si-10%Mg alloy (Table 3) for graphitization.   Then at about 1400°C, 0.4% of inoculation agent which has the composition shown in Tables 4 and 5 was added and simultaneously an oxygen sensor was inserted into the molten metal. The crucible was taken out from the furnace and the electromotive force in the oxygen sensor was measured during cooling down to 1250°C at which temperature the melt was poured into a mold. The cast specimens were used for observation of microstructure and counting of eutectic cells.

Table 2 Chemical composition of base metal.

	C	Si	Mn	P	S
Base metal	3.45	1.53	0.14	0.070	0.006

Table 3 Chemical composition of spheroidizer

Mg	Si	Fe	Al
9.8	45.1	40.8	1.4

Table 4 Chemical composition of inoculant (wt%)

Inoculant	Si	C	P	S	Al	Ti	Mn	Cr	Ca	Cu
Foundry grade Fe-Si (FF)	73.3	0.056	0.040	0.003	0.78	0.091	0.16	0.026	0.17	0.014
Clean Fe-Si (CF)	77.8	0.008	0.017	0.003	0.031	0.030	0.13	0.038	0.02	0.02
Metallic Si (MS)	98.0	0.009	0.008	0.003	0.09	0.034	0.02	0.006	0.21	0.001
Pure Si (PS)	99.999									

Additional amount of inoculant: 0.3% Si increment
Grain size of inoculant: 6 - 10 mesh

Table 5 Non-metallic inclusion in inoculant (wt%)

Inoculant	T.N.M.I.*	SiO_2	Al_2O_3	Fe_2O_3	CaO	MgO	SiC
Foundry grade Fe-Si (FF)	1.30	0.61	0.235	0.038	0.001	0.015	0.15
Clean Fe-Si (CF)	0.11	0.05	0.008	0.020	0.001	0.002	0.02
Metallic Si (MS)	0.17	0.10	0.015	0.009	0.004	0.001	0.02

* Total non-metallic inclusion

3-2. Study of Experimental Results

Figure 6 shows the relationship between the temperature drop and the soluble oxygen (oxygen activity) in the use of various inoculation agents. Figure 7 shows the microstructure and the number of eutectic cells for each specimen. The inoculants were basically a mixture of 75% ferrosilicon and metallic silicon with varying contents of nonmetallic inclusions. Figure 6 shows the change of soluble oxygen with temperature drop after inoculation. It is evident that the soluble oxygen decreases with temperature drop. However theoretically the soluble oxygen or oxygen activity must increase. In practice in contrast to the theory the decrease of soluble oxygen will be a result of the formation of oxides in reaction of the increased soluble oxygen with the active elements present in the molten metal. At first the reaction occurs with Mg and then with Si and others as in the following formula.

$$Mg+O=MgO \dots\dots(5)$$

$$Si+2O=SiO2 \dots\dots(6)$$

The selective oxidation proceeds in the order of higher affinity with oxygen.   In reality, since the content of Si is large, silicate may be formed.

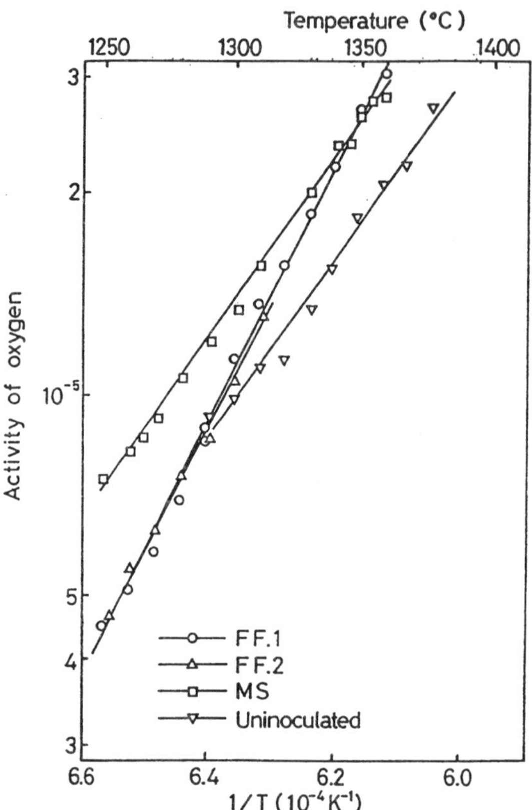

Fig. 6   Relation between temperature drop and soluble oxygen content (oxygen activity)
in molten iron.

On inoculation the soluble oxygen increases for a while but decreases with temperature drop.   The sample was inoculated with FF which contains a large amount of nonmetallic inclusions shows a higher decreasing rate of soluble oxygen than the other samples.   In this sample the inclusions contained in the inoculant disperse in molten metal to provide the graphite nuclei as particles coagulated with inclusions formed by spheroidization or as isolated particles.   As previously mentioned Fig. 7 shows the microstructure and the number of eutectic cells for each sample.   In the figure the shape of graphite appears favorable and the number of graphite or cells is largest for the sample inoculated with FF.   The number of graphite in the sample inoculated by MS containing a small amount of inclusions is nearly the same to the sample without inoculation.

As the result of the present experiment the oxygen activity in the melt decreases with temperature drop after spheroidization.   The decreasing rate is higher for the sample which was inoculated with a larger amount of inclusions.

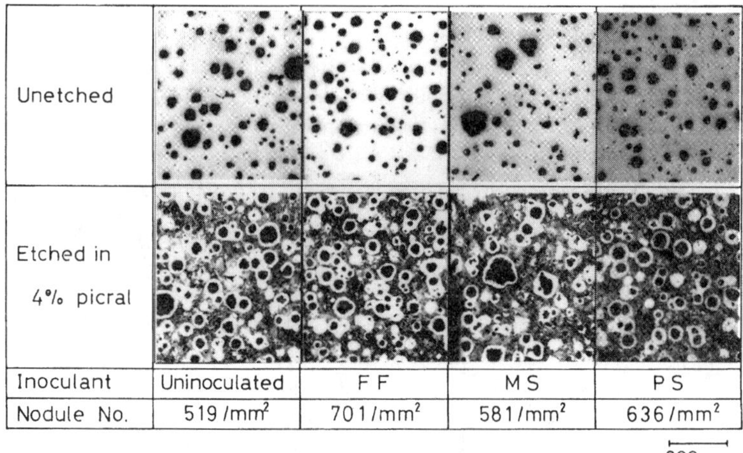

Unetched				
Etched in 4% picral				
Inoculant	Uninoculated	F F	M S	P S
Nodule No.	519 /mm²	701 /mm²	581 /mm²	636 /mm²

200 µm

Fig. 7   Microstructure and eutectic cell number (nodule number of spheroidal graphite) of each spheroidal graphite cast iron inoculated by three kinds of inoculants.

4.   Nucleus formation of spheroidal graphite

The mechanism of graphitization is not yet definite.   As a proposed mechanism the nucleus theory approves the presence of nuclei in the center of graphite.   I have performed experiments in support of the nucleus theory.

4-1   Experimental methods and results

To verify the presence of nuclei in spheroidal graphite, I developed the following two methods.

a)   A method for observation of the core part of spheroidal graphite with EPMA after extracting and crushing down to 5µm thickness.

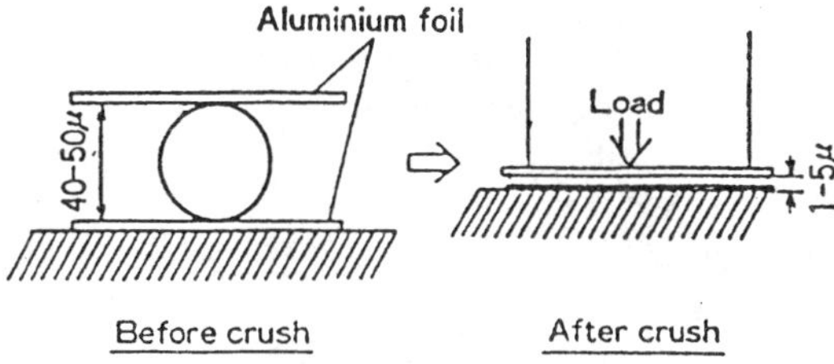

Fig. 8   Making method of test piece for observation of nuclei in the center of spheroidal graphite in spheroidal graphite cast iron.(Crush down method of spheroidal graphite under 0.5)

The sample was dissolved in a condensed solution of hydrochloric acid for extraction of spheroidal graphite.   The extracted graphite was rinsed in a solution of fluoric acid for removing silica adhered on the surface of graphite and further rinsed in warm water followed by desiccation by alcohol.   The graphite particles about 50µm in diameter were wrapped in a aluminum foil and crushed on a glass plate below the thickness of 5µm, as shown in Fig. 8[4].   The crushed specimens were observed with EPMA.   Figure 9 is an example showing the presence of Ca, O, Si, and S in the center of spheroidal graphite.   As expected in many other observations the spheroidization elements such as Mg and Ce are detected.   From these elemental analyses the nucleus material is considered to be silicate but the crystal structure remains to be determined.

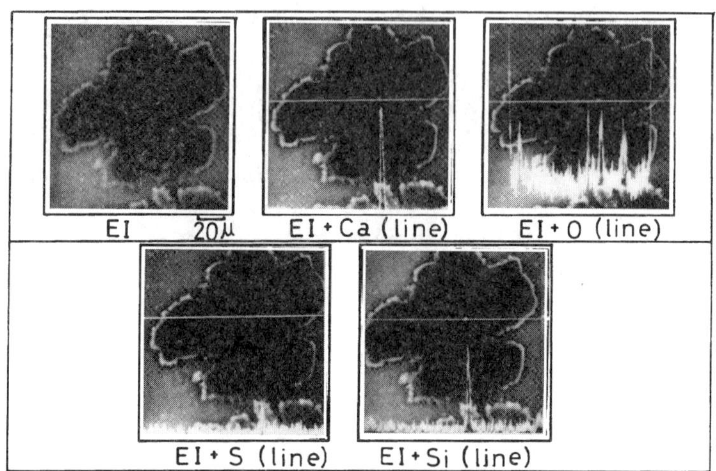

Fig. 9   Observation of spheroidal graphite
extracted from calcium treated spheroidal graphite cast iron by crush down sampling method.

b)   A method for revealing the nuclei of graphite in the center of spheroidal graphite by fracturing and etching with ion milling

We have developed a method, "cathodic vacuum etching process," for revealing clearly the graphite structure by bombardment of ionizing air or nitrogen on the specimen for microstructural observation.   This method enable the electron microscopic observation of the inner structure of spheroidal graphite.   The method is developed into an apparatus, "Ion milling," which is provided with sputtering by bombardment of argon ion beams of 10kV voltage and 5mA current at an angle of 30 degrees on the specimen.   By the sputtering metallic part is etched rapidly, while graphite part remains as it is because of a lower sputtering rate.

The EPMA images of the sputtered surface shown in Figs. 10 and 11 clearly indicate the presence of nuclei in the center of almost all the graphite.   The sample shown in Fig. 10 was spheroidized by the Fe-Si-10%Mg alloy and that in Fig. 11 by a mixture of 85% Ca-Si powder and 15% $CaCl_2$. Both were inoculated by a heated and mixed compound of 75% of Ca-Si powder, 15% $CaCl_2$ and 10% La_2O_3 As evident from Fig. 10 nuclei are found in the core of graphite and the presence of Mg, Si and O are detected by elemental analysis with line scanning.   In Fig.11 the shape of graphite deviate from sphere, but the nuclei are present in the core and they contain the elements of Ca, Si and O as well as La originating from the inoculant. $CaCl_2$ containing water of crystallization easily fuses on heating.   The solution of $CaCl_2$ is small in surface tension and has an effect of dispersing the fine powder of La_2O_3 in the solution without coagulation.   Thus when an inoculant containing $CaCl_2$ is added, the particles of La_2O_3 well disperse in melts and are found in the core of spheroidal graphite in the present experiment.

From this evidence the inoculation is proved to provide the nuclei needed for graphitization when very fine particles of oxides are supplied in a adequately dispersed state.

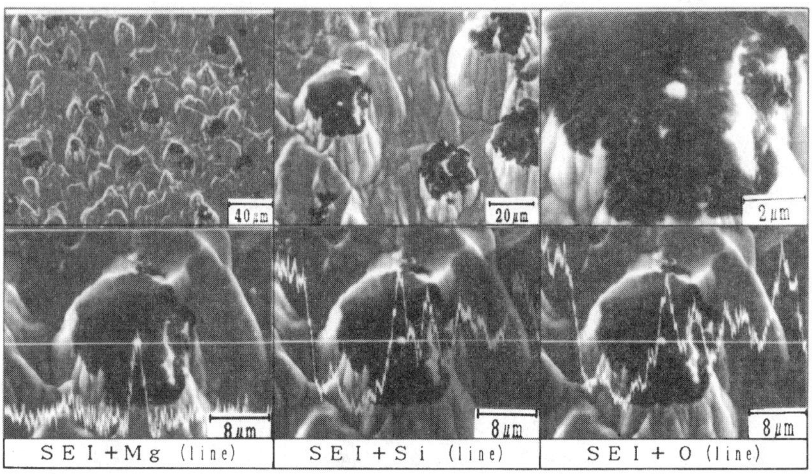

Fig. 10   EPMA image and elemental analysis of spheroidal graphite
in magnesium treated spheroidal graphite cast iron after ion milling etching by argon gas.
Spheroidizer: Fe-Si-Mg(10%)
Inoculant: Ca-Si 75% + La₂O₃ 10% + CaCl 15%.

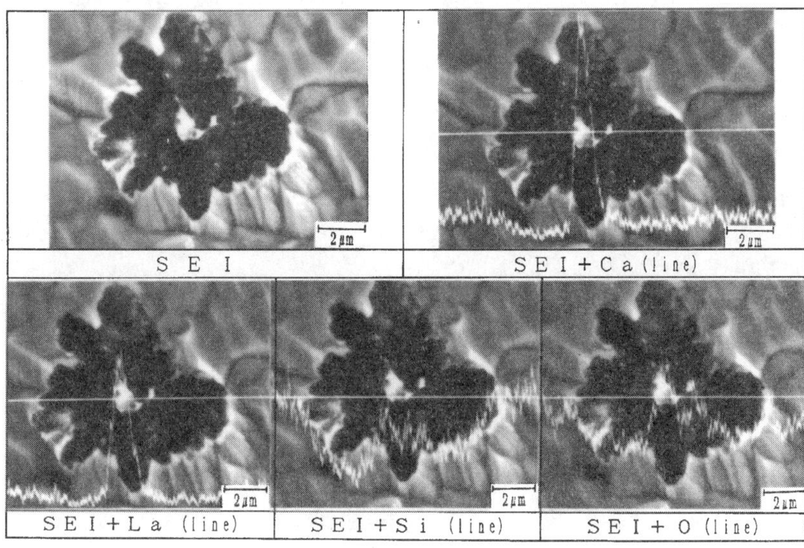

Fig. 11   EPMA image and elemental analysis of spheroidal graphite
in calcium treated spheroidal graphite cast iron after ion milling etching by argon gas.
Spheroidizer: Ca-Si 85% + CaCl 15%
Inoculant: Ca-Si 75% + La₂O₃ 10% + CaCl + 15%.

4-2 Discussion

From the experimental results stated above, the presence of nuclei is indispensable in the core for the occurrence of spheroidal graphite. The materials for nuclei may be silicate containing Mg, Ca, Ce and others. Silicate inclusions in steel is spherical. The melting temperature differs between steel and cast iron: about 1600°C for steel and 1300-1400°C for cast iron. Silicate inclusions is spherical because of melting in steel which has a higher melting temperature. In cast iron because of a higher content of silicon silicate has a higher content of SiO_2. By addition of basic oxides like MgO and CaO the melting temperature of the silicate drops down to about 1300°C and the shape of the silicate will become spherical.

On the other hand as an environment for graphite formation, since the soluble oxygen is low and the total oxygen is also low compared with the flaky graphite formation, graphitization is thought to occur in a supercooled state. Thus in the rapid progress of graphitization, a graphite crystal adheres on a nucleus with its cleavage face (001), grows toward its vertical direction and impinges with the neighboring crystal to continue to grow along the c-axis and to form a spherical form. This concept is not yet completely verified to be a theory, but I would like to propose as an new concept.

5. Conclusion

I proposed a mechanism of graphite spheroidization based on my experiments on the behavior of oxygen, in particular, with respect to the formation of spheroidal nuclei with temperature drop in the solidification process together with discussion on the relationship between supercooling phenomenon and growth velocity of graphite which has been commented until now.

Acknowledgment

I sincerely express my gratitude to Messrs. Lin Hang-Ton and Xu Xing-Cheng who were the visiting researchers, Messrs. Shinnichi Okimoto, Kazuo Ide and Hiroki Okita who were the post graduate students in my laboratory, for their help in performing the experiments reported in this paper.

References

1) W. Pluschkell: Arch. Eisenhüttenwesen, 46(1975), 11.

2) D. Janke and W. A. Fisher: Arch. Eisenhüttenwesen, 46(1975), 12.

3) R. W. Heine: Preprint No. 51-18 and -47 of the Annual Meeting of AFS, (1951).

4) T. Kusakawa, H. Lin and X. Xu: Report of the Casting Research Lab., Waseda Univ., No. 36, Nov., (1985), p.29-34.

5) H. Tsuchikura, T. Kusakawa and T. Okumoto: "Proceedings of International Conference on Electron Microscopy" held in London, July, 1954, p.368-393.

Advanced Materials Research Vols. 4-5 (1997) pp. 73-88
© *1997 Scitec Publications, Switzerland*

Influence of Mould Filling and Natural Convection on the Solidification of Cast Iron

F. Mampaey

WTCM Foundry Centre, Technologiepark 9, B-9052 Zwijnaarde, Belgium

Keywords: Mould Filling, Natural Convection, Solidification, Modelling

ABSTRACT

The influence of mould filling on the final nodule count distribution in a complex thin wall casting is demonstrated. Some aspects of mould filling simulation are reviewed. Simulation results of mould filling are compared with experimental results obtained by pouring liquid cast iron behind a heat resistant glass window. Application of a nucleation model permits to calculate the nodule count distribution. This is in good agreement with experimental nodule counts. The influence of thermal convection has been studied in cylindrical castings. The results illustrate the shift of the thermal centre, the increased heat transfer in the liquid and the thickness variation of the solidified peripheral layer. Convection currents are visualized by a tracer technique. The experiments indicate that graphite nodules are transported by the convection flow in (hyper)eutectic cast iron. There is no evidence that the eutectic cells in lamellar graphite cast iron are moved by the convection flow.

1. INTRODUCTION

The solidification of liquid metals can be influenced by several flow phenomena. The influence of metal flow during mould filling as well as the effect of natural convection are examined in this paper for the case of cast iron poured in sand moulds. Liquid metal flow during mould filling can have a decisive influence on the final casting quality. Moreover, mould filling creates a heterogeneous temperature distribution in the casting and in the mould which directly affects the microstructure of thin wall castings. In more massive castings, natural convection which occurs during liquid cooling interacts with solidification. Increasing efforts on modelling the mould filling of have been carried out. However, only a few comparisons with experimental results have been published, especially with liquid metals. Emphasis is laid on experimental results obtained for mould filling [1–3] and for natural convection [4].

2. MOULD FILLING

Simulation of mould filling requires to solve several equations simultaneously. The flow of liquid metal is governed by the Navier–Stokes and the continuity equations. Heat transport is governed by conduction and convection. During the filling stage, the free surface of the liquid can be located by means of a fluid function F, a step function which represents the fraction of the element filled with liquid. The value of F is 0 for empty elements, 1 for full elements and between 0 and 1 for surface elements. The fluid function F is calculated by

$$\frac{\partial F}{\partial t} + V \nabla F = 0 \qquad (1)$$

where t is the time and **V** the velocity vector. The application of a step function prevents any smearing effects of the fluid function and is in accordance with the discontinuity of the liquid surface.

The surface tension pressure p_{st} can be written as

$$p_{st} = \sigma \cdot \text{sign} \left(\frac{1}{r_i} + \frac{1}{r_j} + \frac{1}{r_k} \right) \qquad (2)$$

where σ is the surface tension of the liquid metal, r_i, r_j, r_k are the three radii of the curvature in the direction of the coordinate axes and sign (+,–) corresponds to a convex or concave surface respectively. The surface tension pressure is added in the Navier–Stokes equations which will change the pressure gradients and affect the velocity field. Details of the numerical scheme to solve the equations have been discribed elsewhere [5]. The present code allows the elements which have been filled to become empty later on so that the formation of waves can be simulated. It can treat an incompletely filled downsprue once the pouring has started. These phenomena have been observed frequently during the experiments. The pouring rate is calculated by maintaining a constant metal level in the pouring cup. At the start of filling, the liquid metal in the cup is stagnant, subsequently it falls due to gravity. In this way, a pouring rate which varies during filling of the mould will be obtained. The variable pouring rate can be demonstrated by Torricelli's law

$$v = \alpha \sqrt{2gH} \tag{3}$$

Here v is the velocity at the ingate and H is the difference in height between the free surface levels of the metal in the pouring cup and in the mould. The discharge–coefficient α depends on the shape of the gating system and on the way of pouring [6]. Since the value of α will be different for each pouring system, this law can only give a rough estimation. Moreover, Torricelli's law is derived from the Bernoulli equation in which the flow is assumed to be steady. This latter assumption is normally not valid during filling of the mould.

3. MOULD FILLING VALIDATION

Direct observation of liquid metal flow during mould filling is relatively difficult which explains the scarcity of the publications on this topic. Movies made by a committee of the Institute of British Foundrymen and at the Naval Research Laboratory have been used to study the influence of the gate design [7]. In these experiments, liquid metal flow was studied in horizontal plate castings which were poured in an open mould. Although only the flow in the casting cavity could be observed, it allowed interesting conclusions on the practical design of gating systems. Ingerslev and Andersen have studied the flow of liquid cast iron entering sand moulds [8,9]. Experimental results of mould filling have been compared with 2–D simulations only [9–11]. Moreover, the gating system was omitted and the inlet velocities simplified [9,10]. In fact, the gating system plays an important role in mould filling. During filling of the mould, the pouring rate and ingate velocities change which will influence the flow pattern. Three – dimensional visualization of the free surface during mould filling can be determined also by using the contact wire method [12,13]. Post processing of the results allows to display the mould filling sequence. An alternative of this technique has been applied by the present author. Quartz thermocouple protection tubes were placed at different locations at the interface between the sand mould and the casting. When the liquid metal contacts the tube, the other end of the tube illuminates and may be filmed. Although this method allows to visualize to some extent the general mould filling sequence, details of the flow pattern are lost by this method even when the number of tubes is relatively high (> 16 tubes / dm^2). Accordingly, this method is not suitable to validate simulation results very accurately. As compared to the contact wire method, superior results can be obtained when pouring liquid cast iron behind a transparent, heat resistant 'glass' window which covers the mould. Moreover, in the pouring temperature range of cast iron, the emitted radiation quickly changes in colour with temperature. This gives additional, qualitative information with respect to the temperature distribution during mould filling.

Fig. 1 shows a characteristic picture during filling of a bottom gated plate casting. The casting measures 250 x 250 mm and has a thickness of 4 mm. When the liquid metal enters the plate cavity, a 'mushroom' is formed which maintains a constant height for some time. The experimental height as measured from the film is compared in Fig. 2 with simulation results. Two kinds of experiments are shown in this figure; one series poured by hand and another series poured with an overflow pouring cup in which a constant height is maintained during mould filling. In the last series the liquid metal in the pouring cup is higher which results in higher values of the mushroom height. Although experiments which are carried out in identical circumstances never produce identical flow patterns during filling of the mould, the dispersion on these results is limited. Fig. 2 gives additional information on the dispersion of the experimental mould filling results. The differences in pouring time when using a constant level in the pouring cup are small. These differences are much larger when pouring by hand. Here the level in the pouring cup constantly changes. Fig. 2 shows that the simulation is in close agreement with the experiments; it produces an average mould filling sequence of the experiments. The simulated mould filling time for pouring by hand is somewhat higher than for the experiments. This is due to different initial conditions in the pouring cup. The metal stream falling down from the ladle spout into the pouring cup will create an irregular downward velocity field. On the contrary, in the simulation model the liquid metal level in the pouring cup is constant and the initial velocities at the free surface are zero. Fig. 2 confirms the experimental observations which show that cast iron and Al–Si produce the same flow behaviour during mould filling. Experiments of mould filling of a stepped side gated plate casting confirm these results [2].

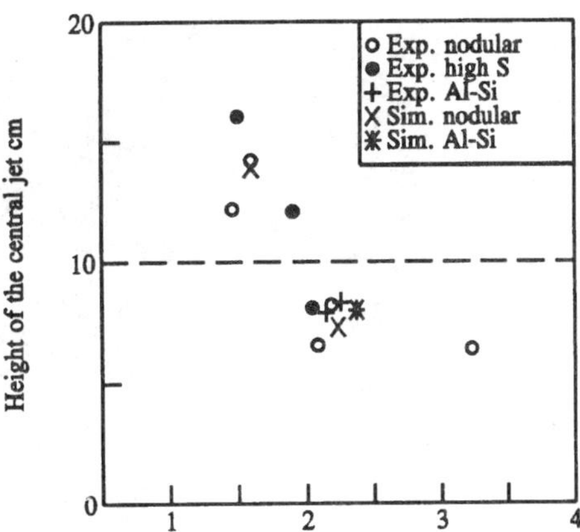

Fig. 1. Typical stage during mould filling of the bottom gated plate casting showing an initial jet in the shape of a mushroom (Experiment)

Fig. 2. Height of the central jet when filling the plate casting with bottom gating: with an overflow pouring cup (upper part) and by hand pouring (lower part).

The influence of surface tension on mould filling has been studied experimentally by comparing the flow of nodular and high sulphur cast iron. The surface tension of nodular irons usually ranges between 1.3 and 1.4 N/m [14]. The surface tension of cast iron with a high sulphur content (0.29 %, experiment) has a value between 0.7 and 0.8 N/m [14]. Despite the high difference in surface tension between both types of cast irons, no differences in flow behaviour could be observed in the experiments (Fig. 2). Simulations confirm that the surface tension has little influence during the mould filling of the vertical plate casting. Indeed, the influence of the surface tension is small as compared to that of the gravity force. This may be illustrated by simulating the mould filling of the vertical plate casting with bottom pouring. The pouring time increases with 0.05 s when surface tension is taken into account. The height of the central jet is 13.5 cm with surface tension and 13.8 cm without.

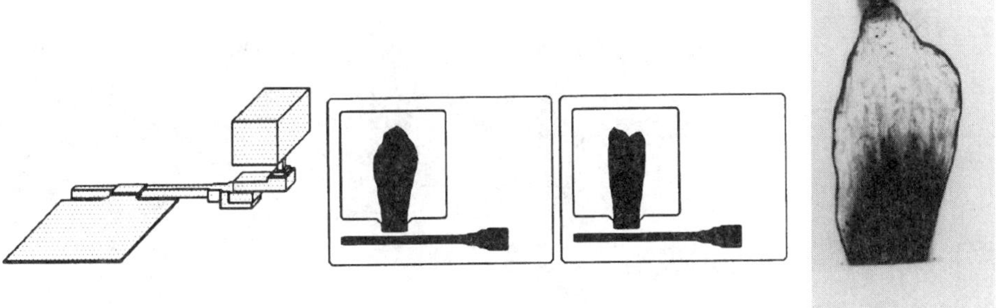

Fig. 3. Form left to right: Geometry model for the horizontal plate casting. Liquid metal entering the plate cavity by simulation with surface tension and without; by experiment.

For horizontal plate castings, surface tension has more effect on the flow. The influence of the surface tension is illustrated for the case of the horizontal plate casting (Fig. 3). This figure shows the simulation results with and without surface tension. Without surface tension liquid metal flows parallel into the cavity. When surface tension is taken into account in the simulation, the flow gradually spreads out as observed in the experiment (Fig. 3). It should be mentioned that if the inlet

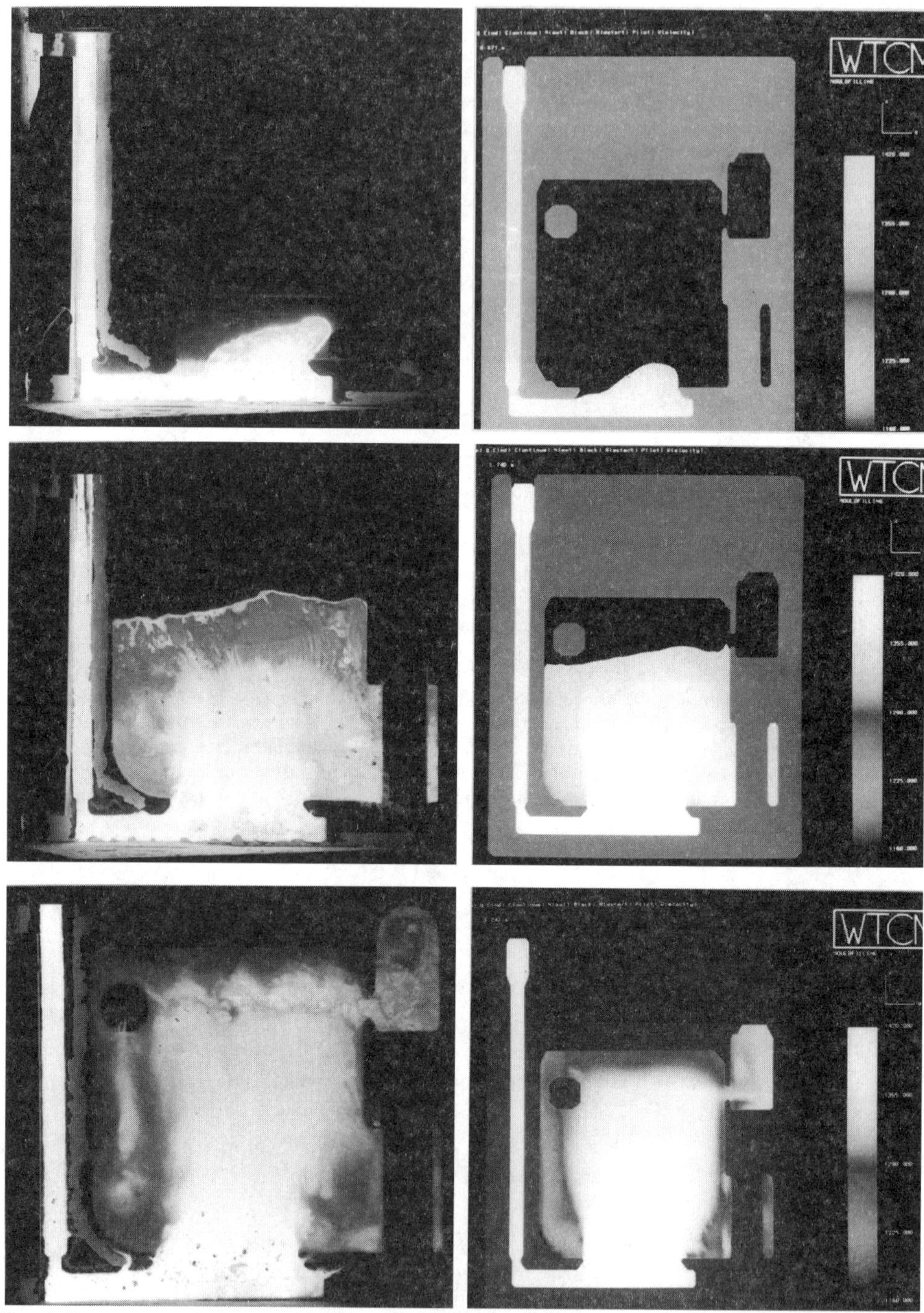

Fig. 5. Some stages during mould filling of a half part of the casting shown in Fig. 4, by experiment (left) and by simulation (right).

velocities are high enough, the surface tension pressure will be relatively small as compared to the internal pressure in the metal. In this case, the influence of the surface tension will be less important.

4. EFFECT OF MOULD FILLING

The effect of mould filling is illustrated for a more complex shaped spheroidal graphite iron casting as shown in Fig. 4. The calculated mould filling sequence has been validated by filming a casting which is composed of nearly one half of the casting shown in Fig. 4; the vertical plate of 3 mm thickness is retained in the mould and covered with the glass window. Some stages during mould filling are shown in Fig. 5. High temperature differences are created in the plate during the filling stage; these differences reach 230 K between the ingate (Fig. 4, C) and the lower left corner (Fig. 4, F). Additionally, the sand in the vicinity of the mould wall is heated during mould filling by the liquid metal. Both metal and sand temperature will have a profound influence on the subsequent cooling rate during solidification which directly influences the grain count. Application of a nucleation and grain growth model allows to simulate the resulting nodule counts.

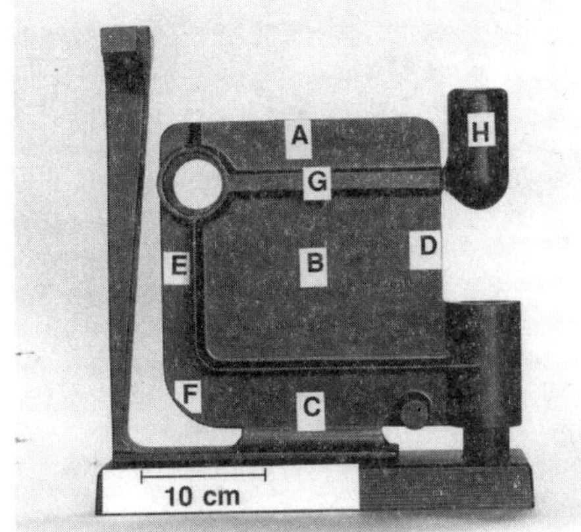

Fig. 4. The spheroidal graphite iron casting. Nodule counts at locations A to H are listed in Table I.

Table I. Nodule count (/mm^2) in various locations of the casting shown in Fig. 4.

Location (a)	Experiment	Simulation (b)
A (3 mm)	3288	3221
B (3 mm)	2375	2613
C (3 mm)	1917	1774
D (3 mm)	3350	3026
E (3 mm)	3525	3274
F (3 mm)	2817	3000
G (20 mm)	542	601
H (50 mm)		
centre	894	903
edge	523	350

(a) Casting thickness or diameter (H).
(b) With mould filling.

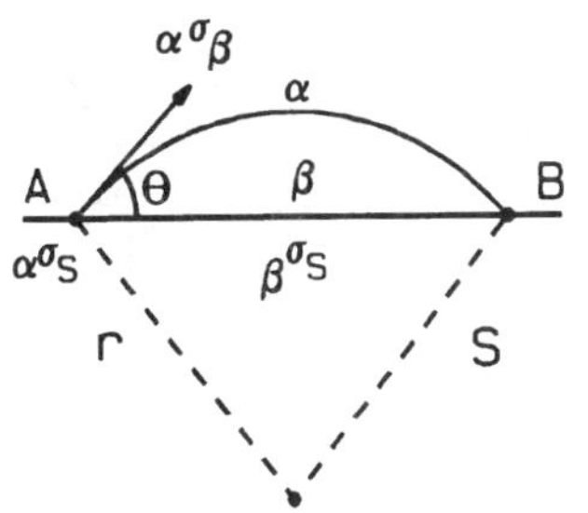

Fig. 6. Model for heterogeneous nucleation.

The model to predict the grain count [15,16] is based on the theory of heterogeneous nucleation. The general model for heterogeneous nucleation is represented as a spherical segment on a flat surface (Fig. 6). Graphite β of the eutectic grows from the melt a on the flat substrate S. This model leads to a formula for the critical radius r*

$$r^* = - \frac{2 \sigma_{\alpha\beta}}{\Delta G_v} \qquad (4)$$

Substituting $\Delta G_v = (\Delta H_S \ \Delta T) / T_F$ in Eq. 4 results in

$$r^* = -\frac{2 \; \alpha\sigma_\beta \; T_F}{\Delta H_S \; \Delta T} \tag{5}$$

$\alpha\sigma_\beta$ represents the interfacial energy between the phases α and β, θ is the angle of contact, ΔG_V the volume free energy change, ΔH_S the enthalpy change, T_F the solidification temperature and ΔT the supercooling.

The critical radius r^* may be related to the width AB of the flat substrate

$$AB = 2\, r^* \sin\theta \tag{6}$$

giving finally

$$AB = -\frac{4 \; \alpha\sigma_\beta \; \sin\theta \; T_F}{\Delta H_S \; \Delta T} = K_2 \frac{1}{\Delta T} \tag{7}$$

The use of other models than the one represented in Fig. 6 might modify the form of Eq. 6. However some relation will exist between the critical radius and the minimal width of the substrate. The nature of the particles influences the minimal width AB by changing the contact angle θ. Since it is not clear if inoculation introduces particles of different nature simultaneously, the present simulations have only taken one into account. Otherwise, multiple values of K_2 should be taken into account. The latter poses no special problems.

The hypothesis put forward now is that a new solid grain with a critical radius r^* can only be formed if a solid particle is present in the melt having a substrate of minimal width AB (Eq. 7). The width of all particle planes available in the liquid is supposed to have a normal distribution. Consequently as the melt becomes more supercooled, the length of the critical radius decreases and hence more nucleation sites will become available. Their number will increase according to $(1-\Psi(t))$, $\Psi(t)$ being the unit cumulative normal distribution function. By introducing a logarithmic scale, the normal distribution may be defined between $-\infty$ ($\Delta T = \infty$; $1/\Delta T = 0$) and $+\infty$ ($\Delta T = 0$; $1/\Delta T = \infty$).

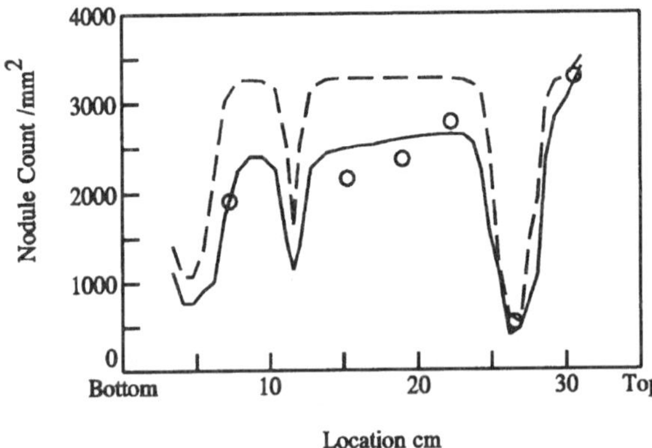

Fig. 7. Nodule count along the vertical line from A to C of the casting shown in Fig. 4. Circles represent experimental measurements. The solid line gives the simulated nodule count based on mould filling (pouring temperature 1420 °C); The dashed line results from an uniform initial temperature distribution of 1320 °C at simulation start.

The present nucleation model has been applied to calculate the nodule count distribution in the casting shown in Fig. 4, starting from the simulated temperature distribution in the sand mould and in the casting at the end of the filling sequence. Experimental and simulated nodule count values are listed in Table I. The nucleation parameters used in the microstructure simulation are: mean 400 K (equal to the supercooling at which half of the nucleation sites would become available); standard deviation 0.35 (expressed as log(1/ΔT)) and total number of nucleation sites 6.0 10^{15}/m^3. Examination of Table I reveals a good agreement between simulated (with mould filling) and experimental nodule counts. The simulated results obtained from a uniform initial temperature field without mould filling are not representative. This is illustrated by Fig. 7 which shows the nodule count along the vertical line from A to C (Fig. 4). Examination of Fig. 7 reveals that the nodule count in the 3 mm thick section of the casting decreases from the top to the bottom of the casting. The simulated nodule count obtained with mould filling is in good agreement with the experimental data. In the lower part of the casting the hot metal heats up the sand of the mould during the filling of the cavity. This results in lower cooling rates and less supercooling and hence in a smaller nodule count. A constant initial temperature distribution creates identical cooling conditions and nodule counts in all the 3 mm thick sections. The differences in the microstructure, i.e. the nodule counts, observed in 3 mm thick sections throughout the casting are only caused by different local temperature histories recorded during the filling sequence. It is obvious that any solidification simulation of thin wall castings which neglects the effect of the mould filling and starts from a constant initial temperature field throughout the casting and mould, has little practical value (Fig. 7). The higher nodule counts in the centre of the riser (Fig. 4, H and Table I) are caused by secondary nucleation. This phenomenon can occur towards the end of the solidification when the supercooling in a particular spot can become larger than at the solidification start.

5. NATURAL CONVECTION

Natural convection is induced in a body of fluid by buoyancy forces due to density perturbations within the fluid. In the liquid state, these density variations result from temperature inhomogeneities set up during filling of the mould and the subsequent cooling. In the case of alloys, buoyant flow can also be driven by mass transfer during solidification. When solute of different density is rejected at the solidification front density inhomogeneities are build up in the liquid. Solutal convection will not be treated in this paper. The existence of thermal convection in a casting is significant because of several reasons. Convection increases the heat transfer from the thermal centre to the mould wall and changes the temperature profile in the casting. Additionally, it influences to some extent the solidification pattern. Little work has been published on convection in real casting conditions.

Cole [17] has measured temperatures ahead of a stationary solidification front in a tin melt under conditions of unidirectional heat flow. These experiments showed that natural convection flow patterns may be deduced from the temperature distribution and the shape of the interface. Szekely and Chhabra [18] carried out similar experiments in liquid lead. They showed that the movement of the melt – solid interface during transient experiments could be predicted from the numerical solution of the unsteady heat conduction equation. These experiments [17,18]show that in the presence of convection, the solidified layer is thicker at the bottom than on the top of the mould cavity. Stewart and Weinberg [19] directly observed thermal convection in liquid tin by using radioactive tracer techniques in combination with quenching. They used a rectangular cavity which was subjected to a steady state heat flux. The overall flow behaviour of liquid metals differs considerably from that of non–metallic fluids due to the differences of the controlling parameters. Calculations show that the convective flow pattern and flow rate of water and liquid tin will be quite different in a system of the same geometry and having equal thermal boundary conditions [20]. The steady state experiments where convection is generated by holding the boundaries to constant temperatures above the solidification temperature indicate that the flow velocity increases with the temperature difference between hot and cold ends and with the average temperature of the melt [19,21].

The steady state experiments where convection is generated by holding one boundary to a constant temperature above the solidification temperature are not good analogues of industrial solidification processes. In real castings there is only a finite amount of superheat is available in the melt which will create a transient convection process. Temperature measurements in cylindrical castings allow Engler and Jansen [22] to conclude that convection currents are always present during liquid cooling in aluminium and copper alloys and cast iron. The intensity of the thermal convection increases with when the time between pouring and solidification start becomes longer. The type of the solidification morphology determines the continuation of the currents. With an exogenous smooth – walled solidification type [23], the convection remains effective long during solidification. However, mushy solidification or dendrite growth obstruct the currents. In additional experiments, Engler and Jansen [24] applied the flow – out method. During the solidification of a cylindrical casting, the mould and bottom of the casting are pierced and the remaining aluminium melt flows out. These experiments show that in the presence of convection, the thickness of the solidified peripheral layer is not parallel to the vertical mould wall. The authors show that the growth direction of dendrites is influenced by convection too. The dendrite tips grow against the direction of the flow, the inclination of the dendrite axes being higher with higher flow rates.

Mampaey [4] studied natural convection in cylindrical castings by combining temperature measurements, a tracer technique, quenching experiments and a simulation model. The tracer technique allows to visualize the convection currents in cast iron. The quenched samples show a thicker solidified layer at the bottom of the casting. As the calculations are a good agreement with the experimental data, the model is reliable and hence can provide much more information than can be obtained by experiments only. Natural convection indeed depends on the geometry of the mould cavity, on the superheat of the liquid metal, on the solidification morphology and on the material properties of the liquid. Mampaey [25] observed different radial nodule count distributions in hypo– and hypereutectic compositions. Equal nodule counts along the casting radius in hypereutectic spheroidal graphite cast iron are explained by convection currents which transport the graphite nodules from the peripheral layer to the centre of the casting. In a hypoeutectic composition, the dendrite network hinders convective flow and nodule transport. This results in higher nodule counts in the peripheral layer because of higher undercooling close to the mould wall.

Finally, natural convection is one of the factors which may have an influence on the columnar to equiaxed transition [26]. Dendrites arms which are detached from the trunks are carried by the convection currents to the centre of the casting which permits the development of the equiaxed zone [27]. The equiaxed growth will be promoted by convection because it lowers the radial thermal gradient.

6. THERMAL INFLUENCE OF CONVECTION

The changes of the melt thermal profile due to convection, depends on the physical properties of the liquid and more specific on the value of a dimensionless number, the Prandtl number Pr, defined as:

$$Pr = \mu\, c\, /\, \lambda \tag{8}$$

with μ the viscosity, c the specific heat and λ the thermal conductivity. For low values of Pr, thermal convection will have less influence on the temperature field. Liquid metals which have high values for the thermal conductivity, have low Pr numbers ; e g 0.012 for Cu and Al. This reduces the effect of the convection on the temperature profile. Unfortunately, the thermal conductivity of liquid cast iron is not available [28]. Its value may be estimated by applying a theoretical relationship, the Wiedemann – Franz – Lorenz law, which exist between the electrical and thermal conductivities of solid metals at high temperatures [29]:

$$\lambda\, \rho_e = 2.45 \times 10^{-8}\, T \tag{9}$$

with ρ_e the electrical resistivity and T the Kelvin temperature. Moreover, measurements for a number of liquid metals are in agreement with this law too. Stefanescu and Craciun [30] have measured the electrical resistivity during solidification of cast iron. They obtained a value of 1.61 μ Ω.m for liquid cast iron close to solidification which allows to estimate the thermal conductivity of liquid cast iron at 21.6 W/m.K. This value is very close to thermal conductivity of white iron at high temperature (19.6 W/m.K at 1073 K) [28]. Accordingly, the Pr number for liquid cast iron can be estimated at 0.42 which is considerably higher than that of Cu and Al. Hence the effect of convection on the thermal profile will be higher in cast iron than in Al or Cu.

Fig. 8 shows the simulated cooling curve, the location of the thermal centre and the history of the velocities after the mould filling of a cylindrical casting. Natural convection is induced as soon as the liquid metal starts to cool at the mould wall, it initiates relatively high velocities in the liquid metal which are in the order of cm per s. However these velocities are about one to two orders of magnitude lower than those introduced by mould filling (order of 1 m/s). The forced flow pattern in the liquid metal resulting from the mould filling quickly disappears once the pouring is completed and is substituted by the natural convection driven flow. For the cylindrical casting of Fig. 8, mould filling takes 3 s, after another 3 s the flow in the casting with mould filling becomes identical to the one which starts from a stagnant liquid. Due to the convection currents, the thermal centre is shifted from the geometrical centre to the top of the casting (Fig. 9, left).

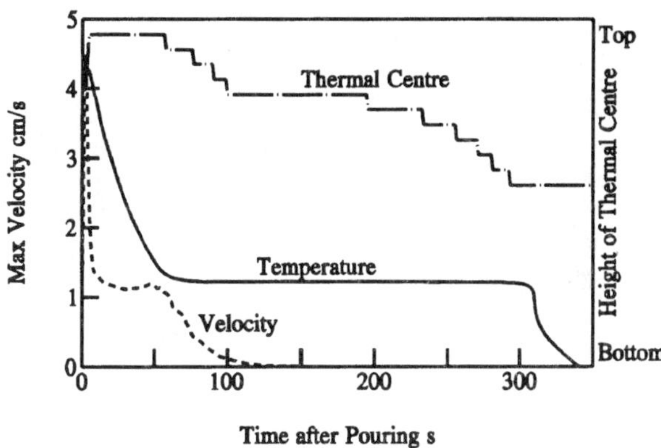

Fig. 8. Simulated maximal absolute velocity in the liquid metal, maximal temperature and height of the thermal centre along the casting axis (bar diameter 38 mm, height 120 mm). The vertical temperature scale ranges between 1050 °C and 1400 °C.

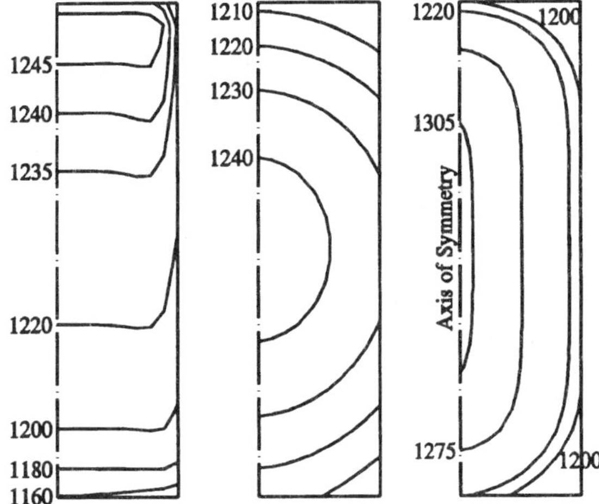

Fig. 9. Simulated temperature profiles 30 s after pouring; with natural convection (left); without convection and enlarged thermal conductivity [167 W/m.K] (middle); with 'normal' thermal conductivity [22 W/m.K] (right) (bar diameter 50 mm, height 120 mm).

During liquid cooling in the mould, the maximal velocities in the casting remain nearly constant and some kind of steady state is reached. Once the solidification starts, a solid shell of uniform temperature is formed around the liquid pool and the superheat in the liquid flows away to the mould. In this stage which makes up most of the solidification, the temperature difference between the liquid interior and the solidification front becomes very low and eliminates the driving force for thermal convection. The velocities gradually become zero and the heat transfer is governed by conduction only. As a result, the thermal centre shifts back from the top of the casting to the geometric centre. The non–symmetric temperature profile in the casting created by natural convection, persists long after the convection currents have disappeared. However, the last spot to solidify in the convection model will only be located 5 mm higher than in the stagnant liquid model.

The simulation results are confirmed by experiments which are based on the examination of cooling curves and quenching. Test castings are cylindrical bars of different diameter (25, 38, 50 mm) and 120 mm height. The castings have been vertically cast; pouring was done through the top surface in the absence of any pouring or gating system. The pouring time varies between 2 and 4 seconds. The relocation of the thermal centre is illustrated in Fig. 10. Cooling curve analysis reveals that the solidification first starts at the bottom of the casting and last close to the top of the casting. This is due to the vertical shift of the thermal centre. On the other hand, the last spot to solidify is located in the middle of the casting, indicating that the thermal effect of convection has disappeared by then.

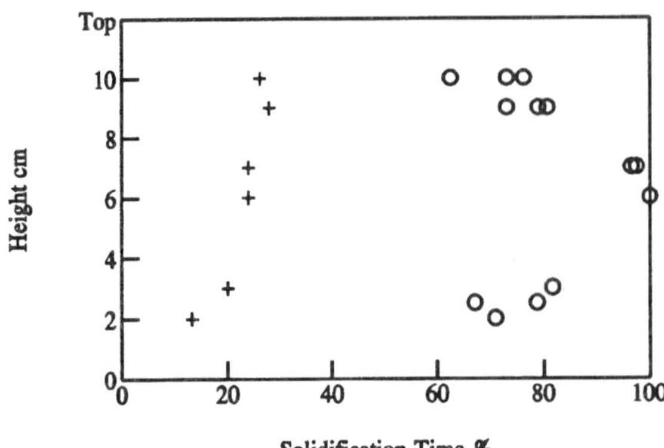

Fig. 10. Time at which solidification starts (+) and ends (o) in a cylindrical casting (Experiment).

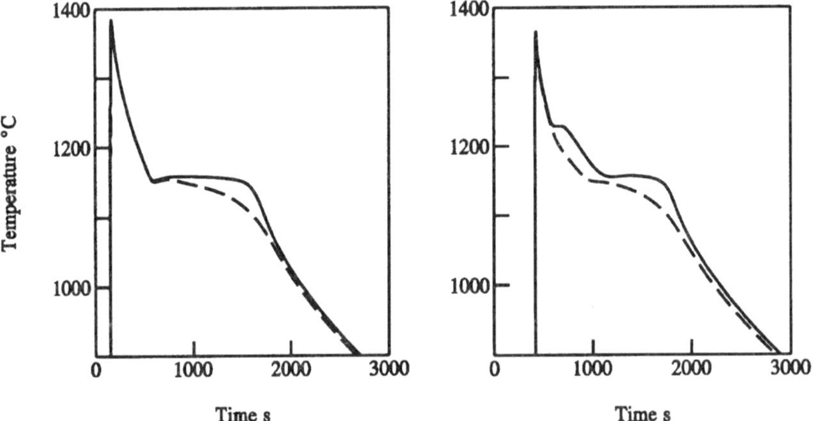

Fig. 11. The effect of natural convection on experimental cooling curves as recorded in the centre (solid line) and in the vicinity of the mould wall (dashed line) in a cylindrical casting (diameter 93 mm, height 185 mm). With natural convection (left) and without any major metal flow during primary dendrite precipitation (right).

Convection substantially increases the heat transfer from the thermal centre of the casting to the mould as compared to conduction only. The increased thermal transfer may be illustrated by another simple experiment. Fig. 11 shows cooling curves measured at mid-height in the centre and close to the peripheral layer in a cylinder with diameter of 93 mm and 185 mm in height. With a eutectic composition, the temperature curves nearly coincide during liquid cooling of the melt; on the average the temperature measured in the centre is only 0.3 K above the one measured at 7.6 mm from the mould wall. This confirms the simulated temperature profile of Fig. 9, left. When a hypoeutectic composition is poured, the dendrite network will hinder and probably obstruct [22], the currents resulting from thermal convection. Consequently, mainly conduction remains and the temperature difference between the centre and the peripheral layer (4 mm from the mould wall) increases from 5.8 K before primary solidification to 35.2 K during austenite growth (Fig. 11, right).

Fig. 12 shows in more detail the temperature differences measured along the casting radius at mid-height of the cylinder. This experiment shows that substantial temperature differences with the centre only exist in the peripheral layer with a thickness of 7.9 mm (casting radius 46.5 mm). The temperature quickly decreases in the vicinity of the mould wall; a max temperature difference of 13 K has been measured at 2.2 mm from the wall. This temperature profile is reflected too by the convection model: no radial temperature differences in the central part of the casting and a steep temperature decrease in the peripheral layer. The small shift between the simulated curve and the experimental data may be due to the presence of the thermocouple protection tubes which may disturb the convection currents to some extent.

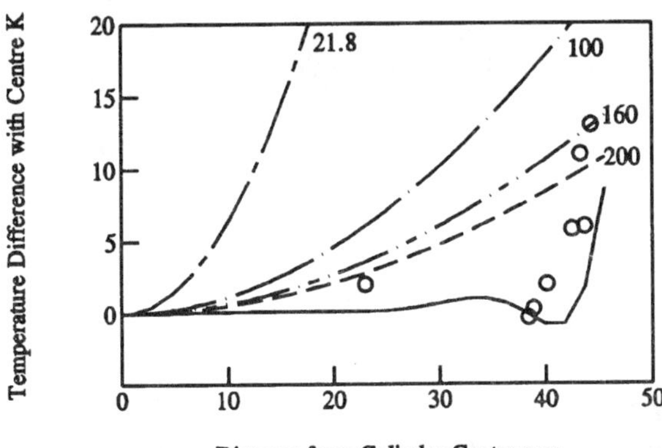

Fig. 12 Temperature difference with the centre along the casting radius at mid–height of a cylinder (diameter 93 mm, height 185 mm). Positive values indicate a higher temperature in the centre.

In solidification simulations which do not treat liquid flow, the effect of convection is approximated by artificially enlarging the thermal conductivity of the liquid metal. Fig. 12 demonstrates that a substantial increase is necessary in order to obtain radial temperature differences which are close the experimental ones. Enlargement of the liquid thermal conductivity reduces the radial temperature difference but cannot account for the vertical shift of the thermal centre in the casting (Fig. 9, middle). Moreover the vertical temperature difference decreases too which is in contradiction with the convection induced temperature field. No enlargement of the liquid metal thermal conductivity leads to erroneous, high radial temperature differences (Fig. 9, right).

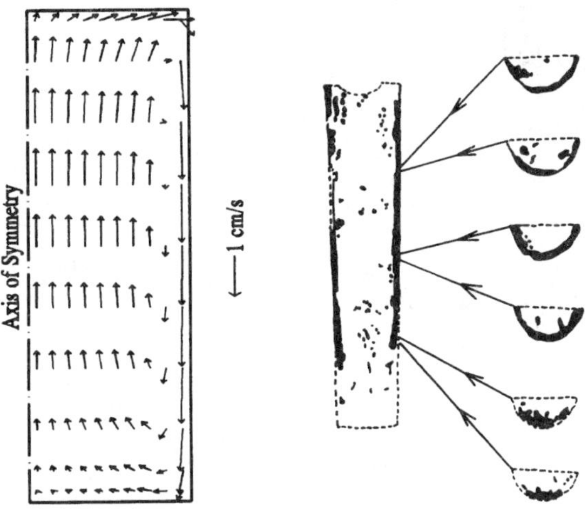

Fig. 13 (left). Simulated flow pattern in a cylindrical bar (diameter 50 mm, height 120 mm) 12 s after simulation start.

Fig. 14 (right). Visualization of the convection currents in a cylinder (diameter 25 mm). The peripheral layer of high sulphur content (dark) moves down along the mould wall.

7. VISUALIZATION OF CONVECTION CURRENTS

The simulated flow pattern which is created by natural convection is shown in Fig. 13. Liquid metal cools at the mould wall, increases in density and moves down in the peripheral layer which has an approximate thickness of 2 mm. This phenomenon may be visualized experimentally by adding sulphur on the liquid top surface of the cylindrical castings. The test castings are either allowed to cool to room temperature or are quenched. This tracer technique does not interfere with the original flow pattern. The advantage of sulphur is associated with the transition from spheroidal graphite to lamellar graphite when the sulphur content of the spheroidal graphite base iron increases. This effect enables to distinguish regions of low and high sulphur content during the subsequent metallographic examination. The macroscopic flow of the convection currents becomes visible after sectioning, grinding and etching the castings. Regions containing high amounts of sulphur are darkened and can be clearly distinguished from the base melt. Fig. 14 shows a macroscopic example of the convection currents in a cylindrical casting and Fig. 15 a microscopic detail at the mould wall for castings normally cooled to room temperature. Fig. 14, 15 reveal a peripheral layer of irregular thickness and several isolated spots of high sulphur content randomly distributed in the interior of the casting. The interfacial layer between lamellar and nodular graphite is relatively thin.

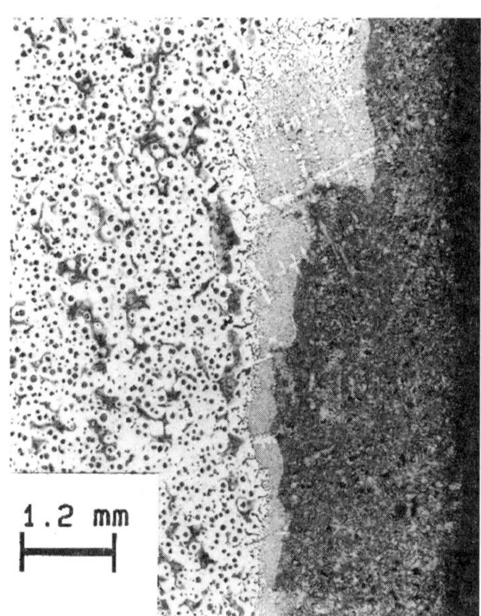

Fig. 15. Peripheral layer of high sulphur content (solidified as a lamellar graphite eutectic) moves down along the mould wall (black). The original melt solidifies as spheroidal graphite cast iron (left).

8. CONVECTION INFLUENCE ON SOLIDIFICATION

In the absence of convection the solidified peripheral layer should be parallel to the vertical casting surface at mid–height of the casting. Convection distorts this solidification profile (Fig. 16). As a result of the shifted thermal centre, the solidified peripheral layer is smaller towards the top of the casting. This effect has been confirmed by experiment. Fig. 17 shows the thickness of the solidified peripheral layer for uninoculated lamellar graphite cast iron with eutectic composition. These results have been obtained by quenching experiments. Fig. 16, 17 illustrate that long after the convection currents have disappeared, their effect lasts.

Experimental cell counts in a plane of the cylinder axis are shown in Fig. 18 for a slightly hypereutectic (CE 4.42) and hypoeutectic (CE 3.82) lamellar graphite cast iron. The presence of a dendrite network in the hypoeutectic iron will inhibit natural convection before the eutectic nucleation starts. Both cell count distributions are quite similar and do not seem to be influenced to a large extent by natural convection. The cell count along the casting radius (Fig. 18, right) in a hypereutectic composition (CE 4.60) shows high cell counts close to the mould wall.

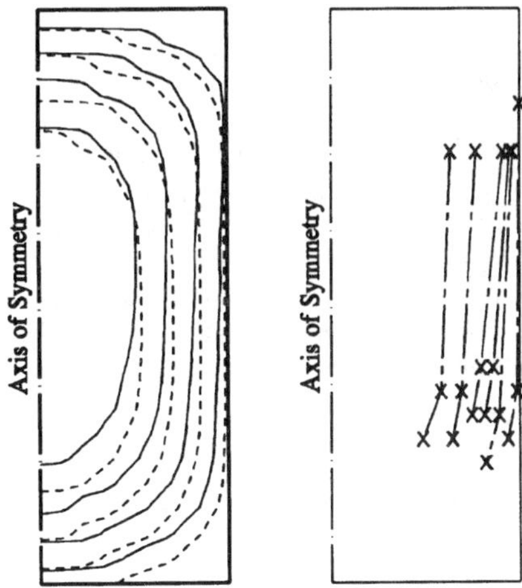

Fig. 16. Simulated thickness of the peripheral layer; with (solid lines) and without (dashed lines) natural convection (bar diameter 38 mm; 120, 160, 200 and 240 s after pouring) (left)

Fig. 17. Thickness of the peripheral layer obtained by quenching experiments during solidification (bar diameter 38 mm) (right)

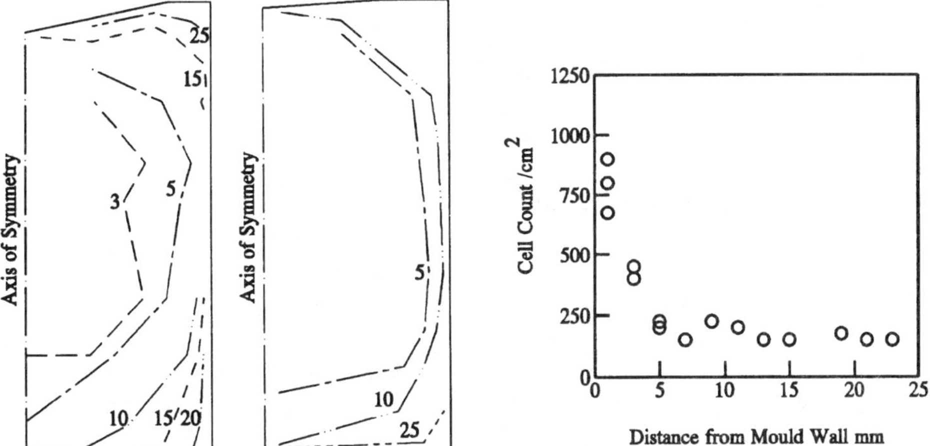

Fig. 18. Experimental eutectic cell counts (/mm^2) in lamellar graphite cast iron, hypoeutectic composition (left) and hypereutectic composition (middle) (bar diameter 38 mm). Cell count along the casting radius at mid–height is shown right (bar diameter 50 mm).

As opposed to lamellar graphite cast iron, hypereutectic (CE 4.60) spheroidal graphite castings show a constant nodule count from the peripheral layer towards the thermal centre of the casting for bar diameters from 10 − 50 mm (Fig. 19). The higher supercooling at the peripheral layer does not result in higher nodule counts. This might be explained by natural convection. Graphite nodules initially grow in direct contact with the melt [31,32]. Since their volume fraction will be very low during this stage, convection currents can remove the nodules from the peripheral layer towards the centre of the casting. This hypothesis is verified by the nodule count distribution in a hypoeutectic (CE 4.09) cast iron. The primary dendrite network obstructs natural convection during eutectic nucleation and results in higher nodule counts at the peripheral layer of the casting (Fig. 19).

Secondary nucleation increases the nodule count in the thermal centre of the casting. It seems reasonable to accept that the graphite nodules which nucleate during the eutectic stage are moved too by the convection flow. If not, higher nodule counts would be expected in the peripheral layer of the hypereutectic composition.

Since the density of graphite (2250 kg/m^3) is considerably lower than that of liquid cast iron (7200 kg/m^3), an upward Archimedes' force acts on the graphite nodules. The upward velocity is given by Stokes' law and is equal to 2.6×10^{-5} m/s for a nodule with a diameter of 5 μm. Wetterfall et al have observed nodules with this diameter without an austenite shell in the liquid [31]. This velocity is about 3 orders lower than the convection flow rate (1 cm/s, Fig. 8). Consequently, it may be expected that the graphite nodules are transported by the convection currents. Flotation of the nodules will become possible when the convection has disappeared.

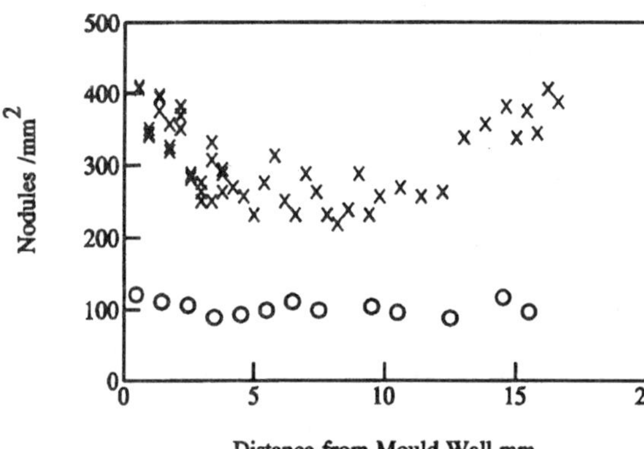

Fig. 19. Nodule count along the casting diameter (38 mm) at half height; composition hypereutectic CE 4.60 (o) hypoeutectic CE 4.09 (x).

The present experiments with cast iron and those of Engler and Jansen [22] with aluminium alloys show that the presence of a dendrite network hinders the convection flow so that its effect cannot be measured any more in the present casting conditions. For the case of cast iron, two reasons let explain this assessment. First, the rejection of carbon during primary dendrite growth decreases the liquid density. The density change may be estimated at 8.0×10^{-3}/%C when applying a simple balance based on the molar weights of iron and carbon. Experimental data [14] give a value of 3.4×10^{-2}/%C. These values may be related to the temperature change by using the liquidus slope which gives -5.7×10^{-5}/K to -2.4×10^{-4}/K (density decrease at lower temperatures). Since the temperature change has an opposite effect ($+1.0 \times 10^{-4}$/K, density increase at lower temperatures) of the same order, the driving force for convection flow in the interdendritic liquid decreases. The second reason is due to the drastic reduction of the permeability of the dendritic medium [33]. According to the Darcy law, the velocity in a porous medium is proportional to its permeability. As a result, the interdendritic flow due to solidification contraction and density differences in the liquid is much smaller than the convection currents present in the melt after pouring. Ohnaka and Matsumoto [34] reported flow rates in the order of 0.1 mm/s, present in the interdendritic liquid of a solidifying ingot.

9. CONCLUSIONS

The mould filling of thin wall castings has a decisive influence on the final nodule count distribution. This results from the inhomogeneous temperature distribution in the liquid metal at the end of the filling and from different temperature histories in the sand layers close to the mould wall. The present nucleation model which is based on the classic theory of heterogeneous nucleation permits to calculate a nodule count distribution in a complex thin wall casting which is in good agreement with experimental measurements.

Natural convection temporary shifts the thermal centre upwards, increases the thickness of the solidified layer from the top towards the bottom of the casting and reduces the radial temperature differences in the melt. These phenomena have been demonstrated by combining a simulation model with temperature measurements and quenching experiments. Convection currents have been visualized by a simple tracer technique. Enlargement of the liquid thermal conductivity is not a good approximation of the higher thermal transport in the melt due to convection. There is experimental evidence that graphite nodules are transported by the convection flow in hypereutectic and eutectic compositions.

REFERENCES

[1] Z.A. Xu, F. Mampaey, 60th World Foundry Congress (1993) paper 18.
[2] Z.A. Xu, F. Mampaey, Trans. AFS (1994) 94–43.
[3] Z.A. Xu, F. Mampaey, Modeling of Casting, Welding and Advanced Solidification Processes VI, T.S. Piwonka et al editors, TMS Warrendale Pensylvania USA 485 (1993).
[4] F. Mampaey, 56th World Foundry Congress (1989) paper 10.
[5] Z.A. Xu, Dr thesis Univ. Ghent Belgium (1993).
[6] H. Nieswaag, H.J.J. Deen, 57th World Foundry Congress (1990) paper 10.
[7] C.W. Briggs, The Flow of Steel in Molds, Steel Founders' Soc. Cleveland, Ohio (1958).
[8] S.T. Andersen, P. Ingerslev, 50th Int. Foundry Congress (1983) paper 7.
[9] P. Ingerslev, S.T. Andersen, 54th Int. Foundry Congress (1987) paper 9.
[10] R.A. Stoehr, C. Wang, W.S. Hwang, P. Ingerslev, Modeling of Casting and Welding Processes III, S. Kou and R. Mehrabian, TMS Warrendale Pensylvania USA 303 (1986).
[11] H. Walther and P.R. Sahm, Giessereiforschung, **38**, 119 (1986).
[12] G. Sciama, D. Visconte, Fonderie **81**, 15 (1989).
[13] S.H. Jong, W.S. Hwang, Trans. Japan Foundrymen's Soc. **11**, 56 (1992).
[14] H.T. Angus, Cast Iron: Physical and Engineering Properties, Butterworths London (1976).
[15] F. Mampaey, Modeling of Casting, Welding and Advanced Solidification Processes V, M. Rappaz et al. editors, TMS Warrendale Pensylvania USA 403 (1991).
[16] F. Mampaey, 55th Int. Foundry Congress (1988) paper 2.
[17] G.S. Cole, Trans. AIME **239**, 1287 (1967).
[18] J. Szekely, P.S. Chhabra, Metall. Trans. **1**, 1195 (1970).
[19] M.J. Stewart, F. Weinberg, J. Crystal Growth **12**, 228 (1971).
[20] M.J. Stewart, F. Weinberg, J. Crystal Growth **12**, 217 (1972).
[21] L.C. MacAuley, F. Weinberg, Metall. Trans. **4**, 2097 (1973).
[22] S. Engler, R. Jansen, Giessereiforschung **29**, 31 (1977).
[23] A. Rickert, S. Engler in The Physical Metallurgy of Cast Iron, ed. H. Fredriksson, M. Hillert North–Holland, New York 165 (1985).
[24] S. Engler, R. Jansen, Giessereiforschung **29**, 59 (1977).
[25] F. Mampaey in Advanced Materials and Processes, ed. H.E. Exner, V. Schumacher DGM Oberursel Germany, 129 (1990).
[26] S.C. Flood, J.D. Hunt in Metals Handbook vol. 15 Casting ASM 130 (1988).
[27] K.A. Jackson, J.D. Hunt, D.R. Uhlmann, T.P. Seward Trans. AIME **236**, 149 (1966).
[28] R.D. Pehlke, A. Jeyarajan, H. Wada, Summary of Thermal Properties for Casting Alloys and Mold Materials, Nat. Science Foundation NSF/MEA–82028 (1982).
[29] T. Iida, R.I.L. Guthrie, The Physical Properties of Liquid Metals, Clarendon Press, Oxford (1993).
[30] D.M. Stefanescu, St. Craciun, Giesserei–Praxis 197 (1973).
[31] S.E. Wetterfall, H. Fredriksson, M. Hillert, JISI **210**, 323 (1972).
[32] M. Hecht, J.C. Margerie, Mém. Sc. Revue Mét. **78**, 325 (1971).
[33] T.S. Piwonka, M.C. Flemings, Trans. AIME **236**, 1157 (1966).
[34] I. Ohnaka, M. Matsumoto, J. Iron Steel Inst. Japan **73**, 1698 (1987).

Advanced Materials Research Vols. 4-5 (1997) pp. 89-104
© *1997 Scitec Publications, Switzerland*

Methodologies for and Performance of Macro Transport-Transformation Kinetics Modeling of Cast Iron

D.M. Stefanescu

Solidification Laboratory, Department of Metallurgical and Materials Eng.,
The University of Alabama, Tuscaloosa, AL 35487, USA

Keywords: Cast Iron, Microstructure Modeling, Transformation Kinetics Modeling, Gray Iron, Spheroidal Graphite Iron, Gray/White Transition, Microsegregation, Deterministic Modeling, Probabilistic Modeling, Nucleation, Eutectic Growth, Dendritic Growth

ABSTRACT

During the last decade, solidification modeling has known a sustained development effort, supported by academic as well as industrial research. The driving force behind this undertaking was the promise of predictive capabilities that will allow simultaneous process and material improvement. The most significant recent progress has been incorporation of transformation kinetics, for both the liquid/solid and the solid/solid transformation, in the macro-transport models. The results of these efforts have materialized in a proliferation of publications and commercial software, some of which have penetrated the industry. Numerous claims are made regarding modeling method accuracy and capabilities. They include prediction of casting defects, of microstructure length scale and composition, and even of mechanical properties. A reality test of these claims is the subject of the present paper.

The methodologies for macro transport - transformation kinetics modeling (MT-TK), and therefore for prediction of microstructural evolution, can be broadly classified as being based on the continuum approach (deterministic), or on the stochastic (probabilistic) approach, or, more recently on a combined approach. These methods and their applicability to modeling of cast iron solidification will be discussed.

Originally the MT-TK analysis was performed for the liquid/solid transformation. Subsequently it has been extended to the solid/solid transformation, thus resulting in prediction of room temperature microstructure. Specifically, for cast iron it has been attempted to predict features like microsegregation, graphite spacing, graphite nodule count, pearlite/ferrite ratio, the gray-to-white transition, hardness, microhardness, and tensile properties. The success of these efforts is critically reviewed in the paper.

INTRODUCTION

With the advent of faster computer paralleled by the rapid development of numerical methods, the metallurgical aspects of microstructure evolution have finally become a quantitative engineering science. Indeed, in the broadest definition of engineering science we know only what we can predict through mathematical models. As late as 1975, to solve the complete macro transport - transformation kinetics (MT-TK) problem was considered a "formidable problem" [1]. Microstructure prediction was strictly an empirical exercise, were elements of the microstructure were correlated with processing and material variables. However, today, as proven by the at least 14 papers that tackle various phenomena occurring during cast iron solidification through mathematical/numerical modeling, the task has lost its reputation for inaccessibility. A multitude of papers boldly claim to predict solidification microstructure at the end of solidification, or even at room temperature. From there it is only a short step to prediction of mechanical properties, and again, many papers have declared victory. The enthusiasm of the academic researchers has infected even skeptical industry based engineers and researchers, and computer modeling of solidification is now accepted "state of the art". But are we really there? This is the main question that will be addressed in the following paragraphs.

METHODOLOGIES FOR MODELING OF MICROSTRUCTURE EVOLUTION

The traditional method of predicting microstructure in castings relies on experimental measurements of some process parameters such as cooling rate ($\dot{T}$), temperature gradient (G), or liquid/solid interface velocity (V) in the casting. Alternatively, classic macro-transport solidification models generate a temporal and spatial description of the movement of the liquid/solid interface. From this description, local $\dot{T}$, V, and G can be obtained throughout the casting. Then, these process parameters are correlated with microstructure length scale (phase spacing λ, volumetric grain density N) and composition (fraction of phases).

A typical example is shown in Fig. 1 for eutectic grain density of cast irons with various carbon equivalents. Apparent discrepancies may result from different methods of evaluating the cooling rate.

The cooling rate results from the coupling of macro heat transfer from the casting to the environment (MT) with the heat evolution during solidification, which is dictated by transformation kinetics (TK). If it is assumed that TK does not influence MT, the two computations can be performed uncoupled. Typically $\dot{T}$ is evaluated from HT, and then λ and N are calculated based on empirical equations as a function of $\dot{T}$. This methodology is presently used by classic macro-transport models that solve the mass, energy, momentum, and species macroscopic conservation equations. They are inherently inaccurate in their attempt to predict microstructure because of the weakness of the uncoupled MT - TK assumption. Indeed, since in effect MT and TK are coupled during solidification, accurate prediction of microstructural evolution revolves around modeling both MT and TK, and then coupling them appropriately.

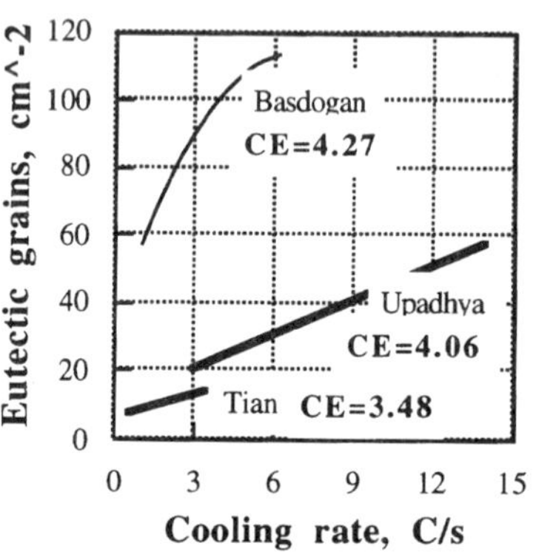

Fig. 1 Variation of volumetric eutectic grain density as a function of cooling rate for cast iron [2,3,4].

Significant progress has been recently made through the use of coupled MT - TK codes. Both deterministic and probabilistic models have been proposed.

Deterministic modeling of TK is based on the solution of the continuum equations over some volume element. First the computational space of the casting is divided in macro-volume elements within which the temperature is assumed to be uniform (Fig. 2). Each of these elements is further subdivided in micro-volume elements, typically spherical, based on some nucleation law. Within each of these elements only one spherical grain is growing at a velocity V, dictated by kinetic considerations. In summary, for the MT-TK models presently in operation, the macro-volume element is an open system in terms of mass transfer and energy, while the micro-volume element is isothermal and closed for mass transport.

The first approach to deterministic modeling of TK relies on the basic simplifying assumption that the solid phase has zero velocity (*one velocity models*), meaning that once nucleated the grains remain in fixed positions. Grain coalescence and dissolution are ignored. Several such models have been applied to cast iron solidification, *e.g.*, [5-11].

While the progress made by implementing the one-velocity MT-TK models is remarkable, their limitations are obvious if one considers for example the role of fluid transport of equiaxed grains during initial stages of solidification on the final microstructure. *Two-velocities models* that concomitantly solve the energy, mass and momentum equations of two-phase systems have been outlined by [12-14]. However, this method is not implemented in an MT-TK code at this time, and such an implementation is not trivial.

Wang and Beckerman [15] have proposed a multiphase solute diffusion model for dendritic alloy solidification. This model accounts for the different length scales in the dendritic structure. Macroscopic conservation equations were derived for each phase, using a volume averaging technique. While thermal undercooling was ignored, the model can incorporate coarsening and was successfully used to predict the columnar-to-equiaxed transition (CET).

The main limitations of the deterministic calculations, that are based exclusively on the continuum model, are that they ignore the discrete and crystallographic aspects of dendritic growth. To include such features in the model it is required to consider the nano-scale. However, to simulate for example 1 mm^3 of metal, 10^{11} atoms must be considered, which is beyond present hardware capabilities. Nevertheless, significant work has already been done to predict microstructure evolution through probabilistic calculations.

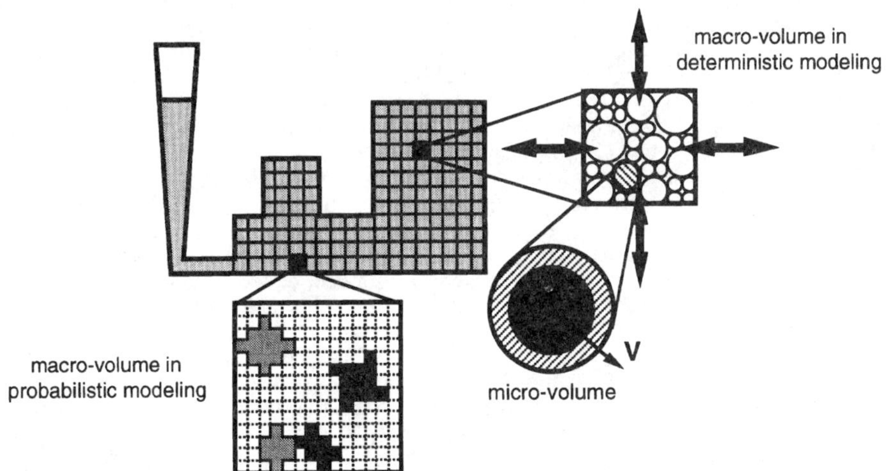

Fig. 2 Division of the computational space for deterministic and probabilistic modeling of macro-transport and transformation kinetics.

Probabilistic models have the advantage that individual grains are identified and their shape and size can be shown graphically throughout solidification. The computational procedure requires the following steps: the domain is divided into a regular network of cells; variables and states are attributed to the cells; a neighborhood is associated with each cell (e.g., the neighborhood of each cell is given by its four neighbors along the North, South, East, and West directions); transition rules that govern the evolution of the possible states and variables of the cells are defined. Phenomena that play an important role in as-cast grain structure formation, such as crystal transport in the melt and crystal remelting, can be simulated. By adapting a Monte-Carlo simulation developed for simulation of grain growth in solids, Spittle and Brown [16] have developed models that give qualitative predictions of the influence of processing variables (e.g., superheat, mold temperature) and of alloy parameters (e.g., m, k, C_0) on grain structure in single phase binary alloys. They have also examined the CET through these models. However, their model ignores the specifics of dendritic growth that include tip kinetics and preferred crystallographic growth direction, and therefore suffers from a lack of physical background. Furthermore, the Monte-Carlo time step is not related to the solidification time.

Xiao et al. [17] simulated microscopic solidification morphologies of binary systems using a probabilistic model that accounts for bulk diffusion, attachment and detachment kinetics, and surface diffusion. An isothermal two-component system contained in a volume element subdivided by a square grid was considered. Diffusion in the undercooled liquid was modeled by random walks on the grid. Through variation of interaction energies and undercooling, a broad range of microstructures was obtained, including eutectic systems and layered and ionic compounds. It was shown that, depending on the interaction energies between atoms, the microscopic growth structure can range from complete mixing to complete segregation. For the same interaction energy, as undercooling increases, the phase spacing of lamellar eutectics decreases. Thus, continuum derived laws, were recovered through a combination between a probabilistic model and physical laws. However, this model can only be used for qualitative predictions, because the length scale of the lattice, a, is several orders of magnitudes higher than the atomic size. Thus, this is not a nano-scale model. The model is limited to micro-scale level calculations, and cannot be coupled to the macro-scale because of computing time limitations. Nevertheless, the main merit of this work is that it has demonstrated that physical laws can be successfully combined with probabilistic calculation for TK modeling.

A significant progress has been made when some simplified TK was incorporated in probabilistic modeling of solidification by Zhu and Smith [18]. Oldfield's [19] continuous nucleation model was used. Grain growth was calculated on the basis of the change of bulk and interfacial free energy associated with the change of state of one cell from liquid to solid. This probabilistic TK model was coupled to an MT model, as shown in Fig. 2, by subdividing the macro-volume element in cells for which probabilistic calculations are run. Note that while the macro-element can be partially liquid and partially solid, the cells can only be either liquid or solid. Thus, the complete model is deterministic in calculations involving heat transfer, but both deterministic and probabilistic when calculating grain growth. It is claimed that the MT model includes both energy transport and solutal field calculations. However, there is no explanation in the paper of how microsegregation is modeled and coupled to the TK model. Two fundamental issues associated with dendritic growth are ignored by the Zhu-Smith model, *i.e.*, dendrite tip kinetics and preferential crystallographic growth direction.

These issues were addressed by Rappaz and Gandin [20]. Their model assumes uniform temperature within the specimen and is not coupled to an MT model. Energy transport is modeled by imposing a cooling rate on the system. A Monte Carlo procedure was used to model the stochastic phenomena associated with the random location and crystallographic orientation of new nuclei. Heterogeneous nucleation at the mold surface and in the bulk of the liquid were taken into account through nucleation site distributions. Grain growth was modeled by defining a probability for a liquid cell to be entrapped by a neighboring solid cell growing with a velocity dictated by growth kinetics. While it is claimed that growth velocity of dendritic grains is calculated with a deterministic growth model, the Péclet number is fixed. This implies that the change in the compositional field (constitutional undercooling) during solidification is simply ignored. In addition, thermal undercooling is also ignored. The model accounts for the role of preferential growth directions.

Complications may arise from length scale differences. Indeed, for models using dendritic tip growth kinetics without mass averaging, when $r/\ell \approx 1$, where r is the dendrite tip radius and ℓ is the size of the cell network, the predicted solidification time is reasonably close to the experimental one. However, when $r/\ell \ll 1$, the calculated solidification time is much shorter than the real one. This is because of the assumption that when the dendrite tip reaches a cell the whole cell becomes solid (see Fig. 3). In the first case much faster evolution of fraction of solid is calculated than in the second case. Nucleation density must also be correlated to the number of cells. The number of cells must be much larger than the number of nuclei in the volume element. Thus, careful selection of cell size is required.

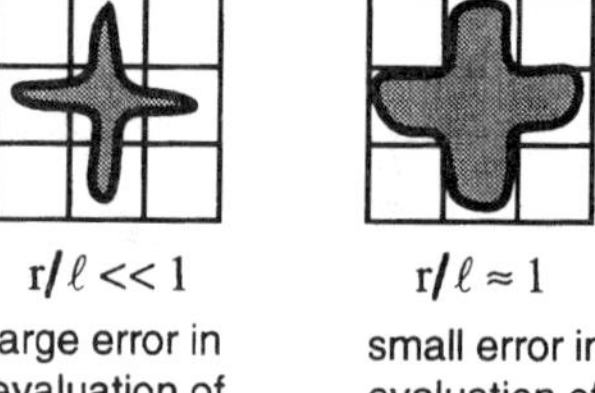

Fig. 3 Influence of length scale on error in calculation of fraction of solid.

The advantage of probabilistic models is that the computed microstructures can be compared with micrographs from various cross-sections of the castings. Thus, the computer output becomes a dynamic metallographic microscope. Since growth competition between columnar and equiaxed crystals, as well as preferred crystallographic growth direction are well described by the model, the CET and crystallographic texture can be predicted quantitatively. Some examples of computer output of probabilistic modeling of TK uncoupled with MT are given in Fig. 4. Nucleation in all samples was triggered at the boundaries of the casting (volume element), as well as in its bulk.

As pointed out in ref. [20], an advantage of probabilistic models is that, unlike FEM or FDM calculations, the CPU time is almost linear with the number of cells. Thus, with the appearance of very powerful workstations on the market, it is believed that this type of calculation will give reasonable computation times while providing an useful tool for the prediction of grain structures in solidification processes.

SPECIFICS OF MT-TK MODELS USED FOR CAST IRON

Nucleation

Two main approaches are used to describe nucleation kinetics in MT-TK modeling as follows: (1) continuous nucleation, and (2) instantaneous nucleation. The continuous nucleation model assumes a continuos dependency of the number of nuclei, N, on temperature (Fig. 5 a). Some mathematical relationship, most of the time empirical, is then provided to correlate nucleation velocity, $\partial N/\partial t$, with undercooling, ΔT, cooling rate, $\dot{T}$, or temperature. The instantaneous nucleation model assumes site saturation, that is that all nuclei are generated at the nucleation temperature, T_n (Fig. 5 b). Then, an empirical relationship must be provided to correlate the final number of nuclei (grains) in a volume element with ΔT or $\dot{T}$. Some of the details of selected nucleation models applied to cast iron will be discussed in the following paragraphs.

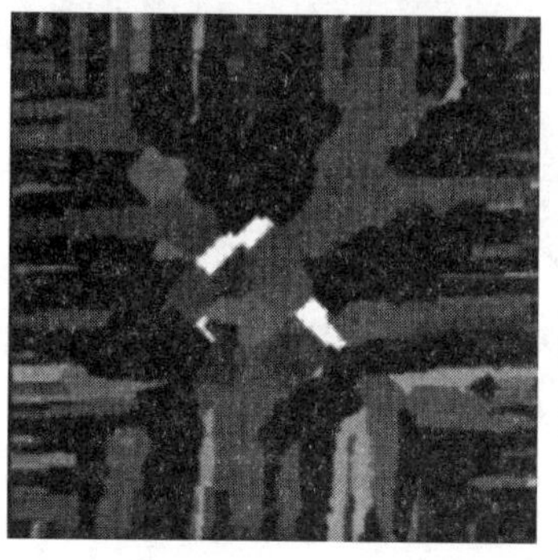
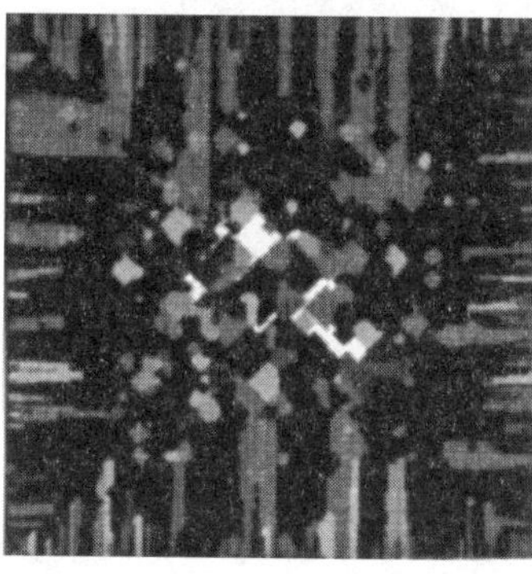

Fig. 4 Results of 2D probabilistic TK modeling of solidification on square configuration. Sample (b) has solidified on more nuclei than sample (a). This has changed the position of the CET, and the distribution of the last regions to solidify, showed in white. The last regions to solidify may be associated with dispersed shrinkage [21].

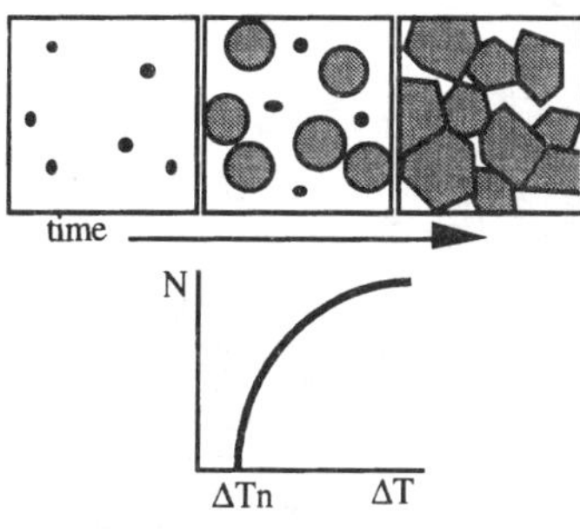
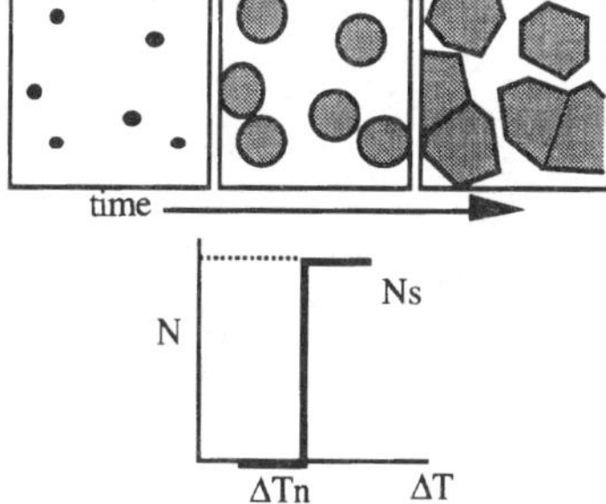

a) continuous nucleation　　　　　　　　b) instantaneous nucleation (site saturation)

Fig. 5 Schematic representation of the assumptions of nucleation models.

In general, for alloys that solidify with narrow solidification interval the instantaneous nucleation model is recommended, since it saves computational time. The use of the continuous nucleation model runs into computation complications, related to the definition of the dimensions of the micro-volume element (diffusion distance), when applied to equiaxed dendritic solidification. A summary of the basic equations and of the parameters that must be experimentally evaluated or assumed is given in Table 1.

Chronologically, the first nucleation model applied to simulate solidification of cast iron was proposed by Oldfield [19]. A power law function fitting the experimental data on cast iron, was used to evaluate the final number of grains:

$$N = K_1 (\Delta T)^n \qquad\qquad (1)$$

where K_1 and n are fitting parameters. A typical experimental and fitted N-ΔT dependency is given in Fig. 6. From Eq(1), an equation for nucleation velocity is derived (Eq 2 in Table 1). A summation procedure is then carried on to determine the final number of nuclei. Such a model was applied to solidification modeling of gray [19,27,28] and ductile [24,29] iron. Some models also assumed that the final grain density corresponds to the onset of recalescence.

Table 1 Summary of nucleation models

Model	Type	Basic equation		Parameters
Oldfield [19]	continuous	$\dfrac{\partial N}{\partial t} = -n\,K_1\,(\Delta T)^{n-1}\dfrac{\partial T}{\partial t}$	(2)	n, K_1
Maxwell and Hellawell [1]	continuous	$\dfrac{dN}{dt} = (N_s - N_i)\mu_2 \exp\!\left[-\dfrac{f(\theta)}{\Delta T^2\left(T_p - \Delta T\right)}\right]$	(3)	N_s, θ
Rappaz and Thevoz [25]	continuous (statistical)	$\dfrac{\partial N}{\partial(\Delta T)} = \dfrac{N_s}{\sqrt{2\pi}\,\Delta T_\sigma}\exp\!\left[\dfrac{(\Delta T - \Delta T_N)^2}{2\left(\Delta T_\sigma\right)^2}\right]$	(4)	$N_s, \Delta T_N, \Delta T_\sigma$
Goettsch and Dantzig [11]	continuous (statistical)	$N(r) = \dfrac{3N_s}{\left(R_{max} - R_{min}\right)^3}\left(R_{max} - r\right)$	(5)	N_s, R_{max}, R_{min}
Stefanescu [7] based on Hunt [23]	instantaneous	$\dfrac{dN}{dt} = (N_s - N_i)\mu_2 \exp\!\left[-\dfrac{\mu_3}{\Delta T^2}\right]$	(6)	$N_s(dT/dt)$

Eq(2) was later modified [22] to include the residual volume fraction of liquid. It appears that for equiaxed solidification this correction is negligible because the nucleation process will cease before a significant portion of the metal has solidified. Indeed, calculations for the Al-Ti system showed that when nucleation was completed the fraction of solid was a mere 10^{-4} [1]. However, it is claimed that such an equation gives a better description of nucleation during directional solidification [53].

A more fundamental nucleation models was developed by Maxwell and Hellawell [1], Eq(3) in Table 1. In this equation N_s is the number of heterogeneous substrates, N_i is number of particles that have nucleated at time i, μ_3 is a constant that includes the interface energy, the entropy of fusion, and Boltzman constant, T_p is the peritectic temperature, and $f(\theta)$ is the classic function of the

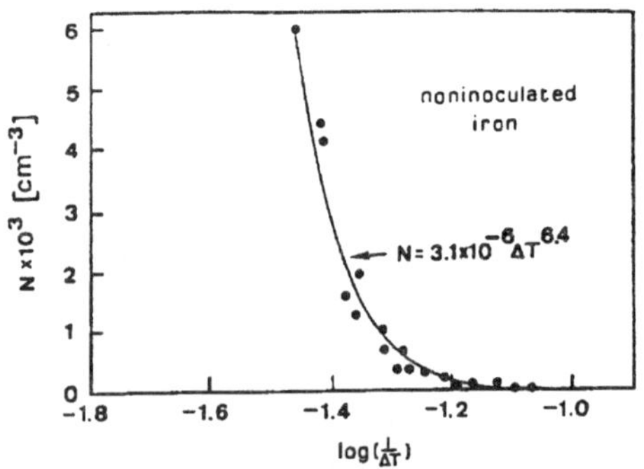

Fig. 6 Correlation between undercooling and volumetric nucleation density of eutectic grains in eutectic cast iron [29].

contact angle. Since θ is unknown, it was used as a fitting parameter to obtain the typical experimental undercooling [1]. The final number of grains (the grain density) was computed as the sum of all nuclei generated at each time step, until all the substrates were consumed, or until the number of nuclei generated decreased to a very small value.

Other continuous nucleation models introduced some statistical functions to help describe the rather large size distribution of grains sometimes encountered in cast iron. A Gaussian (normal) distribution of number of nuclei with undercooling was introduced by Rappaz and his co-workers [8,25], Eq(4) in Table 1, and applied to the solidification of gray iron. In this equation, ΔT_N is average nucleation undercooling and ΔT_σ is the standard deviation.

The same distribution was used by Mampey [26] to model SG iron solidification. However, rather than applying this distribution to the number of nuclei, he applied it to the size of nuclei. To avoid the complication of having to specify θ for heterogeneous nucleation, the width of the substrate was used as a function of undercooling ($K_2/\Delta T$).

Goettsch and Dantzig [11] assume a quadratic distribution of number of grains as a function of their size, $N = a_o + a_1 r + a_2 r^2$. This allows calculation of the number of nuclei of a given radius r,

N(r), as a function of the total number of substrates, the maximum grain size, R_{max}, and the minimum grain size, R_{min}, Eq(5) in Table 1. This equation is then used to calculate the fraction of solid or the solidification velocity, $\partial f_S/\partial t$.

The main assumption used in the instantaneous nucleation model is that all nuclei are generated at the nucleation temperature, T_N. It is based on Hunt's Eq(6) in Table 1. Calculations for eutectic cast iron [7] showed that all substrates became nuclei over a temperature interval of about 2 s. Therefore, Eq(6) can be substituted by:

$$\partial N/\partial t = N_s\,\delta(T - T_N) \tag{7}$$

where δ is the Dirac delta function. Integration gives a total number of nuclei at T_n of $N = N_s$. The nucleation law in Eq(7) predicts instantaneous nucleation, that is that all nuclei are generated at the nucleation temperature T_n. Regardless of the nucleation temperature, if the number of substrates is the same the total number of nuclei is the same. From analysis of Eqs(6) and (7), it becomes obvious that while μ_3 and μ_4 affect the nucleation rate, only N_s will determine the final grain density. Thus, the dependency between cooling rate and grain density reflects a direct correlation between cooling rate and the number of substrates, $N_s(dT/dt)$. This correlation must be evaluated experimentally from data such as shown in Fig. 1.

So, which model works best? In principle they all work, since they are based on fitting experimental data. Thus, the issue is which one fits experimental data better? This is debatable. As shown as an example in Fig. 7, the main difference between Oldfield and Rappaz models is the use of second or third order polynomial, respectively, to fit the data. In other words, they are using two and three adjustable parameters, respectively, to fit the data. However, as found by many investigators, (see for example Fig. 6) two adjustable parameters may be sufficient in most cases. The use of third order polynomials has been proposed for example for hypereutectic SG iron [48]:

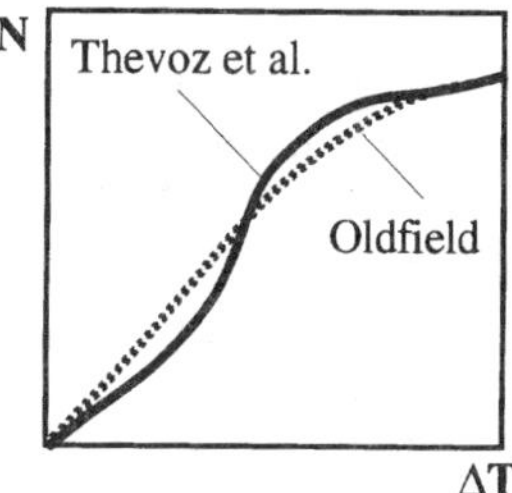

Fig. 7 Schematic comparison between dependency of number of nuclei on undercooling by two continuous nucleation models.

$$N = 8171 + 20.82\,\Delta T + 13.96\,\Delta T^2 - 0.21\,\Delta T^3 \qquad cm^{-2}$$

Growth

When attempting to describe mathematically the microstructure evolution of cast iron, one must model the growth of the primary phase (austenite), irregular eutectic (austenite + lamellar graphite), regular eutectic (austenite + Fe_3C), and divorced eutectic (austenite + spheroidal graphite). The main models proposed to describe solidification of these phases will be summarized in the following paragraphs.

Both deterministic and probabilistic models have been proposed to describe dendritic growth. Deterministic models differ tremendously in complexity based on their ultimate goal. They can be classified as follows:

• tip kinetics models: they attempt to describe solely dendrite tip kinetics, as determined by the thermal and solutal field, and by capillarity (Fig. 8 a);

• simple geometry models: they use tip kinetics models and solutions of the thermal and solutal field to describe growth of simple geometry (1D) dendrites (Fig. 8 b);

• complex geometry models: they attempt to describe growth kinetics of complex geometry dendrites taking into account the thermal and solutal field, and capillarity (Fig. 8 c).

For MT-TK modeling the only deterministic dendrite models that can be used at the present time are the simple geometry models. While they cannot describe the geometric complexities of the dendrite, through the use of averaging techniques and mass balance calculations, they can provide a veridical picture of solidification kinetics. Since the solutal field is the main driving force of dendritic growth, an accurate representation of the liquid composition is required. This is provided by microsegregation models. Thus, a dendritic model is only as good as the microsegregation calculation (see recent reviews of microsegregation models in Ref. [32]).

More recently probabilistic models have made some inroads in dendritic MT-TK modeling. However, there are many problems to solve before it could be really claimed that this models provide an accurate description of temporal and spatial evolution of temperature, fraction of solid, and microstructure scale.

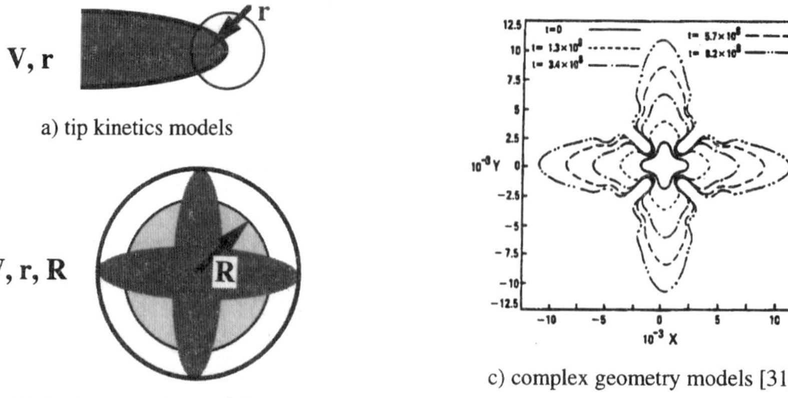

a) tip kinetics models

b) simple geometry models

c) complex geometry models [31]

Fig. 8 Schematics of models for dendrite growth.

A summary of the major assumptions made in selected deterministic and probabilistic models fo
equiaxed dendritic growth is given in Table 2. It is necessary to emphasize that, unless a complet
microsegregation model, the correct dendrite tip kinetics, and coupling with an MT model are used
the chances for correct prediction of solidification events and microstructure are meager.

Typically, for both dendritic and eutectic equiaxed grain growth, velocity is calculated as
function of bulk undercooling:

$$V = \partial R/\partial t = \mu\,\Delta T^2 \tag{8}$$

Various growth coefficients, μ, are used by different researchers, depending on the type of equiaxe
grain described and on their assumptions. A summary of most common growth coefficients is give
in Table 3. The difference between the models described by equations (9) and (10) is not only in th
additional thermal term in Eq(10), but also in the use of more complete microsegregation model i
the later case.

Table 2 Major assumptions used in equiaxed dendritic growth models.

Model	Type	Microsegre-gation model	Growth kinetics	Solutal undercooling	Thermal undercooling	Coupled with MT
Maxwell / Hellawell [1]	deterministic	none	diffusion limited spheri-cal growth	No	Yes	Yes
Dustin / Kurz [33]	deterministic	equilibrium	dendrite tip	Yes	Yes	No
Rappaz / Thevoz [25]	deterministic	no solid dif-fusion + Scheil	dendrite tip	Yes	No	Yes
Nastac / Stefanescu [34]	deterministic	solid and liquid diffusion	dendrite tip	Yes	Yes	Yes
Spittle / Brown[16]	probabilistic	none	arbitrary rules	No	No	No
Zhu /. Smith [18]	probabilistic	none	thermodyna-mic model	No	No	Yes
Rappaz / Gandin [20]	probabilistic	none	dendrite tip	restricted	No	No

Table 3 Growth coefficients used in MT-TK models for equiaxed grains.

Equiaxed grain	Growth coefficient, μ	Reference
star dendrite	$D_L\left[2\pi^2\,\Gamma\,m_L(k-1)C_o\right]^{-1}$ (9)	Rappaz / Thevoz [25]
globulitic dendrite	$\left[2\pi^2\,\Gamma\,m_L(k-1)C_L^*\,D_L^{-1}+\rho L\,K_L^{-1}\right]^{-1}$ (10)	Nastac / Stefanescu [34]
cooperative eutectic (gray iron)	$7.25\cdot10^{-8}$ to $9.5\cdot10^{-8}$ m·s^{-1}·K^{-2} $1.20\cdot10^{-8}$ to $3.1\cdot10^{-8}$ m·s^{-1}·K^{-2}	[35] [30]
divorced eutectic (spheroidal graphite iron)	$D_C^{\gamma}\left[R_\gamma\left(R_\gamma/R_G-1\right)\left(\dfrac{m_S-m_L}{m_S\,m_L}+\dfrac{C_E-C_\gamma}{\Delta T}\right)\right]^{-1}\dfrac{m_S-m_\gamma}{m_S\,m_\gamma}$ (11)	Wetterfall et al. [37], Su et al. [24]

Growth of the cooperative irregular austenite-lamellar graphite eutectic has been mostly modeled through some approximations of the Jackson-Hunt equation (*e.g.*, Eq 8). When Eq(8) is used the major assumption is that the interface undercooling is equal to the bulk undercooling. This is true for moderate cooling rates, in particular during the eutectic arrest [35]. As discussed by Rappaz [36], this assumption implies that the driving force of eutectic solidification is strictly the solutal field, *i.e.*, the thermal undercooling is negligible during eutectic solidification. Experimentally evaluated values for the growth coefficient of the lamellar graphite eutectic are given in Table 3. The lower values correspond to undercooled graphite. Note that the values quoted in ref. [35] were calculated on the basis of experimental results reported in [38].

For the growth of the divorced SG-austenite eutectic a simple diffusion model seems to have been conducive to good results. Eq(8) with the growth coefficient in Table 3 is used to describe graphite and austenite shell kinetics.

PREDICTIVE CAPABILITIES OF MT-TK MODELS

Numerous claims regarding the predictive capabilities of MT-TK models are made in the archival literature. They are broadly summarized in Fig. 9. Most of these features are also of interest in cast iron processing. A short review of the main achievements will be presented in the following paragraphs.

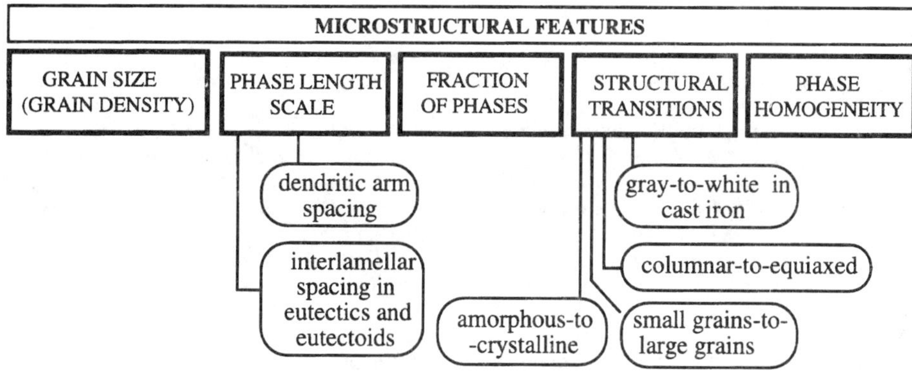

Fig. 9. Microstructural features claimed to be predicted through MT-TK modeling.

Prediction of grain size

All the nucleation models previously described are claimed to give an accurate representation of grain size distribution across the microstructure of the casting, and thus of grain size (Fig. 10 and 11). This should not come as a surprise, since the final grain count measured on the metallographic sample is used as an input for the nucleation law.

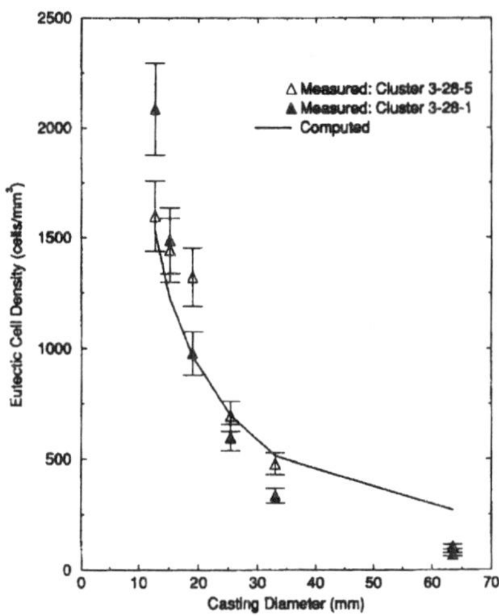

Fig. 10 Predicted and measured eutectic grain densities as a function of casting size [11].

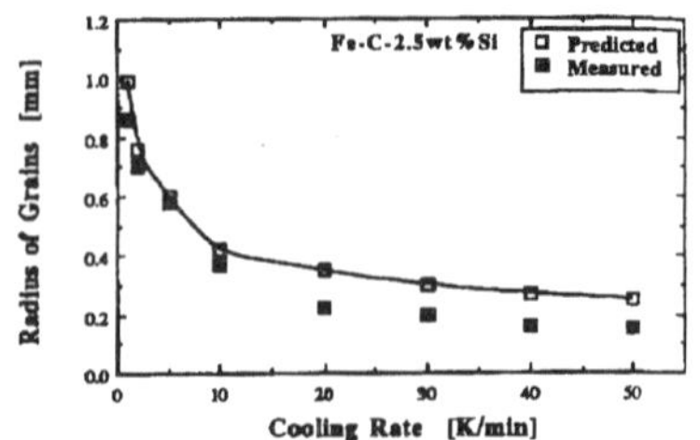

Fig. 11 Predicted and measured eutectic grain densities as a function of cooling rate [8].

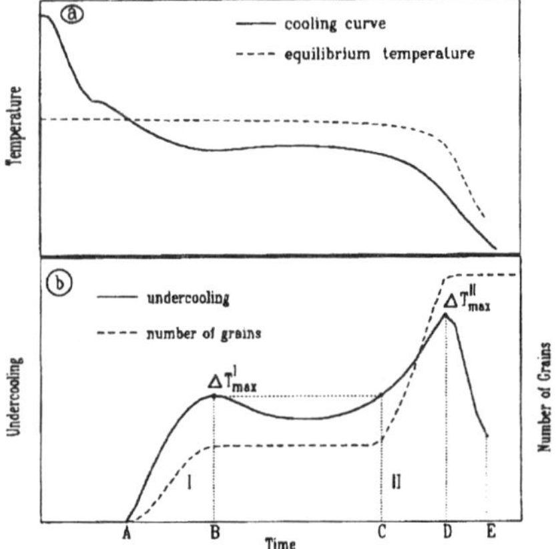

Fig. 12 Correlation between temperature (a), undercooling and nucleation events (b) [52].

An interesting point was raised by Fras et al. [52], who pointed out that in some instances fine eutectic grains are found in the thermal center of the casting. When examining the correlation between undercooling and nucleation events (Fig. 12) it was noted that after the maximum undercooling, ΔT^I_{max}, is reached at time B, nucleation stops until time C, when higher undercooling than ΔT^I_{max} occurs. After that, secondary nucleation starts and ends at ΔT^{II}_{max} (time D). In the time interval B to C the primary nuclei continue to grow and reach considerable size before secondary nucleation starts. The secondary nuclei may have only a short time available for growth, D to E. Thus, a mixed large grains - fine grains structure will occur. Fras and his coworkers demonstrated that such a behavior can be predicted through MT-TK modeling.

The major problem with all nucleation models is that, while experimental evaluation of such correlations is possible, they are only valid under the given experimental conditions. It is likely that the nucleation potential of an industrial melt will never be predicted from first principles. The solution is then to evaluate the nucleation potential of the melt before pouring, and then to bring it to a standard value for which the nucleation constants have been evaluated. For the case of cast iron such evaluation methods include computer-aided cooling curve analysis and the chill test.

In adition, experimental evaluation of heterogeneous nucleation laws has been oversimplified. Typically, the final number of grains at the end of solidification is used to evaluate the parameters required for the nucleation law. However, as shown through liquid quenching experiments [30], the evaluation of nucleation laws from the final grain density may result in inaccurate data, since grain coalescence is ignored. The final eutectic grain density in cast iron was found to be smaller by up to 27% than the maximum number of grains developed during solidification.

<u>Prediction of phase length scale</u>

To the best of this author's knowledge there are no data in the literature comparing predicted and experimental dendritic arm spacings for cast iron. On the contrary, the issue of eutectic spacing has been addressed.

Results by Zou and Rappaz [8] for Fe-2.5% C-Si alloys, shown in Fig. 13, indicate that, when a simple growth law such as that in Eq(8) is used, discrepancies are to be expected at cooling rates higher than 3 to 4 K/min.

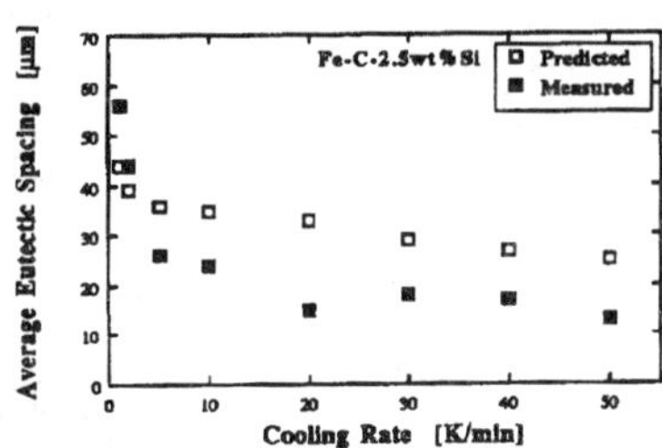

Fig. 13 Predicted and measured average eutectic spacing of lamellar graphite cast iron [8].

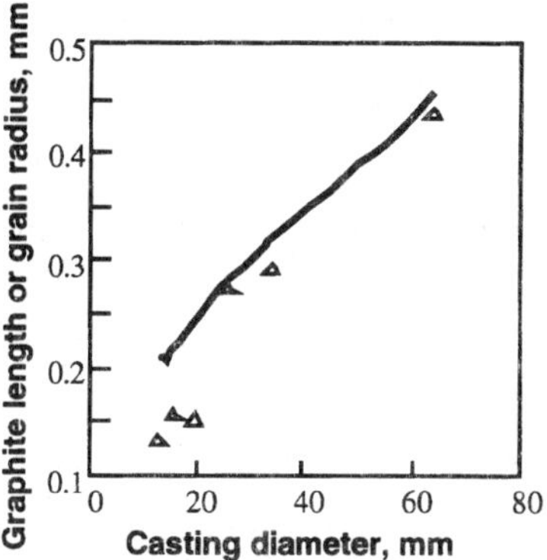

Fig. 14 Comparison between maximum graphite lamella length (Δ) and predicted maximum eutectic grain radius (line) [11].

Since MT-TK models can predict grain size, and since it is reasonable to assume that the maximum length of the graphite lamella is the radius of the grain [11], the maximum graphite length can be predicted (Fig. 14). The maximum graphite length seems to be 75 to 95% of the predicted maximum grain size. This is important because the maximum graphite length seems to be directly related to some mechanical properties [39].

Prediction of spacing of the eutectoid in SG iron has also been attempted [40] starting with the solidification structure obtained from the MT-TK code. The evolution of the pearlite lamellar spacing was obtained by combining the volume-diffusion model (Zener and Hillert) for the transformation of pearlite (V_{pe} λ^2= constant, where V_{pe} is the local growth velocity of the pearlite front, and λ is the local lamellar spacing of pearlite) with the growth velocity of pearlite ($V_{pe} = \mu$ $(\Delta T)^2$). Thus, an equation for the calculation of the evolution of lamellar spacing with temperature was obtained: $\lambda^2 = K_1 (\Delta T)^2$, where K_1 is an experimentally evaluated material constant. Then, an average spacing was calculated from the evolution curve. Typical results are shown in Fig. 15.

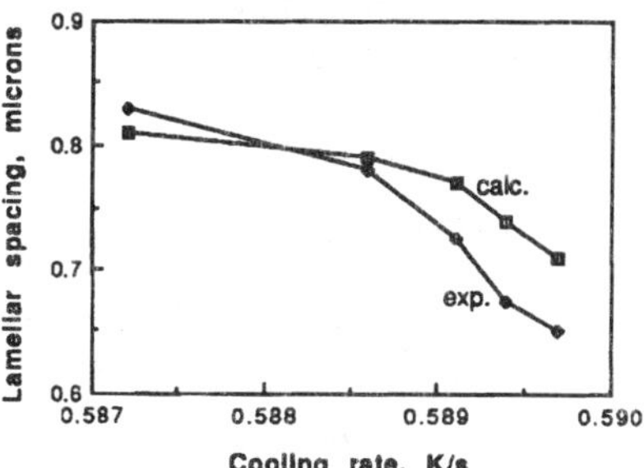

Fig. 15 Measured and predicted average pearlite lamellar spacing as a function of cooling rate in different parts of the casting [40].

Prediction of fraction of phases

Prediction of the fraction of primary and eutectic phases does not seem to pose particular problems. However, additional complications arise if the asymmetric nature of the phase diagram of cast iron is considered. This diagram is conducive to formation of off-eutectic austenite, even under the equilibrium temperature. Such a problem has been tackled by Lacaze et al. [54]. It is claimed that a more exact description of the eutectic reaction for SG iron has thus been developed.

While predicting the fraction of eutectic and primary phase is useful, the real goal of any MT-TK model is to predict the as-cast structure. Thus, for cast iron solidifying with a stable structure one must predict the fractions of graphite, pearlite and ferrite. To achieve this goal it is necessary to develop additional nucleation and growth laws for pearlite and ferrite.

Nucleation data for either pearlite or ferrite in cast iron are not available. However, Stefanescu and Kanetkar [43] proposed the use of a nucleation law developed for eutectoid steel by Mehl and Dube [44]. This law correlates the nucleation velocity of pearlite to temperature:

$$\frac{dN_{pearl}}{dt} = 1.8 \cdot 10^{22} \exp\left(\frac{3.33 \cdot 10^4}{T}\right) \tag{12}$$

This equation was rewritten [11] as a function of eutectoid undercooling:

$$\frac{dN_{pearl}}{dt} = 5.07 \cdot 10^3 \exp\left(\frac{-370}{\Delta T_{etd}}\right) \tag{13}$$

In both equations the nucleation velocity is expressed in mm^{-3} s^{-1}, and the temperature in K. A similar equation was fitted to the experimental data in ref. [44] to describe pearlite growth velocity:

$$\frac{dR_{pearl}}{dt} = 0.168 \exp\left(\frac{-94.8}{\Delta T_{etd}}\right) \quad \text{(mm/s)} \tag{14}$$

For ferrite growth, a diffusion controlled model similar to that used for solidification of the SG iron eutectic, Eq(10), was developed [43]. The number of ferrite grains was specified to be 1.7×10^4 to 2.5×10^6 mm^{-2}.

When predictions with this model were compared with experimental results (Fig. 16), large discrepancies were observed [11]. Subsequently, a simpler power law model was used, based on Eq(1) for nucleation (with $K_1 = 2 \times 10^{-3}$ mm^{-3} K^{-2}), and on Eq(8) for growth (with $\mu = 5 \times 10^{-7}$ mm/s). The constants were fitted to match observed undercoolings.

Venugopalan [41] calculated the isothermal transformation of SG iron based on a model for ferrite growth similar to that developed in Ref. [43]. The complex issue of nucleation of ferrite grains was solved by assuming that each graphite nodule is the nucleus for a ferrite shell (not grain). The model was restricted to the transformation of austenite to ferrite. The number of graphite spheroids was assumed, since this model was not integrated in a complete MT-TK code. The isothermal transformation rates were successfully predicted. The only adjustable parameter in the model was the constant used in the equation for the incubation time for nucleation of ferrite. Good match was obtained between computed and measured ferrite fractions (Fig. 17).

A similar diffusion-controlled growth model was extended to describe also pearlite formation, and was incorporated in an MT-TK code [40]. Thus, the microstructural evolution during solidification and the subsequent solid state phase transformations (eutectoid reaction) during continuous cooling of SG iron was completely described. To avoid the complications related to the data base for nucleation of pearlite, the fraction of pearlite was calculated with Avrami's equation.

The effects of heat treatment can be similarly treated. Skaland et al. [42] were successful in explaining the size increase of graphite nodules after heat treatment. For a 30 mm section size the predicted increase was 8 μm, while the measured one was 13 to 15 μm.

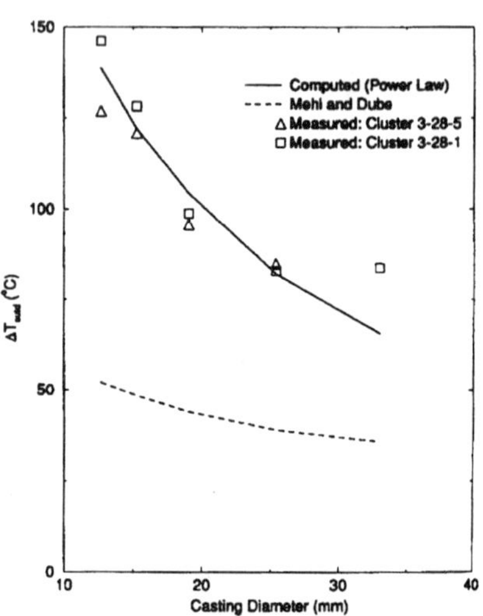

Fig. 16 Measured and predicted maximum eutectoid undercoolings [11].

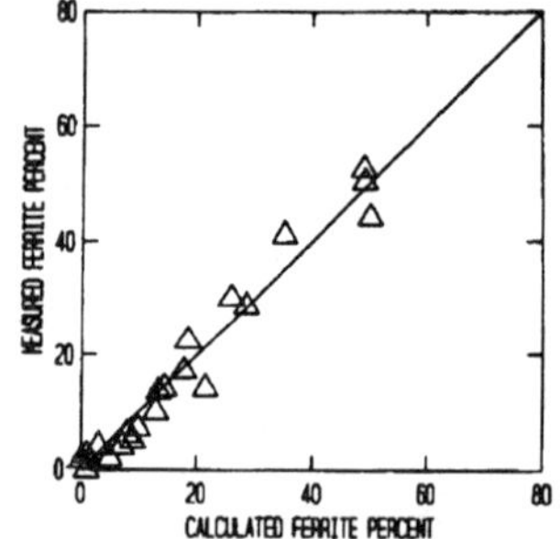

Fig. 17 Calculated and measured fractions of ferrite in SG iron [41].

<u>Prediction of the gray/white transition</u>

The structural gray-to-white transition (GWT) in cast iron is the result of growth and nucleation competition between the stable (gray) and metastable (white) eutectics, as explained by Magnin and Kurz [55]. As shown on Fig. 18, the curves describing the growth velocities of the two eutectics, stable and metastable, intersect at V_{cr}. Since above this velocity the metastable growth velocity is higher than the stable one at any undercooling ($V_{met} > V_{st}$), it can be assumed that above this velocity (or under this temperature) only white iron will solidify, while under V_{cr} the structure is completely gray. Thus, if only growth considerations are invoked, a clear GWT exists.

These arguments have ignored nucleation. It is well accepted that nucleation of white iron is more difficult than that of gray iron. Consequently, additional undercooling, in excess of that predicted from growth velocity considerations, is required for a complete GWT. This is shown in the figure as ΔT_n^{met}, that is the nucleation temperature of the metastable cementite. Thus, a complete white iron is obtained only at growth velocities larger than V_{g-w}. A similar argument holds for the white-to-gray transition, when a complete transition cannot occur unless the undercooling is smaller than ΔT_n^{st}, that is the nucleation temperature of the stable gray eutectic. Thus, a region of mixed structure, gray and white, will exist at growth velocities between V_{w-g} and V_{g-w}. This is the mottled region.

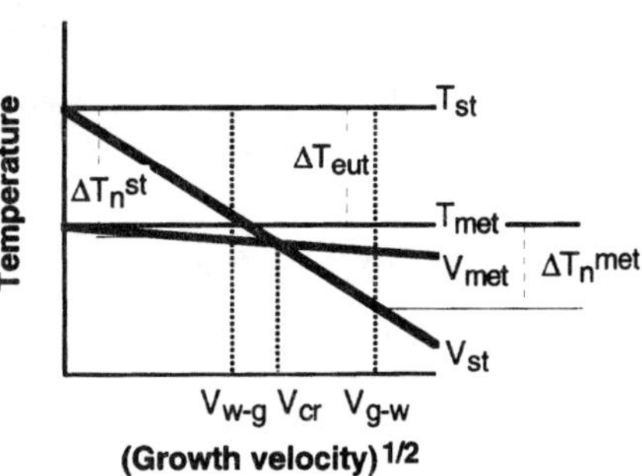

Fig. 18 Structural transitions in cast iron as function of undercooling and growth velocity [55].

On the basis of this discussion it is clear that a model that predicts the GWT must describe nucleation and growth of both the gray and white eutectic. An analytical has been proposed by Fras and Lopez [45]. For eutectic iron the chilling equivalent is:

$$E = \frac{T_{st}^{1.08}}{1.9\,\Delta T_{pour}^{0.5}} \left(\frac{1}{\mu_1 \mu_2{}^3 c_p \left(\Delta T_{eut} + \Delta T_n{}^{met} \right)^{10}} \right)^{1/6} \tag{15}$$

where $\Delta T_{sup\,heat}$ is the superheating above the eutectic temperature, μ_1 and μ_2 are nucleation and growth coefficients, respectively, and c is the specific heat. It is seen that the chilling tendency (chilling equivalent) decreases as the number of eutectic grains and their growth rate increase (μ_1 and μ_2 increase), as the eutectic interval, ΔT_{eut}, and the pouring temperature increase.

Coupled MT-TK models based on growth competition have been proposed almost 10 years ago [35,46]. The major obstacle was and is the unavailability of data for nucleation of the white eutectic. To avoid difficulties with the nucleation data base, Upadhya et al. [47] proposed a fully coupled MT-TK model based on the concept of the existence of a critical cooling rate at which the GW transition can occur. The critical cooling rate is the intersection between the T_{st} and T_{met} curves on a temperature - cooling rate graph. It is a function of chemical composition and nucleation potential and must be experimentally evaluated. Validation of the model against experimental data was satisfactory (Fig. 19). This model predicts chill formation in cast iron, or rather the gray-to mottle transition, but can not describe the mottle-to-white transition and intergranular carbide formation.

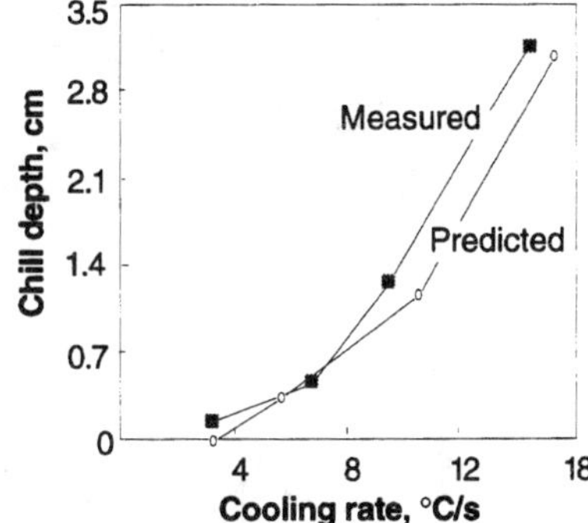

Fig. 19 Measured and predicted chill depth in a step casting as a function of cooling rate [47].

More recently another fully coupled MT-TK has been proposed [49]. It takes into account the nucleation and growth of both the white and gray eutectics, as well as the effects of microsegregation of the third element. The model was incorporated in the commercial software PROCAST ^{TM}. Using a sensitivity analysis of the main process variables, it was found that the % Si of the melt and the nucleation and growth of the gray eutectic have the main positive effects on this structural transition. The model was used to evaluate the influence of cooling rate, silicon content, (Fig. 20) and nucleation potential on the gray-to-mottle and mottle-to-white structural transitions. It was found that both silicon and inoculation increase the mottled and gray zones.

Successful validation of the model was performed against an experimental casting. However, it must be noted that because of the incertitude associated with some of the physical parameters, it is necessary to calibrate the model if useful predictions are desired.

<u>Phase homogeneity (microsegregation)</u>

Assessment of microsegregation occurring in solidifying cast iron is important, since it influences mechanical properties. Also, a comprehensive theoretical treatment of dendritic growth requires accurate tracking of the solutal field during solidification. The major analytical microsegregation models have been recently summarized [32]. Many numerical micro-segregation models have also been proposed. However, the use of numerical segregation models in MT-TK codes increases dramatically the computational time. Analytical models are by far preferable.

Calculation of microsegregation in SG iron is particularly important since, for certain elements such as Mo or Mn, it is not possible to determine the microsegregation

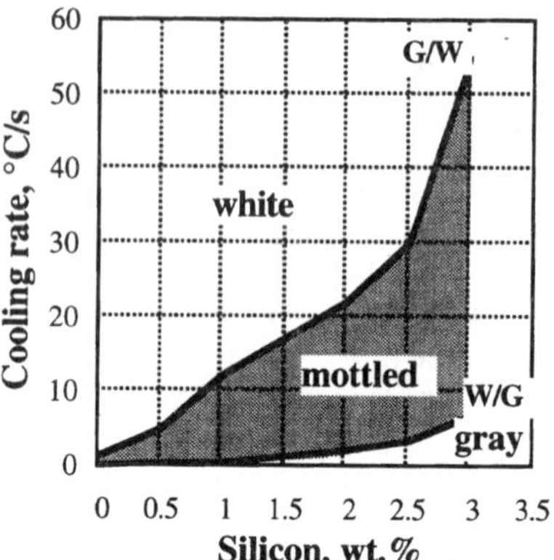

Fig. 20 The influence of Si and initial cooling rate on structural transition in a 3.6 %C cast iron [49].

ratio experimentally, because of carbides formation in the intergranular regions. The classic Scheil model can not be used to predict the microsegregation ratio at the end of solidification. A recently developed analytical model for solute redistribution, that considers diffusion in both liquid and solid [32], was used to calculate segregation of Mn, Mo, Cu, and Si in SG iron. Some typical results are shown in Fig. 21. Note that, in particular for Cu, the correct distribution is predicted to the end of solidification.

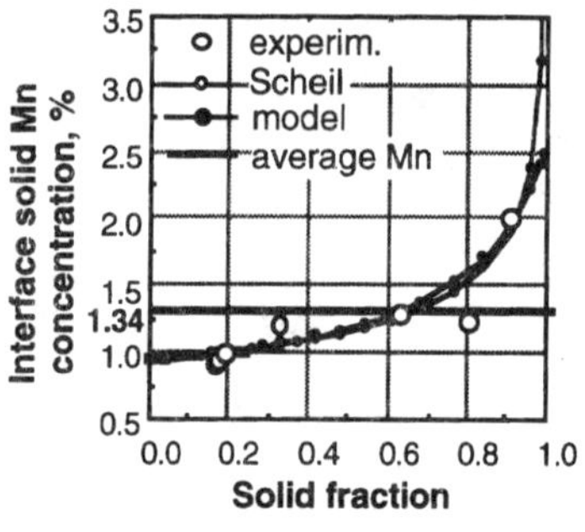

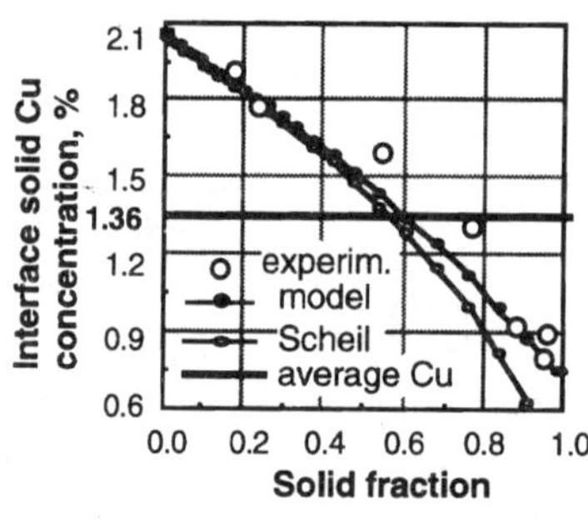

Fig. 21 Measured and predicted microsegregation evolution in SG iron [50]. Experimental data were taken from ref. [51].

CONCLUSIONS

The interest of both industry and academia in modeling of microstructural evolution is obvious from the large number of publications. The progress made over the last 5 to 10 years is nothing short of impressive. However, in spite of increased efforts to validate the models, their accuracy in predicting the microstructural features of interest is still questionable. Validation by authors typically demonstrates that the model <u>can</u> predict the outcome, but not necessarily that it <u>will</u> predict the right one. Adjustable parameters are often unspecified and used in a rather liberal manner. Some progress has been made by incorporating some models into commercial solidification codes. Validation by industry will impart more credibility to the method.

Apparently, accurate prediction of mechanical properties relies heavily upon precise calculation of microstructure length scale and composition. This alone, will continue to be a powerful driving force for the development of models for microstructure evolution.

Acknowledgments: The author would like to thank his students, C. S. Kanetkar, G. Upadhya, D. K. Banerjee, H. Tian, S. Chang and L. Nastac, whose contributions have been used throughout this paper.

Nomenclature

Sym-bol	Definition	Sym-bol	Definition	Sub- or super-script	Definition
C	composition	k	partition coefficient	C	carbon
D	diffusivity	l	length, diffusion distance	E	eutectic
K	thermal conductivity	m	slope of equilibrium line	G	graphite
L	latent heat of transformation	t	time	L	liquid
N	grain density	ΔT	undercooling	N	nucleation
R	radius	Γ	Gibbs-Thompson coefficient	S	solid
T	temperature	λ	phase spacing	f	final
c_p	volumetric specific heat	ρ	density	*	interface
f	fraction of phase	θ	contact angle	γ	austenite

References

1   I. Maxwell and A. Hellawell, Acta Met. **23** (1975) 229-237

2   M. F. Basdogan, V. Kondic and G. H. J. Bennett, Trans. AFS, **90** (1982) 263-273

3   G. Upadhya, D. K. Banerjee, D. M. Stefanescu and J. L. Hill, Trans. AFS, **98** (1990) 699-706

4   H. Tian and D. M. Stefanescu, in *Modeling of Casting, Welding and Advanced Solidification Processes-VI*, T. S. Piwonka, V. Voller and L. Katgerman editors, TMS, Warrendale Pa. (1993) 639-646

5   K. C. Su, I. Ohnaka, I. Yaunauchi and T. Fukusako, in *The Physical Metallurgy of Cast Iron*, H. Fredriksson and M. Hillert editors, North Holland, New York (1984) 181-189

6   C. S. Kanetkar, I. G. Chen, D. M. Stefanescu and N. El-Kaddah, Transactions ISIJ, **28** (1988) 860

7   D. M. Stefanescu, G. Upadhya and D. Bandyopadhyay, Metall. Trans., **21A** (1990) 997

8   J. Zou and M. Rappaz, in *Materials Processing in the Computer Age*, V. R. Voller, M. S. Stachowicz and B. G. Thomas editors, TMS, Warrendale Pa. (1991) 335-360

9   I. L. Svensson, M. Wessen and A. Gonzalez, in *Modeling of Casting, Welding and Advanced Solidification Processes-VI*, T. S. Piwonka, V. Voller and L. Katgerman editors, TMS, Warrendale Pa. (1993) 29-36

10 Z. A. Xu and F. Mampey, in *Modeling of Casting, Welding and Advanced Solidification Processes-VI*, T. S. Piwonka, V. Voller and L. Katgerman editors, TMS, Warrendale Pa. (1993) 485-492

11 D. D. Goettsch and J. A. Dantzig, Metall. and Mat. Trans., **25A** (1994) 1063-1079

12 S. Ganesan and D. R. Poirier, Metall. Trans., **21B** (1990) 173-181

13 M. Rappaz and V. Voller, Metall. Trans., **21A** (1990) 749

14 J. Ni and C. Beckerman, Metall. Trans., **22B** (1991) 349-361

15 C. Y. Wang and C. Beckermann, Metall. Trans., **24A** (1993) 2787-2802

16 J. A. Spittle and S. G. R. Brown, Acta Metallurgica, **37** (1989) 1803-1810

17 R. Xiao, J. I. D. Alexander and F. Rosenberger, Phys. Rev. A, **45**, no. 1 (1992) 571-574

18 P. Zhu and R. W. Smith, in *Modeling of Coarsening and Grain Growth*, S. P. Marsh and C. S. Pande editors, TMS, Warrendale Pa. (1992) 85-99

19 W. Oldfield, ASM Trans., **59** (1966) 945-959

20 M. Rappaz and C. A. Gandin, Acta Metall. Mater., **41**, no. 2 (1993) 345-360

21 L. Nastac, PhD Dissertation, The University of Alabama, Tuscaloosa (1995)

22 J. Lacaze, M. Castro and G. Lesoult, in *Advanced Materials and Processes, vol. 1*, H. E. Exner and V. Schumacher editors, Informationsgesellschaft Verlag (1989) 147-152

23 J. D. Hunt , Mat. Sci. and Eng., **65** (1984) 75-83

24 K. C. Su, I. Ohnaka, I. Yaunauchi and T. Fukusako, in *The Physical Metallurgy of Cast Iron*, H. Fredriksson and M. Hillert editors, North Holland, New York (1984) 181-189

25 M. Rappaz and P. Thevoz, Acta Metall., **35** (1987) 1487-1497

26 F. Mampey, in *Modeling of Casting, Welding and Advanced Solidification Processes-V*, M. Rappaz, M. R. Ozgu editors, TMS, Warrendale Pa. (1991) 403-410

27 D. M. Stefanescu and S. Trufinescu, Zeitschrift Metallkde, **65** (1974) 610-616

28 O. Yanagisava and M. Maruyama, *46th International Foundry Congress*, CIATF (1979) paper 21

29 F. Mampey, *55 th International Foundry Congress*, CIATF, Moscow (1988) paper 2

30 H. Tian and D. M. Stefanescu, in *Modeling of Casting, Welding and Advanced Solidification Processes - VI*, T. S. Piwonka et al. editors, TMS, Warrendale Pa., (1993) 639-64631

31 D. A. Kessler, J. Koplik and H. Levine, in *Computer Simulation of Microstructural Evolution*, D. J. Srolovitz editor, TMS, Warrendale, PA (1986) 95-108

32 L. Nastac and D. M. Stefanescu, Metall. Trans. A, **24A** (1993) 2107-18

33 I. Dustin and W. Kurz, Zeitschrift Metallkde, **77** (1986) 265

34 L. Nastac and D. M. Stefanescu, in *Modeling of Casting, Welding and Advanced Solidification Processes - VI*, T. S. Piwonka et al. editors, TMS, Warrendale Pa., (1993) 209-218

35 D. M. Stefanescu and C. S. Kanetkar, in *State of the Art of Computer Simulation of Casting and Solidification Processes*, Les Edition de Physique, Les Ulis, France (1986) 255-266

36 M. Rappaz, International Materials Reviews, **34**, no. 3 (1989) 93-123

37 S. E. Wetterfall, H. Fredriksson and M. Hillert, J. Iron and Steel Inst., (1972) 323-333

38 H. Fredriksson and S. E. Wetterfall, in *The Metallurgy of Cast Iron*, B. Lux, I. Minkoff, F. Mollard editors, Georgi Publishing Co., Switzerland (1975) 277-293

39 C. E. Bates, AFS Trans., **94** (1986) 889-912

40 S. Chang, D. Shangguan and D. M. Stefanescu, Metall. Trans. **22A** (1991) 915

41 D. Venugopalan, Metall. Trans., **21A** (1990) 913-918

42 T. Skaland, F. Grong, and T. Grong, Metall. Trans., **24A** (1993) 2347-2353

43 D. M. Stefanescu and C. S. Kanetkar, in *Computer Simulation of Microstructural Evolution*, D. J. Srolovitz editor, TMS, Warrendale, PA (1986) 171-188

44 R. F. Mehl and A. Dube, in *Phase Transformations in Solids*, John Wiley , New York (1951) 545

45 E. Fras and H. F. Lopez, Trans. AFS, **101** (1993) 355-363

46 H. Fredriksson, J. T. Thorgrimsson and I. L. Svensson, in *State of the Art of Computer Simulation of Casting and Solidification Processes*, Les Edition de Physique, Les Ulis, France (1986) 267-275

47 G. Upadhya, D. K. Banerjee, D. M. Stefanescu and J. L. Hill, AFS Trans., **98** (1990) 699 - 706

48 D. K. Banerjee and D. M. Stefanescu, AFS Trans., **99** (1991) 747 - 759

49 L. Nastac and D. M. Stefanescu, this conference

50 L. Nastac and D. M. Stefanescu, AFS Trans., **101** (1993) 933-938

51 R. Boeri and F. Weinberg, AFS Trans., **89** (1989) 179-184

52 E. Fras, W. Kapturkiewicz and A. A. Burbielko, in *Modeling of Casting, Welding and Advanced Solidification Processes-VI*, T. S. Piwonka, V. Voller and L. Katgerman editors, TMS, Warrendale Pa. (1993) 261-268

53 G. Lesoult, in *Modeling of Casting, Welding and Advanced Solidification Processes-V*, M. Rappaz, M. R. Ozgu and K. W. Mahin editors, TMS, Warrendale Pa. (1991) 363-375

54 J. Lacaze, M. Castro, C. Selig, and G. Lesoult, in *Modeling of Casting, Welding and Advanced Solidification Processes-V*, M. Rappaz, M. R. Ozgu and K. W. Mahin editors, TMS, Warrendale Pa. (1991) 473-478

55 P. Magnin and W. Kurz, in *The Phyisical Metallurgy of Cast Iron*, H. Fredriksson and M. Hillert eds., North Holland, New York (1985) 263-272

Advanced Materials Research Vols. 4-5 (1997) pp. 105-128
© *1997 Scitec Publications, Switzerland*

The Status in Cast Iron Microstructure Engineering and Interactions with Computer Aided Modelling

R. Weber[1], J.C. Sturm[2] and P.R. Sahm[3]

[1] Eisenwerk Brühl GmbH, Brühl, Germany

[2] MAGMA GmbH, Alsdorf, Germany

[3] Rhine-Westphalian Institute of Technology, Aachen, Germany

Keywords: Cast Iron, Simulation, Product Design, Process Optimization

ABSTRACT

Microstructural engineering in cast iron alloys has come a long way and successfully fulfilled the numerous challenges posed to it in the past. Its microstructural features are extremely variable (graphite shape and size and matrix characteristics), and it still represents the most widely applied casting material. The recent light alloy competitors, i.e. Al-(or even Mg-)base alloys, signify the next challenge. The response to it can only be won by utilizing all available modern tools, the most pre-eminent of which momentarily is given by computer simulation and modelling concerning both the cast iron processing and the microstructural and property (gradient) engineering route. Several examples for the value of interactive optimization between simulation and tasks to be mastered will be presented.

1.   Cast Iron; Quo Vadis?

Cast iron with lamellar graphite is a very old iron base material (500 B.C. in China). In the Europe of the first industrial revolution which was marked by the invention of the steam engine in the 18th century and the railway in the 19th century cast iron with lamellar graphite was the construction material N$\underline{o}$ 1 for any type of machine part. A large fraction of this regime of applications has remained to this day.

Thus, for example, tool machine beds are made of cast iron with lamellar graphite. Cylinders for piston engines, particularly the large sizes, are still based on cast iron. The important materials properties which have kept these areas of application unimpededly strong are above all the damping ability and abrasive stability due to the graphite embedded in the microstructure. Another property, i.e. its heat conductivity, also marks a high potential against the other competitors within its own family, namely the cast steels, ductile iron, and tempered iron. The applications here are, for instance, break disks, permanent molds, glass molds, heating vessels, exhaust-pipes, and other commodities.

In other cases the basis for utilization of cast iron are not only the specific materials properties but the low cost-materials base and -processing approach. However, it must be stated that the ductile cast iron (with nodular graphite) has become an ever advancing competitor to lamellar graphite cast iron. Due to its higher mechanical strength the wall thickness may be remarkably diminished. This then means lower weight and this, in turn, brings economic advantages in that moving masses need less energy, and thus in fuel consuming motors, less carbon dioxide is produced. Although the price of spherulitic cast iron (SGI) is higher than that of the lamellar type the first appears to be constantly winning out against the latter. Also in cases of corrosively engaged castparts SGI is being preferred to the lamellar iron. Due to the fact that corrosion passes along the lamellae, not present in SGI, corrosion, therefore, will occur much slower. This means that also here smaller wall thicknesses are tolerable.

The production numbers of the two last decenia clearly show that SGI is replacing cast iron increasingly. **Figure 1.1** shows this dramatic development. Even if continuously cast shapes (for example centrifugally cast tubes) are exempted from the consideration the statistical trend still prevails.

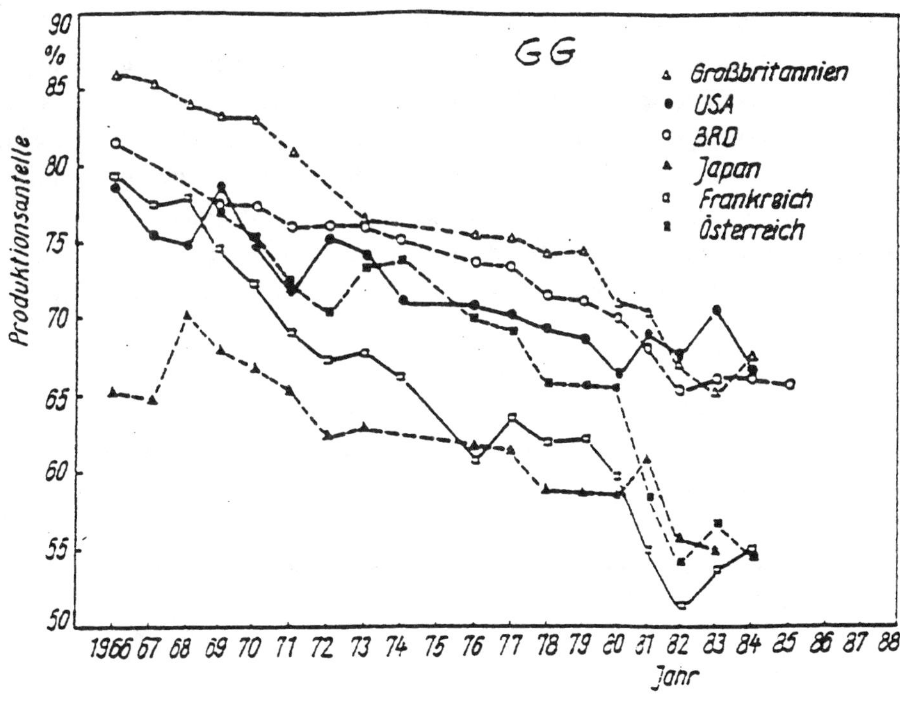

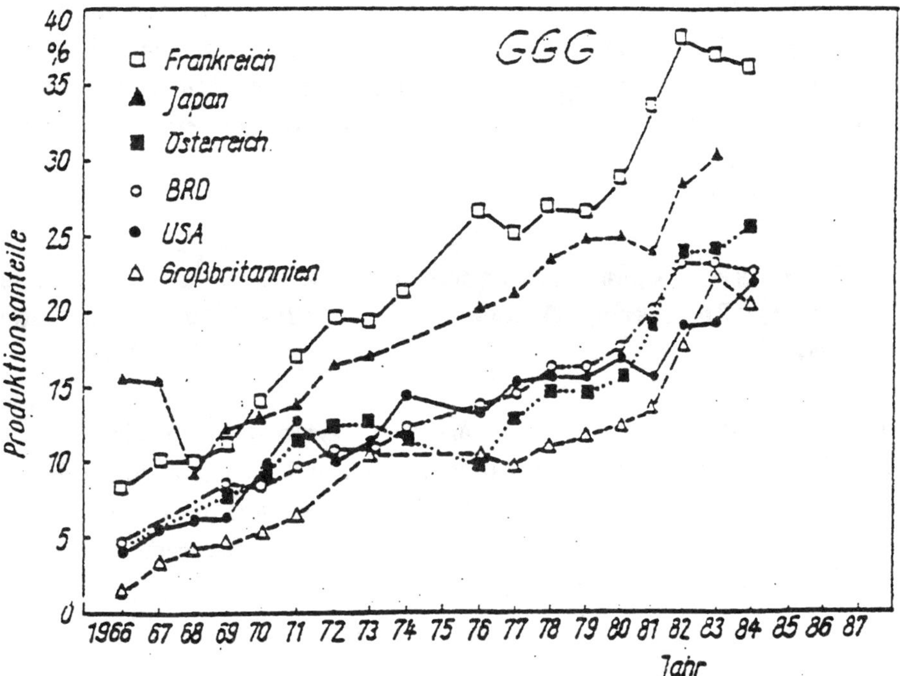

Figure 1.1: Development of LGI and SGI production numbers within the entire Fe-C materials scenario; D. Wittekopf 1987.

The automobile represents an important sink for castparts, **figure 1.2**. **Table 1.1** yields numbers and distinguishes between ferrous and non-ferrous castings. Criteria for the environmentally friendlier car with less CO_2-exhaust, because lighter in construction, see **table 1.2**, appear to ask for aluminum, if not magnesium in the longer run, **figure 1.3**. This trend certainly is noticeable already in the increasing number of Al-castparts for automobiles, **figure 1.4**. As mentioned above, the one advantage on the iron (& steel) side is the price, **figure 1.5**, and it will, therefore, be important to seek novel approaches for cast iron along the lines of modern R&D, i.e.

-   materials related developments:
    o   cast composite parts for multi-leveled approaches,
    o   produce (ceramic) coated cast iron parts for higher performance,
    o   design and manufacture gradient microstructure and properties for narrow tolerance performance,
    o   pursue alloy and microstructural engineering for a higher level performance platform.

-   cast iron processing routes.

The other improvable area is the processing of cast iron parts, so far practiced mostly by utilizing the sand mold technologies in which the environmentally sensible sand reclamation and recycling systems are in competition with Al-base alloy melting and solidification practice into permanent mold (low and high pressure die casting and their various modifications).

2.   Modern Requirements and Expectations on Cast Iron Microstructure and Property Engineering Demonstrated with the Example of Automotive Engines

Till very recently the engine block has been regarded as the classical cast iron component being produced in mass production. The demands on engine blocks were ideally fulfilled by the mechanical and physical characteristics of cast iron, and the complexity of the components geometry has been met by the prime castability of the eutectic cast iron alloy.

Table 1.1: Approximate number of castparts (a) in automobiles and enumeration of essential parts encountered (b) according to eight years' disassembling experience at the Aachen Foundry-Institute*, Rhine-Westphalian Institute of Technology; P.R. Sahm 1991.

Assemble System	Ferrous	Non-Ferrous	Sum
Engine	36	29	65
Wheel/Brake	20	9	29
Gear/Differential	35	9	44
Other	11	49	60
Sum	102	96	198

Table 1.2: Aspirations towards the emission-pauper and energy-thrifty automobile run high worldwide and will determine the R&D concepts of the coming decennium

CO_2-exhaust criterion for the lighter car

till 2005 ⟶ diminish CO_2-exhaust by 25 %

for example, through
lowering

static mass (total mass to be moved on the road)	dynamic masses (within propulsion and transmission systems: accelerating and decelerating masses)

e.g. in

o engine and casings, bodies of various types o brake disc, brake drum	o piston o crankshaft o valves o rocker arm

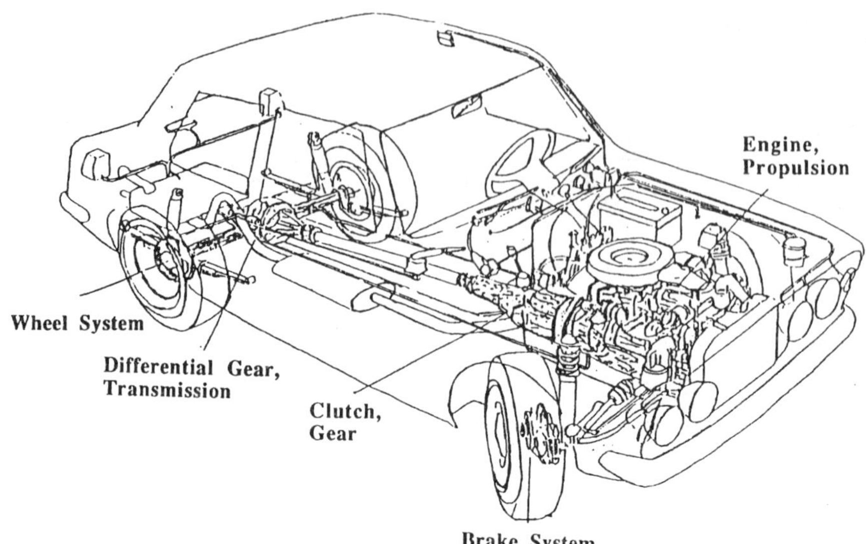

Figure 1.2: The automobile with its various engine and gear requirements linges heavily on the supply of tailered cast parts, P.R. Sahm 1991.

Development of cast iron and aluminium engine blocks in Western Europe

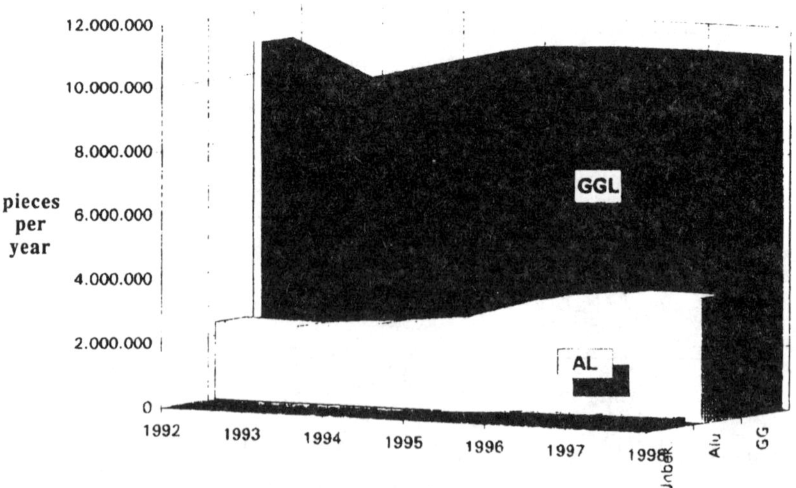

Figure 1.3: Development of cast iron and aluminium engine blocks in Western Europe. Although aluminium engine blocks have demonstrated a strong increase in the recent years a deeper look shows that cast iron engine block will mountain a 70% share of the entire production in this segment (figures from June 1993, taken from Marketing Systems).

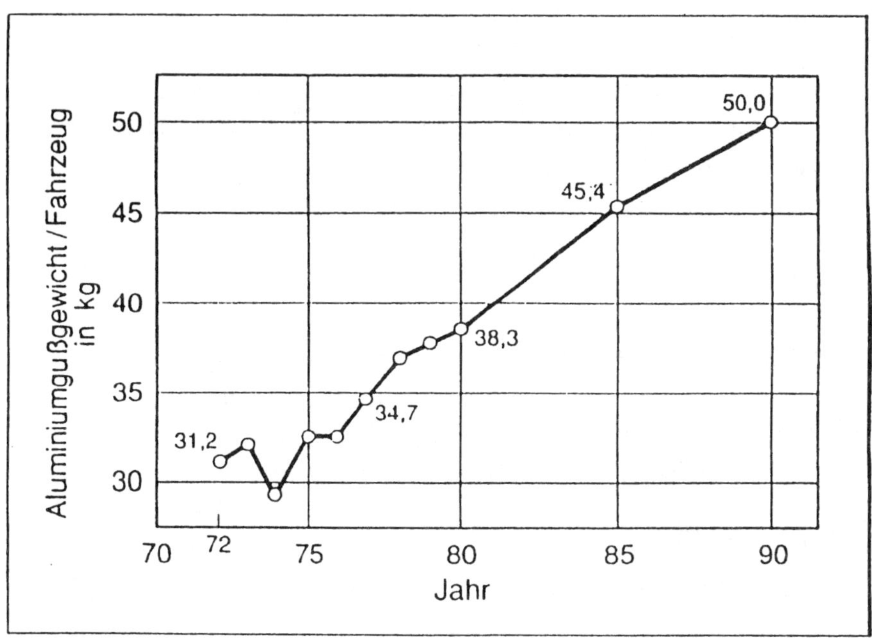

Figure 1.4: Average weight of Al-castparts in French automobiles; M. Santarini 1992.

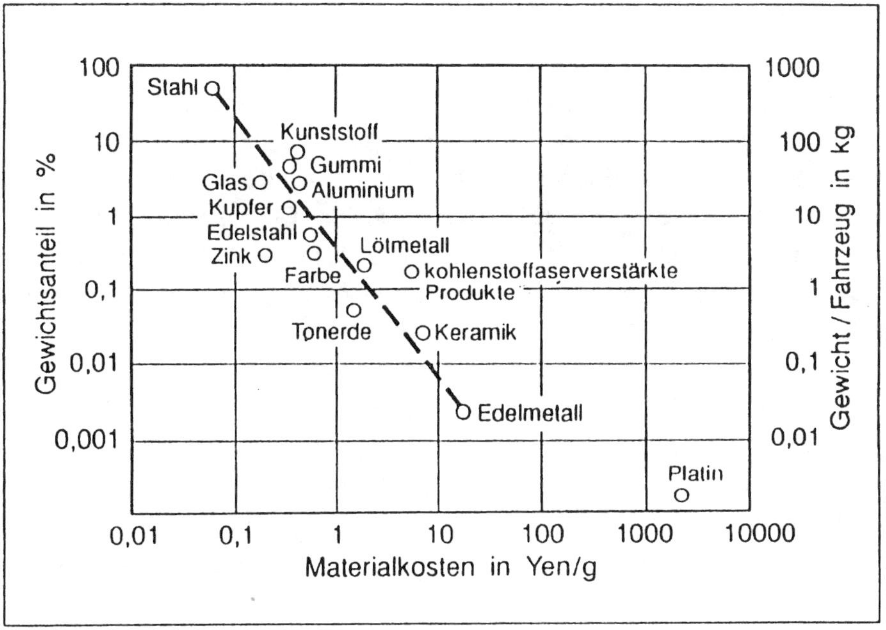

Figure 1.5: Fractions and costs of materials in automobiles; M. Santarini 1992.

Recent developments to substitute the cast iron engine by an aluminum engine block, as contended above, present a strong challenge to keep cast iron competitive. The selection of the cast material and processing routes in the future can no longer be decoupled from the components demands. In the following, the scenario of expectations for the future engine, the demands on the material and the corresponding manufacturing process, and the properties to be achieved will be discussed.

2.1   Development Tendencies and Customer Expectations for the Future Engine

The development of cars is confronted by the growing demands of comfort, safety, and environmental well-behavior. Consequently, the engine has to become more powerful, less noisy, and lighter. For economic reasons tendencies are also towards a maintenance free engine, with less fuel consumption and with a life span of over 200,000 km.

The demands on the future propulsion system can be approached from two sides: injection efficiency and engine construction.

The future engine will have a lighter specific power in any case. By introducing higher compression, turbo-charging and optimization of the combustion, the effective average pressure in the engine and, consequently, the mechanical and dynamic loading of the engine block casting will grow.

The improvement of the specific performance of the burn is aimed at an insulation of the combustion chamber. Less cooling of the engine will lead to improved adiabatic conditions: the ideal "Carnot Process" is attempted of being achieved, i.e. a poor heat conductivity of the material will have to be realized. Combining the performance increase with less cooling work in the engine leads to a substantially higher temperature level of the engine material. Demands on piston rings and liners thus increase accordingly.

These deviating demands enforced a compromise what concerns the materials selection. Whereas the development of powerful and high volume engines is predominated by the aspect "light weight", the decision which cast material will be used for the majority of lower volume engines (less than two liters volume) will be mainly determined by cost. Nevertheless, the customer's demands concerning the cast "engine block" must still be met. **Table 2.1** shows the main aspects and developmental tendencies.

Figure 2.1 shows the potential of weight reduction, obviously a most challenging demand on the successful application of cast iron production in the future.

2.2 Expectations Concerning the Engine Material

The cast material selected has to meet the increased static and dynamic loads within the engine, without distortions and instabilities. Because the engine stiffness plays a major role within the components construction it requires a material with improved mechanical properties. In addition to the mechanical stiffness the thermally induced distortions must also be considered Dynamic instabilities enforce extremely high dynamic stresses and larger noise, which, again, favor cast iron over and above other, for example, Al-based materials.

For gas consuming engines the demands are even higher because the explosion produces stronger pressure waves and flame fronts. So, a material has to be selected which meets the high loading at a high specific performance/mass ratio. **Table 2.2** lists the entire bouquet of requirements a cast material has to fulfill to meet the design goals.

2.3   Current Demands on an Engine Block Material

The statements made above have shown that cast iron still is the privileged material for producing engine blocks. The good mechanical properties of cast iron, even including the higher temperature level, permit a balanced load of the engine block for the performance of the engine.

The low thermal expansion, for example, leads to a decoupling of engine geometry from engine temperatures. Under load the high Young's modulus leads to low components distortions.

The good abrasive resistance of cast iron and its sound running properties advantageously allow a combination of cast iron engine block with aluminum piston

Table 2.1: Expectations for the future engine

- weight reduction

- performance improvement (kW/kg)

- reduction of specific fuel consumption

- 4-valve technology standard technology

- increasing amount of technology engines with most innovative direct injection technology or similar

- latest and future exhaust standards to be fulfilled

- maintenance free life time (> 200,000 km)

- cost reduction

Table 2.2:　Expectations concerning cast engine block materials

1. **Mechanical properties**
   - improved thermal shock resistance
   - improved resistance for static/dynamic loads
   - higher tensile strenghts and bend strenghts in certain functional areas of the engine leading to
     - less distortions in liners
     - improved properties in flange areas, especially for bolts in the oil sump flange and cylinder head flange,
     - high Youngs Modulus

2. **Physical properties**
   - low density
   - heat conductivity
   - heat expansion

3. **Microstructure**
   - homogeneous pearlitic structure
   - avoidance of skin hardness

4. **Casting conditions**
   - reduction of pores and shrinks
   - castability (for minimal wall thickness)
   - controlled mold filling
   - shrinkage

5. **Technological properties**
   - higher ductility
   - improved machinability
     - constant hardness level
     - less volume to be milled away
     - small tool consumption
     - weldability
   - high corrosion resistance
   - wear resistance

6. **General**
   - recycling friendly
   - quality assurance systems by closed loop control
   - interest in compact iron

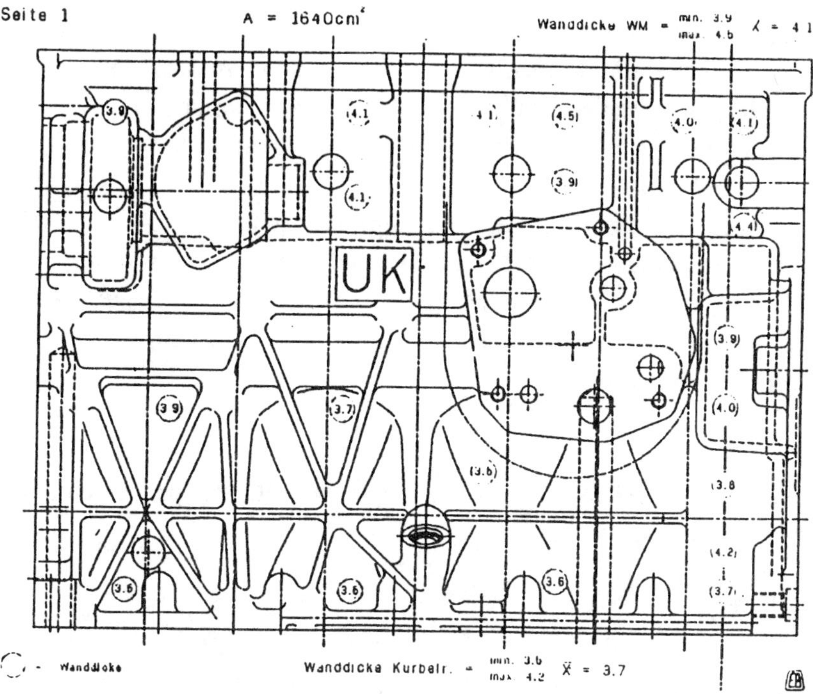

Influence of wall thickness

reduction of wall thckness for
$$a) = 0{,}5 \text{ mm}$$
$$b) = 1{,}0 \text{ mm}$$

A = surface
B = volume
G = weight
g = density of GG —> 7,2 g/cm³

<u>side 1</u>:

a) A = 1.640 cm²
   Va = 1.640 cm² x 0,05 cm = 82 cm²
   Ga = 82 cm³ x 7,2 g/cm³ = 590,4 g

   Ga = <u>0,6 kg</u>

a) A = 1.640 cm²
   Va = 1.640 cm² x 0,1 cm = 164 cm²
   Gb = 164 cm³ x 7,2 g/cm³ = 1.180,8 g

   Gb = <u>1,2 kg</u>

Figure 2.1: Influence on a reduced minimal wall thickness in cast iron engine blocks on the over casting wheight. The sketch demonstrates the wall thicknesses of a selected engine block for one side wall. The engine has an average wall thickness of 4,1 mm. A reduction of the average wall thickness by 0,5 mm leads to a wheight recuction just in this side wall by 600 g, whereas the reduction of the averate wall thickness down to 3,1 mm would reduce the casting wheight by 1,2 kg.

and chromium-molybdenum alloyed of SGI piston rings. Cast iron is especially suitable for low-cooled engines. This is particularly valuable because the exchange of heat to the cooling water system should be kept to a minimum. Lastly, the high damping capacity of cast iron keeps the noise level down.

By further alloying the cast iron may be improved with respect to its physical characteristics. For instance, the heat resistance and high temperature properties are modified by adding chromium and copper in combination with titanium, antimony, nickel and tin. Cu- and Cr-alloyed cast irons shows a stable pearlitic microstructure and subsequently machinability.

3. Strategies and Tools for Cast Iron Engine Block Design Processing and Property Engineering

3.1 Manufacturing

Cast iron also yields itself particularly advantageously to economic production (of 500 and more) cast engine blocks per day. Here, its castability, solidification and cooling characteristics are at an advantage for a sound and reproducible manufacturing route, allowing the production of engine blocks with constant overall quality.

The manufacturing routes for the production of cast iron have been continuously improved, optimized, and automized. The high degree of automation of molding lines, core manufacturing as well as assembling and then summarizing all manufacturing steps (including core making, molding and casting) into an integrated quality assurance system allows to keep and improve very tight materials and dimensional tolerances.

Main target for the future will be to reduce the weight of the cast iron engine block keeping the advantages of the material. Figure 2.1 has shown the potential of weight reduction for a selected example. This challenging goal can only be achieved by controlling the manufacturing process in conjunction with a good quality assurance system, **figure 3.1**.

3.2 Strategies and Modelling Tools for Engine Design

Till recently progress in development of cast iron engine blocks has been mainly made in separate areas, without a global strategy for the engine and its manufacturing processes. The main deficiency was the splitted responsibility of design and manufacturing.

An important new task today for the modern foundry is to regard its product not as a singular casting event but to consider all aspects associated with the entire engine issue. The reason for this is:

1. Latest simulation tools more and more allow to investigate component characteristics of design, performance and manufacturing of an engine during the design stage, without prior production or even laboratory pilot stage experience, **figure 3.2**

2. This permits a change of the R&D schedule from the sequential to a parallel approach, **figure 3.3**.

Overall target of these approaches is the reduction of weight to allow the cast iron engine block so stay competitive to light alloy components on the one hand by stressing its advantages of properties, castability, and costs. The weight reduction will be targeted by a threefold approach:

1. Global reduction of cast wall thickness by improved manufacturing routes and tolerances.

2. Optimization of wall thickness by adjustment of design due to the actual stress and load level.

3. Application of the actual local cast iron properties instead of global standard values to reduce the casting wall design based on the really needed functionality.

Current goals can be summarized as follows: For new construction of engine blocks the weight shall be reduced by 20%. The modification of existing engine designs may

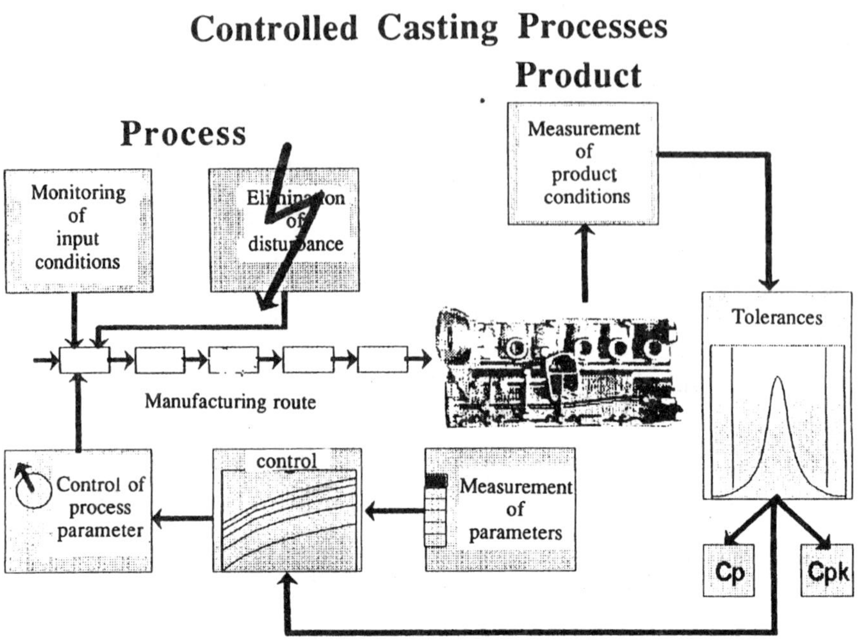

Controlled Casting Processes

Figure 3.1: Controlled Casting Processes: Reduction of casting weight by reduction of wall thicknesses requires a reliable and reproducable casting process route. The process has to be controlled and quality control systems have to be not only fully integrated but also provide a closed loop to adjust the quality determining process conditions.

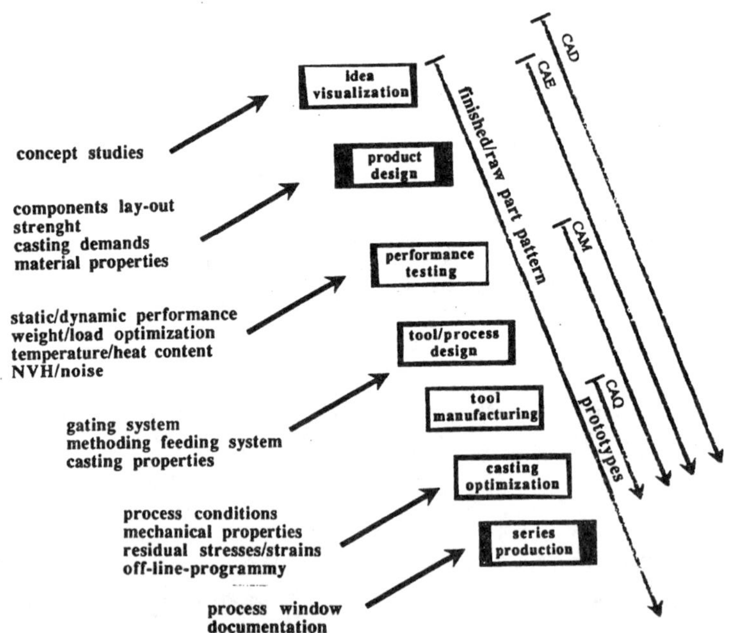

Figure 3.2: Integrated product- and process engineering using simulation tools. The simultaneous engineering of product design and processing route can be decisively supported by an integrated application of simulation tools at all stages of development. Simulation tools shall be applied as soon as possible during the different phases of concept studies, finished part design, and performance testing as well as the lay-out of the raw part and its manufacturing route. Presumption for this approach is the availability on the components geometry or parts of it.

allow a weight reduction of 10 % (basis 60 kg casting). **Table 3.1** shows a selection of measures and tools required to achieve the above goal.

A key area for meeting the above targets will be to couple the performance related design with a process-oriented here casting oriented components design. Together with casting technology, design of the component play a decisive role in determining the specific requirements of cast-part design in terms of defined properties. In this context intensive communication between the design engineer and the supplier is pivotal in order to exploit the available know-how of each party. This applies particularly to the utilization of existing geometry data. The sooner these requirements are met the lower the penalty in time and costs for reworking the design and the necessary tooling.

This is especially true for the introduction of simulation tools to the development and design of new engines and the required cast components which will be the more effective the sooner the different development stages can be supported within the simultaneous engineering strategies, **Figure 3.2.** Even for concept studies first basic information for a as well process as performance oriented component design can be provided. During the product development stages simulation of mechanical strength and simulation of achievable mechanical properties and casting process related demands support the construction of the component. Die and tool design can be supported by a simulation of methoding of the gating and feeding system. The validation of casting tolerances and manufacturing and reject boundaries support the foundryman prior to the production.

3.3 Demand driven design - A case study

In the following this integrated approach will be demonstrated at the example of the optimization of an cast iron engine block. Basic condition for a time and cost effective approach is the availability of a geometrical data base of the component. Here the finished part CAD design has been used as basis for a first evaluation of the stress level and possible distortions of the cast component. By adding the gating system and the machined areas, the solidification and cooling of the first lay-out of the raw part can be done. By coupling the thermal investigations with a fully elastic/plastic stress model the foundry can supply the designer with information about the residual stresses and strains to be expected within the casting, **Figure 3.4.**

Table 3.1: Measures, tools and strategies for weight reduction.

1. Measures
- reduction of process tolerances
- structural analysis for stress/strain
- optimization of castability
- prediction of real casting properties
  establishment of safe process window

2. Functionality of component for
- bosses
- ribs
- nodes
- flanges

3. Innovative principles of construction for the engine block
- removal of walls
- wall thickness reduction
- stress related design of profiles
- contoured walls

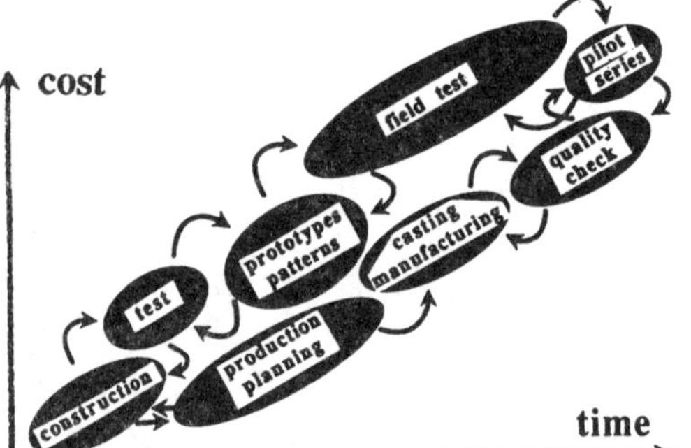

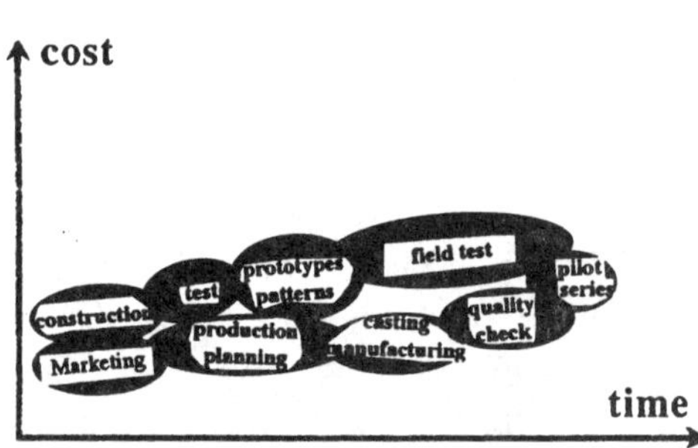

Figure 3.3: Integrated product design and development - traditional and future concepts. The current practise of the traditional concept of sequential solution of different tasks during the development of a new engine requires strict mile stones which may cause delays in the entire time schedule. Within different subtasks iterative procedures may lead to additional delays and cost increase (a). The parallel approach of the future requires a global view on the final component within any of the subtasks and integrated concepts and tools to provide knowledge about any subtasks to other design stages at the earliest time possible. This finally leads to a reduction to a reduction of overall development costs and a gain of time by this more concentrated approach (b).

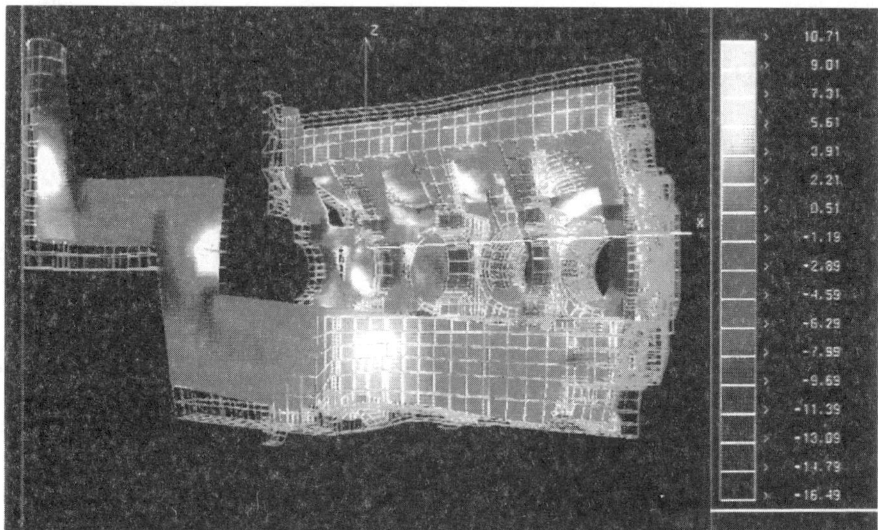

(a)

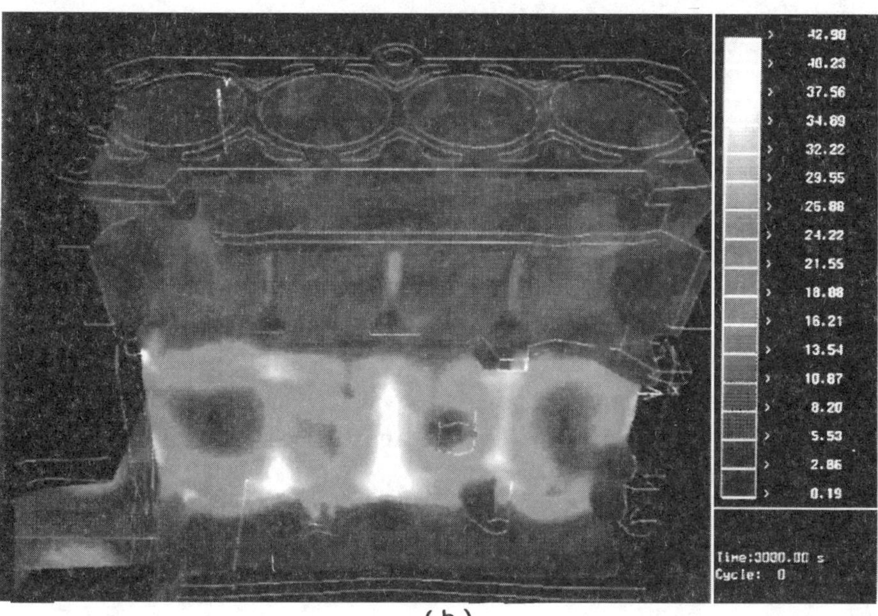

(b)

Figure 3.4: Prediction of the residual stresses in the engine block casting. State of the art casting simulation tools allow the investigation of residual stresses in the casting due to the solid state shrinkage within the mold. The prediction of stresses and strains in the casting done bases on a elastic/plastic stress strain model considering not only the temperature dependent casting conditions but taking into account the stabilty and shrinkage restriction due to the mold. The final residual stresses at room temperature can be used as input conditions for performance simulation. Figure (a) shows the stress level of the casting in z-direction, and as distribution of Mises stresses.

This information can be taken as improved input conditions for the necessary performance simulation of dynamic loads and stress levels during operation to be expected in critical areas within the existing design, **Figure 3.5.**

A detailed investigation of the process conditions for the stress improved design allow to establish the production conditions and provide further feed back of possible mechanical properties within the casting for the designer. State-of-the-art casting simulation tools allow the detailed investigation of the filling and solidification stage. The lay-out of the gating system and the related filling and flow conditions decisively determine the entire process route and the subsequent properties within the casting, **Figure 3.6.** Shrinkage conditions can be evaluated by coupling the thermal map with the evaluation of local austenite shrinkage and graphite expansion, **Figure 3.7.** This simulation enables to establish a first safe process window by investigating alternative process conditions.

Recent developments allow a quantitative prediction of the design and process dependent local properties of the casting. The simulation tools take into account the nucleation, growth and segregation of the specific alloy during solidification on a microscopic scale of the solidification modelling. By calculating the further cooling of the casting down to lower temperatures, the solid state reaction can be taken into consideration. Consequently the final non-isotropic structures of the cast iron component (e.g. eutectic cell size, graphite morphology and distribution, white to grey transition) and mechanical properties (hardness distribution, tensile and yield strength and elongation by fracture) can be predicted. **Figure 3.8 (a)** show the calculation of tensile strength distribution within the casting based on a full filling and solidification simulation including the microscopical prediction of phase formation. **Figure 3.8 (b)** demonstrates the comparison of simulated and measured hardness results for an engine block.

Since the results of these calculations visualize the geometry and process dependent mechanical properties of the casting, the designer can take advantage of the locally varying mechanical characteristics of a casting in a favorable way instead of being dependent on the lowest standardized data for this alloy. This provides a strong potential for further possible weight reduction of the component.

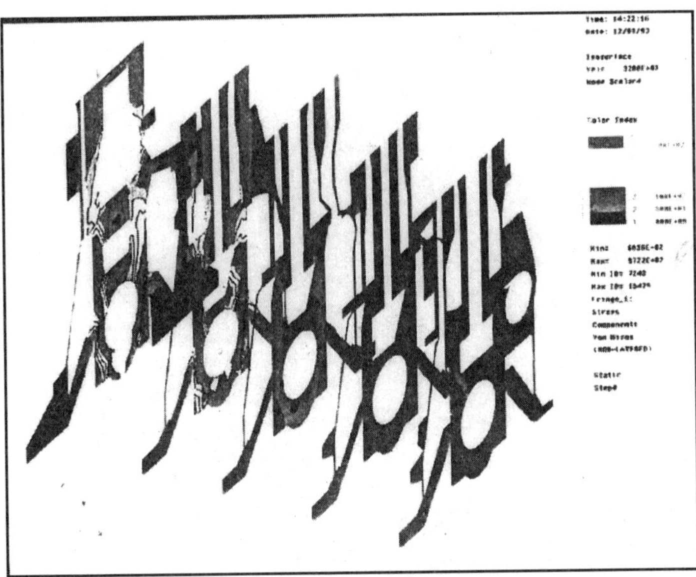

Figure 3.5: Simulation of the dynamic stress distribution of the engine block during performance. The performance simulation of the engine allows the investigation of stresses within critical areas of the component for different conditons , here at igniton time within cylinder 4 at a rotation of 2300 per minute. The figure shows the Mises stress level in MPa in selected cuts for the five bearing areas.

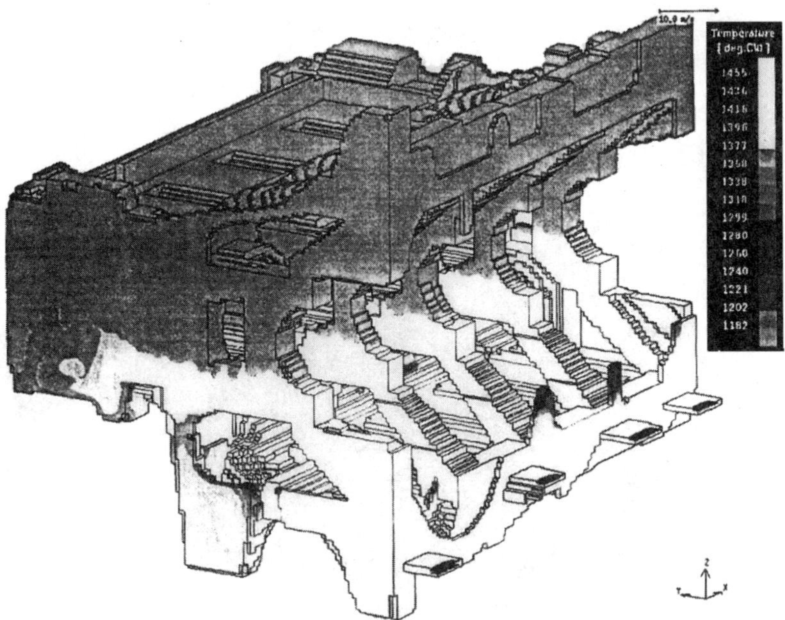

Figure 3.6: Simulation of mold filling of the casting. Latest developments or casting simulation tools allow the investigation of the three dimensio inflow of the liquid superheated melt into the cavity. The complex physics require a coupled simulation of Navier-Stokes equation coupled with energy equation to take the turbulent flow and coupled heat loss of the melt into account. Besides the temperature distribution at any time during fill the velocities and related pressures can be investigated in any area and at any time. State-of-the-art mould filling simulation takes into account vent conditions and mold permeability as well as allows to simulate the influence of filters.

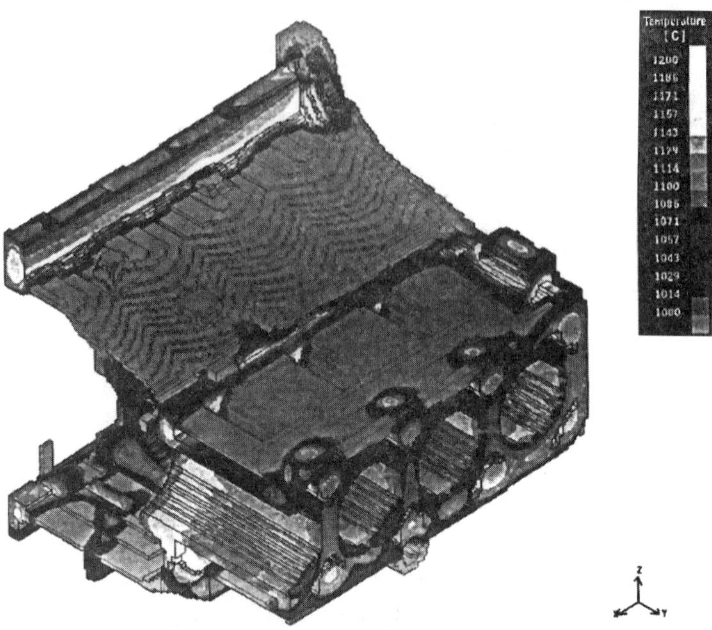

Figure 3.7: Solidification simulation of the casting. Based on the non-uniform temperature distribution the solidification simulation taking into account the detailed composition and innoculation and growth morphology can be investigated. This figure shows the temperature distribution at 2 min 47 s after pour on the surface of the casting and in a cut through the liners.

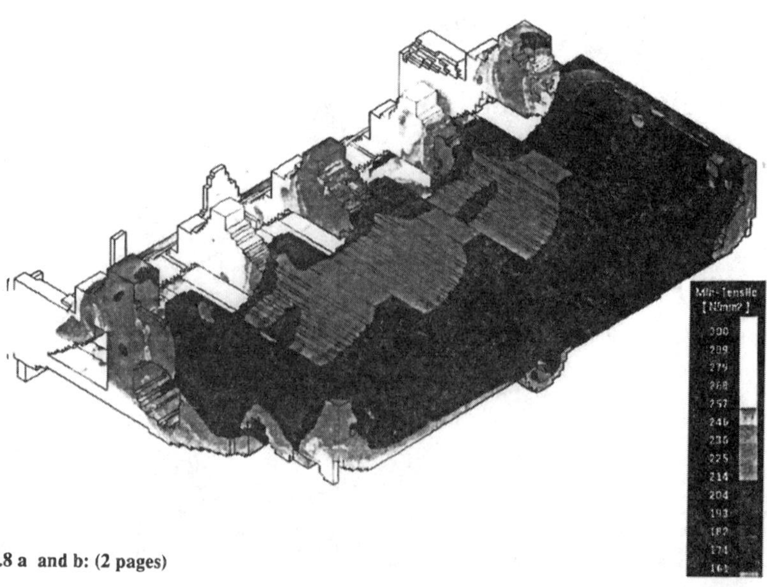

Figure 3.8 a and b: (2 pages)

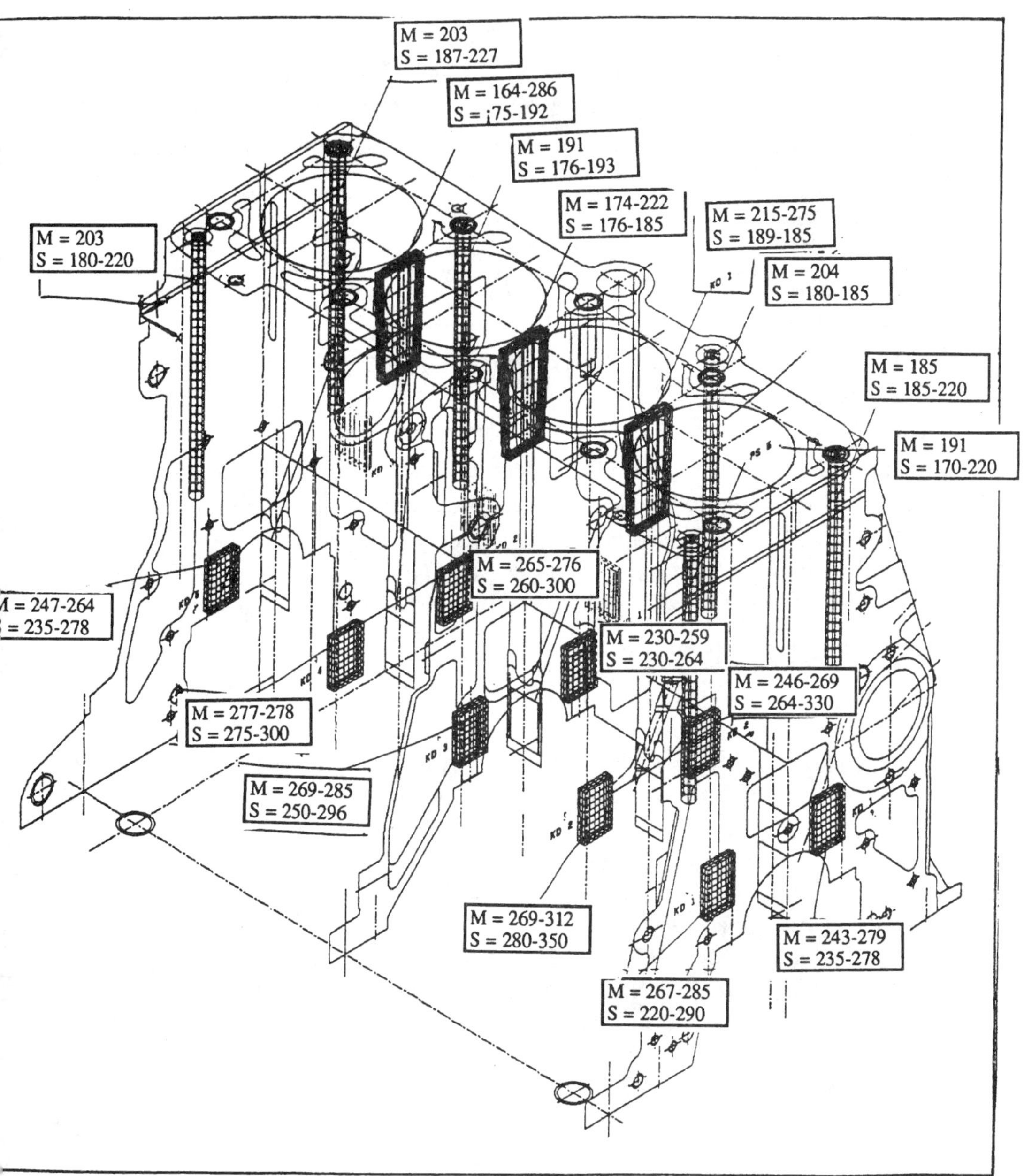

Figure 3.8 : Tensile strength distribution of the casting. The integrated simulation of the entire process comprising mold filling, solidification and cooling simulation using a micro modelling approach allows to investigate the resulting final structures and properties of the casting iron engine block quantitatively. The micromodelling of the solidification takes into account the innoculation and growth condition of the specific alloy, considering undercooling, segregation and morphology of the solidifying structure. So the solidification structure can be prediction, i.g. graphite morphology (White, grey, nodular), structure sized (eutectic cell size, type and average size of lamellae or number of nodules). Calculation the euteciod solid state reaction of the Austenite forming pearlitic or ferritic structures lead to a prediction of final structures and the properties as hardness, yield and tensile strength or elongation by fracture. The figure (a) shows the tensile strength distribution of the engine block in a central cut through the liners and the bearings. Figure 3.8 (b) shows the comparison between experiment and simulation. for the hardness distribution in Brinell units.

Based on this advantageous knowledge the final step of design optimization allows to couple further structural improvements with a noise optimization of the component, **Figure 3.9.** By investigation of different possible geometrical designs with a NVH simulation, the casting weight and at the same time the noise of the engine can be further reduced.

4.   Conclusion

The competition of materials and processes challenged by the future demands on cast components has lead to a change of development strategies and the introduction of latest simulation tools for design and process optimization. The still existent strong advantages of cast iron castings by means of cost competitiveness, castability and dedicated mechanical and physical properties will only prevail their importance if the overall demand on weight reduction can be met. Any locally oriented and separated optimization strategies will fail in future. Integrated concepts providing any available information at the earliest time possible must be introduced to the foundries and the designer - supplier cooperation.

5.   References

E. Flender and J.C. Sturm: Anforderungsgerechte Konstruktion - optimierte Verfahren - Qualitätsgerechte Fertigung, Int. Gießereikongress 1994, Düsseldorf

P.N. Hansen, E.Flender and G. Hartmann: The MAGMA System of Mould Filling and Solidification Modelling" the TMS Detroit Section 18th Annual Automotive Materials Symposium on Numerical Simulation of Casting Solidification in Automotive Applications, May 1-2, East Lansing, USA, 1991

P.R. Sahm and P.N. Hansen: Numerical Simulation and Modeling of Casting and Solidification Processes for Foundry and Cast-House, CIATF, Zurich, Switzerland (1984).

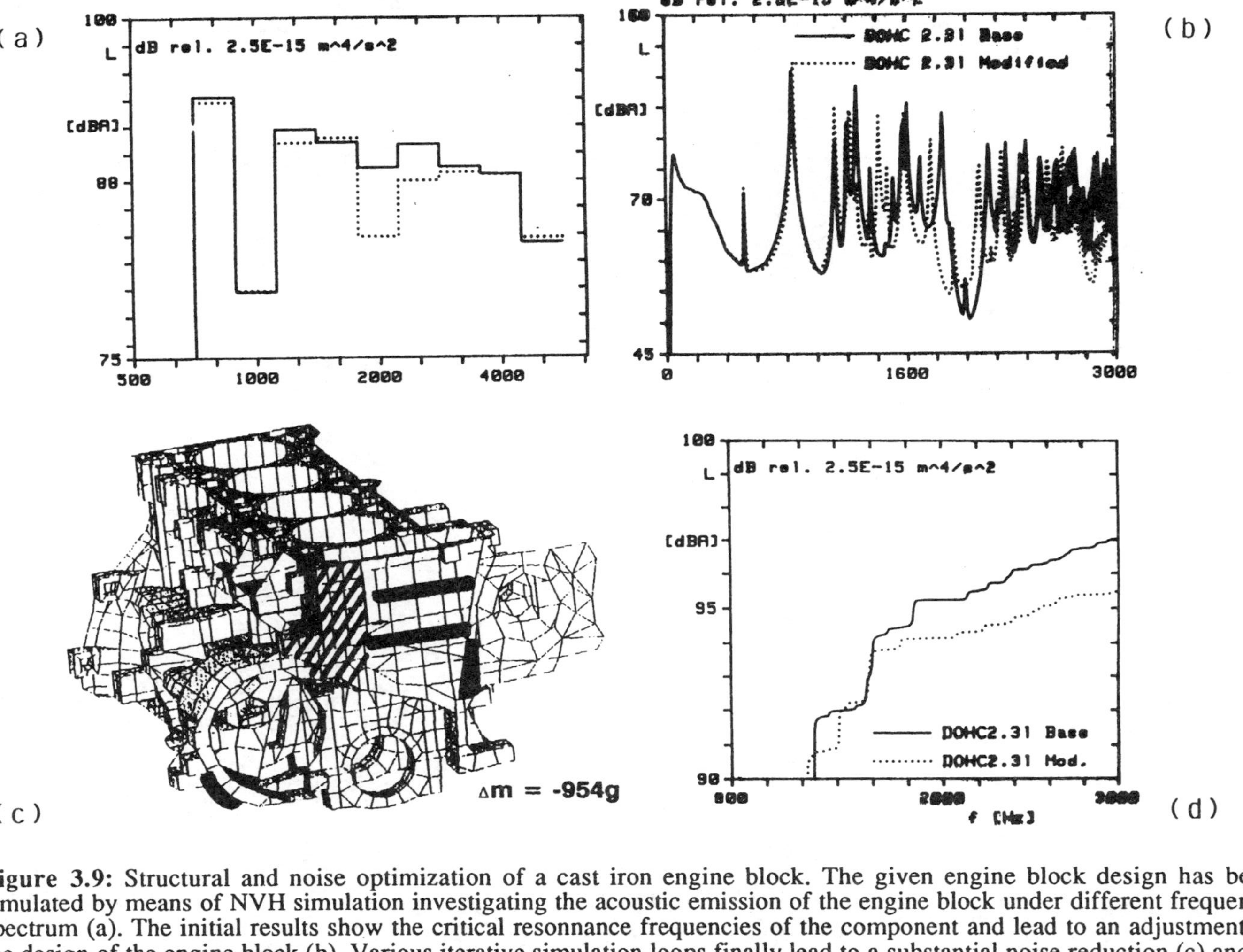

Figure 3.9: Structural and noise optimization of a cast iron engine block. The given engine block design has been simulated by means of NVH simulation investigating the acoustic emission of the engine block under different frequency spectrum (a). The initial results show the critical resonnance frequencies of the component and lead to an adjustment of the design of the engine block (b). Various iterative simulation loops finally lead to a substantial noise reduction (c) and at

Advanced Materials Research Vols. 4-5 (1997) pp. 129-136
© *1997 Scitec Publications, Switzerland*

Trends and Metallurgical Factors Involved in Automotive Cast Iron Parts

J. Le Gal

Renault, France

Keywords: Automotive Cast Iron Parts

Abstract :

The aim is to give some information about the trends in automotive cast iron parts, especially as regards their metallurgy, and to suggest some new areas of research.

First, a brief description of the use of cast iron in the automotive industry is given. The way scientific results can be translated into industrial practice is illustrated. Finally, the areas where extra knowledge is required are described.

What is required is a better description of the automotive industry's future needs.

This information will assist in choosing the best fields of scientific research to meet our future requirements for iron and steel castings.

INTRODUCTION

Every year, more than thirty five million cars are built throughout the world, each one containing more than 100 kg of cast iron parts. Cas iron parts are used both for their specific in-service properties and for their near net shape capabilities, which reduce the need for expensive investments in machining and assembly lines. Moreover, casting is attractive to car manufacturers as the most direct route from liquid metal to a complex machined component. For these reasons, scientific research into cast iron has a great impact upon the automotive industry, both qualitatively and quantitatively.

TRENDS IN THE AUTOMOTIVE INDUSTRY

New priorities are gradually emerging for the future :

- Decrease in exhaust gas pollution,
- Decrease in fuel consumption,
- Recyclability,
- ...

The first two requirements can be partially met by weight-saving, provided that the cost is not too high. The great inertia of the automotive industry, mainly due to the high cost and low flexibility of existing investments, leads to two levels of activity :

Short term work

Better use of existing processes and well-known materials.

Long term work

The potential use of "new" processes and "new" materials. Here, "new" has an automotive meaning : not yet used in mass production.

SHORT TERM WORK

The main ideas are to save weight by the following means :

- geometrical optimization taking into account material properties,
- enhanced reliability of existing processes
- better in-service properties for existing materials without modifying machining or assembly lines.

Given the tight safety margins now current, these goals can only be achieved through better understanding of existing processes and in-service properties.

Car manufacturers need to transfer to their processes the latest scientific results on the following :

- burden quality, melting and solidification processes,
- process contrôl, product quality and reliability,
- undestanding of mechanical behaviour,
- modelling of feeding, solidification and mechanical behaviour so as to get reliable results more quickly.

It is worth noting that this list describes virtually all the papers in this Symposium.

LONG TERM WORK

To get the right answers for future requirement (2000 A.D and beyond) car markers must forecast "break scenarios" which would require the use of new materials and processes.

The next ten years will probably lead to serious competition in the automotive industry between various material/process couples.

Some of them could be big competitors to cast iron. They are described below.

The different components mentionned here after are located on the schemas in the appendix.

1 - <u>Foundry trends</u> (exlcuding cast iron)

<u>High pressure magnesium die casting</u>

High pressure die casting of low-density metals is attractive. Magnesium is already beginning to compete with aluminium in die casting and welded sheet or tube assemblies :
-steering columns, transfer cases, seat frames, steering wheel armatures etc.

Weight saving can be as much as 50% with the same overall cost, thanks to integrated functions.

<u>Aluminium</u>

The most important development concerning aluminium in recent years has been research into better use of high pressure technology, which has led to the following improvements:

- porosity has been reduced dramatically, or even eliminated, ensuring good in-service properties and reliability,

- fine microstructures can be developed by rapid solidification, giving enhanced properties,

- dimensional accuracy has been increased giving near net shape parts with lower machining requirements,

- heat treatment and welding are now possible.

The number of applications of these new processes, such as squeeze casting, thixocasting and HPDC (high pressure die casting) is rising : Brackets, suspension arms, rear knuckles, differential case housing,...

These aluminium parts can easily replace cast iron, provided wear resistance is not required, and the in-service temperature is not above 150°C-200°C.

The required stiffness can be obtained through design geometry (eg : ribs) and the higher cost of raw material can be offset by lower investments in machining.

Steel castings

Steel castings can compete with cast iron in some applications where several of their specific in-service properties are superior.

---> Chassis components   From suspension arms and knuckles.
The higher elastic modulus, fatigue and impact resistance of steel allow attractive weight savings to be made on these highly stressed parts.

--->Engine components   Exhaust manifolds
The lower thermal conductivity and enhanced oxidation resistance of stainless steel, both of which increase the efficiency of catalytic converters, could be attractive if thin-walled parts can be made.

2 - Progress in alternative technologies

Some useful developments have also taken place in the competitors to casting.

- Forged steels new grades have been developed with better machinability.

- Steel plates. High strength low alloy (HSLA) steels have better in-service properties with excellent stamping qualities.

Forged steels could be used for knuckles and conrods, whilst steel plates could be used for the rear axle, including the suspension arms.

- Tubes. The development of hydroforming could allow their use, for instance in engine mountings.

- Welding. Bettter knowledge of process parameters and automation are leading to lower costs and improved reliability. Thus welded steel fabrications of tubes and plates can compete with cast components of complex shape.

- Sintering. The low requirement for machining makes this attractive for parts like conrods, whilst their good wear properties could make it attractive for cylinder liners.

3 - Cast Irons

Despite what has been said, cast irons will probably be used in the future for the following reasons :

- The graphite they contain gives them good friction properties which find a use in brake disks and drums, liners and cylinder blocks.

- Components of complex shape can be machined at low cost.

However, faced with the competitive technologies outlined above, the cast iron foundry industry must improve itself.

Up till now, the foundry industry, and in particular the automotive cast iron foundry, has not made use of its full technical potential, concentrating instead on producing at the lowest price.

The general goals which should be targeted are as follows.

- Enhanced dimensional accuracy.

- Better "skin quality" : lower roughness, removal of inclusions and avoidance of decarburization, all of which are most important with thin walls.

These two requirements suggest that more work is needed on sands, core ramming, filtration, shot blasting, etc.

- "New" grades with much better in-service properties, even if more sophisticated processes are needed.

- Steel castings should also be considered in the light of the same targets.

Below are given some examples showing some potential applications which highlight the need for new scientific knowledge so that reliable mass production can be achieved.

New lamellar cast irons with aluminium replacing silicon

These cast irons were developed in the seventies in Belgium at the C.R.I.F. The results were as follows :

- Die casting is possible, giving excellent dimensional control, less machining and enhanced in-service properties

- Better mechanical properties than usual with the conventional carbon content (3,3% to 3,5%).

- Better thermal properties with mechanical properties equivalent to those of a conventional silicon iron can be achieved by raising the carbon content to 4%.

These properties are most attractive to car makers, since they allow miniaturization and weight saving for parts such as brackets, and for brake disks when the 4%C grade in used.

Unfortunately, reliable mass production has not yet been achieved with conventional techniques due to oxidation of aluminium in the liquid state during melting and pouring.

Research into filtration, into melting and pouring in a neutral atmosphere and their effects on solidification and in-service properties would be of great interest. The improvement in properties and weight saving would probably compensate for the extra production costs.

Compact graphite cast irons

The in-service properties of C.G cast irons, intermediate between nodular and lamellar irons are also attractive to car makers as giving a weight-saving replacement to lamellar iron. Cylinder blocks, brake hub drums and brackets would be good applications for these irons. However, it seems as though extra works is also required in this field to achieve the high reliability needed for mass production. Perhaps the key to success lies in the same studies : melting, treating and pouring under a controlled atmosphere.

Thin wall steel castings of complex shape

As mentioned above, the use of steel castings may also be attractive. Moreover, research on steel castings could also enrich work on cast irons. The potential applications of steel castings suggest a wide range of research topics :

- flowability under vacuum or reduced atmospheric pressure

- mold/metal reactions : skin quality, dimensional accuracy etc

- solidifcation mechanisms, so that complex parts can be designed with, simultaneously, thin walls and large areas.

- in-service propreties of various grades.

CONCLUSION

The mass production automotive industry is being forced today to change its targets to meet new requirements. Low cost is no longer the over-riding priority, but the best compromise between low weight and low cost for the design specification. At the same time, various new materials and processes are appearing which could be used in place of cast iron.

Faced with these competitors, the foundry industry must develop improved techniques.

- Better reliability. New cast irons, such as lamellar aluminium and C.G irons should be considered, using controlled atmospheres. The progress with forged steels over the last twenty years is an example of what can be done.

- Consider the potential uses of thin walled steel castings.

To reach these targets, a great deal of scientific research into the physical metallurgy of cast irons and steels will be indispensable.

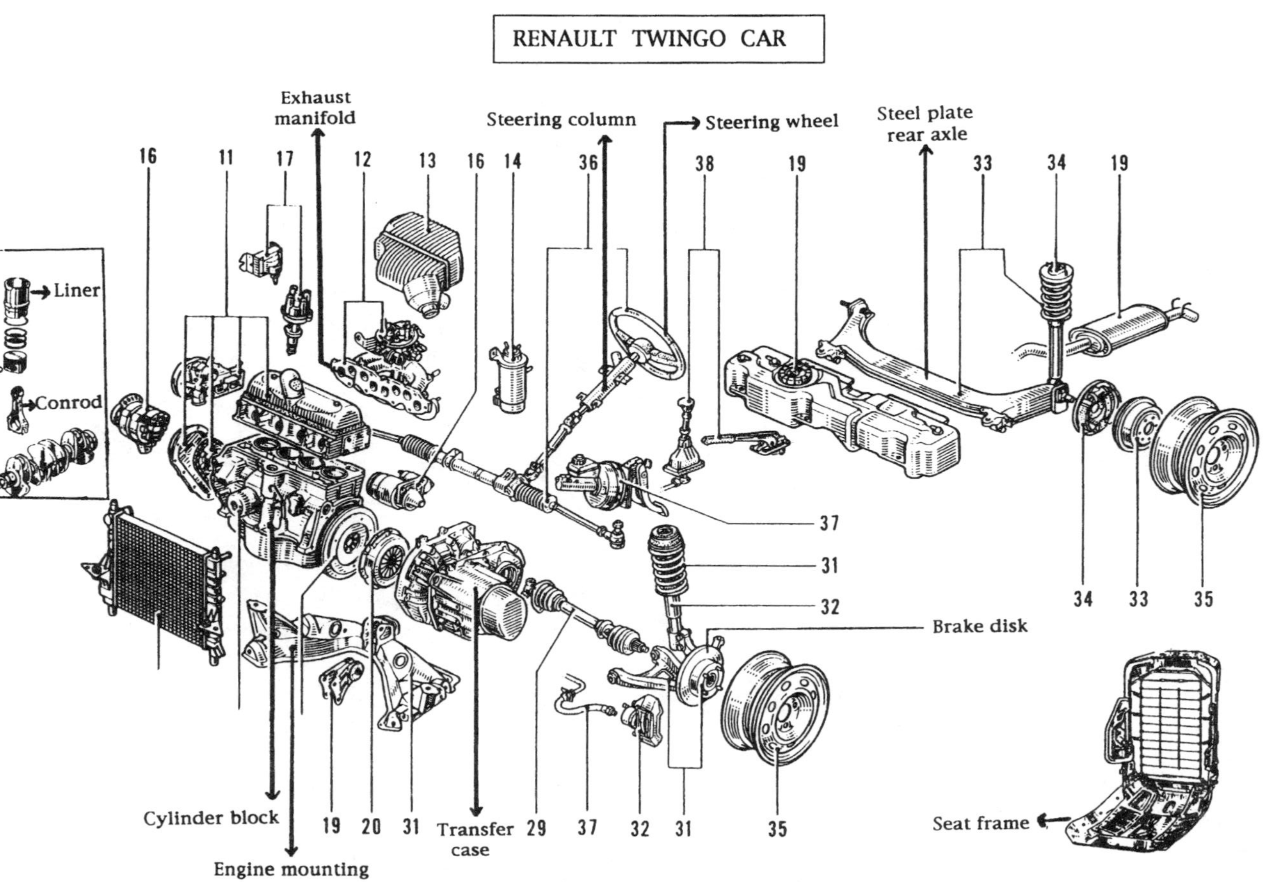

RENAULT TWINGO CAR
Exhaust manifold
Steering column
Steering wheel
Steel plate rear axle
16
11
17
12
13
16
14
36
38
19
33
34
19
Liner
Conrod
37
31
32
34
33
35
Brake disk
Cylinder block
19
20
31
Transfer case
29
37
32
31
35
Engine mounting
Seat frame

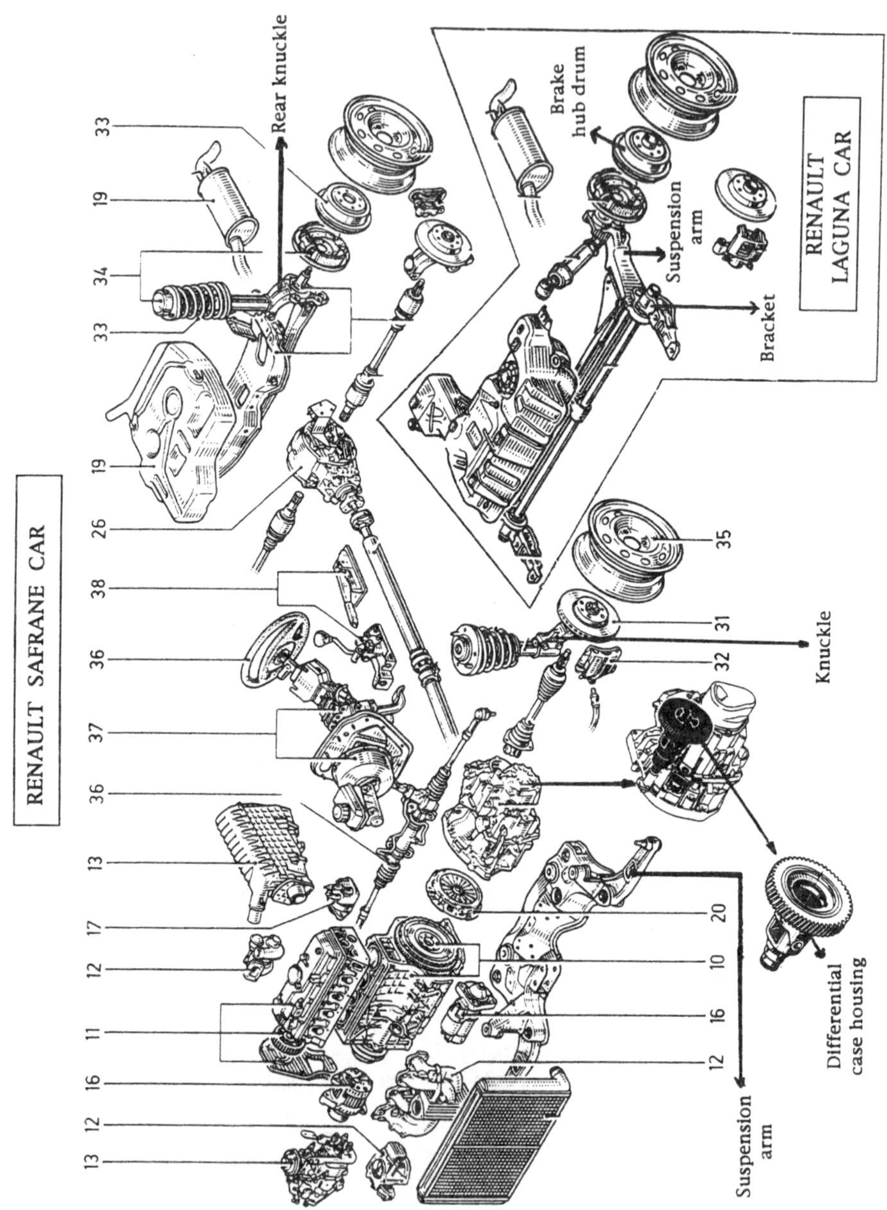
RENAULT SAFRANE CAR
RENAULT LAGUNA CAR
Rear knuckle
Brake hub drum
Suspension arm
Bracket
Knuckle
Differential case housing
Suspension arm
33
19
34
33
19
26
38
36
37
36
13
17
12
11
16
12
13
35
31
32
20
10
16
12

MECHANICAL BEHAVIOUR OF CAST IRON

Microstructure and mechanical behaviour - General

Impact properties

Fatigue and fracture behaviour

Wear behaviour

Advanced Materials Research Vols. 4-5 (1997) pp. 139-146
© 1997 Scitec Publications, Switzerland

Microstructural Evolution of Spheroidal Graphite Cast Iron at High Temperature: Consequences on Mechanical Behaviour

Ph. Bastid[1], Ph. Pilvin[1], Ch. Grente[2] and E. Andrieu[1]

[1] Ecole des Mines de Paris, Centre des Matériaux, URA CNRS 866,
B. P. 87, F-91003 Evry Cedex, France

[2] Renault S. A., Direction de la Recherche,
67, rue des Bons Raisins, F-92508 Rueil - Malmaison, France

Keywords: S.G. Cast Iron, Oxidation, Decarburization, Thermomechanical Behaviour

ABSTRACT

For the last few years, the improvement of car engine performances, as well as the efficiency of emission control by catalytic converters, have lead to an increase of exhaust gas temperature. It is of importance to study the applicability of currently used cast irons (GS50, therefore, SiMo) to more severe service conditions. First of all, an aged manifold has been used to assess the different types of damaging mechanisms. Oxidation, through wall thickness reduction and decarburization, through modification of the mechanical and physical properties of the material, have been identified as the two major damaging processes.

By using fully decarburized testing specimens, a comparison in terms of thermal and mechanical behaviour has been carried out between the as-cast and decarburized materials. The main differences are the following :
- decarburized iron has a lower thermal conductivity
- the α / γ phase transformation disappears in the decarburized material
- low cycle fatigue tests have shown differencies in the elastic and visco-plastic behaviours.

After establishing oxidation and decarburization kinetics and constitutive laws of ferritic, austenitic and decarburized cast iron, a finite elements modelling of local plasticity due to thermal loading of partially decarburized specimens is possible. It thus becomes possible to simulate the behaviour of components subjected to thermomechanical loadings such as exhaust manifolds.

INTRODUCTION

Recent improvements in car engines have lead to an increase in exhaust manifold temperature. This temperature increase is due to :
- wide use of high-performance engines, as turbo-charged or multi-valve engines.
- the use of thermal shields to protect electronic components, confining heat around the manifold.
- the improvement in aerodynamics, leading to a reduction of air flow under the bonnet.
- the use of the catalytic converter, which create another heat source for the manifold.

The thermomechanical resistance of GS50 spheroidal graphite cast iron, generally used in the past by RENAULT is not sufficient to withstand such high temperature level (above 850 °C at the branch pipe connection). Many other materials can be used to prevent damaging of the manifolds. SiMo iron exhibits a good oxidation resistance and a better high temperature mechanical behaviour. Moreover, the allotropic transformation temperature is higher in SiMo than in GS50 iron. Ni-Resist iron and stainless cast steel are also excellent candidates for exhaust manifolds because of its structural stability and interesting high temperature mechanical and oxidation properties [1, 2]. These last two materials seem to fit the technical requirements but are too expensive to be used in this particular industrial application.

Low cost GS50 and SiMo irons can be used as long as the microstructural evolutions and subsequent mechanical behaviour are taken into account in the manifold design. Thus, the approach chosen in the present work consists in three steps :
- examination of damaged manifolds to identify the damaging mechanisms
- study of these damaging mechanisms by isothermal and anisothermal ageing tests
- reproduction of some of the aged microstructures on test specimens and comparative study based on low cycle fatigue tests and dilatometry carried out on aged and as-cast materials.

MATERIALS AND EXPERIMENTAL PROCEDURES

One part of the materials was received as aside cast shafts. Compositions of the two cast irons are given in Table 1. Figure 1 shows the corresponding microstructures, which are similar to those observed in the as-cast manifolds.

	C	Si	Mn	S	P	Mo	Mg
GS50	3,61	2,95	0,011	0,007	0,028		0,035
SiMo	3,20	3,92	0,12	0,011	0,026	0,61	0,043

Table 1 - Chemical compositions of the two cast irons (wt %)

The matrix of GS50 iron is mainly ferritic with an area fraction of pearlite which does not exceed 5 %. The average grain size is 30 μm and the mean diameter of the nodules is 12 μm. In the case of SiMo iron, the matrix is also ferritic, the average grain size is 50 μm and the mean diameter of the nodules is 20 μm. For the two cast irons, the surface fraction of graphite is about 12 %. Differences between GS50 and SiMo microstructures are essentially based on an intense precipitation of lamellar Molybdenum carbides in grain boundaries and at triple points of SiMo iron. These carbides are responsible for the high temperature creep resistance of the material [3]. The precipitation of these carbides is accompanied by the precipitation of thinner particles which have not been indentified in the present work (Figure 1-b).

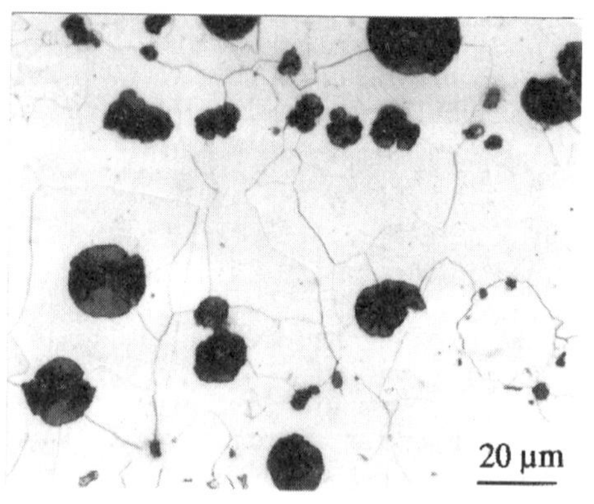

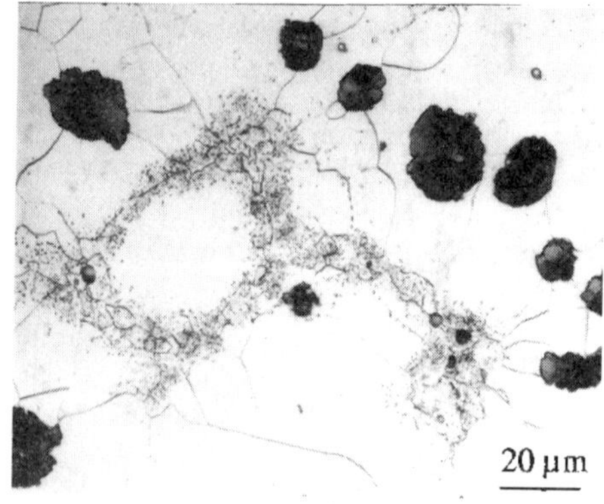

- a -   - b -

Figure 1 - As-cast materials
a - GS50 iron
b - SiMo iron (Molybdenum carbides and precipitates)

Dilatometry has been carried out in order to determine the allotropic transformation temperatures for the two irons. A significant scattering in the results has been observed. This observation might be related to an inhomogeneity of Silicon contents. Mean transformation temperatures are given in Table 2.
Finally, materials were also received as as-cast and aged manifolds.

	Ac_1	Ac_3	Ar_3	Ar_1
GS50	820 °C	890 °C	840 °C	770 °C
SiMo	860 °C	900 °C	860 °C	795 °C

Table 2 - Transformation temperatures ($\alpha \rightarrow \gamma \rightarrow \alpha$)

Strain isothermal low cycle fatigue tests with a symmetric loading ratio have been carried out on a servohydraulic testing machine. A three zone radiation furnace was used to heat the fatigue test specimens. Different strain levels have been applied, namely, 0.2 %, 0.3 %, 0.4 %, 0.5 %, 0.7 % under different strain rates 10^{-3}, 10^{-4}, 10^{-5} s^{-1}.

Thermal expansion properties have been determined with an absolute dilatometer.

RESULTS AND DISCUSSION

<u>Examination of damaged manifolds</u>

A cracked part of GS50 manifold which has experienced a bench test has been observed by optical microscopy. The observation of the crack vicinity shows three aging mechanisms, which are oxidation, decarburization and phase transformation (see Figure 2). The main features of these microstructural evolutions are described below :

- The oxide layer is about 500 µm thick. Consequently, the reduction of wall thickness is close to 20 %. Since oxide scales of similar thickness have been found on non-damaged manifolds, one can conclude that the oxidation process is not responsible in itself for crack formation.
- Phase transformation is revealed by the occurence of pearlite, while structure of the as-cast manifold is quite completely ferritic. Successive allotropic transformations seem to be the cause of the deformation of graphite nodules shown in figure 2-b.
- Decarburization is revealed by disappearance of graphite nodules and formation of a ferrite layer near the surface and around the crack.

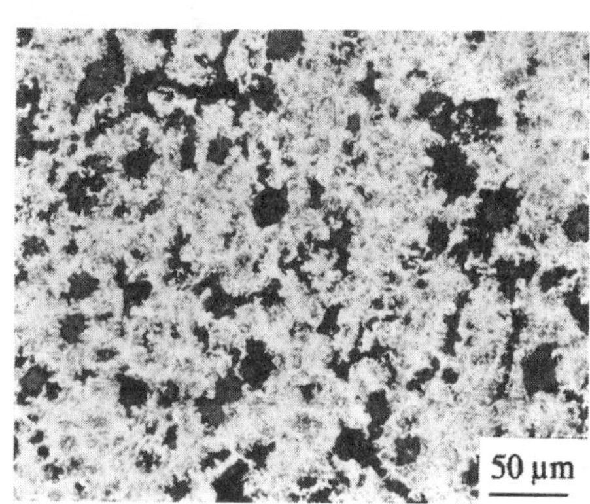

- a -   - b -

Figure 2 - Damaged exhaust manifold
a - Pearlitic matrix and ferrite layer revealing phase transformation and decarburization
b - Degeneration of graphite nodules due to successive phase transformations

The last two ageing mechanisms are certainly involved in the damaging process, because they have not been observed in non-damaged manifolds. Moreover, it is well known that allotropic transformations have to be avoided to prevent stresses due to local dimensional changes.

In fact, phase transformations are likely to reduce the lifetime of the manifold in two different ways. The first is the deformation of graphite nodules. The graphite particles shape change, from spheroid to flake, can promote crack initiation and propagation. The second way is related to the partial decarburization of the wall. Decarburized and non-decarburized cast irons will certainly exhibit differences in their thermomechanical properties. In particular, the absence of Carbon will change the transformation temperature of decarburized iron.

<u>Isothermal ageing</u>

The three aging processes, identified from the examination of manifolds, have been studied by applying high temperature isothermal heat treatments on testing specimens under air environment. The ageing temperatures were 800 °C, 850 °C, 870 °C and 890 °C. This set of experimental conditions allowed us to investigate the effect of phase transformation on oxidation and decarburization processes.

As long as oxidation is concerned, the two materials studied behave similarly in terms of the chemical composition of the external oxide scale (FeO). Nevertheless, differencies appears when considering internal oxide scale. Microprobe analysis has shown that internal oxide structures were different for the two cast irons. The SiMo iron contains much more $Fe_2(SiO_4)$ than does GS50. This oxide, known for its adherence, gives to the iron some additional protection against oxidation.

When dealing with decarburization, it is worth mentioning that this process initiates below 800 °C and is accompanied by the disappearence of graphite nodules, each of them leaving a pore in the ferrite matrix. The final material after decarburization is then a porous steel, with a volume fraction of pores equal to the initial fraction of graphite. A detailed analysis of the pores formation process has shown that the nodules do not disappear entirely . The nodule size decrease induces a loss of coherency between the nodule and the ferritic matrix so that the nodule cannot be anymore considered as a Carbon source. In SiMo iron, the unidentified precipitates observed in as-cast material, disappear if ageing temperature is above 870 °C, either by dissolution or decarburization. Molybdenum carbide precipitation remains unchanged in the grain boundaries of decarburized SiMo iron. This is the main microstructural difference between decarburized GS50 and SiMo irons (see Figure 3).

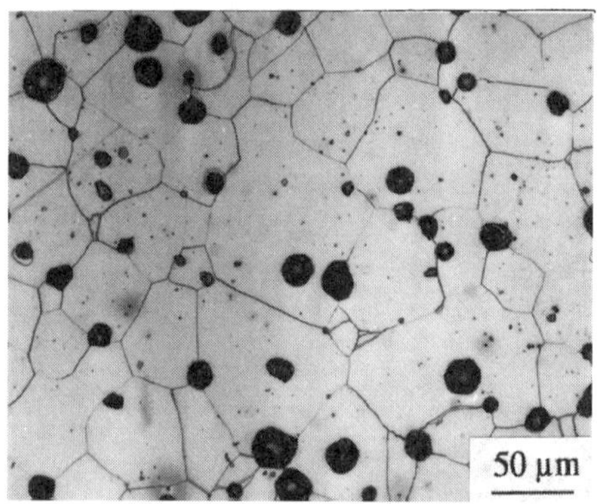
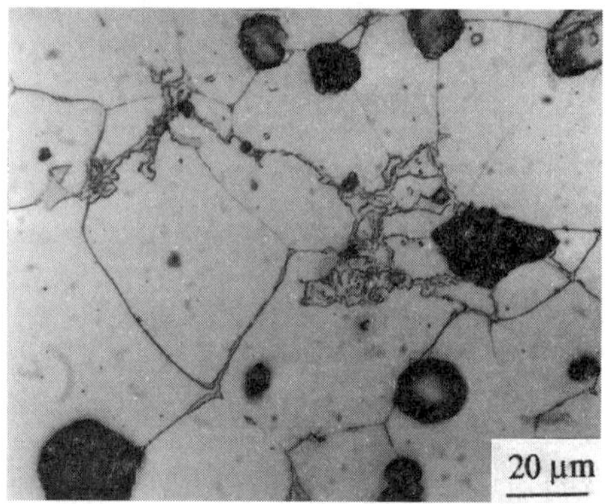

- a -   - b -

Figure 3 - Decarburized GS50 and SiMo irons
a - GS50 iron
b - SiMo iron (steady Molybdenum carbides)

Oxidation and decarburization kinetics have been determined by measuring the oxide layer thickness and the decarburization depth on polished cross sections.

Assuming oxidation and decarburization processes follow the usual Fickian diffusion equations, each thickness evolution can thus be defined by :

$$x_{ox} = K_{ox}.\sqrt{t} \qquad and \qquad x_{dec} = K_{dec}.\sqrt{t} \tag{1}$$

where t is time, x_{ox} and x_{dec} are respectively the oxide scale thickness and the decarburization depth and K_{ox} and K_{dec} are the kinetics constants of the process. The evolution of these two coefficients with respect to temperature is presented on figure 4.

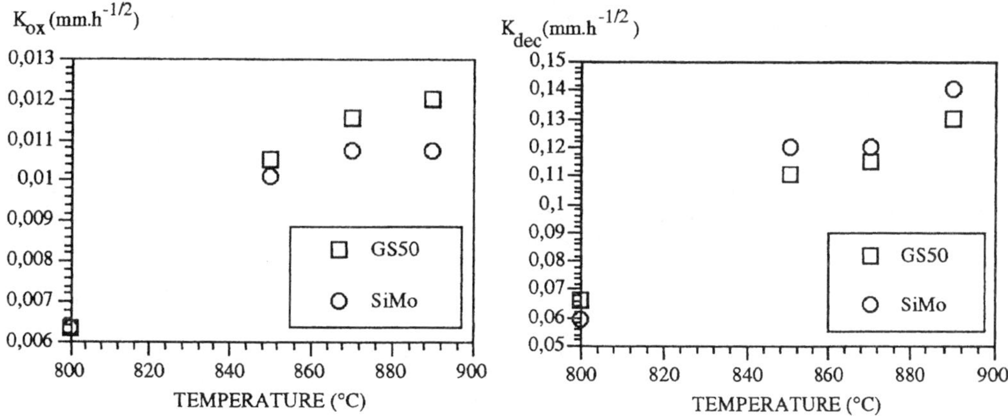

Figure 4 - Oxidation and decarburization kinetics

The main point brought by these results is the surprising continuity of oxidation and decarburization kinetics evolution when the aging temperature exceeds phase transformation temperature.

In the case of oxidation, this can be explained by considering that, despite of having the allotropic transformation deep inside the test specimen, the oxidation front always evolves in a ferritic layer due to decarburization.

In case of decarburization, kinetics depend only on the nature of the material through which Carbon diffuses. Here, Carbon comes from decarburized/non-decarburized interface and diffuses through the ferritic layer and the oxide scale. Oxide scale morphology and thickness seem to play an important role in controlling decarburization kinetics [4].

Moreover, it is worth noting that the formation of the internal oxide in SiMo iron does not produce a significant reduction as expected from the litterature [3].

<u>Anisothermal ageing</u>

The same aging processes are observed when the materials are submitted to thermal cycling. The samples were exposed to two different thermal cycles :
- heat treated 1 hour at maximum temperature, then air quenched during 30 minutes
- heat treated 2 hours at maximum temperature, then air quenched during 1 hour.
The different values chosen for T_{max} were 850°C, 870°C and 900°C.

Even if the quantitative study of oxidation has not been possible because of oxide scale desquamation due to dimensional changes of samples, it is reasonable to suggest that cyclic oxidation is more detrimental than isothermal oxidation.

Concerning decarburization, the kinetics determined by isothermal ageing tests can be used for thermal cycling. Decarburization depth can be calculated by integration of isothermal equations.

The main differences in terms of microstructural evolution between isothermal and anisothermal aging result from the successive phase transformations. Three essential features are presented below :
- circumferential deformation of pores in ferrite ring surrounding non decarburized heart
- cast iron inflation
- degeneration of shape of graphite nodules.
  ı Circumferential deformation of pores in ferrite layer arises from differences in transformation temperatures between decarburized and non-decarburized materials. The external ferrite layer contains mainly Iron and Silicon (about 3 wt% for GS50 and 4 wt% for SiMo). Observation of the Fe-Si phase diagram shows that the decarburized layer is stable up to 1500°C. This point has been partially studied through thermal expansion tests, performed up to 1200°C. For non-decarburized material, Schissler et al. [5] have shown that contraction occuring during allotropic transformation disappears after a few cycles. After this stage has been reached, the effect

of phase transformation is reduced to the effect of differences in thermal expansion coefficient between α and γ phases. Since γ phase has a higher thermal expansion coefficient, the dilatation of the heart of the specimen generates circumferential tension stresses in the ferrite layer which causes the observed pores deformations.

Inflation of non-decarburized cast iron is certainly another source of circumferential deformation for the ferrite layer. It occurs through precipitation of new graphite nodules during $\gamma \rightarrow \alpha$ transformation, as indicated by Schissler et al. [5]. Inflation has been quantified by density measurements. It increases with temperature and number of thermal cycles and tends toward an asymptotic value. The maximum linear expansion obtained in the present study was close to 8 % (GS50, for 320 cycles at $T_{max} = 900°C$). This value is similar to one previously mentioned in [5].

The shape evolution of graphite particles, mentionned during the examination of damaged exhaust manifold, did not appear by isothermal aging tests. In contrast, this degeneration of graphite can occur by anisothermal cycling, especially when T_{max} is inbetween Ac_1 and Ac_3. This phenomenon has to be related to the dual phase (ferrite+austenite) composition of the matrix. During an isothermal exposure between Ac_1 and Ac_3, phase transformation begins first in the vicinity of grain boundaries. Then, during few hours, the sharing of phases evolves to steady state as shown in Figure 5. Quantitative study about this evolution is not yet made, but it seems that it occurs to reduce internal stresses appearing in the beginning of austenite formation. For each heating between Ac_1 and Ac_3, the material is submitted to these stresses producing degeneration of graphite shape (Figure 2-b). The consequences of degenerated graphite on the fracture behaviour has not yet been studied.

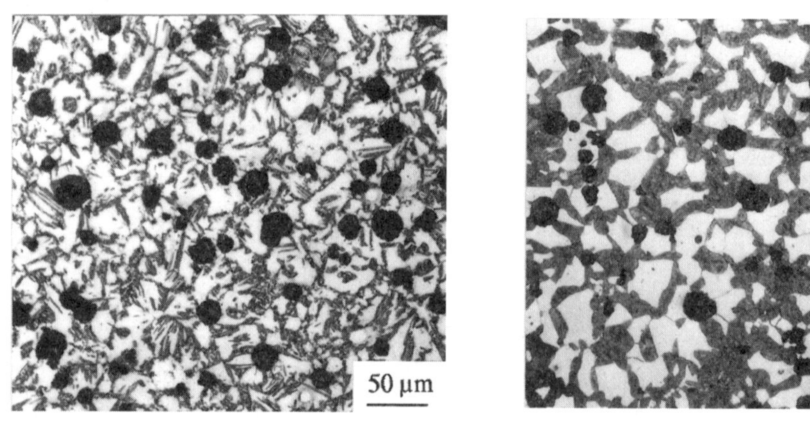

Figure 5 - Evolution of phases distribution during thermal treament of SiMo iron at 870 °C
a - After 20 mn treatment
b - α/γ distribution at steady state

CONSEQUENCES ON MECHANICAL BEHAVIOUR

To analyze the mechanical behaviour, an elastoplastic model developed at ONERA [6], was choosen. This model allows to describe high temperature behaviour of metals under cyclic loading. In the case of monotonic tension, the stress can be devided in four components as :

$$\sigma = R_0 + R + X + \sigma_v \quad (2)$$

where R_0 is the yield stress, σ_v is the viscous stress, R and X describe respectively isotropic and kinematic hardening effects. Expressions of R, X and σ_v, with ε_v is the viscoplastic strain, are given below :

$$R = Q.[1 - \exp(-b.\varepsilon_v)] \quad (3) \qquad X = (C/D).[1 - \exp(-D.\varepsilon_v)] + H.\varepsilon_v \quad (4) \qquad \sigma_v = K.(\dot{\varepsilon}_v)^{1/n} \quad (5)$$

This model has 9 coefficicients : Young modulus E, initial yield stress R_0, viscosity coefficients K and n, linear and non linear kinematic hardening effect coefficients H, C, D, and isotropic hardening coefficients Q and b.

When the number of coefficients included in the model increases, the identification of these coefficients has to be made through a simulation and an optimization code. Identification consists in determining the set of coefficients such as the deviation, for all tests, between the experimental observation and the model simulations is as small as possible. The used code, named SiDoLo, is well adapted for the treatment of any kind of model that may be expressed as ordinary differential equations [7].

In the present study, the numerical identification of the constitutive law parameters has lead to the following results (Figure 6.a). In this figure, the partitioning of the macroscopic stress at two temperatures is presented for as-cast and aged materials.

At 400 °C, it is worth mentionning that the yield stress corresponding to decarburized materials is higher than that of as-cast materials. The physical reasons responsible for these phenomenon are not yet clear. As was expected at the beginning of the study, the viscoplastic component of the macroscopic stress increases significantly when temperature reaches 700 °C. This internal stress is higher in SiMo iron than in GS50, probably because of the grain boundary pinning by Molybdenum carbides precipitation. For the same reason, the effect of decarburization on high temperature properties of SiMo is negligible.

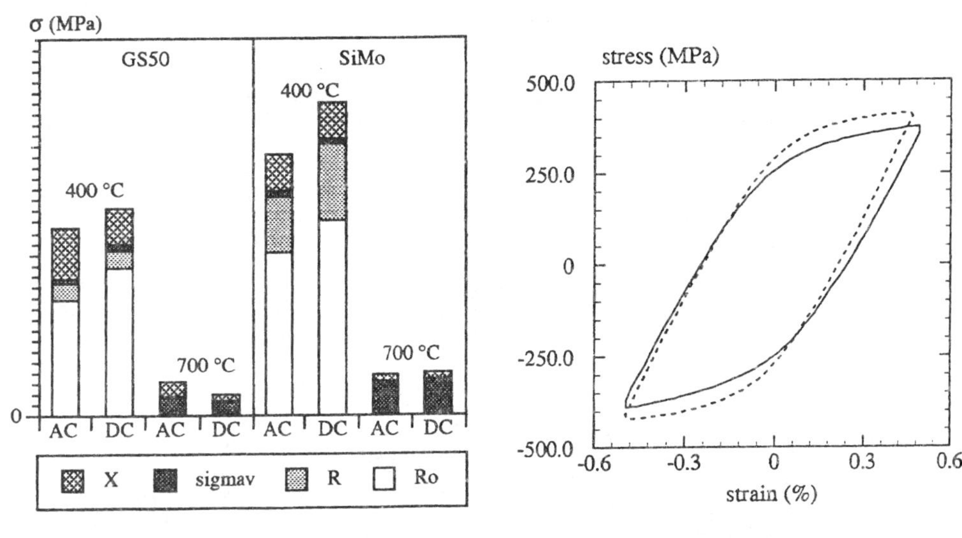

- a -   - b -

Figure 6 - Low cycle fatigue tests
a - Macroscopic stress partitioning for as-cast (AC) and decarburized (DC) materials
b - Fatigue cycle for as-cast and decarburized (dashed line) GS50 iron at 400°C

In figure 6.b, the effect of graphite nodules on elastic properties is highlighted. In the case of as-cast materials, the Young's modulus measured for a tensile loading is lower than the one measured for a compressive loading. This discrepancy strongly suggest that the as-cast material has to be considered as a porous material under tensile loading and a composite material under compressive loading. On the contrary, the porous decarburized material is symmetrical as far as elastic properties are concerned.

CONCLUSION AND PROSPECT

Microstructural evolutions on aged and as-cast exhaust manifold have been observed and the corresponding kinetics measured.

The characterization of the evolution of the physical properties has been done, using representative as-cast and aged microstructures testing specimens.

Low-cycle fatigue tests at different temperatures has lead to the identification of the constitutive equations parameters assuming an elastoviscoplastic model.

The design of structures with account taken of the various points mentionned above seems to be possible.

Finally, ferritic cast iron might be used in more severe working conditions provided that the microstructural evolutions are taken into account in manifold dimensioning.

REFERENCES

[1] K. Akiyama, M. Ike, M. Tsuda, K. Otsuka - Analysis of thermal fatigue resistance of engine exhaust parts - SAE Publication, N° 910430, pp. 63, 71

[2] T. Shimamoto, K. Asano, Y. Shioya, S. Ando - Development of the stainless cast steel exhaust manifold - SAE Publication, N°930621, pp.39, 47

[3] S. Bechet - Les fontes utilisées à haute température - Hommes et Fonderie, février 1981, pp. 3, 16

[4] N. Birks - Mechanism of decarburization - DECARBURIZATION, ISI Publication 133, 1970, pp. 13,33

[5] J. M. Schissler, F. Gross, J. P. Chobaut - Etude du phénomène de gonflement des fontes GS ferritiques par cycles thermiques de type $\alpha \rightarrow \gamma \rightarrow \alpha$ - Hommes et Fonderie, mars 1993, pp. 33, 42

[6] J. L. Chaboche, P. M. Lesne - A non linear continuous fatigue damage model - Fatigue and fracture of Engineering Materials and Structures, Vol. 11, N° 1, p.1, 17

[7] P. Pilvin - Modélisation du comportement des assemblages de structures à barres - Thèse 3ème cycle, Université Pierre et Marie Curie (Paris 6)

Advanced Materials Research Vols. 4-5 (1997) pp. 147-152
© *1997 Scitec Publications, Switzerland*

Characterization and Measurement of the Modulus of Elasticity of Ductile Irons

L. Fang[1], K.E. Metzloff[1], R. Voigt[2] and C.R. Loper Jr.[3]

[1] Research Assistant, Dept. of Materials Science and Engineering, University of Wisconsin-Madison, 1509 University Ave., Madison, WI 53706, USA

[2] Associate Professor, Dept. of Industrial and Management Systems Engr., Pennsylvania State University, 207 Hammond Bld., State College, PA 16802, USA

[3] Professor, Department of Materials Science and Engineering, University of Wisconsin-Madison, 1509 University Ave., Madison, WI 53706, USA

Keywords: Modulus of Elasticity (Young's Modulus), Non-Linear Stress Strain Curve, Hyperbolic Stress-Strain Response, Localized Plastic Deformation

ABSTRACT

The modulus of elasticity (E) or Young's modulus is a measure of the initial slope of the engineering stress-strain curve obtained during conventional tensile testing. For most engineering metals the slope is nearly constant for the elastic portion of the stress-strain curve until gross yielding occurs at values of stress near the yield stress. However, for ductile irons, this "elastic" portion of the stress-strain curve is often highly non-linear. Because of this complex behavior, measured values for the "modulus of elasticity" of ductile irons depends not only on graphite morphology and matrix structure, but also on the techniques used to measure and determine the modulus.

A method has been proposed to use modulus measurement techniques originally developed for aggregate construction materials on ductile irons. Plots of engineering strain vs. strain/stress clearly illustrate the different deformation regimes occurring during "elastic deformation". This method can be used to reproducibly measure the true elastic modulus of ductile irons as well as characterize the non-linearities that occur below the yield stress caused by graphite morphology and matrix effects..

MODULUS OF ELASTICITY

The modulus of elasticity (Young's modulus) is the proportionality constant in equations relating stress to strain during uniaxial tensile testing. Typical modulus values for ductile irons are generally reported as: 165 GPa to 172 GPa (24 to 25 Mpsi) [1], 164 GPa for ASTM A536 grade: 120-90-02 to 169 GPa for ASTM A356 grade: 60-40-18 [2], and 169 GPa (annealed) to 176 GPa (pearlitic) [3]; however, substantially lower values are often recorded in industry, and with great variability. The cause of this modulus variability may be attributed to a few factors: actual variation in modulus due to differences in graphite morphology, the effect of the matrix structure, and the technique used to record the modulus. The basic procedure to determine the modulus of elasticity is to calculate the slope of the stress-strain diagram in the linear region at the beginning of an uniaxial tensile test. However, the slope of the "linear" portion of the stress-strain diagram is not constant for cast irons. This round house behavior presents a severe measurement problem as to which slope is used to characterize the stiffness or modulus of the material.

HYPERBOLIC STRESS-STRAIN RESPONSE

Because of the inherent heterogeneous microstructures present in ductile iron castings, it is evident that the traditional Young's modulus values and modulus measurement techniques developed for steels are not entirely appropriate for ductile irons.

It is necessary to develop a realistic and usable approach by which the modulus of elasticity can be computed and rationalized for these heterogeneous materials. An attempt to do so was made by utilizing a hyperbolic stress-strain response which was first introduced by civil engineers for the analysis of soil mechanics [4]. A hyperbolic plot is developed by utilizing strain (in/in) as the X axis and strain/stress (1/psi) as the Y axis. As shown in Figure 1 (a hyperbolic plot of AISI 1045 cold rolled steel), three regions develop in the diagram. The beginning portion (region A) is an unstable region where the load train and extensometer "seating" produce an initial transient on the hyperbolic plots. The second region (region B) is the horizonal plateau where the modulus is truly constant and strain is perfectly elastic. The final segment (region C) is influenced by plastic deformation; this is usually a straight line with a slope showing a continuous and constant change (decreasing value) in modulus in the plastic region. The point at which the hyperbolic plot increases in slope corresponds to the yield point of the material. The modulus of elasticity from a hyperbolic plot is determined by taking the inverse of the average Y axis values (strain/stress) of the plateau. The start and endpoint of the plateau was estimated and the average of the data that fell between the two points was taken as the value to be the inverse of the modulus. Figure 1 also exhibits the initial part of the engineering stress-strain curve of AISI 1045 cold rolled steel as well as the homogeneous mocrostructure which consists of strain hardened of pearlite and ferrite.

The mechanical properties of the experimental lower carbon heats (alloys #D-1 and #D-2) are summarized in Table I as well as similar tests performed on samples of commercial ductile irons and cold rolled steels for comparison.

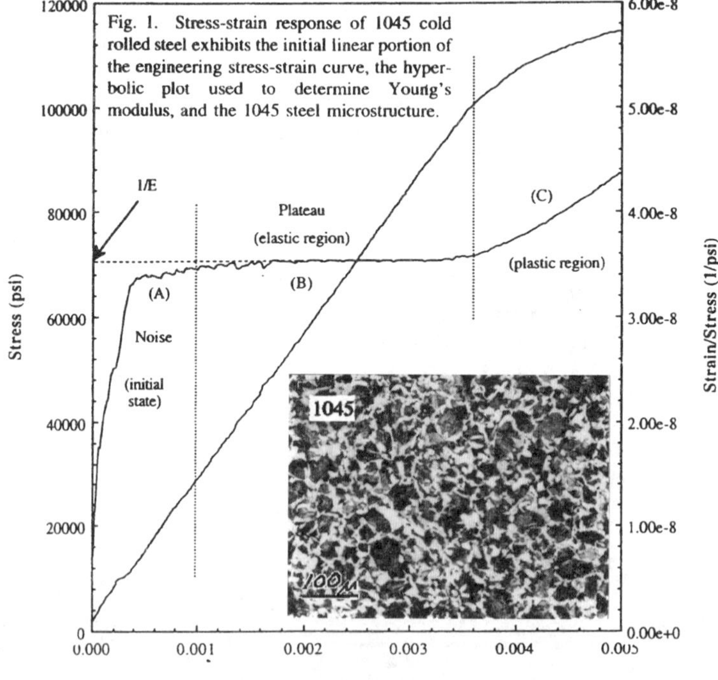

Fig. 1. Stress-strain response of 1045 cold rolled steel exhibits the initial linear portion of the engineering stress-strain curve, the hyperbolic plot used to determine Young's modulus, and the 1045 steel microstructure.

Table I. Mechanical Properties of Experimental Heats.

Sample No.	Tensile Str. MPa (ksi)	Yield Str. MPa (ksi)	Elong. (%)	Hard-ness (HB)	N.C. ($/mm^2$)	Matrix (%)	Hyper Modulus GPa (Mpsi)	Zero Modulus GPa (Mpsi)	Tangent Modulus GPa (Mpsi)	Secant Modulus GPa (Mpsi)
D1-N	903 (131)	679 (98.5)	2.28	341	40	95% P	182 (26.4)	174 (25.3)	181 (26.3)	185 (26.9)
D1-S	710 (103)	435 (63.1)	9.83	207	30	62% P	173 (25.1)	172 (24.9)	171 (24.8)	174 (25.3)
D1-F	506 (73.4)	355 (51.5)	23.92	156	30	7% P	160 (23.2)	157 (22.7)	165 (24.0)	163 (23.6)
D2-N	1000 (145)	670 (97.2)	3.12	321	200	95% P	208 (30.1)	223 (32.3)	185 (26.8)	219 (31.8)
D2-S	636 (92.2)	405 (58.7)	15.45	179	270	57% P	190 (27.6)	181 (26.3)	178 (25.8)	199 (28.8)
D2-F	519 (75.3)	349 (50.6)	25.57	163	300	all F	184 (26.7)	186 (27.0)	172 (24.9)	189 (27.4)
ACDI	758 (110)	459 (66.5)	7.89	235	100	35% F	188 (27.2)	183 (26.5)	174 (25.3)	193 (28.0)
ANDI	602 (87.3)	358 (51.9)	14.14	159	100	50% F	184 (26.7)	175 (25.4)	166 (24.1)	188 (27.2)
NODI	917 (133)	538 (78.0)	5.86	277	100	all P	162 (23.5)	143 (20.7)	158 (22.9)	161 (23.3)
1045CR	875 (127)	820 (119)	9.03	248	---	P + F	196 (28.4)	192 (27.9)	196 (28.5)	199 (28.8)
1045A	703 (102)	433 (62.8)	22.28	201	---	P + F	192 (27.8)	170 (24.7)	171 (24.8)	190 (27.5)

"D-1" = Low carbon equivalent cast iron (1.95% C, 2.45% Si, 0.60% Mn, 0.039% Mg).
"D-2" = High carbon content graphitic steel (1.01% C, 2.33% Si, 0.34% Mn, 0.0028% Mg).
"N" treatment = Normalizing (Austenitized and air cooled).
"S" treatment = Hold at 900°C (1652°F) (12 hrs for #D-1, 1 hr for #D-2). Furnace cooled at 100°C/hr
　　　　　(180°F/hr) to 690°C (1274°F). Furnace cooled.
"F" treatment = Hold at 900°C (1652°F) (12 hrs for #D-1, 1 hr for #D-2). Furnace cooled at 100°C/hr
　　　　　(180°F/hr) to 760°C (1400°F). Cooled to 690°C (1274°F) (at 5°C/hr [9°F/hr] for
　　　　　#D-1, at 20°C [36°F] for #D-2). Furnace cooled.
"ACDI" = Typical as cast ductile iron (3.6% C, 2.7% Si, 0.33% Mn, 0.04% Mg, 0.5% Cu).
"ANDI" = Typical annealed ductile iron (Heat and hold at 900°C [1652°F] for 1 hour.
　　　　　Furnace cooled at 55°C/hr [100°F/hr] to 345°C [650°F]. Air cooled.)
"NODI" = Typical normalized ductile iron (Heat and hold at 900°C [1652°F] for 1 hour, and air
　　　　　cooled.)
"1045CR" = 1045 cold rolled steel (0.43-0.50% C, 0.6-0.9% Mn).
"1045A" = Annealed 1045 cold rolled steel (Heat and hold at 820°C [1508°F] for 1 hour.
　　　　　Furnace cooled at 28°C/hr [50°F/hr] to 650°C [1202°F]. Air cooled.)

A. Effect of Graphite Structure

The modulus of elasticity of the matrix components of cast iron and steel, namely ferrite, pearlite, cementite is approximately constant at 207 Gpa (30 Mpsi). It is the presence of graphite as flakes or nodules that strongly influences the stress/strain response. The "interruption" of the matrix by graphite reduces the effective modulus by an amount dependent upon the form and quantities of these discontinuities [5].

At very low values of stress, nodule decohesion from the matrix can be readily observed in ductile iron. Further loading results in localized plastic strain concentrated in the matrix at the matrix/graphite interface. Considerable localized plasticity can occur around the nodules even though the rest of the matrix is undergoing only elastic deformation. This micro-plasticity at low values of stress in this "composite material" is the primary contribution to the "non-linear" modulus behavior observed for ductile irons. The compactness of the graphite influences the amount of localized strain concentrated at the matrix/graphite interface. Accordingly, the shape, distribution, quantity and size of the graphite particles must be considered in discussing the influence of the graphite structure on the modulus of elasticity. In general the presence of non-spheroidal graphite decreases the modulus.

Localized plasticity occurring at the matrix/graphite interface will be reflected in an upward slope to the strain vs. strain/stress plateau (rather than the plateau being horizontal) and is increased when bull's-eye ferrite is present. This upward change in slope will be observed at low values of strain for cast irons with non-compact graphite morphology and can result in a round house stress-strain diagram with no horizontal plateau in the hyperbolic plot.

Hyperbolic plots of normalized samples of alloys #D-1, #D-2 and ductile iron are illustrated in Figure 2. The difference in modulus values (Table 1) is attributed to graphite morphology and graphite

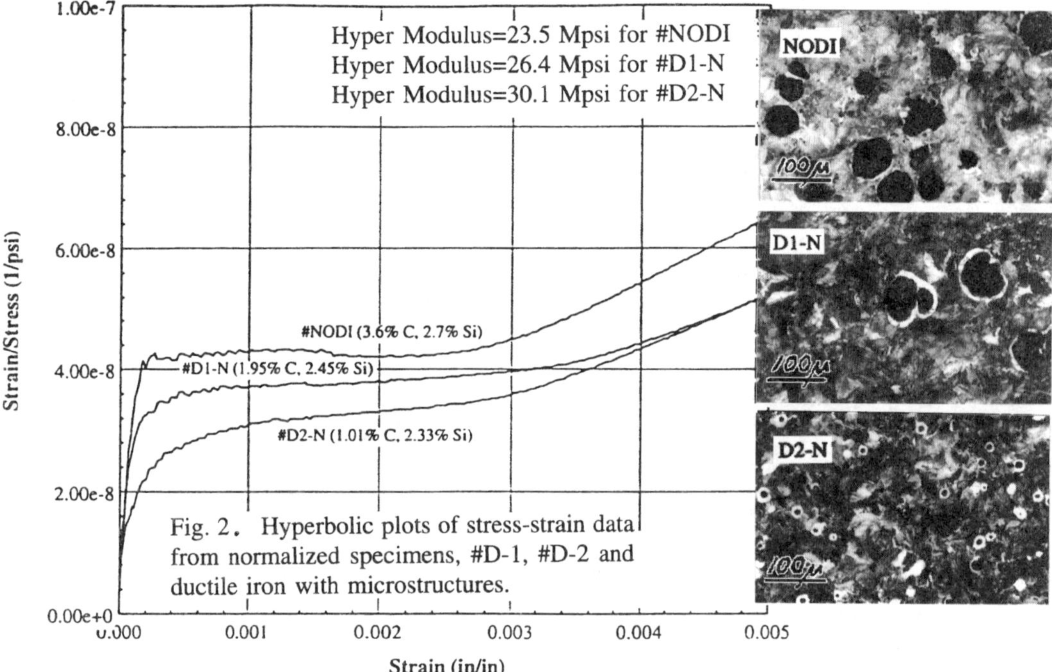

Fig. 2. Hyperbolic plots of stress-strain data from normalized specimens, #D-1, #D-2 and ductile iron with microstructures.

volume fraction (carbon content). The volume of graphite present is controlled by the carbon content (and the matrix structure). The increase in carbon increases the volume fraction of graphite, which in turn decreases the modulus.

B. Effect of Matrix Structure

As previously stated, variation in the matrix structure of steels has little or no effect on the modulus value. However, when graphite flakes or nodules are introduced, small values of strain can cause considerable micro-plasticity at the matrix/graphite interface. If the graphite is surrounded by a constituent with a lower yield strength (i.e., ferrite), micro-yielding occurs well before gross yielding. Therefore, a decrease in effective modulus will be expected as well as a shorter plateau region in the hyperbolic plot. A well-developed long plateau is observed in the normalized specimen which consists of a fully pearlitic matrix. For specimens containing a mixed ferrite-pearlite matrix (#D2-S) or a fully ferritic matrix (#D2-F), there is a less obvious (shorter) plateau region with increasing ferrite content, indicating that the modulus changes gradually in the "elastic" region. This same trend was observed in the hyperbolic plots of commercial ductile irons as exhibited in Figure 4. The flatness and length of the plateau are primarily determined by the matrix structure, while the modulus value is essentially influenced by the carbon content (graphite volume fraction) and graphite nodularity.

SUMMARY

This study has demonstrated the complex stress-strain behavior of ductile irons and the challenges of measuring the Young's modulus for these heterogeneous "composite" materials. A method of analyzing conventional stress-strain information is proposed that clearly characterizes the pseudo linear portion of the curve and allows better measurements of the modulus to be obtained. In summary:

A method is proposed to use a hyperbolic modulus plot (strain vs. strain/stress) to characterize the modulus of ductile irons. This method has clearly demonstrated the non-linearities occurring in the initial "linear" portion of a stress-strain curve and allows more reproducible values of the modulus of ductile irons to be determined.

The influence of graphite morphology, nodule distribution and matrix structure on the modulus of conventional and low carbon equivalent ductile irons has been presented. Micro-yielding at the matrix/graphite interface causes the stress-strain behaviors to deviate from linearity (true elastic behavior) at low values of strain. This is particularly true for ferrite and ferrite/pearlite matrix structures with low nodularity.

REFERENCES

1. W. H. Bowes, L. T. Russell and G. T. Suter, *Mechanics of Engineering Materials,* John Wiley & Sons, Inc., 1984.
2. *ASM Metals Handbook,* vol. 1, ASM International, 1991.
3. *Ductile Iron Handbook,* American Foundrymen's Society, Inc., 1992.
4. R. T. Kondner, "Hyperbolic Stress-Strain Response: Cohesive Soils", *Journal of the Soil Mechanics and Foundations Division, Proceedings of ASCE,* vol. 89, no. SM 1, p. 115 (1963).
5. H. T. Angus, *"Cast Iron: Physical and Engineering Properties",* Butterworths & Co. Ltd., 1976.

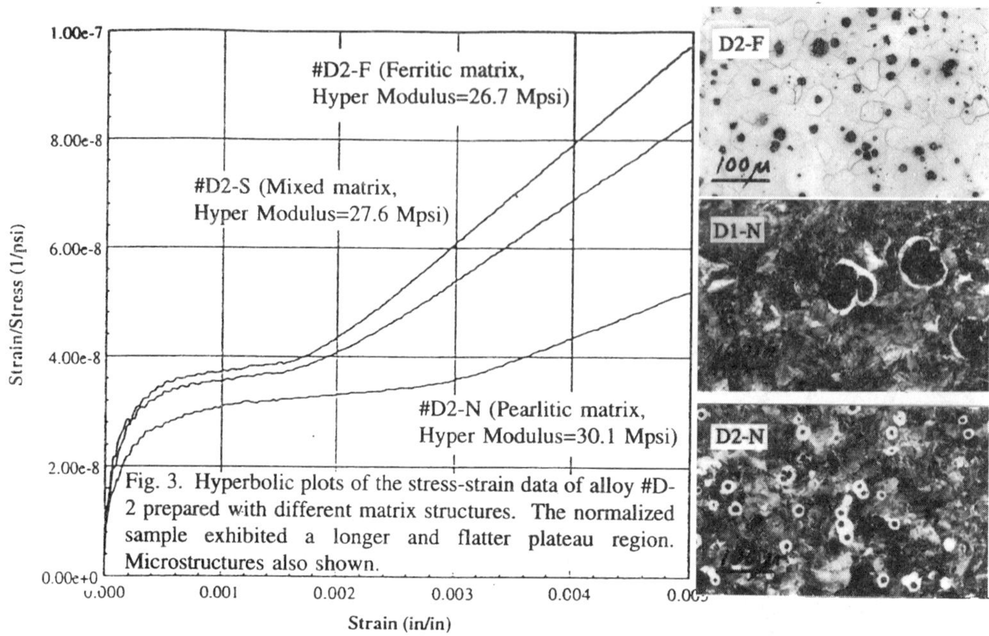

Fig. 3. Hyperbolic plots of the stress-strain data of alloy #D-2 prepared with different matrix structures. The normalized sample exhibited a longer and flatter plateau region. Microstructures also shown.

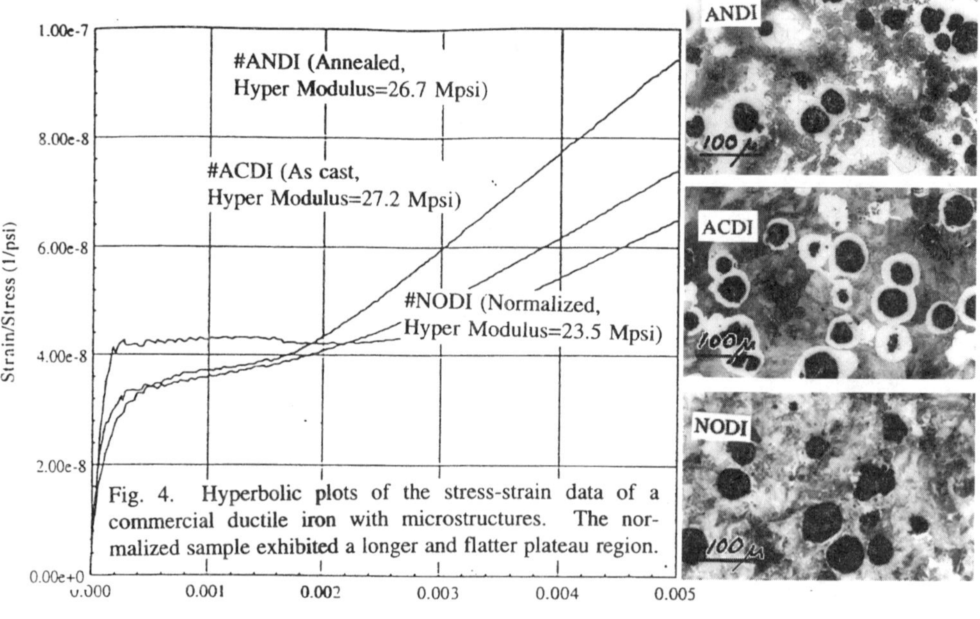

Fig. 4. Hyperbolic plots of the stress-strain data of a commercial ductile iron with microstructures. The normalized sample exhibited a longer and flatter plateau region.

Advanced Materials Research Vols. 4-5 (1997) pp. 153-160
© *1997 Scitec Publications, Switzerland*

Study of Physico-Chemical Mechanisms Responsible for Damage of Heat Treated and As-Cast Ferritic Spheroidal Graphite Cast Irons

P. Dierickx[1], C. Verdu[1], J.-C. Rouais[2], A. Reynaud[3] and R. Fougeres[1]

[1] GEMPPM URA CNRS n° 341, INSA bât 303,
20 av A. Einstein, F-69621 Villeurbanne Cedex, France

[2] CNDRI, INSA bât 303, 20 av A. Einstein, F-69621 Villeurbanne Cedex, France

[3] Centre Technique des Industries de la Fonderie,
44 av de la Division Leclerc, F-92312 Sèvres Cedex, France

Keywords: Cast Iron, Spheroidal Graphite, Ferrite Heat Treatment, Damage Initiation Microstructure

ABSTRACT

The microstructural damage initiation mechanisms in ferritic spheroïdal graphite cast irons before and after different heat treaments of ferritisation are investigated in this work. Two different compositions of ductile cast irons and two different heat treatments of ferritisation were realized. The first step of this study was a microstructural characterization and in a second time SEM and optical microscope "in situ" uniaxial tensile tests were carried out. Different damage mechanisms were identified and their proportions were determined from a statistical point of view. There are three microstructural sites of damage initiation : the graphite/matrix interface, inside the wrapped graphite and the interface between initial graphite and the wrapping graphite. On the one hand, two points appeared to control the damage mechanisms : a good nodularity of the graphite leads rather to interface decohesions and a Si segregation in the wrapping graphite seems to be responsible for within graphite decohesions. There is a competition between these two mechanisms. One the other hand, the heat treatment performed at highest temperature leads to the formation of ferrite inside the wrapping graphite.

INTRODUCTION

Today, ferritic spheroïdal graphite cast irons are more and more used for mechanical applications. The ferritic matrix can be obtained either during the elaboration or after additional heat treatment of ferritisation. The final microstructure seems to be the same but some mechanical characteristics appear to be very different (for example K_{IC} of 80 MPa $\sqrt{m}$ for the as-cast and 30 for the heat treated) [1]. This difference is usually attributed to a matrix-graphite interface brittleness. In fact, two damage initiation sites are reported in the litterature [2-12] : at the matrix-graphite interface and within the graphite. So far, the heat treatment effects on the damage initiation mechanisms are not known. A qualitative and quantitative relationship between mechanical properties and microstructural parameters is missing.

This paper proposes to study this relationship and especially to identify and to understand the microstructural damage mechanisms in ferritic spheroïdal graphite cast irons before and after different heat treatments of ferritisation using the new concept of microheterogeneous material. A statistic study is made in order to evaluate the influence of different important microstructural parameters. A microstructural characterization is performed during a first step and in a second one a mechanical behaviour characterization. Then, decohesion modes can be linked to the microstructural parameters.

EXPERIMENTAL

<u>Materials:</u>
Two ferritic spheroidal graphite (SG) cast irons were elaborated, one, called A1, has a perlito ferritic matrix and the second one, called B2, is almost totally ferritic (Table I).

	C	Si	Mn	S	P	Mg	Ceq
A1	3.38	2.40	0.19	0.007	0.02	0.062	4.19
B2	3.43	3.04	0.03	0.009	0.02	0.055	4.45

Table I : cast irons chemical compositions.

Then, two different heat treatments of ferritisation were performed on the two types of cast irons. The first one, called TT1, consisted of an isothermal treatment at 750°C for 5 hours, cooled to 20°C at 55°/h. The second one, called TT2, was heated at 880°C for an isothermal holding of 2 hours and cooled to 300°C at 15°C/h.

<u>Microstructural characterization:</u>

The metallographic structure analysis has been performed using a Quantimet 570 of Leica. The shape and the size of graphite nodules were determined. It was noted that the computer program allowed the classification of the graphite shape in six classes, according to the ISO 945 standard(class VI = high nodularity and class I elongated shape) [13]. With the reasonable hypothesis that the nodules are spherical, the diameter distribution was deduced from the apparent area distribution. Particles the size of which being lower than 8 µm were not taken into account.

The X-Ray microanalysis pointed out the possible chemical segregations inside the nodule and at the matrix/graphite interface. Line scans were performed from the graphite nodule centre to the matrix to analyse Si, Mg and S.

The influence of chemical segregations on local mechanical properties was characterized by nanoindentation technique (Young's modulus measurements) with a Nanoindenter II from Nano Instrument [14].

<u>Tensile tests:</u>

The second step was a mechanical characterization. SEM "in situ" uniaxial tensile tests were carried out to identify with accuracy, the different damage mechanisms (crack initiation and propagation) [15]. In order to do a statistical study at large scale, the same experiment was performed under optical microscope. Different damage mechanisms were identified and their proportions were determined from a statistical point of view.

<u>Fractographic analysis:</u>

A fractographic analysis was performed on the tensile test fracture surfaces to confirm the damage initiation mechanisms which were observed at the specimen surface during the "in situ" tensile test and to identify the crack growth mechanisms.

RESULTS

<u>Microstructural characterization:</u>
<u>*Microstructure*:</u>

Figure 1 shows the different microstructures. Metallographic observations and the image analysis show evolutions of the nodule size and shape with heat treatments (table II and fig n°2).

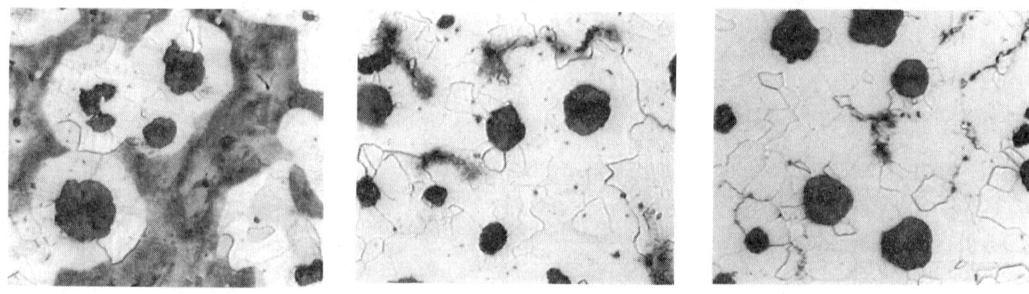

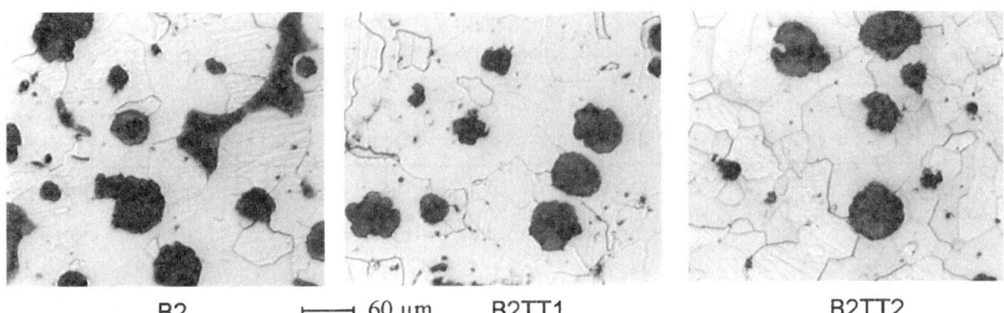

Fig n°1: microstructure: A1: ferrito-pearlitic matrix with ferrite in "bull's eyes", B2 as-cast has more than 95% of ferrite and A1TT1, A1TT2, B2TT1 and B2TT2 almost totally ferritic.

	class VI	Class V	Class IV	Class III	Class II	Class I
A1	60	23.2	10.3	6	0.3	0.2
A1TT1	58.1	24.6	10.1	6.5	0.6	0
A1TT2	52.3	33.4	9.1	4.5	0.2	0.4
B2	34.4	39.7	19.5	5.3	1	0.2
B2TT1	68.3	21.5	6.7	3.1	0.5	0
B2TT2	60.8	25.3	10.3	3.5	0	0.2

Table II : Shape class of graphite (%).

As-cast ferrito-pearlitic sample A1 has a good percentage of class VI, and after both heat treatments only a very slight decrease can be observed. B2 showed a deteriorate shape of graphite nodule compared to A1 and the heat treatments improved its shape. Moreover, a relation between shape and size is observed. The small graphite nodules are rather round and the big ones are rather degenerated.

The repartition of nodule size was studied in all samples (fig n°2) and according to each class of shape. After heat treatments, a slight increase of the large diameters can be observed.

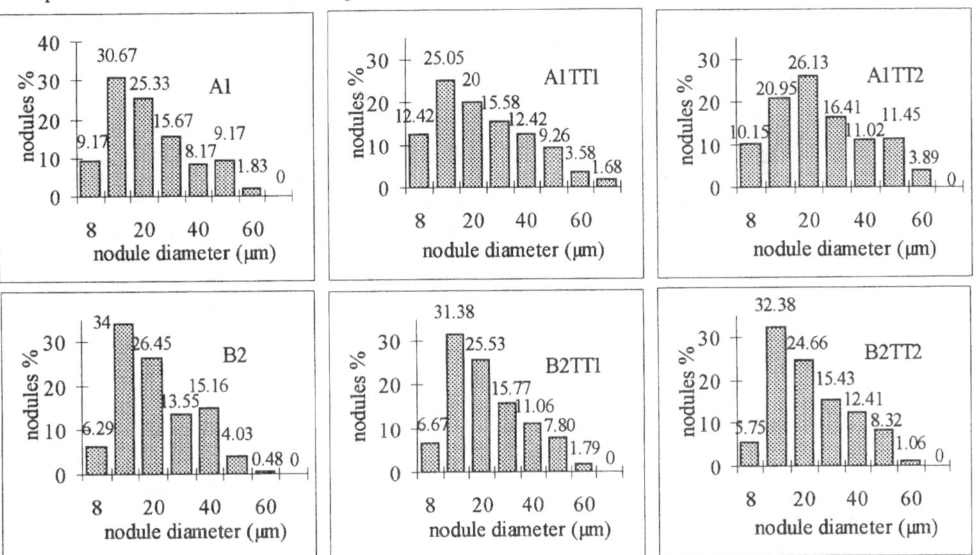

Fig n°2 : Column graphs of nodule size repartitions in all the samples

<u>*Segregation:*</u>

X-Ray microanalysis line scans led to find microsegregations inside the graphite nodules after heat treatments. Indeed, in the graphite, for as-cast samples, a constant percentage of Si and Mg is found. Sometimes, a high value of Si exists at the nodule centre (germination on a inoculant particle). The %Si increases deeply at the matrix-graphite interface (figure 3a).

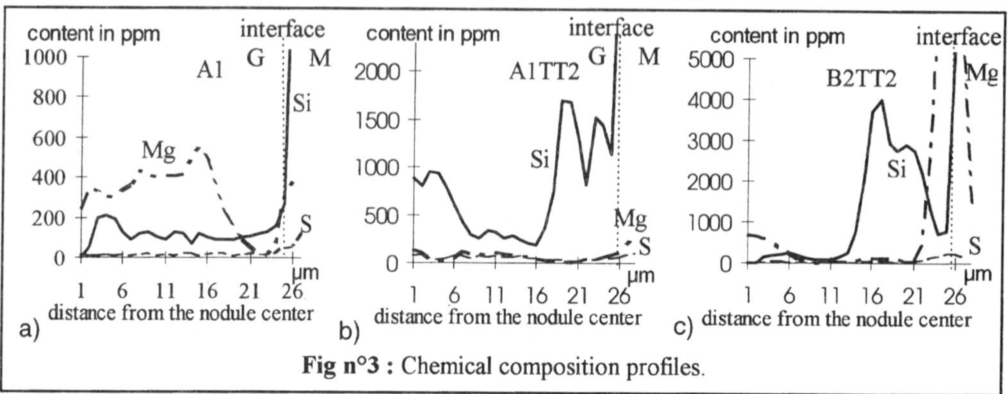

Fig n°3 : Chemical composition profiles.

But after heat treatments :

    - Si content in the wrapping graphite was higher than in the initial one (figures 3b, 3c).

    - Mg increase at the graphite-matrix interface appeared for B2TT1 and B2TT2 (figure 3c).

The variations observed are significant in comparison with the measurement precision.

X-Ray map of as-cast samples have shown microsegregations (Mn and S) in the end solidification areas. It seems that both heat treatments do not eliminate but erase slightly those microsegregations.

<u>*Local mechanical caracteristics:*</u>

Nanoindentation measurements lead to Young's modulus values which increase from 15 ±5 GPa in the graphite to 250 ±10 GPa in the matrix. This transition is more progressive in heat treated TT2 samples than in as-cast ones (figure 4).

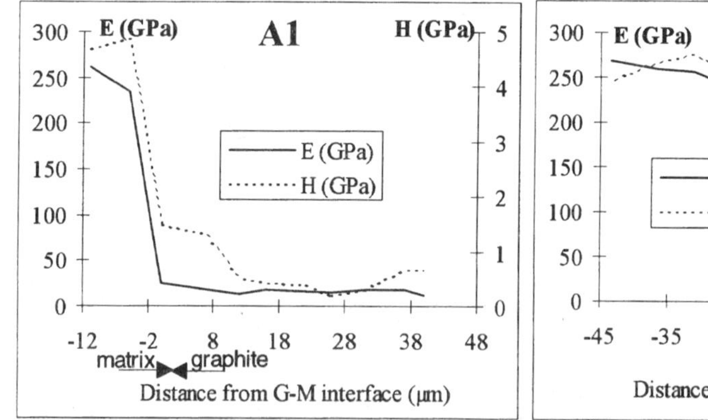
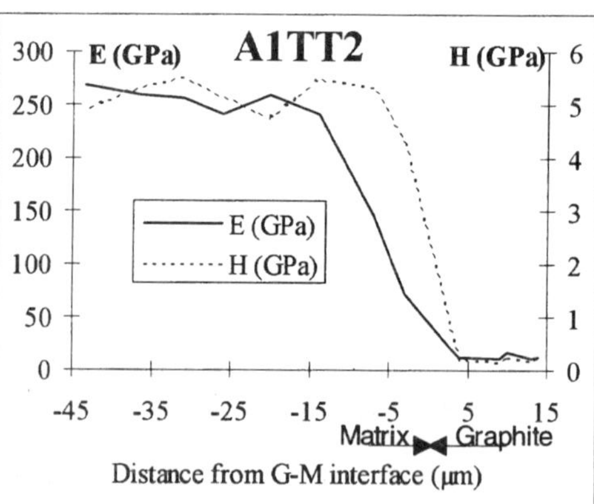

Fig 4: Hardness H and Young's modulus E profiles obtained by nanoindentation for A1 and A1TT2.

<u>Damage:</u>

<u>*Damage initiation:*</u>

SEM observations during "in situ" uniaxial tensile tests showed that there are three microstructural sites of damage initiation (figure 5):

        - at the graphite/matrix interface

        - inside the wrapped graphite

- at the interface between initial graphite and the wrapping graphite

These three types of decohesions exist in different proportions in all samples. Decohesions in as-cast samples are rather at the matrix-graphite interface and in the heat treated samples rather within the graphite.

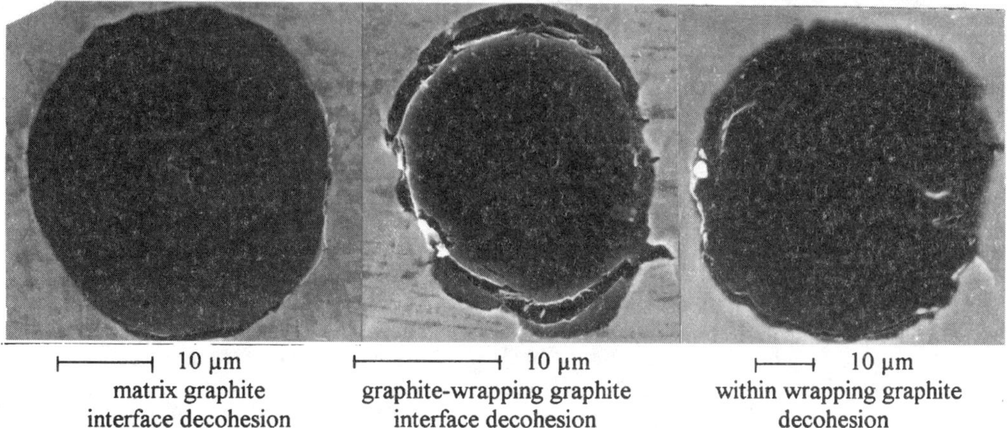

Fig n°5 : SEM micrographies of different damage initiation sites

Moreover, when graphite nodules are spherical, decohesions are located rather at the matrix-graphite interface and when the shape is degenerated decohesions are rather intragraphite.

Under an optical microscope, the low magnification used, do not permit to differentiate the two types of within graphite decohesion. Unfortunately statistic studies cannot be performed on the both within graphite initiation. So, two decohesion modes are considered: the first one at the graphite/matrix interface, and the second one within the graphite. In as-cast ferrito-pearlitic sample A1, most of the decohesions occur at the graphite/matrix interface, and after heat treaments, decohesions are rather intragraphite (see figure 6a and b). In the as-cast ferritic sample B2, decohesions are half intragraphite and half graphite/matrix interface. After heat treatment TT1, decohesions are also half intragraphite and half graphite/matrix interface but after TT2, intragraphite decohesions are mainly observed.

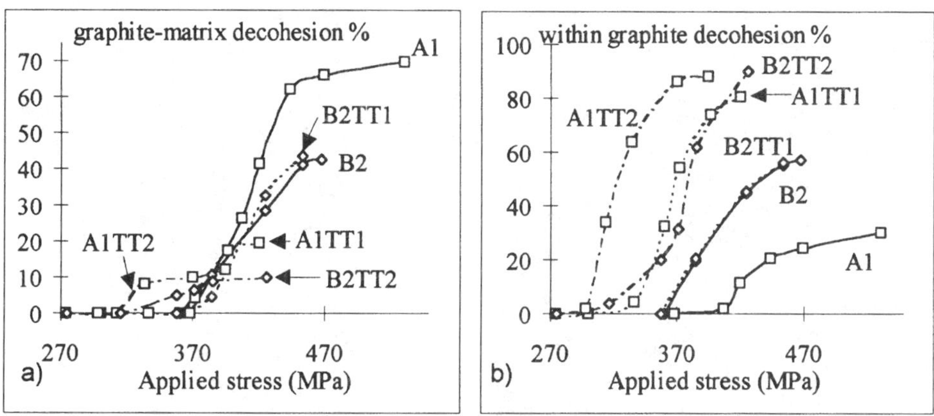

Fig n°6 : Percentage of different damage initiation types vs applied stress.

For all samples, the within graphite decohesions begin before the graphite-matrix decohesions. After the heat treatments decohesions appear earlier. As a matter of fact, tables III and IV give the different threshold decohesion stress σ_{th} for all cases:

	As-cast	TT1	TT2
A1 matrix-graphite decohesion σ_{th}	400	370	335
A1 within graphite decohesion σ_{th}	400	335	300
B2 matrix-graphite decohesion σ_{th}	380	380	350
B2 within graphite decohesion σ_{th}	375	370	315

Table III : Applied stress threshold σ_{th} for the different types of decohesion (MPa)

	As-cast	TT1	TT2
A1 100% decohesion stress	500	410	390
B2 100% decohesion stress	460	440	400

Table IV : Applied stress for 100% decohesion (MPa)

<u>Damage propagation:</u>

In figure 7, the different decohesion modes as well as slip bands in the matrix can be observed. Decohesion and slip bands seems to appear at the same time. From our experiments it is difficult to know if the decohesion is due to an interaction of the slip bands with the graphite nodule or if the slip band is induced by an interface decohesion. The crack propagation occurs by jonction of the microcracks along the slip bands between the nodules and lead to the fracture of the samples.

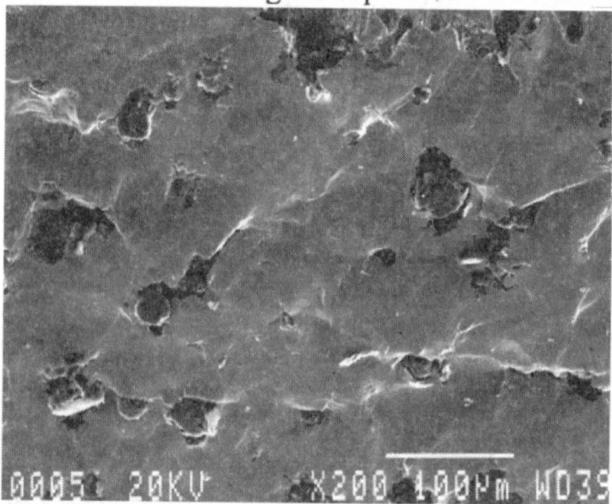

Fig n°7 : A1TT2 micrography of the specimen polished surface, near the fracture surface

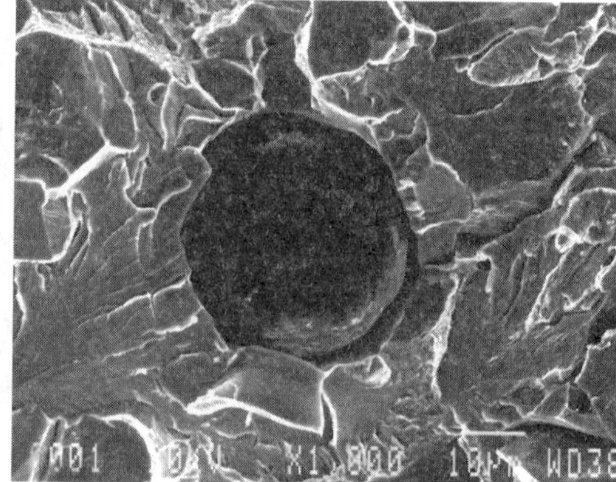

Fig 8: A1. graphite-matrix interface decohesion.

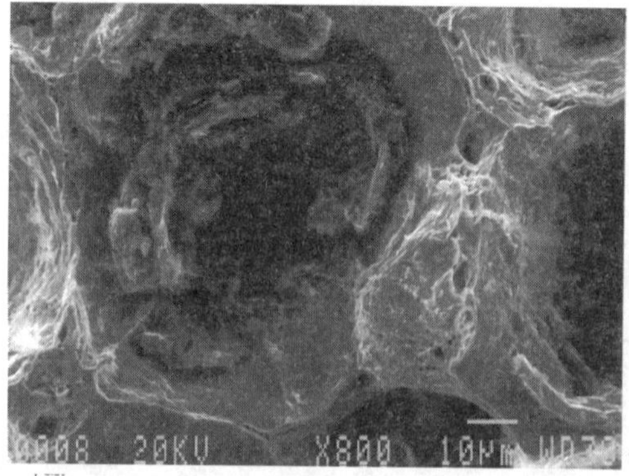

Fig 9: A1TT1. graphite-wrapping graphite interface decohesion. ductile fracture

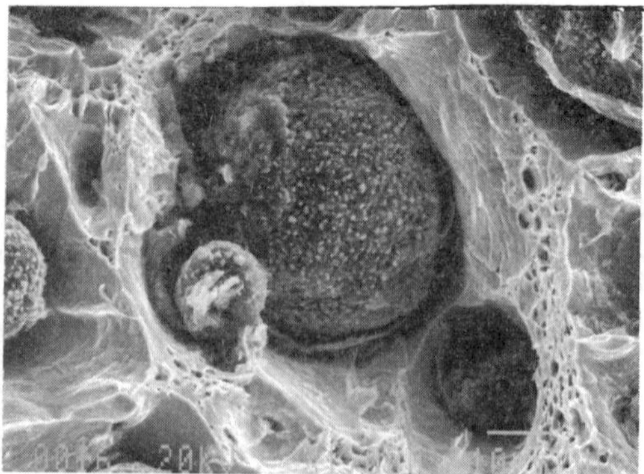

Fig 10: B2TT2. within the wrapping graphite decohesion.

The microfractographic analysis confirms that the previous observations at the sample surface are similar in the bulk sample. Indeed, the three types of decohesion modes were found (Figures 8,9,10) at the fracture surface.

A1 with 50 percent of pearlite has a brittle behaviour. For all the other samples the behaviour is ductile. An other important information appears in this fractographic study: Figure 11 shows a B2TT2 graphite nodule which presents on its surface a white phase. This unknown phase, after X-Ray analysis including carbon element can be assumed to be ferrite and does not content silicon. This white phase is also observed in A1TT2.

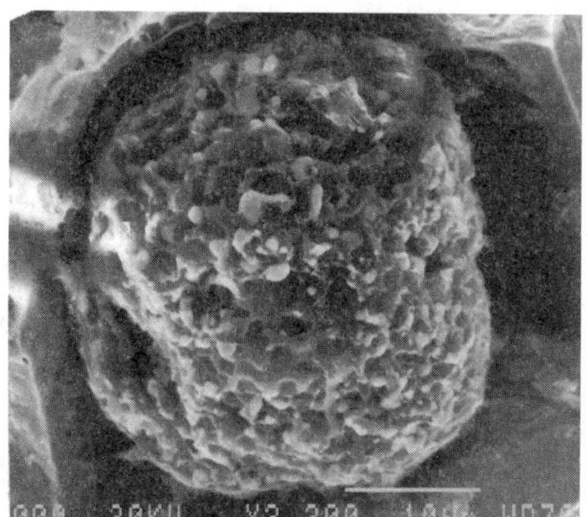 

Fig 11: B2TT2. composite graphite-matrix zone

DISCUSSIONS
<u>Microstructural parameters and damage initiation mechanisms:</u>

As a result, two points appear to control the damage initiation sites : the nodularity (size too) and the microsegregation within the wrapping graphite.

If A1 is compared with A1TT1 and A1TT2, the nodules shape does not change. But Si content within the wrapping graphite is higher than in the initial graphite. At the same time, the 70 % of decohesions at the graphite-matrix interface exhibited in A1, is reduced to 20% after heat treatments. This decrease can be due to Si microsegregation in the wrapping graphite during the heat treatments which embrittle the secondary graphite.

If B2 is compared with A1 the principal difference is the nodule shape which is more degenerated. Moreover, although there is no microsegregation within the graphite: the ratio 70 % of graphite-matrix interface decohesions exhibited in A1, is reduced to 43% in B2. Thus it can be concluded that the nodule graphite degenerated shape in B2 leads to the within graphite decohesions.

If B2 is compared with B2TT1 and B2TT2, the nodules spheroïdicity increases. After TT1, class VI graphite proportion increases from 35 to 68 percent and Mg microsegregation at the graphite-matrix interface appears. The positive effect of class VI graphite proportion improvement is compensated by the Mg microsegregation which yields to a stagnation in the ratio of interface /within graphite decohesions. But, after TT2 the shape improvement is not as high as after TT1 (60% class VI). Mg microsegregation at the graphite-matrix interface appears and also a Si segregation in the wrapping graphite. So, the negative effect of chemical microsegregation is the strongest and as a result 90% of intragraphite decohesion is found.

A good nodularity leads rather to interface decohesions and a Si or Mg segregations in the wrapping graphite seems to be responsible of intragraphite decohesions. A competition between the two mechanisms seems to exist. The nodule graphite size would also be an important parameter to discuss about. Optical microscopy and image analysis showed that small nodules have rather good nodularity and the big ones have a degenerated shape.

The first decohesions appear earlier for heat treated samples and the within graphite decohesions before the matrix-graphite decohesions. This shows the weak strength of the wrapping graphite zone.

Composite zone:

Moreover, with the heat treatment TT2, a composite zone of graphite and ferrite appears in the wrapping graphite. The within wrapping graphite decohesion can be due to the existence of this composite zone. Nanoindentation shows that Young's modulus value increases more progressively from graphite to matrix after TT2 and confirms the mixte zone existence. The composite zone existence can be due to the matrix particules trapped during the graphitization.

CONCLUSION

This study allows the understanding of damage initiation in ferritic (as-cast and heat treated) spheroïdal graphite cast irons. We found two main types of damage sites: one at the matrix-graphite interface and the other within the graphite. This second type of damage can be divided in intra wrapping graphite decohesion and in graphite-wrapping graphite interface decohesion. The nodularity (size too) and the microsegregations within the wrapping graphite appeared to control the damage initiation mechanisms: A good nodularity leads rather to interface decohesions and a Si segregation in the wrapping graphite seems to be responsible for intragraphite decohesions. A competition between the two mechanisms seems to exist. Fractographic analysis shows the existence of a composite graphite-ferrite zone in the wrapping graphite after the heat treatment TT2. Probably, this composite zone is responsible for the within wrapping graphite decohesion. Moreover, the heat treatments lead to a damage initiation at lower stress than in as-cast.

ACKNOWLEDGMENTS

We wish to thank the Centre Technique des Industries de la Fonderie for his financial and scientific supports.

REFERENCES

[1] A. Le Douaron, R. Lafont, D. Poulain, C. Cloitre, Conf Advances in fracture research, Cannes, 29 march-3 april 1981.
[2] R.C. Voigt, 57th World Foundry Congress, Osaka, 23-28 sep. 1990,5.
[3] Y. Iwabuchi, H. Narita, O. Isumura, T. Takenouchi, Trans.of the Jpn Foun. Soc, Oct. 92 Vol 11.
[4] W. L. Bradley, M.N. Scrinivasan, International Materials Rev., 1990, Vol 35, N°3, p. 129-161.
[5] N. Lazardis, R.K. Nanstad, F.J. Worzala, C.R. Loper Jr, A.F.S. Transaction 1977, 51.
[6] M. Ito, T. Umeda, Y. Kimura, K. Kuribayashi, T. Kishi, Jpn. Foun. Soc. april 1987, p. 28-32.
[7] H. Era, K. Kishitake, K. Nagai, Z.Z. Zhang, Mat. Sc. and Tech., march 1992, Vol 8, p. 257-261.
[8] R.K. Nanstad, F.J. Worzala, C.R. Loper Jr, A.F.S. Transactions, 1975, p.245-256.
[9] G. Pusch, Giessereiforschung 44, 1992, N°4. p.138-145.
[10] R. Salzbrenner, Journal of Materials Science 22, 1987, p.2135-2147.
[11] R.C. Voigt, L.M. Eldoky, A.F.S. Transactions, 1986, N°105, p.637-645.
[12] R.C. Voigt, C.R. Loper Jr, Conference Proceedings Cast Iron IV, 1990.
[13] J. Fargues, M. Stucky, Fonderie - Fondeur d'aujourd'hui 129, nov. 1993, p.13-21.
[14] W.C. Oliver, G.M. Pharr, J. Mater. Res; Vol. 7, N°6, June 1992, p. 1564-1583.
[15] C. Verdu, G. Thollet, C. Esnouf, R. Fougères, Six International Aluminium-Lithium Conference, 1991, Garmisch-Partenkirschen, Vol n° 1, p.277-282.

Advanced Materials Research Vols. 4-5 (1997) pp. 161-168
© *1997 Scitec Publications, Switzerland*

Microstructures and Superplasticity of Rapidly Solidified White Cast Iron Materials

G. Frommeyer[1], O.D. Sherby[2], O.A. Ruano[3], W.-J. Kim[2] and J. Wolfenstine[4]

[1] Max-Planck-Institut für Eisenforschung GmbH, Max-Planck-Str. 1, D-40237 Düsseldorf, Germany

[2] Dept. of Mat. Sci. and Engineering, Stanford University, Stanford, CA 94305-2205, USA

[3] Centro Nacional de Investigaciones Metalúrgicas (C.S.I.C.),
Av. de Gregorio del Amo 8, E-28040 Madrid, Spain

[4] Dept. of Mechanical and Aerospace Engineering, University of California, Irvine, CA 92717, USA

Keywords: Superplasticity, Microcrystalline Alloys, Rapid Solidification

ABSTRACT

White cast iron alloys with carbon contents ranging from 3 to 5.2 wt.% were produced by melt-atomization and consolidated by uniaxial extrusion. The dendritic (iron carbide) and partially martensitic (Ni-hard) microstructures of the rapidly solidified powders possess a fine dispersion of $(Fe,Cr)_3C$ particles in the hypoeutectic α- or γ-matrix and microcrystalline γ/α grains in the hypereutectic cementite matrix after compaction. Both equiaxed fine grained materials exhibit structural superplasticity in the temperature regime of 750 to 1000 °C with high strain rate exponents of $.4 \leq m \leq .6$ and tensile elongations up to 600 %. The strain rate in the $n = 2$ region is inversely proportional to approximately the cube of the grain size with an activation energy of 200 or 240 kJ/mol for superplastic flow. It is concluded that superplasticity in the iron carbide material (5.2 wt.% carbon) is due to grain boundary sliding accommodated by slip controlled by iron diffusion along iron carbide grain boundaries (strain rate exponent $m=.5$). For the hypoeutectic Ni-hard material the activation energy was determined to be of the order of 170 kJ/mol. These values give evidence that grain boundary sliding is accommodated by iron diffusion along the boundaries in the fine grained ferrite matrix. Large superplastic elongations of the iron carbide material were observed by the decrease in stress and grain size. This effect is described by a fracture mechanics model that has been developed to predict tensile ductility in superplastic ceramics. In the hypoeutectic Ni-hard alloy cavitation formation along the carbide/ferrite grain boundaries occurred during superplastic deformation in tension. This leads to a premature failure of this material.

INTRODUCTION

Extremely fine grained iron carbide material prepared from rapidly solidified powders exhibits superplastic properties over a wide strain rate and temperature range [1,2]. It was shown that large tensile elongations of about several hundred pct. have been achieved. For the hypereutectic iron carbide material the tensile ductility in the high strain rate sensitivity region, however, is a function of temperature (T) and strain rate ($\dot{\epsilon}$), and increases as the product $\dot{\epsilon}$ exp (Q/RT) is decreasing. Q is the activation energy for superplastic flow. The high m-values of .5 indicate that superductility occurred in this material like in many other oxide ceramics [3]. The hypoeutectic Ni-hard alloy with lower carbon content shows superplastic elongations up to 250 % at somewhat lower temperatures. The present paper describes and discusses the microstructure and specifically the superplastic properties of two selected white cast irons with hypo- and hypereutectic compositions.

MATERIALS AND EXPERIMENTAL PROCEDURE

The compositions of the chosen materials are presented in Table 1.

Table 1.   Chemical Compositions of the Investigated White Cast Irons
(Concentration in wt.%)

Material	C	Si	Mn	Cr	Ni	Fe
Ni-hard W.C.I. (Type-I)	3.26	.38	.45	2.1	4.1	balance
Iron Carbide	5.25	<.3	<.3	1.5	-	balance

Rapidly solidified powders were produced by argon gas atomization. The cooling rate of $5 \cdot 10^4 \leq \dot{T} \leq 3 \cdot 10^5$ K/s was estimated from the secondary dendritic arm spacings of the sieved powder particles with 50 to 80 microns in size. Figs. 1a,b display a typical dendritic microstructure of the rapidly solidified iron carbide (a) and of a martensitic structure of a small Ni-hard powder particle (b).

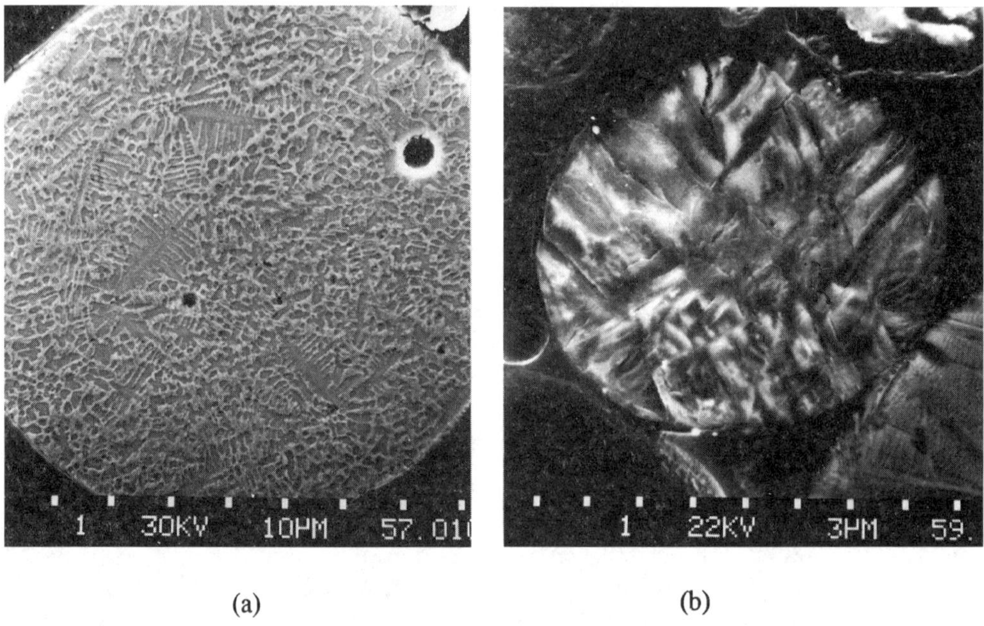

(a)   (b)

Figs. 1a,b: Dendritic (a) and martensitic (b) microstructures of the rapid solidified iron carbide and Ni-hard powders.

The hypereutectic iron carbide powder is characterized by a dendritic microstructure consisting of proeutectic carbides and interdendritic ledeburite. The larger Ni-hard powder particles show also a typical dendritic - quasi eutectic - microstructure whereas the smaller powder particles exhibit martensite. The coexisting phases were determined using X-ray diffraction analysis [4]. The volume fraction of $(Fe,Cr)_3C$ particles in the rapidly quenched state is about 40 vol.% in the iron carbide material, whereas the Ni-hard powders exhibit a cementite portion of about 25 vol.%. Powders of both alloys were encapsulated using mild steel cans for compaction by hot extrusion to a reduction of about 25 to 1 into the shape of round bars of 25 mm in diameter. The extruded bars were then deformed in compression perpendicular to the extrusion direction, with a 6 to 1 reduction at 720 °C. Tensile samples were prepared by spark erosion from the extruded and pressed bars. Strain rate change tests were carried out in the temperature regime from 700 to 900 °C at strain rates between 10^{-5} to 10^{-2} s^{-1} to determined the flow stress-/strain rate-relation as well as the strain rate sensitivity exponent m. Elongation to failure tests were performed to determine maximum superplastic elongations. Microstructural observations were conducted using light optical (LM), scanning electron microscopy (SEM) and transmission electron microscopy (TEM).

RESULTS AND DISCUSSIONS

The microstructure of the extruded iron carbide material shows in Fig. 2 slightly elongated proeutectic carbides which are retained from the rapidly solidified powders with dendritic morphology. After the material was pressed at 720 °C perpendicular to the extrusion direction a dramatic structural change to an equiaxed fine grained iron carbide matrix containing a fine dispersion of α-grains of 20 pct. by volume occurred. The linear intercept grain size of the iron carbide was estimated to $2 \pm .2$ microns and the volume fraction was determined to be 80 %. The high hardness of about 58 HRC is caused by the presence of continuous $(Fe,Cr)_3C$ skeleton, which remains stable after annealing up to 900 °C and somewhat higher temperatures. By quenching the iron carbide material from 850 °C into water hardening to about 70 HRC occurs due to martensitic transformation of the fine dispersed distributed γ-phase.

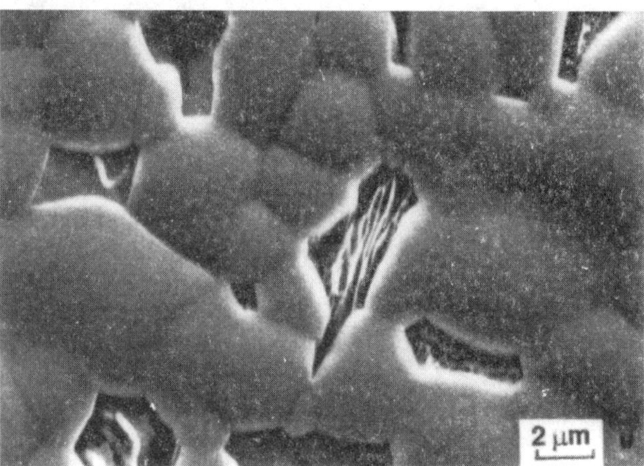

Fig. 2:   SEM micrograph of the consolidated iron carbide material shows small $(Fe,Cr)_3C$ grains and some retained interdendritic ledeburite.

Microstructural observations of the rapidly solidified hypoeutectic Ni-hard powders reveal that particles smaller than 30 microns exhibit an overall martensitic structure, whereas powders with particles sizes of about 50 microns show partially martensitic or dendritic microstructure depending upon the individually experienced cooling rate. The volume fraction of the cementite in the consolidated samples was determined to be about 50 %. The hardness of the compacted Ni-hard material was determined to be 40 $\pm$ 1 HRC. The microstructure consists of small equiaxed ferrite and cementite grains of an average grain interception of 1.5 $\pm$ .3 microns [4,5].

The substructural arrangement of the constituent phases within the consolidated material was studied in greater detail by transmission electron microscopy. Fig. 3 represents e.g. a bright-field TEM image of the Ni-hard material. The cementite grains exhibit a substructure with subgrains of .5 micron in size containing few isolated dislocations. A parallel arrangement of dislocations occurs near the subgrain boundaries. The small ferrite grains of about 2.5 $\pm$ .3 microns are homogeneously distributed within the cementite matrix. The ferrite grains are almost defect free. From this it is concluded that dynamic recrystallization during hot extrusion and pressing occurred, which leads to a substructure of low defect density [6].

Fig. 3:   TEM bright field image shows the equiaxed substructure of the consolidated Ni-hard samples. The ferrite subgrains are virtually free of dislocations, but the cementite subgrains exhibit some dislocations.

<u>Superplastic properties</u>

Mechanical tests were performed at elevated temperature to determine the strain rate stress-relations and to compare the data of both materials. Data analysis were carried out using the constitutive equation:

$$\dot{\epsilon} = K(\bar{b}/L)^p \sigma^n \exp(-Q_c/RT) \tag{1}$$

where $\dot{\epsilon}$ is the steady-state creep rate, σ is the flow stress, n the stress exponent (n = 1/m), Q is the activation energy, R is the gas constant, T is the absolute temperature, $\vec{b}$ is the Burgers vector, p is the grain size exponent, and K is a material constant.

The temperature range of testing was between 650 and 750 °C for the Ni-hard material and from 725 to 1050 °C for the iron carbide base alloy. For both materials a stress exponent of about 2 is observed in the investigated temperature range. Specifically for the iron carbide a decrease in the stress exponent from 2 to 1 at temperatures above 1035 °C was observed. That is due to the fact that Newtonian-viscous flow in the carbide ceramic material occurred. The change in the slopes of the strain rate/stress curves shows evidence for another mechanism of plastic flow at high temperature and low stresses. Generally two possibilities exists: diffusional creep (Nabarro-Herring [7-8], Coble [9]) or Harper-Dorn dislocation creep [10]. In order to establish the specific deformation mechanisms diffusion data, dislocation densities, and the dependence of the creep rate on grain size are required. This topic will be outlined in another paper. Comparing the tension test data of the investigated white cast iron materials, the strain rate vs. flow stress-curves are plotted in Figs. 4a, b in a double-logarithmic plot. In the n=2 region it was observed that the flow stress of the iron carbide at certain strain rates is always larger than that of the Ni-hard material. Here, the flow stress and the m-value are somewhat lower. However, at higher temperatures, above 750 °C, the flow stress of the iron carbide is quite low and does not exceed about 10 MPa at a given strain rate of 10^{-4} 1/s. The determined strain rate exponents (m) are for the Ni-hard material between .35 and .46 and are for the iron carbide slightly higher between .5 < m < .6, respectively. These data imply that for the Ni-hard material, which was tested at lower temperature, grain boundary sliding controlled by diffusion, and for the iron carbide material grain boundary sliding and dislocation creep which is also diffusion controlled, are the dominant deformation mechanisms.

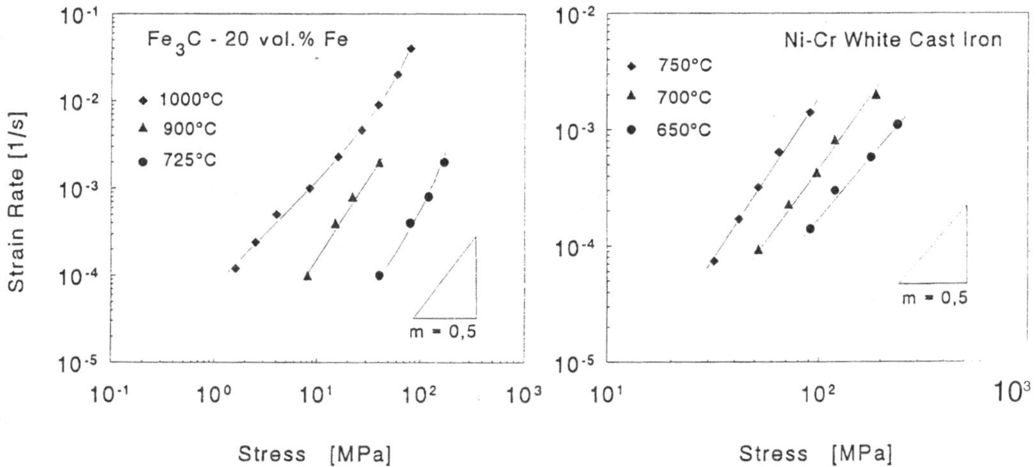

Fig. 4 a-b: Strain rate vs. flow/stress relations for iron carbide and Ni-Cr cast iron. The slopes of the curves are a measure for the strain rate sensitivity of the materials.

Activation energies

Activation energies for superplastic flow were determined from strain-rate change test data at constant stress levels and grain sizes using the following relation:

$$Q = -R \; \frac{d \ln \dot{\epsilon}}{d \, (1/T)} \; \Big|_{\sigma, d} \tag{2}$$

Fig. 5 displays the strain rate versus reciprocal absolute temperature curves in the semi-logarithmic plot at several constant stresses. For the iron carbide material a continuous increase in the creep rate is observed above the A_1 temperature as ferrite transforms to austenite. The weakening effect is attributed to the change in the volume fraction ratio of iron carbide to ferrite during transformation. The slope of the curves is a direct measure of the activation energy. As can be seen, the activation energy increases with increasing temperature and with a decrease in flow stress. The activation energy was determined to 200 kJ/mol below the A_1 and is between 200 and 240 kJ/mol above the eutectoid transformation temperature, where the average strain rate exponent is about .5. For the Ni-hard material the determined activation energy ranges between 150 and 190 kJ/mol at a flow stress of 85 MPa. This is lower than that observed in the iron carbide. The determined activation energies for both materials are related to grain boundary diffusion of iron in the carbide matrix, which is associated with the relative higher activation energy and to grain boundary diffusion of iron in the ferrite matrix in Ni-hard. The latter one is close to the activation energy for grain boundary diffusion of iron in pure α or γ which is of the order of 168 kJ/mol. The slower diffusivity of iron compared to that of carbon is controlling the superplastic flow in iron carbide. Diffusion data for lattice and grain boundary diffusion of iron in iron carbide are currently unavailable. However, carbon diffusion in iron carbide is associated with an activation energy of 170 kJ/mol [11-12].

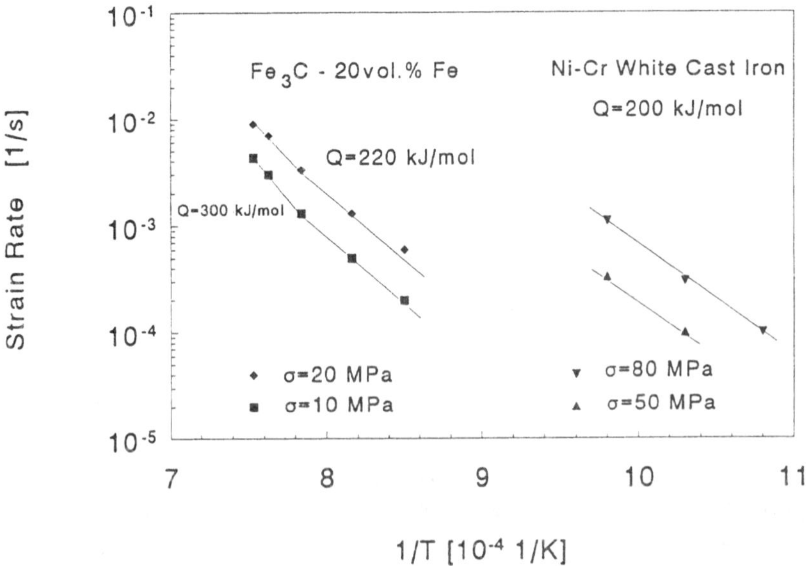

Fig. 5:   Activation energy derived from strain rate vs. reciprocal temperature curves.

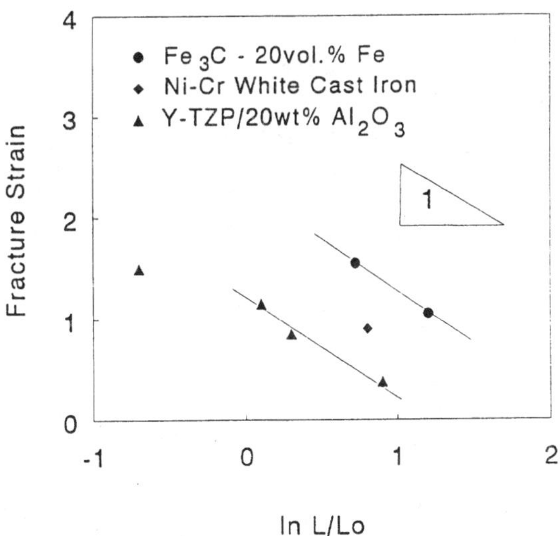

Fig. 6: Fracture strain in dependence on grain size at constant strain rate for iron carbide and selected oxid ceramics.

CONCLUSIONS

From the results of the present investigations it is concluded that

- structural superplasticity occurs in fine grain equiaxed white cast iron materials with carbon contents up to 5.25 %

- the Ni-hard as well as the iron carbide material show large elongations to failure from 210 % up to 600 % which are due to grain boundary sliding accommodated by slip, controlled by iron diffusion along iron carbide grain boundaries. This is derived from activation energy analysis

- the large superplastic elongations of the investigated materials can be described by a fracture mechanics model which has been developed to predict large tensile ductilities of superplastic ceramics. The failure of the material is caused by cavitation formations along carbide/ferrite grain boundaries

- rapidly solidified and compacted white cast iron materials with fine grain sizes can also be shaped into parts with complex geometries by superplastic forming operations.

LITERATURE

[1]   W.J. Kim, J. Wolfenstine, G. Frommeyer, O.A. Ruano, O.D. Sherby, Scripta Metall. **23**, 1515 (1989)

[2]   W.J. Kim, J. Wolfenstine, O.A. Ruano, G. Frommeyer, O.D. Sherby, Met. Trans. A **23A**, 527 (1992)

[3]   T.G. Nieh, J. Wadsworth, in: "Superplasticity in Metals, Ceramics, and Intermetallics", MRS Symp. Proc., M.J. Mayo, M. Kobayashi, J. Wadsworth, eds., MRS, Pittsburgh, PA, 331 (1990)

[4]   G. Frommeyer, Z. Metallkde. **7**, 662 (1985)

[5]   L.E. Eiselstein, O.A. Ruano, O.D. Sherby, J. Mat. Soc. **18**, 483 (1983)

[6]   D.W. Kum, G. Frommeyer, N.J. Grant, O.D. Sherby, Met. Trans. A **18A**, 1703 (1987)

[7]   F.R.N. Nabarro, in: Rep. Conf. on the Strength of Solids, Physical Society, London, 75 (1948)

[8]   C. Herring, J. Appl. Phys. **21**, 437 (1950)

[9]   R.L. Coble, J. Appl. Phys. **34**, 1679 (1963)

[10] J.C. Harper, J.E. Dorn, Acta Metall. **5**, 654 (1957)

[11] B. Ozturk, V.L. Fearing, J.A. Ruth, Jr., G. Simkovich, Metall. Trans. A **13A**, 1871 (1982)

[12] M. Hillert, Metall. Trans. A **15A**, 245 (1984)

[13] Metals Handbook, 9th ed., ASM, Metals Park, OH **1**, 85 (1978)

[14] H.T. Angus, Cast Iron "Physical and Engineering Properties", Butterworths & Co., Ltd., London, Great Britain, 245 (1976)

[15] C. Zener, J.H. Hollomon, J. Appl. Phys. **15**, 22 (1944)

Advanced Materials Research Vols. 4-5 (1997) pp. 169-174
© *1997 Scitec Publications, Switzerland*

Counting Eutectic Grains in S.G. Cast Iron

G. Rivera, R. Boeri and J. Sikora

INTEMA, Faculty of Engineering, National University of Mar del Plata,
Juan B. Justo 4302, 7600 Mar del Plata, Argentina

Keywords: Spheroidal Graphite Cast Iron, Solidification, Grain Count, Metallography, ADI Properties

ABSTRACT

A metallographic study was carried out, aimed to characterize the solidification structure of SG irons and to identify its effects on the microstructure and mechanical properties. Eutectic grains were identified and counted. As the section size increases , the grain count diminishes, and the number of graphite nodules per grain increases.

Tension tests carried out on ADI samples showed that the ductility decreases as the grain structure coarsens. This is attributed to the enlargement of the last to freeze and grain boundary regions. Considering the effect of grain and graphite characteristics, improvement in mechanical properties of SG irons are expected if melting and production practice are designed to obtain high nodule count and the higher possible grain count.

INTRODUCTION

The solidification structure of any casting has direct correlation with its microstructure at room temperature, and consequently with the resulting mechanical properties. Nevertheless this correlation has not been clearly established for SG irons.

The eutectic solidification of SG irons has been examined for many years. Today, most researchers agree that the eutectic solidification involves the formation of solidification units as a result of the interaction of austenite and graphite nodules. These multinodular units of a dendritic substructure have been called "cells" or "grains", and will be referred to as eutectic grains below [1,2,3].

The solidification structure of SG irons is difficult to reveal at room temperature because the austenite is transformed into other phases on cooling. Standard characterization of SG microstructure is based on the examination of features of graphite particles and matrix, neglecting the inspection of the solidification structure. In general practice, when a refinement in the structure is wanted, melting and inoculation processes are modified to obtain high nodule counts. Frequently, high values of nodule count are required for high quality castings. These specifications are generally based on experimental measurements which show that in most cases the mechanical properties improve as the graphite count increases. This experimental result is difficult to explain by considering only the modifications in the distribution of the graphite phase. A more comprehensive interpretation should take into account the solidification structure. Nevertheless, the information available in the literature in insufficient, to the best of our knowledge, to assess the changes in the degree of coarseness of the solidification structure as a result of variations in casting size, cooling speed, or inoculation practice. A complete characterization of the solidification structure must be based on the description of the appearance of graphite, eutectic grains, and microsegregated regions. A schematic representation of the eutectic solidification structure is shown on **Figure 1**. The solidification units (multinodular grains) are separated by areas called grain boundaries. Regions in which three or more grains meet coincide with the location of the last liquid to freeze, and have been called LTF regions [4]. Alloying elements and impurities, such as Mn, Mo, S and P, segregate to the LTF and, in smaller degree, to the grain boundaries. Most microsegregation studies provided useful information by focussing on the precise determination of the microsegregation in LTF regions. Nevertheless, these observations do not provide a larger scale picture, portraying information about the size, distribution and degree of continuity of the less microsegregated regions conforming the grain boundaries. That information is, in fact, needed when the effects of the solidification structure on the mechanical properties are to be determined [5]. Recently, the authors carried out studies aimed to identify efficient methods to reveal the solidification structure of SG irons of commercial quality [6]. Appropriate etching methods have been found to generate a colour map which allows the identification of the bulk and the boundaries of the grains.

The objectives of the present study are: i) To develop a procedure to count eutectic grains in SG iron. ii) To measure changes in grain count as a function of the change in section size, and to correlate it with changes in nodule count. iii) To identify the influence of grain and nodule counts on the microstructure and properties of the iron.

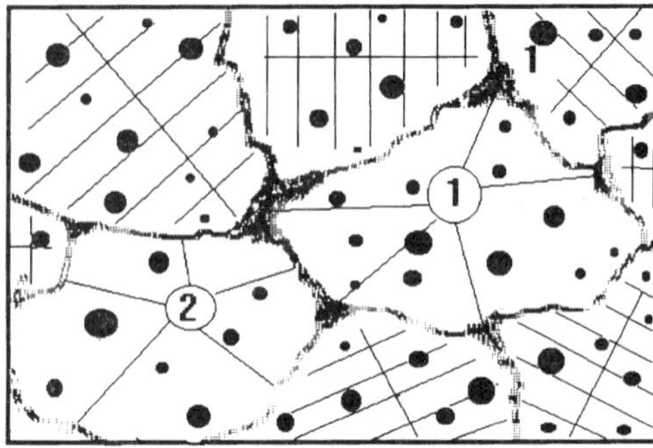

Figure 1: Schematic representation of the solidification structure. (1) LTF regions, (2) grain boundaries.

EXPERIMENTAL METHODS

Spheroidal graphite cast iron samples utilized were taken from Y-blocks (ASTM A395) of 13 and 76mm, and round bars of 20, 40 and 80mm diameter, cast in commercial quality iron. **Table I** lists the chemical composition and details the characterization of the graphite microstructure (ASTM A247) of the specimens.

Table I: Chemical composition and characterization of the graphite phase.

ALLOY	CHEMICAL COMPOSITION (wt%)								SAMPLE ORIGIN	GRAPHITE CHARACTERISTICS		
	C	Si	Mn	Cu	Ni	P	S	Mg		NODULA-RITY (%)	NODULE SIZE	NOD. COUNT NOD/mm²
2F	3.20	2.85	0.32	-	-	0.015	0.014	0.044	BY1/2"	100	6	213
									BY3"	100	5/6	111
									⌀=20mm	80	6	250
									⌀=40mm	90	5/6	187
									⌀=80mm	90	5/6	128
F92	3.11	3.14	0.31	-	-	0.021	0.015	0.034	⌀=20mm	100	6	321
									⌀=40mm	100	6	242
									⌀=80mm	100	5/6	167
1P	3.30	2.70	0.32	0.76	0.56	0.016	0.013	0.036	BY1/2"	100	6	154
									BY3"	80	5/6	90
									⌀=20mm	100	6	203
									⌀=40mm	90	5/6	124
									⌀=80mm	90	5/6	85

Samples of each chemical composition and casting size were sectioned, polished and etched to reveal the solidification structure following a procedure developed in an earlier study[6]. The method begins with the ferritization of the samples, followed by chemical etching with a colour metallography reagent (10g NaOH, 40g KOH, 10g picric acid and 50ml H_2O). The reagent, declared to trace Si microsegregation[7], is toxic and must be handled carefully. The etching generates different colours in coincidence with microsegregated regions, as shown in **Figure 2**. As a result of the etching, LTF regions and most, but not all, grain boundaries are identified. In order to make possible the counting of grains, it was necessary to recognize and mark manually the grain contours. This task was carried out by a qualified operator on colour photographs, as shown in **Figure 2**, for a large number of fields. The nodules inside each grain were then counted for 50 to 100 grains in each sample, and the mean values (Nc) are quoted as a 95 % confidence interval. The number of grains per unit area (**Ca**) was calculated as the ratio between the average number of nodules per unit area (**Na**), and the mean values of nodules per grain (**Nc**). A second average number of

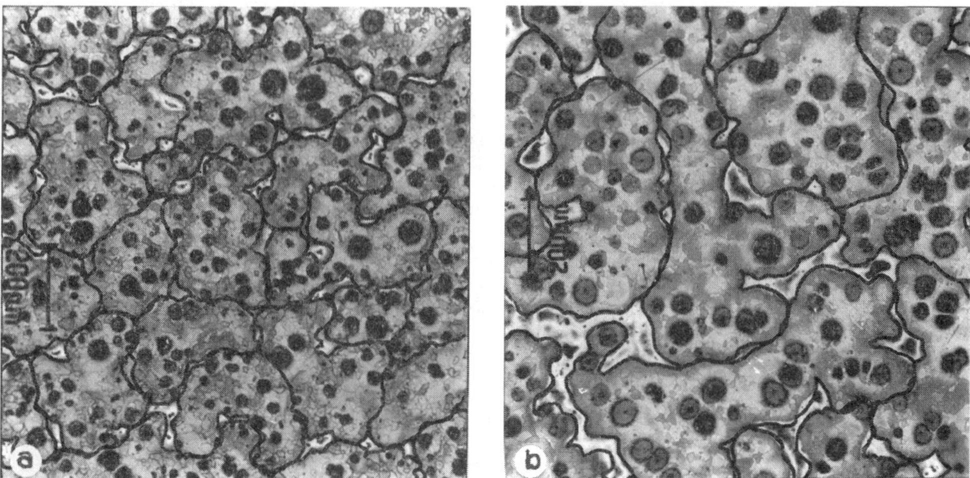

Figure 2:　Black and white reprint of colour metallographies of sample 1P. (a) Bar 80mm diameter, (b) Bar 20mm diameter, grain contour marked manually.

nodules per grain was calculated considering only the 20% of the grains showing the larger count values, called N_{20}. This rule was adopted to obtain a more representative value, by reducing the error induced in the average if the smaller counts, which may correspond to intersections far from the centre of the grains, are considered. It must be noted that the count of nodules per grain, measured following the procedure described above, does not give the actual number of nodules inside the volume of the grain, but is an indicative value.

Additionally, the average distance between nodules **(D)**, and the total graphite/matrix interphase length per unit area **(P)**, was measured by using digital image analysis.

Tensile testing was carried out on samples from Y-blocks of alloys **1P** and **2F**. Round bars of 12mm diameter were first prepared, then austenitized at 910°C, and finally austempered in a salt bath held at 360°C during 90 minutes. Heat treatment was carried simultaneously on all samples. ASTM E8-57T tensile samples were then machined and tested. UTS and elongation values listed are average of six tests. Standard deviation was reported in all cases.

RESULTS AND DISCUSSION

The colour metallography pictures, **Figure 2**, show changes in colour from the bulk of the grains to the grain boundaries and LTF. The colours resulting from the etching may change from sample to sample, since the thickness of the deposited products, which generate the colour, is a function of the time of etch and the reactivity of the substrate. In general, the colour sequence, from bulk to LTF, is yellow, green, red, orange, bright yellow. On occasions, depending on the intensity of the etching, some of this colours may not be evident, or appear only as very thin fringes. The solidification structures revealed in Figure 2a and 2b resemble some features of the schematic representation portrayed in Figure 1. It is evident that the increase in section size gives rise to an enlargement of the LTF and grain boundaries. The graphite phase particles also enlarge and decrease its count.

The results of nodule and grain counts are listed on **Table II**. Typical results are displayed in **Figure 3,** showing the change in the count of nodules, grains, and nodules per grain, as a function of the size, for the cylindrical castings of alloy 2F. As the section size increases, the number of grains and the number of nodules per unit area decrease, while the number of nodules per grain increases. The coarsening of the grain structure as the section size increases, revealed by the decrease in the grain count, is more pronounced than the decrease in the nodule count. This supports the assumption of independent nucleation of austenite and graphite during the divorced growth of the gamma-graphite eutectic. Austenite nucleation appears to be more sensitive to changes in the cooling rate or section size. In foundry practice, inoculation is used as a mean to refine SG iron structure. The effectiveness of inoculation has been always measured in terms of its effect on the nodule count, i.e. on the nucleation rate of graphite. Studies should be carried out to determine the effect of inoculation on the refinement of the grain structure, which appears to be related to the nucleation of austenite.

TABLE II: Nodule and grain counts.

ALLOY	BAR DIAMETER [mm]	Na [Nod/mm²]	Nc [Nod/grain]	N_{20} [Nod/grain]	Ca [Cell/mm²]
2F	20	250	7.74-6.48	13.78	35.16
	40	187	15.08-12.00	24.37	13.81
	80	128	18.32-14.84	27.00	7.72
F92	20	321	12.26-10.40	20.73	28.33
	40	242	12.58-10.24	21.27	21.21
	80	167	20.27-15.89	32.32	9.24
1P	20	203	9.96- 8.24	12.87	22.31
	40	124	14.39- 11.65	19.44	9.52
	80	85	21.26-17.00	28.00	4.44

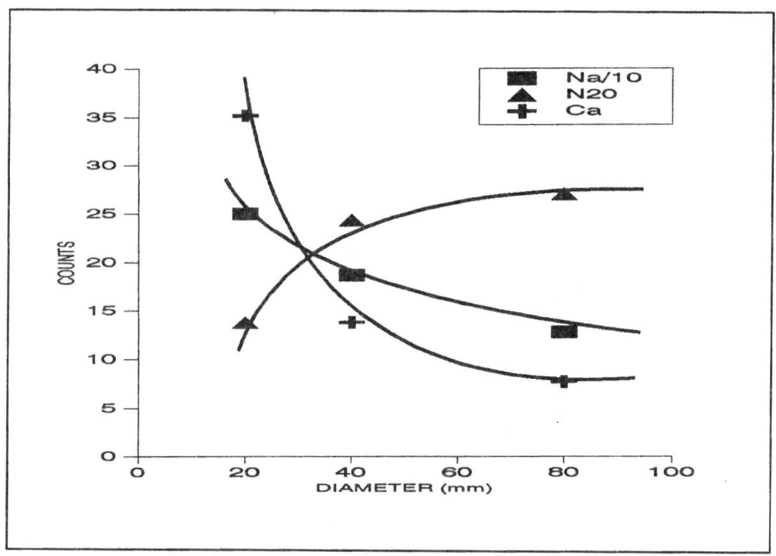

Figure 3:   Change in grain count, nodule count, and nodules per grain, as a function of section size, for alloy 2F.

The results of tension tests carried out on austempered samples of alloys **2F** and **1P** obtained from Y-blocks of different size, are listed on **Table III**. Mechanical properties were measured on high strength conditions, such as those found in ADI structures, in order to maximize the deleterious effect of microstructural defects. The increase in section size causes minor changes in UTS. Elongation, on the other hand, decreases markedly.

TABLE III: Solidification structure characteristics and mechanical properties on ADI samples.

ALLOY	Y-block size	Na [Nod/mm²]	Nc [Nod/grain]	Ca [Cél/mm²]	D [µm]	P [mm/mm²]	UTS (Dispersion) [MPa]	Elongation (Dispersion) [%]
2F	1/2"	213	12.47 9.89	19.05	76.90	11.72	1159.08 (99.52)	12.99 (1.21)
	3"	111	21.43 16.81	5.80	144.70	8.01	1173.23 (20.39)	9.71 (1.83)
1P	1/2"	154	11.34 9.76	14.60	---	---	1124.19 (16.73)	11.85 (0.83)
	3"	90	40.91 30.23	2.54	---	---	1076.30 (53.03)	6.98 (2.26)

The observed decrease in the elongation as the section size increases is in agreement with other studies [5]. Usually, this effect has been attributed to the coarsening of the nodular structure that follows the increase in section size. Nevertheless, recent publications indicate that, within the range of nodule counts usually found in industrial alloys, and provided nodularity ratings remain high, the observed change in elongation cannot be attributed to modifications in the stress distribution resulting from the coarsening of the nodular structure [8]. In reality, the interpretation of this effect can be based on the characteristics of the solidification structure. As shown in **Figure 2**, the enlargement of section size causes the widening of the intergranular regions. These regions are prone to induce deleterious effects on the microstructure, either by yielding to the precipitation of inclusions and carbides or by inducing the retainment of high temperature phases after heat treatment. An example of this is shown in **Figure 4**, where the microstructure of austempered samples of alloy 1P, show white patches of unreacted austenite, located in the intergranular regions. The smaller section size sample, **Figure 4a**, has larger number of smaller grains per unit area, and shows lower tendency for defect formation in the much smaller, and less extensively segregated, intergranular regions. Larger segregated regions found in larger section size samples, **Figure 4b**, show inclusions, unreacted austenite and martensite. The occurrence of these microstructural defects in the larger intergranular regions can be the reason for the observed changes in elongation. Significantly, the decrease in elongation, shown on **Table III**, is more marked in the case of the more alloyed samples of casting 1P. This is probably due to the fact that, as the amount of alloying increases, the segregation to the grain boundaries is more marked.

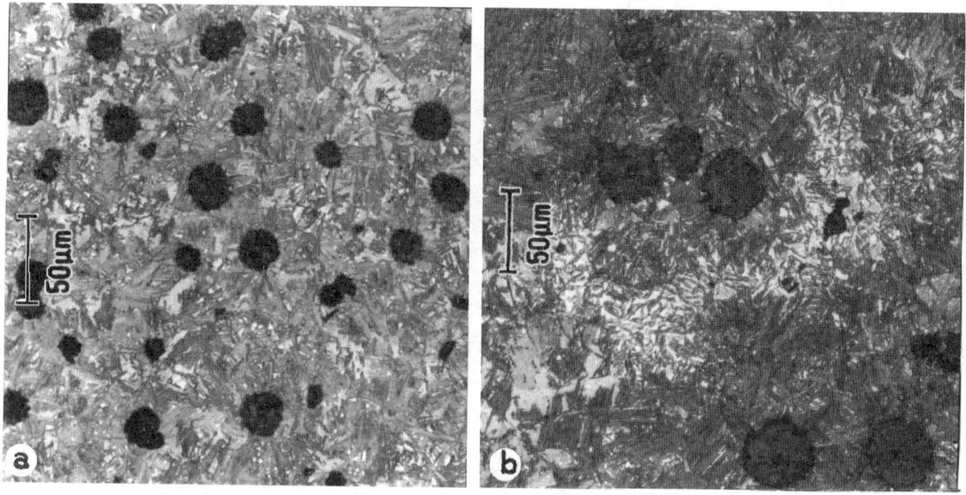

Figure 4: Microstructure of austempered samples of alloy 1P. (a) Y-block 1/2", (b) Y-block 3".

The role of the graphite phase, characterized by its count, size and morphology, can be more directly related to its effect on the phase transformations, than to its influence on the stress concentration. Different nodule size leads to different kinetics during phase transformations. In fact, graphite nodules act as sinks or sources of carbon, and graphite-matrix interphases are preferential nucleation sites for some phases forming in SGI. As **Na** increases, the mean distance for carbon diffusion, given by **D**, decreases, and the interphase area increases proportionally to the change in the value of **P**, as shown on **Table III** for casting 2F. In the case of diffusion controlled transformations, the refinement of the graphite phase leads to faster kinetics and finer products [9].

Considering the discussed effects of grain and graphite characteristics, improvements in mechanical properties of SG irons are expected if melting and production practice are designed to combine high nodule counts, which will conduce to finer matrix microstructure, and the higher possible grain count, to minimize the incidence of defects associated to LTF regions and grain boundaries.

Further studies are necessary to establish the effect of the inoculation practice on the grain count, and the possibility of using **Ca** as a quality control parameter.

CONCLUSIONS

1-   The solidification microstructure of commercial quality SG irons can be revealed and its grain count measured, following a procedure which includes colour metallography.

2-   The increase in the section size of the castings causes a marked decrease in the grain count and a noticeably increase in the number of graphite nodules per grain.

3-   The elongation, measured on ADI samples, diminishes as the section size enlarges. This effect has been attributed to the enlargement of the LTF and grain boundary regions.

4-   Better SG irons should combine higher nodule count with a finer grain structure.

References

[1] J.A.Sikora, G.L.Rivera, H.Biloni, "Metallographic study of the eutectic solidification of gray, vermicular and nodular cast iron", Proceedings of F.Weinberg International Symposium on Solidification Processing, Pergamon Press, 280 (1990) .
[2] R.Boeri and F.Weinberg, "Microsegregation of alloying elements in cast iron", Cast Metals **6**, 153 (1993).
[3] D.K.Banerjee and D.M.Stefanescu, "Structural transitions and solidification kinetics of SG cast iron during directional solidification experiments", AFS Transactions **99**, 747 (1991).
[4] K.L.Hayrynen, D.J.Moore, K.B.Rundman, "Heavy-section ADI: production and microsegregation", AFS Transactions **96**, 619 (1988).
[5] G.P.Faubert, D.J.Moore, K.B.Rundman, "Heavy-section ADI: Tensile and fatigue properties of relatively pure ADI", AFS Transactions **100**, 93 (1992).
[6] G.L.Rivera, R.E.Boeri, J.A.Sikora, "Revealing solidification structure of nodular iron" , Cast Metals **8**, 1 (1995).
[7] J.M.Motz, "Microsegregations- an easily unnoticed influencing variable in the structural description of cast materials", Prac. Met. **25**, 285 (1988).
[8] Y.Iwabuchi, H.Narita, O.Tsumura, T.Takenouchi, "Fracture toughness and its relation to microstructural features in ferritic nodular-graphite iron castings", Transactions of the Japan Foundrymen's Society **11**, 9 (1992).
[9] V.M.Bermont, J.A.Sikora, "Efecto de las morfologías del grafito, aleantes y tamaño de célula en la cinética de los procesos de grafitización y carburación de hierro", Proceedings of VII Congreso Nacional de Metalurgia, III Congreso de la Asociación Latinoamericana de Metalurgia y Materiales, Chile Vol.II, 719 (1994).

Advanced Materials Research Vols. 4-5 (1997) pp. 175-180
© *1997 Scitec Publications, Switzerland*

Influence of Silicon Content and Heat Treatment Conditions on Impact Characteristics in the High Toughened Austempered Ductile Cast Iron with Additions of Nickel and Manganese

M. Aoyama[1] and T. Kobayashi[2]

[1] Department of Mechanical Engineering, Daido Institute of Technology,
Minami-ku, Nagoya, 457 Japan

[2] Department of Production System Engineering, Toyohashi University of Technology,
Toyohashi, 441 Japan

Keywords: Austempered Ductile Cast Iron, Charpy Impact Test, Heat Treatment, Toughenin
Silicon Content

Abstract

The effects of austenitizing, austempering conditions and silicon content on the impact characteristics of austempered ductile cast iron were evaluated using an instrumented Charpy impact testing system.

The optimum condition for heat treatment to improve impact properties was quenching once from the austenite range in order to refine the prestructure, then austempering from near the austenitizing finishing temperature. The special austemper treatment shows no remarkable effect on impact values over an austempering temperature range from 523 to 673 K, and it was possible to obtain high impact values in short austempering periods. A sample containing 2.1% Si exhibited a combination of superior strength and toughness.

1. INTRODUCTION

Recently, austempered ductile cast iron (ADI) has drawn attention for its applicability for various mechanical structural parts due to its very high toughness and superior strength properties in comparison with conventional pearlitic and ferritic type ductile cast irons[1,2].

Kobayashi et al.[2,3] succeeded in developing toughened ADI (TADI) through the use of microsegregation of alloying elements and application of special austemper heat treatment. The purpose of this special austemper heat treatment is to provide ductility through the introduction of refined ferrite and toughness at the fracture initiation sites as graphite-matrix interfaces and eutectic cell boundaries through the stabilization of the retained austenite[2,3] by enrichment of austenite forming elements (Mn, Ni) and refining of grain size. Therefore, the present study is to clarify metallurgical factors to give the optimum impact properties to the developed austempered ductile cast iron.

2. MATERIALS AND EXPERIMENTAL PROCEDURES

The ductile iron used in this study was melted in a high frequency induction furnace having a 500 kg capacity. **Table 1** shows the chemical composition, nodularity of graphite and austenite transformation temperatures of cast irons used in the study. The austemper heat treatments performed in

Table 1. Chemical composition, graphite nodularity and austenite transformation temperature of test irons.

Iron	Chemical composition, wt.%					Trans. Temp.,K			Nodularity, %
	C	Si	Mn	Ni	Mg	A_s	A_f	A_{50}	
I	3.51	1.44	0.58	2.21	0.050	993	1033	1003	76
II	3.33	2.09	0.62	2.15	0.046	1003	1073	1023	81
III	3.31	3.43	0.59	1.97	0.059	1033	1123	1073	82
IV	3.44	2.14	0.19	——	0.055	1043	1103	1073	93

(I) Heat treatment conditions for special austemper
(a) Reaustenitizing holding times   (c) Austempering temperatures and
holding times

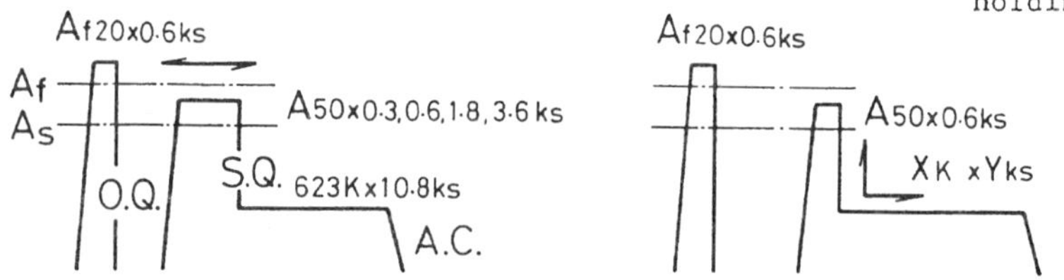

(b) Reaustenitizing temperatures   (II) Heat treatment conditions for
standard austemper

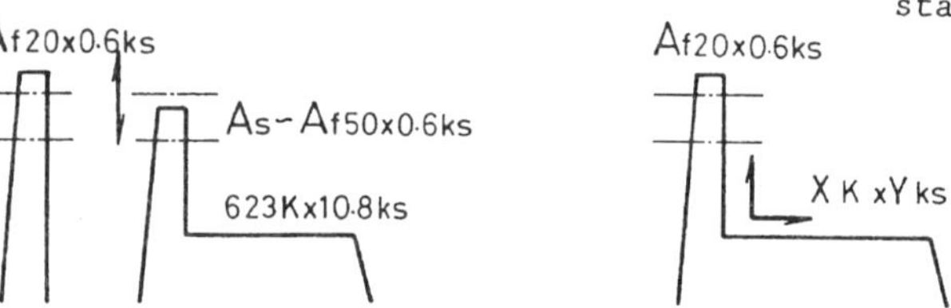

Fig. 1. Schematic representation of austemper heat treatment processes.
Subscripts of A_f temperature show an added temperature. O.Q.: oil-quenching; S.Q.: salt bath quenching; A.C.: air-cooling; A_s: austenitizing starting temperature; A_f: austenitizing finishing temperature; A_{50}: same volume fraction of α and γ coexisting transformation temperature; X: austempering temperature (523,573,623,673K); Y: austempering holding time (0.6,1.8,7.2,10.8,18ks).

this study are shown schematically in **Fig. 1**. Austenitizing treatments were conducted in air by an electric muffle furnace, and austempering treatments were performed in a solution of mixed salts (55% KNO_3 and 45% $NaNO_2$) using a salt bath furnace.

Impact tests were made using an instrumented Charpy impact machine (energy capacity: 9.69 J). Heat treated materials were machined to U-notched Charpy specimens (specimen size: 10x10x55 mm). Tests were conducted at room temperature. Dynamic fracture toughness tests were made using a computer-aided instrumented impact testing (CAI) system developed by Kobayashi et al.[4].

The microstructure were observed by scanning electron microscope (SEM). The volume fraction of retained austenite was determined by X-ray diffraction method using the procedure developed by Miller[5].

3. EXPERIMENTAL RESULTS AND DISCUSSION

The addition of Si serves to prevent carbide formation and it helps stabilize the retained austenite in the austemper heat treatment. It has been reported that addition of Si to steel also produces superior toughness and high strength[6,7].

Figure 2 shows the impact values and retained austenite content for reaustenitizing holding times in the (ferrite + austenite + graphite) range in different Si content irons. The impact values of sample I with a 1.5% Si content increased slightly as the reaustenitizing holding time increased. Iron III was expected to increase toughness because retained austenite was introduced by the addition of a large quantity of Si. The results of iron III, however, indicate low impact values similar to those of iron I, whereas, impact values for iron II tended to increase toughness with a prolonged heating time by the addition of 2.1% Si. The matrix structure on iron II consisted of a mixed structure of refined ausferrite and annealed ferrite at reheating. The ausferrite portion of this structure induced a thorough concentration distribution of alloying elements with a long reaustenitizing holding time at a temperature of A_{50}. The resulting marked improvement in impact values are thought to have been due to carbon, Ni and Mn enrichment in the retained austenite and its stabilization effect[2,8]. The reduced impact vales in high Si content iron III, however, caused coarse ausferrite in a portion of the eutectic cell boundaries, with progressive ferritizing around the graphite nodules.

Figure 3 shows the relationship between impact values and reaustenitizing temperatures. Although reaustenitizing in the (ferrite +

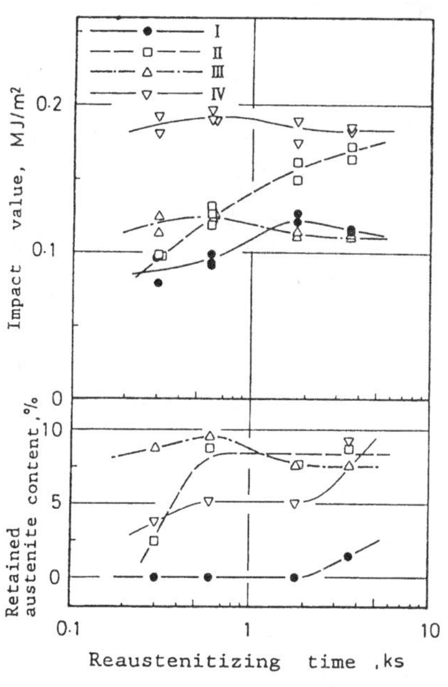

Fig. 2. Change of impact value and retained austenite content with austenitizing holding time. Heat treatment condition of Fig.1(I)(a)

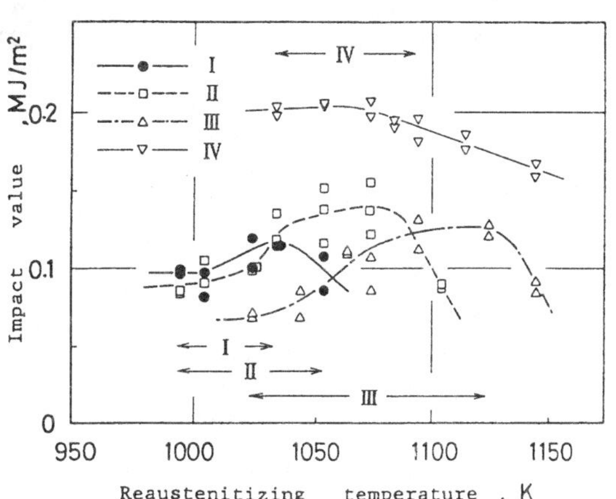

Fig. 3. Relation between impact value and reaustenitizing temperature. Arrows indicate temperature range of (α + γ + Graphite) for each irons.

austenite + graphite) range increased impact values as the heating temperature increased, impact values decreased markedly when the heating temperature entered the austenite phase range. Therefore, reaustenitizing near the A_f range is important to obtain superior impact characteristics. The addition of Si raised impact values in the following order: iron I <iron III <iron II. It is clear that the toughness of the iron II sample with 2.1% Si is optimum.

On the other hand, the reaustenitizing treatment of iron IV in the (ferrite + austenite + graphite) range produced high impact values by growing a coarse ferritic structure through heating at a low temperature. The impact values decreased, however, due to progress of ausferritizing with the elevation of the reaustenitizing temperature.

Figure 4 shows the measurement results of dynamic elastic-plastic fracture toughness values (J_{Id}) in typical iron samples at room temperature, and also gives the results of impact value and maximum fracture load of U-notched Charpy impact specimens. Reaustenitizing at just above A_f temperature produced superior impact values, fracture load and fracture toughness values compared to ones obtained at A_{50} temperature. It is probable that reaustenitizing at just above the A_f temperature improves the efficiency of alloying element redistribution.

Standard austempered iron exhibited superior fracture toughness values and high maximum fracture loads. Special austemper treated iron II exhibited low fracture loads, however, special austempered iron exhibited a stable

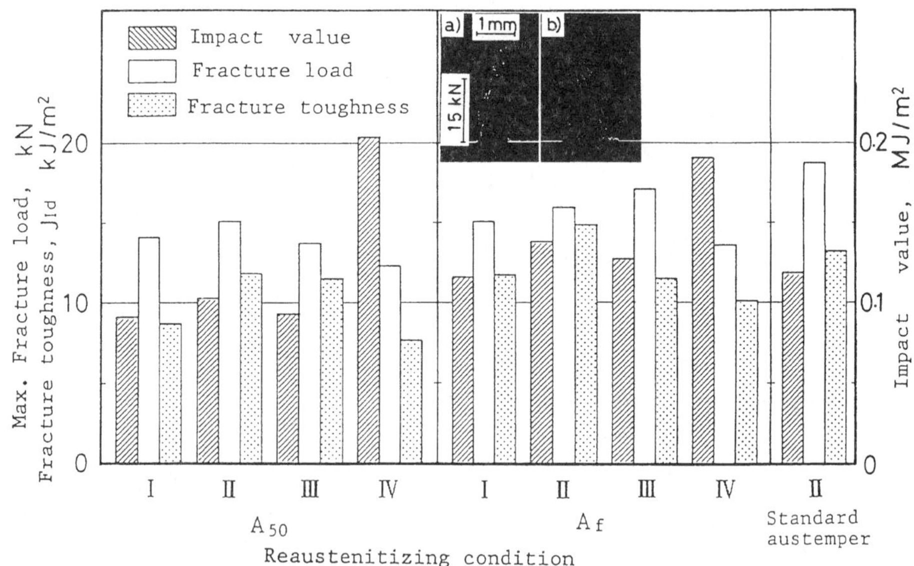

Fig. 4. Fracture properties of various irons at room temperature. Photographs in diagram are load-diffraction curves obtained by Charpy impact test [a)Standard austemper, b)Special austemper; test iron II, A_f].
(reaustenitizing time :0.6ks, austempering condition: 623 K, 10.8ks)

crack propagation process with plastic deformation in the load-displacement curve as shown in the diagram, and the fracture toughness values, especially so-called resistance values, were in no way inferior. The matrix structure with both hard and soft aspects is a useful characteristic for the special austempered iron. In this study, the addition of 2.1% Si produced a suitable balance of strength and toughness. Silicon is thought to have a great effect on the formation of the ausferrite structure.

Figure 5 shows the impact values, hardness, retained austenite contents and austempering holding temperatures for both special and standard austemper heat treatments. Special austemper treated iron produced stable impact values that were unaffected by the highs and lows of austempering holding temperatures. However, standard austempered iron with increased retained austenite produced a maximum impact value at 623 K, and the impact value was reduced at 673 K. These findings agree with reported findings for conventional austempering[9]. The lower impact values in the high temperature range apparently result from the decomposition of retained austenite and precipitation of carbide or secondary graphite due to the austempering treatment[10]. Standard austemper with a short holding time of 0.6 ks produced low impact values and corresponded with the initial stage of ausferrite transformation. These low values were probably caused by no prequenching in the standard austemper treatment. As shown in **Fig. 6**, the incomplete ausferrite reaction was thus the cause.

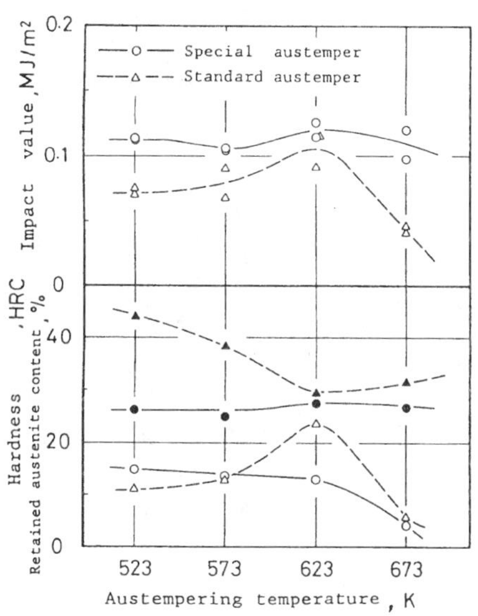

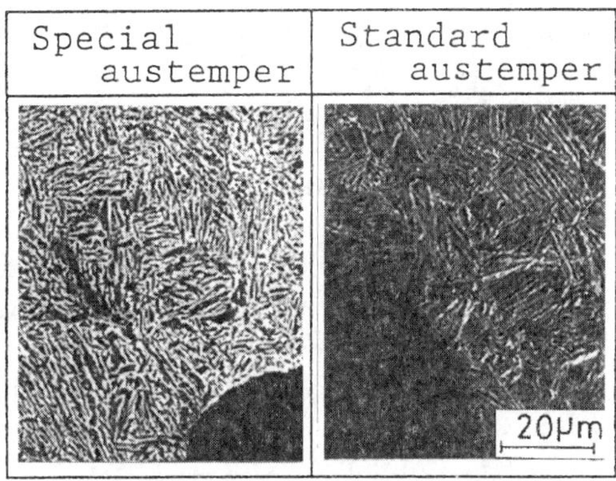

Fig. 6. SEM photographs of microstructure in differently austempered irons at austempering time of 0.6 ks. (test iron II; austenitizing temperature: 623 K)

Fig. 5. Austempering holding temperature dependence of impact value, Rockwell hardness(HRC) and retained austenite content by different austemper treatments. Solid symbols represent hardness.

(test iron II; austempering time: 10.8ks,
 Special austemper; reaustenitizing condition:A_{50} , 0.6ks
 Standard austemper; reaustenitizing condition:A_{f20} , 0.6ks)

4.   CONCLUSION

The influence of austenitizing, austempering conditions and silicon content on the impact characteristics of this iron were evaluated using an instrumented Charpy impact testing system. The findings are listed below.

1. Optimum heat treatment conditions for increasing impact values was obtained by quenching after a short holding time in the austenite range, then conducting an austempering from near the austenitizing fishing temperature. This procedure produces a duplex matrix structure of refined ferrite and ausferrite.

2. Increasing (3.4%) or decreasing (1.4%) of Si content from a proper range did not show the improvement in strength or toughness. The 2.1% Si produced a very satisfactory balance between strength and toughness.

3. Compared to standard austempering processes, special austempered iron showed good stable impact values at austempering temperature range from 523 to 673 K, and it was also possible to achieve a high impact value with a short austempering time.

ACKNOWLEDGMENTS

The authors would like to thank Kurimoto Ltd. for providing the test material.

REFERENCE

[1] S.Nishi, T.Kobayashi and M.Aoyama, Journal of the Japan Foundrymen's Society (Imono), 50(1978), 4, 223

[2] T.Kobayashi and H.Yamamoto, Metallurgical Transactions A, 19A(1988), 2, 319

[3] T.Kobayashi and H.Yamamoto, Transactions of Japan Foundrymen's Society, 8(1989), 4, 30

[4] T.Kobayashi and M.Ninomi, Nuclear Engineering and Design, 111(1989), 27

[5] R.L.Miller, Transactions of the ASM, 57(1964), 892

[6] K.D.Sibley and N.N.Breyer, Metallurgical Transactions A, 7A(1976), 10, 1602

[7] R.G.Davies, Metallurgical Transactions A, 10A(1979), 1, 113

[8] T.Kobayashi and H.Tachibana, Transactions of the Japan Institute of Metals, 24(1983), 281

[9] For example; T.Shiokawa, Technical Bulletin No.56 (Ductile Iron Society of U.S.A.), 56(1983), 11, 129

[10] H.Yamamoto and T.Kobayashi, Physical Metallurgy of Cast Iron IV (MRS), (1989), 243

Advanced Materials Research Vols. 4-5 (1997) pp. 181-188
© *1997 Scitec Publications, Switzerland*

Damage Effect on the Fracture Toughness of Nodular Cast Iron

M.J. Dong[1], B. Tie[1], A.S. Béranger[2], C. Prioul[1] and D. François[1]

[1] Laboratoire de Mécanique, MSS/MAT, CNRS URA 850, Ecole Centrale Paris,
F-92295 Châtenay Malabry, France

[2] Renault Automobiles, Direction de la Recherche Sce 0852,
8-10 av. E. Zola, F-92109 Boulogne Billancourt, France

Keywords: Nodular Cast Iron, Damage, Toughness, Gurson's Model, Pressure Sensitivity Parameter

ABSTRACT

In order to understand the toughness of nodular cast iron, the damaged zone is studied by SEM observations on the prepolished surface of a CT 25 specimen, before and after ductile tearing. The damage is defined as the graphite/matrix interface decohesion. It is shown that the damaged zone is very large in nodular cast iron (through almost the whole ligament ahead of the crack), so the Linear Elastic Fracture Mechanics is not valid in nodular cast iron for small specimens.

The size of the damaged zone is calculated analytically by introducing a damage initiation criterion which is based on observations of the graphite nodules interfaces debonding and a pressure sensitive parameter. To take into account the limit condition, the damaged zone is calculated by numerical modelling, using the modified Gurson's model, for which nodular cast iron is regarded as a porous material. The calculated results give a good agreement with the damaged zone observations. The plastic zone size is nearly the same for plane stress or plane strain calculations, which is opposite to dense materials.

INTRODUCTION

The insufficient understanding of the evolution of the measured fracture toughness of nodular cast iron is a limiting factor for the large development of cast iron applications for safety components in nuclear industry or car manufacturing. For example, the increase of the fracture toughness of nodular cast iron with the thickness of the sample [1-3], opposite to what is usually observed, is not well modelled till now. This behaviour could be attributed to a large damaged zone ahead of the crack tip [4-5].

As observed by SEM, the fracture of the nodular cast iron is a ductile one at room temperature. The initiation of voids at the graphite matrix interface begins in the early stage of the plastic deformation [6]. So, the graphite nodules can be regarded as voids in the numerical or analytical modelling of the constitutive equation [6-7].

The aim of the present work is to get a better understanding and evaluation of the damage which takes place at the crack tip. For this purpose the damaged zone size ahead of the crack tip is determined experimentally after stable crack propagation, by SEM observations of the surface of a pre-polished CT 25 specimen. These observations are compared with the analytical and numerical modelling of the damaged zone size. In the analytical modelling, a damage initiation criterion which is based on observations of the graphite nodules interface debonding and a pressure sensitive parameter are used. The nodular cast iron is regarded as a porous material in the numerical modelling which allows to take into account the real boundary conditions. For this purpose, the modified Gurson's model is introduced into a Finite Element code specially developed at the laboratory to realise an automatic self adapting mesh refinement. This numerical technique strongly enhances the precision of the stress and strain fields determination at the crack tip.

MATERIAL

The nodular cast iron which is studied is labelled GGG40 (German standard). Its chemical composition is presented in table 1.

X Ray microprobe matrix chemical analysis showed that Silicon and Manganese content were 2.6 and 0.2 weight % respectively, with an homogeneous distribution (Figure 2).

Table 1 : Global chemical analysis (weight %)

	C	Si	Mn	S	P	Mg	Cu	Ni	Cr	Fe
GGG40	3.35	2.25	0.30	0.006	0.025	0.039	0.06	0.04	0.02	bal.

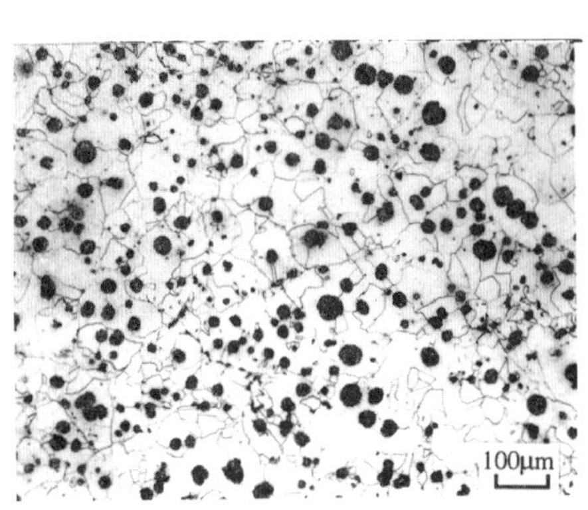

Figure 1. Optical micrography of GGG40

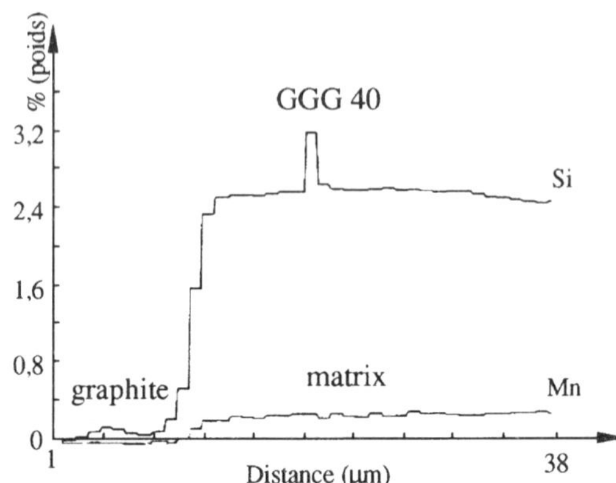

Figure 2. Distribution of Si and Mn in the matrix

EXPERIMENTAL TECHNIQUE AND RESULTS

A CT-25 specimen was used to study the damaged zone ahead of the crack tip. Preliminary experiments showed that the polishing preparation on damaged specimens induced void filling by polishing residues, thus disturbing further damage observations. Therefore, a pre-polished specimen is required for graphite matrix interface damage observation.

At first, the two lateral surfaces of the CT-25 specimen are polished, before straining, by classical polishing techniques with a final polishing using 1/4 µm diamond powder; then reference marks are made at the surface by mean of microhardness indentations to define square grids 4 mm x 4 mm large. This facilitates the observation of nodules interface modification before and after ductile tearing. The precrack is then made by fatigue on a MTS machine, using the following fatigue conditions: loading ratio R= 0.105, maximum load P_{max} = 8100 N, test frequency 20 cycles / s, number of cycles 4570. SEM observations are made before and after pre-cracking in order to select the nodules which will be observed and to verify the absence of damage. Then the tear test is made on an INSTRON machine with a speed of 10^{-4} / s at C point, to have a stable crack propagation (Δu = 0.4 mm). Finally, the SEM observations are made to determine the damaged zone, which is defined as the area on which graphite nodule interface decohesions are observed.

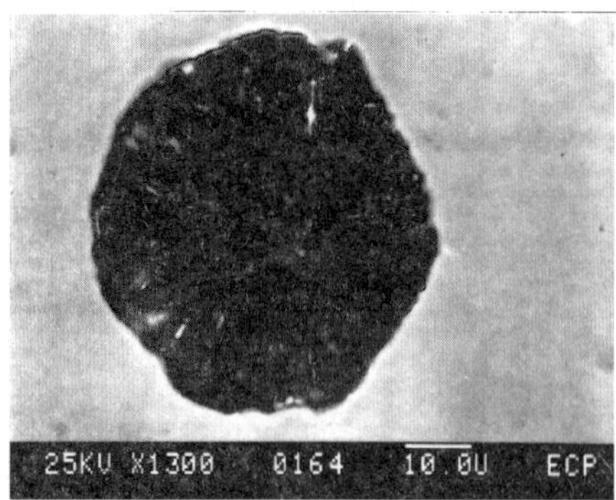

Figure 3. Decohesion of the interface

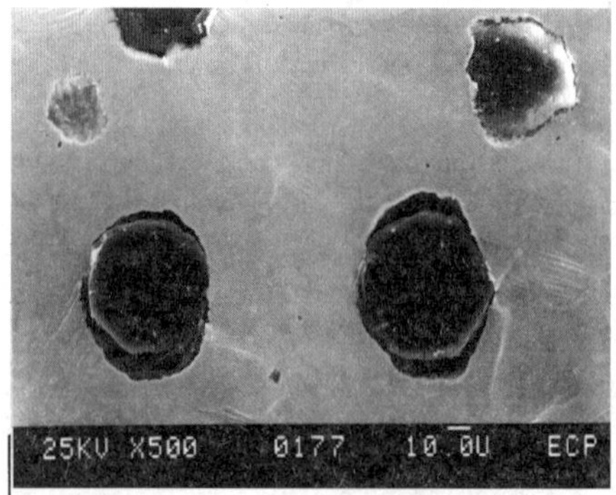

Figure 4. Cavity growth around nodules

From tensile tests performed on smooth specimens we showed (figure 3) that debonding occured at the nodule polar cap (oriented in the direction of tensile loading) as soon as the macroscopic yield stress was reached. Furthermore, when the stress was increased the cavities grew as ellipsoidal voids elongated in the tensile direction (figure 4). Previous modelling of the constitutive equation (figure 5) confirmed that cast iron could be regarded as a porous, so pressure sensitive, material [6-7].

SEM observations, after fatigue pre crack, do not show any void initiation at the graphite matrix interface. After stable crack propagation, the damaged zone extension, which has been reported on figure 6, is very large (through almost the whole ligament ahead of the crack).

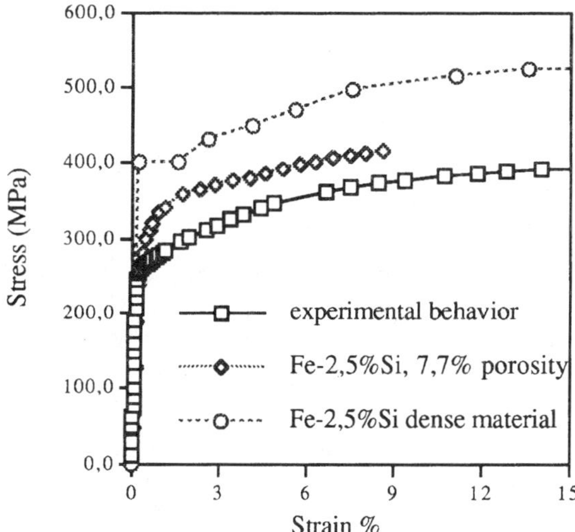

Figure 5. Comparison between experiment and modelling of the elastoplastic behaviour

Analytical damaged zone modelling

After debonding, we must deal with a pressure sensitive material owing to the existence of cavities. For this analytical modelling we have used the results of Li [8] who adopted a simple pressure - sensitive yielding criterion for the pressure - sensitive materials. It contains two invariants: the Mises equivalent stress σ_{mises} and the hydrostatic stress σ_m. The yield criterion is written as:

$$\sigma_{eq} = \sigma_{mises} + \sqrt{3}\,\mu\sigma_m \qquad (1)$$

$$\sigma_{mises} = (3/2\ s_{ij}\ s_{ij})^{1/2} \qquad \sigma_m = \sigma_{kk}\,/\,3$$

σ_{eq} represents the modified equivalent stress.

μ is the hydrostatic pressure sensitivity factor.

For $\mu = 0$, we obtain the HRR crack tip field. A direct measurement of the pressure sensitive factor μ relies on shear experiments under pressure. It can be obtained from the difference between the compressive yield strength σ_c and the tensile yield strength σ_t through the relation [9]:

$$\mu = \sqrt{3}\,\frac{\sigma_c - \sigma_t}{\sigma_c + \sigma_t} \qquad (2)$$

for the material tested, the experiments lead to $\mu = 0.28$.

In our analysis, we assume that nodular cast iron is specified by the Ramberg-Osgood stress-strain relation in shear:

$$\frac{\gamma}{\gamma_0} = \frac{\tau}{\tau_0} + \alpha \left(\frac{\tau}{\tau_0}\right)^n \qquad (3)$$

where τ is the shear stress, γ is the shear strain, n is the strain hardening exponent, α is a material constant, and τ_0, γ_0 are the reference shear stress and the reference shear strain. They can be obtained by the tensile test results. For the material tested we obtain: $n = 7.5$.

Following the procedures used by Li [8] after Hutchinson [10,11], Rice and Rosengren [12], we obtained the dominant asymptotic crack-tip stress, strain fields for nodular cast iron (equations 4,5).

$$\Sigma_{ij} = \Sigma_0 \left[\frac{J}{\alpha \Sigma_0 E_0 I(n,\mu) r}\right]^{\frac{1}{n+1}} \widetilde{\Sigma}_{ij}(\theta,n,\mu) \qquad (4)$$

$$E_{ij} = \alpha E_0 \left[\frac{J}{\alpha \Sigma_0 E_0 I(n,\mu) r}\right]^{\frac{n}{n+1}} \widetilde{E}_{ij}(\theta,n,\mu) \qquad (5)$$

$\widetilde{\Sigma}_{ij}$ and $\widetilde{E}_{ij}$ are the stress and strain dimensionless angular functions depend on the strain-hardening exponent n, the pressure-sensitivity factor μ, and the conditions of plane strain or plane stress. The traction free conditions on the crack faces and the symmetry (mode I) conditions about the crack line provided the necessary boundary conditions for the differential equation. A shooting method based on a combined fourth order Runge-Kutta scheme with error and step size control was employed to generate solutions.

The damaged zone boundary (figure 6) can be found by equating the local stress at the interface of the graphite nodules with a mesured critical stress $\sigma_c = 82$ MPa [13]. This local stress is given as a function of the applied stress tension Σ_{ij} through the Mori Tanaka model. The applied stress level is determined as the one reached at the initiation of crack propagation when J is equal to 27 kJ / m^2 as obtained experimentally using a multi specimen procedure [14].

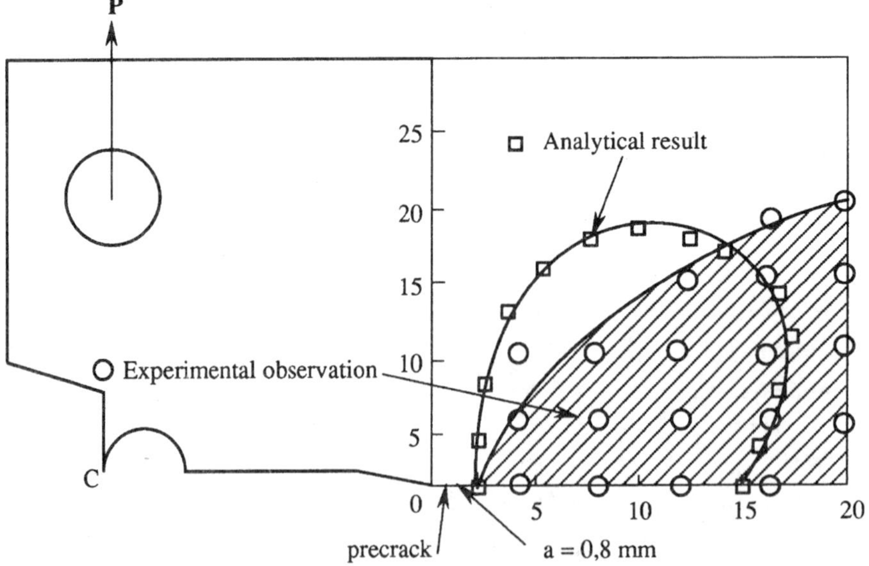

Figure 6. Damage zone of the nodular cast iron

The calculated damaged zone is large but less extended than the observed one, due to semi infinite specimen modelling which does not take into account the existence of a free surface ahead of the crack tip. In order to have a more realistic description of the CT specimen and to verify the validity of this analytical modelling a numerical damaged zone modelling must be performed.

Numerical damaged zone modelling

In this study, the modified Gurson's model [16] has been implemented into a Finite Element code specially developped at the laboratory to realize an automatic self adapting mesh refinement [17]. It has been applied to the modelling of the CT 25 specimen and the study of the damage zone ahead of the crack tip, in order to compare with our experimental observations

Since the different steps of voids evolution, initiation (interface decohesion), growth and coalescence are introduced through different equations in the Gurson's model [15], the different effects, from local damage up to coalescence of voids can be expressed continuously, and used in a numerical modelling. Tvergaard's modification of Gurson's yielding condition is [16] :

$$\phi\,(\sigma) = \frac{\sigma_{eq}^2}{Y^2} + 2\,f^*\,q_1\cosh\left(\frac{3}{2}q_2\frac{\sigma_m}{Y}\right) - [1+(q_1 f^*)^2] = 0 \qquad (6)$$

$$f^* = f \qquad\qquad f < f_c$$
$$f^* = f_c + f_{uc}\,(f - f_c) \qquad f > f_c$$

$$q_1 = 1.5 \text{ and } q_2 = 1$$

σ_{eq} : macroscopically equivalent stress

σ_m : macroscopically hydrostatic stress
Y : flow stress of the material
f_c : critical void volume fraction at void coalescence

At the beginning f^* is identical to f as in the original Gurson's model but is greater than f above a critical volume fraction f_c. By substituting f in the original Gurson's model by f^*, the decrease of load carrying capacity due to the coalescence of voids is modelled. For $f^* = 0$, the plastic potential (equation 6) according to Tvergaard is obviously identical to that of Von Mises. If f^* reaches the limit $1/q_1$ the material looses its load carrying capacity because all stress components have to vanish in order to satisfy equation (6).

Both the nucleation of new voids and growth of existing voids contribute to the increase of the volume fraction of voids, so the growth rate of f can be expressed by:

$$\dot{f} = \dot{f}_{growth} + \dot{f}_{nucleation} \qquad (7)$$

In agreement with our experimental observations, we assume that in nodular cast iron, the nodules can be replaced by voids initially present in the material, and we can neglect the nucleation of new voids. Only the growth rate of the initial cavities were considered.

The graphite nodule volume determined by image analysis was used as the initial void volume in the numerical simulation, $f_0 = 0.077$. The value of f_c, chosen to be 0.08, and the value of the void volume fraction f at fracture, equal to 0.2, were determined from the evolution of the damage parameter given by modulus mesurements. The constitutive equation of the matrix was chosen as the one of a Fe-2.5Si steel, whose composition is close to the one of the matrix as determined by X Ray microprobe quantitative analysis (figure 5).

Figure 7 shows the initial finite element mesh for the CT-25 specimen. 3-nodes or 4 nodes isoparametric elements with two-by-two integration were used. To investigate the influence of the stress state hypothesis on the damaged zone extension, plane strain and plane stress conditions were both tested. No crack propagation was considered in this modelling but the precrack length was taken into account.

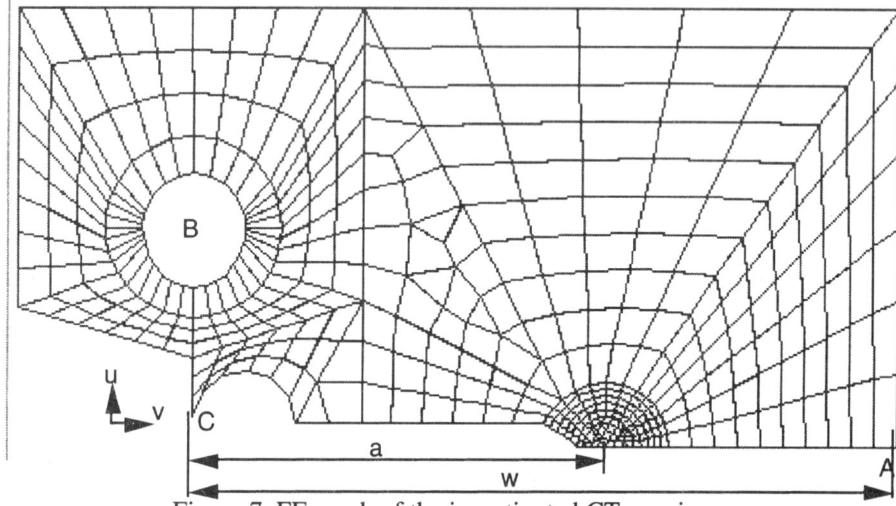

Figure 7. FE-mesh of the investigated CT-specimen

Figure 8 shows for this porous material the incremental plastic zone (with a step of 0.02 mm) for plane strain condition with a crack mouth displacement (measured in C) of 0.4 mm, it is nearly the identical to the damaged zone (figure 9). The calculated damaged zone is very large, almost through the whole ligament. This result is in good agreement to the experimental result. A local damaged zone was observed earlier by using the automatic self adapting mesh refinement than that of the usual mesh, but there is no difference in the global damaged zone.

Figures 10, 11 show the plastic zone and damaged zone for plane stress conditions with the same displacement Δu. The result is nearly the same for both conditions. This is different from the result for dense materials. This result is confirmed by toughness studies in porous materials [18] and also by the analytical study of the influence of porosity on plane strain tensile crack-tip stress fields [19].

Figure 8. Numerical plastic zone for CT25 specimen at $\Delta u = 0.4$ mm **(plane strain)**

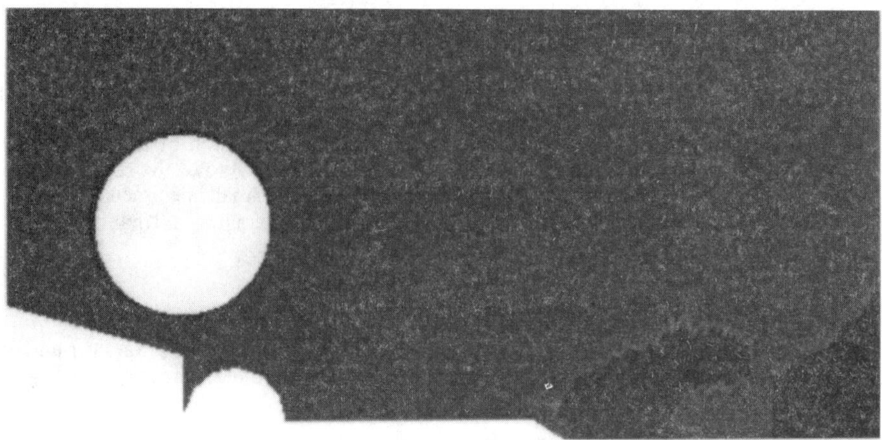

Figure 9. Numerical damaged zone for CT25 specimen at $\Delta u = 0.4$ mm (plane strain)

Figure 10. Numerical plastic zone for CT25 specimen at $\Delta u = 0.4$ mm (plane stress)

Figure 11. Numerical damaged zone for CT25 specimen at $\Delta u = 0.4$ mm (plane stress)

Conclusions

Ductile failure in CT specimen in nodular cast iron can be simulated analytically and numerically. In analytical modelling a damage initiation criterion and a pressure sensitive parameter were used, and the modified Gurson's model was used in numerical modelling.

The observed or calculated damaged zone show that the extension of the damaged zone in nodular cast iron is very large through almost the whole ligament ahead of the crack, and is nearly identical to the plastic zone. So, K_{Ic} measurements are not valid for small specimens (CT 25).

The plastic zone for plane stress conditions is nearly the same as that for plane stress conditions, in contrast with what is observed in dense materials.

Acknowledgements:

The financial support of Commissariat à l'Energie Atomique (DMT Saclay) and Renault Automobiles (Direction des Etudes Matériaux) is greatly acknowledged. The authors are grateful to Dr. D.Moulin and Dr. A.Diboine for their cooperation to this study.

References:

[1] R.K.Nanstad, AFS Transactions, 245 (1975).
[2] W.L.Bradley, International Materials Reviews 35 (3) 129 (1990).
[3] R.Salzbrenner, Journal of Materials Science 22 2135 (1987).
[4] Bradley W. L.,Kinney K. E. M. and Gerhardt.P. C., Fracture Mechanics (ASTM STP 905), 17 75-94 (1986).
[5] F.J.Worzala and R. K.Nanstad, Proceedings of the fifth conference on dimensioning and strength calculations, Budapest, 1 265-276 (1974).
[6] M.J.Dong, G.K.Hu, C.Prioul, D.François, MECAMAT93, International seminar on micromechanics of materials, Moret-sur-Loing, France, 6-8 July, Eds.EYROLLES, 511-521 (1993).
[7] M.J.Dong, G.K.Hu, A.Diboine, D.Moulin, C.Prioul, Journal de Physique IV, Colloque C7, supplément au Journal de Physique III, 3, novembre, 643-647 (1993).
[8] Li F. Z., Pan J., Transactions of the ASME, 57, 40 (1990).
[9] Argon A. S., Met. Trans. A, 6A, 839 (1975).
[10] Hutchinson J. W., J. Mech. Phys. Solids, 16, 13 (1968).
[11] Hutchinson J. W., J. Mech. Phys. Solids, 16, 337 (1968).
[12] Rice J. R., Rosengren G. F., J. Mech. Phys. Solids, 16, 1 (1968).
[13] Dong M.J., C.Prioul, G.K.Hu, D.Moulin, D.Francois, Saclay International Seminar on Structural Integrity, April 28-29, INSTN, France (1994).
[14] Internal report, Saclay - CEA, (1991).
[15] A.L.Gurson, J.Engng. Mater.Tech. 99 2-15 (1977).
[16] V.Tvergaard, International J. Fracture 17 389-407 (1981).
[17] B.Tie, Ph.D Thesis, (1993), Ecole Centrale de Paris, France.
[18]G.Andrews, Powder metall. group meeting, 24/26 Coentry 5 1 (1977)
[19] W.J.Dragan, Y.Miao, J.of Appl.Mech. 59 559 (1992)

Advanced Materials Research Vols. 4-5 (1997) pp. 189-194
© *1997 Scitec Publications, Switzerland*

Relation between Microstructure Size and Ductile-Brittle Transition Behaviour in Fracture Toughness of Ferritic Nodular Cast Iron

S. Komatsu[1], T. Shiota[1], K. Nakamura[2] and H. Kyogoku[1]

[1] Kinki University, Faculty of Engineering, 1 Takayaumenobe, Higashihiroshima 729-17, Japan

[2] Kinki University, Faculty of Science & Engineering,
3-4-1 Kowakae, Higashiosaka, Osaka 577, Japan

Keywords: Ductile Cast Iron, Fracture Toughness, J_{IC}, Ductile-Brittle Transition, Microstructure Size, Graphite Nodule Size, Ferrite, Grain Size, Nodule Count, Temperature, Solidifying Rate, Transition Temperature

ABSTRACT

The relation between the microstructure size (ferrite grain size and graphite nodule size) and the ductile-brittle transition behavior in the fracture toughness of ferritic nodular cast iron was investigated. Four ferritic nodular cast iron samples whose chemical compositions are the same but whose microstructure sizes are diferrent from each other were conducted for fracture toughness testing at every 25 K between 123 K and 293 K. In the J_{IC}-temperature relation, every sample exhibited a horizontal upper shelf region, a ductile-brittle transition region and a lower shelf region. Increasing microstructure size resulted in a moderate increase in upper shelf J_{IC}. This tendency is caused mainly by the increase in the average distance of matrix. In the case of the same chemical composition, increasing graphite nodule size accompanies decreasing graphite nodule count and increasing average matrix distance. The larger the average matrix distance is, the more ductile extension and thus the more energy is required in order to produce a fracture of dimple type. Increasing microstructure size also resulted in a remarkable rising of the transition temperature region. This tendency is caused mainly by decreasing the grain boundary area, decreasing the relieving effects of stress-concentration and plane-strain condition that are produced by increasing the graphite nodule (void) diameter and the average matrix distance.

INTRODUCTION

The relation between temperature and fracture toughness of ferritic nodular cast iron is known to exhibit a remarkable transition behavior being caused by the ductile-brittle fracture mode changing. Nodular cast iron products for engineering use are generally produced in various sizes, and their microstructure sizes (ferrite grain size and graphite nodule size) vary largely because of difference in solidifing rate caused by casting size. This paper aims to clarify how the microstructure size affects the relationship between temperature and fracture toughness of ferritic nodular cast iron.

EXPERIMENTAL PROCEDURES

In this study, four ferritic nodular cast iron samples whose chemical compositions are the same but whose microstructure sizes are different from each other were used. Fig. 1 shows the casting molds and the microstructures of the four samples used in this study. Sample A was produced by the fully metallic casting mold and sample B was produced by the casting mold made of metallic walls and CO_2 sand bottom. Samples C and D were produced by the casting molds made of fully CO_2 sand, but the wall thickness of sample D is much larger than that of sample C. As the sample name changes from A to D, the solidifing rate becomes slower, and thus the microstructure size becomes larger. All the samples were given full annealing heat-treatment in order to get fully ferritic matrixes.

Table 1 indicates the microstructure sizes, mechanical properties and chemical compositions of the specimens. The mean graphite nodule diameters were obtained by averaging the diameters of the graphite nodules that are larger than 5 mm in magnification of X400, and the ferrite grain sizes are indicated by the ferrite grain size numbers according to the JIS-G0552 standard [1].   Fracture toughness testing was carried out by measuring the elastic-plastic fracture toughness J_{Ic} by means of an electric potential method [2,3] using compact tension (CT) specimens 30 mm wide and 15 mm thick which meets to the ASTM-E813 standard [4].   The electric potential method was employed because it is convenient to measure fracture toughness in both the upper shelf region and the transition region in the same manner. Fracture toughness testing was conducted twice at every 25 K between 123 K and 293 K using a chamber of a liquid nitrogen spraying type.

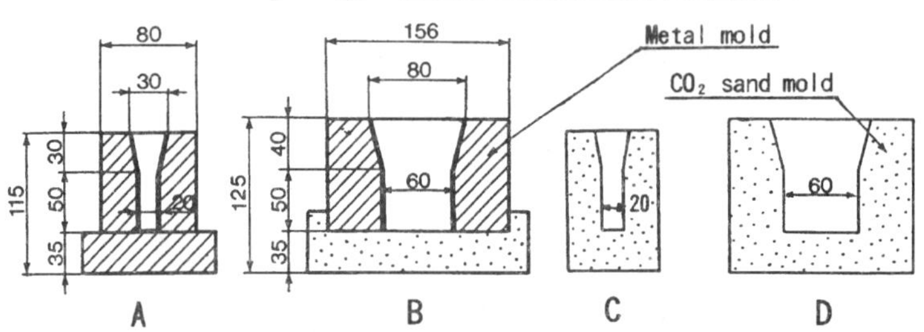

a) The casting molds to modify the solidifying rate

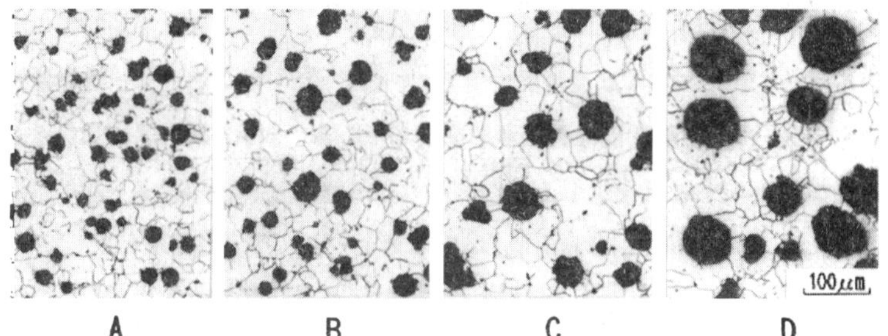

Fig. 1
Casting molds and
microstructures
of the samples

b) The microstructures of the samples

Table 1   Microstructure, mechanical properties and chemical compositions of the samples

Samples	Average graphite nodule dia., μm	Ferrite grain size No.	Tensile strength MPa	Elongation %	Brinell hardness (HB)	Chemical compositions %
A	18	7.0	448	20.0	154	C :3.67 Si :2.42 Mn:0.21 P :0.06 S :0.02
B	25	6.5	426	20.5	151	
C	45	6.0	429	22.0	147	
D	88	5.0	408	22.3	144	

RESULTS AND DISCUSSIONS

Fig. 2 shows the mechanical properties as a function of the microstructure size of the samples.   In this figure, the microstructure size is represented by the average graphite nodule diameter. Therefore, in the figure a sample with a larger graphite nodule diameter does not mean a larger graphite nodule diameter alone but also means a larger ferrite grain size.   Fig. 2 indicates that increasing the microstructure size results in a moderate decrease in tensile strength, yield strength and Brinell hardness,however their decreasing rates are as small as less than 10 % when comparing sample A to sample D. The elongation is influenced little by the microstructure size.

Fig. 3 shows the schematic figures of the load versus load-point displacement curves at CT testing at several temperatures. In the figure, the arrow indicates a continuation of stable fracture or ductile fracture, and the symbol X indicates an occurrence of unstable fracture or brittle fracture. At 293 K all the samples showed ductile fractures regardless of their microstructure size. However at 198 K, while sample A showed ductile fracture, the samples with larger microstructure sizes than sample A showed brittle fractures. At 173 K, all the samples showed brittle fractures, but the samples C and D showed brittle fractures at smaller load-point displacements than the samples A and B whose microstructure sizes are smaller. At 123 K, all the samples fractured in brittle fracture mode during the relation between load versus load-point displacement was keeping linearity.

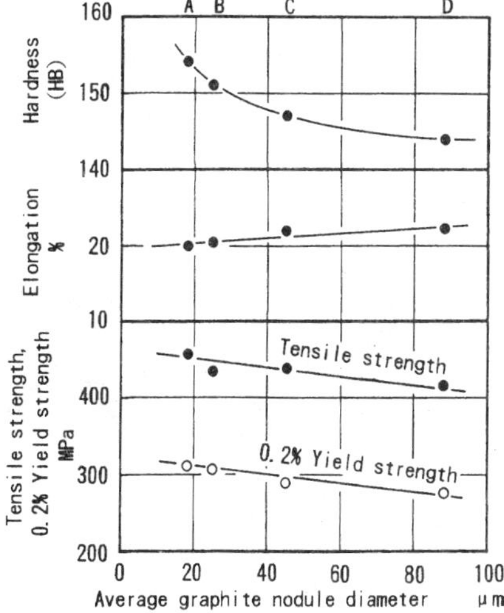

Fig. 2   Influence of microstructure size on mechanical properties plotted versus average graphite nodule diameter

Fig. 3   Schematic diagrams between load versus load-point displacement in CT testing at several temperatures

Fig. 4 shows the J initiation (Jin) data of each sample as a function of temperature. In these figures, the solid points indicate that the fracture mode was ductile at fracture initiation and the circles indicate that the fracture mode was brittle at fracture initiation. All of these Jin data are the ones that qualified all the requirements to be evaluated as valid Jic mentioned in the ASTM-E813 standard. The Jic-temperature relation in the upper shelf region was obtained by averaging the solid data in the region. In the transition region a large scatter was observed in the Jin data. The fracture modes of usual ferritic materials in the transition region were usually ductile, brittle or combinations thereof. The brittle fracture in the transition region is usually due to cleavage fracture, and cleavage fracture is usually initiated by a critical fracture-triggering defect (microcrack, carbide, inclusion, etc) near the pre-crack tip [5]. The fracture-triggering defects vary widely in their nature, size, distance from pre-crack tip, orientation, etc.   Therefore, even at the same temperature, these variations result in variations in fracture mode and load-point displacement at fracture initiation and thus in the measured data of Jin.   From the point of the weakest rink phenomenon theory, it is common to obtain the Jic-temperature relations in the transition region and the lower shelf region by connecting the smallest data at each temperature [6].   In these regions of Fig. 4 this manner was applied to obtain the Jic-temperature relations that are represented by the solid lines.   In Fig. 4 each sample exhibits a horizontal upper shelf region of ductile fracture mode and a transition region where Jic decreases remarkably due to the changing of fracture mode from ductile to brittle.

Fig. 5 shows the comparison of the Jic-t emperature relations of the four samples obtained in Fig. 4. This figure indicates that a sample with a larger microstructure size tends to have a higher

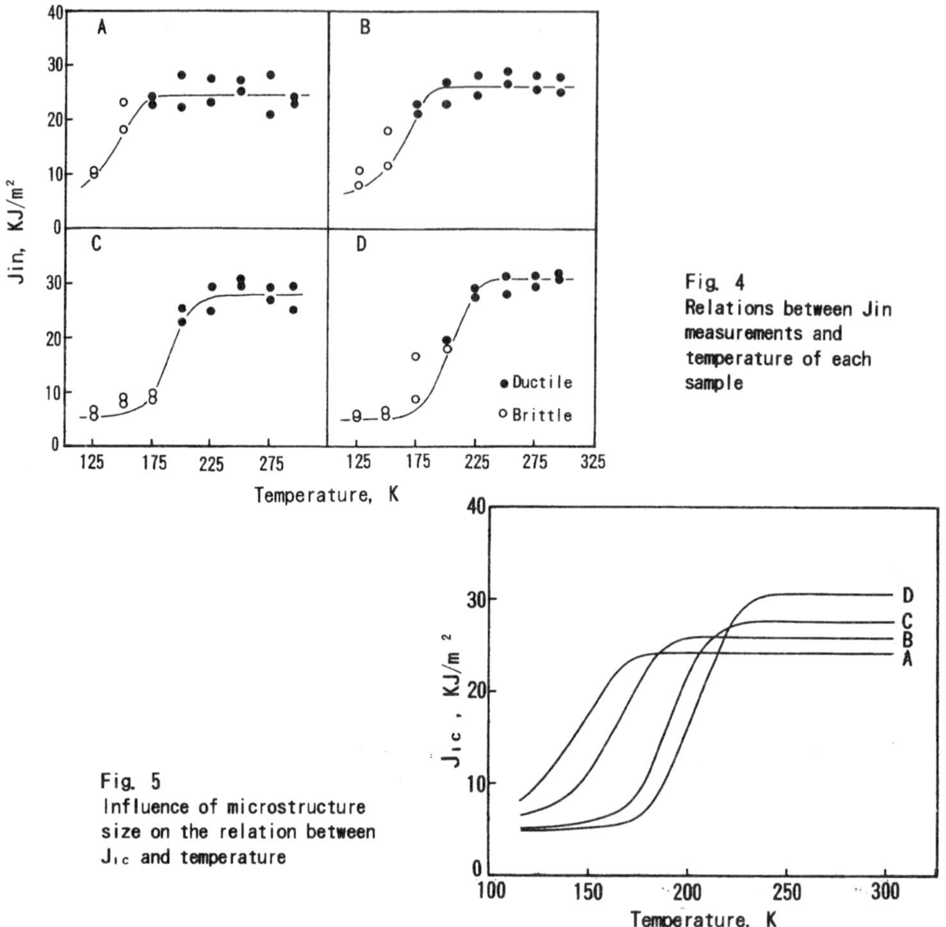

Fig. 4
Relations between Jin measurements and temperature of each sample

Fig. 5
Influence of microstructure size on the relation between J_{Ic} and temperature

upper shelf. This increasing tendency of J_{Ic} with increasing microstructure size agrees well with the results of W. L. Bradley and M. Tanner [7] and R. Salzbrenner [8]. W. L. Bradley and M. N. Srinivasan [9] and R. O. Ritche and A. W. Thompson [10] reported that there is a relatively linear relationship between the square root of the graphite nodule diameter and fracture toughness $K_{Ic}(J)$ which is the converted K_{Ic} value from J_{Ic}.

In the case of fully ferritic nodular cast iron samples with constant chemical compositions, increasing graphite nodule size accompanies inevitably decreasing graphite nodule count and thus increasing average distance of matrix between graphite nodules. As the graphite nodules act as voids and main nucleus of the dimples in a fracture under tensile condition, a larger average distance of matrix requires more extension or ductility and thus more energy in order to produce a fracture of the dimple type. This is a main cause of the increase in the upper shelf J_{Ic} caused by increasing microstructure size. However, this larger extension of the matrix of a sample with a larger microstructure size seems to be contradictory to the results of tensile-test elongation shown in Table 1 where the elongation is affected little by the microstructure size. This can be explained as follows. As the graphite nodule diameter increases and the average distance of matrix between graphite nodules increases, the extension of the matrices around each graphite nodule increases in order to create dimples. However, increasing average graphite nodule diameter also means decreasing the graphite nodule count within the gauge length of the tensile test specimen. Therefore, the positive effect for larger elongation of each matrix around each graphite nodule that is caused by increasing the average distance of matrix is offset by the negative effect that is caused by decreasing the graphite nodule count within the gauge length. This offset of a positive effect and a negative effect is considered to be the main reason for the similar tensile-test elongation values for all the samples regardless the graphite nodule size.

Fig. 5 also indicates that increasing the microstructure size results in considerable shifting of the transition region toward the higher temperature direction. The transition behavior is caused by a change of fracture mode from ductile or dimple fracture type to brittle or cleavage fracture type. Therefore, this figure indicates that a sample with a larger microstructure size tends to begin brittle fracture at a higher temperature than a sample with a smaller microstructure size, i.e., a sample with a larger microstructure size tends to fracture in brittle mode more easily than a sample with a smaller microstructure size at a same temperature in the transition region.

There may be several reasons for this shifting of transition region. The first ones are related to the ferrite grain size. The brittle fracture in the transition temperature region is mostly due to cleavage fracture under a plane-stress condition. T. L. Anderson [5] described that a cleavage fracture initiates from a local discontinuity ahead of the macroscopic crack that is sufficient to exceed the bond strength and a sharp microcrack is one way to provide sufficient local stress concentration. J. F. Knott [11] and others reported that cleavage fractures usually are initiated by grain boundary carbides. As a larger ferrite grain size implies a smaller grain boundary area, a sample with a larger ferrite grain size tends to possess larger grain boundary carbides and more easily initiates cleavage fracture than a sample with a smaller ferrite grain size.

Another critical event may be propagation of the initial microcrack. The ferrite grain boundary interrupts propagation of microcrack because the cleavage plane of the grain with microcrack rarely coincides with that of the neighboring grains. This interrupting effect of microcrack propagation is smaller in the case of a larger ferrite grain size than in the case of a smaller ferrite grain size because of difference in grain boundary area. The other reasons are related to the graphite nodule size. At a temperature low enough to create cleavage fractures graphite nodules act as voids or microcracks that may produce some local stress-concentration which may initiate fracture-triggering microcraking. The combination of a larger average diameter of graphite nodules (voids or microcracks) and a larger average distance of matrix may produce a larger stress-concentration effect compared to the case of the combination of smaller ones. And in tensile condition, the graphite nodules act as voids that may contribute to release the plane-strain condition at the vicinity of pre-crack tip. A sample with a larger graphite nodule size has a smaller graphite nodule count and a larger average distance of matrix between graphite nodules. This means that a sample with a larger graphite nodule size is given a smaller relieving effect for plane-strain condition, and it becomes easier to produce a brittle fracture compared to a sample with a smaller graphite nodule size.

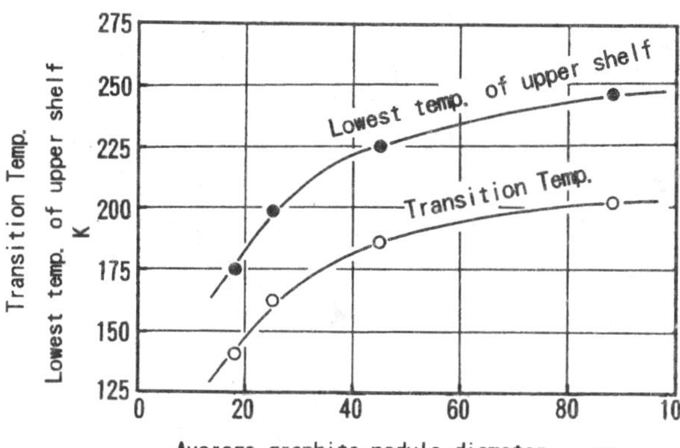

Fig. 6
Influence of microstructure size on the transition temperature and the lowest temperature of the upper shelf plotted versus average graphite nodule diameter

Fig. 6 shows the influence of microstructure size on the transition temperature and the lowest temperature of the upper shelf. The transition temperature was obtained as the temperature at which J_{Ic} becomes the average of the upper shelf and the lower shelf, and the microstructure size is represented by the average graphite nodule diameter in the figure. Both temperatures are observed to rise largely as the microstructure size increases, and the rising rate in the larger graphite nodule region becomes smaller than in the smaller graphite nodule size region. Fig. 7 shows the macro-fractographs of the fracture surfaces of CT specimens after testing at every temperature.   The black portions consist mainly of dimple fracture surfaces, and the white portions consist mainly of cleavage fracture surfaces.   The temperature zone where the color of fracture surfaces changes from black to white indicates the transition region. Fig. 7 shows that this changing temperature zone rises as the microstructure size becomes larger, and this tendency coincides with the tendency mentioned above.

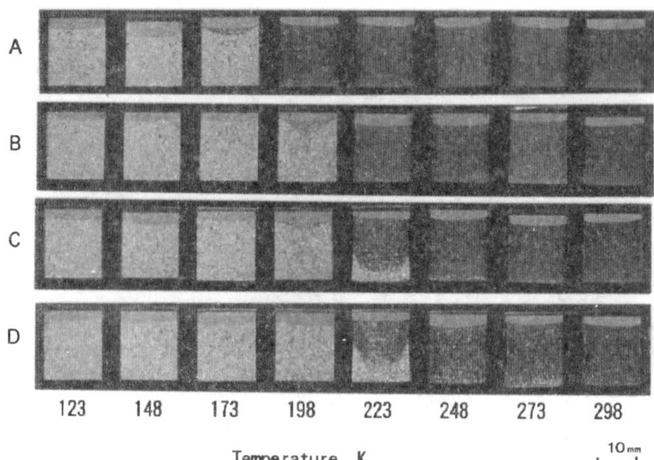

Fig. 7
Macro-fractographs of
the fracture surfaces
of the CT specimens

CONCLUSIONS

Four ferritic nodular cast iron samples whose chemical compositions are the same but whose microstructure sizes are different from each other were subjected to fracture toughness testing at various temperatures, and the following results were obtained.
(1) Increasing microstructure size results in moderate increase in the upper shelf Jic.   This is mainly because a larger average matrix distance requires larger extension of the matrix and thus larger energy to create dimples.
(2) Increasing microstructure size results in a considerable shifting of the ductile-brittle transition temperature region toward the higher temperature side.   The main reasons for this are the decreasing grain boundary area which may arrest microcrack propagation, the larger fracture-triggering particles, the smaller relieving effect of plane-strain condition and the larger stress-concentration effect caused by larger graphite nodules.

REFERENCES

1. Japanese Industrial Standard : JIS-G0552; Method of ferrite grain size test for steel .
2. Japan Society of Mechanical Engineers: JSME-S001-1981; Standard method of test for elastic-plastic fracture toughness Jic.
3. S. Komatsu, T. Shiota and K. Nakamura , in " Physical Metallurgy of Cast Iron IV" MRS Symposia Proceedings , 313 (1989).
4. ASTM-E813-1981.
5. T. L. Anderson ,"Fracture mechanics, fundamentals and applications", CRC Press, 321 (1991).
6. A. Ohtsuka and T. Miyata , J. of The Society of Materials Science, Japan , **33**, 97 (1984).
7. W. L. Bradley and M. Tarnner , in "Fracture Mechanics", STP 945, 18 (1987), ASTM.
8. R. Salzbrenner , J. of Material Science, **22**, 2135-2147 (1987).
9. W. L. Bradley and M. N. Srinivasan , International materials reviews , **35**, No 3, 129. (1990)
10. R. O. Ritche and A. W. Thompson , Metall. Trans., **16A**, 233 (1985).
11. J. F. Knott , Fracture 1977, **1**, ICF4, Waterloo Canada, 61 (1977) .

Advanced Materials Research Vols. 4-5 (1997) pp. 195-202
© 1997 Scitec Publications, Switzerland

Effect of Thermomechanical Treatment on Toughness of Spheroidal Graphite Cast Iron

S. Yamada[1] and T. Kobayashi[2]

[1] Graduate School of Toyohashi University of Technology,
1-1 Tempaku, Toyohashi 441, Japan

[2] Department of Production Systems Engineering, Toyohashi University of Technology,
1-1 Tempaku, Toyohashi 441, Japan

Keywords: Spheroidal Graphite Cast Iron, Austempered Ductile Iron, Thermomechanical Treatment, Toughness, Ausferritic Transformation

ABSTRACT

Effect of mechanical deformation on ausferritic transformation and toughness in spheroidal graphite cast iron was examined. The rolling was performed at austempering temperature, austenitising temperature or both temperatures. The toughness of obtained material was evaluated by instrumented Charpy impact test. Microstructure and fracture surface were observed via scanning electron microscopy, SEM. The rolling at the austempering temperature remarkably refines the matrix microstructure and that accelerated the kinetics of ausferritic transformation. While, the rolling at the austenitising temperature was effective for eliminating the γ-pool and that slightly delayed the kinetics of the transformation. Maximum strength and hardness were obtained by 24.0% rolling at the austempering temperature. Maximum ductility and toughness were obtained by 19.3% rolling at both the austenitising and austempering temperatures.

1. INTRODUCTION

Spheroidal graphite (SG) cast iron has been applied to structural parts because of its good mechanical properties and economical efficiency. Especially, austempered ductile iron, so-called ADI, shows high strength and good wear resistance. It is well-known that coarse unstable austenite (γ-pool) located near the eutectic cell boundary degrades the mechanical properties and machinability in ADI. The refinement of ausferritic matrix (retained austenite + acicular ferrite) and the elimination of γ-pool are, therefore, necessary to improve the toughness of ADI. The refinement of microstructure is, in general, effective for improving the mechanical properties, and that has been achieved by the controlled rolling in steels. However, application of the thermomechanical treatment to SG iron has rarely reported[1,2].

In this study, various thermomechanical treatments, TMT, based on the austempering were applied to SG iron. The rolling was performed at the austempering temperature, austenitising temperature or both temperatures in the austempering process. The effect of rolling on the ausferritic transformation and toughness was examined.

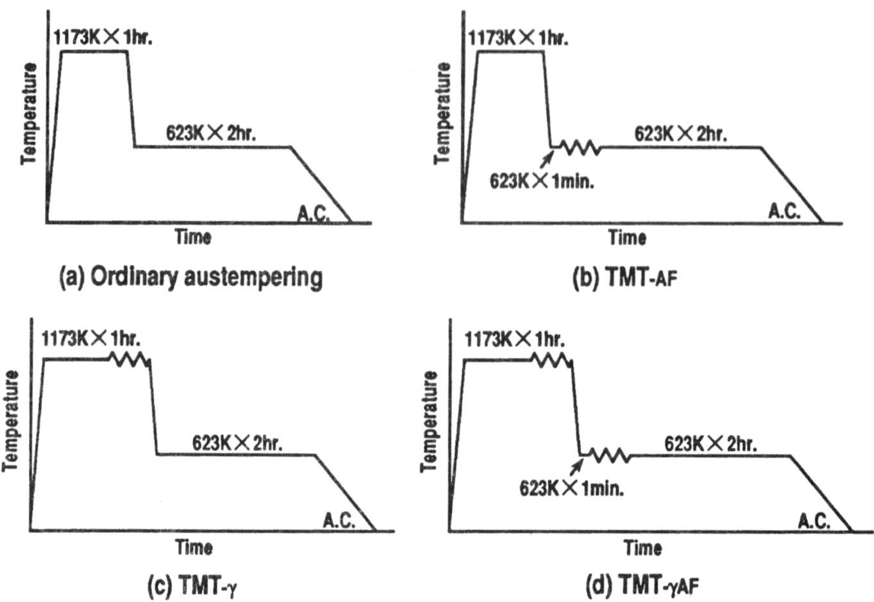

Fig. 1 Schematic illustrations of ordinary austempering and various thermomechanical treatments.

2. EXPERIMENTAL PROCEDURE

A SG iron with the following chemical compositions was used in this study; 3.66% C, 2.82% Si, 0.16% Mn, 1.59% Ni, 0.108% Cr, 0.33% Mo, 0.034% P, 0.003% S, 0.054% Mg. Y-block castings with the thickness of 35mm had almost fully ferritic structure with ordinary graphite distribution.

Ordinary austempering and three types of TMT based on the austempering used in this study are shown in Figure 1. The rolling was performed at the austempering temperature (TMT-AF), austenitising temperature (TMT-γ) or both temperatures (TMT-γAF) in the austempering process. In TMT-γAF, the rolling at each temperature was conducted by 50:50 in the reduction ratio, and a total reduction ratio is regarded as a reduction ratio.

Instrumented Charpy impact test was carried out to evaluate the toughness of the materials conducted with the ordinary austempering and TMT at room temperature. A computer-aided instrumented Charpy impact testing system (CAI system)[3,4] with a 490J capacity developed by one of the authors and unnotched half size Charpy type specimens ($5 \times 10 \times 55$mm) cut from treated pieces were used in all tests. The loading direction was perpendicular to the rolling direction and the loading rate was 2m/s. Microstructures and fracture surfaces at the center of specimen were observed via scanning electron microscopy, SEM.

3. EXPERIMENTAL RESULTS AND DISCUSSION

Typical microstructures around the graphite nodule and the eutectic cell boundary in the material conducted with ordinary austempering and various TMT are shown in Figure 2. The γ-pool with the grain size of around 5μm was observed around the eutectic cell boundary in the material conducted with the ordinary austempering. In the material conducted with TMT-AF, very fine matrix microstructure compared with the ordinary ADI was obtained, and grain size of each phase was in the order of sub-micrometer. Additionally, colonies of acicular ferrites which may be corresponding to previous austenite grain were observed. In the material conducted with TMT-γ, the matrix microstructure was not as fine as that of the material conducted with TMT-AF, although the γ-pool was eliminated. Thus, the rolling at the austenitising temperature contributes to improve the uniformity of the matrix rather than the fineness of that. Intermediate microstructures between those of the materials above-mentioned were observed in the material conducted with TMT-γAF. The γ-pool was almost fully eliminated in the material with 19.3% rolling, and a finer matrix microstructure was obtained in the material with 27.1% rolling in this case.

In order to examine the effect of mechanical deformation on the ausferritic transformation process in SG iron, specimens were water-quenched from various austempering times. Effect of the rolling on the kinetics of the ausferritic transformation in SG iron is shown in Figure 3. The decrease in Vickers hardness after water quench is corresponding to the progress of ausferritic transformation. The rolling at the austempering temperature remarkably accelerates the transformation, because the deformation directly gives a driving force for the transformation or the increase in the precipitation site of acicular ferrite. The incubation time for starting the transformation and the transformation time from a start to an end of the transformation are varied sensitively by the reduction ratio. While, the

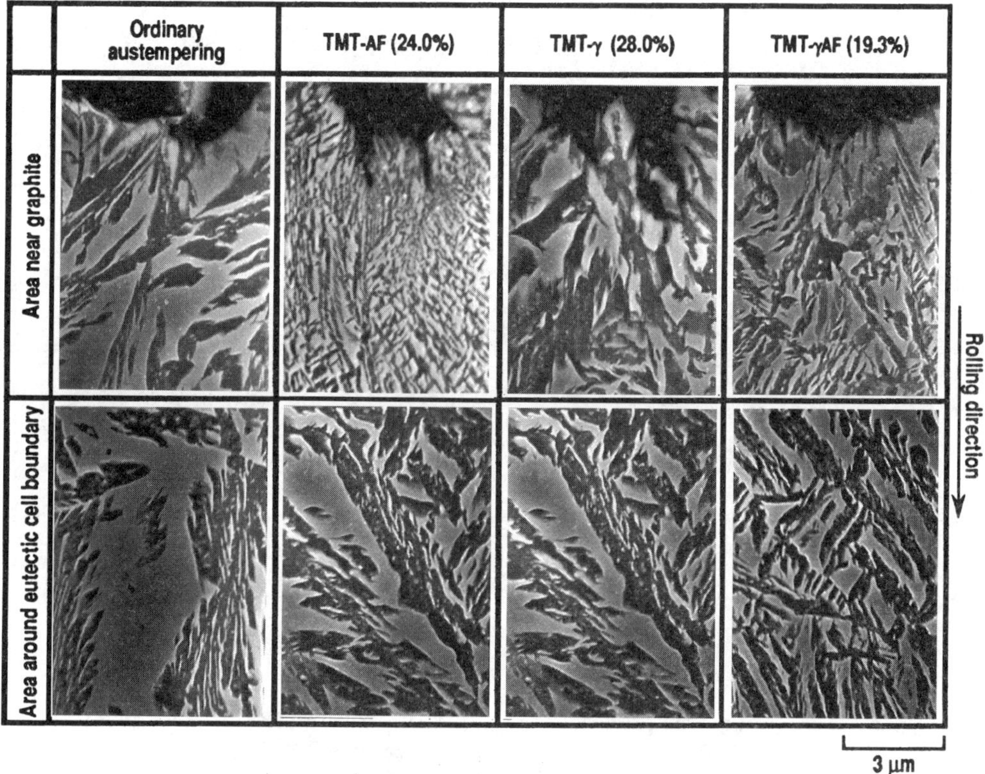

Fig. 2 Microstructure obtained by ordinary austempering and various TMT.

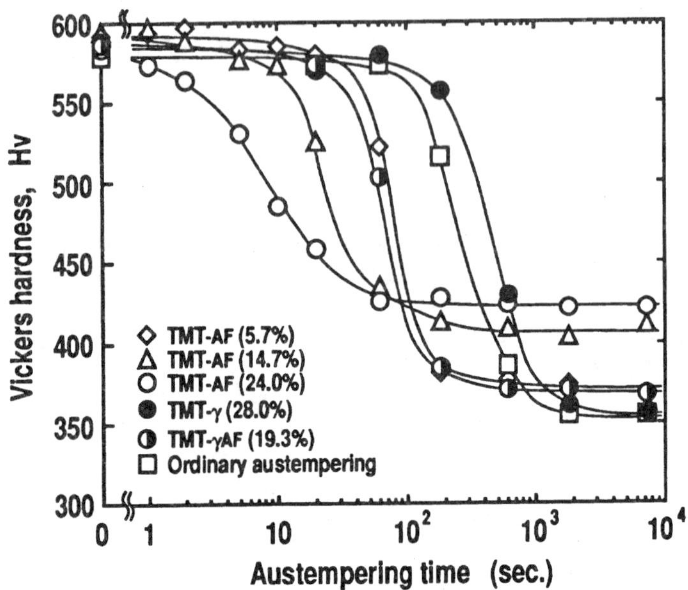

Fig.3   **Relationships between Vickers hardness after water quench, Hv and austempering time.**

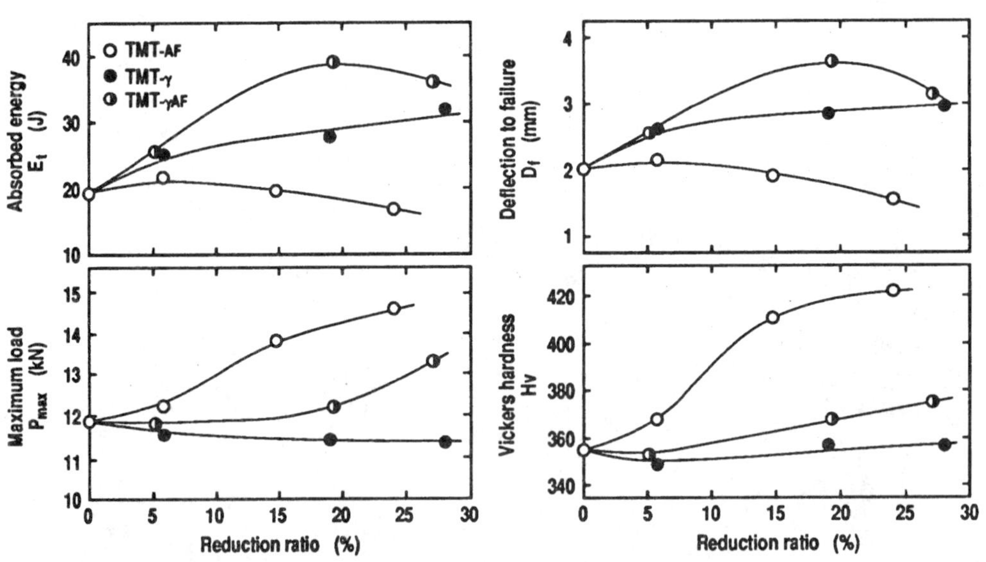

Fig. 4   **Change in various parameters obtained by instrumented Charpy impact test at room temperature and Vickers hardness with reduction ratio.**

rolling at the austenitising temperature brings a little retardation of the transformation. The reason of this phenomenon is that the rolling at the austenitising temperature increases the precipitation site of ferrite, while the length of each ferrite becomes shorter simultaneously.

The toughness and mechanical properties are, in general, related crossly with the microstructure. Therefore, these properties are varied with TMT conditions. The results of instrumented Charpy impact tests and Vickers hardness measurements are shown in Figure 4. In the material rolled at the austempering temperature, the strength and hardness increased with increasing reduction ratio. The ductility and toughness, however, decreased with increasing reduction ratio. By contrast, in the material rolled at austenitising temperature, the ductility increased with increasing reduction ratio.

This is considered to be caused by the increase in the ductility due to the elimination of γ-pool. However, since the degree of refinement of the matrix microstructure is smaller than that in the material rolled at the austempering temperature, little increase in the strength was obtained. In the material rolled at both austenitising and austempering temperatures, furthermore, the strength and hardness increased with increasing reduction ratio. The ductility and toughness of this material show maxima.

Fracture surfaces at the center of Charpy specimen in the material conducted with the ordinary austempering and various TMT are shown in Figure 5. The ductility of SG iron depends mainly on the growth of void nucleated at the interface between graphite nodule and matrix. When neighboring

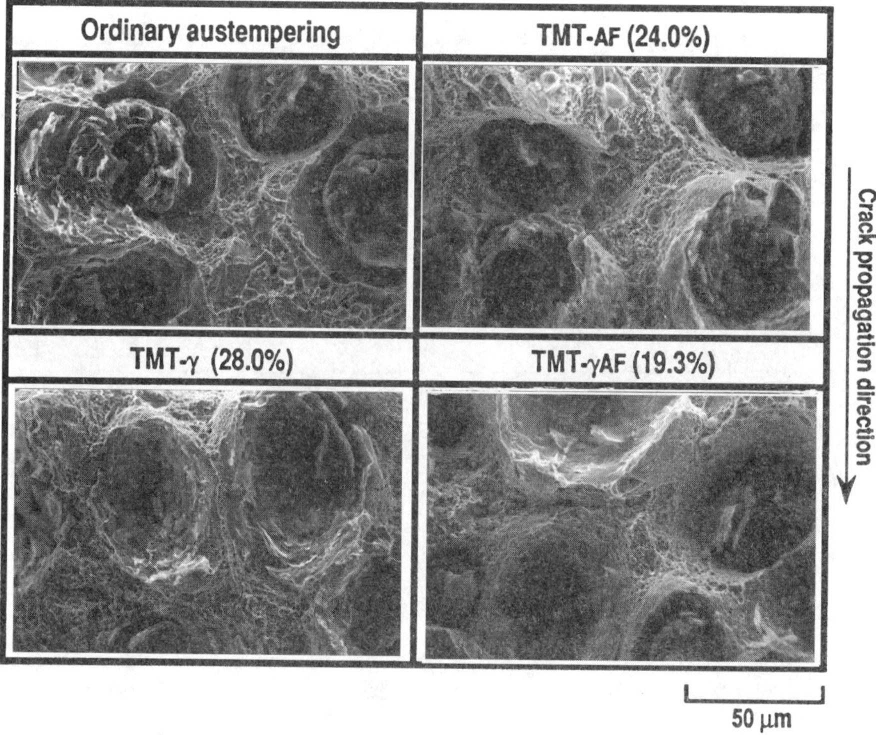

Fig. 5　Fracture surfaces at the center of specimen tested at room temperature in the material conducted with ordinary austempering and various TMT.

voids grow until they connect each other, the material will show high ductility. While, when the voids are connected by intervening quasi-cleavage or flat fracture, lower void growth, the material shows low ductility. In the material conducted with ordinary austempering, the quasi-cleavage fracture surface was observed between voids because of the existence of γ-pool around the eutectic cell boundary. Similarly, in the material conducted with TMT-AF, voids were connected by intervening flat fracture surface with micro-voids corresponding to the fineness of matrix. By contrast, in the material conducted with TMT-γ or TMT-γAF, the void growth was larger than that of ordinary ADI. Especially, little flat fracture area was observed in the material rolled at both temperatures with 19.3% of reduction ratio. Thus, the appearance of the void growth observed on the fracture surface agrees well with the change in the ductility as shown in Figure 4.

TMT widely varies the fineness of matrix microstructure crossly corresponding to mechanical properties and fracture process as mentioned above. Relationship between various properties and fineness of matrix microstructure is summarised in Figure 6. The increase in the strength due to the refinement of matrix and the increase in the ductility due to the elimination of γ-pool can be taken as positive factors for improving the toughness of SG iron conducted with TMT. While, the decrease in the ductility due to the flat fracture surface is regarded as a negative factor. The material with the maximum of toughness, therefore, seems to have a "peak-refined" matrix in a sense of toughness. Additionally, the reason of the maximum of the toughness in the material conducted with TMT-γAF shown in Figure 6 is considered as follows. Since the ausferritic transformation in SG iron tends to start preferentially near the graphite nodule, the rolling at the austempering temperature will give the mechanical deformation mainly to the unreacted austenite region around the eutectic cell boundary. Therefore, the material with "peak-refined" matrix around both graphite nodule and eutectic cell boundary shows the maximum toughness. This is supported by the observations of microstructure and fracture surface.

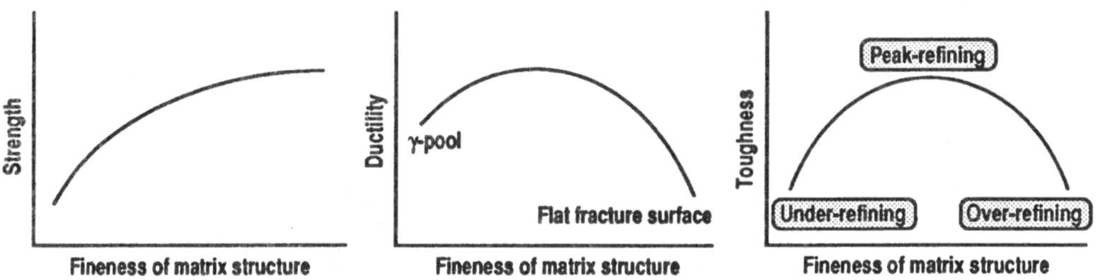

Fig. 6   Schematic illustration of relationship between various properties and fineness of matrix structure.

4. CONCLUSION

The effect of mechanical deformation during the austempering on the ausferritic transformation and toughness was examined in spheroidal graphite cast iron. The following results were obtained.

(1) The rolling at the austempering temperature remarkably refines the ausferritic microstructure and that accelerates the kinetics of ausferritic transformation. While, the rolling at the austenitising temperature eliminates the γ-pool and that slightly delays the kinetics of the transformation.

(2) The refinement of matrix microstructure increases the ductility due to the elimination of the γ-pool and the strength, but that leads to the flat fracture surface simultaneously. Therefore, optimum fineness for toughness exists.

(3) Maximum strength and hardness were obtained by 24.0% rolling at the austempering temperature. Maximum ductility and toughness were obtained by 19.3% (total reduction ratio) rolling at both the austenitising and austempering temperatures.

REFERENCES

1. D. J. Moore, T. N. Rouns and K. B. Rundman, *Journal of Heat Treating*, 4-1, 1 (1985) .
2. K. L .Hayrynen, C. T. Snow, D. J. Moore and H. B. Rundman, *Proceedings of 1991 World Conference on Austempered Ductile Iron*, 195 (1991).
3. I. Yamamoto and T. Kobayashi, *International Journal of Pressure Vessel and Piping*, 55, 295 (1993).
4. T. Kobayashi, I. Yamamoto and M. Niinomi, *Journal of Testing and Evaluation*, 21-3, 145 (1993).

Advanced Materials Research Vols. 4-5 (1997) pp. 203-212
© *1997 Scitec Publications, Switzerland*

The Influence of Austempering and Induction Hardening on Strength and Fatigue of Spheroidal Cast Iron

H. Vetters[1], H. Bomas[1], P. Mayr[1], Y.C. Liu[2] and J.M. Schissler[2]

[1] Stiftung Institut für Werkstofftechnik, Bremen, Germany

[2] Université Nancy I, Nancy, France

Keywords: Ausferritic SCI, Induction Hardening, Fatigue Endurance, Crack Propagation, Microstructure

Abstract

Low alloyed S.C.I. has been austempered into the upper bainitic state. Notched ADI samples have been treated initially by alternating cyclic load until crack propagation conditions have been obtained. Crack propagation under push pull load condition has been realized in a servohydraulic test equipment. Quantitative data of the crack propagation velocity have been determined by measuring the increase of the crack depth with the potential difference measurement method. The influence of structural changes on crack stop has been studied with metallographic and electron beam assisted analyses. Influences on crack propagation by the formation of strain induced martensite have been detected. Stress centres induced by the shape of the notches and the stress horizon due to the loading amplitude determine the local austenite martensite formation. In comparison to the untreated SCI the ADI treated samples showed, as statistically proved, higher resistance against crack propagation.

Introduction

As found in earlier investigations, the presence of unreacted retained austenite affects the mechanical properties of the casting /1-4/. The stability of the unreacted austenite depends on the homogeneity of carbon or other gamma-phase-extending alloying elements, such as nickel and molybdenum /4/. Tempering at elevated temperatures influences strongly the structural dependent properties because of carbide precipitation which induces a decrease in toughness and strength /3,4/. Short tempering times will result in an unstable retained austenite which transforms under the influence of externally applied loads into martensite /11,12/. One can distinguish between stress assisted and strain induced martensite formation /5,8/. Stress assisted formation includes the spontaneous transformation on cooling assisted by the thermodynamic effect of applied stress, while strain induced formation involves the production of new nucleation sites by plastic deformation /6,7/. Bending fatigue tests showed the increased twin formation within the retained austenite, and dislocation crowds in the surrounding ferrite /9/. Fatigue crack propagation is governed by the linkage of small cracks /10/. Within this research work the fatigue properties, fracture range and crack propagation of induction hardened S.C.I. in the ausferritic state have been investigated.

Experimental

The experiments on notched and smooth cylindrical samples have been conducted under push pull loading conditions and the influence of structural changes on crack stop has been studied. Special regard was taken on the role of retained austenite and the possibility of stress induced martensite formation.

One low alloyed type of S.C.I. (contents in mass-%: Mn 0,62; Ni 0,023; Cu 0,33) with a minimum tensile strength limit of R_m = 600 MPa in the ferritic-pearlitic "bull eye" state (SCI 600 P/F) has been treated to the ausferritic state with an ultimate tensile strength of 900 MPa (SCI 900 ADI). Both types have also been induction hardened by induction (fig.1).The case depth of 1,4 mm was determined by hardness profile testing (fig. 2), For fatigue tests cylindrical samples, smooth and radially notched, have been manufactured (fig. 3).

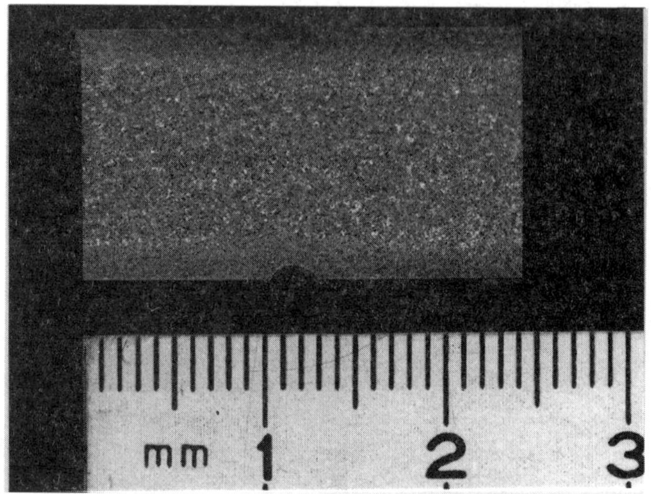

fig. 1   Induction hardened part of SCI 900 ADI (Light micrograph, etchant (nital))

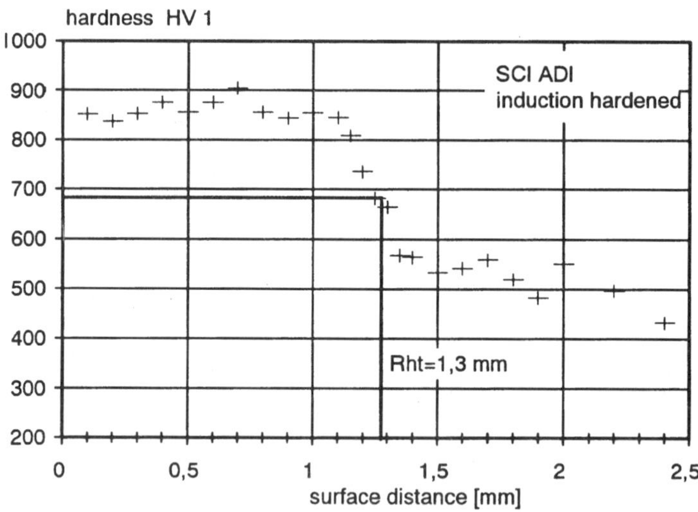

fig.2   Induction hardened SCI 900 ADI: hardness profile and case depth

Fatigue properties

The experiments have been conducted in a servohydraulic test equipment[3] with a 1 Hz sinusoidal load form for alternating stresses. The Woehler characteristics have been determined under stress control with zero mean stress (R = -1), the strain response has been measured by means of a capacitive gauge pin /11/. The number of cycles was limited with $N_G = 10^6$. The Woehler curves (fig.4) are calculated from 30 trials by the arcsin $(P)^{1/2}$ statistics /13/ with a mean fracture probability of 50%. (tab.1). The stress amplitudes (σ_a) have been determined at $N_f = 105$ cycles

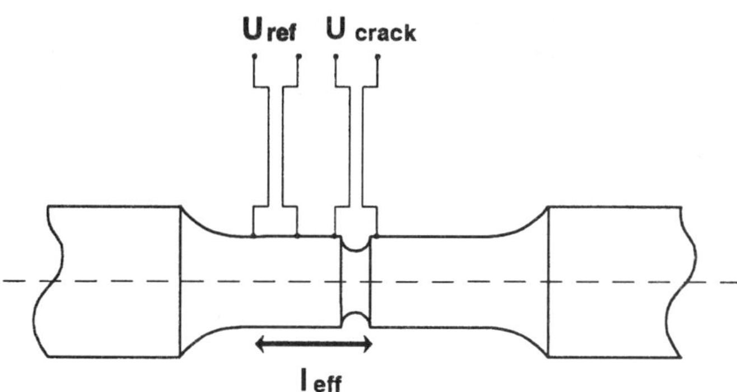

fig.3   Test of crack propagation

Type	SCI 600 P/F			SCI 900 ADI			
state	untreated		ind. hard.	ausferritic		ausferritic + ind. hard.	
geometry	smooth	notched	smooth	smooth	notched	smooth	notched
fatigue limit	278 ± 5	182 ± 8	228 ± 14	359 ± 45	260 ± 45	385 ± 15	316 ± 17
σ_a	335 ± 6	258 ± 6	400 ± 28	526 ± 15	357 ± 20	505 ± 25	358 ± 21

table 1: Push-pull fatigue test results (probability 90%, values in MPa)

Crack propagation

Crack propagation was determined by the load dependent crack area increment (fig 3). This was realized with an alternating current potential drop where the change of conductivity due to the crack propagation is detected /11/. The fatigue crack growth rate (da/dN) was determined by means of the secant method in accordance to the ASTM designation E 647-88 (tab. 2). The linear slope of the curve was determined by the Paris-Erdogan equation with:

$$da/dN = C.(\Delta K)^m \qquad (1)$$

(ΔK : stress intensity range, a: crack length; C, m: material dependent factors). The factors m and C have been determined by linear regression within the growth rate interval of 2.10^{-6} to 2.10^{-3} mm. The threshold value of the stress intensity range (ΔK_{th}) was determined at a of da/dN = 10^{-7} mm.

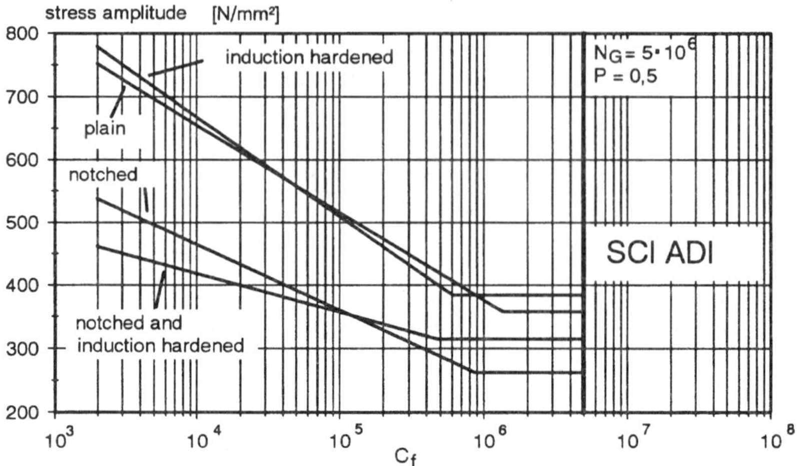

fig.4   Fatigue characteristics of notched and plain samples: ADI, and ADI induction hardened

SCI type	m	C	ΔK_{th} [N/mm$^{3/2}$]	ΔK_C [N/mm$^{3/2}$]
SCI 600 P/F	5,91	$2,4.10^{-22}$	300	1800
SCI 900 ADI	4,41	$2,0.10^{-18}$	500	2800

tab. 2:   cyclic stress intensity (ΔK_c) and threshold value of the stress intensity range (ΔK_{th}) of low alloyed SCI in untreated and ausferritic state

The comparison of the characteristics (fig. 5) indicates the retardation in crack propagation after austempering. The threshold value of ausferritic SCI exceeds the untreated cast with 60% and the crack velocity is reduced with a factor of 1,6. As reported /12/, ausferritic SCI (3,3 % Ni, 0,6% Mn) of extreme toughness extend to a similar threshold value of ΔK_{th}.

Fracture analysis

The fractured samples have been investigated at the marked sites (fig. 6). Crack initiation occurred in the notch root at a stress amplitude of 500 MPa and there local martensite transformation has been observed in some austenite grains (fig. 7). At higher magnifications observed by SEM (fig. 8a-c) the transformed acicular martensite is branching into various directions within the austenite grain and sometimes the needles are cross sectioning each other. By nital etching, the precipitated ferrite and the etch pits indicate the feathery structure. The surface of the fracture (fig.9) indicates in the center the formation of cleavage planes due to the stacking of austenite and ferrite platelets /4/, which indicates the ausferritic state of the alloy. TEM investigations indicated, that the ferrite was enriched with dislocations while the austenite showed the formation of slip planes and martensite formation (fig.10).

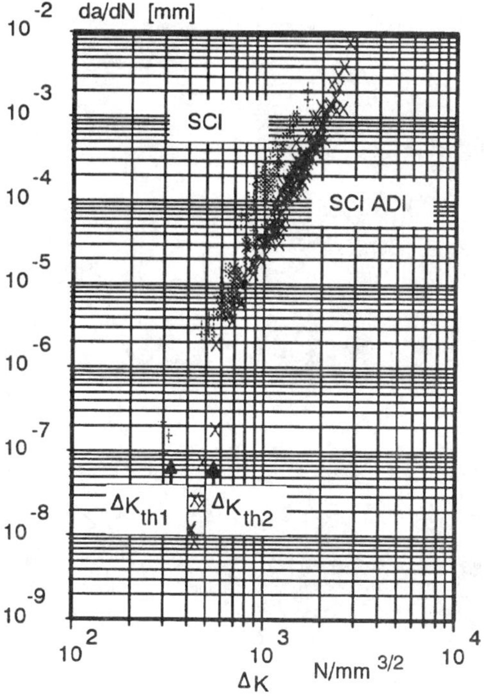

fig. 5   Crack propagation characteristics of: SCI 600 and ausferritic SCI ADI

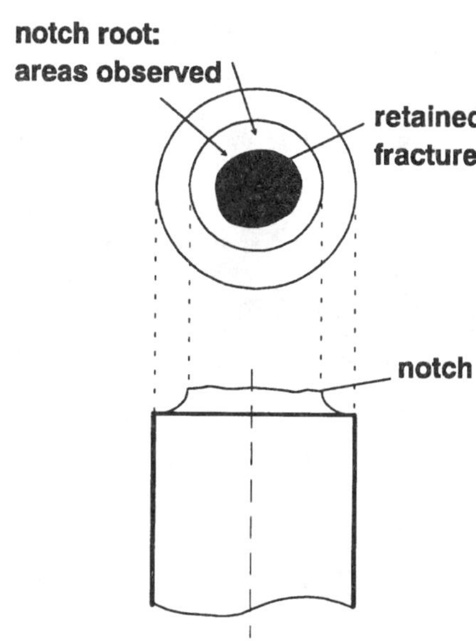

fig.6 Fractured sample
       (examined areas marked)

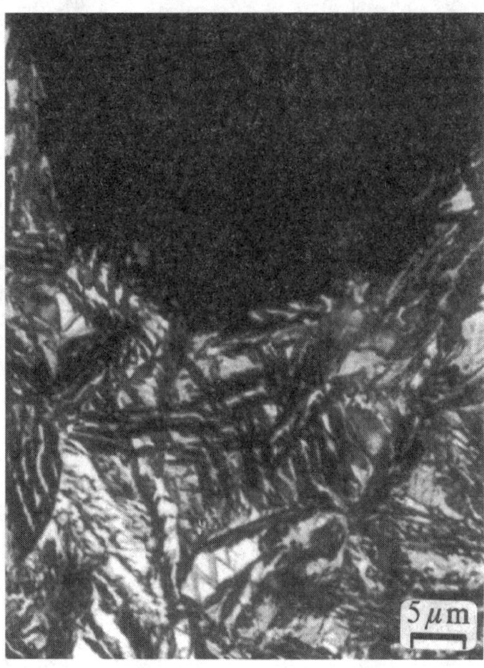

fig.7 Cross section at crack tip (nital etched)

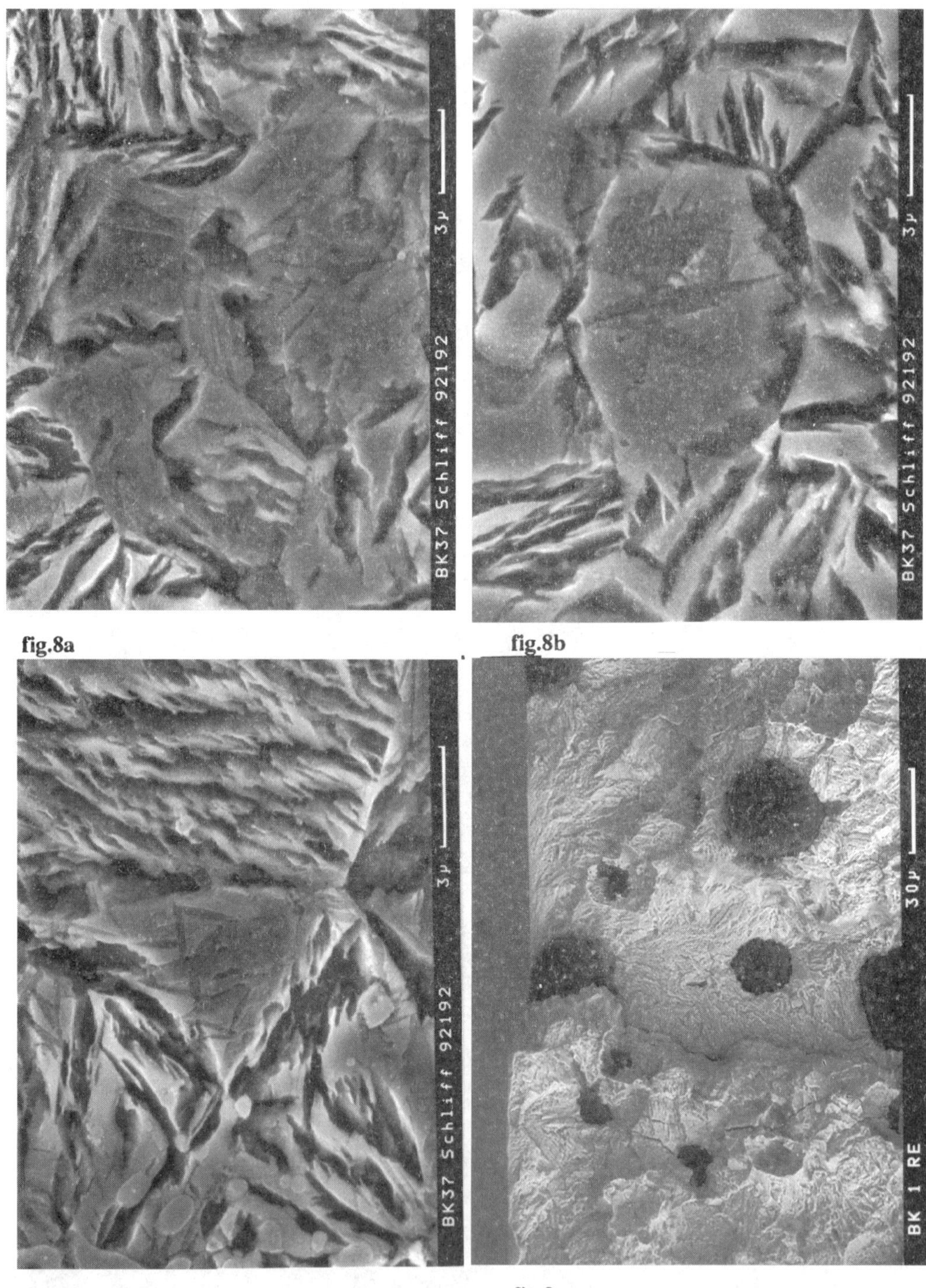

fig.8a   fig.8b

fig.8c   fig.9

Notch root: martensite formation (fig8: SEM:10% nital etched; fig.9: fractured area (SEM: BS))

Discussion

The Woehler results (fig.4) of the ausferritic SCI indicate that the fatigue limit and the fatigue strength at 10^5 cycles (tab.1), are in the same order of magnitude comparing each of the notched and smooth samples. Austempering increased the fatigue limit and the fracture strength at 10^5 cyclesof the smooth samples with 50%. Induction hardening of untreated SCI seems to reduce the fatigue strength at 10^5 cycles for about 30% but the increase of the scatter rate must be regarded. Comparing the results obtained with notched samples, an increase between 40% and 50% after austempering is observed.With push- pull loading trials,the notched samples of ausferritic and induction hardened ausferritic SCI show the same order of magnitude.

The increase of the threshold value of the stress intensity range (tab.2) in tough ausferritic SCI can be explained by the structural influence of the retained austenite on fatigue crack propagation. The retardation is obviously caused by the crack closure effect accompanied with athermic strain induced transformation of retained austenite into martensite /12/. With increasing content of retained austenite due to the alloy composition of stabilizing elements the retardation effect on crack initiation, which is expressed by the threshold value of the stress intensity range, increases. According to the model of Laird /14/, the crack propagates along slip planes. Martensite formation can occur, either by cross slip or by linear slip in defined crystallographic planes, if the induced energy exceeds the threshold value for γ - α transformation (fig. 11). This procedure causes a sharp strain and distortion of the host lattice occurs. As predicted in theory /18/ the plastic deformation at the crack tip is then accompanied by the formation of martensite plates in this volume in agreement with the experimental observation (fig. 8). Regarding the deformation process in the ferrite crystallites, the cell walls contain a large amount of screw dislocations (fig. 10). Within the retained austenite the deformation induces dislocation movement in slip bands. Stacking faults occur in the cross sections of the subsequently formed slip bands which increases the energy content / 15/.

Crack propagation in retained austenite at stage II (Paris- Erdogan- slope) occurs in only two slip planes inclined to each other with an angle of 45° /19/. Then the closing stage activates one of the planes, restoring a very sharp crack tip and a high local stress concentration. In the case of rigid plasticity, dislocation movement by shear stress starts only along the two slip planes /19,20/. Martensite formation occurs then by alternate shear first in the one slip band, then in the other. The typically oriented martensite formation in retained austenite results therefore from the array of the 45° slip planes and the alternate activation of the slip systems. (fig. 12). Thereby the distortion energy exceeds the threshold for γ - α transformation, twist and tilt motion of the lattice occurs in the affected region, and a zig- zag patchwork of martensite platelets is observed if the stress nuclei are concentrated at the slip plane intersections with the grain boundary of the austenite and adjacent ferrite plates (fig.7).

As proved /4,7,11,12/ only a small amount of retained austenite reacts by martensite nucleation. The reduction of the crack propagation rate in ausferritic SCI depends therefore mainly upon the internal structure due to the amounts of ductile components /3,4,16/. In the case of homogeneous bainitisation /1,16/, the increment of the crack propagation rate is kept more constant then in the untempered state. This is indicated by reduced error limits, while the untreated state of SCI (F/P) shows extended error bars and incremental alterations in the slope (fig.5) due to the inhomogeneous ferritic pearlitic structure.

fig.10   TEM micrograph at crack tip
retained austenite (bottom); martensite
(top)

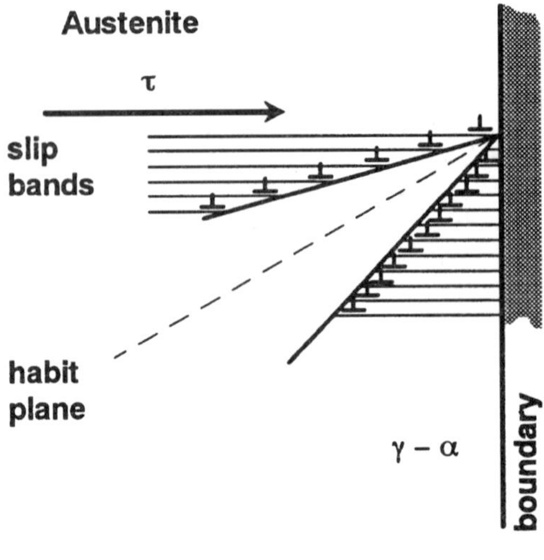

fig.11   Formation of strain induced martensite

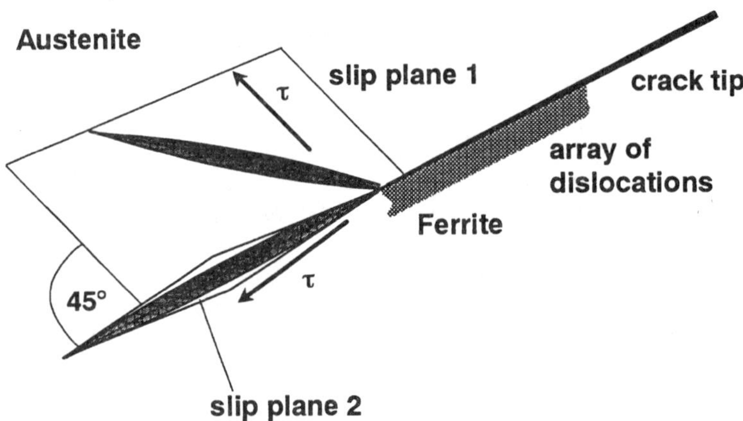

fig.12   Intersection of slip planes, formation of various oriented martensite lenses

Acknowledgement:

The investigations of authors[1] on fatigue and crack propagation studies of ADI have been proposed by the Association of German Foundrymen (Verein Deutscher Gießereifachleute VDG) and financed by funds of the German Association for Industrial Research (AIF) with funds of the Federal Ministry of Commerce wherefore the authors wish to express their gratitude.

Literature

1. J. M. Schissler, J. P. Chobaut, C. Bak, D. Gouvenel: Heterogeneous bainitic transformation observed in unalloyed or alloyed SG cast irons. International Foundry and Heat Treatment Conference, Johannesburg,, 26 (1988)

2. J.M. Schissler, Y.C. Liu, J.P. Chobaut: Etude du comportement de la fonte G.S. bainitique de type ADI au cours du revenu a 360°C 9th internat. Conf. on heat treatment Nice (1994)

3. J.M. Schissler, Y.C. Liu, J.P. Chobaut, P. Brenot: Upper Bainitic Heat Treatment of Manganese, Nickel and Copper Alloyed S.C I., Proc. of Int. World Conf. on ADI (Chicago), 424 (1991)

4. Y.C. Liu: Etude de fontes G.S. bainitiques alliees de type "ADI" . Stabilite de la structure "Ausferritique" soit en cours de revenu, soit en condition abrasive. Doctoral Thesis Nancy Universite I (1994)

5. I. Tamura: Deformation induced martensitic transformation and transformation induced plasticity in steels, Metal Science, 16 245 (1982)

6. H. Vetters, R. Böschen, H. Bomas, P. Mayr: Martensite Formation in Austempered Ductile Iron with Unidirectional and Cyclic Loading. Proceeding of the Conf.."The Martensitic Transformation in Science and Technology" (Ed. E. Hornbogen) Bochum, DGM, (1989)

7. P. Mayr, H. Vetters, J. Walla: Investigations on the Stress Induced Martensite Formation Proceedings on the 2nd International Conference on ADI, Ann Arbor, 171 (1986)

8. E. Hornbogen: Martensitic Transformation at a Propagating Crack. Acta Metallurgica 26, 147 (1978)

9. R. Boeschen, H. Bomas, P. Mayr, H. Vetters: Strength and fatigue of austempered ductile iron (ADI). Proceedings on the World Conf. of Austempered Ductile Iron , Chicago, IL. AFS spec. publ., 468 (1991)

10. R. C. Voigt, L. M. Eldoky, H.S. Chou: Fracture of ductile cast irons with dual matrix structures, AFS transaction 100, 645 (1986)

11. R. Boeschen: Schwingfestigkeitsverhalten von randschichtgehärtetem Gußeisen mit Kugelgraphit, Doctoral thesis, Univ. Bremen (to be published as VDI-monograph) (1993)

12. T. Kobayashi, H. Yamamoto, S. Yamada: On the toughness and fatigue properties of austempered ductile iron,.Proceedings on the World Conf. of Austempered Ductile Iron, Bloomingdale Chicago, IL. AFS spec. publ., 567 (1991)

13. D. Dengel: Empfehlungen für die statistische Abschätzung des Zeit- und Dauerfestigkeitsverhaltens von Stahl, Mat. Wiss. und Werkstofftechnik 20 75 (1989)

14. C. Laird, In: Fatigue Crack Propagation. Philadelphia/Pa. (ASTM spec. tech. publ. No 415 131 (1967)

15. K.P. Staudhammer, L.E. Murr, S.S. Hecker: Nucleation and evolution of strain induced martensite embryos and substructure in stainless steel, a transmission electron microscope study, Acta Metall. 31 267 (1983)

16. J.M.Schissler, J.P. Chobaut: Stabilität des Gefüges von bainitischem Kugelgraphitguß bei Raumtemperatur. Härterei Tech. Mitt., 42 205 (1987)

17. O. Voehringer, E. Macherauch: Struktur und mechanische Eigenschaften von Martensit, Härterei-Tech. Mitt. 32 153 (1977)

18. U. Dehlinger: Präformierte Keime bei Umwandlungen und Ausscheidungen, Zeitschr. f. Metallkunde 51, 353 (1960)

19. R. Pelloux: in D. Hodgson and A.S. Tetelman: Fracture (ed. P.L. Pratt London), 731 (1969)

20. P. M. Kelly, The Cristallography of Lath Martensite in Steels, In: C. M. Wayman and J. Perkins: Proc. of the int. Conf. on Martensite (ICOMAT-91), 299 (1992)

Advanced Materials Research Vols. 4-5 (1997) pp. 213-218
© *1997 Scitec Publications, Switzerland*

Notch Strength of Cast Iron under Static and Fatigue Loading

T. Noguchi[1], K. Shimizu[2] and M. Fujita[1]

[1] Dept. of Mech. Eng., Faculty of Eng., Hokkaido University,
North 13, West 8, Kita-ku, Sapporo 060, Japan

[2] Dept. of Information and Control Eng., Oita National College of Technology,
1666 Maki, Oita 870-01, Japan

Keywords: Notch Effect, Flake Graphite Iron, Spheroidal Graphite Iron, Static Strength, Fatigue Strength, Stress Distribution, Finite Element Method, Elasto-Plastic Theory, Fracture Criteria

Abstract

To examine the effect of notches on the strength of cast irons, static and fatigue notch tests were perfoemed on several flake graphite and ductile irons with various notch configurations. The results were analyzed based on the stress distributions at the notch section considering the non-elasticity of the stress-strain relation, and a fracture criteria with an overstressed depth, δ, was derived. A variety of notch effects, very limited effect in flake graphite iron bars, clear strength decrease in plate specimens although the amount is smaller than presumed from the form factors, marked decrease in low temperature strength and the high cycle fatigue, are described well with δ of 0.1 - 3 mm depending on material and the fracture mode. Strength increase in ductile irons is due to tri-axial stress state induced by the notches.

1. Introduction

In strength design of structural components, notch sensitivity is an important material characteristic. Cast iron is known for its low notch sensitivity. Occasionally, the fracture strength of notched specimens of cast iron is little different from unnotched specimens with the same cross section[1-3]. This has been attributed to an microscopic or internal notch effect of graphite flakes, which overshadows the effect of the macroscopic machined notches[3]. The authors have shown that the notch effect of cast iron depends strongly on material, notch configuration, and fracture mode[4-6].
This article summarizes the results of notch tests of various cast irons both in static and fatigue loading, and discusses the reasons for the notch characteristics of cast irons based on the stress distribution at the notch section considering non-elasticity in the stress-strain relations of this material.

2. Notch strength of various cast irons in static loading

To examine the notch effects under static loading, tension tests with smooth and notched specimens with various configurations were performed on two spheroidal and three flake graphite cast irons. FCD450 and 600 are spheroidal graphite irons with tensile strength level of 450 and 600 MPa, and FC250A, B and C are flake graphite irons all with strength level of 250 MPa. The test pieces were circumferentially notched bars and edge notched or holed plates as shown in Fig.1. The notch radius were from 0.1 to 30 mm, and the form factors were from 1.1 to 10. The notch strength, σ_{nf}, is defined by the maximun load P_m divided by the smallest cross sectional area. The effect of the notch is expressed by the Notch Strength Ratio, NSR = σ_{nf}/σ_f, where σ_f is the strength of the smooth specimen.

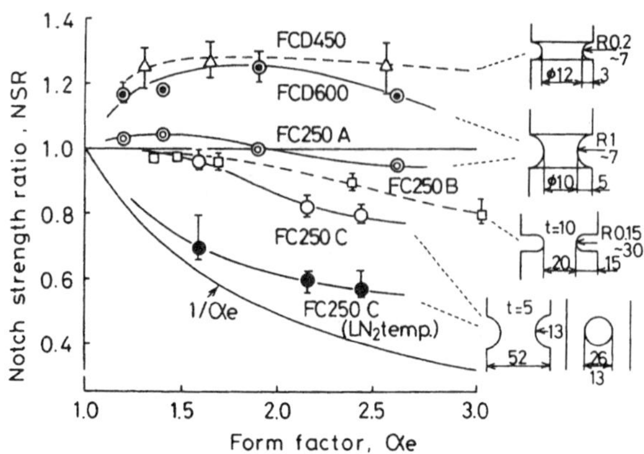

Fig.1 Effect of notches on the static strength
of various cast iron specimens.

Fig.1 shows the experimental results.   The abscissa is the form factor,
the elastic stress concentration factor α_e, of the notch.   The strength change
due to notches in FC250A is only within about 5 %, indicating that the notch
has little effect on strength.   In FC250B, however, strength clearly decreases
with increasing α_e, to about NSR = 0.85 at R = 3 mm, and at α_e = 5, the NSR was
about 0.75 [7].

At liquid nitrogen temperature, 77 K[4], the NSR of FC250C decreased to 0.6
with α_e = 2 - 2.5 notches, while it is about 0.8 at room temperature.   The
notch effect at liquid nitrogen temperature is close to $1/\alpha_e$, the maximum
strength decrease in elastic theory.   Similar results were obtained in
circumferentially notched bars[8].   High notch sensitivity at the low
temperature can not be explained by microscopic notch effects of the graphite
flakes, because the notch effects of the graphite would be more pronounced at
the low temperature where the matrix becomes very brittle and notch
sensitive[9].

In the spheroidal graphite(ductile) irons, FCD600 and FCD450, NSR is larger
than 1 for all notches, indicating strength increases due to notches.   These
results can not be explained by micro notch effects either.

3. Discussion of the stress distribution at notch sections

3.1 Stress analysis

To discuss the experimental results, the authors noticed macroscpic stress
distribution at the notch section rather than the microscopic ones around the
garaphite flakes.   Fig.2 shows examples of stress strain(σ-ε) curves of cast
irons.   Both flake graphite iron and ductile iron exhibit non-elasiticity,
although the fracture strain of the flake graphite iron is less than 1 %[4,7,9]
at room temperature.   To consider the non-elasticity, the finite element
method(FEM) together with elasto-plastic theory[10] was applied.   The stress
strain curve was approximated by $\sigma = C(A + \varepsilon_p)^n$ (ε_p: plastic strain = ε -
σ/E, E: elastic modulus), and the factors C, A and n were determined
experimentally.   In the plastic state, strain incremental theory and Mises's
equivalent stress σ-ε_p relation were assumed to hold[10].

3.2 Stress distribution at the notch section

Fig.3 shows the calculated distribution of axial stress, σ_z, at the notch

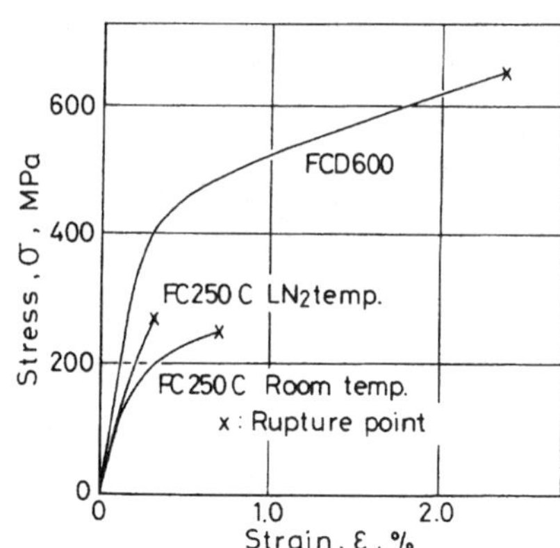

Fig.2 Examples of stress-strain
       curves of cast irons.

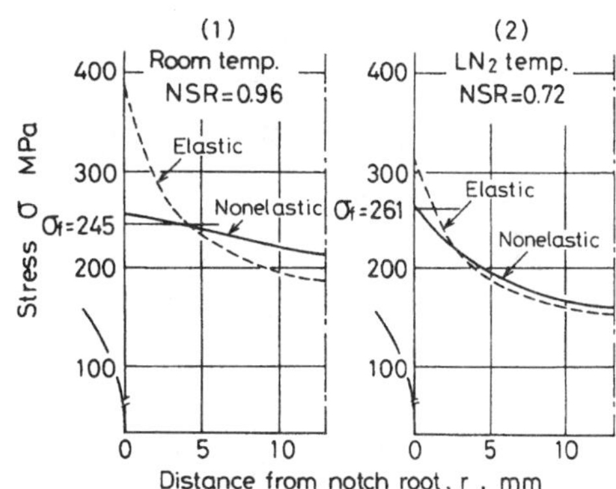

Fig.3 Stress distribution at fracture
       at the notch section of semi-
       circular edge notches.

section of semi-circular edge notches in Fig.1.   Fig.3(1) is for fracture
stress at room temperature, NSR = 0.96.   The solid line is the non-elastic
distribution and is much flatter than the elastic one, shown by the broken
line.   The maximum stress at the notch root is 255 MPa, only 0.65 of the
elastic estimation, 390 MPa.   The stress concentration factor is decreased to
1.1 from α_e = 1.7.   This is the main reason for the low notch sensitivity of
cast iron.

The stress distribution in Fig.3(1) also shows that the maximum stress at
the notch root is higher than the tensile strength, 245 MPa.   This indicates
that the second reason for low notch sensitivity is the existence of a region
where the working stress is above the smooth specimen strength[9].   This
overstressed depth, δ, is about 3 mm.

At liquid nitrogen temperature, in Fig.3(2), the effect of non-elasticity
is smaller than at room temperature, but the maximum stress is lower than the
elastic value and is close to the tensile strength, 261 MPa.   The overstressed
depth, δ, is 0.2 mm at this temperature.   Similar results were obtained in
haled plates, where δ values were 3 - 3.5 mm at room temperature and about 0.5
mm at liquid nitrogen temperature.   Circumferentially notched bars of FC250A
showed similar values[8].

Fig.4 shows δ values obtained from
notch tests with various materials and
notch configurations at room and liquid
nitrogen temperatures, together with
the results of beam bending with various
depth[11].   At room temperature, δ
varies from 1 to 5 mm depending on the
notch radius, ρ, and beam depth, h,
showing small values at small ρ and h.

At liquid nitrogen temperature, δ is
almost constant and less than 1 mm
independent of root radius.

The concept of a minimum region for
fracture has been proposed by Neuber[12]
,Sibel[13] and others, and is related to
crystal grains or other microscopic
structures.   The minimum value of δ in
cast iron for small ρ or h, and at
liquid nitrogen temperature, coincides
with the graphite eutectic cell size,

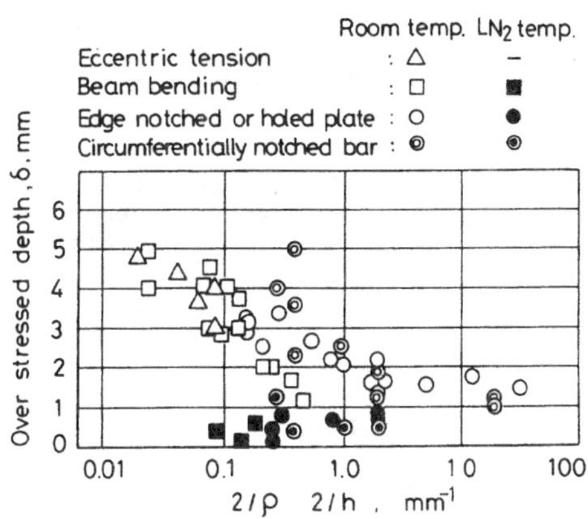

Fig.4 Over stressed depth δ for
       various notches and beames.

0.4 - 1 mm[4].   The δ value at room temperature and larger P is 3-5 times the cell size.   The smallest δ is interpreted as an initiation condition for cracks, and the larger δ the propagation and the final fracture[8,9,11].

3.3 Stress distributions in notched bars of ductile iron

Fig.5 shows the stress distributions with circum-ferential notches in ductile iron, FCD600.   In the R = 5 mm specimen, the axial stress σ_z exceeds the tensile strength throughout the section, and reaches the maximum value at the center. The distribution of tangential stress σ_θ and radial stress σ_r are also convex.   These results show clearly that the fracture strength increased by the tri-axial stress state induced by the notch, this is known as the plastic constraint effect in ductile materials[14].   The σ_z is higher than σ_f in R = 2 and 1 mm specimens, also due to the tri-axial stress states, where the maximum values of σ_r and σ_θ reach 1/3 - 1/2 of σ_z.

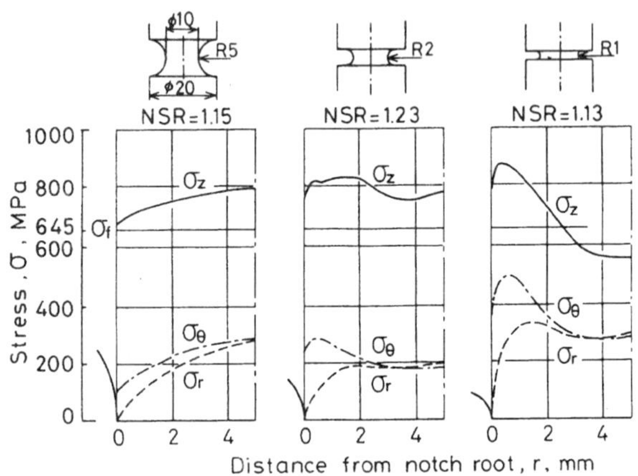

Fig.5   Stress distribution at fracture in circumferentially notched ductile iron.

4. Notch effect in fatigue

To examine the notche effects in fatigue, cyclic tension fatigue tests were performed on five cast irons: a pearlitic(FCP) and a ferritic(FCF) flake graphite irons, a bull's eye(FCDP) and a ferritic(FCDF) ductile irons, and an austempered ductile iron, ADI.   The specimens were bars with 8 mm in smallest cross section, the notches were circumferential with root radius 2(R2) or 0.7 mm(R07), both 2 mm deep.   The form factors are 1.6 and 2.8[5].
The top graphs in Fig.6 are the stress(S) vs. number of cycles to failure(Nf) curves for FCP, FCDP and ADI.   With these curves, the notch effect at the Nf are defined as:
Notch factor, Kf = (Fatigue strength of smooth specimen)/(Fatigue strength of notched specimen), the reciprocal of NSR.
The bottom graphs in Fig.6 are the Kf vs. Nf relations.   In flake graphite FCP, Kf increases with Nf from 1.0 in static fracture to around 1.2 at Nf>10⁶ region, indicating that notch sensitivity increases with fatigue life, although the maximum value is still much lower than the form factor, α_e.   In ductile iron, FCDP, the increase of Kf with Nf is more pronounced.   The Kf at the fatigue limit region is 1.3(R2) and 1.6(R02).   In ADI, the Kf values at Nf=10⁶ region are 1.6 and 2.1, close to the form factors, 1.6 and 2.8, indicating that ADI is very notch sensitive in high cycle fatigue[6,15].
Fig.7 shows the stress distribution at the notch sections for different Nf.
With the corresponding fatigue strength of the smooth specimens, expressed by the horizontal lines, the overstressed depths are determined as in static fracture.   The value of δ is 0.5 - 0.8 mm in FCP, 0.3 - 0.7 mm in FCDP, and 0.13 - 0.8 mm in ADI, all showing the smallest values around the fatigue limit.
Similar experiments on a ferritic flake graphite iron, FCF, gave 0.7 - 2 mm, and in a ferritic ductile iron, FCDF, 0.4 - 2 mm, as summarized in Table 1.

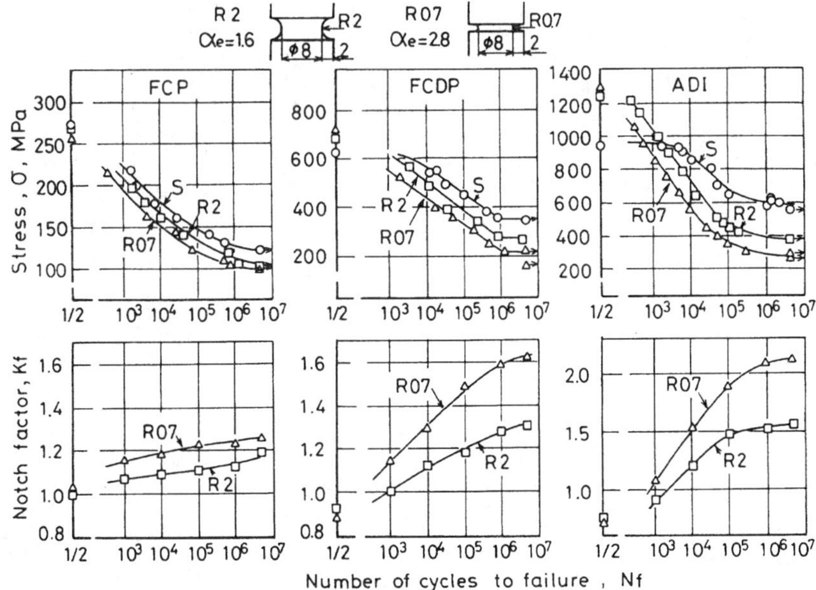

Fig.6 S-Nf curves and notch factors Kf in various cast iron specimens.

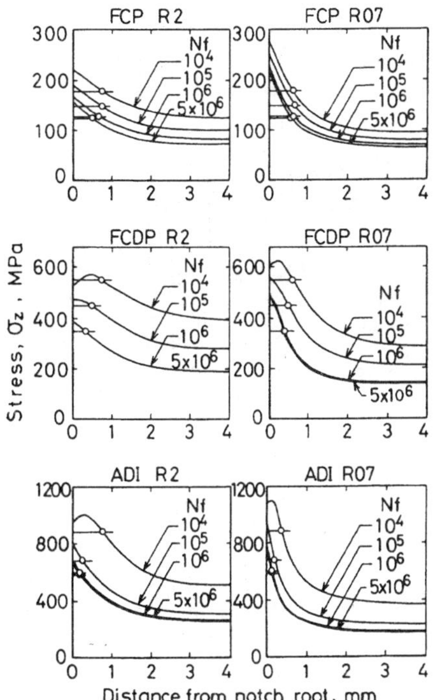

Fig.7 Calculated stress distributions at the notch section.

Table 1 Estimated overstressed depth δ in fatigue(mm)

Specimen		Nf=10^4	10^5	10^6	5×10^6
FCP	R2	0.78	0.72	0.67	0.54
	R07	0.59	0.63	0.62	0.57
FCF	R2	1.4	1.1	0.75	0.73
	R07	1.1	0.83	0.79	0.86
FCDP	R2	0.74	0.52	0.36	0.36
	R07	0.59	0.48	0.40	0.37
FCDF	R2	2.30	1.26	0.96	0.96
	R07	1.00	0.60	0.43	0.43
ADI	R2	0.78	0.25	0.20	0.18
	R07	0.38	0.19	0.14	0.13

As in static fracture, the minimum δ is interpreted as a necessary region for fatigue to initiate, and the value in flake graphite irons coincides with the eutectic cell size.   In ductile irons, the smallest δ in ADI, 0.1-0.2 mm is on the order of the graphite nodule size and the mean nodule distance.   The larger δ values are considered a condition for initiated cracks to grow to macroscopic cracks[16.17].   The δ values of FCDP in $Nf>10^6$ is about twice ADI, and larger with FCDF, while the nodule distributions are similar.   This indicates that δ depends on matrix ductility, showing smaller values in the brittle matrix.   Among flake graphite irons, δ in pearlitic FCP is smaller than ferritic FCF, too.

5. Conclusions

The effects of notches on the strength of cast iron varies widely depending on material, notch configuration and type of fracture.   In static loading and with low cycle fatigue, there is little effect of notches in round bars.   In plate specimens, the strength clearly decreases but the amount is much smaller than that suggested by form factors.   Low temperature strength of flake graphite irons and the fatigue limit of ductile irons decrease markedly with notches.   Circumferentially notched ductile irons show strength increas. This variety of notch characteristics can be described by two factors.   One is the non-elastic stress-strain relation, that reduces stress concentration and causes a tri-axial stress state.   The other is a fracture criteria with an overstressed region, δ, at the notch root.   The variety of δ values were related to the initiation and propagation conditions of the fracture in each type, small in brittle matrix and fracture mode, and large in ductile ones. With this concept, the strength characteristics such as notch strength and the bending of beams and plates of various cast irons are considered unified as a fracture problem of non-elastic materials under a stress gradient.

Referances

[1] AFS, Cast Metals Handbook, Des Plaines, IL.p92(1957)
[2] Ishibashi,T., Trans.JSME, 18,68(1952),87
[3] Nakanishi, F., Okamoto, S., Trans.JSME,17,61(1951),103
[4] Noguchi,T., J.Matrelials Science,Japan,28(1979),306
[5] Noguchi,T., Trans.AFS,97(1989),389
[6] Noguchi,T., Proc.Int.Conf. on Mechanical Behavior of Ductile
    Cast Irons & Other Cast Metals, Kitakyushu, Japan(1993),139
[7] Noguchi,T., J.Matrelials Science,Japan,22(1973),366
[8] Noguchi,T., J.Matrelials Science,Japan,29(1980),387
[9] Noguchi,T., Cast Metals,6(1993),146
[10] Yamada,Y.;Computer Structural Analysis Series II 2A,Baifukan,
    Tokyo,(1987),173
[11] Noguchi,T., J.Testing & Evaluation,18[1](1990),70
[12] Neuber,H., Kerpspannungslehle,Berlin(1937),144
[13] Siebel,E.,VDI-Z,97(1955),121
[14] Hill,R.,The Mathematical Plasticity,Oxford at the Clarendon
    Press(1964),248
[15] Noguchi,T., Trans.JSME,60-A(1994),1524
[16] Nishitani,N., Trans.JSME,51-A(1985),1442
[17] Takao,K., Nishitani,N., J.Soc.Materials Science,Japan,36(1987),1060

Advanced Materials Research Vols. 4-5 (1997) pp. 219-226
© 1997 Scitec Publications, Switzerland

Estimation of the Tolerable Defects in Brake Disk Made from Spheroidal Graphite Cast Iron

M. Biel Gołaska, L. Gołaski and A. Kowalski

Foundry Research Institute, 73 Zakopiańska Street, PL-30-418 Cracow, Poland

Keywords: Spheroidal Graphite Cast Iron, Brake Disk, Fracture Toughness, Impact Strength, Low Temperatures, Tolerable Defect, Fracture Models

ABSTRACT

The spheroidal graphite cast iron assigned for the brake discs and designed for service at low temperatures has been tested. To estimate the tolerable crack length, the methods of fracture mechanics including FAD have been applied. The relationship between the results of impact strength and fracture toughness to estimate the tolerable defect size in castings has been determined. Employing two fracture models an attempt has been made to explain the influence of graphite precipitations on fracture mechanism of spheroidal graphite cast iron.

INTRODUCTION

In Foundry Research Institute in Cracow a technology of the manufacture of cast brake disks for rail vehicles has been elaborated along with the testing procedure and requirements of technical acceptance. Castings were made in spheroidal graphite cast iron, grade 70002. This kind of material is characterised by high strength and low plasticity, and because of this is defects sensitive. The rail vehicles were to be used in different climate zones and low temperatures up to -60°C. The risk of fracture of metallic materials usually increases when temperature decreases. Therefore it was advisable to estimate the tendency to failure at low temperatures, tolerable defect size and possibility of damage of disks under consideration. These targets were realized utilizing fracture mechanics methods. To estimate the risk of failure of disks when temperature is changing, Failure Assessment Diagram has been applied.

To make possible the estimation of tolerable crack size in industrial laboratories equipped with impact pendulum only, an empirical relationship between impact strength and fracture toughness has been developed. Beyond utilitarian there are also cognitive aspects of this paper where an attempt is made to explain the role of graphite precipitation in fracture mechanism of spheroidal graphite cast iron. For this reason two fracture models developed by Osborne-Embury [1] and Czoboly-Havas [2] were applied. An agreement process zone size and mean free path between graphite inclusions was obtained. This shows that the graphite inclusion number is essential for failure processes.

MATERIAL, EXPERIMENTAL PROCEDURE AND RESULTS

One of the aims of this work was estimation of the tolerable defects size and assessing the failure probability of brake disks of railway cars containing crack and working at low temperatures. The second aim was to establish a relationship between the results of fracture toughness and impact strength in temperature range from +20 to -60°C. The third aim was an explanation of the role of graphite precipitates in the fracture process of ductile iron on the basis of two mathematical fracture models.

The spheroidal graphite cast iron was applied for brake disks for rail vehicles. The shape and dimensions of brake disk are shown in fig.1. The rail vehicles were to be used in different temperatures, so it was necessary to test the yield strength, the tensile strength, the impact strength and fracture toughness of ductile iron in the range of performance temperatures of rail vehicles, i.e. between +20 and -60°C. The changes in the impact strength and fracture toughness of ductile iron in function of temperature are presented in sequence in fig. 2. Along with the drop of temperature from +20 to -60°C , the impact strength decreased by 36% and fracture toughness decreased by 46%. The results of the fracture toughness test at various temperatures were used for computation of the length of tolerable defects in brake disk. The dimensions of the tolerable defects in a brake disk for performance at different temperatures are shown on the bar chart in fig.3. In the tested brake disk the length of admissible surface defect is 3 mm at temperature 20°C and 0,98 mm at temperature -60°C. If the defect is smaller than the computed tolerable defect, it is advisable to assess the reliability of the examined brake disk under the changing conditions of load, temperature and crack length. This evaluation is done using Failure Assessment Diagram (fig 4). In the case of brake disk under the assumed load of 500MPa and crack length a=2mm, the critical temperature of operating is -60°C.

To estimate the role that graphite precipitates plays in the fracture mechanism of the examined ductile iron, the metallographic analysis were made. The graphite precipitates are shown in fig 5.

The stereological examinations for the size and distribution of graphite precipitates were performed. It was stated that the mean free path between the graphite precipitates is equal to 0,34mm.

The analysis of the role of graphite precipitates in fracture toughness can be done utilising fracture models in the tested material. Two models proposed by Osborne-Embury [1] and by Czoboly-Havas [2] were taken into account. In these models the dependence of strength and microstructure on fracture toughness is given. The applied models are described with the help of equations (1) and (2).

Model Osborne-Embury

$$K_{Ic} = (\sigma_c * E * \varepsilon_c * s * \ln\frac{\varepsilon_c}{\varepsilon_o})^{1/2} \tag{1}$$

where: K_{Ic} - fracture toughness

σ_c - true fracture stress

E - Young's modulus

ε_c - elongation at fracture of smooth sample

s - process zone size

ε_o - elongation corresponding to yield stress

Model Czoboly-Havas

$$J_{Ic} = W_c * L_o \qquad (2) \quad \text{in which} \quad W_c = (\sigma_c + \sigma_f) * \ln\frac{d_o}{d_f} \qquad (2a)$$

where: J_{Ic} - critical value of J-integral

    W_c - absorbed energy per volume unit

    L_o -length of hypothetical tensile specimen situated in

        the vicinity of the crack tip and equal to process zone size

    σ_o - tensile yield stress

    σ_f - ultimate tensile strength

    d_o - initial diameter of the specimen

    d_f - diameter of the specimen at fracture

The calculated process zone size at different temperatures is shown in fig.6. As can be seen in bar chart the values of the process zone size calculated in agreement with both models are similar. The process zone size was compared with the mean free path between graphite precipitates. It was found that both values are similar in the range of temperature from -20 to -60°C. At temperature of +20°C the process zone size is about twice the values of the mean free path. Thus, in agreement with the RKR critical cleavage stress model , the fracture initiation will take place if the stress reaches the critical value in the distance equal to the mean free path between graphite precipitates.

The technical acceptance requirements for castings include the estimation of fracture toughness basing on the results of an impact strength. A relationship between the fracture toughness K_{Ic} and impact strength KCU has been derived in the form of the following equation:

$$K_{Ic} = 40,5 * KCU^{1/2} - 27,5 \qquad (3)$$

where: K_{Ic} - fracture toughness expressed in [MPa*m$^{1/2}$]

      KCU- impact strength expressed in [J/cm^2]

The relationship (3) gives the possibility of calculating the fracture toughness K_{Ic} on the basis of the results of impact strength KCU, and next of estimating the length of tolerable defects in brake disks.

CONCLUSIONS

1. Application of fracture mechanics methods allowed a quantitative estimation of the tolerable defects in brake discs made from ductile iron.

2. The analysis of reliability carried out according to FAD has proved that brake disks can be operating at low temperature of up to - 60°C.

3. The relationships between fracture toughness and impact strength can be used for estimation of the length of tolerable defects using the results of impact strength.

4. The role of graphite precipitates in the fracture process of ductile iron was explained on the basis of fracture models.

REFERENCES

1. D.E.Osborne and J.D.Embury, Metall. Trans. 4 , 2051 (1973).

2. E.Czoboly and J.Havas, Proc.Symp. on Absorbed Specific Energy and Strain Energy Density, Budapest, 107 (1981).

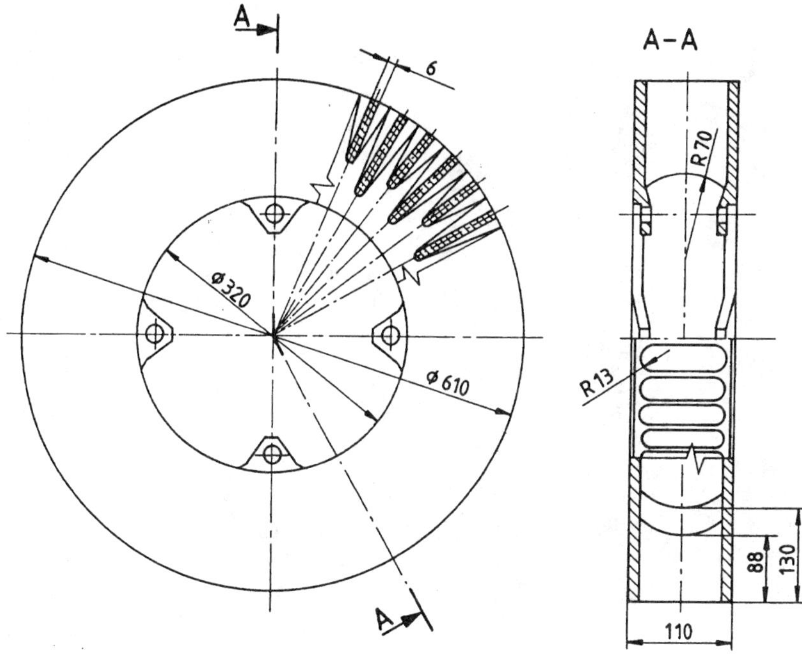

Figure 1 : The brake disk for rail vehicles.

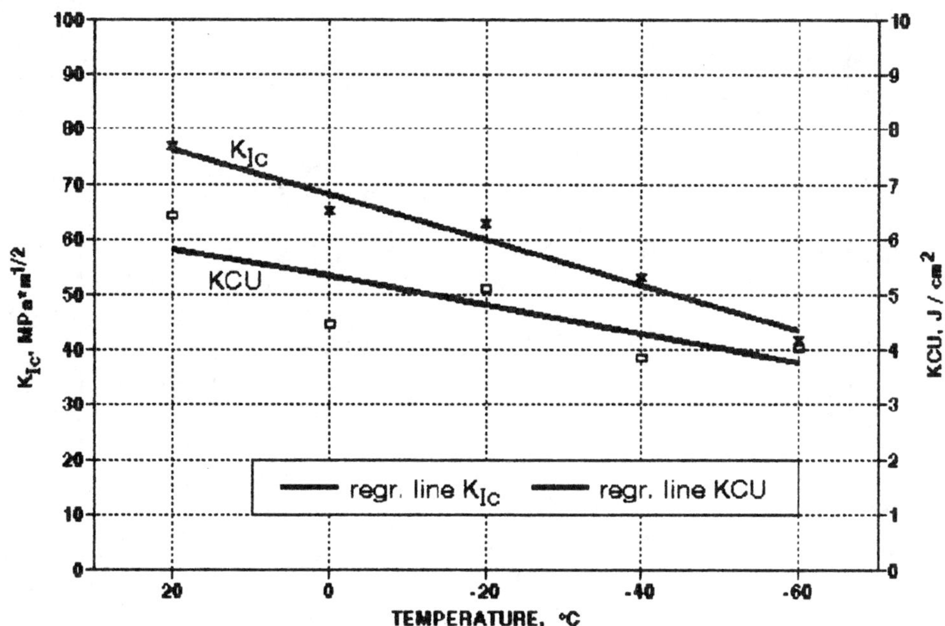

Figure 2 : Fracture toughness and impact strength versus temperature.

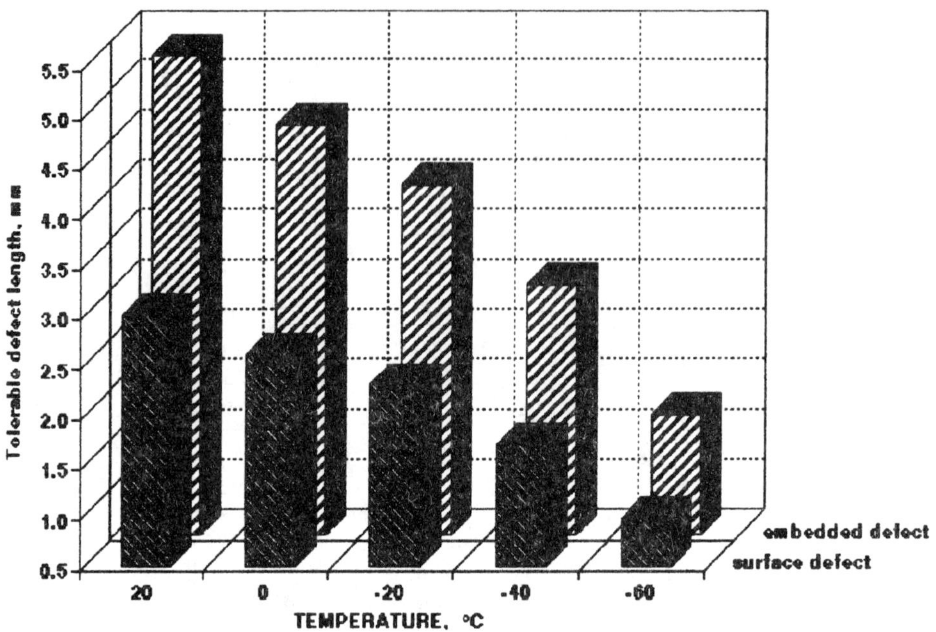

Figure 3 : Tolerable defects length at various temperatures.

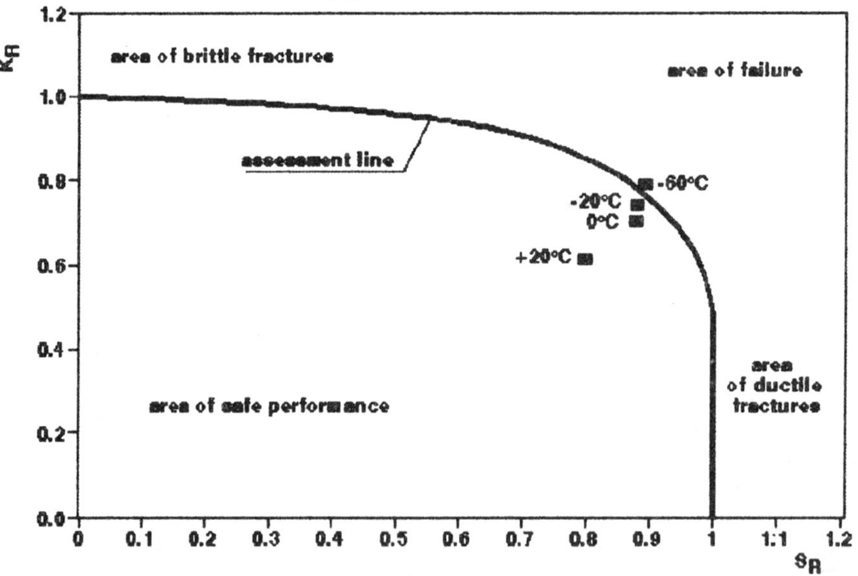

Figure 4 : Failure assessment diagram (FAD)

$$K_R = \frac{\text{stress intensity factor}}{\text{fracture toughness}} = \frac{K_I}{K_{Ic}} \qquad S_R = \frac{\text{applied load}}{\text{collapsed load}}$$

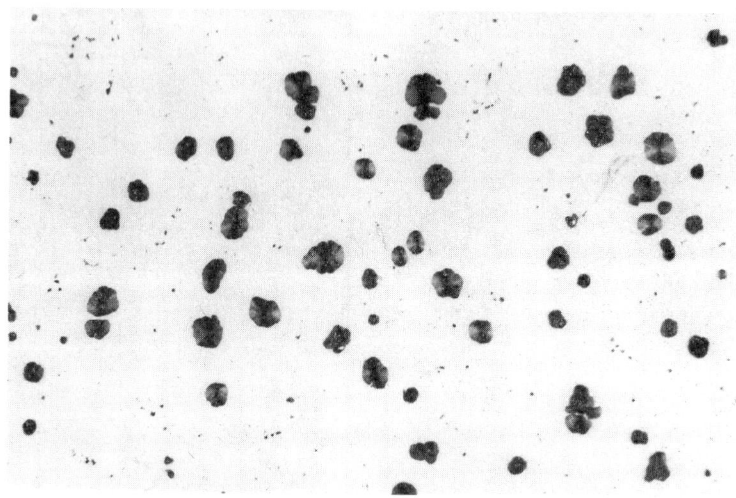

Figure 5 : Graphite precipitates in ductile iron.

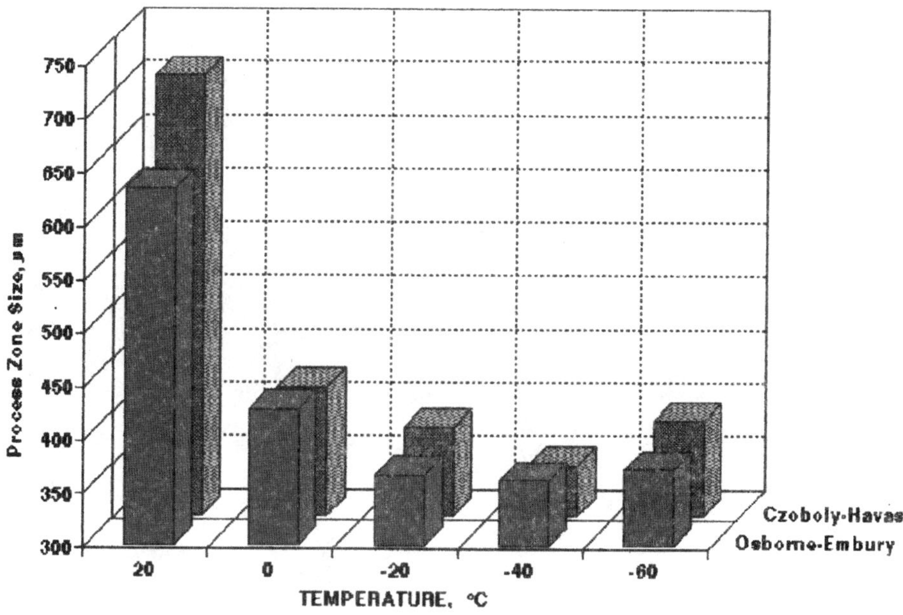

Figure 6 : The process zone size at different temperatures.

Advanced Materials Research Vols. 4-5 (1997) pp. 227-232
© *1997 Scitec Publications, Switzerland*

Study on the Fatigue Properties of Austempered Ductile Irons

C. Cheng and J.J. Vuorinen

Helsinki University of Technology, Lab. of Foundry Engineering,
Puumiehenkuja 3A, FIN-02150 Espoo, Finland

Keywords: Austempered Ductile Iron, Fatigue Properties

ABSTRACT

This paper presents a fatigue study on different grades of austempered ductile irons. The rotating bending fatigue tests with staircase method were conducted to compare the fatigue properties for different ADI materials with optimized chemical composition, heat treatment and surface treatment parameters. The microstructures were studied and the austenite contents were measured. The results from the fatigue, compression, tensile and hardness tests are also discussed.

INTRODUCTION

With the achieved combination of high strength, toughness, fatigue strength and wear resistance, austempered ductile irons (also known as ADI) have already found many applications in gears, railroad wheels, crankshafts, connecting rods, support wheels and other subjects to dynamic stresses and wear. Therefore, it has been an attractive alternate material to replace forged steel components in many applications.

The appropriate mechanical properties of austempered ductile iron are controlled with the final microstructure of the ADI components which results from the austempering reaction. The most critical constituent of the ADI microstructure is austenite. The morphology, distribution, carbon content and volume fraction of this phase can be strongly affected by the heat treatment parameters and metal composition. These structural characteristics will determine the strength and toughness of ADI materials.

The information regarding the fatigue properties of austempered ductile iron is very limited in the literature [1-5]. The mechanical properties of ADI material could be very different because of the wide variations of its composition, heat treatment and surface treatment processes. Therefore, the data on the mechanical properties, especially the fatigue properties of different ADI materials are still much needed for its design applications.

The aim of this study is to investigate the fatigue properties of ADI materials with optimized chemical composition, heat treatment and surface treatment processes.

EXPERIMENT

On the basis of the early study [3], the optimized chemical composition and heat treatment parameters were chosen and only the austempering temperature (for different grade of ADI) and surface treatment process were adjusted.

The chemical composition of the material used in this study which was optimized from the orthogonal experiment is presented in Table 1.

Table 1. The chemical composition in percentage of test material.

C	Si	Mn	P	S	Mg	Cu	Mo	Ni
3.46	2.48	0.39	0.020	0.010	0.043	0.80	-	-

The test specimens were austenitized in an electric furnace at 910 °C for two hours. After the austenitization the test specimens were directly transferred to a salt bath with different austempering temperatures, i.e. test codes A1, A2 and A3 for three hours in order to get three grades of austempered ductile iron. After the austempering the test specimens were air cooled to room temperature. The heat treatment parameters are shown in Table 2.

Table 2. The heat treatment parameters.

Parameters for heat treatment	Test code		
	A1	A2	A3
Austenitizing temperature (°C)	910	910	910
Austenitizing time (h.)	2	2	2
Austempering temperature (°C)	350	320	290
Austempering time (h.)	3	3	3

The tensile test was carried out in a Losenhausen tension machine having a load capacity of 400 kN. The testing temperature was +20 °C. The Brinell hardness testing was performed using a Alpha hardness tester, with a 10 mm steel ball and a load of 3000 kg.

In the compression testing the dimension of specimen was ϕ 5 x 10 mm. The specimens were compressed 5 %, 10 %, 20 % and 30 % from their original height, 10 mm in a Losenhausen tensile test machine having a load capacity of 400 kN with a speed of 1 mm/min at + 20 °C. The specimens were submerged in oil during compression in order to be deformed evenly in each cross section.

After compression the austenite contents of the test specimens for different ADI materials and deformations were measured with the Saturation Magnetization Analyzer (SATMAGAN). Subsequent to the austenite measurements, the Vickers hardnesses were measured from the same specimen on the surface at the end of the specimen.

The shot peening surface treatment is one of the mechanical surface treatments in order to improve the fatigue strength of dynamically loaded components. In the shot peening for austempered ductile iron, there is the benefit of the material responding favourably to work hardening. In the present study a part of the fatigue test specimens were shot peened and others only polished.

The shot peening of specimens for rotating bending fatigue testing was performed in a Hunziker machine (test code CK). The optimized parameters for shot peening procedure were used.

The rotating bending fatigue testing was applied to compare the fatigue behaviours for different ADI materials in this study. In the rotating bending fatigue testing machine a specimen with a round cross section is subjected to a dead-weight load while bearings permit rotation. A given point on the circular test section surface is subjected to a sinusoidal stress amplitude from tension on the top to compression on the bottom with each rotation.

In the present study, the rotating bending fatigue testing was performed in a Carl Schenck PUN rotating bending fatigue machine having a maximum stress level of 900 N/mm^2. The rotating speed in the tests was 12000 rpm. The testing temperature was +20 °C.

The microstructural observations were made from polished and etched specimens by optical microscope and scanning electron microscope (SEM). The fracture surface after rotating bending fatigue test was also studied by scanning electron microscope.

RESULTS AND DISCUSSION

The tensile strength and elongation vs. austempering temperatute is presented in Figure 1. The values from Figure 1 indicate that the test material austempered at lower temperature would have higher tensile strength and slight reduced ductility.

The austenite content and Brinell hardness vs. austempering temperature are presented in Figure 2. From Figures 1-2 it is clear that the ADI material austempered at higher temperature will have higher austenite content, elongation, and reduced tensile strength and hardness.

The austenite contents and Vickers hardness before and after the deformation in the compression tests for ADI material austempered at 350 °C are shown in Figure 3.

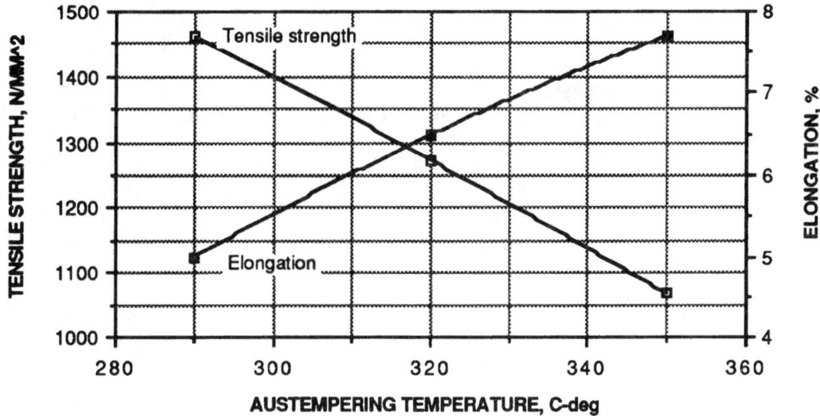

Figure 1. Tensile strength and elongation vs. austempering temperature.

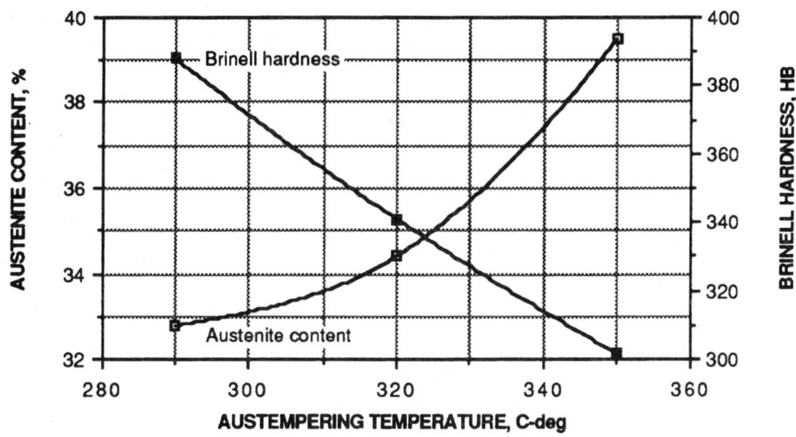

Figure 2. Austenite content and Brinell hardness vs. austempering temperature.

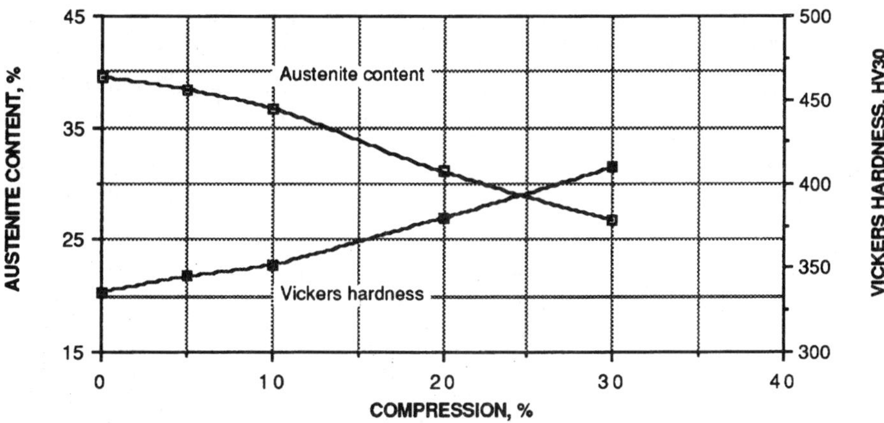

Figure 3. Austenite content and Vickers hardness vs. deformation in compression for test code A1.

It can be seen from the data for the compression test in Figure 3 that with increasing the deformation, the austenite content will drop and hardness increase. At least a deformation of about 10 % is needed to cause a significant change in the austenite content and hardness. A deformation of 30 % has caused a significant drop of about 13 % units for austenite content, corresponding to a hardness increase of 74 HV, that is, the deformation of 30 % has decreased the austenite content to two-thirds of its original one for the ADI material austempered at 350 °C.

In present study on rotating bending fatigue behaviours for different ADI materials a staircase method was applied because only the median fatigue resistance was required to be estimated and only a limited number of specimens were available.

The statistical analysis of the data with staircase method is straightforward for producing estimates of the mean fatigue strength at a specific fatigue life value, i.e. 2×10^6 cycles in present study.

The optimized shot peening intensity was 0.44 A according to the rotating bending fatigue test. Fatigue strength with different surface treatment vs. tensile strength is shown in Figure 4. It is clear that the higher the austempering temperature, the better the fatigue property in this study. After shot peening surface treatment the fatigue strength can increase about 5 - 21 %. Austempered at higher temperature the difference for fatigue strength between shot peened specimen and only polished one is bigger. Therefore, the shot peening is more effective for the ADI material austempered at higher temperature.

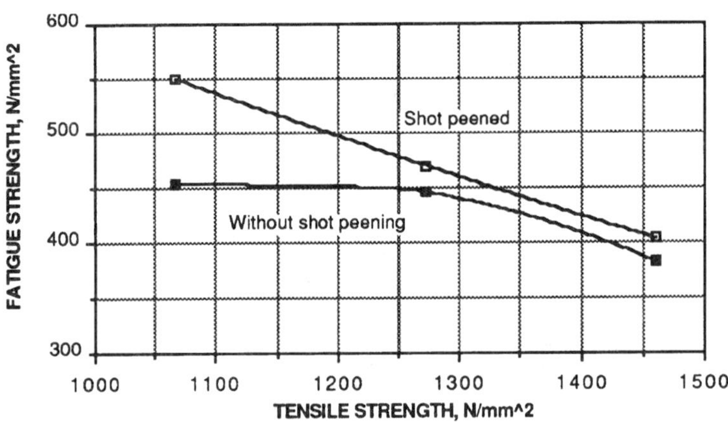

Figure 4. Fatigue strength with different surface treatment vs. tensile strength.

The microstructures of ADI material austempered at 350 °C from rotating bending fatigue test specimen in the central area and near the shot peened surface examined by the scanning electron microscope are shown in Figures 5-6. The fracture surface of fatigue test specimen for same material are presented in Figures 7-8.

It should be noticed that austenite has partially transformed to martensite in the shot peened layer near the surface in Figure 5. The general fractographical micrographs, as shown in Figures 7-8, reveal that the fracture surface contains faceted brittle areas and more dimpled ductile areas in the central part of the specimen for ADI material austempered at higher temperature but only the faceted brittle mode of fracture is clearly visible in the shot peened layer near the surface.

The surface treatment with shot peening that will improve the fatigue strength of components is based on the principle of the introduction of favourable compressive residual stresses in the surface layer. Also the micrographs reveal that a fraction of austenite has transformed to martensite in the shot peened layer near the surface. Silicon and manganese could decrease the stacking fault energy, i.e. SFE and lowering SFE can promote the martensitic transformation [6-9]. When the austenite transforms to martensite, the expanding tendency could occur in this shot peened layer because of the bigger specific volume of martensite than that of austenite. Therefore, the additional compressive stresses will be introduced in the shot peened layer. Then the initiation and propagation of small crack can be strongly retarded by surface compressive stresses and the ADI material containing optimum manganese and silicon contents will achieve better fatigue behaviour.

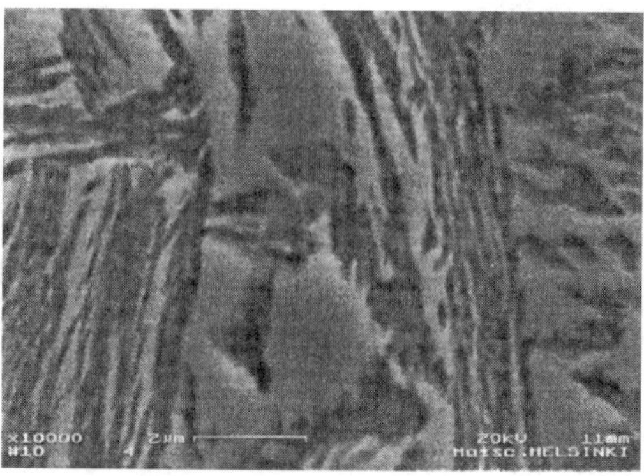

Figure 5. The microstructure of ADI material A1CK from rotating bending fatigue test specimen near the shot peened surface. SEM. 10000 X

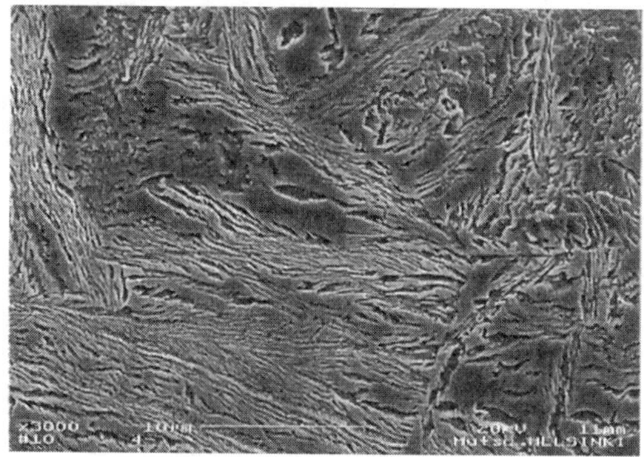

Figure 6. The microstructure of ADI material A1CK in the center of rotating bending fatigue test specimen. SEM. 3000 X

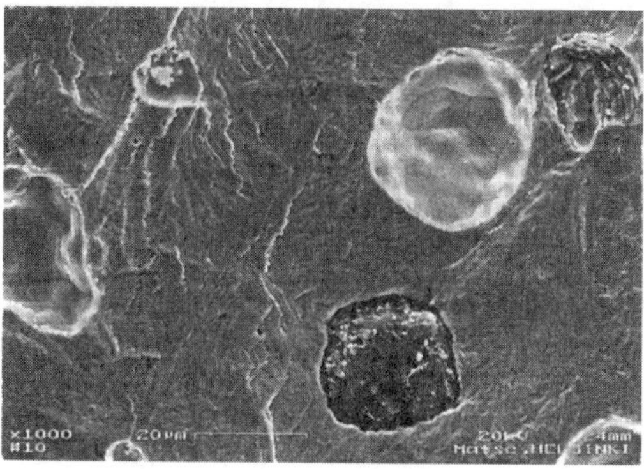

Figure 7. The fracture surface of ADI material A1CK from rotating bending fatigue test specimen near the shot peened surface. SEM. 1000 X

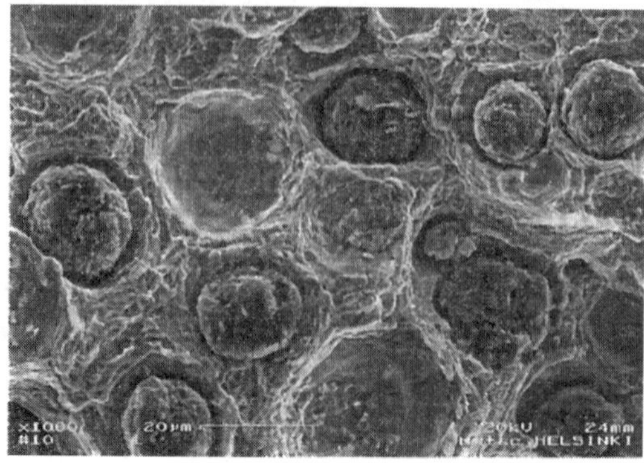

Figure 8.   The fracture surface of ADI material A1CK in the center of rotating bending fatigue test specimen. SEM. 1000 X

CONCLUSIONS

In this study the ADI material austempered at higher temperature will have higher austenite content, fatigue strength and ductility.

Shot peening offers a controllable means of selectively hardening certain parts of a finished casting to produce improvements in fatigue properties. The shot peening is more effective for the ADI material austempered at higher temperature in this study.

Silicon and manganese can promote the martensitic transformation in the shot peened layer when the casting is surface treated by shot peening process. Therefore, the ADI material containing optimum manganese and silicon contents will achieve better fatigue property.

ACKNOWLEDGMENTS

The authors would like to express our thanks to TEKES and JOT-Companies Ltd. for the financial support.

REFERENCES

[1]   M. Johansson, AFS Trans. **85**, 117 (1977).

[2]   Cast Metals Development Ltd., Materials & Design, **13** (5), 285 (1992).

[3]   C. Chen, Report (Dec. 1992) (unpublished).

[4]   R. Böschen et al., ADI Conf. Proc., 468 (1991).

[5]   S. Yoshino, ADI Conf. Proc., 337 (1986).

[6]   J. J. Vuorinen, Ph. D. Thesis. HUT, Espoo, Finland (1981).

[7]   R. E. Schramm et al., Met. Trans. **6** A (7), 1345 (1975).

[8]   P. M. Kelly, Acta Metall. **13**, 635 (Jun. 1965).

[9]   Z. Nishiyama, "Martensitic transformation" Academic Press, New York, 467 (1978).

Advanced Materials Research Vols. 4-5 (1997) pp. 233-238
© 1997 Scitec Publications, Switzerland

Effect of Initial Flaws on Fatigue Crack Initiation of Austempered S.G. Cast Iron

A.S. Béranger[1], R. Billardon[2] and F. Hild[2]

[1] Renault, Direction de la Recherche, Service 0852,
860 quai Stalingrad, F-92109 Boulogne Billancourt, France

[2] Laboratoire de Mécanique et Technologie, E.N.S. Cachan, C.N.R.S., Université Paris 6,
61 Avenue du Président Wilson, F-94235 Cachan Cedex, France

Keywords: Initial Flaws, Failure Probability, Austempered Spheroidal Graphite Cast Iron, Cyclic Fatigue

Abstract. Fatigue failure of austempered nodular graphite cast iron is analyzed by accounting for the presence of initial flaws within the material. An expression of the cumulative failure probability is derived in the case of cyclic loading conditions. An identification procedure is developed to determine the flaw distribution as well as the crack growth law.

1. INTRODUCTION

Due to its good properties, S.G. cast iron is widely used in automotive industry for safety components. For example, it is utilized in ground link elements such as steering knuckle holder, suspension arms. With the current production technology, flaws (e.g. pin–holes, shrinkage, cavities) are unavoidable. The fatigue strength of components may be reduced by the presence of initial casting flaws randomly distributed within the material. Nevertheless, conventional design procedures to assess the structural integrity of a component use deterministic crack initiation criteria and ignore micro–inhomogeneities (e.g., flaws) within the material. Consequently an evaluation method that accurately calculates the effect of casting flaws is required. It is proposed to model the presence of these inhomogeneities, their possible evolution with the number of cycles, and if needed their statistical distribution.

2. RELIABILITY OF STRUCTURES CONTAINING FLAWS

In this paper, it is assumed that initial flaws are randomly distributed within a structure and that the flaw distribution is characterized by a probability density function f. The function f depends upon the flaw size a. Other parameters such as orientation are not considered, since only flaws in pure mode I are considered (i.e. with their orientation perpendicular to the maximum principal stress).

In the case of cyclic loading conditions, the evolution of the flaw size leads to the evolution of the flaw size distribution. After N cycles, it is assumed that the flaw distribution is described by a function f_N. At this stage, it is useful to introduce a function ψ that relates the initial flaw size a_0 to the flaw size after N cycles a_N

$$a_0 = \psi(a_N) \tag{1}$$

If the flaw size evolution is deterministic, the probability of finding a flaw of size a_N after N cycles is equal to the probability of finding an initial flaw of size $\psi(a_N)$. Therefore the cumulative initiation probability P_{I0} can be written as [1]

$$P_{I0} = \int_{\psi(a_c)}^{+\infty} f_0(a)\, da \tag{2}$$

where $\psi(a_c)$ denotes the initial flaw size that, after N cycles of loading reaches the critical flaw size a_c. The flaws are supposed to be described by cracks of size a. whose geometry is taken into account by a dimensionless factor Y such that a general stress intensity factor K is given by

$$K = Y\, \sigma\, \sqrt{a} \tag{3}$$

where σ stands for an equivalent uniaxial stress (for instance the maximum principal stress). It is worth noting that the values of the parameter Y depend upon the geometry of the initial defect and the fact that this flaw intersects or not a free surface. Under monotonic and cyclic loading conditions, local failure can be described by a criterion referring to a critical value of the stress intensity factor

$$K \geq K_c \tag{4}$$

In the case of ductile materials subjected to cyclic loading conditions, stable crack propagation can be described by a generalized Paris law [2]

$$\frac{da}{dN} = C \left(\frac{K_{max}\, g(R) - K_{th}}{K_c - \dfrac{K_{th}}{g(R)}} \right)^n \quad \text{with} \quad R = \frac{K_{min}}{K_{max}} \tag{5}$$

where K_{max} (resp. K_{min}) stands for the maximum (resp. minimum) stress intensity factor over one cycle, K_{th} denotes the threshold stress intensity factor under which $(K_{max}\, g(R) < K_{th})$ no propagation occurs, and N the number of cycles. C and n are material parameters, and the function g models the influence of the load ratio R. Eqn. (7) corresponds to a modified version of an Elber law [3,4]. When the load ratio R is negative, g(R) remains close to unity. The approximation $g(R) \approx 1$ is usually sufficient to model effects of negative load ratios. From Eq. (5) the following closed form solution can be derived

$$\varphi\left(\sqrt{\frac{a_c}{a_M}}\right) - \varphi\left(\sqrt{\frac{\psi(a_c)}{a_M}}\right) = C^* \left(\frac{g(R)}{1 - \dfrac{K_{th}}{K_c\, g(R)}} \right)^n \left(\frac{\sigma}{S_u}\right)^n N_F \tag{6}$$

where a_M denotes the maximum flaw size in the structure, the dimensionless constant C^* is equal to C / a_M, and S_u the maximum value of stress σ is to be defined by Eq. (7.1). When the maximum flaw size is bounded, a monotonic threshold stress, S_u, may be defined as the lowest value of the stress level below which the monotonic failure probability has a zero value so that

$$S_u = \frac{K_c}{Y\sqrt{a_M}} \tag{7.1}$$

In the case of cyclic loading, a cyclic threshold stress may be defined as the lowest value of the stress level below which the cyclic failure probability has a zero value. It is worth noting that this cyclic threshold stress depends upon the load ratio R. For identification purposes, the cyclic threshold stress, S_{th}, is defined for R equals 0 and is related to the threshold stress intensity factor by

$$S_{th} = \frac{K_{th}}{Y\sqrt{a_M}} = S_u \frac{K_{th}}{K_c} = S_u\,k \qquad (7.2)$$

If it is assumed that the interaction between flaws is negligible, an independent events assumption can be made. The expression of the cumulative initiation probability, P_I of a structure Ω of volume V can be derived in the framework of the weakest link theory. The expression of P_I can be related to the cumulative initiation probability, P_{I0}, of a single link by

$$P_I = 1 - \exp\left\{\frac{1}{V_0}\int_\Omega \ln(1 - P_{I0})\,dV\right\} \qquad (8)$$

In the case of global unstable propagation, the structural failure corresponds to the initiation and the expression of the cumulative failure probability P_F is given by

$$P_F = P_I \qquad (9)$$

It is worth noting that in the case of high cycle fatigue, the propagation stage tends to become negligible when compared, in terms of number of cycles, with the initiation stage. Since the propagation stage is neglected when Eq. (9) is used, this equation corresponds to a *lower bound* to the cumulative failure probability of the structure. Hence, in the following, 'failure' refers to local failure i.e. macroscopic initiation or lower bound to macroscopic failure.

3. ANALYSIS OF FATIGUE TESTS

In this section, a series of experiments reported in [5] are analyzed in details. These experiments have been carried out at different stress levels on Kymenite, which is an austempered ductile iron, grade K–10005 [5]. The specimens were subjected to rotating bending with a test area diameter equal to 7.5 mm. The ratio between the threshold stress intensity factor and the critical stress intensity factor is on the order of 1/3.

When the fatigue limit is known, the identification can be performed in two different stages. The first stage consists in identifying the flaw size distribution. A minimization scheme is used to determine the minimum error between all the available experimental data on fatigue limits. Such an error can be defined as

$$Err = \sum_{i=1}^{N_e}\left\{P_{Fi} - P_F\left(\alpha;\beta;\frac{V}{V_0};\frac{\sigma_{Fi}}{S_{th}}\right)\right\}^2 \qquad (10)$$

where N_e denotes the number of experimental data (here 7), P_{Fi} is the experimental cumulative failure probability and $P_F(\alpha;\beta;V/V_0;\sigma_{Fi}/S_{th})$ is the predicted cumulative failure probability.

If we assume that the maximum flaw size is bounded by a_M, the flaw size distribution f_0 can, for instance, be given by a beta distribution

$$f_0(a) = \frac{a_M^{-1-\alpha-\beta}}{B_{\alpha\beta}}\,a^\alpha\,(a_M - a)^\beta\ ,\quad 0 < a < a_M\ ,\quad \alpha,\beta > -1 \qquad (11)$$

where α and β are the parameters of the beta function, and $B_{\alpha\beta}$ is equal to $B(\alpha+1,\beta+1)$, where $B(.,.)$ is the Euler function of the first kind. The parameters to identify are the powers α and β of the beta distribution, the volume ratio V/V_0, and the cyclic threshold stress S_{th}. When the number of cycles to failure tends to infinity, the cumulative failure probability can be rewritten as

$$P_F\left(\alpha;\beta;\frac{V}{V_0};\frac{\sigma_{Fi}}{S_{th}}\right) = 1 - \exp\left\{\frac{1}{V_0}\int_\Omega \ln\left(1 - \int_{a_{th}}^{+\infty} f_0(a)\,da\right)dV\right\} \tag{12}$$

with

$$a_{th} = \left(\frac{K_{th}}{Y\sigma(\sigma_{Fi},M)}\right)^2$$

where $\sigma(\sigma_{Fi},M)$ is the applied equivalent stress level at a point M of Ω when the maximum equivalent stress level is σ_{Fi}. When applying these relations to the experimental results, the following values are identified: $\alpha = 1.8$, $\beta = 17.5$, $V/V_0 = 112$, and $S_{th} = 170$ MPa. The identification is shown in Fig. 1 in terms of cumulative failure probability versus stress.

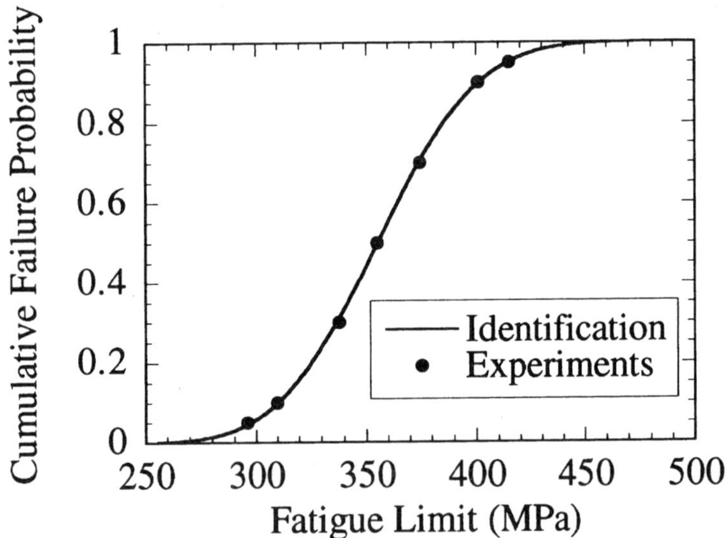

Figure 1: Experiments and predictions of the fatigue limits
for different cumulative failure probabilities in rotating bending of Kymenite K–10005.

The second step of the identification concerns the crack growth law (parameters C and n). In tension, this identification would be straightforward by studying one iso–cumulative failure probability (e.g. 50%) since an iso–cumulative failure probability is described by a constant cumulative failure probability P_{F0}. In bending however, because of the stress heterogeneity in the specimen, the previous result is not valid. One needs to compute the cumulative failure probability of the structure and to identify the constants that best fit the experiments. One single iso–cumulative failure probability is however still sufficient to identify the two material parameters. The iso–cumulative failure probability 50% is used to minimize an error similar to that defined in Eq. (12).

The following values were obtained: $n = 2.34$, and $C / a_M / (1-k)^n = 1.3 \ 10^{-4}$. In Fig. 2 the predictions of the number of cycles to failure are compared with the experimental observations. Three points were used for the identification and the remaining points are predictions. Out of 19 experimental data, 15 points exhibit a difference between experimental and predicted number of cycles to failure less than 15%, and the maximum difference is 75%.

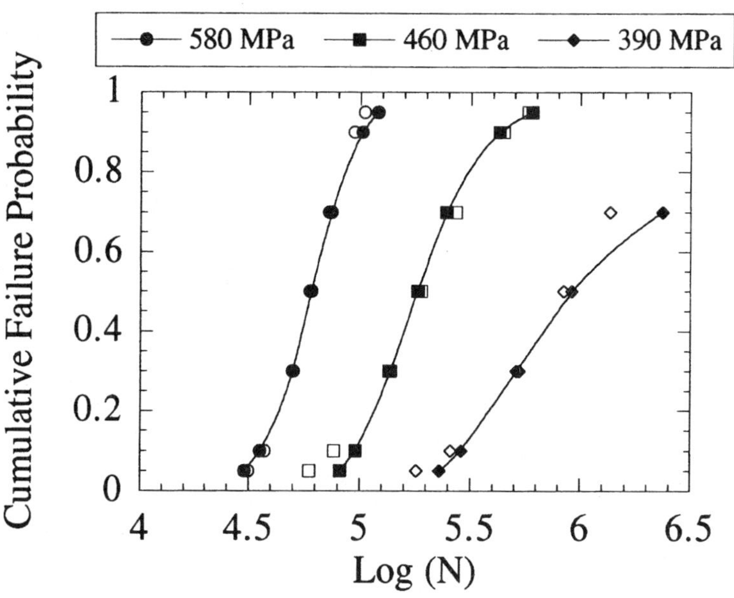

Figure 2: Comparisons between experimental (solid symbols) and predicted (open symbols) values of the cumulative failure probability for the three different stress levels in rotating bending of Kymenite K–10005.

4. CONCLUSIONS

A reliability analysis taking account of flaw size distributions has been developed for components subject to cyclic loading conditions. Emphasis is put on the initiation stage, which is directly related to the evolution of initial flaws. An expression of the cumulative initiation probability is derived in the framework of the weakest link theory and by assuming that the flaws do not interact.

Experimental data on nodular graphite ferritic cast iron in rotating bending are analyzed within this framework. The predictions of the whole set of data is in good agreement with the experimental number of cycles to failure. This last result shows that the expression of cumulative failure probability proposed herein is able to model fatigue data obtained on nodular graphite ferritic cast iron.

To model the effect of initial flaw distributions on the fatigue behavior of heterogeneous materials, fatigue failure maps can be introduced. They consist in plotting contours of fatigue limits in a space representative of the initial flaw distributions. Such maps are drawn for a given cumulative failure probability. When the initial flaw distribution is modeled by a beta distribution of exponents α

and β, the maps are drawn in an α–β plane. The fatigue failure maps allow to extrapolate the results obtained for a given flaw size distribution to other flaw size distributions (Fig. 3).

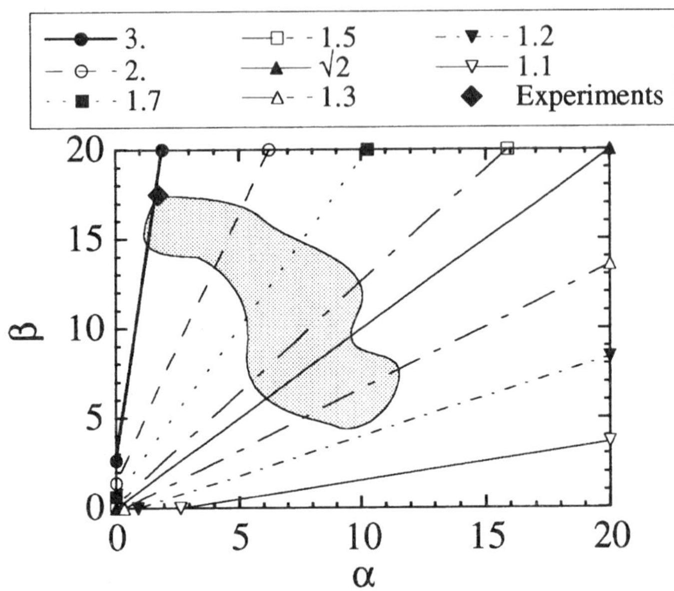

Figure 3: Fatigue failure maps giving the normalized fatigue limit σ/S_{th} in tension

for different values of the parameters α and β when $P_{F0} = 50\ \%$.

      Typical application of this approach concerns the reliability analysis of casted components. More and more tools are available to predict the different flaw size distributions from one point of a part of a cast to another, whereas the fatigue behavior of the material is in general merely derived from experiments performed on so–called flawless specimens. A fatigue failure map representative of the sensitivity of the material to flaws will enable to predict the reliability of the whole component under cyclic loading conditions; in other words, it will enable to predict the number of cycles to macrocrack initiation, or the probability of reaching a certain number of cycles without failure at every point of the component.

ACKNOWLEDGMENTS

The authors gratefully acknowledge the financial support of Renault through contract CNRS/109 (H5–24–12) with the Laboratoire de Mécanique et Technologie, Cachan.

REFERENCES

[1] F. Hild and S. Roux, Mech. Res. Comm., **18**, 409 (1991).
[2] J. Pellas et al., Recherche aérospatiale, **3**, 191 (1977).
[3] W. Elber, Eng. Fract. Mech., **2**, 37 (1970).
[4] W. Elber, Damage Tolerance in Aircraft Structures, ASTM, STP 486, 230 (1971).
[5] K. Jokipii, European Meeting on Castings in Austempered Spheroidal Graphite Cast Iron, Sèvres, France (1992).

Advanced Materials Research Vols. 4-5 (1997) pp. 239-244
© 1997 Scitec Publications, Switzerland

Basic Study on Erosion of Ductile Iron

K. Shimizu[1], T. Noguchi[2], T. Kamada[2] and S. Doi[3]

[1] Oita National College of Technology, 1666 Maki, Oita 870-01, Japan

[2] Hokkaido University, North 13, West 8, Kita-ku, Sapporo 060, Japan

[3] Oita University, 700 Danoharu, Oita 870, Japan

Keywords: Wear, Erosion, Ductile Iron, Impact Angle, Hardness

Abstract

Erosion tests were performed on austempered ductile iron (ADI), and ferritic ductile iron and pearitic ductile iron (F/PDI in the follow), using a shot blast machine. Erosion damage was measured by the removed material volume at impact angles between 10 and 90 deg. The surface metal flow in the vertical section was also observed. The mechanism of erosive wear, the effect of impact angles, and the difference in wear features of the specimens were discussed. After an initial stage, the eroded volume increases almost linearly with blasting time in both ADI and FDI, PDI, and the erosion rate in ADI is about 1/10-25 of FDI, PDI, showing that ADI has superior erosion resistance.

The surface hardness of eroded ADI specimens increased up to $HV700$ from the initial $HV\,350$ after 600s of blasting. The amount of retained austenite was measured about 40% before the test, but it decreased by transformation of retained austenite to martensite, hardening the surface and eliminating the erosion rate.

1. Introduciton

Surface damage caused by the impact of dispersed particles in gas or liquid flow is called *"erosion"*. Recently, this phenomenon is receiving much attention as a serious problem, particularly, at pipe-bends or valves in pneumatic conveying system. For instance in steel making, where the injection of coal dust and iron ore pellets into molten iron and steel has been improved secondary refining and new steel making. Here, the measures against erosion and predictions of the life span of pipe-lines are important problem to be solved as soon as possible. The erosive wear of pipe-lines caused by particles is complicated, and involve parameters of velocity, diameter, hardness, impact angle, material and mechanical properties of particles. We have studied the erosion of pipe metal made of SS 400 and investigated the damage behaviours and mechanism. In our experiments, ferritic ductile iron(FDI), pearlitic ductile iron(PDI), and austempered ductile iron(ADI) which is widely used due to its strength, were used. The results are used to consider the erosion behaviors and mechanism, and the potential of ADI as an erosion resistant material is examined. [1]-[9]

2. Test specimens and experimental method

Three kinds of specimens were used: FDI, PDI (pearlite ratio 70%), and ADI. The chemical compositions and mechanical properties are reported in Table 1 and the process of austempered heat treatment is shown in Figure 1. The microstructure of the specimens are in Figure 2. The specimen is a 50x50x10mm plate with ground surface(Fig.3). A sketch of the blast machine is shown in Figure 4. This machine can accelerate particles using compressed air.[1]-[5]

The specimen is fixed on a stage which enables the impact angle to vary from 0° to 90°. The grit has the following characteristics: average diameter 660 μm, and Vickers Hardness Number(HV) 420, spherical ratio 96%, and total volume 5kg. The grit was changed after every test. The test is carried out at ambient temperatures. The impact angle is changed by 10° increments from 0° to 90° and the test time is 7200s. The

Table 1　Chemical composition (mass%) and mechanical properties of specimens

	C	Si	Mn	P	S	Cu	Mg
F D I	3.75	2.08	0.34	0.011	0.007	0.04	0.037
P D I	3.65	2.34	0.48	0.020	0.008	0.75	0.043
A D I	3.66	2.15	0.35	0.021	0.030	0.53	0.049

	σB	ϕ	HB	ρ
F D I	400	25.8	140	7.0
P D I	750	7.2	229	7.0
A D I	1015	10.5	293	7.2

σB : Tensile strength (MPa)
ϕ　: Elongation (%)
HB : Hardness
ρ　: Density

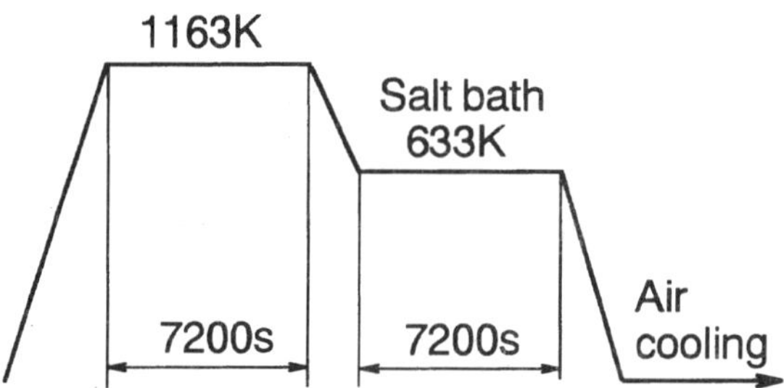

Figure 1　Heat treatment diagram for ADI

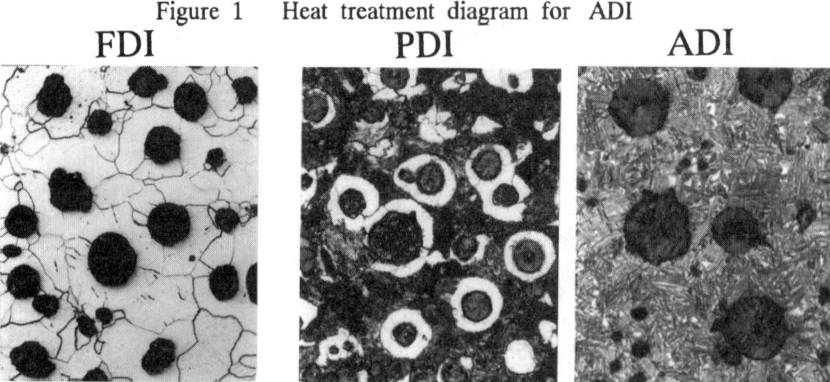

Figure 2　Microstructure of the specimens

eroded volume is measured by an electronic micro weight scale (0.01mg sensitivity) after the impact experiment, and the eroded volume is calculated from the reduced mass. After the test, the eroded surface and the metal structure of the cross section are observed.   1)-5) 10)-15)

3.Experimental results

In discussing the erosion resistance of materials of different densities, it is not appropriate to simply compare the eroded weight of material. Therefore, by employing the average density of each material, the eroded mass can be calculated and the erosion ratios compared. Figure 5 shows the relation between impact angle and the erosion ratio after blasting FDI, and PDI for 7200s. All the cases show the maximum erosion ratio with an impact angle of 60°. The erosion ratios of PDI is about one third of FDI. This is explained by the fact that the Brinell hardness(HB) of both FDI is about 150, while PDI is much harder, at 229. This allows the conclusion that harder materials have lower damage rates.

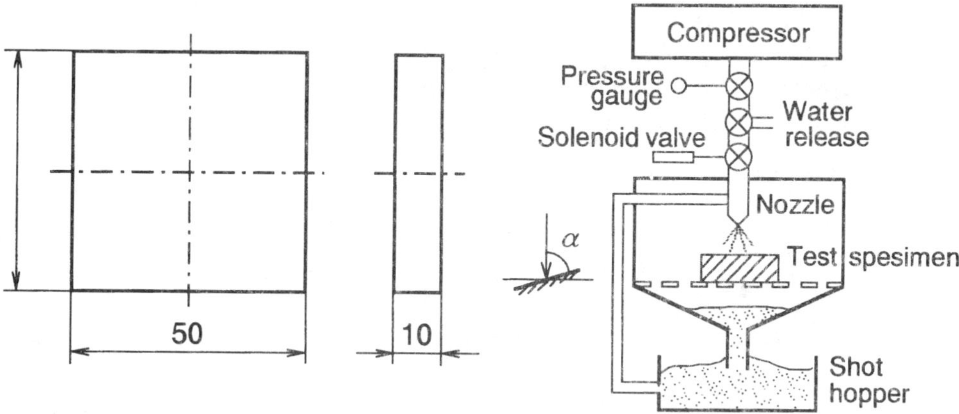

Figure 3   Test   specimen   dimensions   Figure 4   Sketch of the shot blast machine

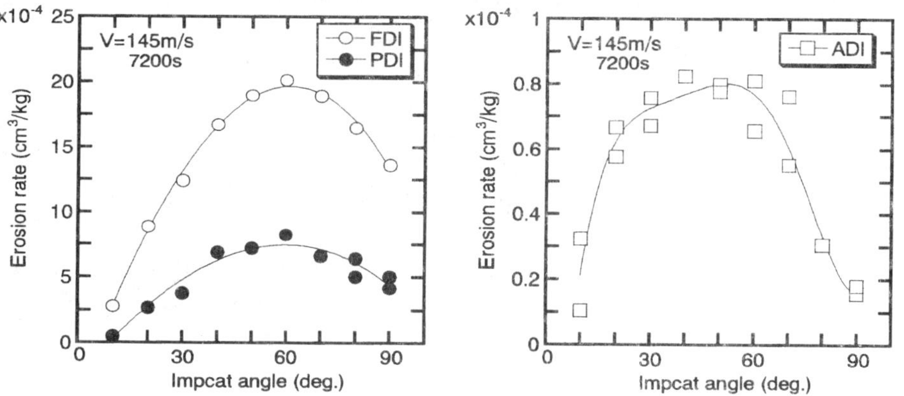

Figure 5   Erosion rate vs. impact angle,   Figure 6   Erosion rate vs. impact angle,

for   F/PDI   for   ADI

The relation between impact angle and erosion rate of ADI blasted for 7200s is shown in Figure 6.   The erosion rate reaches the maximum when the impact angle is 40° ~60° and the this behaviour is different   from that of FDI and PDI.   The erosion rate of ADI is about 1/25 of that of FDI, and 1/10 of that of PDI.

4. Observation of the Eroded Surfaces

To evaluate the above results in relation to specimen surface changes, the eroded surface was observed macroscopically. Figure 7 shows the eroded surface of specimens blasted for 7200s. All specimens had craters as large as the impact areas. With FDI and PDI, there are lamellar patterns at right angles to the flow of the eroding material at 20° ,40° but not at 60° , 80° . The craters become deep from 40° .

The ADI surface is quite different from that of FDI, and PDI. At   20° the surface is almost smooth. At 40° , there is an indistinct ripple pattern. At 60° and 80° ripples are conversed into small dimples. At 60° and 80° ,   the erosion does not   occur in the direction of depth.

Figure 8 shows cross sections of specimen with impact angle 60° and impact velocity 145m/s, after 7200s of blasting.   All specimens have ridges caused by plastic deformation near the surface. The largest ridge is about $150\,\mu$ m in FDI, $200\,\mu$ m in PDI (70), and $25\,\mu$ m in ADI. This indicates the very limited erosion on ADI.

The above shows that erosion causes big ridges due to plastic transformation near the surface owing to the impact.   The corrosion proceeds with the growth and breaking off of ridges.   The size of ridges depends on the kind of material, and this causes differences in erosion rates.

To confirm the superior erosion resistance of ADI over FDI and   PDI, the hardness

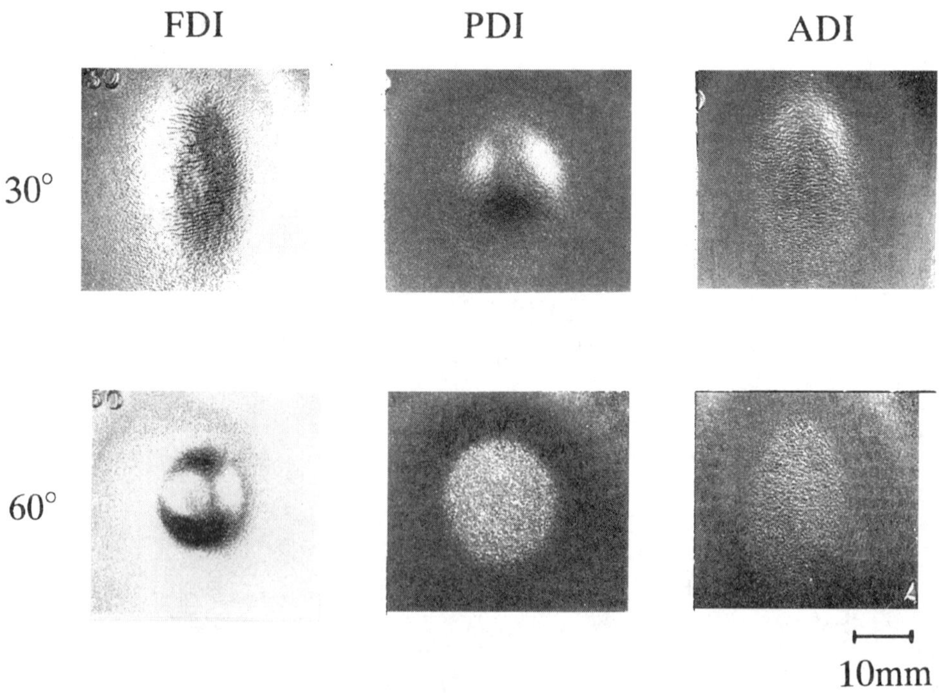

Figure 7 Microscopic appearance of eroded surface, F/PDI and ADI

near the surface was measured. Figure 9 shows the hardness distribution of FDI and ADI in the direction of depth for impact angles of 30° , 60° , 90° with 7200s of blasting. The initial Vickers hardness (*HV*) of FDI is about HV180, but on the eroded surface it is HV350, showing the hardening caused by the impacts.

The value of hardness for initial ADI is around HV350 but it reaches HV750 on the eroded surface. To establish the reason for this remarkable hardening the volume of the retained austenite was established by X-rays, as shown in Figure 10. Before the test, there is about 40% and after the test (7200s), only 3% to 5%, for impact angles of 30° , 60° , and 90° . From hardness test, the surface phase is almost the same at 600s as at 7200s, and the transformation of the retained austenite to martensite may be considerd to have started already at 600s. The results lead to the assumption that, the austenite on the surface transforms into martensite by the impacts, and that the erosion rate is slowed by the hardening. As erosion proceeds, the top martensite layer is soon away and the austenite in the underlying surface transforms into martensite maintaining the hardness to some depth below the surface.

5.Conclusion

In order to investigate the potential of spheroidal graphite cast iron as an erosion resistant material, blast tests were made and the erosion behaviors were observed. The following results were obtained:

1) The erosion rate of ADI is approximately 1/25 of FDI and 1/10 of PDI. ADI is clearly superior in erosion resistance. The relation between impact angle and erosion rate is similar for FDI and PDI but quite different for ADI. Both FDI and PDI reach the maximum value of erosion rate near 60° , while ADI presents a maximum at is 40° ~ 60°

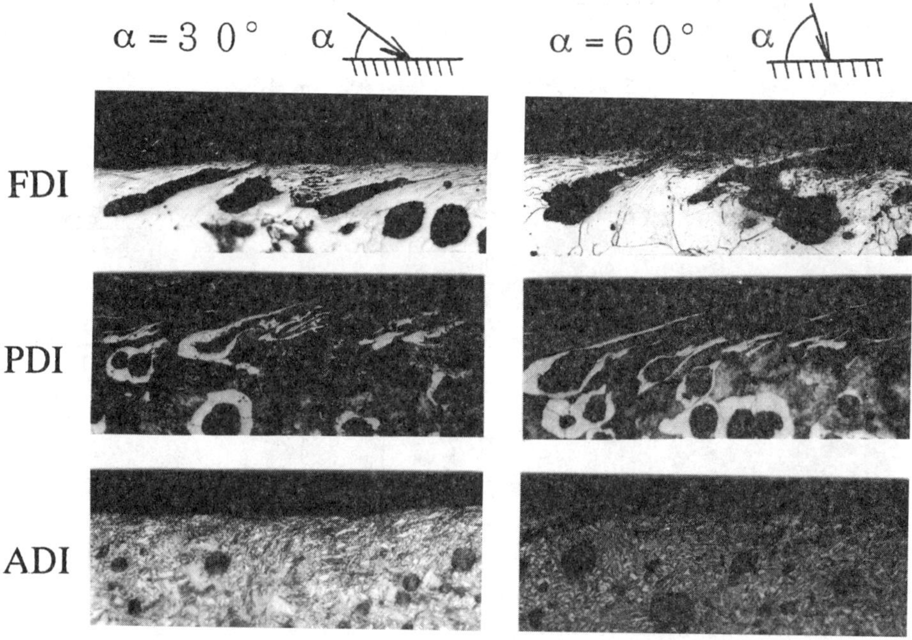

Figure 8   Microstructure of the eroding surfaces of F/PDI and ADI

2) In erosion of FDI and PDI, ridges are produced and erosion proceeds by their growth and breaking off. Near the high impact angle, at 90° , the erosion proceeds with formation of ridge in every direction.

3) With ADI, at impact angles around 60° (20° to 70°), small 5~25 μm ridges are formed in the direction of the particle impact and the erosion proceeds by their breaking off.

4) With ADI, the impact by the particles transforms the retained austenite on the surface into martensite, and this hardening decreases the erosion rate. The above shows that ADI has excellent erosion resistance and that it can be used to replace spheroidal graphite cast iron.   This extraodinary material with its superior erosion resistance can be expected to find applications in various fields.

6. Acknowledgement

The autors wish to express their gratitude to Dr. Masato Goka (Hitachi Metals, L.t.d) for his suggestions and advice in performing the experiments.

7. Reference

1) K.Shimizu & T.Noguchi, AFS, 93-078 (1993)
2) K.Shimizu ,T.Noguchi and Others, JSME, 920-78 (1992), 483
3) K.Shimizu & T.Noguchi, Wear, A2751 (1994) in print
4) K.Shimizu & T.Noguchi, JFS, 66-7 (1994)
5) K.Shimizu ,T.Noguchi and Others, AFS, 94-063 (1994)
6) I.Finne, Wear, 3 (1960), 87-103
7) J.G.A. Bitter, Wear, 6 (1963), 5-21
8) J.G.A. Bitter, Wear, 6 (1963), 169-190
9) Neilson.J.H and Gilchrist.A, Wear, 11 (1968), 111
10) K.V.Pool, C.K.H.Dharan & I.Finne, Wear, 107 (1986), 1-12
11) T.-H.Tsiang, ASTM STP 1003 (1989), 55-74
12) A.W.Ruff & L.K.Ives, Wear, 35 (1975), 195-199
13) S.Okazaki & K.Hasegawa, JSME, A , 56-527 (1990), 1668-1671
14) R.C.Tucker, Wear Failures, Metals Handbook, 11 (1986), 145-162
15) R.C.Tucker, Wear Testing, Metals Handbook, 11 (1986), 602-607

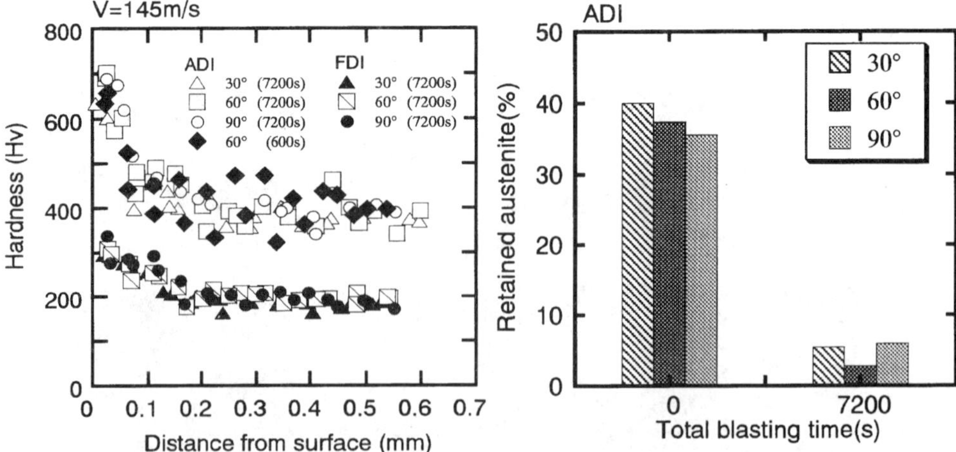

Figure 9 Hardness vs. distance from surfaces for FDI and ADI

Figure 10 Retained austenaite (%) without and with 7200s blasting for ADI

Advanced Materials Research Vols. 4-5 (1997) pp. 245-250
© *1997 Scitec Publications, Switzerland*

Study of the Wear Resistance of Gray, Vermicular and Ductile Iron under the Oxidational Condition

Y. Liu[1,2], S. Ren[3], J.M. Schissler[1,2] and J.P. Chobaut[2]

[1] LSG2M CNRS-UA 159 Ecole des Mines de Nancy, Parc de Saurupt, F-54042 Nancy, France

[2] CRITT METALL-2T, Parc de Saurupt, F-54042 Nancy, France

[3] Harbin University of Science and Technology, Xuefu Lu, 150080 Harbin, P.R. China

Keywords: Wear, Surface Oxidation, Cast Iron

Abstract

In this paper, the influence of surface oxidation on the wear-resistance of ductile iron, grey iron and vermicular iron during dry sliding friction and the mechanisms of wear have been studied. The results show that the effect of surface oxidation (formed under normal atmospheric conditions) on the wear rate depends on graphite morphology and matrix structure in a complex manner. Generally the presence of surface oxidation make the wear rate of gray iron decreasebut the wear rate of ductile iron and vermicular iron increase when the cast iron has high hardness. This trend is reversed for low hardness cast iron.

Introduction

When rubbing takes place in an oxidising environment, such that in air, the surface reactions and oxide films are found to form on the rubbing metals. The surface films are subsequently removed by further rubbing. This process is continually repeated and results in oxidational wear[1,2,3]. The original wear of the metal surface may be due to adhesion, however the worn spalling iron oxides become debris which may cause abrasion on the surface. Therefore the metal wear under atmospheric conditions usually results from a complex interaction between various wear mechanisms.

Some references have noted that surface oxidation of metal is not always detrimental [4,5,6]. For example, it has been shown that metallic wear may be increased under conditions of vacuum where oxide films were not able to form. This is due to some oxide films preventing the adhesion of metal asperities and thereby reducing the effect of the original process of wear. Successful industrial applications using the above principle have been reported. For example, plain stainless steel bearings which are used at elevated temperatures in nuclear reactors [1] have tough and tenacious oxide films, which offer excellent protection from wear. Despite the fact that most published research is chiefly concerned with steel, a systematic study of the oxidational wear of cast irons is lacking. Some wear mechanisms based on assumptions have been put forward but they lack experimental data. From factors affecting wear such as graphite morphology, mechanical characteristics, thermal properties and corrosive environment, this paper mainly studies the influence of surface oxidation on the wear-resistance of ductile iron, grey iron and more specially, vermicular iron, under normal atmospheric conditions and a argon protective atmosphere.

Experimental Approach and Conditions

To study the effects of different graphite morphologies on the wear-resistance of cast iron, various spheroidizers were used to produce ductile iron and vermicular irons. Selected samples of vermicular iron and ductile iron were normalized to compare the effects of different matrix structures.

The form of the graphite was defined by the graphite morphology coefficient which was calculated from the equation $K=4\pi S/L^2$, where S and L are respectively the area and circumference of graphite[7]. The chemical composition of studied cast irons, the values of K, the hardness and the quantities of spheroidizer are given in Table 1.

Dry sliding friction tests were carried out on a Sliding Wear Tester. A schematic diagram of samples tested are given in Figure 1. During the experiment, the upper samples were fixed and loaded vertically onto the curved surface of a rotating ring (lower specimen) of quenched grey cast iron (Rockwell hardness of 41 HRC), 40 mm in diameter, 10 mm wide, and the turning speed of the ring was maintained at 200 turns per minute. The applied load was always 15 kg. Upper specimens

Table 1 Chemical composition, hardness, quantities of spheroidizer and graphite morphology coefficient of experimental irons

Type	C%	Si%	Mn%	P%	S%	Cr%	K	HB	QS%
Grey iron	3.2-3.4	1.8-1.2	<0.7	<0.07	0.12	0.3	0.095	295	
Ductile iron	3.6-4.0	2.4-2.6	<0.7	<0.07	<0.04		0.801	197	1.7
Re-Si-Ca CV iron							0.245	179	1.4
Re-mg CV iron	3.4-3.6	2.0-2.2	<0.7	<0.07	<0.04		0.305	197	0.7
Re-Mg-Sn CV iron							0,389	213	0.75

Note: $K=4\pi S/L^2$;   Re--Rea earth;   QS--Quantities of Spheroidizer;
CV iron--Vermicular Graphite iron.

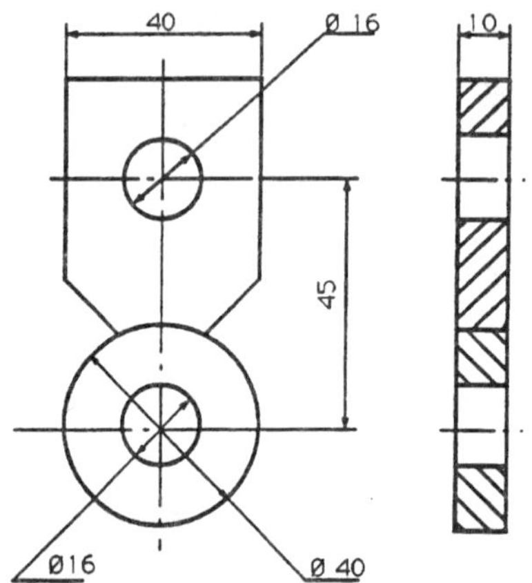

Fig. 1 *Shape and size of wear test sample*

consisted of grey iron, vermicular iron and ductile iron. The loss of material due to wear was determined by interrupting sliding after 40 minutes and weighing the samples. The value of wear "W" is given by a decrease in sample weight. By comparing the relative wear resistance, ε is defined by the relation: $\varepsilon=W_1/W_2$ where W_1 is the weight loss of gray iron sample, chosen as the reference wear quantity and W_2 is the wear quantity of other samples. All wear rates reported here are those measured after equilibrium wear had been established.

The wear tests under a protective argon atmosphere were carried out within a sealed cabin. In addition to measurements of wear rate, selected wear surface and wear debris were examined in detail using scanning electron microscopy and X-ray diffraction.

Results and Discussions

Under normal atmospheric conditions, dry sliding wear of cast iron is a complex process. It is mainly generated by the mechanisms of adhesion, corrosion and microcutting. There is usually complex interactions between various mechanisms, which occur in varying degrees. In this study, the purpose and function of the use of argon is to eliminate the effect of oxidational wear. Figure 2 shows the results from tests carried out under normal atmospheric conditions and under an argon atmosphere.

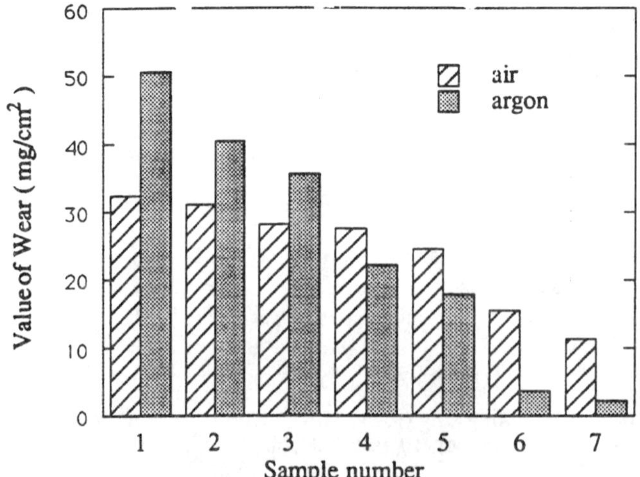

Fig. 2 *The wear values of different cast irons in air and in argon atmosphere (sliding length: 1 km, applied load: 15 kg)*

As shown in Fig. 3 curve a, Fe_2O_3 can be observed from the X-ray diffraction pattern of the wear debris which was formed in normal atmospheric conditions. Oxides were not found under argon atmosphere, as shown in Fig. 3 curve b. These results clearly show that argon can effectively stop samples from oxidising. Hence the method designed is satisfactory.

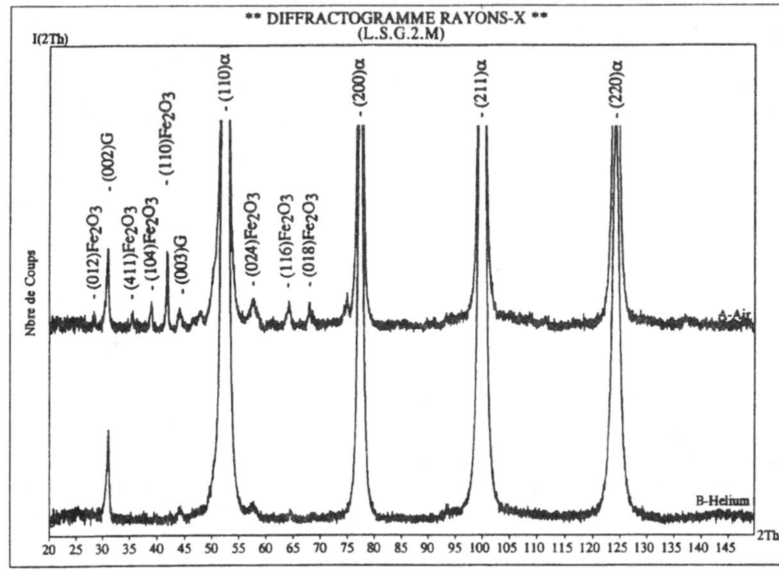

Fig. 3 X-ray diffraction pattern of the wear debris of ductile iron formed under normal atmospheric conditions (a) and under an argon atmosphere (b) (sliding length: 1 km, applied load: 15 kg)

From Fig. 2, it can be seen that the wear rates change significantly between normal atmospheric conditions and argon protection conditions. This variation is strongly related to the graphite morphology and the matrix structure of the cast iron.

The test results indicated that the wear rate of vermicular iron was between that of grey and ductile iron. The disparities of wear-resistance effected by the three graphite morphology types were mainly caused by the different degrees of stress concentration near the edges of graphite as a result of the surface deformation occurring during the friction of cast iron. Certainly, the higher the stress concentration, the easier the micro-crack occurrence and expansion. The latter finally resulted in wearing. Figure 4 is a S.E.M. micrograph of three graphite morphologies. It is shown [see Fig. 4(a)] that the interconnected graphite flake of gray iron was sharp and thin. The micro-cracks were easily formed and spread at the tips of graphite flakes where there were high stress concentrations. Figure 4(b) shows that the chunky graphite of vermicular iron took the form of blunt edged stubby

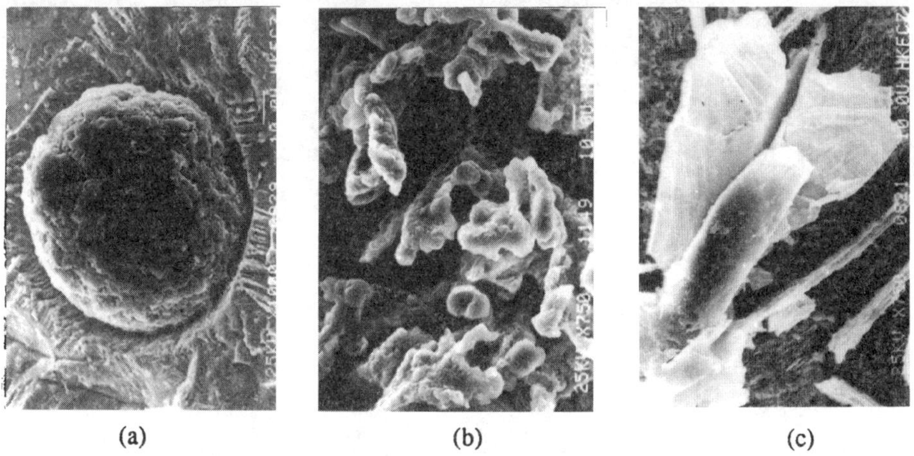

(a)   (b)   (c)

Fig. 4 SEM micrographies showing the various graphite morphologies
(a) flake graphite   (b) vermicular graphite   (c) ductile graphite

flake that decreased the stress concentration compared to flaky graphite. Furthermore, isolated spheroidal graphite [see Fig. 4(c)] not only kept its original good mechanical properties to the greatest extent, but also greatly promoted the resistance to deformation of surfaces of the metal matrix. The above-mentioned was the main reason why the wear-resistance of vermicular and ductile iron were better than that of gray iron.

In this experiment, the wear value of the gray iron increased remarkably in argon because its flake graphite practically cut and/or isolated the matrix[8]. Taking the loss of sample weight during the test under normal atmospheric conditions as W_{air} and under the argon atmosphere as W_{Ar}, ΔW was defined as the amount of variation of wear by the relationship: $\Delta W = W_{air} - W_{Ar}$. Thus, the ΔW value of gray iron reached -18.24mg/cm^2.

The wear value of ductile iron, both under as-cast and normalized conditions, decreased under an argon atmosphere. Furthermore, the higher the hardness of its matrix, the lower the average wearing value. ΔW of the normalized ductile iron was 9.14mg/cm^2.

The test results of vermicular iron were divided into two groups based upon the different effects of the argon atmosphere. Under the protection of argon, the wear rate increased for both vermicular irons with added Re-Si-Ca and ReMg as spheroidizer. The value of ΔW of ReMg vermicular iron reached -7.4mg/cm^2. On the other hand, the wear rate decreased for the Re-Mg-Sn and normalized vermicular irons under the same conditions. As the matrix hardness got higher, ΔW was up to 11.76mg/cm^2.

From observations and analysis of the debris and of appearance of the wear surface during the study, the oxidational wear process may be described by the following sequence:

Under normal atmospheric conditions, the surface of cast iron reacts chemically with oxygen, forming oxide films. A clean metal surface will acquire such a film of 5 to 50 molecules thick within a few seconds[1]. In the real area of contact, oxide islands or plateau are developed which are preferentially oxidized due to high friction heating[9,10,11]. When an oxide film covering a sliding asperity is removed by rubbing, the clean metallic surface will almost instantaneously be recovered by a monomolecular layer. Consequently, if the oxide is removed from an asperity by rubbing, it will almost certainly be reformed before the asperity makes another contact with the opposing surface. When the oxide film reaches a certain thickness, its rate of growth on the friction surface will decrease exponentially with time, and therefore, unless the oxide film is removed by rubbing, the metal-to-oxide reaction will rapidly become negligible [1].

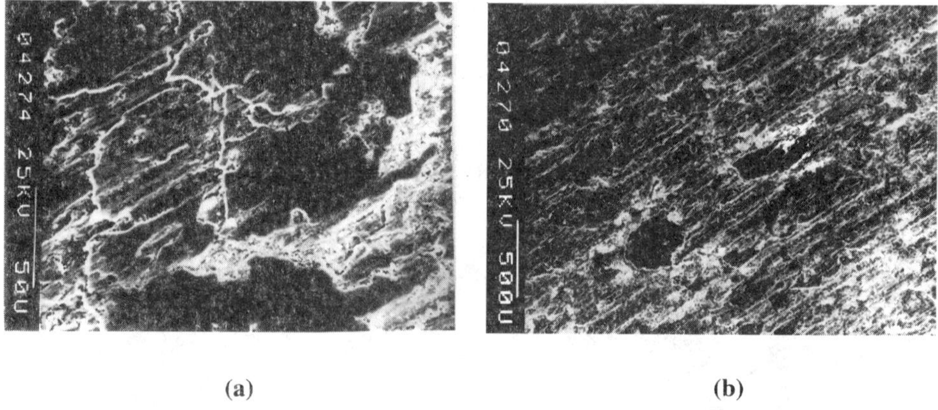

(a)   (b)

Fig. 5 Morphology of wear surface and the debris formed during oxidational wear (a) and under argon atmosphere (b) (Vermicular graphite iron, sliding length: 1km, applied load: 15kg)

Two different results may take place after the formation of oxide films during the rubbing:

If the cast iron has a high strength and hardness, and thus good properties for a hindering adhering and microcutting, the oxide film will have enough time to grow to a critical thickness and then separate from the underlying metal and break up to form flakes and/or powder wear debris (Fig. 5a). This is caused by normal and tangential stresses[12,13]. In view of what is written above, the wear under these conditions occurs chiefly by a corrosive mechanism, oxidation being a major cause of this wear.

In our experiment, the hardness, strength and toughness of Re-Mg-Sn and normalized vermicular irons, as-cast and normalized ductile irons were relatively higher than the other irons. These irons had better properties to hinder adhesive wearing and microcutting. Oxidational wear greatly increased the wear rate in a normal atmospheric condition. However, for the rubbing process in argon, the wear occured only by adhesion and microcutting. It is kown that adhesive wear occured easily on a clean interface under the non-oxidation environment, and that wear rate is inversely proportional to the strength and hardness of the matrix. Thus, the advantage of the combination of the matrix and graphite morphology of the cast irons mentioned above appeared significantly. This is why their wear resistance under argon conditions were greater than that found in air. More simply, oxidation will promote the wear rate of cast iron if it has good adhesion-resistance and microcutting-resistance.

Secondly, for the metal which has low hardness and strength, wear occurs by mechanisms of adhesion, microcutting and corrosion, in varying degrees. In this case, adhesion is a major cause of wear and plays the most important role during the wear procedure[14]. Thus, the presence of the oxide films can be very beneficial in this situation. As a "lubricating layer", oxide films can, to a certain extent or even completely, actually prevent intermetallic contact. They can decrease the normal stress which acts directly on the metal matrix and which results in a reduction of the adhesion. In this case, surface oxidation decreases wearing.

Although the gray iron selected for the experiment had relatively higher hardness, its coefficient of graphite morphology (K) was less and the graphite flakes cut apart the metal matrix. Furthermore, internal stresses of the gray iron was large and graphite flakes themselves easily split along the interface of layers which caused micro-cracks to expand. The capability of adhesion resistance and microcutting resistance of gray iron was much lower. In this case, oxide films isolated the rubbing surfaces from contact with each other and reduced adhesive wearing under normal atmospheric conditions. Consequently oxide films decreased the wear rate of gray iron.

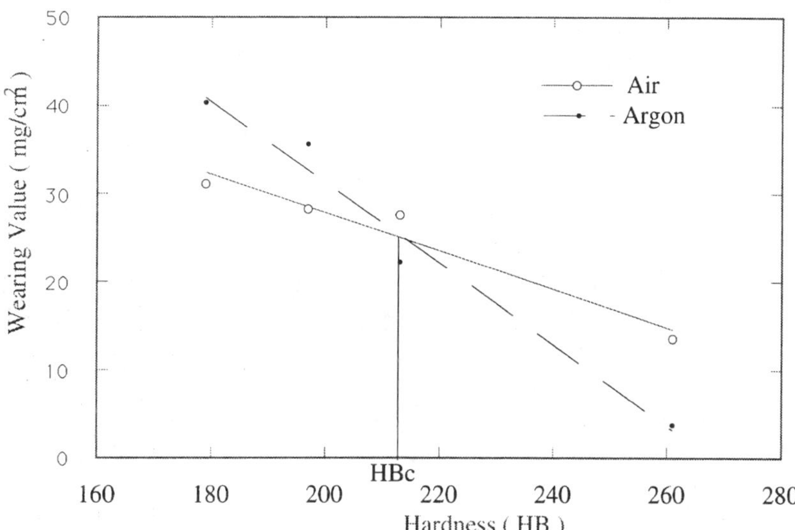

Fig. 6 *Linear regression curves of the wear value and hardness of vermicular graphite irons in air and in argon*

In contrasted to the gray iron, the vermicular graphite morphology in Re-Si-Ca vermicular graphite iron and ReMg vermicular graphite iron was better than that in gray iron. However, great quantities of ferrite existed in their matrix structure. They had lower hardness, adhesion and scratch resistances. Therefore, when the rubbing took place under argon atmosphere, the serious adhesion and microcutting occured because the load stress acted directly on the matrix (see Fig. 5b). Thus the wear rate was high. On the contrary, the wear rate of the irons mentioned above under the normal atmospheric condition was significantly reduced, because the "lubrication" role of oxide films increased their adhesion resistance. It is clear that surface oxidation may decrease wearing if the cast iron has a low adhesion resistance and a low microcutting resistance.

In addition, from the analyses of the experimental data of four types of vermicular graphite iron, it was found that, for the vermicular irons whose graphite morphologies were between flake and spheroid (the coefficient of graphite morphology was from 0.245 to 0.389), the effect of surface oxidation on the wear resistance depended on the hardness of matrix. Linear regression was used to obtain two equations describing the wear data of vermicular irons:

under an argon atmosphere $W=66.333-0.19164HB$, $R=0.947935$
under normal atmospheric condition $W=126.36-0.46054HB$, $R=0.98809$

As shown as Fig. 6, it is obvious from the regressive curves that the wear rate of vermicular graphite iron under an argon atmosphere is higher than that under normal atmosphere when the hardness was less than a critical value (HB_c); whereas if the hardness was greater than HB_c, the wear rate under argon atmosphere decreased compared to that in normal atmospheric conditions.

Conclusions

1. Under a normal atmospheric conditions, surface oxidation may either significantly decrease or increase the wear resistance of cast iron. This effect depends on both the characteristics of graphite morphology and the metal matrix.

2. Generally, surface oxidation may decrease the rate of wear for gray iron and increase wear rate of ductile iron. The experimental values of wear of grey iron and normalized ductile iron produced in this study under normal atmospheric conditions were 0,64 and 5,04 times those obtained in argon.

3. With an approximately similar graphite morphology, the surface oxidation of vermicular iron which has high strength and hardness, increases wear rate but decreases wear rate if the vermicular iron has lower hardness and strength.

References

[1] J. Halling, Principle of Tribology, The Macmillan Press LTD, London, (1975)
[2] K. Okabayashi, IMONO, Japan, (1974) No.4
[3] K. Okabayashi, Report of the 101th Regular Grand Lecture Meeting, IMONO, Japan, (1982)
[4] H. Czichos, Tribology, Elsevier Scientific Publishing Co., New York, (1978)
[5] Karl-Heninz Zum Gahr, Microstructure and Wear of Materials, Elsevier Scientific Publishing Co. New York, (1987)
[6] A. Iwabuchi, Wear of Materials, 1991 Vol. 1, 49-54, USA (1991)
[7] N. Tsutsumi, Analyze of Structure on the CV Graphite Cast Iron, Report of the 101th Regular Grand Lecture Meeting, IMONO, Japan, (1982)
[8] R.C. Voigt, S.D. Holmgren, AFS Trans. Vol.98 (1990) 213-225
[9] T. F. J. Quinn, The Origins of Oxidational Wear, Tribology Int., October (1983) Vol.16 No.5
[10] T. F. J. Quinn, Recent Developments and Future Trends in Oxidational Wear Research, Tribology Int., December (1983) Vol.16 No.6
[11] T. F. J. Quinn, D. M. Rowson and J. L. Sullivan, Application of the Oxidational Theory of Mild Wear to the Sliding Wear of Low Alloy Steel, Wear, 65(1980)
[12] J.J. Liu, Y. Chen, Y.O. Cheng, Wear, May (1992),154 (2) 259-267
[13] S.T. Dktay, N.P. Suh, J. Tribology, Apr. (1992), 114 (2) 379-393
[14] R. Okumoto, IMONO, Japan, (1974) No.4

Advanced Materials Research Vols. 4-5 (1997) pp. 251-258
© *1997 Scitec Publications, Switzerland*

Wear and Fatigue Properties of Austempered Ductile Iron

J. Tartera, J.M. Prado and À. Pujol

Dpt. Ciència dels Materials i Enginyeria Metallúrgica, Universitat Politècnica de Catalunya,
Diagonal 647, E-08009 Barcelona, Spain

Keywords: ADI, Wear, Fatigue

ABSTRACT

A series of austempered ductile iron samples were prepared at austempering temperatures of 370, 340, 300 and 270 °C. Their sliding dry wear behaviour was studied under applied loads of 110, 300, 500 and 800 N. The results obtained showed that the wear resistance is independent of austempering temperature for low applied loads, but it shows a strong dependence on austempering temperature reaching a maximum value for 340 °C. Rotary bending fatigue test were also carried out. Fatigue limit showed again a maximum value for an austempering treatment at 340 °C. Wear and fatigue are considered as related phenomena being the crack growing stage the controlling mechanism.

INTRODUCTION

Ductile cast iron has been evolving during the last decades giving place to better combinations of mechanical strength and toughness. A further and recent advance has been the introduction of the thermal treatment of austempering. The resulting microstructure, consists in a mixture of bainitic ferrite and carbon enriched austenite. Mechanical properties obtained in this way are very attractive with high strength (1000-1500 MPa), acceptable ductility (3-15% elongation) and toughness (60-90 MN.m$^{-3/2}$) [1,2]. Thus, the use of austempered ductile iron (ADI) has been increasing steadily in the last few years and consequently many mechanical pieces such as connecting rods crankshafts and even gears are being made out of this material with energetic and economic advantage.

Several reviews have been published to evaluate the response to thermal treatment of the mechanical behaviour of ADI [3,4]. But, up to present, very little of this work has been focused onto the study of its dry wear and fatigue characteristics. Therefore, in this work the wear and fatigue behaviour of ductile iron under different austempering treatments are studied.

EXPERIMENTAL PROCEDURE

The chemical composition of the ductile iron used in this work is given in Table 1. Ductile iron Y-blocks were cast in sand moulds. Blanks cut out from the Y-block legs were subjected to heat treatment to obtain four different matrix structures.

Ct	Si	S	P	Mn	Cr	Ni	Cu	Mo
3,52	2,60	0,01	0,01	0,20	0,03	0,41	0,73	0,20

Table 1 Chemical composition of the ductile iron, (wt-%)

The austempering heat treatments consisted in austenitising for 45 minutes at $900°C$ followed by isothermal holding for 60 minutes in a nitrite salt bath, at four different temperatures, namely 370, 340, 300 and $270°C$.

The microstructures obtained after these thermal treatments were studied by means of scanning electron microscopy (SEM) and have been shown elsewhere [5]. In all cases the microstructure consisted in bainite ferrite and retained austenite, but thinner ferrite plates and lower amounts of austenite were obtained as holding temperature decreased.

Mechanical properties, for the different thermal treatments, such as, yield stress (σ_o), ultimate tensile stress (σ_u), hardness and K_{IC} were previously determined [2] and are given in Table 2. Also included are the amounts of retained austenite for each austempering temperature, which were obtained by X-ray diffractometry [6].

	$270°C$	$300°C$	$340°C$	$370°C$
σ_u (MPa)	-	1230	1108	942
σ_o (MPa)	-	1189	1063	761
K_{IC} (MPa·$\sqrt{m}$)	57	65	70	64
Hardness (HV)	450	440	394	322
$\gamma(\%)$	10	18	29	36

Table 2. Mechanical properties for the different treatments.

Wear tests were carried out in a Shimadzu-Nishihara machine, in which two ring-shaped specimens were made to roll-slide against each other. This tests were stopped when the total weight lost was approximately two grams. Consequently, the sliding distance for each test was different. Details of the experimental arrangement have been given elsewhere [7]. Specimens were 30 mm. in diameter and 8 mm. wide. Both upper and lower specimens were of the same material and thermal treatment. The sliding velocity was held constant at a value of 0,11 ms^{-1}.

The variation of weight lost with sliding distance, for both upper and lower specimens, under four different applied loads (110, 300, 500 and 800 N) was measured. In this way the dependence of wear rate with applied load for each austempering temperature and with austempering temperature for each load could be studied.

Rotary bending fatigue tests were carried out by using a Shenck PUPN machine. Specimens had a gage length of 96 mm and 7,52 mm in diameter. Fracture and worn surfaces were studied by means of SEM.

EXPERIMENTAL RESULTS

Wear

The variations in the weight lost with sliding distance under the four different applied loads and for the five specimens families are shown in Fig. 1. Four of them were obtained with different isothermal treatments (a, b, c, d) and the other without isothermal treatment. In all cases we have represented the total weight lost between the two specimens in contact. The weight lost shows a linear relationship with sliding distance, for all the thermal treatments and applied loads, with no indication of an initial transient stage of severe wear. As it was expected the effect of the applied load, for the different thermal treatments, is to increase the weight lost at a given slid distance. This was particularly marked for applied loads of 800 N.

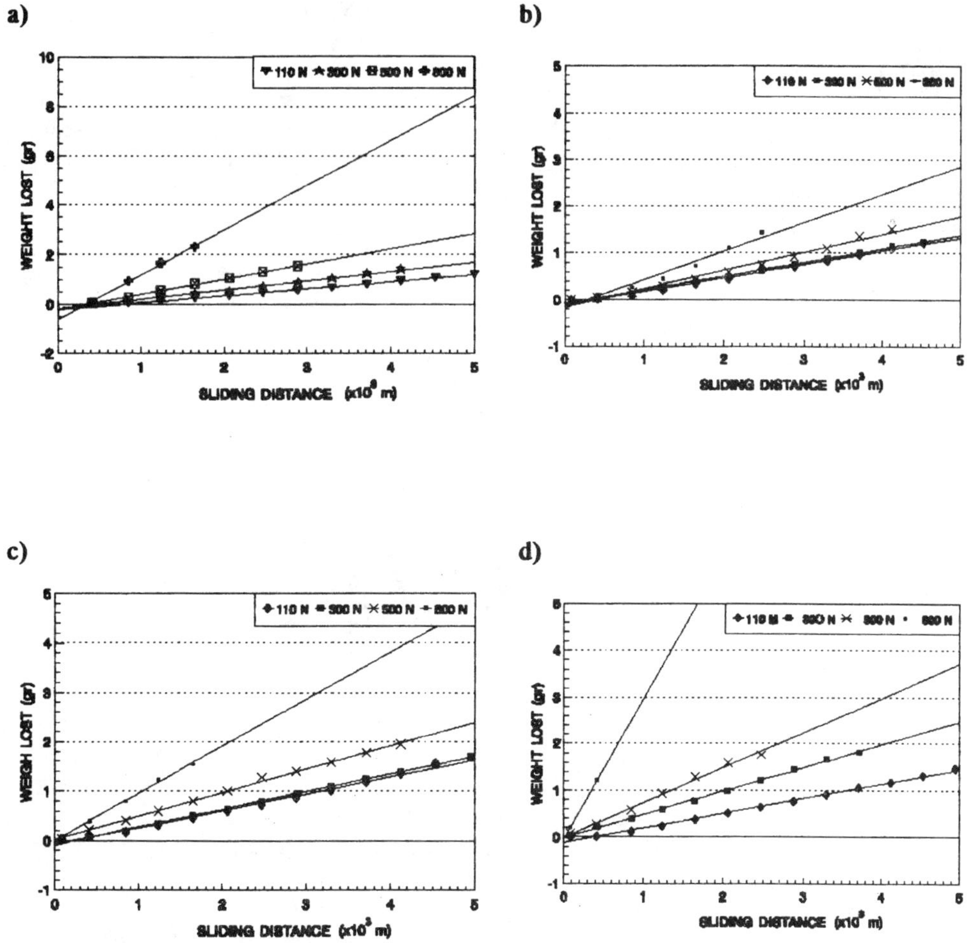

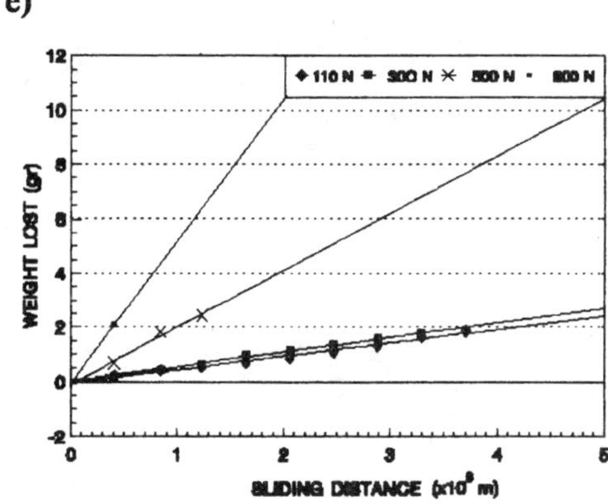

Fig. 1 Variation of weight lost with sliding distance for the different applied loads with thermal treatments at 370°C (a), 340°C (b), 300°C (c), and 270°C (d); and without thermal treatment (e)

A classical way of representing these results is shown in Fig. 2, where the variation of wear rate with applied load for each of the five specimens families is depicted. Each point of this figure corresponds to the slope of one of the straight lines in Fig. 1 (a, b, c, d, and e). It is clear from Fig. 2 that increasing loads below 300 N have a small effect on wear rate. But above this load wear rate increases quite rapidly for all thermal treatments. From the results in Fig. 2 it can be stated that nodular iron is less resistant to wear than austempered ductile iron. When a load of 800 N was applied, the wear rate of nodular iron was an order of magnitude higher than that of ADI at the austempering temperature of 340°C.

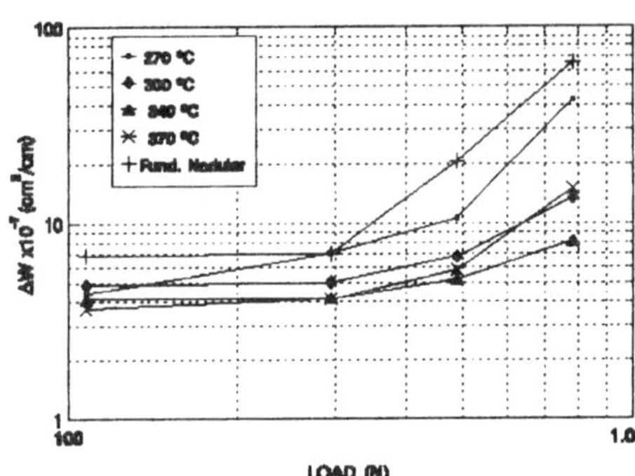

Fig. 2. **Variation of wear rate with applied load**

The effect of austempering temperature on wear rate is better shown in Fig. 3, which shows the variation of this parameter with holding temperature. Four curves, one for each applied load, have been represented in this figure. The smaller applied loads (110 and 300 N) produce what can be considered as "mild wear" for all thermal treatments with very little change between them. On the other hand, wear rate for the two higher loads shows a clear dependence on austempering temperature which is better marked for the applied load of 800 N. In this latter case it seems to exist a holding temperature of 320-340°C for which wear rate has a minimum value. The decrease in wear resistance is also steeper on the lower austempering temperatures side than on the higher one. This has the unusual meaning that a higher bulk hardness can result in a greater wear rate.

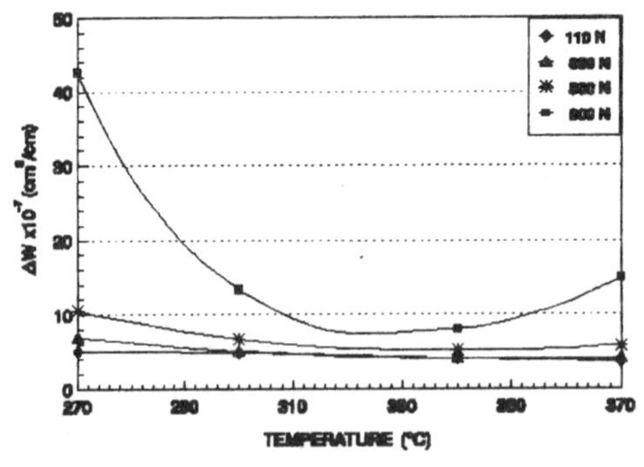

Fig. 3. **Variation of wear rate with austempering temperature**

It is now important to study the microstructural changes that could have occurred in the subsurfaces of the samples as the TRIP effect (transformation induced plasticity) has been frequently invoked to explain the wear behaviour of steels with high contents of retained austenite and even that of ADI [8].

Fig. 4 shows the variation of hardness with distance from the surface for two representative samples after a wear test. Very high increases in superficial hardness can be observed but affecting only a very narrow layer, thinner than 100 μm, below the surface.

a)   b)

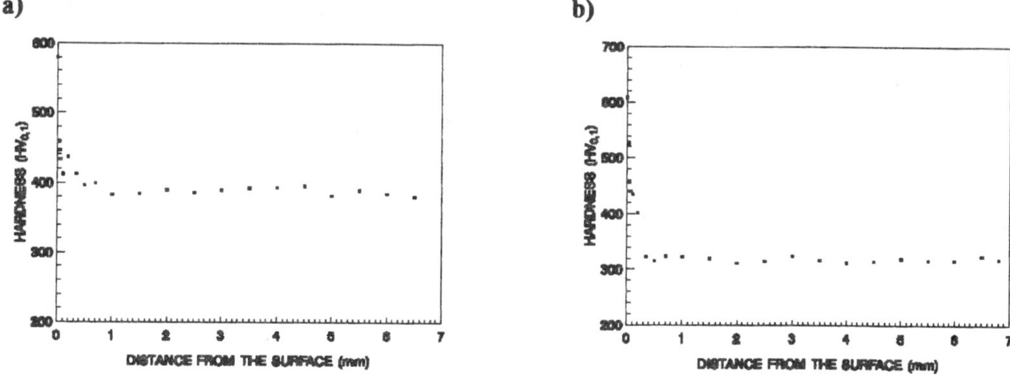

Fig. 4. **Variation of hardness with distance from surface**

Fig. 5 shows a SEM micrograph of the microstructure below the surface of a sample treated at 340°C. No TRIP phenomenon is distinguishable, but what seems only to have happened is a strong plastic deformation of the retained austenite. This plastic deformation is also patent in the graphite nodules close to the surface which deformed and acquire an ellipsoid shape oriented according to the imposed deformation. The depth of the layer plastically deformed is of about 50 μm.

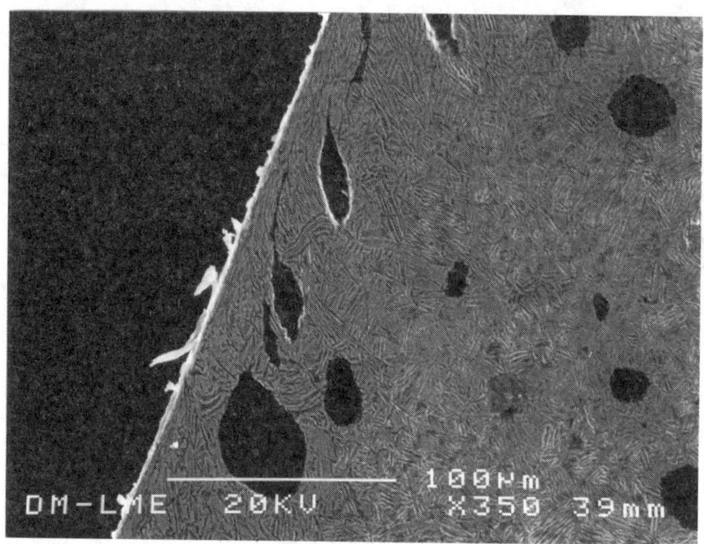

Fig.5
Microstructures of the plastically deformed layer below the surface for a specimen

The experimental results presented here seems to indicate that the wear behaviour of ADI is governed by a somehow different mechanism of that acting in high hardness steels.

Rotary bending fatigue

The S-N fatigue curves corresponding to the different thermal treatments are shown in Fig. 6. Specimens treated at 370, 340 and 300°C present a similar low-cycle behaviour, but those treated at 270°C have shorter fatigue lifes.

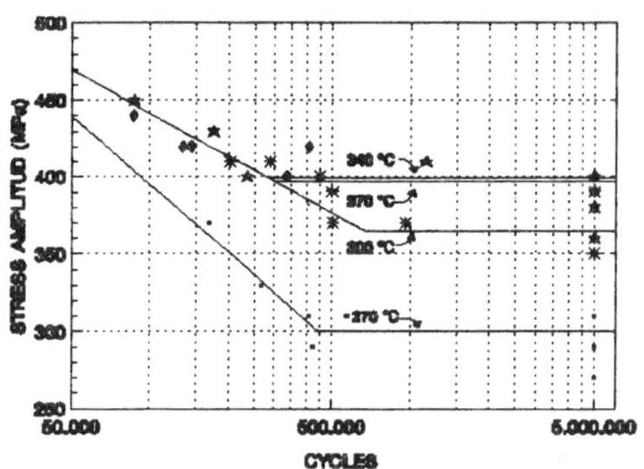

Fig. 6. The S-N fatigue curves for the different thermal treatments

The fatigue limit reaches a maximum for an austempering temperature of 340°C and decreases for lower and higher isothermal holding temperatures. This dependence of the fatigue limit with both, austempering temperature and hardness, is better shown in Figs. 7 and 8 respectively.

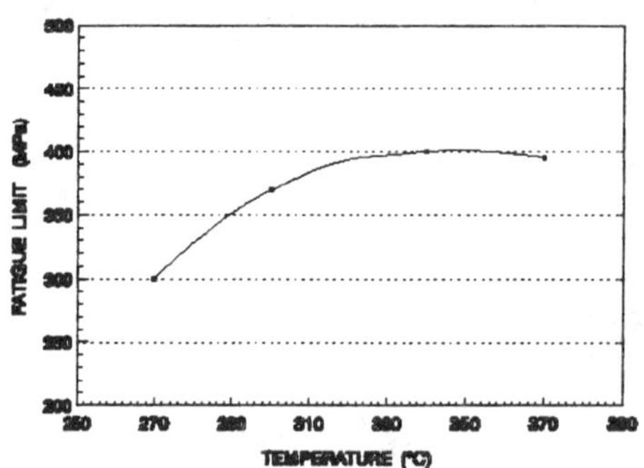

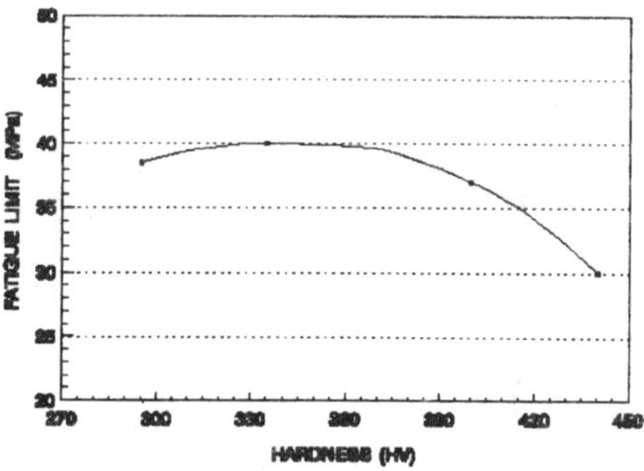

Fig. 7. Dependance of the fatigue limit with austempering temperature

Fig.8. Dependance of the fatigue limit with hardness

The values of the fatigue limit obtained for ADI with austemperings at 370, 340 and 300°C are comparable with those of steels treated between 800 and 1000 MPa, and clearly higher than those of conventional perlitic nodular cast irons.

The observation, by means of SEM, of the fatigue crack nucleation sites shows that they mainly correspond to the graphite nodules of bigger size situated at or near the specimen surface (Fig. 9). Occasionally, internal defects such as shrinkage microvoids can also nucleate cracks fatigue. It is, therefore, very important, in order to improve the fatigue properties of ADI to control the size and shape of graphite nodules and also the soundness of the cast.

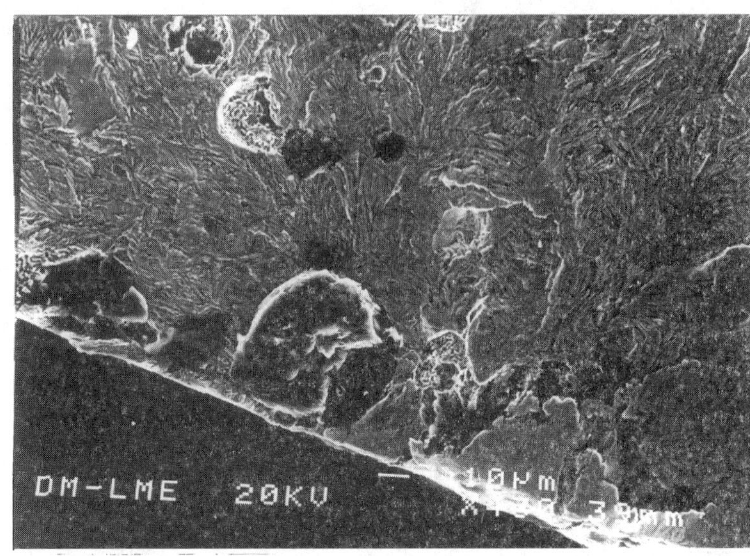

Fig. 9.
Microstructure of fatigue crack

DISCUSSION

In agreement with many authors [9,10], wear can be considered, for the levels of applied loads and sliding velocity of this work, as a fatigue phenomenon. Hence, it consists in the nucleation of cracks below the surface, growth of these cracks and final breaking of a metal chip.

In steels nucleation of cracks is the controlling mechanism of wear being the "delamination theory" the more generally accepted one to explain this stage of wear [11]. In ADI the crack nucleation is easier than in steels due to the presence of graphite nodules; as it is shown in Fig. 5 graphite spheroids close to the surface deform readily becoming ellipsoid-shaped. Therefore, they soon act as cracks just below the surface of the specimens. As a consequence of this facility for crack nucleation this stage is no longer the controlling wear mechanism. Consequently this role can be taken over by crack growth.

In sliding wear cracks grow parallel to the surface under mode I or II of loading (or more probably a mixed mode of both) [9] until they reach a length for which the complex state of stress at the crack tip changes its path towards the surface; final breaking of a metal chip follows immediately.

It is logical to imagine that, also in this case, crack growth rate can be described with the well-know Paris equation:

$$\frac{\delta a}{\delta N} = C \cdot (\Delta K)^m \qquad\qquad (Eq.\ 1)$$

where $\delta a/\delta N$ represents the crack grown rate and C and m are materials constants. The ΔK term correspond to the stress intensity range, which is the difference between the maximum and minimum stress intensity factors and is given by:

$$\Delta K = K_{max} - K_{min} \qquad\qquad (Eq.\ 2)$$

According to Murakami et al [11]. the stress intensity factor K can be expressed, for a defect or a crack near the specimen surface, as:

$$K = 0,650 \cdot \sigma \, (\pi \cdot \sqrt{Area})^{1/2} \qquad\qquad (Eq.\ 3)$$

the area corresponds to that of the defect projected onto the plane perpendicular to the maximum tensile stress σ. Because during a cycle the applied stress on each point of the rings surfaces varies between σ (nominal applied stress) and zero, depending on whether they are or not under contact, the value of K_{min} is zero and the stress intensity range ΔK coincides with K_{max} becoming a function of applied stress and crack size of the type of expression (Eq.3).

As is shown in Table 2 the value of K_{Ic} increases with increasing amounts of retained austenite up to a value corresponding to a holding temperature of $340^{\circ}C$, from which it stays constant or even shows a slight decrease. We think, that this behaviour of K_{Ic} together with the fact that cracks nucleate readily at graphite nodules at a rate independent of applied load, for the levels of load here used, can explain the wear results obtained in this work.

The values of m and C in Paris equation have not been sufficiently studied for the case of ADI, but it seems that microstructure has only a small influence on them [12]. Hence, the extension of the crack grown stage depends mainly on the value of K_{Ic}, because when $K_{max} > K_{Ic}$ unstable propagation of the crack takes place. Small values of K_{Ic} will make easier for cracks to reach their critical length; but small applied loads means also small initial values of K_{max} and cracks will have to grow longer before they reach a critical length. Thus, both K_c and σ, a material characteristic and an external factor respectively, determine the dry wear behaviour of ADI.

When applied loads are small $K_{max} < < K_{Ic}$ at the beginning of the crack growth stage. Integration of equation (1) indicates that the number of cycles needed to reach the crack its critical length, such that $K_{max} > K_{Ic}$, is very high because crack growth rate, $\delta a/\delta N$, is small during most of the crack growing stage. In this situation the influence of the value of K_{Ic} is of secondary importance and consequently also the thermal treatment has a small influence in the dry wear behaviour.

High applied loads mean initial values of K_{max} close to K_{Ic}. Now crack growing stages are short because crack growth rates, $\delta a/\delta N$, are high from the start. The K_{Ic} becomes an important parameter to determine the length of the crack growing stage. The dry wear behaviour is now highly dependent on austempering temperature.

The results of rotary bending fatigue can be explained in a similar way to those of wear. This statement implies that graphite nodules act as cracks that propagate from the beginning of the fatigue test. In this case fatigue life is again controlled by the crack growing stage together with the value of K_{Ic}. If graphite nodules had to act as cracks the initial ΔK_{app} should be equal or greater than ΔK_{th} for each thermal treatment. Values of ΔK_{th} quoted in literature for ADI are very scarce, but according to Bartusiewicz et al. [12] they decrease with decreasing retained austenite, with values between 6 and 4 MPa$\sqrt{m}$ in agreement with the statement that $\Delta K_{app} \geq \Delta K_{th}$.

REFERENCES

1. T.N. Rouns; K.B. Roundman; D.M. Moore. A.F.S. Trans., 1984, **92**, 815.
2. J.Tartera, J.M. Prado, E. Roig, M. Marsal. Tecnica Metalurgica, 1989, **44**, 11.
3. R.C. Voight, C.R. Loper. First International Conference on Austempered Ductile Iron., 1984, 83.
4. R.B. Gundlach, J.F. Janowak. First International Conference on Austempered Ductile Iron., 1984, 1.
5. J.M. Prado, A. Pujol, J. Cullell, J. Tartera. Dry sliding wear of ADI, Materials Science and Technology, to be published.
6. H. P. Klug and L. E. Alexander, X-ray Diffraction Procedure for Polycrystalline and Amorphous Materials, 2nd edn., Wiley, New york, 1974, 554.
7. J.L. Arques, J.M. Prado. Wear, 1985, **103**, 321-331.
8. W.S. Zhou, Q.D. Zhou, S.K. Meng. Wear 1993, **696**, 162-164.
9. J.P. Hirth, D.A. Rigney. Strength of Metals and Allogs, 1979, 3,1883.
10. S.C. Lim, M.S. Ashby. Acta Metalurgica. 1987, **35**, 1.
11. Y. Murakami, S. Kodama, S. Koduma. International J. Fatigue, 1989, 11, 291.
12. L. Brartosiewicz, A.R. Krause, B. Kovacs, S.K. Putatunda. A.F.S. Trans., 1992, **100**, 135.

Advanced Materials Research Vols. 4-5 (1997) pp. 259-266
© *1997 Scitec Publications, Switzerland*

Shaping the Structure and Properties of ADI with High Mn Content through Changes in Chemical Composition and Heat Treatment

A. Kowalski and M. Biel-Gołaska

Foundry Research Institute, 73 Zakopiańska Street, PL-30-418 Cracow, Poland

Keywords: ADI, Chemical Composition, Microstructure, Mechanical Properties, Heat Treatment, Hardenability, Abrasion Resistance

ABSTRACT

Basing on Hartley's plan of experiments, eleven melts of ductile iron with high Mn content (0.9%) and various content levels of Ni (0–2.8%), Cu (0–1.4%) and Mo (0–0.8%) were made. Tensile strength, elongation, impact strength and hardness were tested. The microstructural examinations of cast iron in as–cast state as well as after normalizing and austempering were carried out. The hardenability and abrasion resistance of ductile iron from the individual melts were determined. The mathematical and graphical relations between some chosen characteristics of ADI and its chemical composition were established. The ductile irons of high abrasion wear resistance, high strength and good "through" hardenability were chosen.

INTRODUCTION

It was the aim of the present research to deterrmine the effect of Ni, Cu and Mo in a selected content range on the microstructure and some mechanical and technological properties of ductile cast iron with a high manganese content (about 1%).
The effect of the above mentioned elements on the structure and mechanical properties of ductile cast iron was discussed in numerous publications. [1–8].

Manganese in cast iron makes it much more susceptible to the formation of chills and reveals a notable tendency towards the ordinary segregations. With Mn content in cast iron of about 0.4% Mn, in the boundary regions of grains the content of this element may reach even 4% Mn. The network of the formed carbides precipitates is very stable; it is not decomposed even in the course of prolonged annealing at a temperature of 900 – 1000°C [2,4]. When the content of manganese is in the range of 2–3%, in the structure may appear bainite or ever martensite. Further increase in the content of manganese (possibly also in that of other alloying elements) results in the formation of a martensitic or bainitic structure. With 2.55% Mn in the cast iron containing 3.2% C, 0.5% Si, up to 1% Cu, 0.2% Mo, the point M_s is slightly above room temperature, and the matrix has already got some martensite [1].

Copper gets dissolved in the liquid cast iron to an unlimited extent. The alloying elements do not affect in a significant way the solubility of Cu in cast iron; the only exception is nickel which makes the solubility of Cu increase strongly [2]. Copper reduces the point of eutectoid transformation by about 10° C per every 1% Cu. Along with the decreasing point of eutectoid transformation A_1, favouring the formation of fine pearlite, copper raises the temperature of the beginning of eutectoid cementite decomposition, raising in this way the stability of pearlite [4].

Like copper, nickel also reveals a tendency towards the negative microsegregation [2–4]. It reduces carbon content in austenite by about 0.05% per every 1% Ni. Nickel increases the stability of austenite in the range of temperatures of both pearlitic and bainitic transformations restricting, at the same time, the range of temperatures of a pearlitic transformation. With Ni content increasing appears the tendency towards the formation of bainitic (especially in the presence of Mo) or martensitic matrix [3].

Molybdenum hinders graphitisation during the solidification of eutectic. Although molybdenum extends the range of Fe, it also strongly impedes the eutectoid transformation of austenite, its effect being much stronger than that of Cu and Ni [4]. An important advantage of molybdenum is that it strongly affects the bainitic transformation of austenite and improves the hardenability.

METHODS OF OWN INVESTIGATION

The study undertaken at FRI had as its main aim to obtain different structures in ductile iron, depending on the limited content of alloying elements, like manganese, nickel, copper and molybdenum, and on the heat treatment, i.e. normalising and austempering. This paper is but only a fragment of a more extensive study [9].

It was decided to examine the effect of alloying elements added in the amount of:
Mn =~1%, Ni = 0 – 2.8%, Cu = 0 – 1.4%, Mo = 0 – 0.8%.

Eleven melts of ductile cast iron with various contents of Mn, Ni, Cu and Mo were made. The contents of above mentioned elements were chosen basing on Hartley's plan of experiments PS/DS – P: Ha3/hs [10], enabling obtaining of cast iron microstructure characterised by the highest abrasion wear resistance, i.e. the one composed of pearlite, bainite and/or martensite.

Table 1 gives the codes as well as the computed and obtained values of the content of alloying elements in the individual melts determined by spectroscopy.

The content of remaining elements was in the range: C = 3.4 – 3.7%, Si = 2.4 – 2.8%, Mn = 0.8 – 1.0%, P = 0.06 – 0.08%, S = 0.015 – 0.03%, Mg = 0.04 – 0.08%.

Melting procedure.

The cast iron was melted in an electric, medium–frequency, acid–lined furnace of 60 kg crucible capacity. The charge was composed in about 87% of steelmaking pig iron and in about 13% of Armco iron. The required silicon content was obtained by addition of an appropriate amount of 75% ferrrosilicon. The spheroidising treatment was carried out using FeSiMg7 master alloy containing 48% Fe, 45% Si and 7% Mg. The addition of master alloy to the crucible was 2%. The required content levels of Ni and Cu were made up introducing these metals in the form of commercially pure elements; Mn and Mo were added in the form of ferroalloys.

From each melt the Y2 test pieces for mechanical testing and structure examinations were cast; samples for the determination of chemical composition were also taken. The temperature of pouring test pieces, measured by a Pt–RhPt thermocouple, was 1360 – 1380 ° C.

Heat treatment.

Some of the ingots were heat treated, i.e. normalised and austempered .

Austempering. Austenitising of all the specimens was carried out in an electric resistance furnace at a temperature of 900 ° C for 2 h. To protect the specimens from oxidising, they were covered with charcoal.

The isothermal process was carried out in a quenching tank with salt bath (a mixture of 50% KNO_3 and 50% $NaNo_3$ at a temperature of 330 °C for 2h.

Normalising. Austenitising was carrried out at a temperature of 950 °C for 2 h; then the specimens were cooled in the air.

Metallographic and structure examinations.

The hardness of cast iron was determined in as–cast state and after heat treatment using a Rockwell test (scale HRC); the microstructure was also examined. The results of these examinations were compiled in **Table 2**.

The hardenability of cast iron was determined by Jominy test according to Polish Standard PN–79/H–04402. The austenitising temperature of 900 °C was adopted along with the following parameters of hardenability:
– maximum hardness on specimen surface,
– depth of hardening up to the hardness of 45 HRC,
– depth of hardening up to the hardness of 35 HRC.
These data are compiled in **Table 3.**

The abrasion wear resistance in oil lubrication was tested on an Amsler machine according to Polish Standard RN–60/MPC–22201. The counter specimen was made of a 55 grade steel, quenched and tempered, and characterised by a hardness of about 60 HRC. The counter–specimen diameter was 40^{+1} mm, the load applied during the test was 50 kG. The test was conducted up to 560 revolutions of the counter–specimen. The loss in weight of the specimen was measured, and the coefficient of friction was calculated. The results of measurements are given in **Table 4**.

The as–cast tensile strength, elongation and notched impact strength were measured. These characteristics were determined by traditional methods on the specimens consistent with Polish Standard PN–91/H–83123. The results of the tests are given in **Table 5.**

Following the adopted Hartley's plan of experiments, the approximation multinomials, determining the effect of alloying elements on the tensile strength, impact strength and hardness, were computed. It is to be remembered that these relationships are valid only for and in the examined range of the elements content. The following equations were obtained:

Rm = 555 + 31*Ni − 43*Cu + 122*Mo + 30*Ni*Cu − 86*Ni*Mo..............(Eq.1)

HRC = 42 − 7*Ni + 10*Ni*Mo + Cu − 15*Mo...(Eq. 2)

KVU = 7.2 − 8.3*Cu − 6*Mo + 1.7*Ni*Cu − 3*Ni*Mo + 14.7*Cu*Mo.......(Eq. 3)

Basing on the equations, the three–dimensional graphs were plotted *(Figs. 1– 3)*.

DISCUSSION OF RESULTS

In Table 1 the <u>chemical composition</u> of cast iron from the individual melts was given. It can easily be noted that the obtained composition does not stray very much from the assumed mean composition, not exceeding – as a matter of fact – +/–10%, which means the value of the scatter admissible and unavoidable under the existing conditions of melting.

As regards the as–cast <u>structure</u>, from Table 2 it follows that the graphite from individual melts assumed the shape of regular and irregular spheroids (Gf9, Gf8) with small amounts of vermicular graphite (Gf5).

In the metallic matrix are present carbides or the precipitates of cementite, caused by an elevated content of Mn and the presence of Mo. The latter element is particularly active in promoting the precipitation of cementite.

In melt no. 3 containing Cu only (1.35%) the martensite is present, while melt no. 9 contains a mixture of martensite and bainite (the melt with 1.55% Ni and 0.79% Cu). Martensite was also present in melts nos. 5 and 10 but together with ferrite. These structures were not obtained in previous studies [22,23] carried out on melts containing about 0.3% Mn. Hence it follows that the addition of 1% Mn combined with Cu (about 1.3%) and Ni (1.5%) gives even in as–cast condition a martensitic structure.

After normalising, all the melts contained martensite with austenite combined, as a rule, with carbides, and sometimes with cementite. In melt no. 2 with the sole nickel, the precipitates of pearlite and ferrite i.e. a mixture of all the phases, were also noted to be present. A mixture of this type is not favourable to the abrasion resistance, or the tensile strength, or – finally – the corrosion resistance.

Here, it should be emphasized that the martensitic–austenitic cast iron has not been up to now discussed in detail in the literature.

After austempering only in melts no. 3 and no. 9 the optimum bainitic–austenitic matrix was obtained. In other melts the precipitates of martensite, of carbides, or of cementite were noted. Those phases do increase the hardness but reduce the ductility.

<u>The hardenability.</u> As it follows from Table 3, included in this study, the cast iron from all melts has a hardness of more than 35 HRC along the whole specimen length, which is 95 mm. On the other hand, the hardness of more than 45 HRC along the entire length of the specimen was obtained in melts nos. 1 (0.73 Mo), 3 (1.35% Cu), 4 (maximum content of Ni, Cu, Mo) and in 8 and 10 (various contents of Ni, Cu and Mo). Hence it follows that the cast iron with an addition of 1% Mn and 1.35% Cu, that is the cheapest one out of all the examined grades, is characterised by a very good hardening capacity and due to that can be used for heavy castings.

This result should be verified by another method, e.g. using Grossman's procedure.

The results of <u>abrasion wear resistance</u> tested in oil in the as–cast condition reveal quite a notable scatter (the loss in weight from 1.8 G up to 28 G). From Table 4 it follows that the lowest abrasion wear resistance has the cast iron which contains nickel only; the highest resistance has the cast iron with Ni+Cu+Mo. This fact was confirmed by tests carried out on the cast iron after normalising and austempering; the highest abrasion wear resistance was revealed by the cast iron from melt no. 7 (1.2% Ni and 0.37% Mo), i.e. by the typical ADI used most commonly

Normalising improves many times the abrasion wear resistance ; that is , its effect is even more pronounced than that of austempering. Hence, the logical implication is that the latter treatment is of no major significance as regards the abrasion wear resistance of ductile iron.

The coefficient of friction was comprised in the range of from 0.10 up to 0.12, and it was comparable with the result obtained in one of the earlier studies [9].

<u>Mechanical properties.</u> The tensile strength of cast iron in the as–cast state was comprised in the range of 565 – 645 MPa, the elongation was minimal from 0 up to 0.9%, the impact strength was from 2.4 up to 3.3 J/cm^2, while hardness assumed the values from 25 up to 45 HRC. While the tensile strength is corresponding to grade 600–3, the plastic properties are very low. This is probably due to a high content of Mn. The cast iron characterised by such properties can be used for elements exposed to abrasion wear only under low static–dynamic loads.

The above given <u>equations,</u> computed according to Hartley's plan of experiment [21,22], allow for the effect of Ni, Cu and Mo in the presence of an elevated Mn content on the tensile strength, impact strength and hardness.

a). As regards Rm it can be noted that its increase is most prominent when Mo is added together with Ni or with Cu. The lack of coefficients for other expressions of the multinomial suggest that the effect of single elements is much less pronounced than when they are combined together.

b). The value of impact strength is definitely reduced by an addition of Cu and increases when Ni is added.

c). The effect of alloying elements on the hardness is of a much more complex nature; in general terms it can be stated that it is improved most effectively by an addition of Mo and Ni+Mo.

SUMMARY.

The cast iron with a martensitic matrix and a martensitic – bainitic matrix in the as–cast state was obtained. Both grades are characterised by a very good hardenability (above 50 HRC along the specimen length of 95 mm) and by a high abrasion wear resistance.

After normalising the above mentioned grades have a hardness of 55 HRC. The obtained grades of ductile iron are much cheaper in production than the grades of martensitic cast iron known until now. They can also replace the high–chromium cast iron with the cost of manufacture cut down considerably.

REFERENCES

1. J.C.T. Farge et al.: Foundry Tr. J. ,<u>131</u>, 319 (1971)
2. K. Rohring et al.: Giess.–Verlag, Dusseldorf (1970,1974)
3. D.M. Stefanescu: Giess.–Prax. , 419 (1974)
4. A. Kosowski, C. Podrzucki: Żeliwo stopwe, AGH, Kraków (1977)
5. J.A.Sikora et al.: The British Foundryman, 180 (1986)
6. .R.H. Juneja et al.: Foundry m&t, 64 (1989)
7. J.–M. Schissler et al.: Metallurgical Science a. Technology, 71 (1987)
8. M.Nili Ahmadabadi et al.: Cast Metals_2, 62 (1992)
9. A.Kowalski: Research studies. FRI Cracow (1986 – 1993)
10. Z.Polański: Planowanie doświadczeń w technice, Warszawa (1984)

Table 1

Schedule of Hartley's plan of experiment PS/DS–P:Ha3 (hS); content of alloying additions

| n | Code marks | | | Allloying additions, % | | | | | |
| | X1 | X2 | X3 | Assumed | | | Obtained | | |
				Ni	Cu	Mo	Ni	Cu	Mo
1	1	–1	1	0	0	0,8	–	–	0,73
2	2	–1	–1	2,8	0	0	2,65	–	–
3	–1	1	–1	0	1,4	0	–	1,35	–
4	1	1	1	2,8	1,4	0,8	2,65	1,30	0,65
5	–1	0	0	0	0,7	0,4	–	0,73	0,36
6	1	0	0	2,8	0,7	0,4	2,65	0,74	0,36
7	0	–1	0	1,4	0	0,4	1,20	–	0,37
8	0	1	0	1,4	1,4	0,4	1,50	1,35	0,42
9	0	0	–1	1,4	0,7	0	1,55	0,79	–
10	0	0	1	1,4	0,7	0,8	1,45	0,65	0,72
11	0	0	0	1,4	0,7	0,4	1,55	0,70	0,45

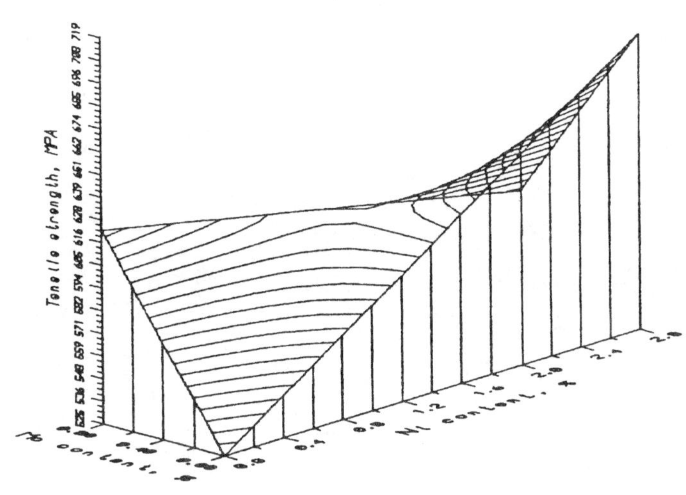

Fig. 1. Influence of Ni and Mo on tensile strength, Cu = 0.7%

Table 2

Microstructure and hardness for ductile iron of successivee melts

No of melt	Condition					
	As–cast		Normalised		Austempered	
	Structure	HRC	Structure	HRC	Structure	HRC
1	P85,Fe,C4,C w2000	31	M,C,A,Crb	50	B,tr.M,A,C	39
2	P45,Fe,F2,Fw 2000	22	M,B,P,F, C,Crb,A	39	B,tr.M,A,C	35
3	M,Crb	45	M,C,Crb,A	52	B,A	45
4	P96,,Fe, C10,Cw2000	25	M,C,Crb,A HRV=723	56	B,C,A	40
5	M,Fe,B,Crb Cw2000	38	M,C,Crb,A	53	B,M,A,C	43
6	P92,Fe F4,Fw2000	29	M,B,C, Crb,A	56	B,tr.M,A,C	40
7	P96,Fe C2,Cw2000, F4,Fw2000	34	M,C,Crb,A	55	B,M,A,C HRV=488	42
8	P96,Fe C4–ślady	31	M,C,Crb,A	55	B,tr.M,A,C	29
9	M,B,Crb	44	M,C,Crb,A	50	B,A	44
10	Fe,M,B, Crb, Fe–traces	39	M,C,Crb,A	57	B,M,A,C	37
11	P85,Fe C4,Cw2000 F4–tr	25	M,B,Crb,A	49	B,M,A,C HRV=523	36

M – martensite, C – cementite, A – austenite, B–bainite, Crb–carbide phases
P–pearlite, Fe–ferrite, F–phosphorus eutectic

Table 3

Hardenability of ductile iron from succcessive melts

No of melt	Maximum hardness, HRC	Deph of hardening till 45 HRC hardness, mm	Deph of hardening till 35 HRC hardness, mm
1	55	95	95
2	58	40	95
3	53	95	95
4	58	95	95
5	52	30	95
6	57	95	95
7	43	40	95
8	55	95	95
9	45	30	95
10	57	95	95
11	47	20	95

Table 4

Results of abrasion   wear resistance tests in oil for ductile iron from succesive melts

No of melt	Condition					
	AS–cast		Normalised		Austempered	
	Weight loss,G	Coeffic-tien of friction,f	Weight loss,G	Coeffic-tien of friction,f	Weight loss,G	Coeffic-tien of friction,f
1	13,4	0,11	2,0	0,11	2,4	0,12
2	28	0,11	1,6	0,12	2,4	0,11
3	7,2	0,11	1,0	0,12	2,2	0,12
4	5,2	0,11	1,0	0,11	4,0	0,11
5	5,2	0,11	1,2	0,11	4,0	0,12
6	5,8	0,11	1,4	0,12	1,6	0,12
7	4,0	0,12	0,6	0,11	0,6	0,12
8	14,0	0,11	1,4	0,11	1,4	0,12
9	7,5	0,1	1,6	0,11	2,7	0,12
10	1,8	0,12	1,2	0,11	1,0	0,12
11	8,2	0,10	1,0	0,10	2,6	0,11

Table 5

Mechanical properties of ductile iron from successive melts in as–cast condition

No of melt	Rm, MPa	A_5, %	KVU,J/cm2-5
1	610	0,5	3,3
2	688	0,0	3,3
3	480	0,0	2,7
4	614	0,9	3,4
5	568	0,0	3,7
6	622	0,2	3,5
7	560,3	0,2	2,5
8	556	0,0	2,4
9	645	0,0	3,0
10	611	0	3,2
11	644	0,0	3,9

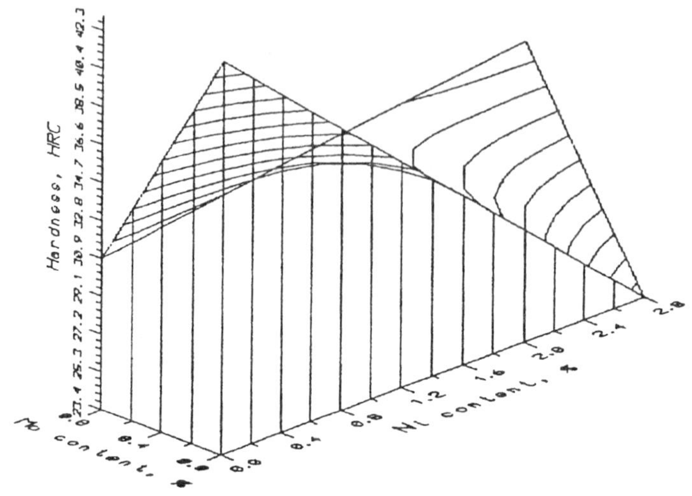

Fig. 2. Influence of Ni and Mo on hardness, Cu = 0.7%

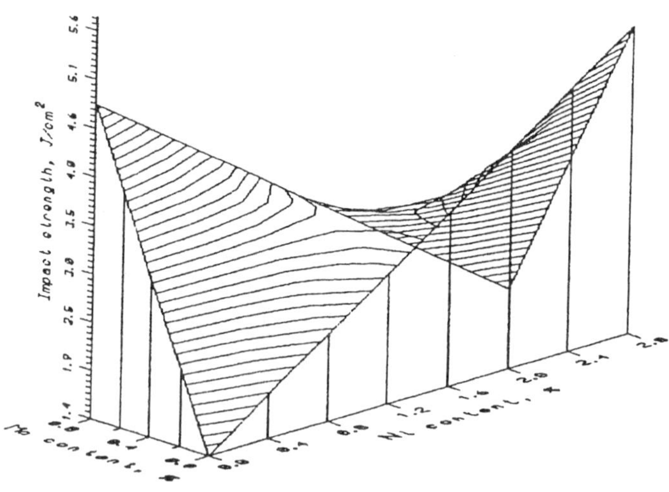

Fig. 3. Influence of Ni and Mo on impact strength, Cu = 0.7%

MELT, SOLIDIFICATION AND PRODUCT PROCESSING

Melt processing and related properties of cast iron

Solidification and product processing

Advanced Materials Research Vols. 4-5 (1997) pp. 269-276
© *1997 Scitec Publications, Switzerland*

Oxygen Activity of Ductile Iron Melts - Its Relationship to the Solidification and Shrinkage Behaviour

R. Hummer

Austrian Foundry Research Institute, Parkstrasse 21, A-8700 Leoben, Austria

Keywords: Oxygen Activity, Graphite Formation, Shrinkage Tendency

ABSTRACT

The oxygen activity measurement is a recognized measurement technique which reached industrial maturity about 15 years ago. It is now being used in steel plants to evaluate the degree of deoxidation in the production of high quality steels. The present report shows that this measurement technique can also be used to great advantage in the production of cast iron with spheroidal graphite. It permits a quick evaluation for the treatment condition of melts which have had a magnesium treatment and have been inoculated. It not only provides data on the tendency to form spheroidal graphite [1], but also on the shrinkage tendency. It is especially appropriate for industrial control of melts.

SIGNIFICANCE OF THE OXYGEN CONTENT

Before more is said about the measurement technique, the significance of oxygen in the production of cast iron with nodular graphite should be underlined. There are two processing steps in the formation of nodular graphite. First magnesium is added whereby, according to the following reaction equation, extensive deoxidation and desulphurization take place and in addition some magnesium is dissolved.

$$[FeO] + \{Mg\} \Rightarrow (MgO) + [Fe]$$
$$[FeS] + \{Mg\} \Rightarrow (MgS) + [Fe]$$
$$\{Mg\} \Rightarrow [Mg]$$

Only after the magnesium is dissolved nodular graphite can be formed. After that an inoculation agent based on FeSi is added which contains singulary or in combination Ca, Ce, Al, Bi, Ba, Sr. This provides crystalization nuclei for the graphite and ensures grey solidification.

As soon as magnesium is dissolved an equilibrium between the dissolved magnesium and dissolved oxygen is valid. The line of equilibrium for 1427°C [2] in Fig. 1 shows that the contents of dissolved magnesium and dissolved oxygen are indirectly proportional so that conclusions can be drawn from one element to the other. As dissolved oxygen and oxygen activity are proportional there is also a close connection between the oxygen activity and the content of dissolved magnesium, but not with the entire content of magnesium, which consists not just of the dissolved magnesium but also of the portion which is bonded (oxides, sulphides).

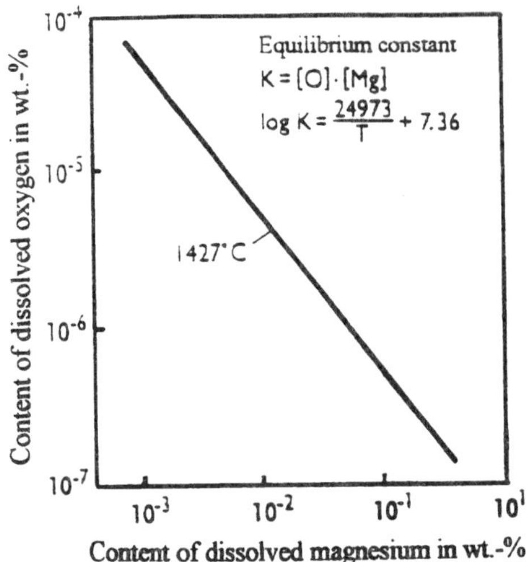

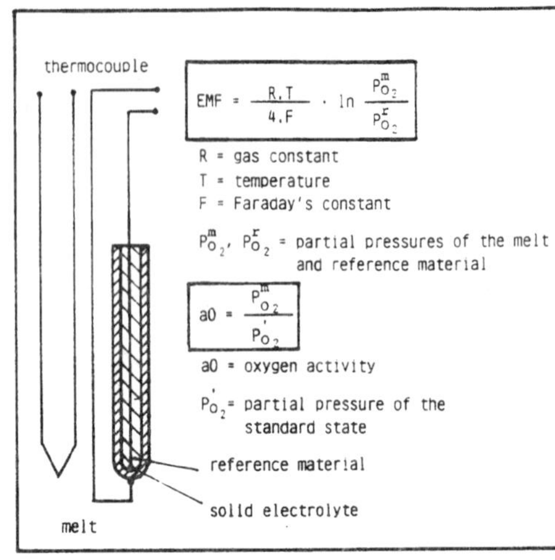

Fig. 1: Relationship between the dissolved magnesium and dissolved oxyen in iron melts at 1427°C [2].

Fig. 2: The principle scheme of the electrochemical cell

MEASUREMENT PRINCIPLE AND THE EXPERIMENTAL PROCEDURE

The principle of oxygen activity measurement is based on the use of an electrochemical cell whose principle scheme is shown in Fig. 2 in a simplified manner. The cell is essentially established by the melt, the solid electrolyte and the oxygen reference material. The solid electrolyte of the cell consists of magnesium stabilized zirconia in the form of a closed end tube which is filled with the oxygen reference material $Cr+Cr_2O_3$. If the solid electrolyte is submerged into the melt a voltage is created between the inside and the outside surface of the tube, as the reference material has a much higher oxygen content than the melt and an ion current runs from the reference material to the melt. The created EMF obeys the Nernst`s law shown in Fig. 2 and results in the oxygen partial pressure of the melt as the other parameters of the formula are known. The oxygen partial pressure of the melt related to the standard state leads, subsequently, to the oxygen activity. Therefore, the oxygen activity can be estimated from the EMF and the also measured temperature by the following formula given from the probe manufacturer [3].

$$\log a\,[O] = 1.36 + 0.0059*[E + 0.54*(t - 1150) + 2*10^{-4}*E*(t - 1150)]$$

a [O] = oxygen activity in ppm

E   = EMF in mV

t   = temperature in $^{\circ}C$

The measurements were carried out in the experimental research foundry of the Austrian Foundry Institute and in industrial foundries with measuring probes CELOX Al from the Electro–Nite n.V. Company, which are made for measurements from 1 to 20 ppm oxygen activity.

OXYGEN ACTIVITY AND THE FORMATION OF SPHEROIDAL GRAPHITE

As expected, the oxygen activity is closely related to the formation of spheroidal graphite. If you plot, as in Fig. 3, the spheroidal graphite portion over the oxygen activity, the transition from spheroidal graphite to flake graphite occurs at 10^{-7} oxygen activity, which is also indicated by decreasing of the tensile strength and elongation. For the evaluation of the tendency to form spheroidal graphite, it is not necessary to convert the EMF into oxygen activity [1].

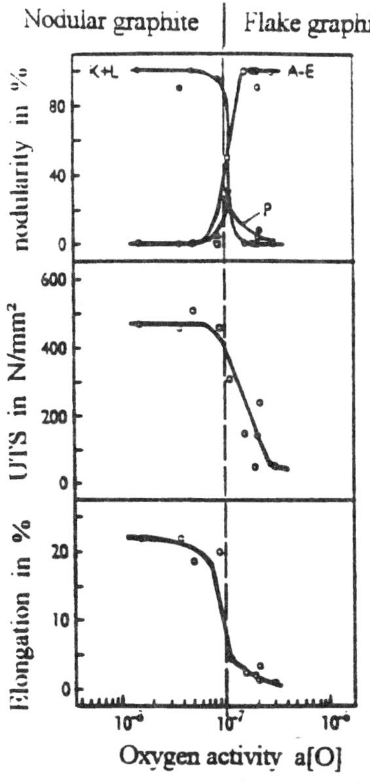

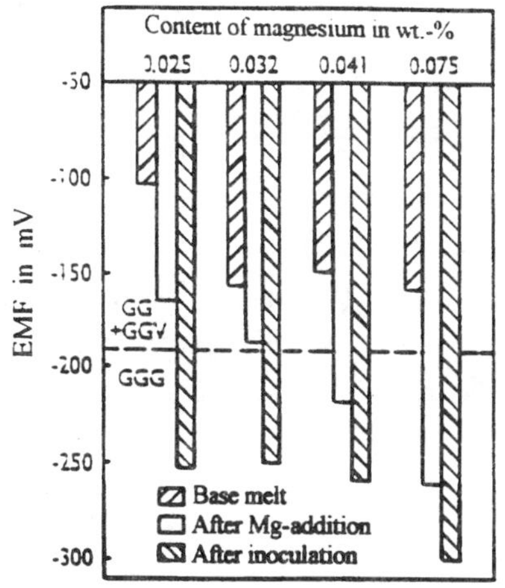

Fig. 4: Influence of the magnesium addition and inoculation of the EMF and spheroidal graphite formation at different magnesium contents [5]

Fig. 3: The transition from spheroidal graphite at 10^{-7} oxygen activity [1], (Y2–sample, 25 mm wall thickness)

The change in the graphite formation can be qualitatively followed very well using the EMF. In Fig. 4 several melts of pig iron with different magnesium content from 0.025 to 0.075 % are presented in the form of a bar diagram. The first bar shows the EMF of the initial condition, the second the EMF after the addition of magnesium and the third the EMF after inoculation. In addition, the boundary between flake graphite and spheroidal graphite is indicated by a broken line. As expected, the EMF sinks due to the magnesium treatment. And the subsequent inoculation also causes a considerable additional decrease of the EMF. This effect is the stronger, the lower the magnesium content is. At the low magnesium contents of 0.025 and 0.032 %, the area of the formation of spheroidal graphite is just reached by inoculation. This underlines the importance of inoculation. It is not only essential for the grey solidification, but in borderline cases, it is also necessary for the formation of spheroidal graphite. This fact is often given too little attention in practice.

One could raise the argument against the process that elements, such as Ti and Al, which disturb the formation of spheroidal graphite and at the same time lower the oxygen activity, limit the truth of this statement. This is theoretically correct but practically of no importance since these elements would only be disturbing at higher content. This would hardly ever occur and would be discovered nowadays by the chemical analysis which is carried out in every foundry with an awareness for quality. It should only be taken into consideration that strong deoxidizing elements like Ti, Al, Ce and so on could slightly move down the boundry between flake and nodular graphite cast iron.

OXYGEN ACTIVITY AND THE TENDENCY TO SHRINKAGE

The oxygen activity does not only indicate the tendency to form spheroidal graphite, it also predicts the shrinkage tendency which, in due practical simplification, is described by the feeding requirement and the feedability. The feeding requirement of spheroidal graphite cast iron depends, to a great extend, on mould strength. The feedability, against it, is connected strongly with the "metalurgical quality" of the melt [4].

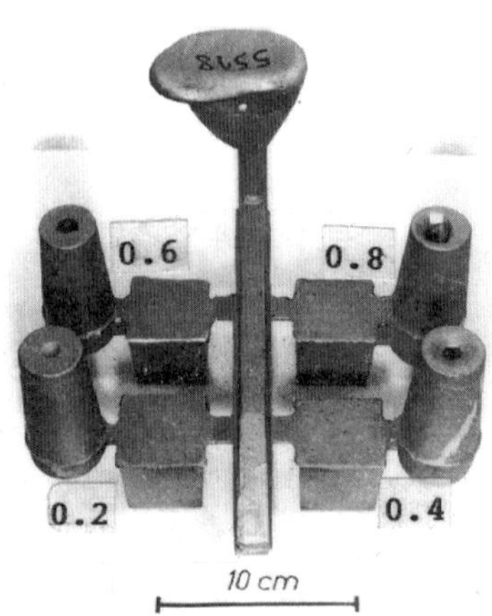

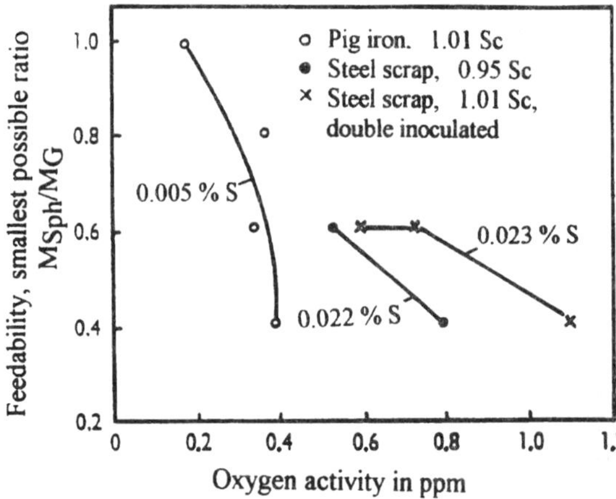

Fig. 5: Feeder–neck–sample with four cubes and "feeder–neck–ratios" M_{Sph}/M_G from 0.2 until 0.8 [5,6]

Fig. 6: Relationship between oxygen activity and feedability [5]

The term, feedability means the ability to feed through narrow cross sections. It is measured by the feeder neck sample cast in CO_2 sand moulds and shown in Fig. 5 [5]. The sample consists of four cubes each having side dimensions of 40 mm, cast from a common sprue and runner, each cube having a feeder of equal size, 40 mm diameter and 60 mm high. The feeder necks are of square cross–section and were dimensioned in such a way that the ratio of the feeder neck modulus M_{Sph} to the casting modulus M_G varied between 0.2 and 0.8 (The Chvorinov's solidification modulus M distinguishes the freezing time of the casting and is determined by the ratio of the casting volume to the heat releasing casting surface). The smallest ratio, a dense microstructure can still be achieved with, is serving as a measure of feedability.

The feedability plotted above the oxygen activity in Figure 6 is better the higher the oxygen activity is. The advantage of the pig iron appears by moving the curve to the left. The same is valid for the sinking tendency [5].

The results from EMF measurements, carried out in three foundries have shown that the optimum range in which under normal industrial conditions the best casting properties can be expected lies between −200 and −250 mV EMF. The magnesium alone is not sufficient criterion for the feedability.

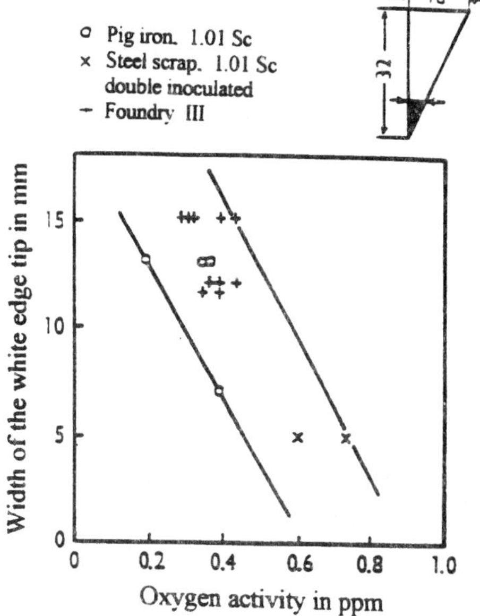

Fig. 7: The chilling tendency as function of oxygen activity

A "better more than less" dosage of treatment agent does not increase safety, but does increase cost and causes more shrinkage. A particularly "nice" spheroidal graphite formation is purchased with the occurance of porosity and empty spaces with their well known bad influence on the properties. In addition the excessive Mg additives connected with low oxygen activity values cause a stronger tendency to chill (Fig. 7) and an increasing tendency to pearlite formation. This has a very disturbing effect in the production of ferritic ductile iron. And last but not least at the same time the tendency to form dross defects increases.

Finally it must be mentioned that in spite of the strong influence which oxygen activity has on feedability, other factors such as charge materials, degree of saturation, melt processing, casting temperature etc. also have an influence. Best results have been achieved when all the influential factors are optimized, whereby the oxygen activity can make a valuable contribution.

PREPARATION OF THE BASE IRON

It should be shown by the following two examples, whereby optimal conditions should be obtained by using pure sorel pig iron, choosing a just hyper eutectic chemical composition with 1.03 and 1.05 degree of saturation, adding only so much magnesium from the master alloy FeSiMg3.7 (with 42 % Si, 1.02 % Ca, 0.58 % Al, 1.39 % RE and 3.7 % Mg) as for the nodular graphite formation is necessary and reducing the magnesium requirement by deoxidizing the basis melt with the deoxidant FeSiCe2 (with 60 % Si, 4.5 % Mn, 1 % Ca, 1 % Al, 4.5 % Zr and 2 % Ce) and adding the inoculant

FeSiBi (with 60 % Si, 1 % Ca, 1 % Al, 1 % Bi and 1 % RE) which produces a low chill tendency and a high nodule count [5].

$$\text{Degree of saturation } S_C = \% \text{ C. } [4.23 - 0.3 \cdot (\% \text{ Si} + \% \text{ P})]^{-1}$$

The course of the treatment of the first example, shown in Fig. 8, corresponds to the normal mode of work. The melt contains 0.007 % S. First 2.5 % FeSiMg3.7 was added, which caused the oxygen activity to drop from 3.04 ppm to 0.38 ppm with which the area of nodular graphite formation was reached. The following inoculation with 0.5 % FeSiBi was leading to further sinking the oxygen activity to 0.33 ppm (−221.7 mV EMF).

In the second example the melt, which had 0.008 % sulphur, was first deoxidized with 0.25 % FeSiCe2 and inoculated with 0.25 % FeSiBi. Thanks to this "preparation" of the initial melt, the oxygen activity was reduced from 2.43 to 2.09 ppm. For the nodular graphite formation it was only necessary to add 1.5 % FeSiMg3.7. The oxygen activity then amounted to 0.40 ppm (−215.6 mV EMF), as can be seen in Fig. 9.

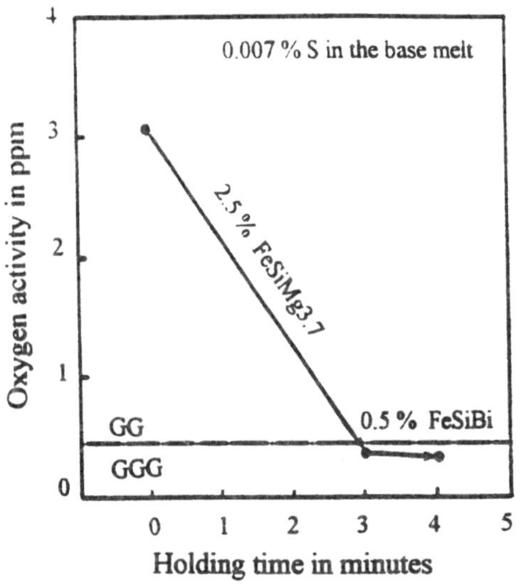

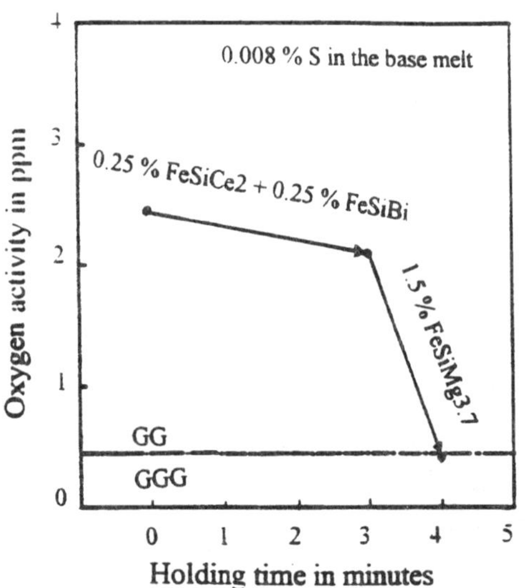

Fig. 8: First example: Normal mode of work

Fig. 9: Second example: Preparation of the base melt by deoxidation with 0.24 % FeSiCe2 and inoculation with 0.25 % FeSiBi before the magnesium treatment

Since in each case the limit to nodular graphite formation was only undershot by very little, very good properties were achieved as shown in Table 1. The mechanical properties, especially the high

matrix.The tendency to chill is low. The widths of the white tips of the wedge test pieces amount only to 4.5 and 5.0 mm. The feedability is excellent and amounts 0.2 M_{Sph}/M_G ratio. The nodule count lies at 228 and 245 nodules/mm^2.

Table 1: Results of the examples No. 1 and 2 [6]

Example No.	1	2
EMF mV	– 221.7	– 215.6
aO ppm	0.33	0.40
Tensile strength *) N/mm²	477	472
Yield Point *) N/mm²	318	327
Elongation *) %	19.9	23.1
Feedability M_{Sph}/M_G	≥ 0.2	≥ 0.2
Width of white tip of the wedge test piece mm	5.0	4.5
Nodule count/mm *)	228	245

*) Y2–sample, 25 mm wall thickness

Finally it may be referred to an advantage of today's special meaning. Only few smoke is developed when FeSiMg3.7 with low magnesium content is added to the melt. Much more smoke could be evoluted during the addition of master alloys with higher magnesium content, as the smoke evolution is the stronger the higher the magnesium content of the alloy is [6].

SIMPLE EXECUTION OF THE MEASUREMENT

The measurement apparatus (Fig. 10) is simple and consists of the oxygen probe, the immersion lance, the interface and the computer with its screen and the printer. The oxygen probe is placed on the immersion lance and submerged into the melt. Already during the measurement the temperature and the EMF appear on the screen. After 15 seconds the result is available and can be printed out if desired.

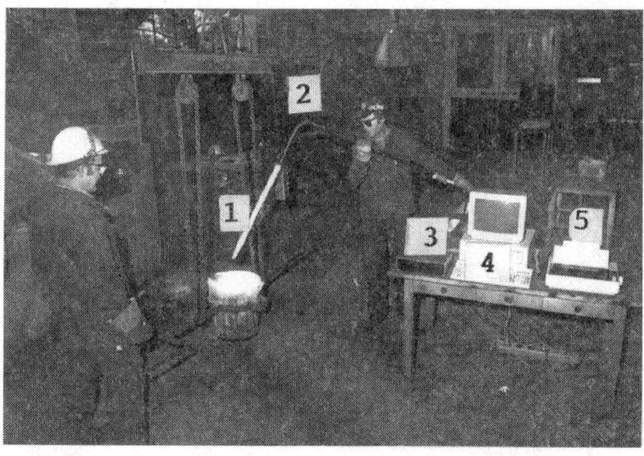

Fig. 10: Simple procedure: The oxygen probe (1) on the immersion lance (2) is connected the interface (3), the computer and the printer and is dipped into the melt

CONCLUSIONS

In conclusion it can be said that EMF measurement is a very useful process to assure sufficient magnesium treatment and additional it can predict the casting behaviour. Technically seen, its use can be highly recommended. A disadvantage is the high price for the oxygen probes which cost presently about DM 30 per measurement.

However considerations of economic efficiency should take the following advantages into consideration. The measurement is simple and takes only one minute. The measurement result is clear and no special experience is needed for its interpretation. Incorrect charges can be identified immediately and be corrected or eliminated. Waiting times for release through the normal microstructure controls are eliminated, the stream of production becomes more fluid. A quality improvement is to be expected because not only under–treatment but also over–treatment with the well known unfavourable effects can be recognized quickly.New metallurgical treatment techniques deviating from the common procedure can be developed as the knowledge of the principal process has been increased. For the quality assurance it is important that the results can be picked up, evaluated and recorded by a computer which can then be made available when needed for subsequent operational evaluations, controls in case of complaints, etc. .

ACKNOWLEDGEMENTS

The main part of this work was done within the scope of the European Concerted Action project on castings technology COST 504–2; the author would like to thank the Austrian Industrial Research Promotion Fund and the Austrian working group "Nodular Iron" for continued financial support.

REFERENCES

1.   Hummer, R., W. Meyer, R. Schlüsselberger jun.,
     48th , Int. Foundry Congress, Okt. 1981 in Varna, report No.1a, or
     Gießerei–Rundschau 28 (1981), No. 12, p. 13–21

2.   Schlüsselberger, R., sen., unpublished

3.   Electro–Nite n.V., "Celox" brochure

4.   Hummer, R., Cast Metals, Vol. 1., No. 2, 1988, p. 62–68

5.   Hummer, R., Gießerei 78 (1991) No. 24, p. 884–889

6.   Hummer, R., BCIRA International Conference 1994, Warwick, report No. 25

Advanced Materials Research Vols. 4-5 (1997) pp. 277-284
© 1997 Scitec Publications, Switzerland

Inoculation of Spheroidal Graphite Cast Iron

V. Cochard[1], R.A. Harding[1], J. Campbell[1] and R. Hérold[2]

[1] IRC in Materials for High Performance Applications, The University of Birmingham,
B15 2TT Birmingham, UK

[2] Péchiney Electrométallurgie, Laboratoire Central de Recherche,
Chedde, F-74190 Le Fayet, France

Keywords: Ductile Iron, Spheroidal Graphite, Inoculant, Nucleus, La, Ca

ABSTRACT

An assessment has been made of the effects of La and Ca in a pure ferrosilicon when used to inoculate SG cast iron. As well as chill tests, nodule counts and thermal analysis, particular emphasis has been placed on examining the nuclei found in the graphite nodules. Inoculation with FeSiLa gave nuclei composed of S, Mg and La, whereas the FeSiCa inoculant gave nuclei mainly consisting of Mg and Si with traces of Ca, S and Al. The FeSiLaCa inoculant gave nuclei consisting either of S, Mg, La and Ca or of Mg and Si with traces of La, S and Ca. In contrast to the commonly-made assumption, no trace of oxygen has been found in the nuclei consisting of Mg and Si.

INTRODUCTION

The homogeneous nucleation of graphite in cast iron is considered to be impossible [1-2] because of the huge undercooling that would be required (~230°C). It is widely accepted that graphite appears in the melt by a heterogeneous nucleation process. The driving force is the minimisation of the free energy of the system consisting of the nucleus, the graphite and the melt [3]. The value of the critical free energy is not available but is related to the lattice disregistry between the nucleus and the graphite. This disregistry can therefore be used to assess the ability of a surface to initiate the graphite nucleation.

The widely accepted theory considers that the graphite nucleates on oxide, sulphide or oxysulphide inclusions [4-13]. Depending on the treatment of the melt, the inclusions have been found to contain Mg, Si, Ca, S, Al, Sr and Ce. However, the only authors who have positively established the presence of oxygen [8, 14, 15] actually analysed huge particles which do not look like the usual kind of nuclei found in graphite [10, 11, 16]. The oxygen was found in round particles of ~8μm diameter [14, 15], with a layer of only 3 to 4μm of graphite surrounding it, or embedded in a hole [8] in the centre of graphite spheroids. Yamamoto [6] showed the presence of oxygen in the graphite without connecting it with any other elements.

Studies of the dissolution of inoculants in quenched samples of cast iron melts [17] have shown the presence of inclusions such as CaMgS, Rare Earth-CaMgS, and SiAl(Ca)MgS close to the remainder of the inoculant grain around one second after inoculation. As the same kinds of particles were also identified in the graphite, it shows that the nodularizer and the inoculant are directly at the origin of the nuclei on which the graphite nucleates.

The purpose of the present research is to understand how the inoculant affects the nucleation of the graphite and, in particular, to locate the active elements in the cast iron. Two active elements were chosen: Ca which is a common element in inoculants and a rare earth, La, which is sometimes present in inoculants. The effect of rare earths as a nodularizer is well established, but their use in inoculants is less well known. Lanthanum was preferred to cerium because it has been shown [16, 18] that it has a slightly superior inoculating power. As interaction between active elements is a key toward the production of powerful inoculants, it was decided to study the effect of La, Ca, and La with Ca introduced via the inoculant.

EXPERIMENTAL PROCEDURES

Design of the treatment alloys

Simplified treatment alloys were used in order to be able to determine the origin of elements found in the nuclei. They were produced in an induction furnace by melting pure elements in a graphite crucible. The nodularizer produced had a composition of 31.5 at.% Fe, 59 at.% Si, 9.5 at.% Mg. The inoculants were designed to correspond to an average of classical compositions found in commercial inoculants and nominally contained 20 at.% Fe, 78.7 at.% Si and 1.3 at.% of active elements. The following alloys were produced:
- FeSi (20 at.% Fe, 80 at.% Si)
- FeSiLa (19.3 at.% Fe, 79.4 at.% Si, 1.28 at.% La)
- FeSiCa (19.6 at.% Fe, 78.8 at.% Si, 1.36 at.% Ca)
- FeSiLaCa (19.6 at.% Fe, 78.8 at.% Si, 0.64 at.%La, 0.58 at.% Ca).

The structures of the inoculants were observed with a SEM to identify the different phases, especially those in which La and Ca were located.

Test castings

All experimental melts were based on Sorel iron with additions of mild steel, silicon, and treatment alloys. The analyses of individual melts after nodularization and inoculation are given in Table 1. Nodularization was carried out at 1570°C by the sandwich method with 2.5% of FeSiMg. Inoculation was carried out at 1420°C by adding 0.6% of the inoculant in the melt stream during transfer to a second preheated ladle. The test casting comprising an array of square bars and chill test fins was poured 15 s later at 1350°C in a phenolic urethane resin bonded silica sand mould.

Inoculant	C	Si	S	Mn	P	Mg
None	3.69	2.36	0.005	0.09	0.02	0.051
FeSi	3.68	2.58	0.008	0.09	0.02	0.058
FeSiLa	3.79	2.68	<0.01	0.09	0.02	0.047
FeSiCa melt 1	3.71	2.69	<0.01	0.09	0.02	0.047
FeSiCa melt 2	3.65	2.57	<0.01	0.11	0.02	0.059
FeSiLaCa melt 1	3.70	2.61	0.005	0.09	0.02	0.051
FeSiLaCa melt 2	3.68	2.68	0.012	0.11	0.02	0.054

Table 1: Melt analysis after inoculation, in weight percentages, balance Fe.

The observations

Chromel/alumel thermocouples were used to measure cooling curves at the centre of 20 mm square bars and the curves were digitized and differentiated. The peak of the first derivative was used to define the eutectic temperature and the first and second peaks of the second derivative were used to define the recalescence and end of solidification temperatures respectively. The nodule count and the chill depth were measured to assess the inoculating power of the different inoculants.

Selected specimens were examined with an SEM with the particular aim of locating the La and Ca introduced from the inoculant. Examinations were made of the nuclei inside the graphite spheroids, of the interface between the graphite and the matrix and of particles in the matrix. Elements heavier than Na (Z = 11) were quantitatively analysed and those from B (Z = 3) to Na were analysed qualitatively. The probe diameter was around 0.3 µm so that the volume analysed was smaller than a 2.5 µm side cube. The detection limit for all elements was around 0.5% and it was therefore possible to locate only the very high concentrations of Ca and La. Indeed only around 0.0078 weight percent of these elements were added to the cast iron. The quantitative results must be considered cautiously as the accuracy is difficult to evaluate. Different tests appeared to show that the relative errors were approximately +/- 5%.

RESULTS

Metallographic examination of the inoculants

A silicon phase (>99.5 at.% Si) and a phase containing iron and silicon (25 - 30 at.% Fe, 75 - 70 at.% Si) were found in all inoculants. In addition, in the FeSiLa and the FeSiCa inoculants,

La and Ca were present as $LaSi_2$ and $CaSi_2$ phases respectively. In the FeSiLaCa inoculant, La and Ca were in a phase consisting of 22 at.% Fe, 75 at.% Si, 2.6 at.% Ca and a $LaCaSi_4$ phase.

Cooling curves
     Typical cooling curves are shown in Fig. 1 and characteristic temperatures measured from such curves are reported in Table 2. It can be seen that the cooling curves of the uninoculated iron and of the iron treated with the pure ferrosilicon were similar and were the only ones not to show any recalescence. Their eutectic temperatures were approximately 1105°C.

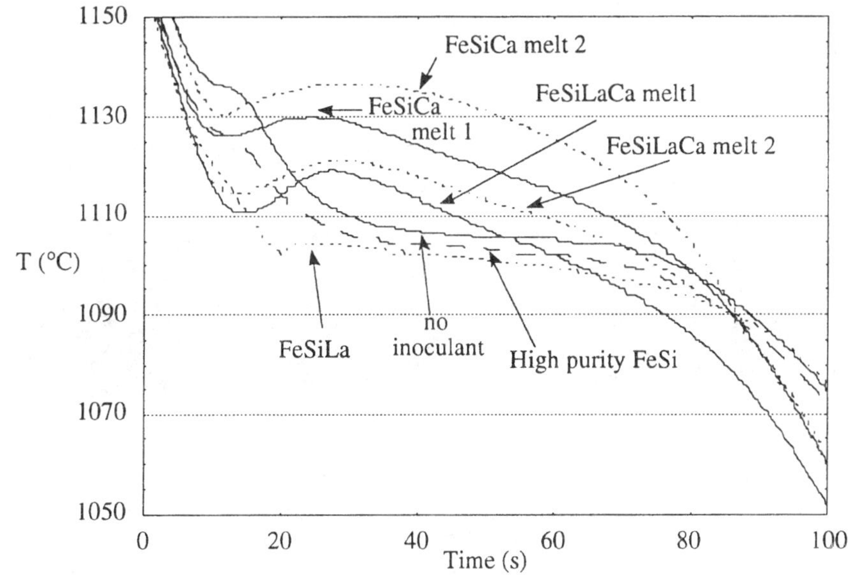

Figure 1: Cooling curves corresponding to the different treatments.

Inoculant	Nodule count (mm^{-2})	Chill depth (mm)		T (°C) eutectic	T (°C) recalescence	T (°C) solidification
		in 6 mm fin	in 9 mm fin			
None	14	>30	>30	1106	none	1040
FeSi	34	>30	>30	1103	none	1050
FeSiLa	133	>30	>30	1102	1104	1082
FeSiCa melt 1	310	19	7	1126	1130	1094
FeSiCa melt 2	330	16	5	1131	1136	1107
FeSiLaCa melt 1	350	23	12	1118	1124	1092
FeSiLaCa melt 2	370	13	4	1115	1121	1081

Table 2: Nodule count, chill depth and characteristic temperatures for the different inoculation treatments.

     The cooling curves also showed that the eutectic transformation in the sample inoculated with FeSiLaCa occurred at a temperature between that for the eutectic transformation in the sample inoculated with FeSiCa (~1030°C) and that in the sample inoculated with FeSiLa (~1102°C).

Metallographic examination of the cast irons
     The results of chill depth and nodule count measurements are also shown in Table 2. The uninoculated sample and those treated with FeSi and FeSiLa were fully carbidic but significant differences were noted. In the uninoculated sample the nodule count was extremely low and moreover the graphite nodules were not even surrounded by ferrite but were embedded in primary cementite. The samples inoculated with FeSi and with FeSiLa had higher nodule counts of 34 and 133 per mm^2 respectively.

The use of FeSiCa or FeSiCaLa led to a significant reduction in chill depth but scatter in the results prevented a clear distinction to be drawn between these two alloys. The samples inoculated with FeSiCa and FeSiLaCa were free of carbide in the 20 mm section bars and had a far higher nodule count. The FeSiLaCa showed a slightly higher inoculating power than the FeSiCa. (Table 2)

<u>Particles found in the graphite</u>

Inoculation with FeSiLa

Fig. 2 shows a typical particle found at the centre of the graphite spheroids in the sample inoculated with FeSiLa. They were spherical with a diameter of 1 to 2 µm and were found to contain S, La and Mg, sometimes with Si and traces of Al and Ca. The analyses of 26 particles from the same sample are summarised in Fig. 3 which shows that the proportions of the different elements fluctuated over a wide range. For example, the La content varied from 5 to 50% and the S content from 5 to 40%.

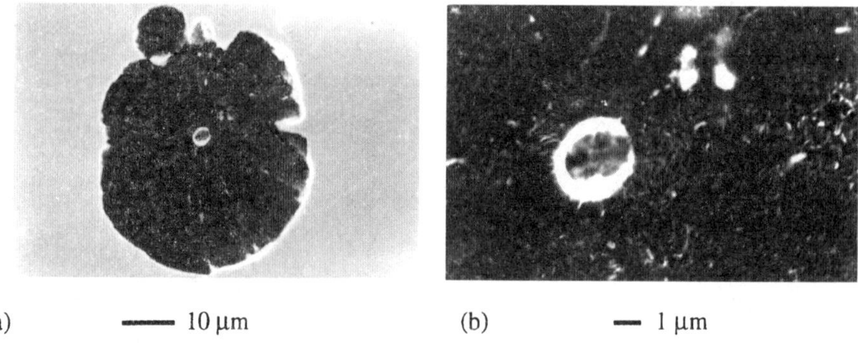

(a) ——— 10 µm   (b) — 1 µm

Figure 2: Graphite nodule and nucleus found in ductile iron inoculated with FeSiLa. Analysis of nucleus (elements heavier than Na in atomic percentages): 1.7% Si, 9.1% Mg, 47.4% S, 41.8% La.

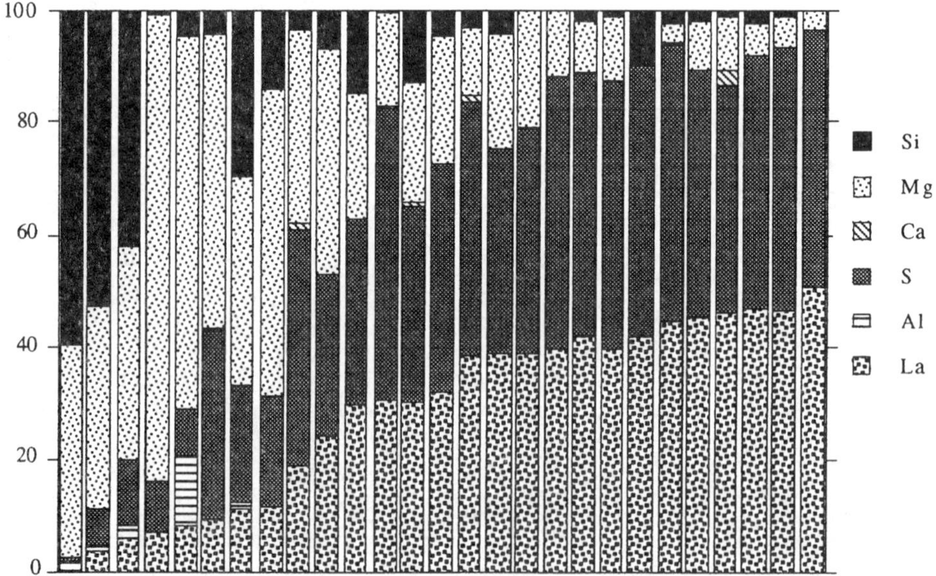

Figure 3: Chemical analysis, in atomic percentage, of particles found in graphite spheroids in a ductile iron inoculated with FeSiLa.

Inoculation with FeSiCa

In the sample inoculated with FeSiCa most of the nuclei were rectangular particles with dimensions 0.5 x 1 µm up to 1 x 2 µm: a typical example is shown in Fig. 4. Analysis showed that they consisted mainly of Si and Mg with small amounts of Ca, S and Al. The analyses of 14 particles are summarised in Fig. 5 and again show considerable variations. In addition to Si and Mg-based nuclei, two nuclei containing Mg, Ca and S and one containing little other than Ca and oxygen were also found.

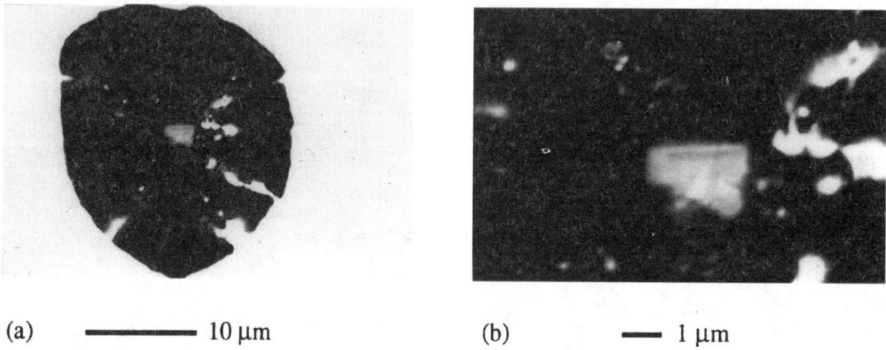

(a) ———— 10 µm (b) — 1 µm

Figure 4: Graphite nodule and nucleus found in ductile iron inoculated with FeSiCa. Analysis of nucleus (elements heavier than Na in atomic percentages): 50% Si, 41.7% Mg, 7.3% S, 1% Ca.

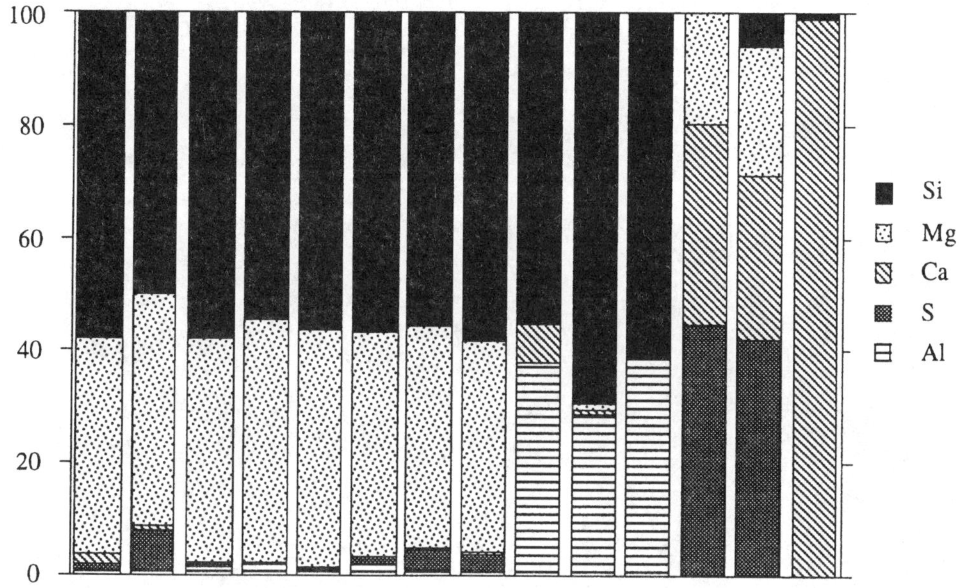

Figure 5: Chemical analysis, in atomic percentage, of particles found in graphite spheroids in a ductile iron inoculated with FeSiCa.

Inoculation with FeSiLaCa

Two examples of nuclei found in the sample treated with FeSiLaCa are shown in Figs. 6 and 7. As shown in Fig. 8, analysis of 16 particles established that there were two main families of particles. One family consisted of S, Mg, La and Ca. One of the particles having a low sulphur content was found to have a detectable amount of oxygen. The other main family of particles contained Mg and Si with traces of La, S and Al.

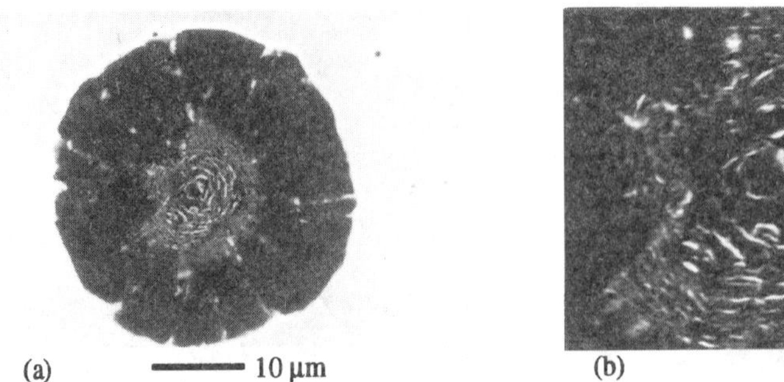

(a) ——— 10 µm　　　　(b) — 1 µm

Figure 6: Graphite nodule and nucleus found in ductile iron inoculated with FeSiLaCa. Analysis of nucleus (elements heavier than Na in atomic percentages): 59.6% Si, 35.7% Mg, 0.8% S, 1.9% Al, 0.6% La.

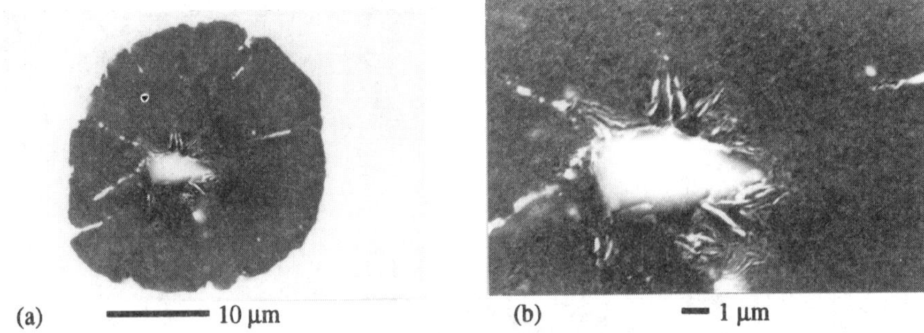

(a) ——— 10 µm　　　　(b) — 1 µm

Figure 7: Graphite nodule and nucleus found in ductile iron inoculated with FeSiLaCa. Analysis of nucleus (elements heavier than Na in atomic percentages): 50.6% Si, 13.7% Mg, 17.5% Ca, 16.9% Al, 1.3% Ti.

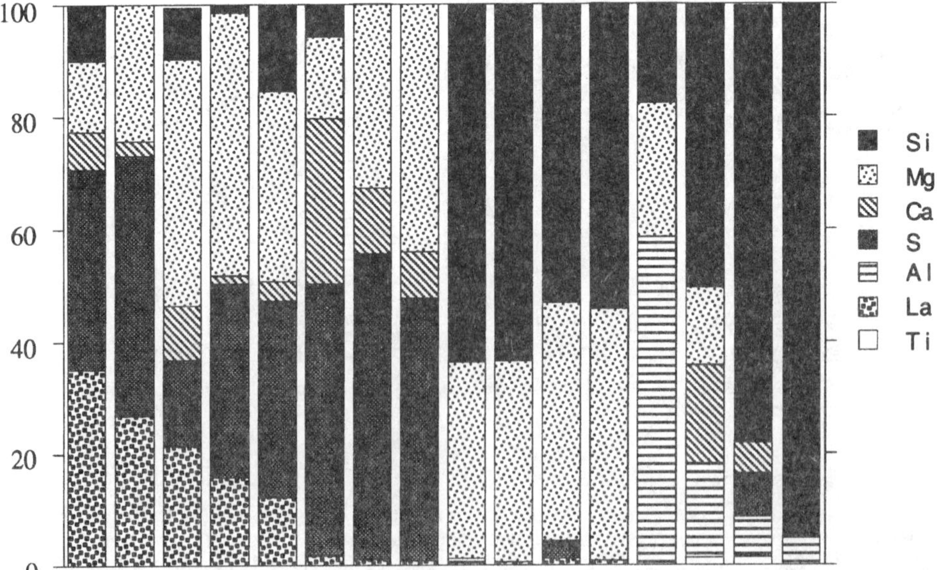

Figure 8: Chemical analysis, in atomic percentage, of particles found in graphite spheroids in a ductile iron inoculated with FeSiLaCa.

<u>Particles found in the matrix</u>
Most of the particles found in the matrix were very different from the nuclei and consisted of MgO, Mg-P-Mn-S and Ti-V-C. However numerous rectangular particles consisting mainly of Si and Mg with the same trace elements as in the nuclei, and roughly the same size as the nuclei, were also found in all the samples.

DISCUSSION

<u>Nuclei composition</u>
Although there were large variations in the compositions of nuclei found in the iron inoculated with FeSiLa, it is considered that this would not cause a very large change in the lattice parameter since MgS and LaS are both FCC with lattice parameters of 5.19 and 5.86 Å respectively. The difference of lattice parameter with the graphite (5.71 Å along the prismatic direction) will therefore not exceed 10% and may remain acceptable for the nucleation of graphite. As CaS is also FCC with a lattice parameter of 5.69 Å, the same comment is valid for the nuclei consisting of S, Mg, La and Ca found in the sample inoculated with FeSiLaCa.
Comparisons of the nuclei found in the samples inoculated with FeSiLa and FeSiCa, and of the inoculating power of FeSiCa which is far higher than FeSiLa, suggest that the nuclei consisting mainly of Mg and Si are more efficient than the nuclei containing S-La-Mg. In the sample inoculated with FeSiLaCa, which has an inoculating power as good as FeSiCa, the predominance of nuclei consisting of S-Mg-La-Ca (similar to those found in the sample inoculated with FeSiLa) suggests that their higher Ca content increased their efficiency.
Skaland [10] established that the modification of the surface of Si-Mg particles by elements such as Ca, Al, Sr and Ba enable them to nucleate graphite. The present observations effectively show the presence of trace amounts of Ca and Al in the Si-Mg nuclei found in the FeSiCa inoculated cast iron. However no Ca has been detected in the Si-Mg nuclei found in the FeSiLaCa inoculated cast iron. The presence of La in the inoculant, or the fact that Ca is not present as $CaSi_2$ in the inoculant, appears to prevent Ca from taking part in the modification of the Si-Mg nuclei.

<u>Oxygen in nuclei</u>
As a result of the high affinity of Si and Mg for oxygen, some authors [9, 10] have considered it reasonable to assume that nuclei contain oxygen. The presence of oxygen was also assumed for nuclei containing cerium [19] which has an even higher affinity for oxygen than Mg or Si. However, the present results have established that the family of nuclei containing Mg and Si were not found to contain oxygen. In fact, oxygen was only found rarely, for example, in the family of nuclei consisting of S, Mg, La and Ca and also in particles which appeared to be CaO or SiO_2.

CONCLUSIONS

1. Two main types of nuclei exist:
- those consisting mainly of Mg-Si, with traces of Ca, S, Al and La. They are rectangular, sized from 0.5 x 1 μm up to 1 x 2 μm. All of them have roughly the same composition;
- those consisting mainly of S and Mg with Ca or La. They are spherical with diameters from 1 to 2 μm. The presence of La seems to induce a great spread of composition.

2. When La is the only active element in the inoculant, S-Mg-La nuclei are formed.
When Ca is the only active element in the inoculant, Si-Mg nuclei containing traces of Ca, Al and S are formed.
When both Ca and La are present in the inoculant, nuclei are either Si-Mg particles containing traces of La, S, Al or S-Mg-La-Ca particles.

3. Although it is commonly assumed that Mg-Si nuclei contain oxygen, no evidence has been found to support this.

ACKNOWLEDGEMENTS

The authors would like to thank Péchiney Electrométallurgie for funding this research and the IRC for providing laboratory facilities. The collaboration of R.D. Forrest and G. Nussbaum is fully acknowledged.

REFERENCES

[1] B. Lux, A.F.S. Cast Met. Res. J., **8**, 25 (1972).
[2] D.M. Stefanescu, A.S.M. Metals Handbook, **15**, 168.
[3] M. C. Flemings, Solidification Processing, McGraw Hill, 295 (1974).
[4] B. Miao, Acta Met. et Mat., **38** (1), 2167 (1990).
[5] M.J. Lalich, A.F.S. Trans., **84**, 653 (1976).
[6] S. Yamamoto, Metal Sci., **9**, 360 (1975).
[7] P.C. Liu, A.F.S. Trans., **91**, 119 (1983).
[8] H. Itofuji, A.F.S. Trans., **98**, 429 (1990).
[9] J.C. Mercier, Fonderie, No. 277, 191 (1969) (in French).
[10] M.H. Jacobs, Metals Tech., **3**, 98 (1976).
[11] T. Skaland. Doktor Ingeniøravhandling, Metallurgisk Institutt. Universitetet i Trondheim, Norway (1992).
[12] R.J. Warrick, A.F.S. Cast Met. Res. J., **2**, 97-108 (1966).
[13] J.C. Margerie, in "The Metallurgy of Cast Iron", Georgi Publ. Co, 545 (1975).
[14] H. Horie, Cast Metals, **3**, No.2, 73 (1990).
[15] H. Horie, Cast Metals, **1**, No. 2, 90 (1988).
[16] M.J. Lalich, in "The Metallurgy of Cast Iron", Georgi Publ. Co, 561 (1975).
[17] M. Hecht, Fonderie, Fondeur d'aujourd'hui, **91**, 27 (1990). (in French).
[18] T. Kusakawa, Waseda University, Casting Res. Lab. Report No. 39, (1988).
[19] K.M. Fang, J. of Rare Earths, **10**, No. 3, 208 (1992).

Advanced Materials Research Vols. 4-5 (1997) pp. 285-292
© *1997 Scitec Publications, Switzerland*

An Investigation of the Influence of Chemistry on Metal Penetration in Gray Iron Castings

F.J. Bradley[1], E.J. Kubick[2] and A.K. Mirle[1]

[1] Department of Materials Science and Engineering, University of Wisconsin-Madison
1509 University Avenue, Madison, WI, USA

[2] Grede Foundries, Liberty Division, 6432 West State Street, Wauwatosa, WI 53213, USA

Keywords: Gray Iron, Metal Penetration, Chemistry Variables, Expansion Related Defects, Solute Redistribution, Specific Volume

ABSTRACT

A 2^{5-1} fractional factorial experiment was conducted to investigate the influence of carbon, silicon, manganese, phosphorus, and sulfur on metal penetration in gray iron castings. Carbon and phosphorus were found to be the dominant chemistry main effects on metal penetration, with metal penetration increasing with increasing carbon content and decreasing with increasing phosphorus content. In a prior study using similar test castings, Levelink and Julien also observed the strong effect of phosphorus on decreasing metal penetration, but reported that metal penetration increased with increasing carbon equivalent (rather than carbon), indicating that the silicon main effect ought to be one of increasing penetration with increasing silicon content. In this study the silicon effect on metal penetration was found to be insignificant in comparison to the carbon and phosphorus effects, with increasing silicon content being associated with a slight reduction in metal penetration.

INTRODUCTION

Metal penetration is a common defect in gray iron castings characterized by the penetration of metal into the voids between sand grains to various depths yielding a phase of sand grains surrounded by metal. As discussed in reference [1], metal penetration may occur due to various mechanisms. The present study is concerned with penetration resulting from exudation pressure caused by expansion associated with graphite precipitation during the eutectic solidification. This paper presents results of an experimental study concerning the influence of carbon, silicon, phosphorus, manganese, and sulfur on metal penetration in gray iron. In addition, the experimental results are interpreted using a simple solute redistribution/specific volume model.

EXPERIMENTAL DESIGN

A 2^{5-1} fractional factorial experiment involving the five chemistry variables was conducted at a sponsor's short run foundry. Table 1 gives the experimental design in standard order as well as the randomized run order. Target maximum and minimum levels of the final iron chemistry variables were as follows: carbon (3.00-3.60%), silicon (1.70-2.50%), manganese (0.45-0.80%), phosphorus (0.030-0.16%), and sulfur (0.05-0.15%). Table 1 also contains analyzed final iron chemistries and measured metal penetration for each run. Carbon and sulfur were determined using a combustion method with the remainder of the elements being determined spectrographically.

In addition to the standard 16 runs (A-P) in the 2^{5-1} design, four extra runs were conducted. As the 2^{5-1} design includes a run with all variables at the '+' level, but not one with all of the

Table 1
Experimental Design, Final Iron Chemistries, and Metal Penetration (mm^2)

Des	Run	Experimental Design					Final Iron Chemistries						Penetration	
		C	Si	Mn	P	S	%C	%Si	%Mn	%P	%S	CEV	Ave	Max
A	3	-	-	-	-	+	3.06	1.68	0.46	0.04	0.14	3.62	353	450
B	13	+	-	-	-	-	3.66	1.74	0.46	0.04	0.059	4.24	458	575
C	14	-	+	-	-	-	3.09	2.43	0.48	0.05	0.057	3.90	217	295
D	5	+	+	-	-	+	3.54	2.43	0.47	0.04	0.15	4.35	377	450
E	18	-	-	+	-	-	3.13	1.71	0.79	0.04	0.062	3.70	223	335
F	8	+	-	+	-	+	3.58	1.68	0.83	0.04	0.13	4.14	528	600
G	2	-	+	+	-	+	3.05	2.37	0.89	0.04	0.12	3.84	132	250
H	17	+	+	+	-	-	3.63	2.41	0.78	0.03	0.058	4.43	218	260
I	9	-	-	-	+	-	3.11	1.72	0.49	0.14	0.054	3.68	2	5
J	12	+	-	-	+	+	3.69	1.75	0.48	0.13	0.14	4.27	126	205
K	16	-	+	-	+	+	2.99	2.39	0.40	0.14	0.13	3.79	1	5
L	6	+	+	-	+	-	3.58	2.43	0.51	0.13	0.073	4.39	152	260
M	15	-	-	+	+	+	3.07	1.66	0.78	0.12	0.15	3.62	9	15
N	4	+	-	+	+	-	3.62	1.65	0.84	0.14	0.049	4.17	161	230
O	7	-	+	+	+	-	3.10	2.48	0.84	0.13	0.036	3.93	10	30
P	19	+	+	+	+	+	3.59	2.42	0.83	0.13	0.14	4.40	102	170
Q	20	-	-	-	-	-	3.09	1.66	0.43	0.04	0.064	3.64	289	365
R	1	0	0	0	0	0	3.39	2.19	0.67	0.08	0.091	4.12	245	305
S	10	0	0	0	0	0	3.46	2.15	0.65	0.07	0.11	4.18	252	295
T	11	0	0	0	0	0	3.47	2.20	0.66	0.08	0.11	4.20	290	345

variables at the '-' level, run Q was added. The design was further augmented with an intermediate point '0', which was replicated three times (runs R, S, and T). This provided a means of assessing experimental error and the reproducibility of the results. Levels of the chemistry variables for these runs were chosen to correspond more closely to typical foundry chemistries.

For a 2^{5-1} fractional design, each of the main effects is confounded with a four factor interaction, and each of the two factor interactions is confounded with a three factor interaction. Due to production constraints at the foundry, it was necessary to conducted the experiments over two days spaced one week apart. To minimize the effect of changing processing conditions, the runs were blocked into two groups using the sulfur/phosphorus interaction as the blocking variable ('+-' or '-+' for the Day 1 runs and '--' or '++' for the Day 2 runs). The blocking effect was found to be insignificant.

EXPERIMENTAL PROCEDURE

A constant charge mix of 50% foundry returns and 50% steel scrap was used for all of the runs with the target chemistries being achieved by trimming with appropriate alloy additions (ferrosilicon, ferrophosphorus, iron pyrite, ferromanganese, and graphite). Melting was carried out in a 340 kg coreless induction furnace. The furnace was argon shrouded to minimize oxidation of the melt. Following alloy additions, the bath was reheated to 1510 C to ensure solution of alloy additions and allowed to cool to 1482 C. When the bath reached 1482 C, the heat was tapped into a pouring ladle with stream inoculation commencing after tapping 25% of the heat. A foundry grade 75% FeSi in the of 0.25% was used for inoculation. The target pouring temperature for each heat was 1400 C.

A centrally-cored cylindrical test casting similar in design to one employed by Levelink and Julien (see reference [2]) was utilized to enable direct comparison of results with previous work. Chemically-bonded silica sands molds were used in the present study in contrast to the green sand molds employed in the Levelink and Julien experiments. Duplicate test casting were poured for each run. The test castings were vertically sectioned and the area of metal penetration observed on each half section was measured using a transparent grid. Table 1 contains both the average metal penetration as well as the highest value observed for two test castings.

RESULTS OF STATISTICAL ANALYSES

A series of regression analyses was conducted. In each case both forward and backward stepwise regression was used to determine the final model. The F-ratio for entering and removing variables was set at a value of 4.0. The R^2(adj) value (adjusted for degrees of freedom in the model) and the significance level of each independent variable are reported. The significance level represents the probability that a larger absolute t-value would occur if there were no marginal contribution from that variable. A low significance level indicates that the coefficients provide useful predictive information.

The initial regression analysis yielded the following predictive model for average metal penetration (MPave)

$$MPave = -246 + 255 \cdot \%C - 2570 \cdot \%P - 91.9 \cdot \%Si \tag{1}$$

for which R^2(adj)=78.62%. From magnitude of the significance levels of the variables: C (0.0007), P (0.0000), and Si (0.0611), it is seen that carbon and phosphorus are the dominant main effects on increasing and decreasing metal penetration, respectively, with the effect of increasing silicon on decreasing metal penetration being somewhat less significant in comparison.

In the initial analysis, run H was flagged as an unusual residual. Removing this data point from the data set yielded the following result

$$MPave = -562 + 305 \cdot \%C - 2912 \cdot \%P \tag{2}$$

for which R^2(adj)=85.88%. A significance level of 0.0000 was obtained for both carbon and phosphorus. Similar results were obtained from an ANOVA analysis for which the qualitative variables 'high' and 'low' were used to represent the two levels of the silicon factor. Silicon level was found to be a small yet significant effect on <u>reducing</u> metal penetration when the result for run H was included, but not when this data point was excluded.

When maximum metal penetration (MPmax) was considered as the response variable, the following result was obtained with run H included

$$MP\max = -455 + 292 \cdot \%C - 3057 \cdot \%P \tag{3}$$

for which R^2(adj)=76.48% and the significance levels are C (0.0011) and P(0.0000). No significant silicon effect was observed, but run H was again flagged as an unusual residual. With run H removed, the following result was obtained

$$MP\max = -635 + 360 \cdot \%C - 3445 \cdot \%P \tag{4}$$

for which R^2(adj)=90.44% and the significance levels for both the C and P terms is 0.0000.

Figure 1 is a plot of observed versus predicted metal penetration for the regression model corresponding to Eqn. (4). It is seen that highest metal penetration is associated with the high carbon-low phosphorus runs B, D, and F. The unusual residual (H) is also a member of the high C-low P group, but a relatively low value of MPmax=260 mm^2 was obtained for this run. Lowest metal penetration is associated with the low C-high P group (I, K, M, O). For these runs, microstructural examination revealed microporosity at the metal/core interface.

It is noted that greater metal penetration was generally associated with the low carbon-low phosphorus group (A, C, E, G, Q) than with the high carbon-high phosphorus group (J, L, N, P). Within the low carbon-low phosphorus group, the highest metal penetration is associated with the lowest carbon equivalent irons (A, E, Q). Regression analysis of MPmax versus CEV revealed no correlation between metal penetration and carbon equivalent.

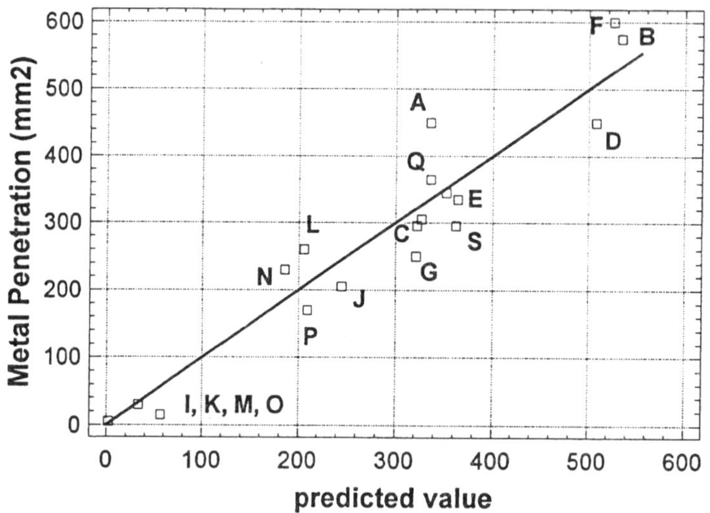

Figure 1 Observed Versus Model Predicted Maximum Metal Penetration

SIMPLE MODEL FOR ESTIMATING METAL PENETRATION TENDENCY

A simple solute redistribution/specific volume model has been developed as a starting point from which to gain insight into the relative impact of metallurgical factors on metal penetration, particularly with respect to the influence of chemistry variables on the volume changes occurring during primary and eutectic solidification. An overview of preliminary work to date on modeling metal penetration is presented below.

Solute Redistribution Model

Simple equilibrium and Scheil equation models for the case of a constant mass volume element of specified initial carbon, silicon, and phosphorus content are considered [3]. Carbon solute redistribution is assumed to occur according to the equilibrium solidification model, and silicon and phosphorus solute redistribution according to the Scheil equation. The initial step in the algorithm is to estimate the mass fraction of primary phase and the composition of liquid at the end of primary solidification, which is used to estimate the average eutectic liquid concentrations of carbon, silicon, and phosphorus and the mass fraction of eutectic graphite and eutectic austenite. These results are used in conjunction with specific volume models to estimate solidification contraction or expansion associated with eutectic solidification.

In the case of hypoeutectic alloys, the mass fraction of primary austenite at the start of eutectic solidification is determined by solving the following system of equations:

$$C_\ell^{eo} = \frac{C_\ell^o - f_{p\gamma} C_{p\gamma}}{1 - f_{p\gamma}} \tag{5}$$

$$Si_\ell^{eo} = Si_\ell^o \left(1 - f_{p\gamma}\right)^{k_s - 1} \tag{6}$$

$$P_\ell^{eo} = P_\ell^o \left(1 - f_{p\gamma}\right)^{k_p - 1} \tag{7}$$

$$C_{p\gamma} = 2.1 - 0.20\, Si_\ell^{eo} - 0.35\, P_\ell^{eo} \tag{8}$$

$$C_\ell^{eo} + \frac{1}{3} Si_\ell^{eo} + \frac{1}{3} P_\ell^{eo} = 4.26 \tag{9}$$

with all concentrations being in weight percent. Equations (5)-(7) are the solidification models for C, Si, and P; equation (8) represents the effect of Si and P on the solubility of carbon in the austenite [4,5]; and equation (9) represents the composition constraint on the eutectic liquid [13] which is assumed to define the end of primary solidification and start of eutectic solidification. The subscripts and superscripts are defined as follows: (ℓ) liquid phase, ($p\gamma$) primary austenite phase, (o) initial concentration, (eo) concentration at start of eutectic solidification.

In the case of hypereutectic melts, a simple lever rule calculation is used to estimate the mass fraction of primary graphite. Precipitation of primary graphite is assumed to have negligible effect on the changing the initial Si and P concentrations at the start of eutectic solidification, which are taken as the initial melt concentrations.

Assuming that only austenite and graphite form during solidification of the liquid present at the start of eutectic solidification, the mass fractions of eutectic austenite ($f_{e\gamma}$) and eutectic graphite (f_{eG}) in the solidified eutectic are obtained from a carbon mass balance:

$$f_{e\gamma} = \frac{100 - C_\ell^{eo}}{100 - C_{e\gamma}} \tag{10}$$

$$f_{eG} = \frac{C_\ell^{eo} - C_{e\gamma}}{100 - C_{e\gamma}} \tag{11}$$

where the concentration of carbon in the eutectic austenite ($C_{e\gamma}$) is obtained using

$$C_{e\gamma} = 2.1 - 0.20\, Si_\ell^{ea} - 0.35\, P_\ell^{ea} \tag{12}$$

The superscript 'ea' denotes average concentration in the liquid during eutectic solidification. The precipitation of graphite during eutectic solidification is ignored in determining the average concentration of Si and P in the liquid, which is estimated as the integrated average obtained using the simple Scheil models, so that

$$Si_\ell^{ea} = \frac{Si_\ell^{eo}}{k_{Si}} \tag{13}$$

$$P_\ell^{ea} = \frac{P_\ell^{eo}}{k_P} \tag{14}$$

Data for equilibrium partition coefficients k_{Si} and k_P are obtained from reference [6]. The average concentration of carbon in the eutectic liquid is estimated using

$$C_\ell^{ea} = 4.26 - \frac{1}{3} Si_\ell^{ea} - \frac{1}{3} P_\ell^{ea} \tag{15}$$

Specific Volume Models

Only interactions involving the main component Fe with the remaining components are considered. As the composition of cast iron melts is generally expressed in terms of weight percent, the following expression is used as the basis of the liquid specific volume model:

$$v_\ell = \sum_{i=1}^{n} m_i v_i + \frac{\sum\sum x_{Fe} x_i I_{Fe-i}}{M_\ell} \tag{16}$$

where v_ℓ is the specific volume of the melt in cm^3/g, m_i and v_i are the mass fraction and specific volume of component i, and M_ℓ is the molar mass of the solution. The first term in Eqn. (16) represents the specific volume of an ideal solution containing n components. The second term represents the deviation from ideal mixing associated with the interaction of components i and j, where I_{ij} is the interaction parameter in units of cm^3/mole.

Jimbo and Cramb [7] recently reviewed data in the literature for the density of Fe-C alloys and reported new results of their own obtained using the sessile drop method. As considerable deviation exists in Fe-C melt density values, the Jimbo and Cramb regression relationship was modified as follows to best represent the average of all previously reported values:

$$\rho_{Fe-C} = (6.993 - 0.0732 \cdot \%C) - (12.92 - 0.874 \cdot \%C)(T - 1550)10^{-4} \tag{17}$$

The density of liquid silicon is taken from reference [8]

$$\rho_{Si} = 2.524 - 0.0003487(T - 1410) \tag{18}$$

Fe and C is treated as a single component and any non-ideal mixing contribution is assumed to be accounted for in Eqn. (17). Olsson's expression for the temperature dependence of the interaction parameter for iron and silicon is used [9]

$$I_{Fe-Si} = -6.20 - 0.0083 (T - 1550) \tag{19}$$

where the units of the interaction parameter are cm3/mole. The density in the above expressions is in units of g/cm3 and T is in °C.

Phosphorus in the solid state exists in several allotropic forms with varying densities [10]: white (1.82 g/cm^3), red (2.20 g/cm^3), and black (2.25-2.69 g/cm^3). A value of v=0.42 cm^3/g was arbitrarily chosen, which corresponds to a density of 2.38 g/cm^3. The approach to determining the specific volume of a melt of given composition and temperature is to use Eqns. (17) and (18) to first compute the densities of Fe-C and Si, invert the density to get specific volume, calculate the first term in Eqn. (16) including the P contribution, and then add the Fe-Si interaction correction.

The specific volume of austenite is estimated using

$$v_\gamma = \frac{0.6023\, a^3}{n_{Fe} A_{Fe} + n_C A_C + n_{Si} A_{Si}} \tag{20}$$

where a is the lattice parameter, and A and n represent the atomic weight and number of atoms of each component per unit cell. The following relationship reported by Palermo and Flinn [11] for the temperature and carbon dependence of the austenite lattice parameter is used

$$a = 3.5814 + 1.709(10^{-2}) \cdot \%C_\gamma + 7.97(10^{-3}) \cdot \%C_\gamma^2$$
$$+ 6.7(10^{-5}) \cdot T + 6.2(10^{-9}) \cdot T^2 \tag{21}$$

where a is in angstroms and T is in °C. It is noted that the effect of silicon on influencing the number of atoms is accounted for in the present model, but not the effect of silicon on the lattice parameter.

A value of v_G=0.51 cm^3/g is used for the specific volume of graphite.

<u>Metal Penetration Index Parameter</u>

Based on the results of the designed experiment, it was postulated that metal penetration tendency was proportional to the product of fraction of eutectic phase and average solidification expansion during eutectic solidification. This led to the definition of a metal penetration index parameter (MPI) which is computed as

$$MPI = f_e \cdot \beta_e^{ave} \tag{22}$$

where $f_e = 1-f_{p\gamma}$. Average eutectic expansion is determined as

$$\beta_e^{ave} = \frac{v_{se} - v_\ell^{ea}}{v_\ell^{ea}} \cdot 100 \tag{23}$$

where the v_{se}, the specific volume of the solidified eutectic, is computed using

$$v_{se} = f_{e\gamma} v_{e\gamma} + f_G v_G \tag{24}$$

and v_ℓ^{ea} is the specific volume of the eutectic liquid of average composition determined at a temperature corresponding to the mean of the undercooling and recalescence temperatures obtained from thermal analysis specimens poured at the time of the experiments.

DISCUSSION OF RESULTS

Figure 2 is a plot of observed maximum metal penetration versus MPI. It is apparent from a comparison of the multiple regression results shown in Figure 1 and the theoretical results in Figure 2 that similar trends are evident. The solute redistribution/specific volume model also predicts significantly higher metal penetration for the unusual residual run H. In this model the effect of carbon on increasing metal penetration is primarily attributed to the increase in fraction of eutectic

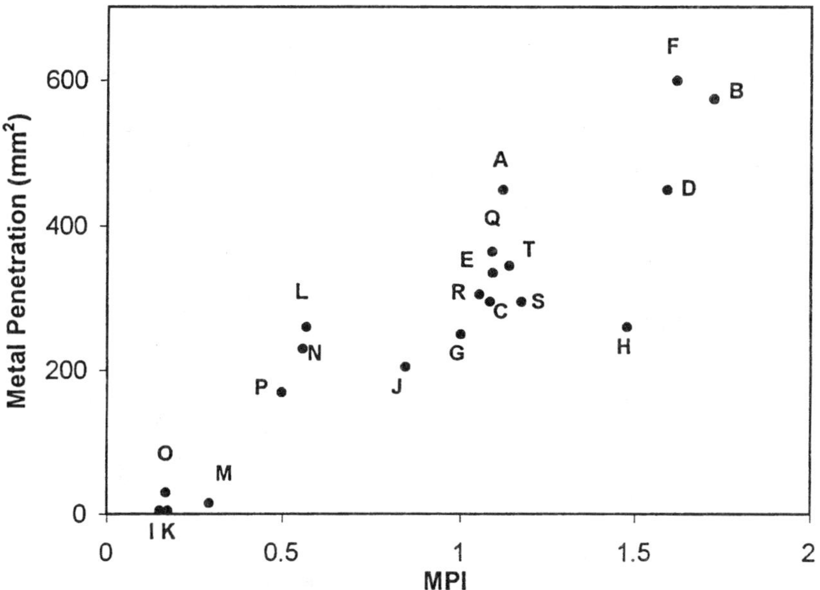

Figure 2 Maximum Metal Penetration Versus Metal Penetration Index Parameter

phase associated with increasing carbon content. The effect of phosphorus on decreasing metal penetration is related to the significant increase in the specific volume of the eutectic liquid that occurs with increasing phosphorus content. This is associated with the increase in phosphorus content in the liquid due to segregation effects (k_P=0.14) in conjunction with the low density of phosphorus. Model predictions of the influence of P on the specific volume of the liquid are in agreement with experimental results presented by Hamaker et. al. [12], who observed that the specific volume of the liquid dramatically increases with increasing phosphorus content.

The model does not account for the influence of casting geometry on metal penetration related to delayed solidification associated with heat saturation in cores or at reentrant corners. In another set of penetration test castings poured during the experiments for which the heat saturation conditions were less severe, metal penetration was only observed in the case of the high carbon - low phosphorus runs (B,D,F,H). Finally, it is emphasized that the model presented above represents initial efforts in this area. Research is ongoing to refine the solute redistribution model and incorporate it within a finite element stress model to account for the geometry effect. The present focus is on the development of a eutectic solidification model that better accounts for the influence of segregation on the distribution of carbon between the graphite and austenite phases throughout the course of eutectic solidification.

SUMMARY

Results of a designed experiment conducted to investigate the influence of C, Si, P, Mn, and S on metal penetration in gray iron castings have been presented and interpreted using a simple solute redistribution/specific volume model. Both regression analysis and model predictions indicate that the dominant main effects on penetration are C and P, with penetration increasing with increasing carbon content and decreasing phosphorus content. The carbon effect is associated primarily with the fraction of eutectic phase precipitated. The P effect is attributed to the low solubility of P in the austenite phase and the resultant increase in the specific volume of the liquid over the course of eutectic solidification. Silicon is observed to have a neglibible effect on metal penetration.

ACKNOWLEDGMENTS

Suport for this research provided by the Cast Iron Research Consortium at UW-Madison and NSF grant DDM-9102682 is gratefully acknowledged.

REFERENCES

[1]   A.B. Draper and J.L. Gainghar, AFS Transactions, Vol. 85, 163-199 (1977)
[2]   H.G. Levelink and F.P. Julien, AFS Cast Metals Research Journal, June, 105-109 (1973)
[3]   M.C. Flemings, Solidification Processing, McGraw-Hill, NY, 33-35 (1974)
[4]   F. Neumann, Recent Research on Cast Iron ASM Seminar Detroit, 659-705 (1964)
[5]   M. Hillert, Physical Metallurgy of Cast Iron, Vol. 34, 233-238 (1985)
[6]   ASM Metals Handbook, 9th Edition, Vol. 15, Casting 66 (1988)
[7]   I. Jimbo and A.W. Cramb, Metallurgical Transactions, Vol. 24B, 5-10 (1993)
[8]   A.F. Crawley, International Metallurgical Reviews, Vol. 19, 32-48 (1974)
[9]   A. Olsson, Ph. D. Thesis, Royal Institute of Technology, Stockholm (1987)
[10] Handbook of Chemistry and Physics, 53rd Edition, The Chemical Rubber Co. B-24 (1973)
[11] H.F. Palermo and R.A. Flinn, AFS Transactions, Vol. 83, 503-518 (1975)
[12] J.C. Hamaker Jr., W.P. Wood, and F.E. Rote, AFS Transactions, Vol. 60, 401-431 (1952)
[13] R. Elliott, Cast Iron Technology, Butterworths, London, 66-67 (1988)

Advanced Materials Research Vols. 4-5 (1997) pp. 293-300
© *1997 Scitec Publications, Switzerland*

Sulphur Inoculation of Mg-Treated Cast Iron - An Efficient Possibility to Control Graphite Morphology and Nucleation Ability

M. Chisamera and I. Riposan

Politehnica University of Bucharest, 313 Splaiul Independentei, RO-77206 Bucharest, Romania

Keywords: Sulphur Inoculation, Graphite Nucleation and Morphology, Mg-Treated Cast Iron, Compacted and Nodular Graphite, Titanium, Aluminium and Oxygen Inoculation, Denodularising Elements

ABSTRACT

Mg-treated irons were inoculated with Ti, O, Al or S, in ladle or in the mould, and the nucleation and the morphology of the graphite were taken into account.

More type of graphite (CG + LG + Degenerated Nodules Graphite) are associated, without a clear delimitation of the front of solidification under Aluminium, Titanium or Oxygen influence.

Sulphur has the clearest influence, facilitating the transition: SG → SG + CG → CG → Coral Graphite → B-type Graphite → A-type Graphite, when S- additive increases. So, sulphur can replace Ti, Al, Sb etc. in order to obtain different values of SG/CG ratio, thus the iron returns are no more dangerous because of their antinodularising elements content - the highest problem in CG Cast Iron production.

The lower sulphur addition after Mg-treatment under severe control conditions can also be the solution to improve the graphite nucleation process in ductile iron, especially in the presence of some silicon-base inoculants, such as SiCa.

INTRODUCTION

Cast iron with mixed structures of spheroidal (SG) and compacted (CG) graphite forms are often used in Romania. In previous works [1, 2, 3], the authors have suggested two main categories of cast irons: a) Standard CG Iron (max. 20 % SG) and Standard SG Iron (max. 20 % CG) on the one hand and b) Conventional CG Iron (50 to 80 % CG + 20 to 50 % SG) and Conventional SG Iron (50 to 80 % SG + 20 to 50 % CG) on the other. Every one of these four types of cast iron has particular properties making it suitable for several applications.

Contrary to Standard SG Iron, making CG Iron (Standard and Conventional) and Conventional SG Iron involves the use of antispheroidising (denodularising) elements (AE) to increase structural stability, in order to extend the range of CG formation. For this purpose Ti is especially used, but good results have also been obtained by using other elements (Al, Sn, Sb etc.).

When Mg and Ce are used as spheroidising elements and Ti and Al as antispheroidising elements (AE), the stability of the production of standard CG Cast Iron (max. 20 % SG) is higher than in the situation when only magnesium is used [4]:

$$Mg : Mg + Ce : Mg + AE : Mg + Ce + AE = 1 : 1.3 : 1.5 : 1.8$$

Useful in controlling the morphology of the graphite, denodularising elements are present in the return scraps and they may have a negative influence on the new obtained SG and CG Irons [1, 2, 5-11]. A previous work [11] has highlighted that CG returns containing Ti and Al were not to be used to make Ferritic Ductile Iron of high ductility. As for thin-walled castings in SG and CG Cast

Irons, the controlled use in charge of CG Iron Returns with Ti (Al) content has a positive influence due to the increase of eutectic cells number, the decrease of formation of carbides for both types of cast irons and the limitation of SG in CG Cast Iron [12].

In the present work, Mg-treated irons were inoculated with general (O, S) denodularising elements and specific (Ti, Al) ones and the nucleation and the morphology of the graphite were investigated.

TITANIUM, ALUMINIUM, OXYGEN OR SULPHUR INOCULATION OF Mg-TREATED CAST IRONS DURING SOLIDIFICATION

<u>Experimental procedure</u>

Irons were produced in a 10 kg/8.000 Hz induction crucible furnace. The charge included pig iron, scrap iron and ductile iron returns, at a sulphur content of max. 0.03 % at the end of the melting process.

The irons were treated with FeSiCaMgCe-type nodulariser by the Sandwich technique, followed by ladle inoculation with 0.5 % FeSi75. At a temperature of 1350 °C, the liquid iron was poured in test-pieces of ϕ 90 x 90 mm in graphite mould. Final chemical composition of iron is: 3.6-3.8 % C, 2.5-2.7 % Si, 0.4-0.45 % Mn, 0.1-0.15 % P, 0.01-0.022 % S, 0.035-0.05 % Mg_{res}, Equivalent Carbon = 4.35 to 4.65 %.

In the liquid iron, at the middle of the test-piece denodularising elements were separately introduced 60 seconds after pouring it in the mould, during the beginning of solidification respectively (time predetermined by drawn cooling curves). They were used: Titanium (90 % Ti), Aluminium (99 % Al), Sulphur (FeS_2) and Oxygen (Fe_2O_3).

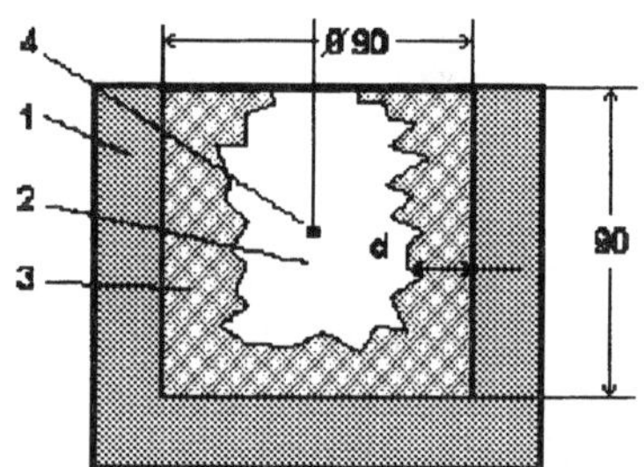

Fig.1 Test-piece used to study the influence of denodularising elements
(Ti, Al, S, O) introduced in Mg-treated irons, during solidification
1- graphite mould; 2- liquid iron; 3- solidified superficial layer; 4- denodularising additive.

Structural and fracture analyses of the cross section of the test-pieces were carried out and it was observed (Fig.1):

a) a superficial layer which was solidified before adding the denodularising agent (d- Fig.1);

b) a central area where the denodularising agent was introduced;

c) the front of solidification in the contact area of the previous zones.

<u>Results and Discussions</u>

An electronic microscope (JXA - 5A) was used to detect the inoculating elements (Ti, Al, O, S) in the front of solidification and the central areas of the test-pieces. The action of the denodularising additives on the nucleation and the morphology of the graphite have been specially investigated.

The inoculation with Ti, Al, O or S of Mg-treated irons during solidification in graphite moulds brings about important changes in the nucleation ability of the graphite.

All these four elements determine an increase by 2-4 times of the number of graphite pieces on surface unit, aluminium and oxygen having the greatest influence.

The analysis of the Ti content has emphasised its uneven distribution along the cross section of the test-pieces, due to the low diffusion coefficient of Ti in liquid iron. Some areas very rich in Ti (over 20 % Ti) and which contained carbides were observed, along with other areas with a low Ti content (less than 0.2 % Ti), that contained compacted graphite and degenerate graphite nodules.

Aluminium has mainly a similar influence to that of titanium but the specific denodularising effect was observed due to the easier formation of lamellar (flake) graphite of small dimensions (maximum 40 μm), compacted graphite and degenerate graphite nodules well.
In those areas where compacted graphite was present, aluminium content did not exceed 0.25 % Al, as compared to 20...25 % Al in the lamellar graphite zones.

As for the graphite morphology, the association of several types of graphite, exploded, compacted and lamellar graphite, was observed under oxygen influence, but the front of solidification was not precisely delimited. Oxygen has a high capacity to interact with many elements present in the melt, so that it is very difficult to use it in order to control the SG/CG ratio. A variable and uncontrollable part of the oxygen introduced in the Mg-treated liquid iron may interact with other elements beside the nodularising elements in excess.

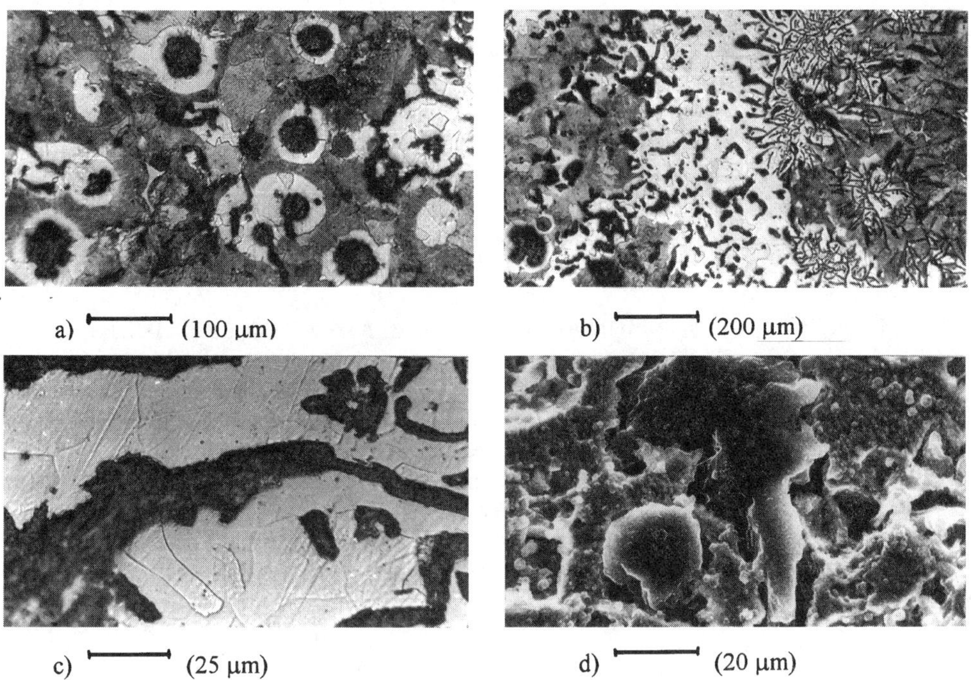

a) ⊢——⊣ (100 μm)   b) ⊢——⊣ (200 μm)

c) ⊢——⊣ (25 μm)   d) ⊢——⊣ (20 μm)

Fig.2. Influence of sulphur inoculation after Mg-treatment on the graphite morphology
a) shell; b) transition zone; c)"hybrid" graphite; d)"hybrid" graphite (SMP).

Sulphur inoculation of Mg-treated iron had a strong influence on the morphology of the graphite and transition from nodular to compacted graphite was very clear and precisely delimited (Fig.2).

Another specific influence of sulphur was the transition from compacted to lamellar graphite, having some characteristics of the growth mechanism of compacted and the way of action of sulphur.

Thus, graphite shapes (" hybrids " called) became visible at the limit of demarcation between the area containing compacted graphite and that containing coral or lamellar graphite. In the initial phase, those grew as compacted shapes, while later, under the influence of sulphur, they continued to grow mostly in the sense of the " a " axis, according to a mechanism characteristic of lamellar graphite.

The analysis of the dimensions of the graphite shapes slowed that "hybrid" graphite presents a more developed lamellar area, it has a length 35 % bigger than of the compacted area. This shows a rather checked increase of the "hybrid" graphite shapes in the lamellar area in comparison with the neighbouring area (towards the centre of the test-piece) where the lamellar graphite shapes were approximately twice longer. The checked increase of the "hybrid" shapes in determined by a higher activity of the nodularising elements (Mg, Ce). However, this activity is below the critical limit necessary to compact the graphite, yet higher than in centre area of the test-pieces, where the sulphur content is bigger.

The dualism of the growth mechanism of compacted graphite is also promoted by the layout of eutectic cells. While in the superficial area of the test-piece, each eutectic cell contains only a single graphite shape (nodular or compacted graphite), in the transition area, graphite shapes grew in colonies, which contain coral or lamellar graphite (A and B type).

Thus, it results that sulphur inoculation in Mg-treated iron produces a transition similar to:
SG $\rightarrow$ SG + CG $\rightarrow$ CG $\rightarrow$ Coral Graphite $\rightarrow$ B-type Graphite $\rightarrow$ A-type Graphite.

A comparative analysis of the influences of the four denodularising elements (when they were used to inoculate Mg-treated iron) leads to the conclusion that the diffusion coefficient of sulphur in liquid iron is high at low temperatures of the melt.

As regards the influence of sulphur, a clear and controllable transition from nodular to compacted graphite can occur more frequently than in the case of the other denodularising elements. Thus, sulphur inoculation after Mg-treatment of the melt may be used in order to settle the SG/CG ratio in the structure of irons.

SULPHUR LADLE INOCULATION AFTER Mg-TREATMENT OF THE MELT

<u>Laboratory experiment</u>

In order to obtain CG Iron or irons with certain SG/CG ratio the use of sulphur inoculation of Mg-treated iron was studied.

According to the program of experiments, the irons were treated by the Sandwich - technique with FeSiCaMgCe alloy, followed by ladle inoculation with variable quantities of sulphur (FeS$_2$) and FeSi75. Thus treated irons were cast in stepped-wedges sample (g = 5...20 mm thickness) and in round pieces (ϕ 10...30 x 100 mm) in sand and metal moulds. Holding time between Mg-treatment and pouring in moulds did not exceed 5 minutes.

The variation of the SG/CG ratio under the influence of the sulphur inoculation of the Mg-treated iron (0.024...0.030 % Mg$_{res}$) and of the wall thickness of the samples is shown in fig.3.

After the Mg-treatment a structure containing more than 80 % SG (up to 20 % CG) was obtained, thus corresponding to standard SG Iron (Ductile Iron). In sulphur-ladle inoculated irons SG/CG ratio decreases, when addition of sulphur increases.

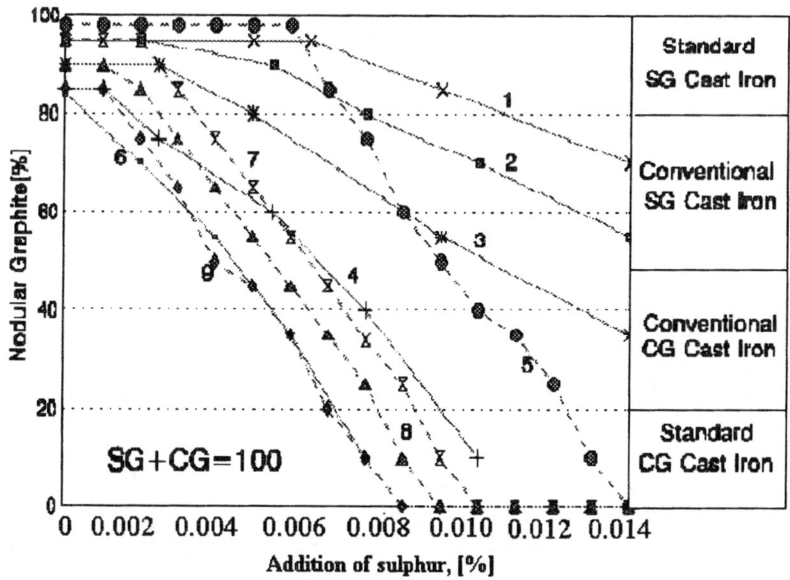

Fig.3 Graphite Nodularity according to the Sulphur-addition after Mg-Treatment
1, 2, 3, 4 - Stepped Wedge Samples - Sand Mould (**1**:5 mm - 0.03 % Mg_{res}; **2**:5 mm - 0.027 % Mg_{res}; **3**:20 mm - 0.03 % Mg_{res}; **4**:20 mm - 0.027 % Mg_{res}); **5** - Round Sample - Metal Mould (17 mm - 0.03 % Mg_{res}); **6, 7, 8, 9** - Round Samples - Sand Mould (**6**:20 mm - 0.024 % Mg_{res} ; **7**:10 mm - 0.03 % Mg_{res}; **8**:17 mm - 0.03 % Mg_{res}; ; **9**:30 mm - 0.03 % Mg_{res}).

The amount of sulphur needed for inoculation of Mg-treated irons depends on the residual magnesium content after Mg-treatment, holding time prior to pouring, wall thickness of the casting and type of the mould (sand or metal mould) [4, 13].

In order to obtain standard grade CG Cast Iron (max. 20 % SG), it is recommended to use Mg or Mg-Ce treatment of iron, getting ratio of 0.025...0.04 % Mg in it. So, the formation of lamellar graphite, on the one hand and the graphite nodularity, on the other hand, are avoided. Low addition of sulphur (usually up to 0.015 % S) makes possible to obtain reproducible Standard Grade CG Cast Iron.

Plant experiment

A schematic technological process of obtaining Standard CG Iron by melting irons in 3.5 t-capacity acid induction crucible furnace in plant conditions is shown in fig.4.

The irons were treated with FeSiMgCaCe-alloy (48-52 % Si, 8-12 % Mg, 0.5-2.5 % Ca, 0.5-1.0) % Ce, Fe-Balance) by Tundish - Cover technique (so that residual magnesium content after this treatment keeps up to 0.04 % Mg) followed by ladle inoculation with 0.01 % S (FeS_2) and 0.4 % FeSiCaBaAl (62-66 % Si, 0.5-1.2 % Ca, 1-3 % Ba, max. 1.2 % Al).

After sulphur inoculation of irons, graphite nodularity was changed and the structure was modified from nodular to compacted irons (P1 to P2 samples - fig.4). This structure keeps a long holding time after Mg-treatment of irons (13...18 min.) at a high stability of the effect of Mg-treatment (fig.4). For a variation of wall thickness of the casting between 5...45 mm, Standard CG Cast Iron made in this way has a distinctive structural stability [4; 13].

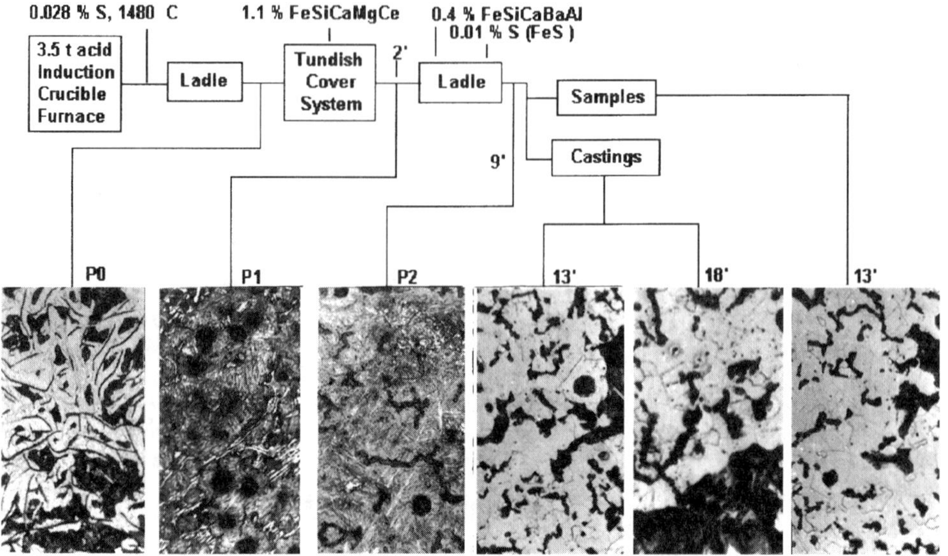

Fig.4. Scheme of technological process of CG Cast Iron
obtaining by S-Inoculation after Mg-Treatment in plant conditions

SULPHUR INOCULATION OF DUCTILE IRON

It is known that after the Mg-treatment, liquid iron becomes low in oxygen, sulphur and nitrogen so that the efficiency of inoculation in reduced, especially either on high content of residual magnesium or high cooling rate processes.

Inoculation efficiency of ductile iron depends on occurrence of compounds (oxides, sulphides, nitrides, carbonitrides etc.) which have sustaining part in graphite nucleation and their stability in liquid cast iron. The literature analysis shows the presence of sulphur components as primary nucleation support both in gray and ductile irons, so it can remark the importance of sulphur in the inoculation process of ductile iron.

Ductile iron having slight hypereutectic composition has been produced in a 10 kg/8,000 Hz induction crucible furnace. The spheroidising treatment was made by the Tundish - Cover procedure, using FeSiCaMg alloy, to ensure the high spheroidising capacity (Mg_{res}= 0.055...0.065 wt %). The inoculation treatment has been made by different procedures(in ladle, pouring basin, pouring gates) using conventional inoculants (FeSi, SiCa), associated with sulphur (FeS), Ce or Bi, test-pieces poured either in conventional sand moulds or chill test.

Experiments have been tried to find out the influence of inoculants on the graphite nucleation capacity (number and size of graphite nodules) and chill tendency of cast irons (the ratio of carbides as against to the metallic chill area) (fig. 5, 6).
The experimental tests pointed out the following aspects:

- sulphur addition as FeS in Mg-treated cast iron has a small inoculanting influence;

- in association with FeSi, sulphur has also a low inoculant effect, but the addition as SiCa+FeS leads to increase the graphitising potential (the greatest increase of nodules number and decrease of the size of chilled area);

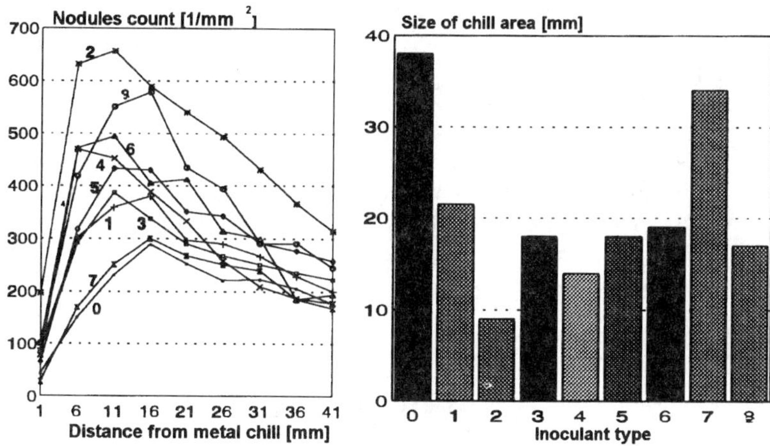

Fig. 5. Variation of nodules count
with the height of chill test

Fig. 6. Influence of inoculant type on
chill size area

0-Base; **1**-0.35 % SiCa; **2**-0.35 % SiCa+0.015 % FeS; **3**-0.35 % SiCa+0.015 % FeS+0.01 % Ce;
4-0.35 % FeSi; **5**-0.35 % FeSi+0.015 % FeS; **6**-0.35 % FeSi+0.015 % FeS+0.01 % Ce;
7-0.015 % FeS; **8**-0.35 % FeSi+0.005 % Bi

- Bi+FeSi inoculation leads to a similar graphitising capacity to SiCa+FeS addition, but it has a lower effect on the chill tendency;

- S- inoculation of ductile iron can be a means of increasing the efficiency of SiCa as inoculant having the advantage of not contaminating the charge materials.

The experiences pointed out the necessity of sulphur employment in association with such elements able to form stable sulphides. It is noted that in the case of singular FeS addition a very low effect was produced. This suggests about low graphite nucleation capacity of MgS in absence of elements able in increase carbon activity, such as silicon.

The experiment has shown that the lower sulphur addition (0.002...0.01 wt % S) after Mg-treatment under severe control conditions can be the solution to improve the graphite nucleation process in ductile iron. This   effect is significant in the presence of the silicon-base inoculants, especially SiCa. The efficiency of SiCa+FeS addition as inoculant was much more effective in the mould inoculation.

GENERAL CONCLUSIONS

1. The inoculation with Ti, Al, O or S of Mg-treated iron during solidification in graphite moulds brings about important changes in the nucleation and the morphology of the graphite:

a) All these four elements determine an increase by 2 - 4 times of the number of graphite shapes on surface unit, aluminium and oxygen having a greatest influence;

b) Ti and Al exert the same influence on the graphite morphology determining an unhomogeneous structure - CG + SG + Degenerate Graphite Nodules; in the case of aluminium, less than 40 μ m lamellar graphite (LG) also occurs.

c) Under the influence of oxygen, more types of graphite are associated (CG + LG + Exploded Graphite), without a clear delimitation of the front of solidification.

d) Under the influence of sulphur, a clear transition from SG $\rightarrow$ SG + CG $\rightarrow$ CG $\rightarrow$ Coral Graphite $\rightarrow$ B-type Graphite $\rightarrow$ A-type Graphite takes place. The transition from nodular to vermicular/compacted graphite is exempt of other types of graphite. The transition from compacted to lamellar graphite is made through "hybrid" graphite shapes.

2. Sulphur ladle or mould inoculation of Mg-treated iron makes possible to obtain different values of SG/CG ratio, without using classical denodularising elements (Ti, Al). So, sulphur can replace these elements in order to control graphite nodularity, thus that iron returns are no more dangerous because their antinodularising elements content.
3. The amount of S-additive in Mg-treated iron needed to obtain a certain SG/CG ratio depends on the residual magnesium content after Mg-treatment, holding time prior to pouring, wall thickness of the casting and type of the mould.
4. The lower sulphur addition (up to 0.01 wt. % S) in association with silicon-base inoculants (especially SiCa) after Mg-treatment can be the solution to improve the graphite nucleation process in ductile iron.

REFERENCES

1. I. Riposan and L.Sofroni - **Fonta cu grafit vermicular** (Vermicular Cast Iron), Editura
       TEHNICA, Bucuresti (RO), 1984, 334 pages.
2. I.Riposan et al, 53rd World Foundry Congress, Prague, 1986
3. I.Riposan et al, Metalurgia (RO), 5 (1986), 233-238
4. I.Riposan and M.Chisamera, Giesserei-Praxis, 9/10 (1991), 155-162
5. L.L.Fosbinder and S.A.Sauer, 51st World Foundry Congress, Lisbon, 1984.
6. J.B.Nesselrode, Giesserei, 16 (1985), 388-391
7. S.Kigucki et al IMONO, 5 (1983), 279-284
8. W.Simmons and J.Brigges, AFS Transactions, 90 (1982), 21-32
9. M.Costes, Hommes et Fonderie, 128 October (1982), 21-32
10. M.Chisamera et al, Metalurgia (RO), 2 (1988), 98-103
11. I.Riposan et al, Cast Metals, Volume 2, No.4 (1990), 207-213
12. M.Chisamera et al, Romanian patents No. 95564, 95565
13. I.Riposan and M.Chisamera - Casting, Forging and Heat Treatment (Japan), 547 (1993), 17-25.

Advanced Materials Research Vols. 4-5 (1997) pp. 301-306
© *1997 Scitec Publications, Switzerland*

Synthetic Cast Iron Melted in an Induction-Plasma Furnace

W. Lybacki, Z. Ignaszak, A. Modrzynski and A. Baranowski

Poznan Technical University Foundry Laboratory,
3, Piotrowo, PL-60-965 Poznan, Poland

Keywords: Grey Cast Iron, Recarburization, Induction-Plasma Melting, Solidification, Properties

Abstract: The article presents preliminary results of the application of an induction-plasma furnace for melting of synthetic grey cast iron. High efficiency of recarburization process exceeding 90 % was obtained. Grey cast iron was obtained with strength from 100 up to 340 MPa.(without inoculation). It was shown that synthetic grey cast iron melted in an induction-plasma furnace is cheaper by over 45 % than foundry iron.

INTRODUCTION

At present crucible induction furnaces are often applied for melting of cast iron. The essential disadvantages of these furnaces is low temperature of slag which limits the reaction between slag and metal. Melting in an induction furnace consists of remelting of a suitably chosen charge of materials and of overheating of the liquid alloy until the required temperature is reached.

As a result of co-operation between the Foundry Laboratory and the Institute of Electrical Engineering (both of the Poznan Technical University) an induction-plasma installation was developed.

Laboratory tests showed a positive influence of the induction-plasma melting on the mechanical and plastic properties of cast ferrous alloys. For instance improved the impact strength of cast ferrous alloys such as grey cast iron, cast carbon steel and cast stainless steel. The tests results were confirmed in a foundry using an induction furnace with capacity of 300 kg. Moreover, it was shown that the efficiency of the induction-plasma melting doubled and the consumed electrical energy dropped about 25 % [1,2,3].

Low temperature plasma as a source of additional energy in an induction furnace has the following advantages:
- it does not pollute the liquid alloy with undesirable impurities,
- its high temperature (up to 14000 $^{\circ}$C for argon plasma) activates refining processes and intensifies the melting process,
- it allows one to melt at any atmosphere depending on plasma gas employed.

World production of foundry iron dropped lately and cheap raw materials have been sought. Steel scrap is the cheapest charge material used for the production of cast iron, called the synthetic cast iron in this case. Winning of synthetic cast iron consists of enriching iron with carbon and silicon and of applying a high temperature treatment in order to change the structure of the charge materials.

EXPERIMENTS

Research on the kinetics of iron and steel recarburization has been reported in numerous papers [4-9]. They show that recarburizing of liquid ferroalloys depends on the following factors:
- temperature of the process,
- chemical composition of the liquid metal,
- mixing of liquid metal,
- carburizer.
- furnace atmosphere.

Two main processes of recarburization are used, namely strewing the liquid metal surface with carburizer or blowing on powdered carburizer with nitrogen or air fluxes. The first method assures the efficiency of the process from 30 to 50 %, the latter one from 50 to 70 %.

The purpose of the preliminary research was to evaluate the usefulness of an induction-plasma furnace for melting cast iron from cheap charge materials such as steel scrap.

Research on obtaining synthetic cast iron in an induction-plasma furnace was carried out in the Foundry Laboratory. The installation consisted of induction furnace PI50/8, plasma torch, plasma supply, and control systems. A scheme of the induction-plasma furnace is presented in Fig.1.

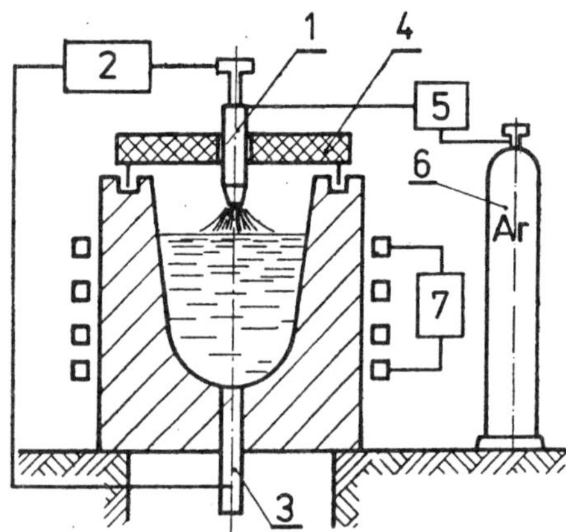

Fig.1 Scheme of the induction-plasma furnace
1-plasma torch, 2-DC-power supply, 3-electrode,
4-furnace, 5-rotameter, 6-argon cylinder, 7-inductor

The following are the characteristic features of induction furnace PI/50/8 and the plasmatron
- maximum capacity of a crucible   - 50 kg
- active power   - 56 kW
- frequency   - 8000 Hz
- cooling of inductor   - water
- furnace lining   - quartzite.

Arc Plasma torch of direct current:
- overall length   - 700 mm
- inside diameter of nozzle   - 10 mm
- plasma gas   - argon

Melting of synthetic grey cast iron in the induction-plasma furnace followed research on efficiency of recarburizing of Armco iron and cast steel. Armco iron or steel was used as a charge material which chemical composition is given in Tab.1. Carburizers and their chemical composition are presented in Tab.2 (carbon and sulphur contents were analysed using automatic analyser METALYT CS ELTRA Gmbh).

All induction-plasma melts were carried out in a similar way. The furnace was loaded with either Armco iron or steel and with carburizer. During all tests liquid metal was held at the temperature 1460-1500 $^{\circ}$C for 15 minutes, and samples were taken to define the final carbon content. Samples for testing carburization efficiency were encapsulated into quartz tubes. Carbon content was determined using Leco analyser CS244. Formulas given in Tabl.3 were used for calculation of the carbon assimilation ratio and carburization efficiency.

Samples for the testing of mechanical properties were cast according to the ISO standard 185:1988. Testing machine INSTRON 4260 and hardness tester WOLPERT were used for the testing of strength and hardness, respectively.

The computer aided thermal and derivative analyses (CA-TDA) were obtained with the application of the SCHLUMBERGER automatic data acquisition system (for recording) and the in-house software for the analysis. The shape and the size of a sample for the   TDA analysis were similar to the QUICK-CUP (Electronite) ones. The first test was carried out after the recarburization of the Armco Iron or steel, and the second one after Si addition.

Table 1: Chemical composition of Armco iron and steel

Chemical composition, %					
C	Si	Mn	P	S	Cr
0,012	0,020	0,015	0,0065	0,006	0,0005
0,21	0,360	0,600	0,0140	0,020	0,1700

Table 2: Chemical composition of carburizers

Carburizer		Chemical composition , %			
Kind	Marking	Carbon	Sulphur	Ash	Volatile matter
Graphite A	I	96,88	0,136	1,03	1,95
Graphite B	II	94,78	0,556	4,44	0,22
Anthracite A	III	75,85	0,799	15,61	7,74
Anthracite B	IV	88,45	0,853	4,38	6,32

RESULTS AND DISCUSSION

Table 3 shows the results of the recarburizing efficiency tests. Table 4 shows the rate of Armco iron recarburization at the temperature 1460-1480 °C. The initial tests results on efficiency of Armco iron recarburization in the induction-plasma furnace using different carburizing materials allowed us to select as the most effective carburizer for further research graphite carburizer marked I in Tab.3. Using different fractions of the carburizer we were able to determine its optimal size with respect to the point of carbon assimilation ratio and efficiency of the process.

Results of recarburizing of Armco iron and steel show the high efficiency of recarburizing in the induction-plasma furnace especially for the carburizer I. It should be underlined that about 90 % of total carbon content was assimilated during overheating of the liquid metal at 1460°C .

Mechanical properties of the obtained synthetic cast iron depended on the final chemical composition of the alloy. Results given in Tab.5 show that grey cast iron with strength of 340 MPa was obtained without any inoculation.

The pearlitic matrix structure of the synthetic cast iron was mainly observed. The shape of the graphite was quasi flake with interdendritic distribution.

Cooling plots after recarburization and after Si addition are presented in Fig.2. The shape of cooling curves immediately after recarburization of the melt evidently indicate high efficiency of the process. It was possible to estimate the CE (carbon equivalent) value from the liquidus temperature T_l. The less pronounced undercooling and recalescence effects as well as the eutectic arrest temperature (solidus metastable) prove that the white structure of the synthetic iron was obtained at this stage.

The cooling curve after Si addition indicated changes in the nature of carbon crystallisation by:
- presence of undercooling and recalescence eutectic effects,
- increase of eutectic temperature arrest.

These results were confirmed by Remi [10], and Chen [11]. The plasma torch application for the recarburization did not cause any new changes in the cooling curve and its derivative. It was observed for both of the synthetic cast irons.

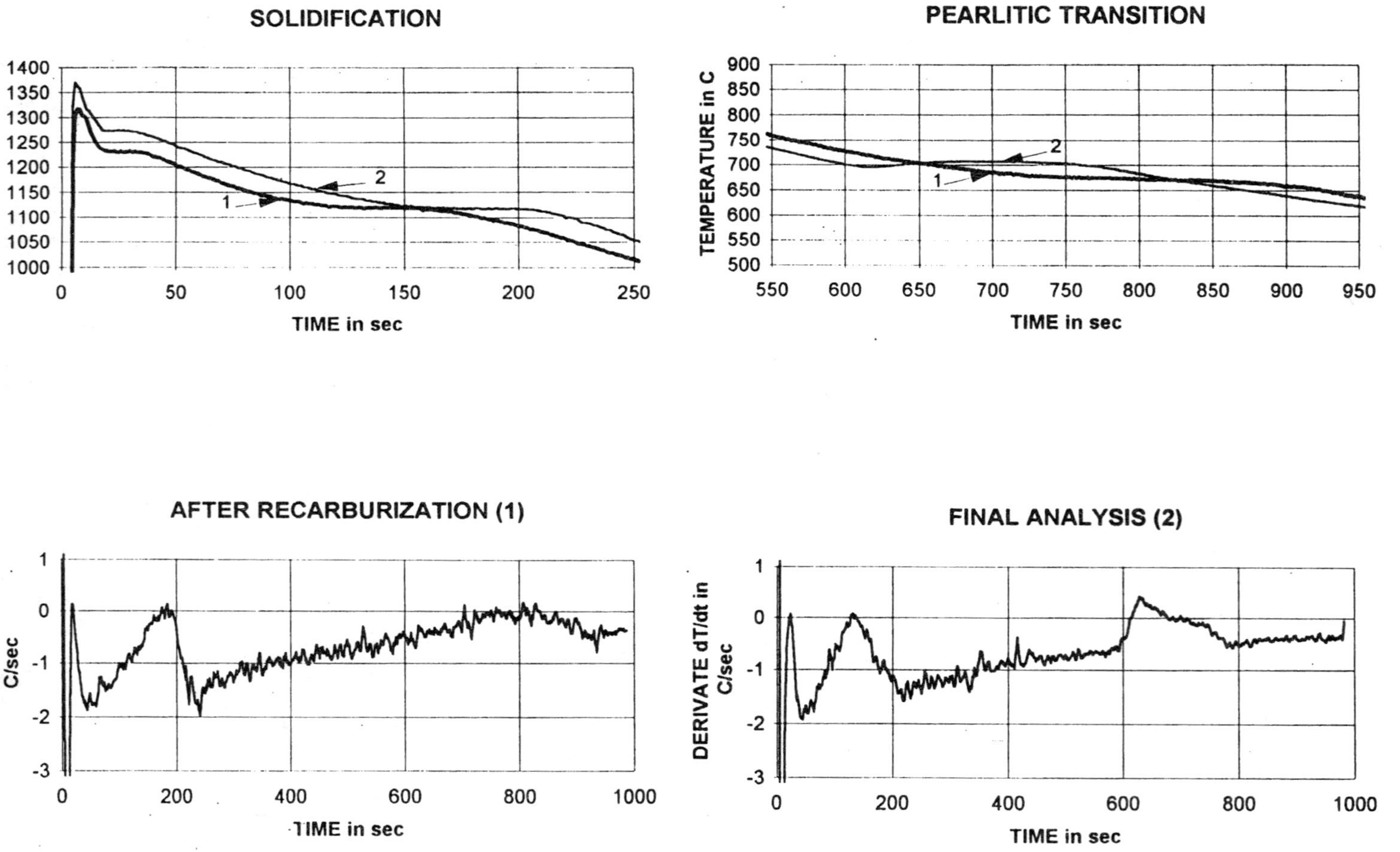

Fig.2. Cooling curves and their derivatives for synthetic cast iron samples

Table 3: Results of recarburization of the Armco iron and Steel

Carburizer	Mass of charge, [kg]	Mass of carburizer [kg]	C1 [%]	C2 [%]	B [%]	E [%]	Charge material
I	32	1,4	0,012	3,95	92,9	90,0	Armco
II	30	1,6	0,012	3,22	63,5	63,5	Armco
III	30	1,6	0,012	2,71	68,6	68,6	Armco
IV	30	1,6	0,012	3,26	68,9	68,9	Armco
I	30	1,4	0,012	4,28	94,5	91,5	Armco
I	30	1,4	0,012	4,35	96,3	92,9	Armco
I	30	1,4	0,012	4,27	94,3	91,2	Armco
I	30	1,2	0,21	3,89	94,9	92,0	Steel
I	37	1,3	0,21	3,54	97,8	94,8	Steel
I	30	0,9	0,21	3,05	97,7	94,7	Steel

C1 - initial content of carbon, C2 - final content of carbon,
B - carbon assimilation ratio E - efficiency of carburization,

$$B = 100 * m_c \frac{(C2 - C1)}{a * m_{cr}} \qquad E = m_c * \frac{(C2 - C1)}{m_{cr}}$$

where:

m_c - mass of charge, kg
m_{cr} - mass of carburizer, kg
a - pure carbon content. %

Table 4: Recarburizing rate of the liquid Armco iron at the temperature 1460 - 1480 °C

Time [min]	0	5	10	15	20
Carbon [%]	0,012	1,67	2,99	3,12	3,56

Table 5: Mechanical properties of the synthetic grey cast iron

Stress at Max.Load	Maximum Percent Strain	Hardness
MPa	[%]	HB
105	0,798	186
194	0,722	205
267	0,659	200
275	0,662	212
315	0,729	226
336	0,525	232
340	0,759	235

COSTS

The results of carburizing show that 0,036 kg of graphite carburizer per 1 kg of steel is needed to raise the carbon content from 0,2 up to 3,9 %. Cost was calculated using prices given in Tabl.6.

Table 6: Charge material prices

Foundry iron	230 DM/t
Steel scrap	100 DM/t
Carburizer	410 DM/t
Synthetic grey cast iron	115 DM/t

Economic analysis shows that synthetic cast iron melted in the induction plasma furnace is cheaper almost 50 % than foundry iron (without costs of ferroalloys)

CONCLUSIONS

Preliminary research on the induction-plasma furnace application for the melting of synthetic cast iron confirmed that this new way of obtaining synthetic cast iron is very effective. The neutral atmosphere, and high temperature create favourable conditions of melting of good quality cast iron .

The modified CA-TDA method will be used for future investigation of synthetic inoculated and SG iron. The authors plan to expand their study on correlation between the characteristic points of cooling curves and the shrinkage/expansion phenomena. The investigations will be continued using the results reported in references [12,13].

REFERENCES

1. W.Lybacki and others : Proceedings of First Conference on Plasma Metallurgy, Czestochowa, Poland, pp.234-246, (October 1987).
2. S.Idziak and others : Proceedings of First Conference on Plasma Metallurgy, Czestochowa, Poland, pp.235-257, (October 1987).
3. W.Lybacki and others : Przeglad Odlewnictwa t.39, No 3, pp.91-94, (1989).
4. A.Kosowski : Przeglad Odlewnictwa ,t.32, No 1-3, pp.11-13, (1982).
5. C. Podrzucki, C.Kalata: Metalurgia i odlewnictwo zeliwa, Slask, Katowice (1971)
6. R. Krzeszeski: Prace Instytutu Odlewnictwa, z. 3-4, pp.175-197, Krakow, (1957).
7. R.V.Riley : Foundry Tr. J., vol. 91 , No 1829, pp. 331-339, (1951).
8. J.Czikel , H.Miketta : Freiberger Forschungshefte Reihe B, H. 3, pp.56-59, (1953).
9. L.Krol, J.Latusek: Archiwum Hutnictwa t.28, z.1. pp. 101-116, (1983).
10. A.Remy and others : Fonderie-Fondeur d'Aujourd'hui, 68, pp.11-18, (Octobre 1987).
11. I.G.Chen , D.M.Stefanescu : AFS Trans. , 30, pp.947-964, (1984).
12. Z.Ignaszak and others : Intern. repport ,Foundry Laboratory of Poznan Techn.Univ. (1993).
13. Z.Ignaszak : "Solidification de la fonte GS", Rapport interne FERRY-CAPITAIN (aout 1994).

Advanced Materials Research Vols. 4-5 (1997) pp. 307-312
© *1997 Scitec Publications, Switzerland*

Development of New Container Manufacturing Process by Vertical Type Centrifugal Casting Method and its Solidification Analysis

S.G. Kim[1], T. Umeda[1], K. Murata[1], D. Sakurai[2] and M. Minami[2]

[1] The University of Tokyo, Hongo 7-3-1, Bunkyo, Tokyo 113, Japan

[2] Nippon Steel Corp., Otemachi 2-6-3, Chiyoda, Tokyo 100-71, Japan

Keywords: Centrifugal Casting, Recycling of Low Level Radioactive Metallic Wastes, Nodular Cast Iron, Critical Solid Fraction for Fluid Flow

Abstract
The experimental and theoretical works are performed to develop the process for producing a container by centrifugal casting without cores. The production of cylindrical container castings can be made by dropping the metal volume below the critical solid fraction for the fluid flow with reducing / stopping rotation of castings. The bottom thickness can be mainly controlled by a falling time. Coating materials and their thickness greatly influence the bottom thickness of castings. In this method the critical solid fraction for the fluid flow of drop is estimated as 0.3~0.4.

1 INTRODUCTION
Centrifugal casting is widely applied as a pipe forming process, and this process can be easily automated, and also has many characteristics such as fewer production steps and less amount of wastes through manufacturing.

In these days, the necessity of decreasing or recycling of machinery or structural parts wastes becomes more and more important, especially in case of large amount of hazardous wastes like as radioactive ones the demand is very strong.

In this situation, a recycling idea for low level radioactive metallic wastes of the atomic power generation is proposed, that is, melting of wastes gives the reduction of the their volume and the uniformity of nuclides, the centrifugal casting which is operated with almost full automation and least environmental restraints makes storage containers for newly produced radioactive wastes. This production system of storage containers exerts not only the utilization of wastes but also an easy watch of both containers and contaminated materials.

During solidification under a centrifugal force to decrease / stop the rotation of the mold makes metal less than the critical solid fraction for the fluid flow drop to form the bottom of the container. This report proposes both the experimental and theoretical development of the process for manufacturing the container with Centrifugal Casting without 'Cores'.

2.EXPERIMENTAL PROCEDURE
2.1 Manufacturing Principles and Test Castings
The principle of manufacturing storage containers by centrifugal casting is shown in Fig. 1. Once pipe shape is formed by pouring metal into the mold rotating with constant speed, side wall metal less than the critical solid fraction for the fluid flow should drop to form the bottom without a core by reducing / stopping the rotation of the mold.

To confirm this idea, nodular cast iron (FCD 370 /JIS G 5202) and stainless steel castings (SCS 5 /JIS G 5121) were made experiments. A carbon steel mold whose inner diameter is 196 mm

Table 1 Chemical composition of castings   (mass %)

Alloy	C	Si	Mn	P	S	Ni	Cr	Mo	Mg
SCS5	0.04	0.48	0.46	0.010	0.006	3.98	12.45	0.45	-
FCD370	3.58	1.85	0.17	0.015	0.005	0.74	-	-	0.048

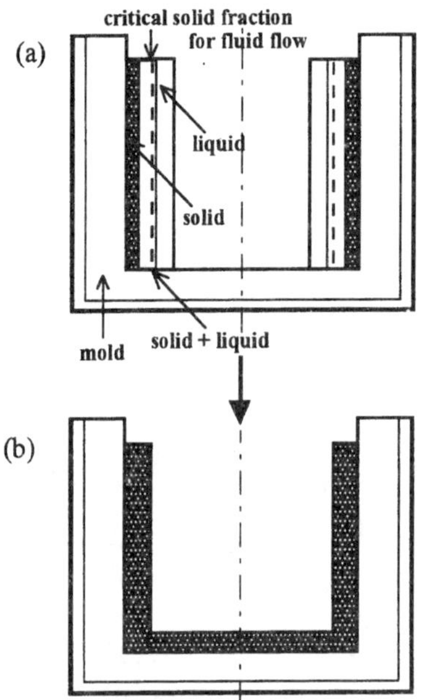

Fig.1 Principle of manufacturing a container by centrifugal casting   Fig.2 Size of castings (unit in mm)

was coated with zirconia dressings . Molten metal was poured into the mold after the rotation of the mold reached constant speed (Gravity Number is 120). After the constant speed was kept for the prescribed time, the speed was reduced and then the rotation stopped. The size of the castings was 300 mm in height and 190 or 176 mm in outer diameter according to the thickness of the coating. The thickness of side wall and bottom was determined by the conditions of the reduction of centrifugal force. Size of casting in case of 3 mm coating thickness is shown in Fig. 2 where circled area at the bottom of a side wall is an important part to weld between falling metals and the side end. Chemical compositions of castings are shown in Table 1.

In this report, time from the start of pouring to the stop of rotation is called "Falling Time". The real falling time should be the time of the falling of metal less than the critical solid fraction for the fluid flow during deceleration of the speed. But pouring was within a few second and falling of metals was confirmed just before the stopping of rotation. Then we defined the time from the start of pouring to the stop of rotation as a falling time.

2.2 <u>Simulation Method</u>

It is necessary to solve the equations of fluid flow and heat transfer, because molten metal even the solid can move under the centrifugal and the gravitational forces. But it is very complicated, therefore the following simplifications are assumed. 1) Solidification proceeds under the constant rotation. 2) Even when decreasing of the rotating speed, the balance of centrifugal force and gravity is kept, shape of molten metal is maintained as a pipe , and solidification progresses. The balance is broken suddenly, and molten metal falls down to the

bottom instantaneously. 3) Molten metal less than the critical solid fraction for the fluid flow falls down. 4) Solidification is analyzed on the fall part (bottom) and the side wall.
 Solidification analysis of symmetric pipe castings was made according to the next 2 dimensional unsteady heat conduction equation.

$$\rho c_p \frac{\partial T}{\partial t} = \lambda \left[\frac{1}{r}\frac{\partial}{\partial r}\left(r\frac{\partial T}{\partial r} \right) + \frac{\partial^2 T}{\partial z^2} \right]$$

where T; temperature, t; time, ; density, Cp; specific heat, and K; heat conductivity. Boundaries between the castings and the coating and between the coating and the mold are perfect contact, and those between the casting and air and between the mold and air are followed by the Newton's cooling law. Thermal properties and boundary conditions used for calculations are shown in Tables 2 and 3. The liberation of a latent heat is treated as the progress of equilibrium solidification.
At first, temperature and solid fraction at each mesh are calculated for the shape shown in Fig. 1(a). Next, according to the for mentioned simplification 1) ~4), the amount of fallen metal was calculated from the total meshes that are less than the critical solid fraction for the fluid flow at a given falling time. Then the amount of the fallen metal are converted to the bottom

thickness, and the calculation was continued for the shape of Fig. 1(b).

Table 2 Thermal properties used for calculation (K; conductivity, ρ; density, C_p; specific heat, L; latent heat, T_i; pouring temperature, T_L; liquidus temperature, T_s; solidus temperature, T_e; eutectic temperature)

	K W / m K	ρ kg / m³	C_p kJ / kg K	L kJ / kg	T_i K	T_L K	T_s, T_e K
SCS 5	21.0	7700	0.71	251	1873	1721	1601
FCD 370	21.0	7100	0.59	239	1573	1493	1403
steel mold	41.9	7100	0.59				
sand mold	1.30	1500	1.13				
coatings	0.8 ~ 3.4 (1.7, 2.1)	1500	1.13				

Table 3 Heat transfer coefficients between various interfaces and calculation treatment of solidification

heat transfer coefficient (W / m² K)				release of latent heat
castings - coatings	coatings - mold	mold - air	castings - air	
perfect contact	perfect contact	8 4	8 4	
				equilibrium solidification

Fig.3 Shapes and macro-structures of centrifugally cast SCS 5.
Falling time: (a) 110 sec, (b) 140 sec and (c) 170 sec

3. EXPERIMENTAL RESULTS

Fig. 3 shows shapes and macrostructures of centrifugally cast SCS-5 with changing of falling time. From these figures sound containers are produced by a vertical type centrifugal casting even without cores to adjust the falling time and at same time to control the falling volume. A condition of a short falling time such as Fig. 3(a) makes a falling metal volume large and a container bottom thick. A shrinkage cavities are also present in the upper part of a central bottom. Fig. 3(c) of a long falling time results in uneven walls and fails to weld between a bottom and a side. Therefore it is very important to select the suitable falling time.

4. SOLIDIFICATION ANALYSIS

Factors to control the solidification are a kind of coatings, their thickness and casting conditions such as, casting temperature, mold temperature, and rotating speed, when the size of the container and the mold is given. Only the coating effect is discussed because it is essentially important to decide the critical solid fraction for fluid flow from the point of view of solidification analysis.

The analytical and experimental (•) results of FCD 370 for the condition that the coating thickness was 3 mm or 10 mm are shown in Figs. 5 and 6.

The analytical results show the relationships between the falling time and the bottom thickness which are directly obtained from the experiments, when the critical solid fraction for the fluid flow (which is shown as FSOL in the figures) was given as 0.2, 0.4, 0.6 and 0.8. For example, the curve, FSOL=0.40 in Fig.5(a), shows the relationship between the falling time and the bottom thickness which was formed by falling elements whose solid fraction was less than 0.4. And this curve shows the shorter falling time gives the larger bottom thickness, and also shows that the falling time should become shorter to get a constant bottom thickness when the critical solid fraction for the fluid flow is small. The effect of a kind of coatings which is expressed as the difference of heat conductivity of coating is shown in Fig. 5(a)~(c). In Fig. 5(c) the heat conductivity is 2.5 W/(mK) which is 50 % larger than that in Fig. 5(a), solidification speed becomes larger and the bottom thickness decreases.

Comparing the result of Fig. 6, in which the coating thickness is 10 mm, with that of Fig. 5, for example, in which the heat conductivity is 1.7W/ (mK), the falling time to obtain rather larger bottom thickness such as about 50 mm is about 80% longer, and one to obtain 20 ~40 mm bottom thickness is also about 60% longer. This result shows that the tolerance control of the falling time to obtain a certain bottom thickness is easier in case of slower solidification with 10 mm coating thickness.

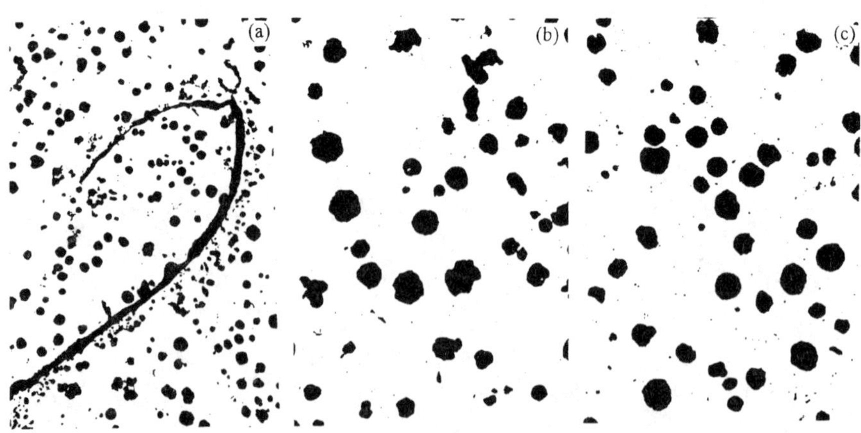

Fig. 4 Microstructures (a) Bad Welded Area (SCS-5), (b) Noduler Graphite (Wall), (c) Noduler Graphite (Bottom)

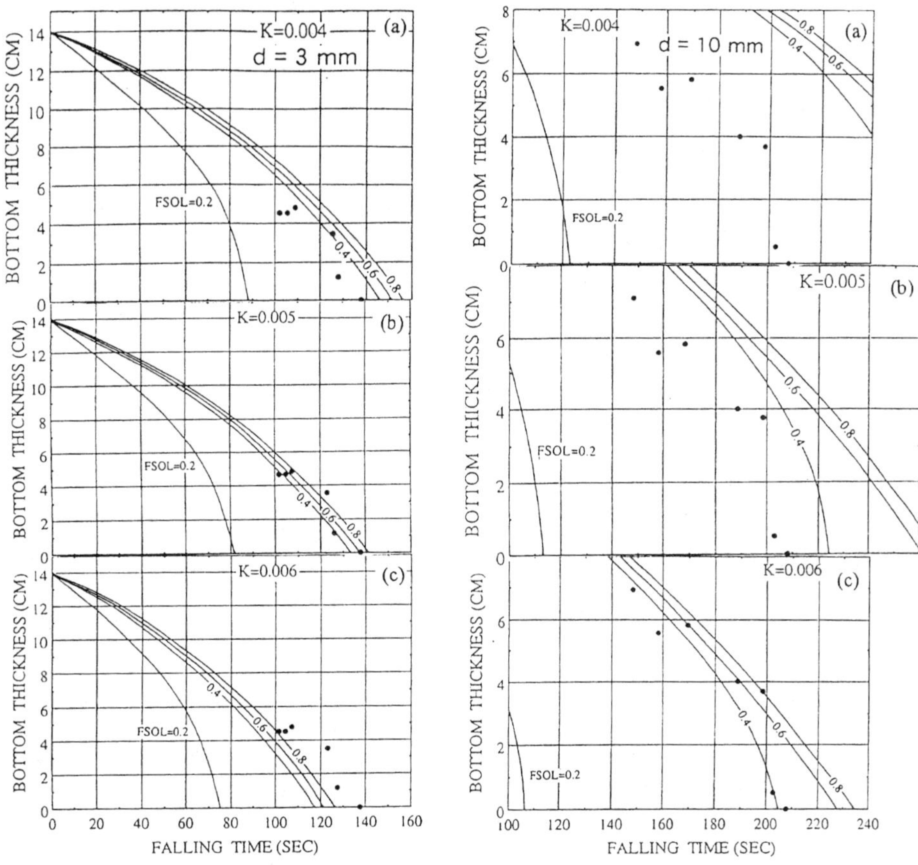

Fig. 5 Relationship between Falling Time and
Bottom Thickness
(FCD370, Coating Thickness: 3mm)
(a) Thermal Conductivity; 1.7W/m•K
(b) Thermal Conductivity; 2.1W/m•K
(c) Thermal Conductivity; 2.5W/m•K

Fig. 6 Relationship between Falling Time and
Bottom Thickness
(FCD370, Coating Thickness: 10mm)
(a) Thermal Conductivity; 1.7W/m•K
(b) Thermal Conductivity; 2.1W/m•K
(c) Thermal Conductivity; 2.5W/m•K

Comparing the analytical results with the experimental ones of 3 mm coating thickness for the used coating material of which heat conductivity is 1.7W/(mK) , the experimental results are quite similar to the simulation results in which the elements whose solid fraction is less than 0.35 , should fall down, that is, the critical solid fraction for the fluid flow is about 0.35. In case of slow solidification like as 10 mm coating thickness, the critical solid fraction for the fluid flow tends to be small as 0.30.

The critical solid fraction for the fluid flow is estimated as 0.30 ~ 0.40 for SCS-5. The thicker coating and the slower solidification brings the smaller the critical solid fraction for the fluid flow in case of SCS-5.

These values of the critical solid fraction for the fluid flow is the same as the value of "q_2 zone" [1, 2] by Takahashi et. al. and "solidification front" [3, 4] in solidification studies of steel ingot and continuous casting. Then falling of metals during reducing the speed in vertical type centrifugal casting can be thought to be controlled by the same mechanism as flow mechanism.

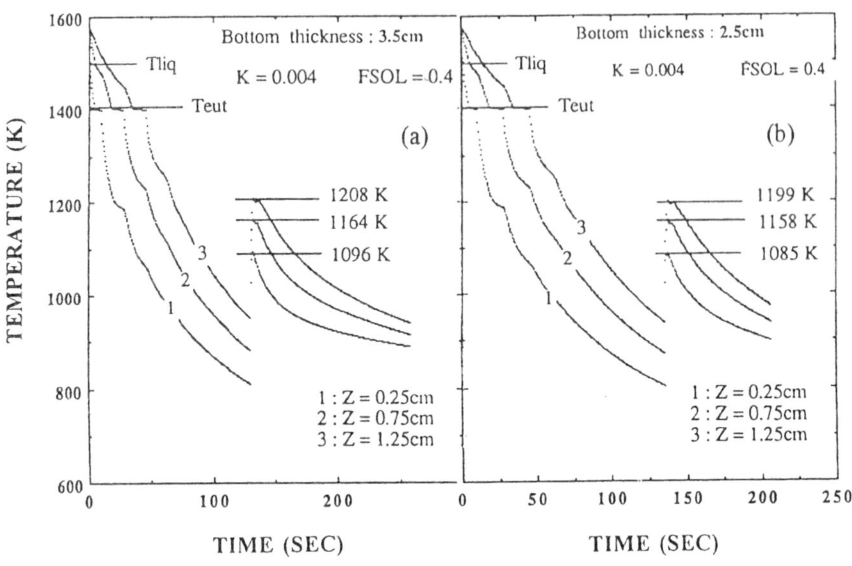

Fig.7 Cooling Curve (FCD370)
(a) Bottom Thickness: 25 mm
(b) Bottom Thickness: 35 mm

The analytical results of cooling curves at the bottom area (the circurated area in Fig. 2) of FCD 370 are shown in Fig. 7. Temperature rise by falling of metals at the bottom of the wall is large, when the bottom thickness is large. These results are approximately the same as the experimental results in Figs. 3 and 4.

5.CONCLUSIONS

The following conclusions are obtained through the experiments and theoretical analysis to manufacture cotainers without "cores" by applying centrifugal casting method.
1) The manufacturing of cylindrical castings with bottoms, by forming pipe shape by centrifugal casting and then by falling the metal less than the critical solid fraction for the fluid flow with reducing the rotation, was performed successfully.
2) The suitably controlled falling time gives sound bottom formation.
3) The coating thickness control brings the suitable falling time control.
4) The falling metal volume is one of the liquid-solid metal less than 0.3 ~0.35 of the critical solid fraction for the fluid flow.

REFERENCES

1. T. Takahashi, I. Hagiwara : Jpn. Inst. Metals. 29, 1152 (1965).
2. T. Takahashi, M. Kudo, S. Nagai : Tetsu to Hagane, 68, 623 (1982).
3. K.Suzuki, T. Miyamoto : Tetsu to Hagane, 63, 45 (1977).
4. S. Ogibayashi, M. Kobayashi, M. Yamada and T. Mukai : ISIJ International 31, 1400 (1991).

Advanced Materials Research Vols. 4-5 (1997) pp. 313-320
© *1997 Scitec Publications, Switzerland*

Production of As-Cast Austenitic-Bainitic Ductile Iron Rolls

Y. Niu[1], S. Long[1] and G. Zheng[2]

[1] Chongqing University, China

[2] Anshan Iron & Steel Co-operative, China

Keywords: As-Cast Austenitic-Bainitic Ductile Iron, Heavy Section Casting, Cast Iron Roll

ABSTRACT

The working condition of a steel mill requires that a cast iron roll should have a hard surface working layer and a core with high strength and toughness. The work presented here aimed to produce an as-cast roll with a combination of the high hardness of chilled cast iron and the high strength and toughness of an austenitic-bainitic ductile iron. A technique has been developed, which is characterised by medium alloying stabilisation of the austenite and a controlled cooling of the roll during casting. Optical microscopy observation and mechanical testing show that a satisfactory microstructure distribution and a significant improvement of mechanical performance of the as-cast austenitic-bainitic roll have been achieved.

INTRODUCTION

Austenitic-bainitic ductile iron has an outstanding tensile strength, wear resistance, contact and bending fatigue strength and hardness, as well as reasonable toughness. This combination of mechanical performance is superior to that of other well-known ductile iron grades. From its origin at the beginning of 1970s[1, 2], it has attracted considerable research interest and has been gradually used to produce high duty castings, such as gear wheel, and crank-shaft, etc., in which the combination of high mechanical properties is required.

The working condition of a steel mill requires that a working roll should have a hard surface working layer with high contact fatigue strength, and a core with high toughness and strength. Despite the usage of the chilled ductile cast iron the failure of the roll in its early service life frequently happens due to wear of the working layer, roll breakage, spalling of the working layer from the core, etc. Analysis of these failure modes shows that improvement of the service life requires to increase the hardness and fatigue strength of the working layer, and the tensile strength and toughness of the whole roll[3, 4]. Obviously, a ductile iron roll with a combination of the high hardness of chilled cast iron and the high strength and toughness of an austenitic-bainitic matrix is desirable, provided that the working temperature of the roll is in the thermally stable temperature of the matrix.

The austenitic-bainitic matrix is

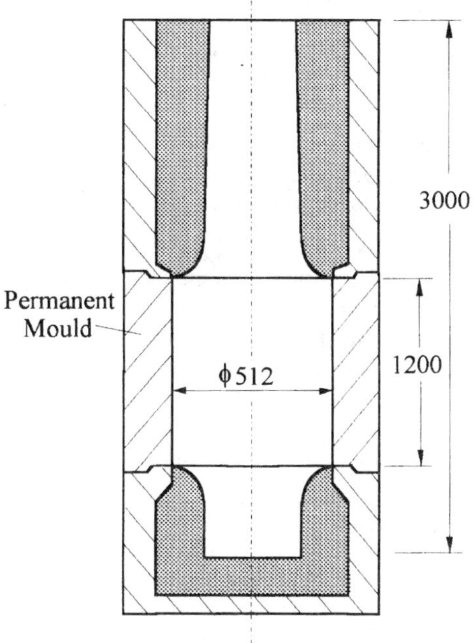

Fig. 1 Schematic of a φ512 roll in the mould during casting Process (unit: mm)

usually obtained by austempering heat treatment of the casting alloyed with austenite stabilising alloying additions to prevent the supercooled austenite from transforming into pearlite. However, austempering heat treatment is not suitable for a ϕ512 roll (512mm in diameter and ~3000mm in length) shown in Fig. 1 because the heat treatment not only induces serious thermal and phase transformation stresses, but also softens the working layer during the austenitisation which also involves a considerable energy consumption. Therefore, it is more attractive to develop a technique that can produce an as-cast austenitic-bainitic ductile iron roll, the mechanical performance of which can meet the requirement of its working conditions.

EXPERIMENTAL RESEARCH

To obtain an as-cast austenitic-bainitic matrix in ductile iron casting two requirements from the phase transformation kinetics must be met:

1. For a hypothetical alloyed casting it must be cooled rapidly enough to avoid the nose of austenitic-pearlitic transformation in the relevant TTT diagram. Alternatively, for a hypothetical cooling rate during the casting process, a certain amount of alloying elements is required to produce sufficient pearlite hardenability.

2. A casting must be held long enough into the austempering temperature range to allow the full development of the desirable austenitic-bainitic phase combination in the matrix.

The presence of austenite stabilising alloying elements, such as, Ni, Mo, Cu, Cr, etc., in the cast iron can increase the pearlite hardenability of the matrix during cooling, but their effectiveness is different[5-8]. Ni is the favoured element for this purpose and, therefore, the effects of the other alloying elements are evaluated as an equivalent Ni addition expressed as Ni_E. The recommended formula for this evaluation in present research is[9]:

$$Ni_E = \%Ni + \%Cu + 2\%Mo + 0.81\%Cr$$

Statistic analysis of the published data about the effect of Ni_E on the pearlite hardenability in Fig. 2 clearly shows, albeit with considerable scatter, the increasing tendency of the pearlite hardenability as the Ni_E increases, but no data about the Ni_E more than 3 are available for ductile iron, especially, for a roll in the as-cast condition[9~12].

The Ni_E required for the formation of the as-cast austenitic-bainitic matrix in a roll has been empirically determined in Qinghuan University, China. The results show that to obtain an as-cast austenitic-bainitic matrix in a casting cooled at a rate of about 0.4~0.6 °C/min, the cooling rate of a ϕ512 Roll at the position ~120mm away from the chilled working surface, the Ni_E should be within 8.7~9.4% (5.0~5.5% Ni, 1.0~1.2% Mo, 1.3~1.5% Cu).

However, the amount of alloying elements required to get an as-cast austenitic-bainitic matrix is considerably high, and this amount

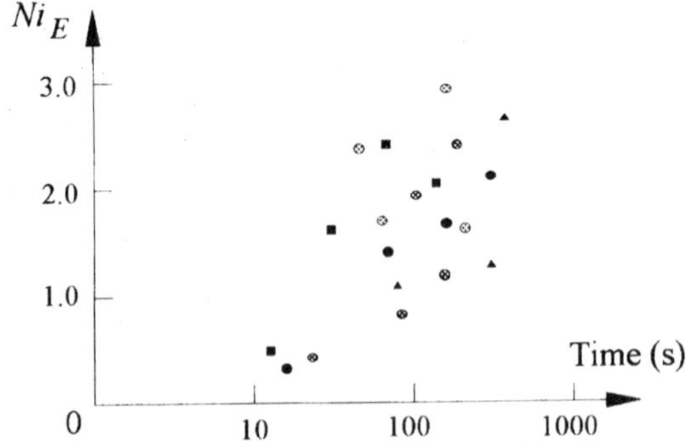

Fig. 2 The relationship between Ni_E of alloying and the pearlite hardenability

of alloying addition will not only cause dramatic cost increase but also ruin the mechanical performance of the roll because the slow solidification of a roll results in very severe macro and micro segregation of the alloying elements.

To produce an as-cast austenitic-bainitic ductile cast iron roll at an affordable cost and with acceptable mechanical properties measures have to be taken to enhance the cooling rate at the temperature range of austenitic-pearlitic transformation and the corresponding Ni_E has to be determined.

1. Relationship between Ni_E and Pearlite Hardenability

The dependence of pearlite formation on cooling rate correlates with the position of the pearlite nose on TTT diagrams. To determine the relationship between pearlite hardenability and Ni_E requires evaluation of the effect of the Ni_E on the incubation time required for austenitic-pearlitic transformation at its nose temperature. It is widely accepted that the nose temperature is almost a constant of about 650 °C despite the presence of austenite stabilising elements. Therefore, the relationship between Ni_E and the pearlite hardenability within a Ni_E range of 3%~9% has been determined as follows.

Six series of ductile iron specimens with different Ni_E have been cast under the same conditions, and all the specimens in any one series have the same chemistry. After austenitisation at 1000 °C for 1 hour the specimens are air cooled to 650 °C and, then, retained at that temperature in a furnace. After different periods of time, the specimens are water quenched one after another and their microstructures are investigated by optical microscopy to determine the incubation time (t) when pearlite first occurs in the matrix, i.e., the pearlite hardenability. The whole heat treatment process is shown in Fig. 3. The chemistry and investigation results of the specimens in every series are given in Table. 1, and the obtained pearlite hardenabilities in each series are plotted in Fig. 4, which shows that the Ni_E--t relationship can be can be statistically expressed as

$$log\ t = 1.516 + 0.33 Ni_E$$

Assuming that the A_{c1} temperature of the alloyed cast iron is 850 °C[10], the obtained Ni_E--t relationship is consistent with the result obtained by Qinghua University under as-casting condition.

Table 1   **Chemical Composite and Pearlite Hardenability (t)**

Experiment No.	%C	%Si	%Mn	%S	%P	%Cu	%Ni	%Mo	%Ni_E	t (min)
AB-1	3.4	2.7	0.57	0.036	0.04	---	1.5	0.75	3.0	5
AB-2	3.4	2.7	0.57	0.036	0.04	1.75	1.0	0.5	3.75	10
AB-3	3.4	2.7	0.57	0.036	0.04	---	3.0	0.75	4.5	15
AB-4	3.4	2.7	0.57	0.036	0.04	1.75	3.0	0.7	6.15	70
AB-5	3.4	2.7	0.57	0.036	0.04	1.75	4.0	0.7	7.15	105
AB-6	3.4	2.7	0.57	0.036	0.04	1.75	5.0	0.7	8.15	260

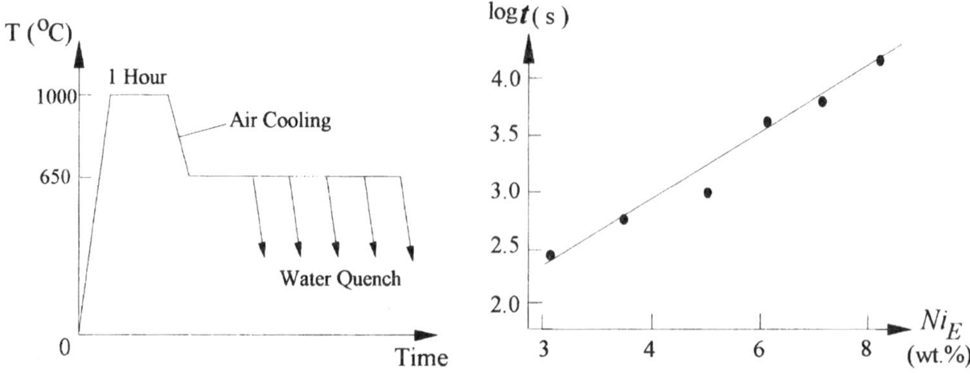

Fig. 3 The austenitisation and austempering
conditions of the specimens

Fig. 4 The relationship between Ni_E and
pearlite hardenability

However, the obtained Ni_E–t relationship indicates that to obtain an as-cast austenitic-bainitic matrix in a roll it is necessary to increase its cooling rate if Ni_E is to be reduced.

2. Cooling Rate Control

To increase the cooling rate of the roll after solidification to reduce Ni_E, various cooling enhancement measures have been taken after the solidification of the roll. The temperature variation of the roll have been measured by three thermocouples pre-placed in the roll 25mm, 125mm and 250mm away from the roll surface, as shown in Fig. 5. As Fig. 5

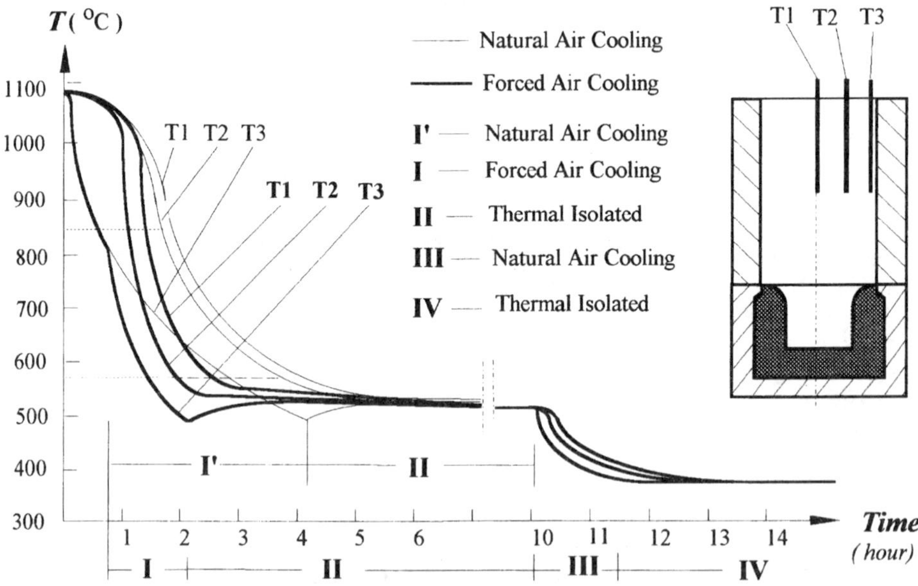

Fig. 5 Schematic of temperature measurement and temperature variations of the
roll under natural air cooling and forced air cooling conditions.

shows, if the permanent mould is taken away as the roll temperature drops down to about 820 ^oC after solidification, the cooling rate of the roll under natural air convection is 2~2.5 ^oC/min at the temperature range between 850~570 ^oC, and if a forced air cooling is added the cooling rate can be further increased to 6~8 ^oC/min. Obviously, this increase of the cooling rate can effectively reduce the requirement on the pearlite hardenability and, thereby, the Ni_E of alloying additions. More severe cooling methods are impractical because they will induce serious thermal stress in the roll although the amount of alloying addition can be further reduced.

As alloying significantly prolongs the incubation time and the duration of the austenitic-bainitic transformation, full development of an as-cast austenitic-bainitic matrix requires the roll to be retained in the austempering temperature range for a significant time. However, the cooling rate of the roll experiencing the enhanced cooling is still high in this temperature range and, therefore, thermal isolation precautions have to be taken to slow it down. Experiments showed that if surrounded by a thermal isolation layer the cooling rate of the roll can be controlled in 1~2 ^oC/h in the temperature range around 350~400 ^oC. That means, there is enough time for the austenitic-bainitic transformation.

Assuming that A_{c1} is 850 ^oC and the nose temperature for austenitic-pearlitic transformation in TTT diagram is 650 ^oC again, the proper Ni_E to obtain adequate pearlite hardenability in a roll cooled at an average cooling rate of 6~8 ^oC/min is 5.0~5.4%Ni_E according to the obtained Ni_E--t relationship. The reults indicate that an as-cast austenitic-bainitic matrix can be obtained with a combination of a controlled cooling process (a forced air cooling followed by holding at 360 ^oC) and a medium alloying of about 5% Ni_E.

3. Empirical Investigation of Microstructure

To ensure a proper Ni_E value for a roll cooled at 6~8 ^oC/min and to investigate the obtained microstructure, another series of specimens with a geometry of ϕ30mmX300mm and three levels of Ni_E have been cast under a cooling process similar to the controlled cooling process of the roll. The results of the optical microscopy observation of the specimen microstructures are given in Table 2.

Despite the appearance of the martensite in the matrix caused by the employed short holding time in austempering temperature range, the results clearly indicate that as the $Ni_E >$ 4.75% the transformation of the austenite into pearlite can be prevented, and as the Ni_E is increased to a value around 5.25% the desirable austenitic-bainitic matrix will be obtained, which is consistent with the prediction of the Ni_E--t relation.

Table 2   **Chemistry of Specimens and Their Matrix Components**

Experiment No.	%C	%Si	%Mn	%Cu	%Ni	%Mo	%Ni_E	Matrix Components
AB-11	3.52	2.4	0.55	1.75	2.0	0.5	4.75	B + M + Small Amount of P
AB-12	3.52	2.4	0.55	1.75	2.5	0.5	5.25	B + M
AB-13	3.52	2.4	0.55	1.75	3.0	0.5	5.75	B + M

N. B. : 1. The content of S and P in the above three specimens are 0.036% and 0.04%, respectively.
     2. Where, B, M and P represent bainite, martensite and pearlite, respectivelv

Hence, all the necessary requirements for the production of an as-cast austenitic-bainitic ductile roll have been empirically determined.

PRODUCTION OF AUSTENITIC-BAINITIC ROLLS

Based on the above empirical investigations, the final process parameters for production of two experimental $\phi512$ rolls have been determined as

1. <u>Chemical composition</u>: C: 3.0~3.2%, Si: 1.7~2.0%, Mn: 0.55~0.65%, P: <0.03%, S: <0.01%, Ni_E: 5.0~5.2%.

2. <u>Parameters of the enhanced cooling after solidification</u>: As the temperature of the roll at 20mm away from its surface reaches 820 °C, the permanent mould is taken away and, then, a forced air cooling at a rate of about 7 °C/min is performed until the average temperature of the roll reduces to ~550 °C. Following a thermal isolation peroid to even the temperature differnce inside the roll, the roll is exposed in a natural air convection environment. As the temperature is further reduced to 350 °C a surrounding thermal isolation layer is immediately put on the roll to keep cool rate of the roll less than 2 °C/hour.

Optical microscopy reveals that a desirable microstructure has been obtained, which consists of a surface layer with alloyed eutectic carbide distributed in an austenitic-bainitic matrix and a core of austenite-bainite decorated with graphite nodules. With the increase in depth from the surface of the rolls, the area fractions of the eutectic carbide and austenite decrease dramatically, concommitant with the increase of the bainite and graphite phases. In the central area of the rolls, a small fraction of pearlite and ferrite can be observed. The microstructures of the rolls at 20mm and 120mm away from roll surface is given in Fig. 6.

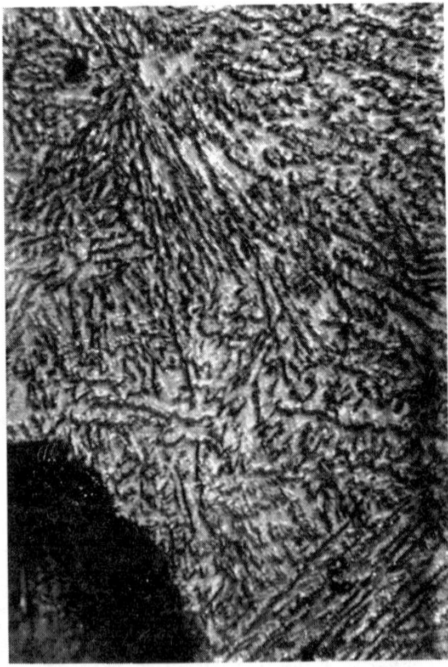

<table>
<tr><td>a. 20mm away from surface</td><td>b. 120mm away from surface</td></tr>
</table>

Fig. 6   Typical microstructures of the rolls. Note the austenitic-bainitic matrix. (X500)

The results of the mechanical testing of the specimens cut beneath the working layer of the rolls are: σ_b=845N/mm^2, δ=2% and H$_s$=59 for 1671 Roll, and σ_b=890N/mm^2, δ=5% and H$_s$=58 for 1571 Roll, showing a significant improvement of the mechanical performance compared with the properties of common ductile iron roll with pearlite matrix: σ_b=500N/mm^2, δ=3~5%, and H$_s$=55.

CONCLUSIONS

The relationship between Ni_E of the alloying addition and pearlite hardenability of the ductile iron has been determined as *log t =1.516 + 0.33Ni$_E$*. The cooling rate of the roll after solidification can be controlled by forced air cooling and thermal isolation to reduce Ni_E and to meet the requirements of the desirable austenitic-bainitic phase transformation.

An as-cast austenitic-bainitic matrix can be obtained in a ϕ512mm ductile iron roll by medium alloying and controlled cooling. The mechanical performance of the produced roll has been significantly improved.

Due to extended service life of the roll and the increase of the surface quality of the milled materials, as well as the recycling of the used roll, there is no cost barrier for the commercial application of the as cast austenitic-bainitic ductile iron roll.

REFERENCE

1. J. Vuorinen, Y. Ingman, M. Johansson, and M. Kurkinen, Pat. US: 3860457 (Kymin Osakeyhtio-kymmene Aktiebolag, Finland), Applied 1973, Issued. 1975.

2. M. Johansson, **AFS Trans.**, Vol. 85, p117~122, (1977).

3. S. Long, Y. Niu, and L. Lee: *'Distribution of Fracture Toughness in a Cast Hot-Milling Roll under Working Condition'*, **Research Report,** November, 1991.

4. S. Long, Y. Niu and L. Lee: *'Temperature and Stress Distribution in a Hot-Milling Roll under Working Condition'*, **Research Report**, November, 1991.

5. S.-C. Lee, and C.-C. Lee, **AFS Trans.**, Vol. 96, p827-838, (1988).

6. S. K. Yu, and C. R. Loper, Jr., **AFS Trans.**, Vol. 94, p557-576, (1986).

7. A. L. Muralidhara, **AFS Trans.**, Vol. 96, p387-394, (1988).

8. T. N. Rouns, K. B. Rundman, and D. M. Moore, **AFS Trans.**, Vol. 92, p815-840, (1984).

9. C. A. Siebert, D. V. Doane, and D. H. Breen: *'The Hardenability of Steels'*, American Society for Metals, 1977.

10. **Atlas of Continuous Cooling Transformation Diagrams for Engineering Steels**, British Steel Corporation, 1978.

11. **Source Book on Heat Treatment**, American Society for Metals, 1977.

12. A. L. Muralidhara, **AFS Trans.**, Vol. 96, p387-394 (1988).

Advanced Materials Research Vols. 4-5 (1997) pp. 321-326
© 1997 Scitec Publications, Switzerland

Microstructure of Semi-Solid Fe-C Alloy Billet for Thixocasting by Powder Compaction

M. Tsujikawa[1], K. Tanaka[2], C. Ushigome[2], S. Nishikawa[2] and M. Kawamoto[1]

[1] Dept. Metallurgy & Materials Science College of Engineering, University of Osaka Prefecture, Gakuen-cho, Sakai-shi 593, Japan

[2] KOGI Corporation, Kanbee-cho, Ohtsu-ku, Himeji-shi 671-11, Japan

Keywords: Semi-Solid, Thixocasting, Eutectic Fe-C Alloy, Powder Compaction

ABSTRACT

The powder compaction route was investigated for making a sound and fine semi-solid slurry and billet for thixocasting. Usually semi-solid slurries with spherical solid particles are made by stirring in the semi-solid temperature range. In such a conventional method, solid particles collide with each other and coalesce. These slurries with coarse microstructure affect mechanical properties of the final product. The powder compaction method without stirring was thought to be a way to get slurries and billets with a fine microstructure. Two powders with different melting point, pure iron and eutectic Fe-C alloy, were mixed and compacted into a cylindrical shape. These samples were heated up to just above the eutectic temperature rapidly. Fine solid particles of prior austenite were observed in the fine ledeburite matrix. This fine microstructure was stable under isothermal holding condition above the eutectic temperature.

INTRODUCTION

From the 1970's, there have been many researches about semi-solid processing[1,2]. For example, the application of rheocasting to magnesium alloys, which are very reactive, was one of the best achievements of these researches[3]. However, although many merits have been presented, the use of this process is still limited.

Two main problems are not resolved yet. One is the coarsening of primary solid during stirring. In the stirring method, alloys are cooled with stirring to the semi-solid temperature range. The size of the primary solid mainly depends on the cooling rate. The cooling rate is usually limited by the stirring to break down the dendrite into spherical solid particles. Also, the primary solid particles grow with the collision of deformed primary solid with each other. The other reason is the difficulty of homogeneously reheating of the billet for thixocasting. The former degrades the quality of the final thixocasted products and the latter prevents the use of the process.

To solve these problems, a way to producing a semi-solid billet was considered. A compact of two different powders with different melting points are rapidly heated up to the temperature range between the two melting points. The powder with the lower melting point liquidates and the powder with the higher melting point remains as solid particles in the liquid matrix. In this method, the probability of collisions is much smaller than that in the stirring method, and the solid crystals are not deformed by stirring. Then the size of solid particles is expected to be smaller than that of particles in the stirring method. If rapid heating is achieved and the difference in melting point of the two powders is larger, the latitude for temperature deviation in the billet to obtain a certain solid fraction may be wider. This is suitable for thixocasting.

In this paper, semi-solid Fe-C alloy slurries were made by the rapid heating of mixed compacts. The compacts consisted of pure iron powder and eutectic Fe-C alloy powder. The microstructures of slurries were investigated.

EXPERIMENTAL PROCEDURE

The chemical composition and estimated melting points of the two powders used in this experiment are shown in **table 1**. Iron powder was made by reduction of magnetite ore particles. This powder was almost impurity free. Therefore, the melting point was estimated as 1807 K. Eutectic Fe-C alloy powder was made by crushing of chilled eutectic cast iron. A small amount of chromium was added to avoid graphitization of carbon in the alloy powder. The melting point of this eutectic cast iron powder was measured as 1421 K. These powders were sieved before the experiments.

Features of the powders are shown in figure 1. Average particle size of pure iron powder was 150 μm, and had irregularity of surface as shown in **figure 1(a)**. Cross section view of particles, **figure 1(c)**, reveals that the interior of this powder was porous and that they consisted of flaky fragments. Eutectic alloy powder was crushed from cast bulk so that they have no pores inside each particle as shown in **figure 1(b)** and **(d)**. Pure iron powder with smaller diameters were sieved and used to get slurry of finer microstructure.

Table 1 Chemical composition and melting point of powders

mass%	C	Si	Mn	P	S	Cr	Melting point
Iron powder	<0.01	-	-	0.004	0.005	-	1807 K
Eutectic Fe-C alloy powder	4.24	0.64	0.21	0.10	0.011	0.46	1421K

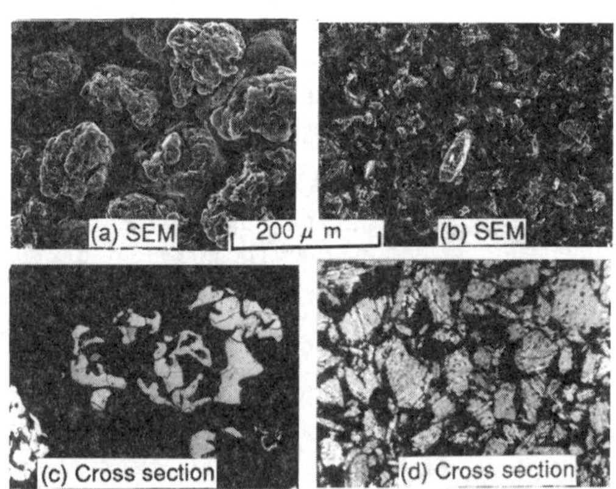

Figure 1 Features of the powders. (a)(c); Iron powder, (b)(d); Eutectic Fe-C alloy powder.

Pure iron powder and eutectic Fe-C alloy powder were mixed in various ratio. Mixed powders were compacted in the air under 20 kN with a die into pellets with 30 mm diameter and 10 mm height. These compacts were heated in an induction furnace up to the temperature just above eutectic point of alloy powder under atmospheric pressure in argon gas. Specimens were isothermally held at the temperature for prescribed period, then they were quenched in water for microstructure observation.

RESULTS AND DISCUSSION

<u>Microstructure of slurries</u>

Table 2 shows the relationship between ratio of mixing and anticipated solid fraction of slurry at eutectic point. Mixed powders were compacted and heated up to 1443 K and isothermally held for 120 s. The samples were water quenched, and the solid fraction was measured by microscopy.

The microstructure of obtained heat 2 slurry is shown in **figure 2**. Spherical particles are in the eutectic matrix. This is the ideal microstructure with fluidity for thixocasting. The shape of the solid phase was as spheroidal as acquired by conventional stirring method. Besides, the size of solid particles was much smaller than that of particles obtained by stirring method. Usually it produces solid particles from 200 μm to around 1000 μm in size[4,5]. The size of solid particles obtained by this powder compaction method is around 40 μm. Furthermore, the value is smaller than 150 μm, that is the average diameter of material pure iron powder. This is because the iron powder were very porous as shown in **figure 1(c)**. Therefore, they shrank in heating up to the solid-liquid range.

Matrix was liquid phase at elevated temperature and now it is fine ledeburite. Spheroidal particles were solid phase floating in former liquid phase. Interiors of solid particles are martensite with retained austenite. Because of the high rate of carbon diffusion, the solid particles were austenite at elevated temperature.

Table 2 Proportion of mixing powders and estimated volume fraction of solid particles at eutectic point

Heat	Iron powder (mass%)	Fe-C alloy powder (mass%)	total carbon content (mass%)	anticipated solid fraction (vol%)
1	40	60	2.55	80
2	30	70	2.97	60
3	20	80	3.39	40

Figure 2   Microstructure of water quenched slurry.
Heat 2: Iron powder; 30mass%,
Alloy powder: 70mass%.

The measured solid fractions are shown in **figure 3** with their microstructure. At the contrast of anticipated solid fraction value, the volume fraction of solid was smallest at heat 2, 30 mass% in pure iron mixture. That is, solid fraction of heat 3 is higher than that of heat 2. **Figure 3 (c)** reveals the microvoids among solid particles. They look like micro shrinkages. This is because, the liquid runs out easily from the slurry of heat 3, which had the smallest calculated solid fraction 40 vol%. Under a certain solid fraction, solid particles are isolated by liquid phase. In heating without crucible as in this case, such a slurry cannot keep its shape. Then, liquid phase can leak out from the slurry.

Microvoids were also found in the matrix of heat 1 specimen. High content of pure iron powder made the volume fraction of liquid too low to complete the consolidation of such a compact under atmospheric pressure. The volume fraction of heat 2 specimen with 30 mass% pure iron was 67%. As confirmed with Al-Cu alloy, slurries with solid fraction of these grade have most intense thixotropy[6,7]. They recovered fluidity with application of shear. From here, experiments carried out with mixture of powder 30 mass% pure iron and 70 mass% eutectic cast iron powder.

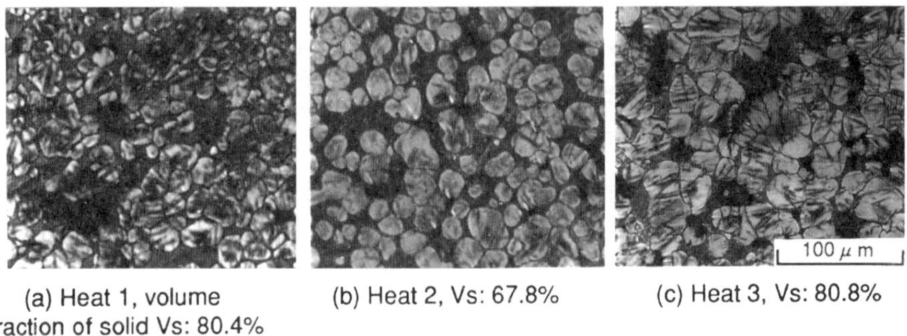

(a) Heat 1, volume
fraction of solid Vs: 80.4%
  (b) Heat 2, Vs: 67.8% (c) Heat 3, Vs: 80.8%

Figure 3 Influence of powder mixture on microstructure of slurry

<u>Stability of microstructure</u>

A change in microstructure during isothermal holding at 1443 K is shown in **figure 4**. Isothermal holding time is the time elapsed at 1443 K without heating time. At the holding time 15s,

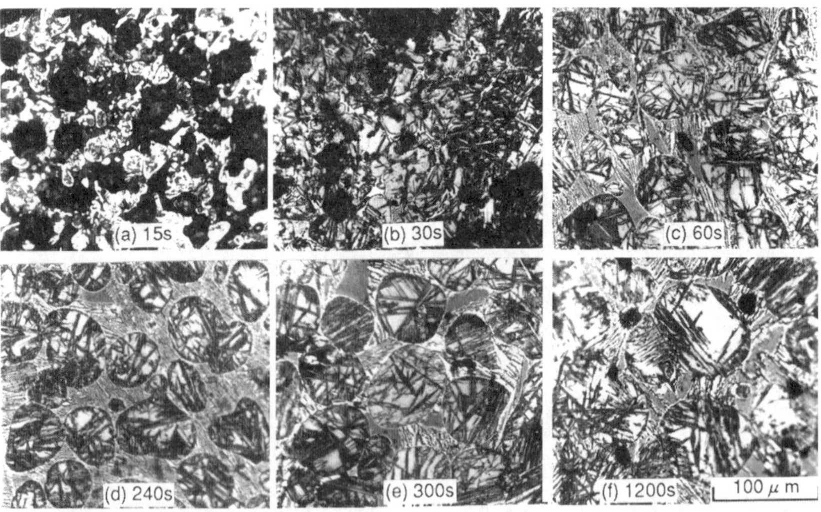

Figure 4 Change in microstructure of compacted powders during isothermal holding at 1443 K

compact was not consolidated. Flaky fragments of pure iron powders are still able to be observed and a lot of large voids exist. After 30s, area of void was reduced, but many small voids exist. The particulated solid did not appear yet, although, acicular microstructure was observed.

By 60s, **figure 4(c)**, voids had disappeared and solid particles clearly appeared. This is the process of consolidation and, at the same time, appearance of solid particles from aggregation of flaky pure iron shown in **figure 1(c)**. At longer isothermal holding time, microstructures of the slurry were basically the same as that of 60s specimen (**figure 4(d)(e)(f)**).

Figure 5 shows relationship between isothermal holding time and diameter of solid particles. The growth of solid particles was observed, but the rate of growth was not high. Within 100s, diameter of solid particles remained around 50μm. Even after 1200s, average diameter did not exceed 80 μm. The stability of fine solid particles in the slurries of powder compaction method at elevated temperature is thought to be enough for the thixocasting process.

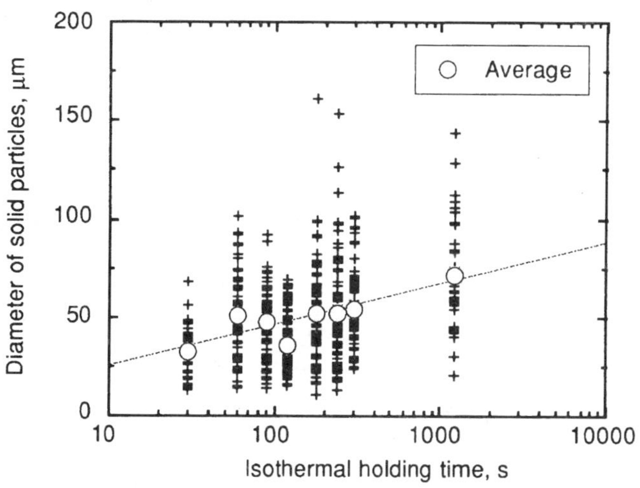

Figure 5 Stability of fine microstructure during isothermal holding

Effect of powder size on microstructure

Another series of experiments carried out to obtain a slurry of finer solid particles. The size of solid particles was thought to be affected mainly by the properties of the pure iron powder, for example, size and extent of hollowness. The powder of low melting point does not affect the particle size of the solid directly because they liquidate at the elevated temperature.

Pure iron powder with average particles size of 50 μm was mixed with eutectic Fe-C alloy powder which was the same powder as in the 150 μm experiments. The mixture of iron powder was 30 mass% and isothermal holding time was 60s.

As presented in **figure 6**, the frequency of smaller particles was high at 50 μm compact although the mode values of these slurries were almost the same. That was, the diameter of solid particles was 30μm for the finer powder and 40 μm for the coarser powder. The resulting difference in solid particle size between the two was much smaller than the difference in diameter between the two materials. The reason why the finer microstructure was not achieved is mainly due to the inner structure of the pure iron powder particles characterized by its processing as shown in **figure 1**. The unit structure that constitutes powder particles did not vary with decreasing diameter. Flaky fragments governed the size of solid particles. It is important to obtain finer microstructure of slurry by powder compaction method that the powder particles have finer interior structure.

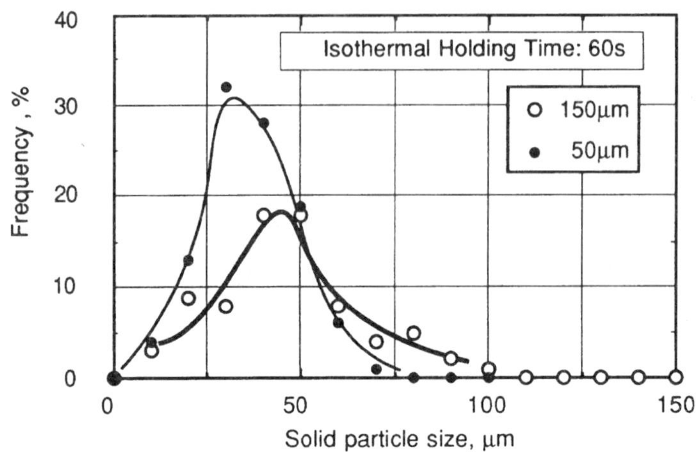

Figure 6 Effect of iron powder size on size distribution of solid particles in slurry.

CONCLUSIONS

By the rapid heating powder compaction method, slurries with particulated solid phase have been obtained for Fe-C system alloys. The microstructures of these slurries were much finer than that of slurries produced by conventional stirring methods. This is because the probability of collision or contact of solid particles is smaller in this method. Furthermore, the solid particles size was three times smaller than that of material pure iron powder particles. The size of solid particles was mainly affected by substructure of reduced iron powder particles.

The fine microstructure remained after long duration of isothermal holding enough to be operated by injection moulding. The solid particles did not grow fast so this specimen has the required nature of billet for thixocasting.

REFERENCES

[1] D. B. Spencer, R. Mehrabian, M. C. Flemings, Met. Trans. **3**, 1925 (1972)

[2] R. Mehrabian, M. C. Flemings, Aluminium **51**, 71 (1988)

[3] P. S. Frederick, N. L. Bradley, S. C. Ericson, Metal Progress 53 (1988)

[4] M. Tsujikawa, M. Matsumoto, M. Kawamoto, Trans. Japan Foundrymen's Soc. **11**, 65 (1992)

[5] J. A. Cheng, D. Apelian, R. D. Doherty, Met. Trans. **17A**, 2049 (1986)

[6] M. Tsujikawa, M. Kawamoto, IMONO **64**, 626 (1992)

[7] M. Tsujikawa, M. Matsumoto, T. Tomonaga, M. Kawamoto, Proc. 57th World Foundry Congress, Osaka 95 (1990)

Advanced Materials Research Vols. 4-5 (1997) pp. 327-334
© *1997 Scitec Publications, Switzerland*

Measurement of Permeability on the Mushy Zone of Lamellar Graphite Cast Iron

S. Hiratsuka and D.M. Stefanescu

Solidification Laboratory, Department of Metallurgical and Materials Eng.,
The University of Alabama, Tuscaloosa, Al 35487, USA

Keywords: Permeability, Lamellar Graphite Cast Iron, Mushy Zone

ABSTRACT

The permeability of the mushy (solid-liquid) region to liquid flow, for gray cast iron, was measured by the seepage of interdendritic liquid into a cylindrical hole formed in a test sample. The hole was produced in the mushy region by means of an alumina cylinder, which was extracted from the sample after various degrees of solidification. The relationship between fraction of solid, permeability and cooling rate was examined. The dendrite morphology in the mushy zone was observed.

For the cast iron studied (3.33 %C, 1.64 %Si, 0.61 %Mn), the critical fraction of solid at which dendrites begin to form networks (the coherency threshold) was found to be 0.24. Both specific permeability, and the effective permeability, which includes density, viscosity and specific permeability, were considered. At the critical fraction of solid, measured values of the effective permeability and the specific permeability were 7.9 $\times 10^{-9}$ m/s , and 8.0 $\times 10^{-13}$ m^2, respectively. On the other hand, the fraction of solid at which the liquid cannot flow through the dendrites was found to be 0.51 when the cooling rate was of 0.067 °C/s, and 0.50 for 0.25 °C/s. At an intermediate value of the fraction of solid of 0.48, the specific permeability was 1.8 $\times 10^{-14}$ m^2 for a cooling rate of 0.25 °C/s.

First, the effective permeability, K , was determined as a function of the fraction of solid, f_S, and of the cooling rate, R. Then, the specific permeability, k, was calculated from the effective permeability as:

$$\log k = -11.4\, R^{0.012} + 9.2\, R^{0.021} \log (1 - f_S).$$

INTRODUCTION

Fluid flow in the liquid metal affects the solidification rate, the solid-liquid interface morphology and casting defects, in particular shrinkage porosity and microsegregation. One of the major factors contributing to macrosegregation and shrinkage in castings is fluid flow through interdendritic channels during solidification. It has been shown that interdendritic fluid flow during solidification of a casting is the main mechanism responsible for occurrence of macrosegregation [1,2] and shrinkage porosity [3]. The flow is caused by solidification contraction, gravity acting on a fluid of variable density, or both. The fluidity in the mushy zone is affected by the dendrite morphology which depends on the cooling rate. Theoretical analysis verifying experimental observations of macrosegregation have shown that the nature and extent of segregation is strongly influenced by the resistance of the solid dendritic network to the interdendritic fluid flow. In order to clarify the flow of interdendritic liquid contributing to the formation of porosity or macrosegregation in an ingot, various types of test method have been used by several investigators [4-7] to measure permeability. The seepage method [7] was adopted in this study because it allows to measure the effective permeability, which is influenced by the viscosity and the density of the liquid. The effective permeability can be measured for different dendrite morphologies occurring at various cooling rates. Then, the specific permeability can be evaluated from the effective permeability obtained by this experimental method.

EXPERIMENTAL PROCEDURE

Experimental samples were prepared by melting gray cast iron in an induction furnace and pouring it into cylindrical sand molds having 40 mm ID and 80 mm height. The chemical composition of the cast iron used was as follows (mass %): 3.33C, 1.64 Si, 0.61 Mn, 0.018 P, 0.029 S. After cooling to room temperature, 10 mm holes were machined in the middle of the cylindrical samples, and they were cut to a height of 55 mm. Then, the samples were remelted in a crucible resistance furnace. The inner diameter of the crucible was 40 mm and the height was 80 mm. Fig. 1 shows a schematic diagram of the experimental equipment. An alumina tube, having a diameter of 10 mm, was inserted into the sample hole. At the bottom of the tube a graphite plug was inserted to prevent molten metal flow in the hole before the alumina tube is pulled out. Melt temperature was controlled by Pt/Pt-Rh thermocouples, which were positioned 5 mm above the bottom of the crucible.

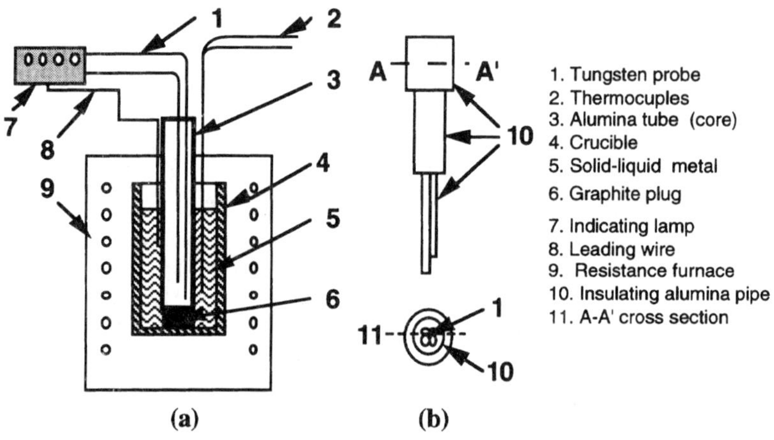

Fig. 1 Experimental apparatus for measuring the velocity of the rising liquid flowing into the mushy zone: (a) furnace assembly; (b) the probes used for velocity measurement.

The molten cast iron was cooled from 1350 °C with cooling rates of 0.067 or 0.25 °C/s to one of six pre selected temperatures, below the liquidus temperature. This resulted in two levels of dendrite arm spacing and six levels of fraction of solid. Then, the solid-liquid metal was held at that temperature for 5 minutes, after which the alumina tube was pulled out to open the hole. This allowed the liquid from the mushy zone to flow into the hole.

The velocity of the rising liquid flowing into the hole was measured using tungsten leading wires which were protected by alumina insulators, and were set at several positions in the hole. Another tungsten wire was inserted in the molten metal. The velocity was calculated from the height reached by the liquid flowing into the hole, and the time elapsed from the removal of the alumina tube. Once the flow of liquid into the hole stopped, the metal was cooled in the furnace.

The dendrite morphology in the mushy region was examined by optical microscopy. Measurements of the secondary dendrite arm spacing were performed.

EXPERIMENTAL RESULTS

The fraction of solid varies with the temperature in the solidification interval. According to the Scheil model [8]:

$$f_S = 1 - \left(\frac{T_M - T}{T_M - T_L} \right)^{1/(k-1)} \tag{1}$$

where f_S is the fraction of solid, T_M is the melting point of pure solvent (1539 °C), T_L is the liquidus temperature (1271 °C), k is the equilibrium partition coefficient (0.467), and T is the temperature.

The relationship between the height from the bottom of the liquid flowing into the hole and the time elapsed after the formation of the hole, at several fraction of solid, is shown in Fig. 2(a) and (b) for cooling rates of 0.067 °C/s and 0.25°C/s, respectively. The rising velocity is high immediately after tube pull-out, after which it decreases. The last experimental point on each line corresponds to the time at which flow stopped because all the available liquid has drained. It can also be seen that the rising velocity decreased with increasing fraction of solid. At a fraction of solid higher than 0.5, the liquid could not flow into the hole. At the higher cooling rate of 0.25 °C/s, at the same fraction of solid, the flowing velocity was smaller and the flow decreased. This is considered to be due to the more complex network formed by the dendritic channels at high solidification rates.

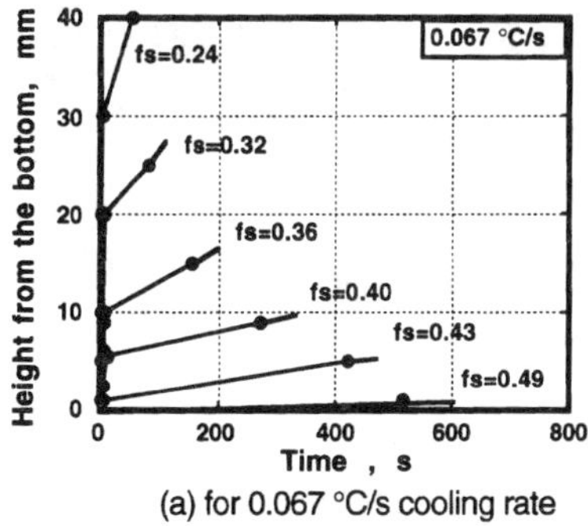

(a) for 0.067 °C/s cooling rate

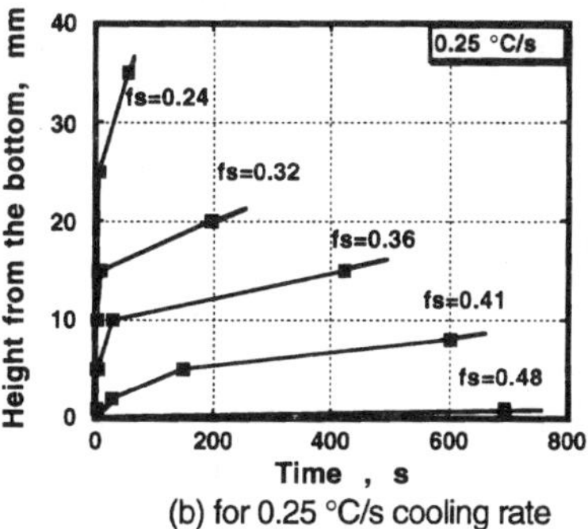

(b) for 0.25 °C/s cooling rate

Fig. 2 Relation between the height from the bottom of the liquid flowing into the hole and the time after the formation of the hole, at different fractions of solid, f_S.

From the results of Fig. 2(a) and (b), an average rising velocity of liquid flowing into the hole, from time zero until flow stopped, was calculated for given fractions of solid. The relationship between the rising velocity and the fraction of solid is shown in Fig. 3. The velocity of the liquid flowing decreased with increased fraction of solid. At fractions of solid above *0.35* velocity decreased suddenly. This tendency is more notable as the cooling rate is higher.

The fraction of solid above which the liquid cannot flow through the interdendritic channels decreased from 0.51 to 0.50 as the cooling rate increased from 0.067 to 0.25 °C/s. On the other hand, at the fraction of solid of 0.24 the depends little on the cooling rate. It was not possible to measure the velocity below the solid fraction of 0.24 because it was too high. This is because the dendrite network in the mushy zone was not yet developed, so that solid and liquid can flow into the hole together.

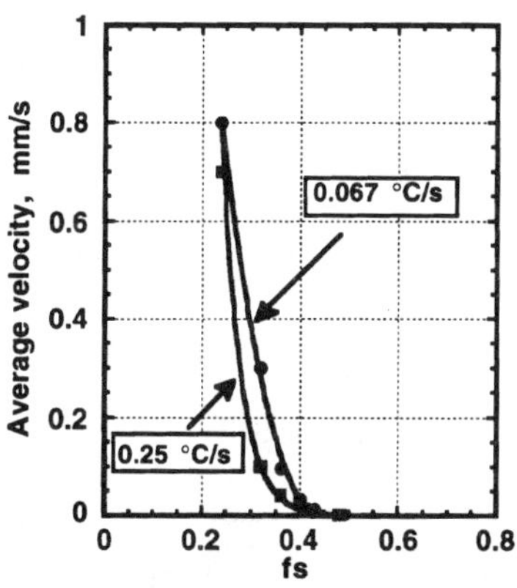

Fig. 3 Correlation between the rising velocity of the liquid flowing into the hole and the fraction of solid, at different cooling rates.

EVALUATION OF THE EFFECTIVE PERMEABILITY

Interdendritic fluid flow in the mushy zone is affected not only by the dendrite morphology and by the fraction of solid, but also by the density and the viscosity of the liquid. A method for the evaluation of the specific and effective permeability based on viscosity and density was considered. Using fluid physics theory [9] and similarity with water flow into a cylindrical hole, the fluidity of the interdendritic liquid was evaluated first. Then, the effective permeability was obtained based on the present experimental data. The final equation is:

$$(t_2 - t_1) = \pi a Q / 8K \tag{2}$$

where:

$$Q = T \ln \frac{\tan \pi (1 + l_2/h)/4}{\tan \pi (1 + l_1/h)/4} + U(\sin \pi l_2/2h - \sin \pi l_1/2h) + W(\sin^3 \pi l_2/2h - \sin^3 \pi l_1/2h)$$

with $T = \left[B(1) - B(3)/3 - B(5)/5 \right] / \left[B(1) \right]^2$

$U = 4 \left[B(3)/3^2 + B(5)/5^2 \right] / \left[B(1) \right]^2$

$W = 16B(5) / 75 \left[B(1) \right]^2$

and $B(n) = \dfrac{-K_1(n\pi d/2h) \bullet I_1(n\pi a/2h)/I_1(n\pi d/2h) + K_1(n\pi a/2h)}{K_1(n\pi d/2h) \bullet I_0(n\pi a/2h)/I_1(n\pi d/2h) + K_0(n\pi a/2h)}$

where I_n is the modified Bessel function of the first kind, and K_n is the modified Bessel function of the second kind (h: height of the hole, d: radius of the ingot, a : radius of the hole, l : height of the liquid in the hole, t: time).

The effective permeability, K, in Eq. (2) includes the specific permeability, k, the viscosity, μ, the density, ρ, and the gravitational acceleration ,g. Note that $K=k\rho g / \mu$..

THE PERMEABILITY OF GRAY CAST IRON

The relationship between the calculated effective permeability and the fraction of solid at cooling rates of 0.067 and 0.25 °C/s is shown in Fig. 4. The effective permeability decreased with increased fraction of solid and with higher cooling rate. The effective permeability at solid fraction of 0.36 and cooling rate of 0.067 °C/s and 0.25 °C/s was 1.4 x10^{-7} m/s and 5.5 x10^{-8} m/s, respectively. The relation between the effective permeability, K, the fraction of solid, f$_S$, and cooling rate R derived from experimental results is:

$$\log K = -10.56\, R^{0.036} - 6.89 R^{0.059} \log f_S \tag{3}$$

In addition, the specific permeability, k, was calculated from the relation between effective permeability, density and viscosity. It was assumed that the viscosity of the liquid at fractions solid between 0.2 and 0.5, was close to that reported for cast iron[10] , *i.e.*, 0.7 to 2.2 poise. The density of cast iron was taken as 7.1 g/cm^3.

Fig. 5 shows the relation between the permeability and the fraction of liquid, 1-f$_S$. The permeability increased with increased fraction of liquid. At same liquid fraction, the specific permeability was smaller for the higher cooling rate. The specific permeability at solid fraction 0.36 and cooling rate 0.067 °C/s was 2.1 x10^{-13} m^2 . The specific permeability, k, was derived from experimental results in a similar way as described for effective permeability to give:

$$\log k = -11.4 R^{0.012} + 9.2 R^{0.021} \log (1- f_S) \tag{4}$$

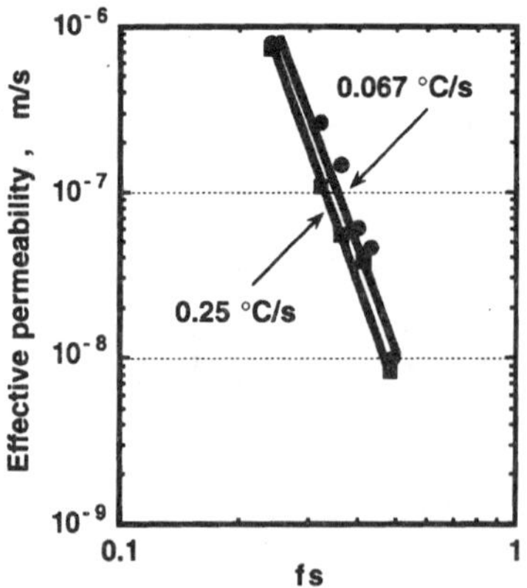

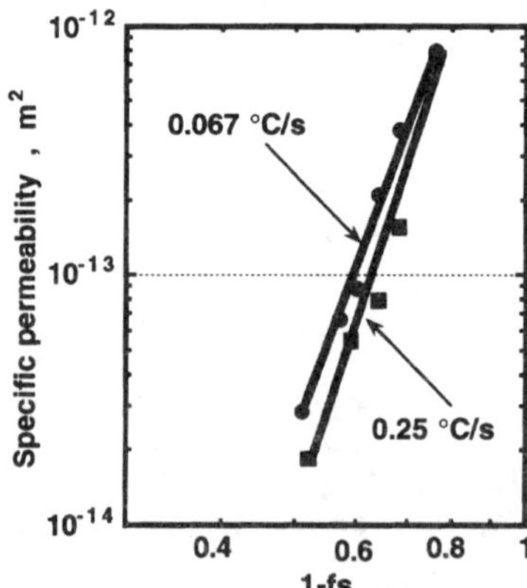

Fig. 4 Relation between effective permeability and the fraction solid at different cooling rates.

Fig. 5 Relation between permeability and the fraction liquid at different cooling rates.

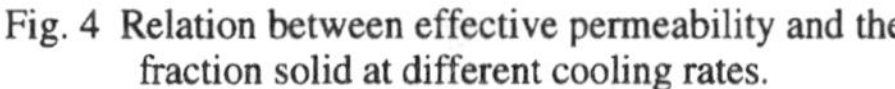

A comparison between the permeability values obtained in this investigation and three literature results from work by Piwonka and Flemings on Al-4.5 %Cu [4], by Apelian *et al.* on Al-4% Si [5], and by Takahashi *et al.* on Fe-0.86%C [7] is given in Fig. 6. The present experimental results on cast iron are very close to those of Takahashi *et al.*. This is not surprising in light of the fact that both the experimental approach and the material tested are similar. However, results on aluminum based alloys were quite different, showing higher permeability than for cast iron. Beyond the difference in materials, it is important to note that in Piwonka and Fleming's experiments, the data were obtained by measuring the pressure exerted on one side of the sample by flowing molten lead or flowing nitrogen gas injected through partially solidified Al-4.5% Cu alloy. A similar approach was used in Apelian's *et al.* experiments, but the fluid injected in the sample was distilled water. The pressure of the fluid injected may break portions of the week dendrite network, thus giving higher values for permeability.

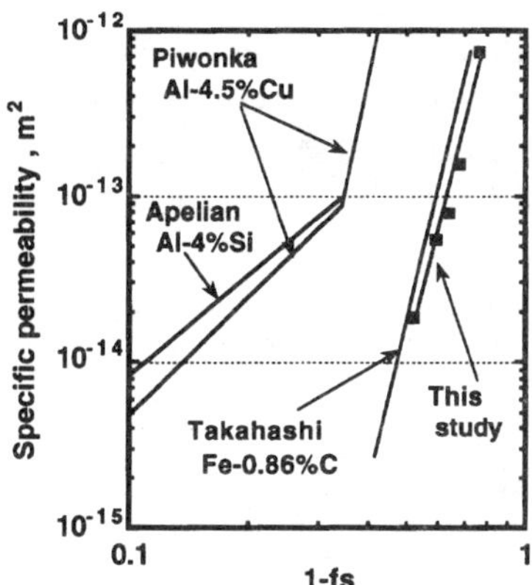

Fig. 6 Comparison between the specific permeability obtained in this study and literature data.

CONCLUSIONS

- The velocity of liquid flowing into the hole decreased with increasing solid fraction. Over a solid fraction of 0.35, the velocity decreases suddenly. This tendency was larger at higher cooling rate.

- The effective permeability, K, was evaluated as a function of the fraction solid, f_S , and of the cooling rate, R, and is given by:

$$\log K = -10.56\ R^{0.036} - 6.83R^{0.059} \log f_S$$

- The specific permeability, k (m^2), also depends on the fraction solid and on the cooling rate, and can be described by:

$$\log k = -11.4R^{0.012} + 9.2R^{0.021}\log (1- f_S)$$

Acknowledgments: The authors wish to thank the Japanese government for supporting Mr. S. Hiratsuka during his work at The University of Alabama. The useful advice of Professor E. Niyama of Tohoku University and Professor H. Horie of Iwate University are greatly appreciated.

References

(1) R, Mehrabian and M. C. Flemings : Met. Trans., **1**(1970) p445
(2) R, Mehrabian, M. Keane and M. C. Flemings : Met. Trans., **1**(1970) p1209
(3) J. Campbell : Cast Met. Research J. : **4**(1969) p1
(4) T. S. Piwonka and M. C. Flemings : Trans. Met. Soc. AIME, **236**(1966) p1157
(5) D. Apelian, M. C. Flemings and R. Mehrabian : Met. Trans., **5**(1974) p2533
(6) N. Streat and F. Weinberg : Met. Trans. B, **7B**(1976) p417
(7) T. Takahashi , M. Kudoh and S. Nagai : Iron and Steel, **68**(1982) p623
(8) E. Scheil : Zeitschrift Metallkde, **34**(1942), p70
(9) D. Kirkham and C. H. M. Van Bavel: Soil Science Soc., **13**(1948) p75
(10) A. Shibuya, K. Arihara and Y. Nakamura : Iron and Steel, **66**(1980) p1550

APPENDIX I

Evaluation of the Effective Permeability

The following derivation of the relative permeability is based on the work by Takahashi *et al.*
[7]. Fig. I.1 shows a schematic illustration for the analysis of the effective permeability for a liquid
flowing into a hole.

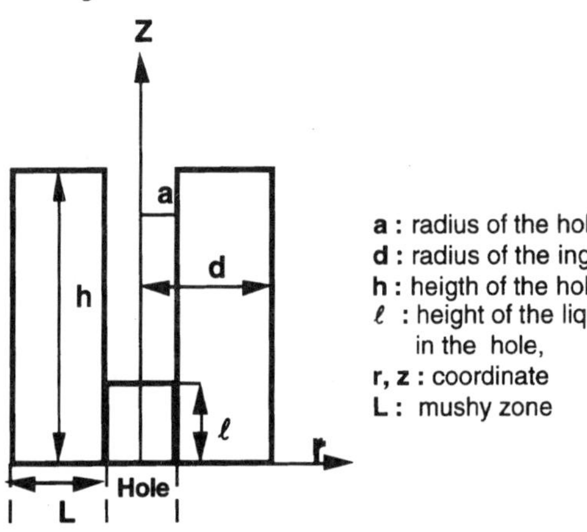

Fig. I.1 Schematic illustration for the analysis of the effective permeability for a liquid flowing into a hole.

The origin of cylindrical coordinates was taken at the bottom center of the hole. If the cylindrical coordinates and Z are considered to extend perpendicularly downward, the continuity equation is:

$$\frac{1}{r}\frac{\partial}{\partial r}\left(r\frac{\partial\varphi}{\partial r}\right)+\frac{\partial^2\varphi}{\partial z^2}=0 \tag{I.1}$$

where φ is head. The angle term was omitted because of symmetry. A velocity potential, ϕ, is given using φ as follows:

$$\phi=\frac{k\rho g}{\mu}\left(\frac{P}{\rho g}+z\right)=\frac{k\rho g}{\mu}\varphi=K\varphi \tag{I.2}$$

where k is the specific permeability, ρ is the density, g is the gravitational acceleration, μ is the viscosity, P is the pressure, K is the effective permeability. Note that $K=k\rho g/\mu$.

The following boundary conditions were used:

(1) z=0 to z=l $\varphi = l$
(2) z=l to z=h $\varphi = z$
(3) z=h $\varphi = h$
(4) r=d $\partial\varphi/\partial r = 0$
(5) z=0 $\partial\varphi/\partial z = 0$

Thus, the solution of equation (I.1) is:

$$\varphi=h-\frac{8h}{\pi^2}\sum\frac{1}{n^2}\cos\frac{n\pi l}{2\pi}R_0\left(\frac{n\pi r}{2h}\right)\cos\frac{n\pi z}{2h} \tag{I.3}$$

where n=1,3,5,..., and R_0 is given by the following the equation:

$$R_0\left(\frac{n\pi r}{2h}\right)=\frac{K_1(n\pi d/2h)\bullet I_0(n\pi r/2h)/I_1(n\pi d/2h)+K_0(n\pi r/2h)}{K_1(n\pi d/2h)\bullet I_0(n\pi a/2h)/I_1(n\pi d/2h)+K_0(n\pi a/2h)} \tag{I.4}$$

where I_n is the modified Bessel function of the first kind, and K_n is the modified Bessel function of the second kind. When r=0, the denominator of equation (I.4) is 1.

The volume of flowing liquid into the hole per unit time, dQ / dt , is given as follows:

$$\frac{dQ}{dt} = 2\pi a K \int_{z=0}^{h} \left.\frac{\partial \varphi}{\partial r}\right|_{r=a} dz \tag{I.5}$$

Therefore the rising velocity of the flowing liquid into the hole is:

$$dl/dt = (dQ/dt)/\pi a^2 = 16hKS/a\pi^2 \tag{I.6}$$

S in equation (I.6) is given as follows:

$$S = cos(\pi l/2h) \bullet B(1) - cos(3\pi l/2h) \bullet B(3)/3^2 + cos(5\pi l/2h) \bullet B(5)/5^2 - \cdots\cdots$$

$$B(n) = \frac{-K_1(n\pi d/2h) \bullet I_1(n\pi a/2h)/I_1(n\pi d/2h) + K_1(n\pi a/2h)}{K_1(n\pi d/2h) \bullet I_0(n\pi a/2h)/I_1(n\pi d/2h) + K_0(n\pi a/2h)}$$

When the liquid moves from l_1 to l_2, time elapses from t_1 to t_2. By integrating equation (I.6), while neglecting the terms over n=7, one obtains:

$$(t_2 - t_1) = \pi a Q/8K \tag{I.7}$$

Where Q in equation (I.7) is given as follows:

$$Q = T \ln \frac{tan\,\pi(1+l_2/h)/4}{tan\,\pi(1+l_1/h)/4} + U(sin\,\pi l_2/2h - sin\,\pi l_1/2h) + W\left(sin^3\,\pi l_2/2h - sin^3\,\pi l_1/2h\right)$$

with　　　$T = [B(1) - B(3)/3 - B(5)/5] \, / \, [B(1)]^2$

$$U = 4[B(3)/3^2 + B(5)/5^2] \, / \, [B(1)]^2$$

and　　　$W = 16B(5) \, / \, 75[B(1)]^2$

The effective permeability, K, in equation (I.7) includes the specific permeability, the viscosity, and the density.

APPENDIX II

Development of an approximate equation of permeability

The relationship between the logarithm of permeability and that of fraction of solid can be approximately expressed by a linear equation, such as:

$$log\,K = G\,log\,f_S + K_o \tag{II.1}$$

This equation was obtained by a least squares approximation method. The gradient, G, of the fitting line and the intercept, K_o, were evaluated for the two cooling rates used in these experiments. Then, it was assumed that the gradient and the intercept depend on cooling rate, through exponential functions:

$$G = C_2 R^{C_1} \tag{II.2}$$

$$K_o = C_4 R^{C_3} \tag{II.3}$$

where R is the cooling rate, and C_1, C_2, C_3 and C_4 are constants. Then combining Eqs. (II.1), (II.2) and (II.3), the constant s are evaluated, and an approximate linear equation of permeability is obtained.

Advanced Materials Research Vols. 4-5 (1997) pp. 335-340
© *1997 Scitec Publications, Switzerland*

Castability Test Applied to Grey Iron

B. Garda and F. Durand

INP de Grenoble et CNRS, EPM /Madylam, ENSHMG BP 95,
F-38402 Saint Martin d'Hères, France

Keywords: Grey Iron, Solidification, Fluid Length

ABSTRACT

A simplified castability test was applied to grey iron, in order to determine the influence of superheat and hydrodynamic conditions on the fluid length and the fluid time. The results are analysed by applying the thermal resistance model in order to take into account the variation in geometry (the tube radius r_0), in casting conditions (initial velocity V_0, casting temperature T_c), and in melt composition.

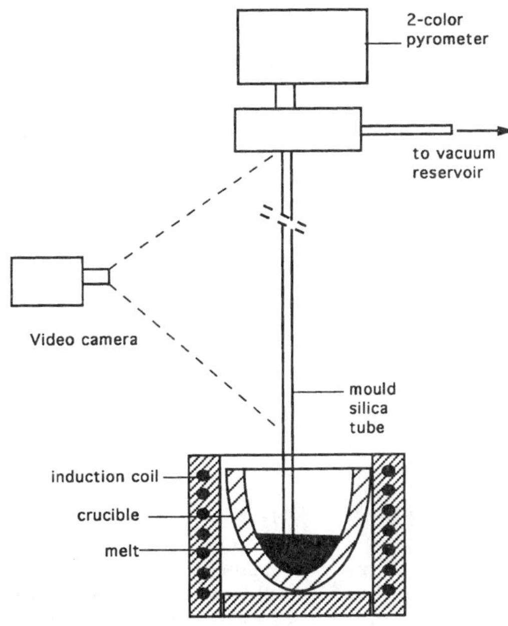

Fig. 1- Casting system

EXPERIMENTAL

The charge (approximately 1.4 kg) was prepared from pure pig iron and pure iron. The sulphur and phosphorus contents were low (less than 0.010 and 0.020 wt%). The charge was melted at 1550°C, then the temperature was lowered to 1350°C and ferrosilicon powder (400 μm size) was added.When the casting temperature T_c was obtained, the mould-tube (fused silica) was dipped into the liquid, which was sucked by connecting the tube to a vacuum reservoir (ΔP = 0.8 to 0.9 bar)(Fig. 1). A video camera recorded the movement of the liquid, giving in particular the initial velocity V_0, and the fluid time t_{fl}. The length covered by the melt before it stops (fluid length Z_{fl},) is a measure of the foundry property called castability

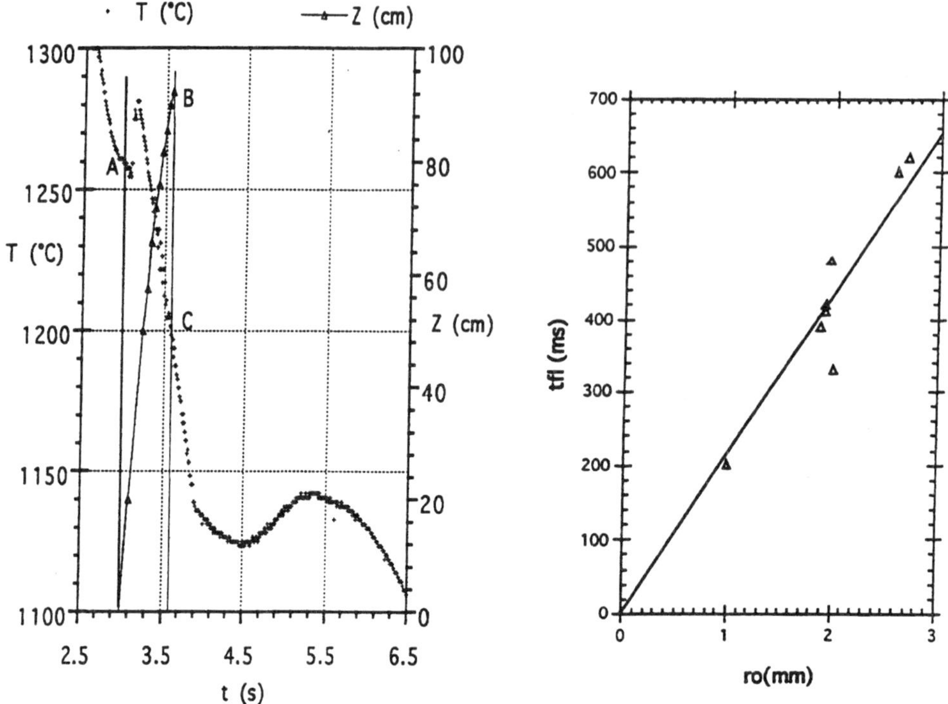

*Fig 2- Position (Z), and temperature (T)
of the liquid front, as functions of time.
Test C21.*

*Fig. 3- Effect of the tube radius (r_o)
on the fluid time.
Parameter S = 0.127±0.002*

(better than fluidity). Moreover, in the test C20, the temperature of the liquid front was measured using a 2-color pyrometer, and recorded during all the solidification process. Alloy composition, casting conditions and results are summarized in table 1.

FLOW AND TEMPERATURE MEASUREMENTS

On fig. 2, two curves are reported (triangles and crosses). The triangles give the position of the liquid front as a function of time. The front velocity is practically constant (V_o) during the initial half part of the curve, then it slows down until the front stops (time t_{fl} = 0.6 s). During this time, the front temperature (crosses) cools down at a constant rate, even after the front is stopped. Then, 0.6 s after the stop, the cooling rate changes, a minimum appears, the curve taking a shape similar to the classical recalescence effect.

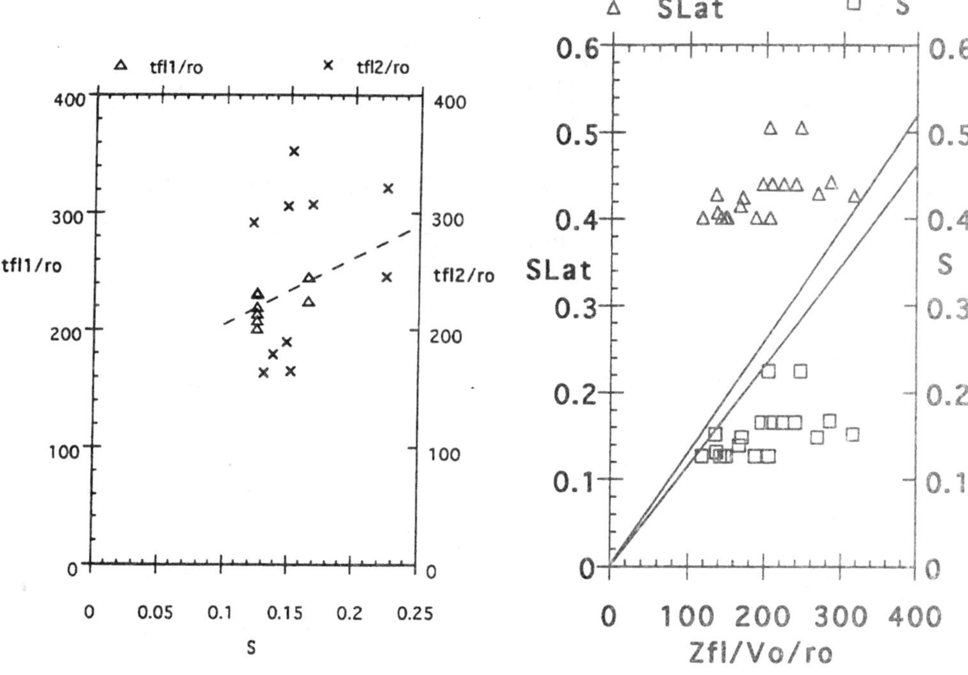

Fig. 4- Effect of the superheat,represented by S on the fluid time.
Units of tfl/ro are s m⁻¹.

Fig 5- Fluid length, effect of superheat. .According to eq.8, each experiment is represented by 2 points on a vertical lever: a square (S), and a triangle (Lat) . Units of Zfl/Vo/ro are s m⁻¹.

VARIATION OF THE FLUID TIME

The fluid time t_{fl} is measured directly from the video records. In order to separate the effect of the different parameters, we analyse the measured value by applying the thermal resistance model, initially proposed by Portevin and Bastien [1]. The fluid time t_{fl} is the sum of 2 heat release times, the first for the superheat, the second one for the latent heat. If the heat extraction is controlled by the metal-mould heat transfer coefficient h, the 2 times can be expressed (with classical notations) :

$$t_{fl} = t_1 + t_2 \tag{1}$$

where

$$t_1 = \frac{r_0 \, \rho C}{2h} S \quad \text{with} \quad S = \ln\left(\frac{T_c - T_a}{T_M - T_a}\right) \tag{2}$$

and

$$t_2 = \frac{r_0 \, \rho C}{2h} Lat \quad \text{with} \quad Lat = \frac{\Delta H_F}{C(T_M - T_a)} \tag{3}$$

For grey iron, 3 equilibria are to be considered: liquid+austenite, liquid+ graphite, and eutectic liquid+austenite+graphite. The corresponding equilibrium temperatures T_{La}, T_{Lg}, T_{Eg} were calculated as functions of the carbon and silicon percentages, by applying the relationships selected by Lacaze [2]. T_M is the temperature at which the solid is formed. The application to grey iron involve some difficulty. By considering fig. 2, T_M is calculated as follows :

$$T_M = \max\,[T_{Lg} - 144°C, T_{La} - 18°C] \tag{4}$$

The effect of the mould radius is represented on fig. 3. All experiments involving the same superheat ($S = 0,127 \pm 0.002$) are placed along a straight line, results which support the above model, and consequently dismisses the concurrent model of heat transfer controlled by the accumulation into the mould (t_1 and t_2 proportional to r_o^2).

The second application was to test the effect of the casting temperature. On Fig. 4, the experimental values of t_{fl}/r_o should be proportional to $S + Lat$. The large scatter of the points does not support this conclusion. Flemings and others [3]. determined h not from this overall linear relationship, but from the local effect of the S variation :

$$\frac{\partial(\theta_{fl}/r_o)}{\partial S} = \frac{\rho C}{2h_1} \tag{5}$$

The dotted "tangent" on fig.4 gives a first estimate of h:

$$h_1 = 5400 \ W \ m^{-2} \ K^{-1} \tag{6}$$

VARIATION OF THE FLUID LENGTH

As revealed by the video records (Fig. 2), the front velocity is not a constant, so 2 flow regimes must be distinguished. First the melt is entirely liquid ($0 < t < t_1$, flow rate $V_o \, \pi \, r_o^2$), then a columnar crust is formed at the tube wall, and restricts the flow cross section. If the mean velocity is assumed constant (V_o), the total fluid length is then [5] :

$$Z_{fl} = V_o \left(t_1 + \frac{1}{2}t_2\right) \tag{7}$$

More generally the fluid length can be written

$$\frac{Z_{fl}}{V_o \, r_o} = \frac{\rho C}{2h} \, [(1-\alpha)\,S + \alpha(S + Lat)] \tag{8}$$

The non-dimensional parameter "α" should range from 0 to 1. If $\alpha = 1$, then all the fluid length results from the time for extracting the latent heat, which is the case for a casting at very low superheat. Physically α depends on the way the solidification stops the liquid flow. It varies with the casting conditions.

According to eq. 8, on Fig.5 each experiment is represented by 2 points, the lower one for S, the upper one for S + Lat noted SLat on the figure, on a vertical "lever". For a given value of h (slope of the straight line),the lever arms give a measure of α and 1-α. Fig. 5 shows that eq. 8 can be applied to all experiments only if the value of the heat transfert coefficient is in the range :

$$h_2 = 3700 \pm 200 \ W \ m^{-2} \ k^{-1} \tag{9}$$

Moreover, Fig. 5 shows that with a smaller superheat, the contribution of the latent heat to the fluid length becomes larger.

The difference $V_o \, t_{fl} - Z_{fl}$ ranges from 2 r_o (experiment C8) to 247 r_o (experiment C21). It is a measure of the more or less long delay of the flow constriction. No relationship was found with S or V_o or chemical composition.

DISCUSSION

Quite evidently the concept of "h" strongly depends on the instantaneous process conditions. "h_1" theoretically represents the heat transfer between the flowing liquid and mould initially cold. "h_2" comes from a heat balance integrated on the complete filling and solidification process. Even if there is no contraction of the solid crust for grey iron, during the solidification step the mould is preheated and the heat flow rate is lowered, so "h_2" is theoretically lower than "h_1".

Besides, the temperature of the melt is not uniform, even during the filling step. Due to the convective contribution to heat transfer the axial gradient is relatively low (Peclet number $Vr_o \rho C/\lambda_{th}$ ranging from 200 to 500). With $h = 3700$ Wm^{-2}K^{-1}, the Biot number hr_o/λ_{th} is about 0.30, indicating that the radial temperature gradient is non negligible. It means that the solid crust can be formed at a distance lower than $V_o t_1$.

CONCLUSION

The simplified castability test can be a source of data for validating numerical models concerning the fluid life and the fluid length in casting.

The "thermal resistance model" represented by eq. 1, 2, 3, even completed by eq. 8 should not be considered as a predictive model, but it can be used as a first approximation to discuss the effects of changes in geometry (r_o), in filling conditions (V_o), in casting temperature (T_c), and in alloy composition (T_M).

Essai	ro (mm)	C (wt%)	SI (wt%)	Tc (°C)	Tla (°C)	Teg (°C)	Tlg (°C)	TM	tfl (ms)	Vo(mm/s)	ZII (mm)
C1	1.95	3.80	3.10	1300	1137	1168	1294	1150			550
C2	1.90			1300	1137			1150			560
C3	1.72			1300	1137			1150			580
C4	1.72			1400	1137			1150			800
C5	1.87			1300	1137			1150			500
C6	1.87			1300	1137			1150			510
C7	1.96			1300	1137			1150			525
C8	1.88			1300	1137			1150	390	1240	480
C9	1.70			1350	1137			1150			400
C10	1.97			1350	1137			1150	480	1875	725
C11	1.96			1300	1137			1150			610
C12	2.00			1300	1137			1150		2500	750
C13	2.00			1300	1137			1150			800
C14	1.92			1300	1137			1150	420	1250	450
C15	1.92			1300	1137			1150	410	1250	450
C16	1.92			1350	1137			1150	430	1250	500
C17	1.92			1350	1137			1150		1870	860
C18	1.92			1350	1137			1150		1750	750
C19	2.70			1300	1137			1150	620	2050	788
C20	1.00			1300	1137			1150	200	1550	230
C21	2.60			1300	1137			1150	600	2187	670
B1	1.96	3.80	2.77	1300	1144	1167	1257	1126	600	1150	605
B2	2.90	3.80	2.77	1400	1144	1167	1257	1126	930	1200	855
B3	2.93	3.80	2.77	1400	1144	1167	1257	1126	720	1650	990
A1	2.95	3.32	1.91	1290	1210	1163	973	1192			920
A2	1.92	3.22	4.03	1300	1173	1173	1173	1155	560		660
A3	1.92	3.81	3.00	1295	1138	1168	1287	1143			990
A4	1.87	3.51	3.05	1330	1166	1168	1175	1148	660	1220	720
A5	1.92	3.51	3.05	1350	1166	1168	1175	1148	590	1460	800
A6	1.95	3.55	3.02	1322	1163	1168	1188	1145	370	1690	560
A7	2.01	3.55	3.02	1300	1163	1168	1188	1145	330	2080	575
A8	1.90	3.57	2.94	1310	1163	1167	1186	1145	340	1830	580
A9	2.00	3.57	2.94	1327	1163	1167	1186	1145	330	1870	510
A10	1.94	3.57	2.94	1310	1163	1167	1186	1145			550

Table 1- Alloy composition, casting conditions and results of the castability tests.

DATA

$$\rho \quad = 7200 \text{ kg m}^{-3};$$

$$\Delta H_F \quad = 256 \ 10^3 \text{ J kg}^{-1}$$

$$C \quad = 840 \text{ J kg}^{-1} \text{ K}^{-1}$$

$$\lambda_{th} \quad = 30 \text{ W m}^{-1} \text{ K}^{-1}$$

REFERENCES

[1] A.Portevin, P.Bastien,"Castability of ternary alloys", J.Inst. Metals 54,45-48 (1934)
[2] J.Lacaze quoted by C.Selig,"Developpement des microstructures et des microségrégations lors de la solidification des fontes...", Thèse de Docteur, I.N.P. de Lorraine (1994)
[3] M.C.Flemings,"Fluidity of metals. Techniques for producing ultrathin section casting",Brit. Foundrymen 312-325 (1964)
[4] B.Garda "Essais de coulabilité en fonderie: aspects thermiques, hydrodynamiques et structure de solidification" Thèse de Docteur, I.N.P. de Grenoble (1993).

AKNOWLEDGEMENTS

This work was a contribution to the european program COST 504 "Advanced casting and solidification technology". It was funded by the French contract MRT 90 A 101, which is gratefully acknowledged.

Advanced Materials Research Vols. 4-5 (1997) pp. 341-346
© *1997 Scitec Publications, Switzerland*

Spheroidal Graphite Cast Iron Hard-Faced by TIG Arc Remelting Process and its Fatigue Properties

T. Hiraoka[1] and Y. Tanaka[2]

[1] Nippon Piston Ring Co., Ltd., Nogi, Tochigi-Pref. 329-01, Japan

[2] Muroran Institute of Technology, Muroran, Hokkaido 050, Japan

Keywords: S.G. Cast Iron, Surface Treatment (Remelting), Fatigue Properties

ABSTRACT

Properties of the chill structure of the spheroidal graphite cast iron processed by the Tungsten Inert Gas-arc as a heat source, and fatigue properties of piston rings remelted and chilled are described in this investigation.

The depth p, the aspect ratio b/p and surface hardness HV of chilled zone are proportional to a parameter $f(J) = (I^4/E^2v)^{1/3}$, *i.e.,* $p = 0.038\ f(J) - 0.3,$ $b/p = -0.043\ f(J) + 7.1,$ $HV = -2.33\ f(J) + 910$
Because these equations depend on chemical composition and graphite shape of the irons, different equation should be used for different iron. The fatigue test results showed that the fatigue limits of TIG arc remelted and chilled s.g.iron piston rings can be improved by ten percent or more compared with ones not treated.

The TIG arc remelting hardening process is expected to be further used for sliding components which are strongly required to be of more resource-saving, of higher performance and of lower cost.

INTRODUCTION

The remelting hardening process [1] , by which cast iron surface is partially remelted and rapidly self-cooled to get the chilled structure, offers prospect in terms of wear and scuffing resistance [2], because of many advantages such as its economical alloying elements, easy operation and local spot hardening [3].

Fig.1 shows the floating seals which are used in construction machines or tractors. These floating-seals were processed by use of TIG (Tungsten Inert Gas) arc to have sealing, wear resistance and corrosion resistance against earth and sand, or muddy water. Chilled zone of cast iron treated by the TIG arc contains a mixed matrix of fine cementite and martensite with complete ledeburite.

It has been acquired in previous paper [4] that the cast irons chilled by TIG arc remelting process had higher wear resistance and scuff-resistance. It was also shown that the corrosion resistance of the iron in NaCl and H_2SO_4+HCl dilute

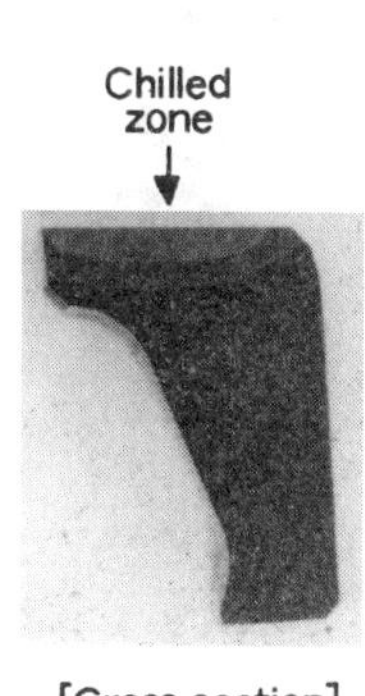

[Cross section]

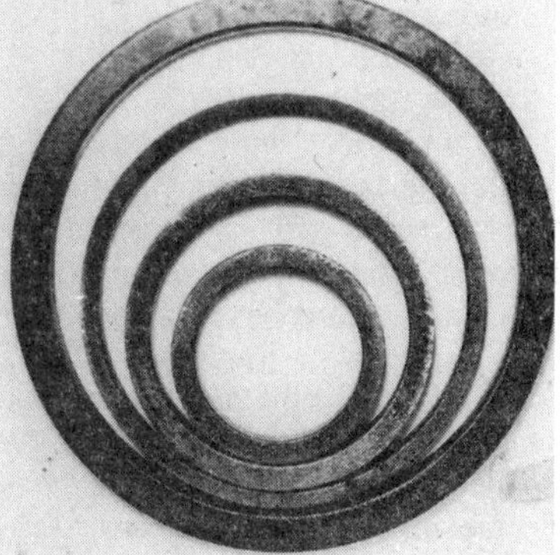

Fig. 1   Appearance of floating seals hardened by TIG arc remelting process

solution was greatly improved compared to the iron which is not surface treated [5]. The TIG arc remelting process easily makes it possible to provide the mechanical and chemical properties required sliding cast iron parts because of fine structure and high-hardness sub-surface layer [2] .

This paper discusses properties of the chill structure of the spheroidal graphite cast iron (s.g.iron) processed by the TIG arc remelting process and fatigue properties of piston ring remelted and chilled.

PROFILE OF CHILLED LAYER AND ITS PROPERTIES

Apparatus for surface remelting

The base equipment of a TIG arc welder is used for surface remelting of s.g.iron. Its remelting conditions are given in Table1. The remelting pass is made by moving the specimen under welding gun by means of a variable carriage in order to attain as constant and reproducible conditions as possible.

Table 1   Remelting conditions

Arc Voltage	V = 24 V
Distance between Electrode–Work	3 mm
Protective Gas	Argon
Gas Flow Rate	8 liter / min
Travel Speed	v = 0.33 – 6.67 mm / s

Profile of chilled layer

Fig.2 shows the cross-section of s.g.irons with ferritic and pearlitic matrix remelted at arc current of I = 200 A. Fig.2 indicates that 1) the cross-section of the chilled zone is semi-oval ; 2) the chill depth and width increase with the decrease in melting rate or the increase in heat input ; 3) the different matrix has no influence on the profile of the chilled zone.

Relationship between parameter $f(J) = (I^4/E^2v)^{1/3}$ which is correlated with melting depth of submerge- arc welding and profile of chilled zone is shown in Fig.3. Where, W is arc voltage, I is arc current and v is travel speed. The chill depth p and the aspect ratio of the cross-section of the chilled zone b/p are proportional to parameter $f(J)$, respectively :

$$p = 0.038 f(J) - 0.3 \quad \cdots \quad (1)$$
$$b/p = - 0.043 f(J) + 7.1 \quad \cdots \quad (2)$$

Fig.3 shows also the same relation for the flake graphite cast (FC) iron. The s.g. iron melts extremely less in the perpendicular direction, when compared with the flake graphite cast iron. This is due to difference in graphite type and in chemical composition. The slower melting of the graphite nodule into the matrix takes more time to reach the eutectic composition and the more silicon content in the s.g.iron raises the eutectic temperature. Therefore, it is considerable that the melt volume, *i.e.*, the chilled zone depth reduces for s.g.iron. The remelting condition of cast irons having different chemical composition

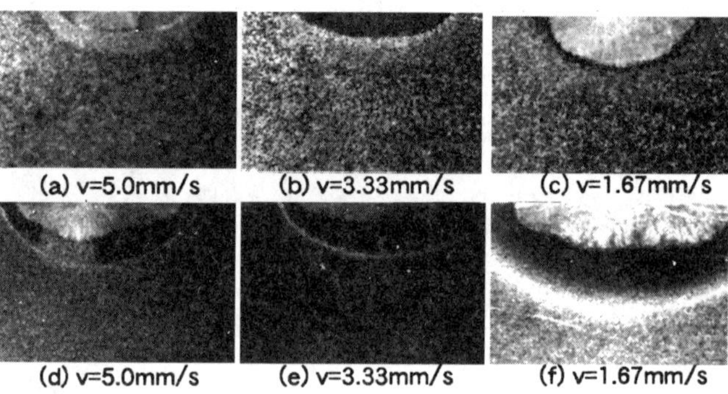

Fig.2 Transverse sections of remelted zone for various irons [(a),(b) and (c):ferritic; (d),(e) and (f):pearlitic]

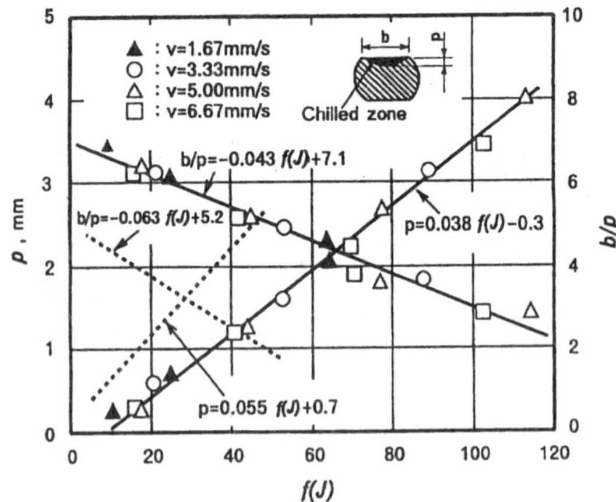

Fig.3 Relationship between depth and aspect ratio of chilled zone and f(J) for ferritic s.g.iron

[Full line: s.g. iron, dotted line: FC iron]

and graphite type should be determined individually for getting the proper profile of chilled zone.

Hardness of chilled zone

Fig.4 shows hardness distribution at the center (A - A' in the figure) of the cross- section of ferritic and pearlitic irons remelted at the current of I = 200 A. The irons of both type have almost constant high hardness levels in the range from the surface to a certain depth. Then, the hardness of the former rapidly decreases to that of the base iron, while the latter shows another step at Vickers Hardness Number (HV) of 750 to 700 and then decreases to the hardness of the base iron.

Fig.5 shows photo-micrographs taken at points (a), (b), (c) and (d) as shown in Fig.4. The chilled zone, (Fig.5 (a) and (c)), is a uniform chill structure and the heat affected zone, (Fig.5 (b) and (d)), just under the remelted zone is a martensitic structure. The sub-surface part with constant high hardness and the 2nd step part with relatively high hardness, in particular, in pearlitic iron, correspond to the chilled zone and the martensite structure zone, respectively. On the other hand, in the ferritic s.g.iron, part near the heat affected zone has an mixed structure of martensite and ferrite because the diffusion of carbon into matrix, which is requisite to the martensite transformation, is not sufficient.

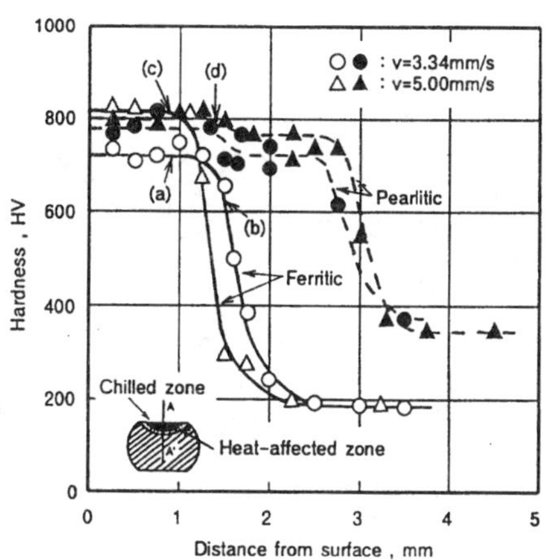

Fig.4 Hardness distribution of remelted s.g. iron

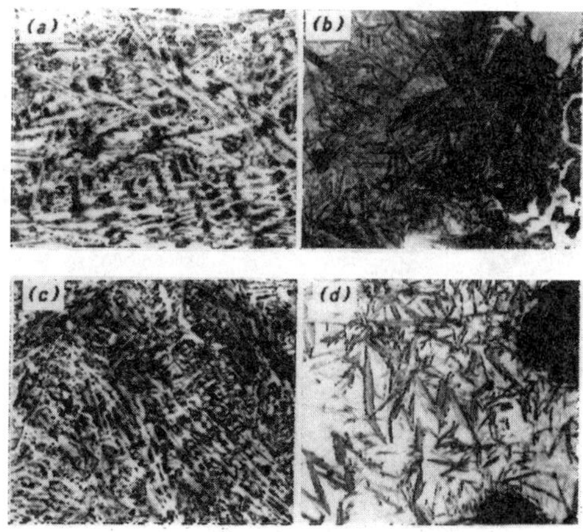

Fig.5 Microstructure of chilled zone and heat affected zone for s.g. iron
[(a), (b), (c) and (d) correspond to positions (a), (b), (c) and (d) in Fig.4 .   I = 200 A, v = 3.33 mm/s]

As seen in Fig.4, the chilled zone has almost constant hardness at a certain depth. Supposing that this constant hardness is the surface hardness, the curves in Fig.4 can be arranged by the parameter $f(J)$ and turned to those in Fig.6. The surface hardness HV is not influenced by the matrix structure and deceases linearly with increase in $f(J)$. It can be indicated by the following equation :

$$HV = -2.33\, f(J) + 910 \quad\text{-------- (3)}$$

As $f(J)$ increases, the ledeburitic eutectic structure is turned coarse and the primary austenite is transformed from the martensite to pearlite. The less the $f(J)$ value is, the more ledeburite is fined, and primary austenite is turned to martensite and the hardness rises, because the small $f(J)$ value equals the low heat input at the same melting rate, and thus lessens the melt volume and raises the cooling rate.

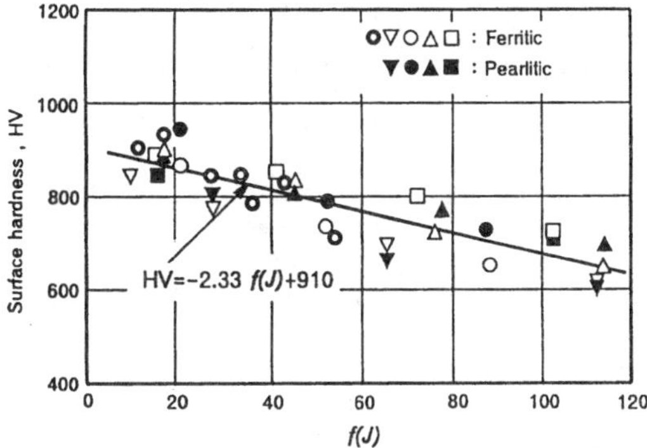

Fig.6 Relationship between surface hardness and f(J)

Cracking at chilled zone

The cracking is observed in the chilled zone and the heat affected zone in a certain melting condition as shown in Fig.7. It is considered that such cracks result from the shrinkage stress caused by solidification of the chilled zone, because cracks in the chilled zone appeared at the center part of the melted zone with their few number and had large width.(Fig.7 (a)) Further, fine cracks were observed in the chilled zone (Fig.7 (b)) near the boundary between the remelted zone and heat-affected zone and in the heat-affected zone (Fig.7 (c)). Such cracking may be caused by the local stress accompanying with the martensitic transformation in particular in the heat-affected zone and the difference between the thermal shrinkage in the chilled zone and that in the heat affected zone. In ferritic s.g.irons, the martensite formed was a minimum and so no cracking is observed in the heat-affected zone in any remelting conditions. On the other hand, in pearlitic iron, the cracking is formed in most of specimens independent of the remelting condition.

Fig.8 is the result of arranging the conditions for the crack formation in the chilled zone by the heat input H = 60 IE/v and the parameter $f(J)$. Fig.8 ndicates that the less heat input per unit time and the larger $f(J)$ lead to the more cracking.

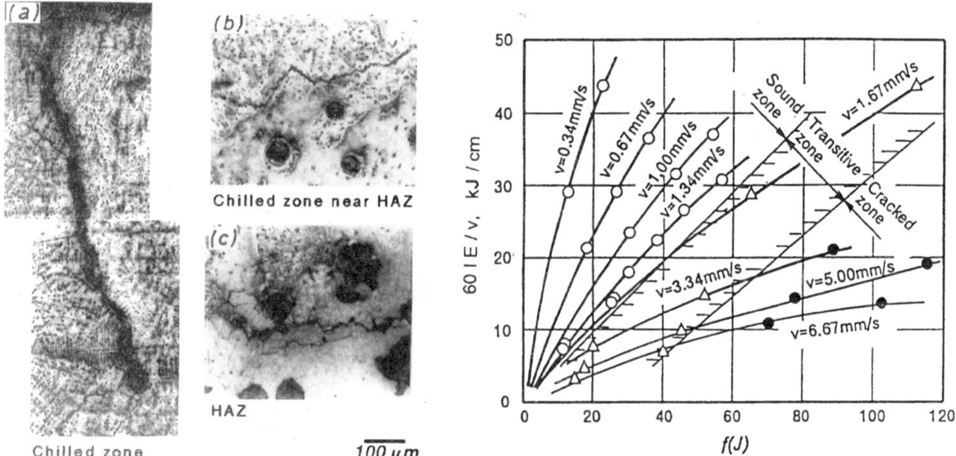

Fig.7 Cracking in chilled zone and HAZ of pearlitic s.g. iron [I = 200 A, v =1.67 mm/s]

Fig.8 Effect of remelting condition on soundness of remelted and chilled zone of pearlitic s.g. iron

To prevent the cracking in the chilled zone, $f(J)$ should be reduced by reducing the arc current, and the melting rate should be lowered to increase the heat input.

FATIGUE PROPERTIES OF PISTON RING REMELTED

Fatigue test of actual piston ring

Schematic representation of a fatigue tester for actual piston rings used is shown in Fig.9. The load is applied on one end of a piston ring which is remelted on its peripheral surface, with the other end fixed. The stress amplitude S is adjusted by determining the difference between the free gap X and the closed gap L and by cutting the piston ring end to change the free gap. The adjusting is according to the following equation :

$$S = 4TE'(X-L)/3(D-T)2 \qquad \text{------- (4)}$$

where E ' is the elastic modulus, and E ' = 176.5 GPa for untreated specimen is used.

Fatigue limit of piston ring remelted

S-N curve for remelted piston ring is shown in Fig.10, comparing with untreated s.g.iron and hard chromium plated s.g. iron piston rings. It is clear that the piston ring remelted has the highest fatigue strength. Fig.11 shows a comparison of fatigue limits which determined from the S-N curves. The fatigue limit of piston ring remelted and chilled can be improved by 10% or more as compared with ones not treated. The fatigue limit decreases when hard-chromium plated, and the thicker the chromium plating is, the lower the

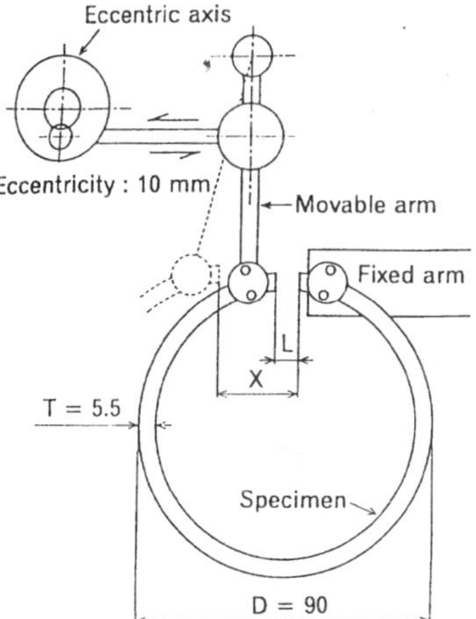

Fig.9   Schematic of fatigue tester for
piston ring

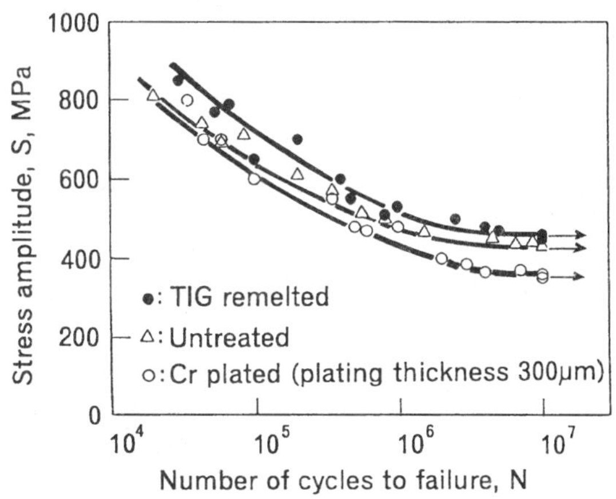

Fig.10   S – N curves of various

[Specimen : Cast iron of 3.63%C, 2.81%Si, 0.77%Mn,
0.42%Cu and Bal.Fe]

fatigue limit becomes. The fatigue cracking on all piston ring started on the peripheral surface of chilled zone 180° apart from the gap. This starting point of crack is a position where the maximum tensile stress is applied.

It is well known that the fatigue strength of s.g.irons is higher as their graphite nodule size is smaller and the matrix hardness is harder. The fatigue strength of chilled iron, therefore, is more advantageous than that of non-treated cast irons.

The influence of graphite nodules, non-metallic inclusions or micro-casting defects which may exist in cast iron, on the fatigue strength can be considered as follows :

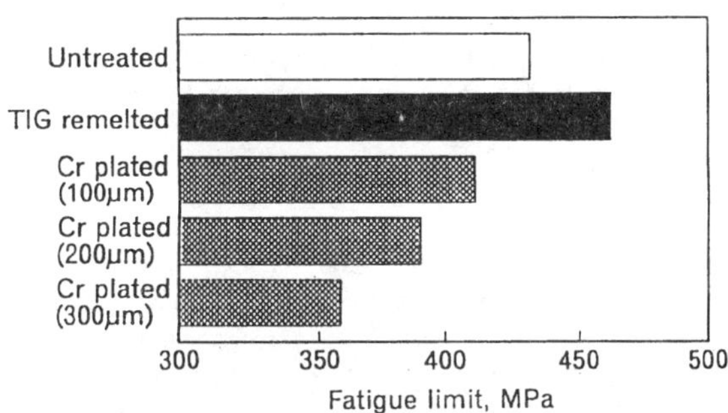

Fig.11   Comparison of fatigue limit of remelted cast
iron and hard chromium plated cast iron

Supposing that graphite nodules or surface defects are round with radius of ρ, the stress gradient χ is given as the following equation :

$$\chi = 2.3 / \rho = 4.7 / d \quad \text{------ (5)}$$

Defining that diameters of defects are d_1 and d_2 ($d_1 < d_2$), the following equation is expressed :

$$\chi_1 (= 4.7 / d_1) > \chi_2 (= 4.7 / d_2) \quad \text{----- (6)}$$

The equation (6) indicates that if large defects exist, the stress gradient around the defect edges is small, while the stress-concentration in both is the same. Therefore, the specimens that have larger defects may be easily cracked.

The strength of a material with cracks is depended on stress intensity factor K_I :

$$K_I = \sigma \sqrt{a\pi} \cdot F \quad \text{------- (7)}$$

where F is a parameter depending on shape and dimension of specimen and crack. Once initiation of cracks occurs, effective notch effect increases with inceasing the hole size, the stress intensity factors increase and the crack propagation resistance decreases more than that occurring at micro-defects, because the larger hole has a cause of the large "effective crack length" a of hole diameter plus crack length and thus the stress intensity factor grows. As a result, the fatigue limit decreases [6] .

The hardness of the chilled zones is practically as high as HV 800-900. This causes increase of the notch sensitivity of micro-defects, decrease of the fatigue limit, the advantages from the graphite-free structure, which extremely improve the fatigue limit by the chilling, may have offset, and thus the fatigue limit improves only a little.

Fatigue limit of hard chromium plated piston ring

The test result was such that hard chromium plated piston rings had a lower fatigue limit. The observation of surfaces and cross-sections of specimens before the fatigue testing showed that specimens had granular holes and micro-cracks on the chromium plating surface and its inside. When peripheral surface are subjected to the repeated tensile stress during the fatigue testing, these micro-cracks may grow during the fatigue testing and spread easily to base metal surfaces. This means that the cracks lengthen in proportion to the thickness of hard chromium plating. Fig.12 is the case of the hard chromium plating thickness of 300 μm.

The stress concentration factor K_t at crack end is given by the following equation :

$$K_t = 1 + 2 (c/\rho_c)^{1/2} ------ (5)$$

where 2 c is length of crack and ρ_c is radius of curvature of crack end. The equation (5) denotes that cracks grow with the hard chromium thickness, and the fatigue limit reduces with the increase the thickness.

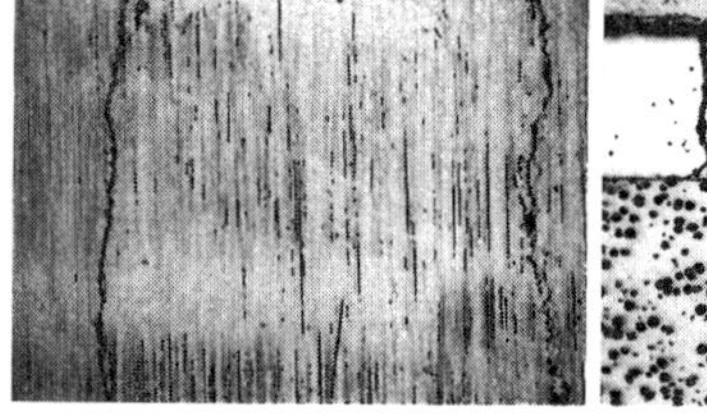
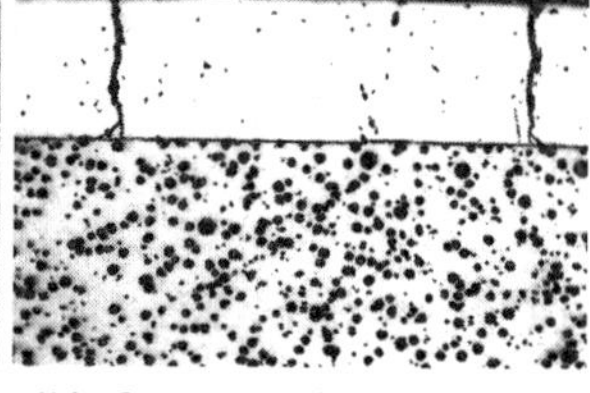

(a) Surface   (b) Cross section

$\overline{300\,\mu m}$

Fig.12   Crack growing in Cr–plated layer during fatigue test

CONCLUSION

This paper describes the consideration of properties of hardened sub-surface layers of TIG remelted spheroidal graphite cast irons and the comparative consideration of results of fatigue tests of actual piston ring remelted. The conclusion obtained as follows :

1) The depth p, aspect ratio b/p of the TIG arc remelted zone and surface hardness HV are correlated fairly closely with parameter $f(J)$; i.e.,

$$p = 0.038\,f(J)\, - 0.3 , \quad b/p = -\,0.043\,f(J)\, + 7.1 , \quad HV = -\,2.33\,f(J)\, + 910$$

2) The fatigue limit of remelted cast iron piston rings is higher than that of non- treated ones by 10 % or more.

3) The fatigue limit of hard chromium plated iron piton rings is lower than that of non- treated ones. The crack extends in the thicker chromium plating, in which the stress concentration factor increases and thus the fatigue limit extremely decreases.

REFERENCE

[1]   F.Emde : Elektrowärme International, **37** (1979), B3, B140

[2]   T.Hiraoka, K.Nakamura and Y.Tanaka : IMONO (Journal of Japan Foundrymen's Society), **66** (1994), 8, 573

[3]   Y.Tanaka and T.Hiraoka : IMONO , **62** (1990), 6, 412

[4]   Y.Tanaka, M.Kon and T.Hiraoka : IMONO, **60** (1988), 5, 332

[5]   Y.Murakami, H.Kohno and T.Endo : Journal of Japan Materials Society, **29** (1980), 988

[6]   M.Sofue, S.Okada and T.Sasaki : AFS Transaction, **78** (1978), 173

Advanced Materials Research Vols. 4-5 (1997) pp. 347-352
© *1997 Scitec Publications, Switzerland*

Rim Zone Metallurgy:
A Coating System to Improve the Microstructure in the Surface-Affected Zone of Permanent Mold Cast Spheroidal Ductile Iron

V. Helling and P.R. Sahm

Foundry Institute, Aachen University of Technology,
Intzestr. 5, D-52072 Aachen, Germany

Keywords: Rim Zone Metallurgy, Coating System, Spheroidal Ductile Iron in Permanent Mold, Gradient Microstructure

Abstract

This paper introduces a rim zone metallurgy to improve the microstructure of spheroidal ductile iron casts into permanent mold is presented. Decisive process parameters are identified, and a coating system based on an insulating and a metallurgical coating is developed. The necessary binder disintegration is worked out so that inoculants as a part of the metallurgical coating are released into the melt and the nucleation rate in the surface affected zone of cone specimens is promoted. The influence of the metallurgical coating on the solidification shows evidence of nucleation efficiency. Additionally, a method to examine the microstructure quantitatively is introduced. Finally, typical results of specimen solidification are presented and the metallurgical potential of producing, for example, gradient microstructures are being shown.

Introduction

Detrimental effects on the environment play an important role today in the development of production methods, whereby in Germany special emphasis is being placed upon the avoidance of waste materials. For this reason permanent mold casting will become increasingly important in the foundry industry because the waste product 'used sand' can be eliminated.

In addition to environmental advantages permanent mold cast spheroidal ductile iron has technical benefits, Table 1. Nevertheless, it represents an innovative challenge for the foundry engineer, as high cooling rates typically lead to chill tendencies. Thus, the costs and the competition within the European market require a ferritic/perlitic as cast microstructure without any annealing process, whereas advanced process know-how is necessary. Figure 1 shows the relevant parameters which can influence the solidification of castings. In addition to the base metallurgy, the heat transfer coefficient fixed by the coating between casting and permanent mold, as well as the present state of the nucleation of the melt, is especially important.

Methods

Due to the decisive position of the heat transfer coefficient and the state of nucleation in the melt, a coating system was developed with which both parameters could be optimized, figure 2. It consists of an insulating and a metallurgical coating, which are sprayed successively on the surface of the permanent mold in a layer technique, figure 3. The base material and the binder, as parts of the coating system, make it possible to decrease the heat transfer coefficient, figure 2, left. The binder of this coating is thermally stable, and high durability is guaranteed.

The metallurgical coating influences the solidification tendency of the melt in the permanent mold. In addition, inoculants are sprayed together with a binder on the insulating coating of the permanent mold just before pouring, figure 2, right. During the contact with the melt, the thermally unstable binder of the metallurgical coating dissolves and the inoculants are released into the melt. Rim zone metallurgy thus helps to improve the nucleation conditions in thin-walled castings, because the diffusion-paths of carbon are decreased and the chill tendency is reduced significantly.

Results

Figure 4 shows the influence on the solidification of a specimen with 12 mm diameter with and without a coating system. With a metallurgical coating the specimen solidifies without any chill, and a ferritic microstructure is noticeable. Without it, a significant chill tendency can be observed. Thus, the comparison shows evidence of nucleation efficiency.

The decisive parameter to optimize the effectiveness of the metallurgical coating is an appropriate binder. Due to the extremely high solidification rate of the casting inside of the permanent mold, the binder must rapidly release the inoculants upon contact with the melt. Therefore a fast binder disintegration under thermal stress is necessary.

Figure 5 compares different binders to specimens etched with ammonium persulfate. Thus, a differentiation between a ferritic/perlitic and zemenitic/ledeburitic microstructure is possible. During the use of a thermally stable binder the chill tendency of the specimen is considerably higher than it is when an unstable binder is used. The coating system also extends the metallurgical potential of the

Table 1: 'Pros and cons' of permanent mold casting of spheroidal ductile iron

- Advantages
  - no molding sand
  - separated circulation of core sand
  - increased output (~ 20%)
  - improved surface quality
  - improved product tolerance
  - refined structure leads to improved tensile strength
  - high pressure tightness

- Disadvantages
  - chill tendency
  - permanent mold costs
  - limited life time of the permanent mold

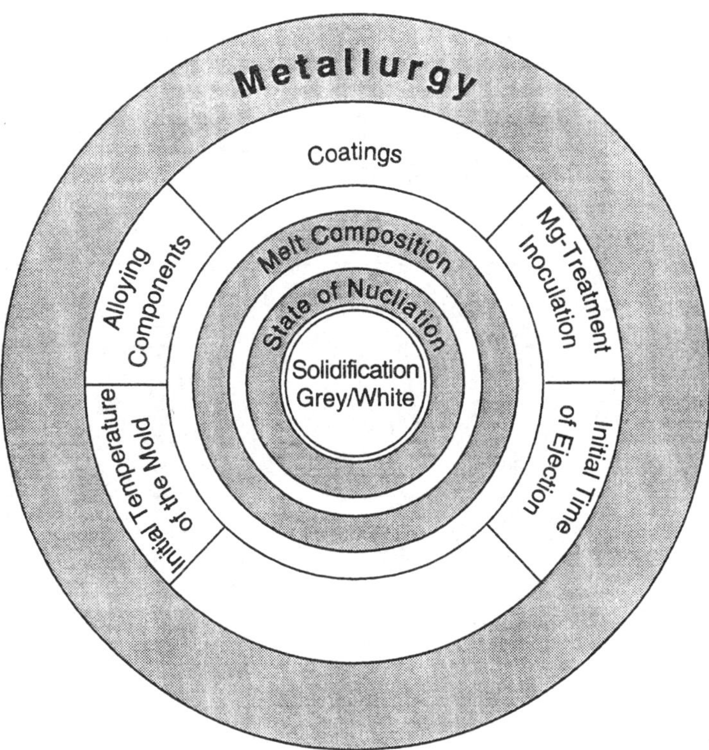

Figure 1: Problem area 'Metallurgy' with focused question 'Solidification' and relevant parameters.

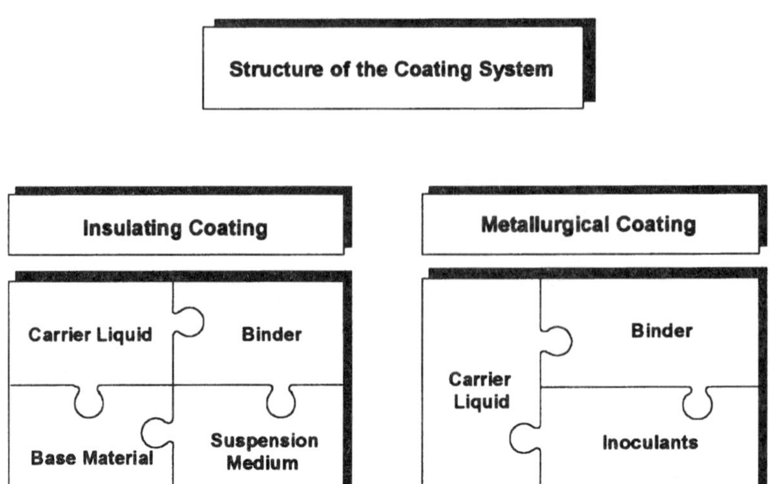

Figure 2: Rim zone metallurgy with an efficient coating system consists of an insulating and a metallurgical coating.

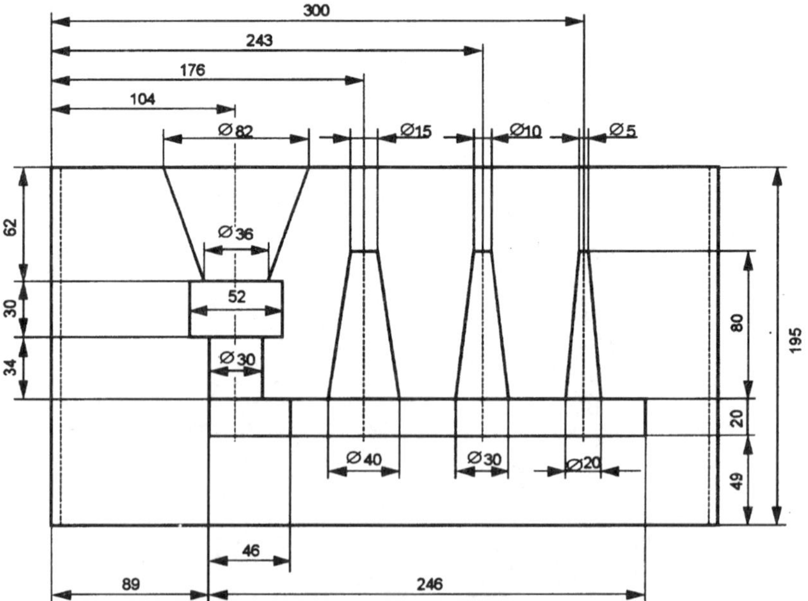

Figure 3: Representation of the mold cavity with cone geometries for developing the inoculation efficiency on the solidification with varying wall thickness'.

production method by modifying the surface-affected zone as compared to the casting's center. The result is a gradient microstructure. In addition, the ammonium persulfate etching makes it possible to use an interactive picture-analysing system to examine the microstructure quantitatively so that the effectiveness of the coating system is improved, figure 6.

Conclusions

Rim Zone Metallurgy based on an advanced coating system is a tool which can help foundrymen to minimize chill effects in spheroidal ductile iron in permanent mold. The surface affected zone may be modified and this offers the engineer to consciously produce gradient microstructures, whereby the production spectrum can be greatly expanded.

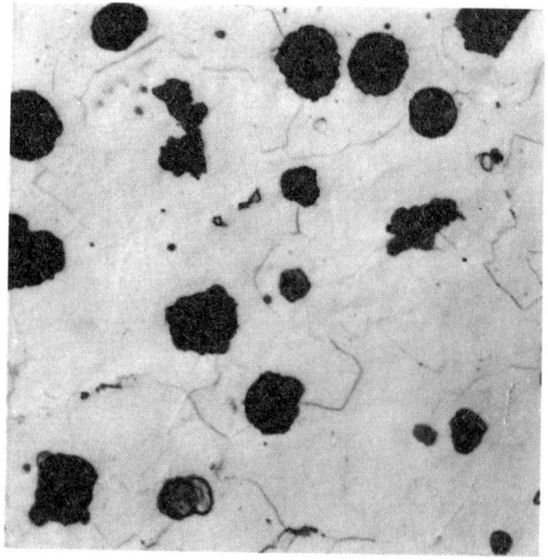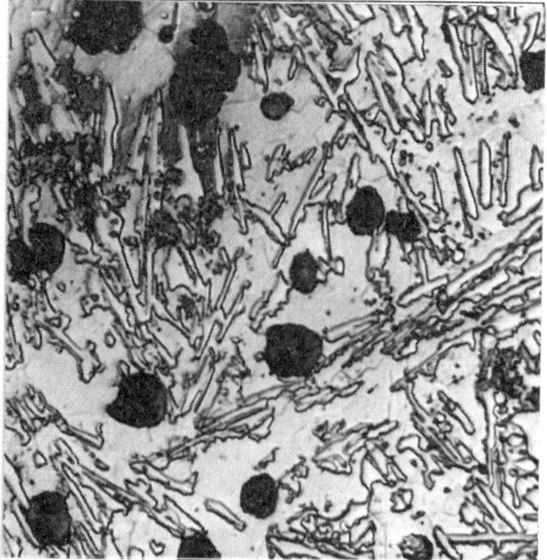

Figure 4: Effects of a metallurgical coating on the solidification of the specimen. The solidified specimen with the metallurgical coating is chill free (left), the part without the metallurgical coating contains chill microstructure (right).

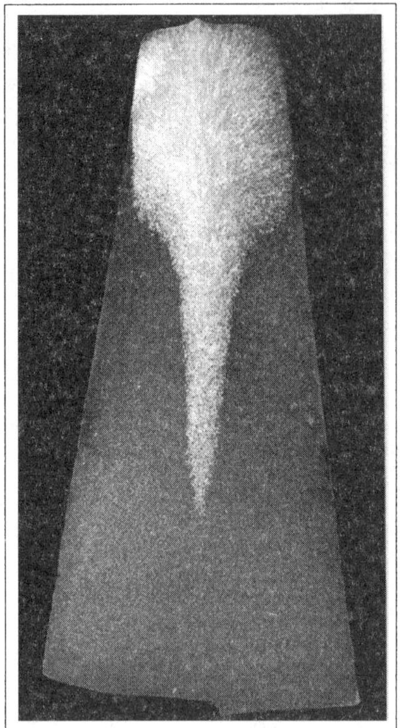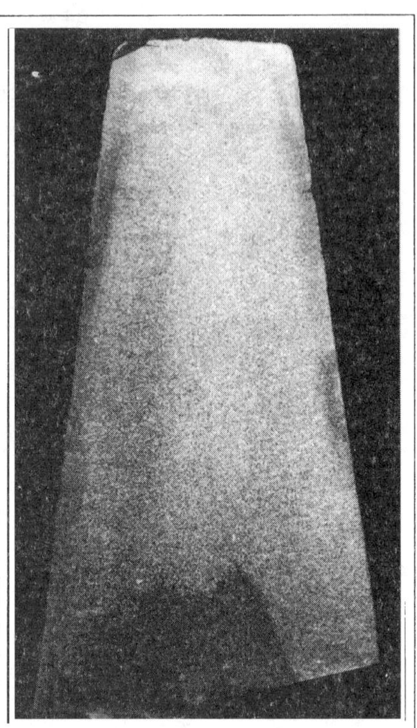

Figure 5: Influence of the binder on the solidification by using a thermally unstable binder, left. The area of the grey solidification depends on wall thickness, as can be clearly seen. When using a thermally stable binder the metallurgical coating is ineffective, and the specimen solidifies white. The specimen on the right shows the influence.

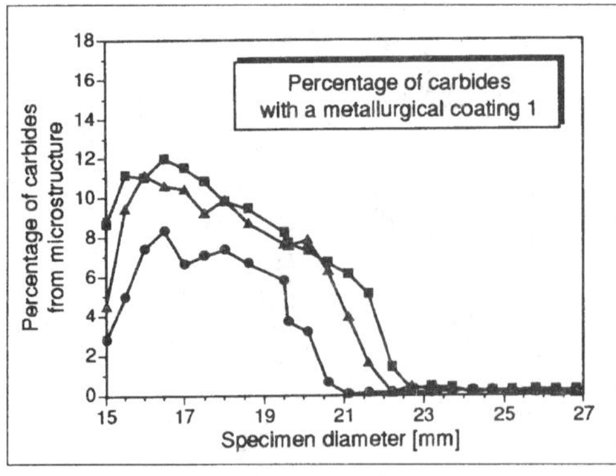

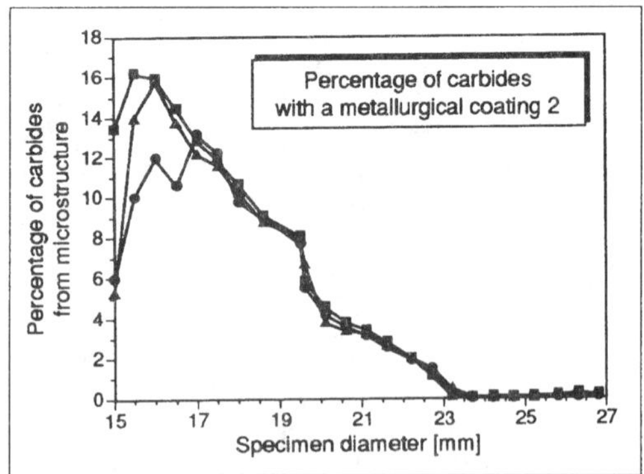

Figure 6: Quantitative effect of the influence of different inoculants on the microstructure: In comparison to figure 5 a smaller effect is noticeable.

STRUCTURAL ANALYSIS

Solidification microstructures

Solid state transformations

Quantitative analysis

Advanced Materials Research Vols. 4-5 (1997) pp. 355-360
© 1997 Scitec Publications, Switzerland

The Effect of Grain Growth upon Eutectic Carbide Particle Coarsening in H.S.S.

M. Durand-Charre, J.M. Chaix and P.H. Yang

Institut National Poytechnique de Grenoble, LTPCM Associé au CNRS (URA 29),
BP 75, F-38402 Saint Martin d'Héres, France

Keywords: High Speed Steels, Ostwald Ripening, Dendritic Growth, M_6C Carbides, Annealing, Liquid Phase (Supersolidus) Sintering

ABSTRACT: The growth of dendrites and M_6C carbides during isothermal annealings was studied on a (Fe,Mo,C) HSS-type alloy. The experiments were performed using an original DTA controlled process which enables accurate control of temperature and duration, this latter ranging from 10 to 180 minutes. Size distributions were measured by image analysis on polished cross sections. The growth rate was analysed using LSW-type models adapted to 3-component systems with large solubility in both phases and large volume fractions of particles. The dendrite growth is consistent with Mo/Fe diffusion limited growth. Two M_6C populations must be distinguished. The INTRA-dendritic carbides with a growth rate far higher than expected from LSW: their apparent growth is due to continuous entrapping by growing dendrites of larger carbides from the liquid phase. The INTER-dendritic carbides with a growth rate in the liquid phase far lower than expected from LSW. A model, proposed to take into account the actual geometry of carbide particles immersed in the liquid and the interaction with dendritic coarsening, leads to a more reasonable growth rate.

INTRODUCTION

The solidification of high carbon iron-base alloys involves a step in which a liquid phase is present, together with solid γ dendrites and eutectic carbides. In the liquid-solid zone, often referred as the mushy zone in castings, solute exchanges are very effective. Liquid phase sintering of alloyed steels in supersolidus sintering results in a similar state at a finer scale. During this step, both dendrites and carbides grow and the final microstructure controlling the mechanical properties is related to their respective growth rates. This latter kinetic aspect is investigated here, on the basis of isothermal annealing experiments, performed at temperatures ranging in the 3-phase field.

EXPERIMENTS

The alloy

The alloy is a model of M type HSS limited to three elements. Its composition (12.4wt%Mo,1.3wt%C,Fe bal.) was chosen by reference to the previously reassessed ternary system so as to form M_6C eutectic carbides [1]. The eutectic temperature T_e, *i.e.*, the temperature of appearance of the first M_6C carbides on cooling from melt in the DTA furnace is 1513K. A previous thermodynamic study [2] provided the phase composition data of Table 1 used in the forward calculations.

The typical microstructure of the quenched alloy is illustrated on Fig. 1. The dendritic pattern can be observed, with three different populations of M_6C carbides, namely:

-the INTRA-dendritic carbides, inside the dendrites.
-the fine eutectic carbides formed on quenching from the liquid phase
-the large INTER-dendritic carbides, located inside the quenched liquid phase, in the interdendritic groove et supposed to have grown in the liquid during all the annealing time.

The finer eutectic carbides are not liable to exhibit different morphologies for different annealing times; then our interest is only focused on dendrites, INTRA and INTER-dendritic carbides.

Table 1. Phase composition data on the studied HSS [2]

	γ dendrites	liquid	$M_6C = Fe_3Mo_3C$
atomic fractions	$x^{0\gamma}_{Fe} = 0.902$ $x^{0\gamma}_{Mo} = 0.053$	$x^0_{Fe} = 0.64$ $x^0_{Mo} = 0.25$	$x^{0M6C}_{Mo} \approx \dfrac{3}{7}$ $x^{0M6C}_{Fe} \approx \dfrac{3}{7}$
molar volumes	$\Omega^\gamma = 7.25\,cm^3\,mol^{-1}$	$\Omega = 8.16\,cm^3\,mol^{-1}$	$\Omega^{M6C} \approx 6.94\,cm^3$ / mole of $M_{6/7}C_{1/7}$

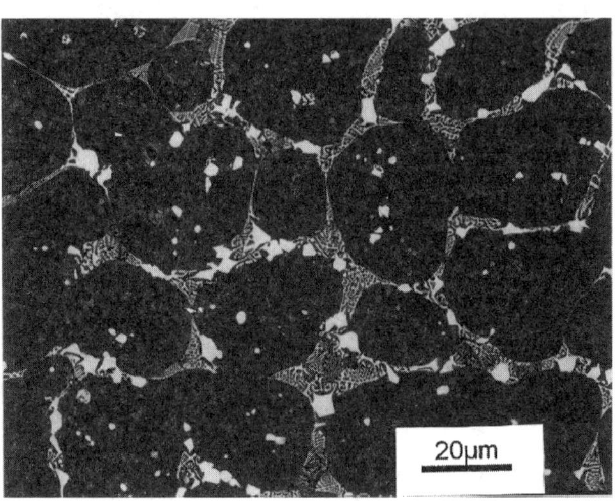

Figure 1. Microstructure of quenched samples

The annealing experiments

Experiments are performed in a DTA furnace. The determination of the appropriate temperature is made by direct observation of a first reference sample. Then the temperature of the furnace (Te-δT) is held for a series of experiments consisting in setting and removing new fresh samples for different durations. There is a constant control of temperature in the sample by reference to the running-on thermogram.

The small 2g samples are quickly put or removed and readily set at the annealing temperature or quenched. This procedure enables experiments with short annealing times from 10 up to 180 minutes.

Image analysis

Numerical 512X512 pixels SEM backscattered electron images are acquired at magnifications of 150 to 200 for dendrites and 400 to 500 for carbides. An automatic segmentation process is used to separate the three populations of interest [3]: dendrites, INTRA and INTER-dendritic carbides.

The image treatment leads to three separate binary images constituted by individual objects of the considered phase. 9 images per sample are used for dendrites, 16 images for carbides. The area A of each object is measured, then translated in equivalent diameter $d_2 = \sqrt{4A/\pi}$. Histograms $N_A(d_2)$ are transformed in 3D diameter distributions $N_V(d_3)$ using the classic Schwartz-Saltykov correction [4] and in volume weighed size distributions $V_V(d_3)$. The 3D number weighed mean diameter, used for comparison with models, and the 3D volume weighed mean diameter, more convenient to relate microstructural features to properties, are deduced.

ADAPTATION OF THE LSW MODEL TO THREE ELEMENTS SYSTEMS

The LSW model [5,6] is the reference model to analyse particle growth in a matrix. This model however deals with a <u>two element</u> system with a <u>vanishing volume fraction</u> of <u>single element</u> solid particles and a <u>small amount of this component dissolved</u> in the matrix. For γ dendrites as well as for M_6C carbides, the system is a three element one, with all three elements in both the particles and matrix. In addition, for γ dendrites, a high volume fraction of particles is present. We therefore adapted the model to account for these features.

In both cases, the diffusion of Mo (and therefore Fe) is assumed to be the rate limiting process, the diffusion of carbon being rapid.

To account for the high content of Mo in both the liquid and solid, a correct mass balance was used, similar to the one recently proposed by Calderon et al. [7]. In a three component system, the 3 local equilibrium conditions and the mass balance in the two phases are not sufficient to determine the 6 concentrations [8], and an additional relation is needed. We used the hypothesis that the diffusivity of carbon in the liquid is very high, and that the chemical potential in the liquid phase can be assumed constant. With the additional hypothesis that all phases behave as ideal solutions — which is a very rough approximation for carbides — and using the equilibrium equations from [8] the particle growth can be described by :

$$d^3 = K\,(t+t_0) \qquad (1)$$

$$\text{with } K = 4 \;\; \frac{D(\Omega^p)^2 \gamma_{pm}}{\Omega^m RT} \;\; \frac{1}{X_{Mo}^{0p} - X_{Mo}^{0m}} \;\; \frac{1}{\dfrac{X_{Mo}^{0p}}{X_{Mo}^{0m}} - \dfrac{X_{Fe}^{0p}}{X_{Fe}^{0m}}} \qquad (2)$$

in which the superscripts p and m refer to the particles and matrix respectively. The symbols and corresponding data used in the numerical applications of the formula are in Table 1.

In the case of dendrites, to take into account the low volume fraction V_V of liquid matrix and INTER-dendritic carbides ($V_V \approx 0.2$ [2]), we used the Sarian and Weart's corrective factor $\dfrac{4\,(1 - V_V)}{3\,V_V}$ [9]

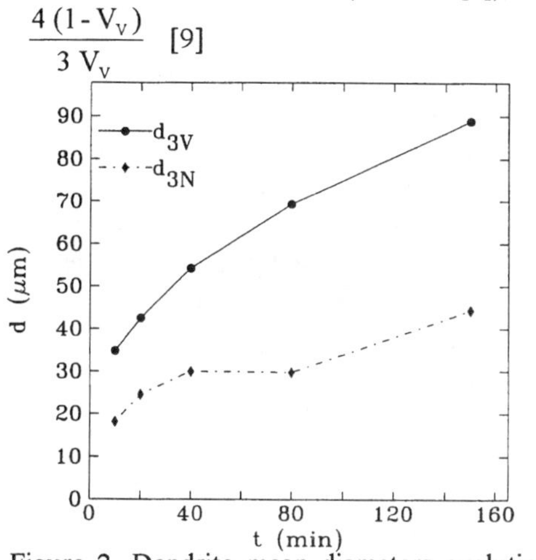

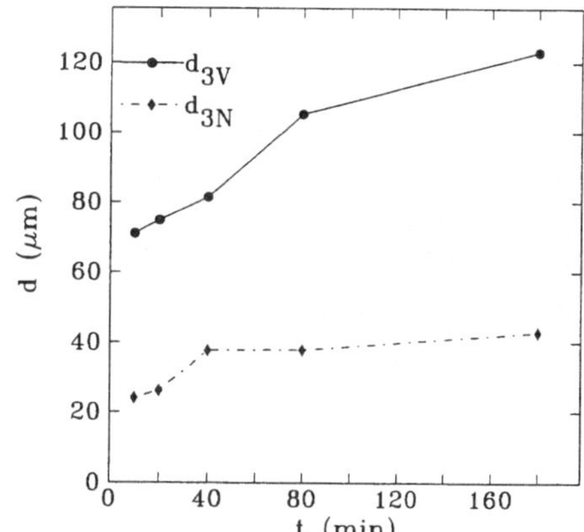

Figure 2. Dendrite mean diameters evolution with time for a presintered alloy at Te-7K

Figure 3. Dendrite mean diameters evolution with time for as cast alloy at Te-14K

RESULTS AND DISCUSSION

Dendrites

The number (d_{3N}) and volume (d_{3V}) weighed mean diameters evolution with time is plotted on Figs. 2 and 3 for an as cast and a presintered HSS (water atomised powder) at Te-14K and Te-7K respectively. The volume weighed diameter is more than twice larger than the number weighed one, which is due to the rather large size distributions.

The estimation of the cubic growth rate in the liquid phase with $D_{Mo}=3\ 10^{-5}$ cm^{2}sec^{-1} [10] and $\gamma^{\gamma L}=0.1$ Jm^{-2} lead to :

$$K_\gamma=800\ \mu m^3\ min^{-1}$$

The comparison of the calculated and measured number weighed diameters with time (Fig.4) shows a very good agreement. This confirms the Mo/Fe diffusion limited growth of dendrites.

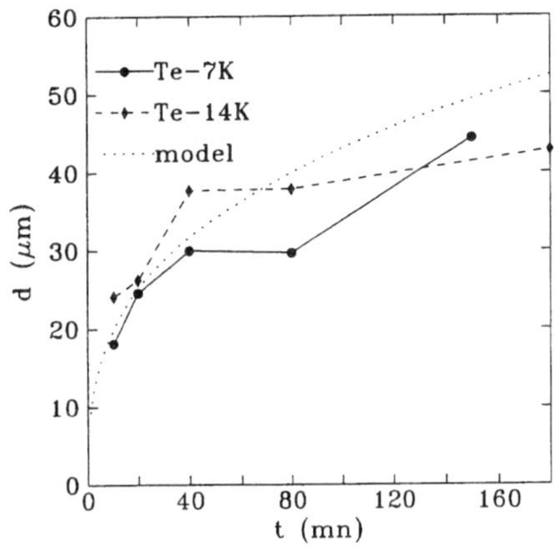

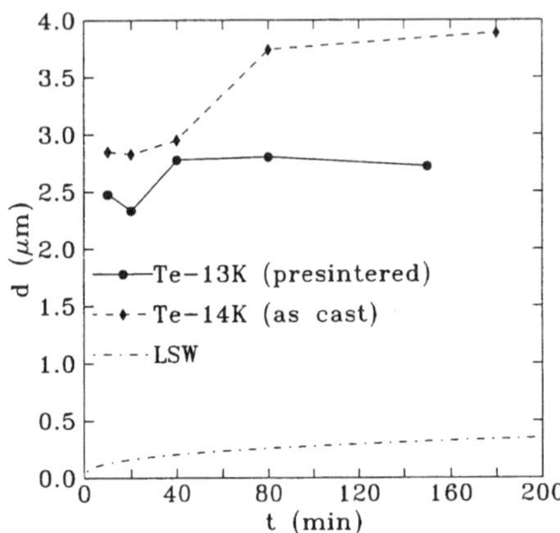

Figure 4. Comparison of model and experimental number weighed diameters for dendrites

Figure 5. Comparison of model and experimental number weighed diameters for INTRA-dendritic carbides

INTRA-dendritic carbides

The estimation of the cubic growth rate inside the γ dendrites with $D_{Mo}=1.6\ 10^{-10}$ cm^{2}sec^{-1} (effective diffusivity) and $\gamma^{M6C-\gamma}=0.3$ Jm^{-2} leads to :

$$K=4.5\ 10^{-4}\ \mu m^3\ min^{-1}$$

Two series of experiments have been performed to analyse the carbides growth, on a presintered alloy and an as cast one at T_e-13K and T_e-14K respectively. The comparison of the calculated and measured number diameters with time (Fig. 5) reveals that the observed diameters are far higher than expected from LSW. The main "apparent growth" mechanism is the entrapping of large carbides grown in the liquid phase by the moving boundaries of growing dendrites.

INTER-dendritic carbides

The number and volume weighed mean diameters evolution with time is plotted on Figures 6 and 7 for the two alloys. As for dendrites, the volume weighed is more than twice larger than the number weighed one, which is due to the rather large size distributions.

The estimation of the cubic growth rate in the liquid phase with $D_{mo}=3\ 10^{-5}$ cm^2 sec^{-1} [10] and $\gamma^{M6C-liq}=0.3$ J m^{-2} lead to :

$$K=550\ \mu m^3\ min^{-1}$$

The comparison of the calculated and measured number diameters with time (Fig.8) indicates that the calculated growth rate is too high. This could be explained by the geometry of the liquid phase between the dendrites, which makes the spherical symmetry assumed in the LSW model out of application.

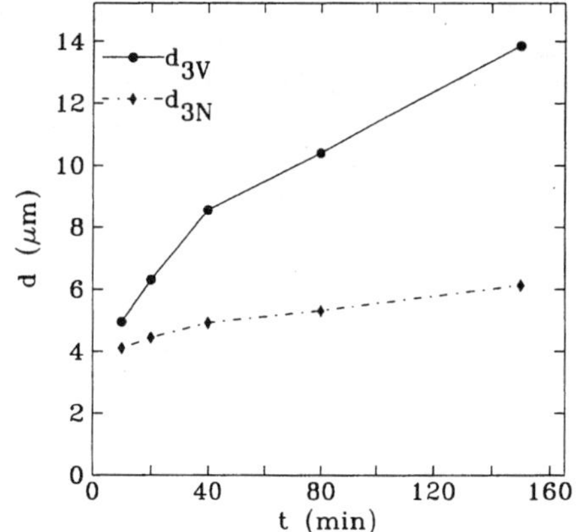

Figure 6. INTER-dendritic carbides mean diameters evolution with time for a presintered alloy at T_e-13K

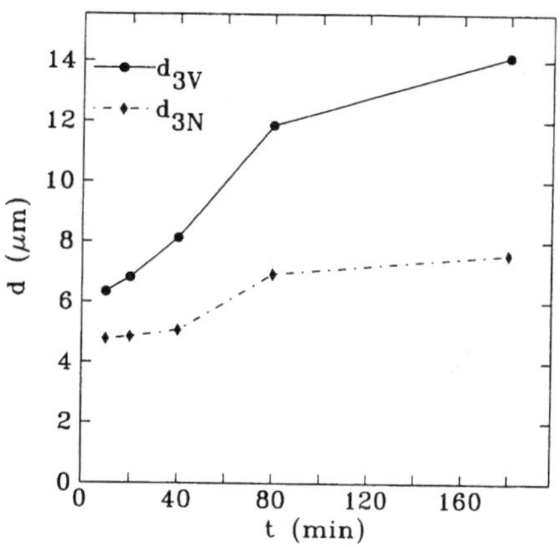

Figure 7. INTER-dendritic carbides mean diameters evolution with time for as cast alloy at T_e-14K

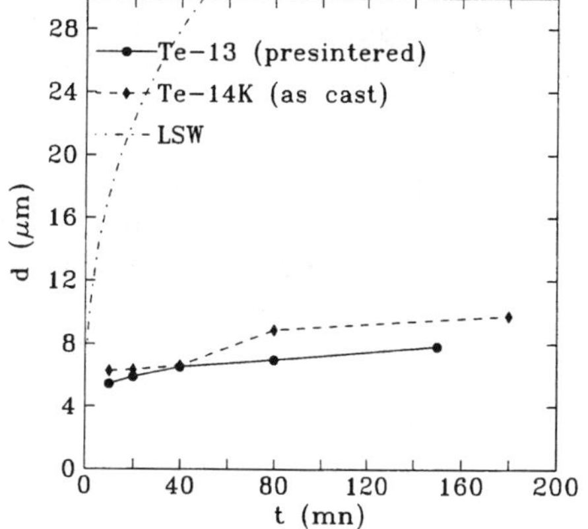

Figure 8. Comparison of LSW model and experimental number weighed diameters for INTER-dendritic carbides

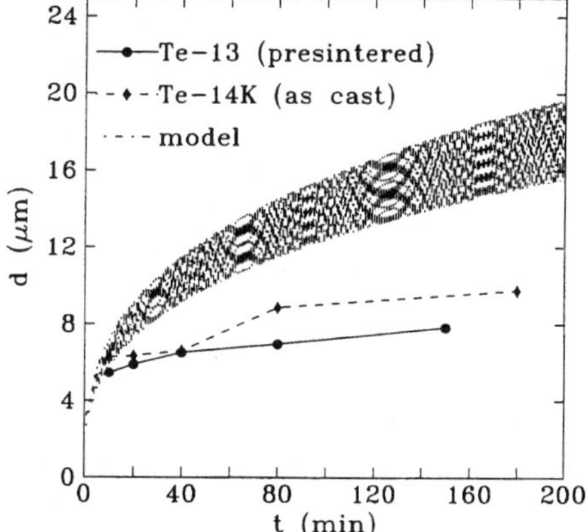

Figure 9. Comparison of liquid layer model and experimental number weighed diameters for INTER-dendritic carbides

A model for INTER-dendritic carbides growth

To describe the INTER-dendritic carbides growth, we have derived a model in which the carbide particles grow inside a liquid layer of width W, which is a more realistic description of the geometry of the system. In that case, and for a constant layer width W, a power 4 growth law is obtained, as in the case of grain boundary diffusion limited growth in solids [11]. The growth rate is proportional to the layer width. In practice, W is directly proportional to the dendrite diameter dγ, and therefore obeys a cubic growth law. taking into account this coupling of the two growing mechanisms through the layer width, the growth law derived for INTER-dendritic carbides is:

$$d^4 - d_0^4 = A\,(t + t_{0\gamma})^{4/3} \quad (3)$$

$$\text{with}\quad A = \left(\frac{3}{4}\right)^4 \frac{D(\Omega^{M6C})^2 \gamma^{M6C-liq}}{\Omega RT} \frac{1}{x_{M_0}^{M6C} - x_{M_0}} \frac{1}{\dfrac{x_{M_0}^{M6C}}{x_{M_0}} - \dfrac{x_{Fe}^{M6C}}{x_{Fe}}} \left(\frac{W}{d_\gamma}\right)(K_\gamma)^{1/3} \frac{1}{\ln(R_2/R_1)} \quad (4)$$

In this equation, the ratio R_2/R_1 is the ratio of the diffusion distance to the particle size, which depends on the volume fraction V_V^{M6C} of INTER-dendritic carbides (V_V^{M6C} =0.07 from image analysis). An upper value can be estimated taking R2 as the mean free path between carbide particles inside the liquid phase ($R_2/R_1=1+\pi(V_V-V_V^{M6C})/2V_V^{M6C}$), a lower limit from the liquid volume associated to each particle($R_2/R_1=(V_V/ V_V^{M6C})^{1/2}$). Using the above calculated K_γ, a range of values is therefore proposed for the growth constant A:

$$50 \leq A \leq 130 \; \mu m^4 \; min^{-4/3}$$

The comparison of the calculated and measured number diameters with time (Fig.9) shows an unperfect but better agreement than LSW, showing that the geometry of the system widely explains the discrepancy between the LSW type model and experiments. The uncertainties on the kinetic data may partly explain the remaining difference, but the main point is probably the assumption of ideal solution for the carbide phase.

CONCLUSIONS

The experimental **dendrite** growth is well predicted by the LSW-type model using diffusion limited by the Mo/Fe mass balance; Sarian and Weart geometric correction factor for small volume fraction of liquid; local equilibrium and mass balance written for 3-component solutions with the assumption of high, and the very high diffusivity of carbon in the liquid phase.

The experimental **INTRA-dendritic M_6C carbides** growth rate predicted by the LSW-type model is far higher than the observed one. In fact, the more effective mechanism is not the pure Ostwald ripening but the entrapping of M_6C particles, immersed in the liquid, by the coarsening dendrites.

The experimental **INTER-dendritic M_6C carbides** growth rate predicted by the LSW-type model is far lower than the observed one. A model is proposed which takes into account the liquid distribution in corridors surrounding the γ dendritic grains and the interaction with the γ grains growth. This modified model gives much more realistic predictions, though again over-estimates growth rates.

Our description of particles and grain growths involves coupled mechanism only through geometric factors (liquid layer width). Diffusion of molybdenum is treated independently for carbide and dendrite growths ; this is an arbitrary uncoupling of two phenomena of small amplitude occurring at different scales. The description of this 3-element, 3-phase non-equilibrium system is very complex and probably could not be completed analytically.

REFERENCES

[1] G. Giron, M Durand-Charre : Z. Metallkd. **86**, 15 (1995)

[2] P. H. Yang : DEA report, INP Grenoble, (1992)

[3] P.H. Yang, J.M. Chaix, M. Durand-Charre : Proc. PM'94 Powder Metallurgy Congress, vol.III, 2291 (1994)

[4] R.T. DeHoff, F.N. Rhines : Quantitative Microscopy, McGraw-Hill (1966))

[5] I.M. Lifschitz, V.V. Slyosov : J. Phys. Chem. Solids **19**, 35 (1961)

[6] C. Wagner : Z. Elektrochem. **65**, 581 (1961)

[7] H.A. Calderon, P.W. Voorhees, J.L. Murray, G. Kostorz : Acta Met. Mater. **42**, 991 (1994)

[8] J.M. Chaix, C.H. Allibert : Acta. Met. **34**, 1593 (1986)

[9] S. Sarian, H.W. Weart : J. Appl. Phys. **37**, 1675 (1966)

[10] P.J. Alberry, C.W. Haworth : Metal Science **8**, 407 (1974)

[11] R.D. Vengrenovitch : Acta Met. **30**, 1079 (1982)

AKNOWLEDGEMENTS

This work is a part of a BriteEuram program (BREU 0165-C). The authors wish to thank their partners Dr M. Oliveira (INETI Lisboa) for powder atomisation and Dr S. Wright for preparation of presintered samples. Helpful discussions with them resulted in choosing a relevant annealing temperature such that it was in the sintering window of the alloy investigated.

Advanced Materials Research Vols. 4-5 (1997) pp. 361-368
© *1997 Scitec Publications, Switzerland*

Solidification of High Speed Steel Type Cast Iron

K. Ogi, H. Miyahara, Z. Hong and N. Murai

Department of Materials Science and Engineering, Faculty of Engineering, Kyushu University,
Hakozaki 6-10-1, Higashi-Ku, Fukuoka 812, Japan

Keywords: White Cast Iron, Solidification, Carbide, Wear, Abrasion, High Speed Steel

ABSTRACT

Phase diagram and solidification processes are investigated on high speed steel type cast irons containing 5%Cr, 5%Mo, 5%W and 0~11%V. The diagram obtained for liquidus surfaces shows the compositional extent of each primary phase and the eutectic lines for (γ +MC) and (γ + M3C). M2C and M7C3 crystallize as eutectic carbides because of segregation of Mo, W and Cr to the residual liquid though they don't appear in the diagram. The species and amount of carbides of this alloy system are estimated on the basis of the diagram and the analysis of the redistribution of alloying elements.

INTRODUCTION

High speed steel type cast irons, which contain carbon and vanadium much higher than usual tool steels, have been developed as abrasion wear resistant materials. They are widely applied to rolls for hot steel strip mill[1,2,3]. Since the control of the distribution of primary and eutectic carbides is essentially important for the improvement and assurance of quality of products, the phase diagram and the solidification processes are investigated on a series of alloy specimens containing 5%Cr, 5%Mo, 5%W, 0~11%V and 1~4.4%C. δ , γ , MC and M3C appeared as primary phases, and (γ + MC) and (γ + M3C) eutectic crystallize at boundary compositions between liquidus surfaces of the constituent phases. The compositional extent of each liquidus surface is similar to those of Fe-V-C and Fe-5%Cr-V-C alloy system[4,5]. As the partition coefficients of Mo and W to γ + MC eutectic are less than 1, both elements are concentrated to residual liquid during γ + MC eutectic solidification and cause the crystallization of eutectic M2C. The partition coefficients of Cr are also less than 1 to primary γ , (γ + MC) and (γ + M2C) eutectic, the chromium content of residual liquid increased to 20% at the fraction solid of 95% in Fe-5%Cr-5%Mo-5.3%V-2.5%C, and chromium carbide crystallizes as M7C3 + γ eutectic.

EXPERIMENTAL PROCEDURES

The experimental specimens, shown in Table 1, were prepared from electrolytic iron, electrode graphite and ferro-alloys. The charge materials were melted in an alumina crucible under argon atmosphere and cast into an iron mold. Each alloy specimen of 30g was remelted and cooled at 0.17K/s in electric furnace with an argon atmosphere. Recording the cooling curves of specimens, they were quenched into water just after the completion of solidification. The crystals in each quenched specimen were identified by combined analysis of color etching, X-ray diffraction and EPMA. The compositional extent of each primary phase was expressed in relation to carbon and vanadium contents.

A series of specimens were quenched at several solidification stages for typical alloys and the changes in quenched structure were observed to clarify how the solidification proceeded. The alloy contents in both crystals and quenched residual liquid were also analyzed with EPMA to get the partition coefficients of each alloying element to primary and eutectic phases.

RESULTS AND DISCUSSION

Identification of carbides

MC, M_3C, M_2C, and M_7C_3 crystallized as primary or eutectic in this alloy system. Since the structure and amount of carbides changed depending on the chemical composition of alloy, color etching, X-ray diffraction and EPMA analysis were applied to identify the carbides. Each carbide was uniquely etched by Murakami reagent and picric acid alkali solution, as shown in Table 2. X-ray diffraction pattern was taken on carbide powder which was prepared by

Table 1 The chemical compositions of specimens.

Sample number	Chemical composition(mass%)					Carbide
	C	Cr	V	W	Mo	
No.1	1.01	4.32	—	5.20	4.98	M_6C
No.2	1.29	4.57	—	5.18	4.95	$M_6C+M_7C_3$
No.3	1.39	4.23	3.03	4.77	4.90	$MC+M_2C$
No.4	1.87	4.53	6.45	5.35	4.80	$MC+M_2C$
No.5	1.93	4.41	8.52	5.34	4.90	$MC+M_2C$
No.6	1.99	4.16	3.06	4.30	4.94	$MC+M_2C$
No.7	2.02	5.09	6.24	3.78	4.98	$MC+M_2C$
No.8	2.07	4.82	11.13	4.62	4.77	$MC+M_2C$
No.9	2.11	5.00	8.48	5.09	4.64	$MC+M_2C$
No.10	2.48	5.17	5.31	4.73	5.27	$MC+M_2C+M_7C_3$
No.11	2.50	4.49	3.13	5.05	5.21	$MC+M_2C+M_7C_3$
No.12	2.57	4.38	6.93	5.02	4.83	$MC+M_2C$
No.13	2.91	5.09	7.16	4.42	5.28	$MC+M_2C$
No.14	3.27	4.06	3.42	3.85	5.06	$MC+M_2C+M_7C_3$
No.15	3.76	4.64	3.26	5.10	4.96	$MC+M_2C+M_7C_3$
No.16	3.81	4.54	3.24	4.84	4.80	$MC+M_2C+M_7C_3$
No.17	4.13	4.68	—	4.81	4.88	$M_3C+M_7C_3$
No.18	4.16	4.69	3.44	5.02	4.50	$MC+M_2C+M_7C_3$
No.19	4.16	4.61	3.77	4.64	4.82	$MC+M_2C+M_7C_3$
No.20	4.31	4.57	—	4.77	4.90	$M_3C+M_7C_3$
No.21	4.31	4.63	1.22	5.39	4.83	$M_3C+M_7C_3$
No.22	4.35	4.62	3.40	5.60	4.83	$MC+M_2C+M_7C_3$

Table 2 The results of color etching.

Carbide	Murakami reagent	Picric acid alkali solution
MC	—	—
M_2C	black	—
M_3C	—	brown
M_6C	dark brown	light brown
M_7C_3	brown	—

dissolving matrix with a solution of phosphoric acid, and the X-ray diffraction spectra for carbides in Fe - 5.2%Cr - 5.3%Mo - 4.7%W - 5.3%V - 2.48%C alloy was shown in Fig.1. Characteristic peaks from MC, M2C and M7C3 were clearly recognized. A strong peak from M6C of Fe-5%Cr-5%W -C alloy was detected at 70° . EPMA analysis was also useful in identification of carbides, since each carbide showed the characteristic composition, as shown in Fig.2. By combined use of these three methods each carbide was accurately identified.

<u>Phase diagram and solidification sequence</u>

A primary crystal of each alloy in Table 1 was identified and the results were expressed in relation with the C and V contents of alloy, as shown in Fig.3. The number in the Figure indicates the specimen number, and △, ○, □ and ▽ represent that the primary crystal was δ, γ, MC and

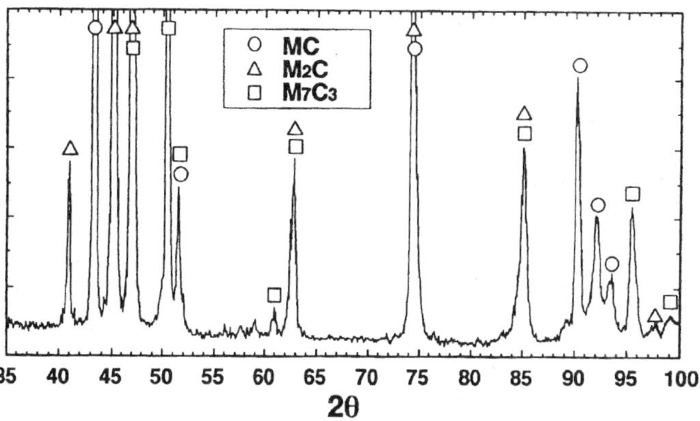

Fig.1 X-ray diffraction spectra of carbides in Fe-5.2%Cr-5.3%Mo-4.7%W-5.3%V-2.48%C alloy.

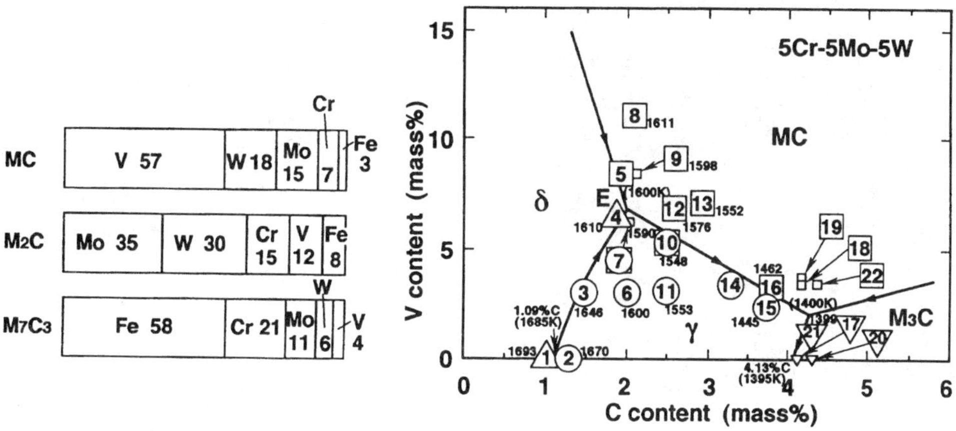

Fig.2 EPMA analysis of carbides.

Fig.3 The diagram for liquidus surfaces of Fe-5%Cr-5%Mo-5%W-V-C alloy system.

M3C, respectively. δ and γ crystallized in dendrites, M3C in coarse plates and MC in dendrites as shown in Fig.4. Since a small amount of dendritic γ and MC appeared simultaneously in the specimens No.7 and No.10, the compositions of these alloys should be close to the eutectic. Based on detected species and amount of primary crystals the compositional extent of liquidus surface of each primary phase was determined as shown in Fig.3. The diagram was rather similar to those of Fe-V-C ternary alloy system and Fe-5%Cr-V-C quadri-alloy system [4,5], though the extent of γ became smaller (Fig. 5). The alloys at boundaries between liquidus surfaces of δ / MC, γ / MC and γ / M3C started the solidification by the eutectic reaction of $L \rightarrow \delta + MC$, $L \rightarrow \gamma + MC$ and $L \rightarrow \gamma + M3C$, respectively. Peritectic reaction $L + \delta \rightarrow \gamma$ occurred at boundary between liquidus surfaces of δ and γ.

Fig.3 can be used to evaluate the relation between alloy composition and solidification structure. As an example the solidification process was estimated on hypo-eutectic alloy specimens No.3, 6, 11, and 14 which contained 1.4~3.3%C and 3 ~4%V. To estimate the changes in residual liquid during solidification of primary γ, the partition coefficients (km) of alloying elements to primary γ were evaluated as the ratios of alloy contents at the core of dendrite (Cc) to the average contents of alloy. Cc was determined with EPMA. km of V, Cr, Mo and W were less than 1, and they decreased with increasing the carbon content of alloy, as shown in Fig.6. These coefficients were similar to the equilibrium partition coefficients (ke) obtained on Fe-C-Cr-M(M = V, Mo etc.) [6]. Though km was a little larger than ke for the element with ke less than 1, the difference was small. The partition coefficients of C to primary γ was also dependent on the composition of melt, it was assumed to be constant at 0.35[7]. By using these partition coefficients the changes in C and V contents of residual liquid were estimated assuming complete diffusion of C and no diffusion of V in dendrites. The calculation [8] was made at every fraction solid 0.005 and the results were given in Fig.7. The composition of residual liquid changed in the direction shown by arrow. The eutectic reaction $L \rightarrow \gamma + MC$ started when the composition of melt reached the boundary between liquidus surfaces of γ and MC. Therefore more primary γ crystallized in the alloy whose

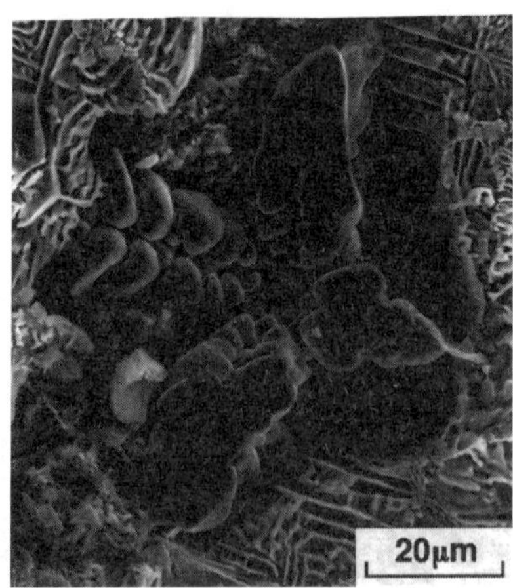

Fig.4 Morphology of primary MC.

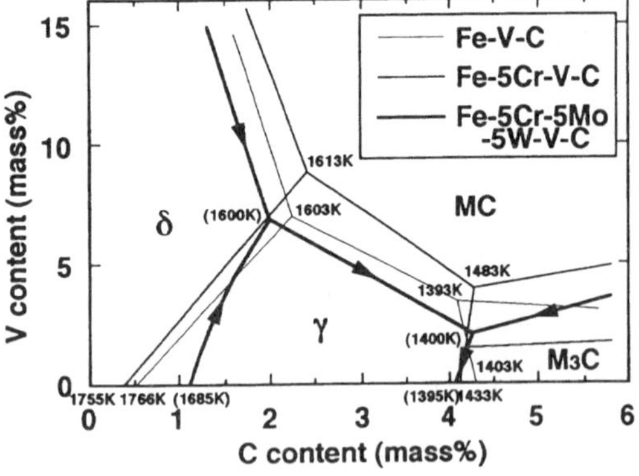

Fig.5 Influence of Cr, Mo and W on the diagram for liquidus surfaces of Fe-X-V-C alloys (X = Cr, Mo, W).

composition was further from the eutectic line and more (γ + MC) eutectic crystallized as the alloy composition became close to point E in Fig.3. In hyper-eutectic alloys with higher V content, dendritic MC crystallized as primary and the composition of residual liquid changed toward (γ + MC) eutectic line.

However, Cr, Mo and W contents varied during eutectic solidification and the diagram could not be applied to the later stage of solidification. As an example, the solidification sequences were studied on a series of alloys containing 3~4%V and the results were expressed in Fig.8 as a Fe-C system diagram which well corresponds to H. F. Fischmeister's results on Fe-4%Cr-5%Mo -6%W-2%V-0~1.4%C [8]. Eutectic reactions L→ γ + M_2C and L → γ + M_7C_3 occurred in the alloys with higher carbon content. The influence of V content on the solidification sequence was

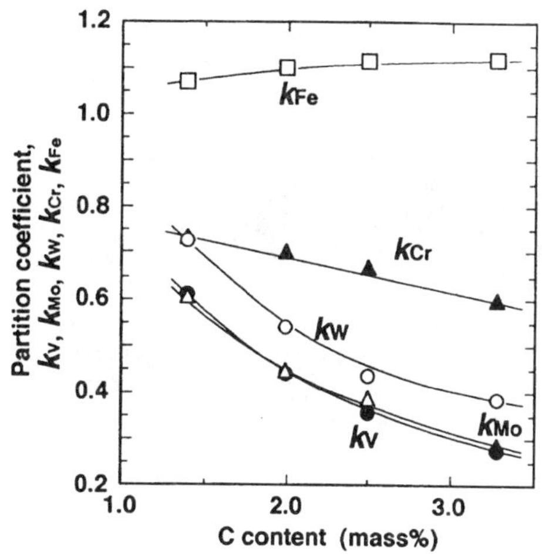

Fig.6 Partition coefficients of alloying
elements to primary γ vs. C content.

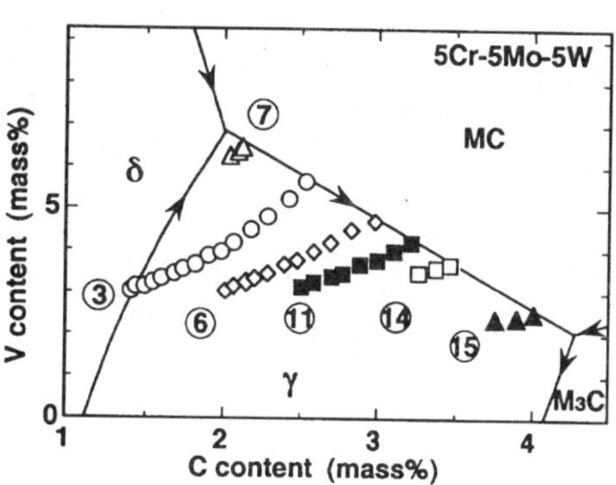

Fig.7 The changes in V and C contents
of residual liquid during the solidification of primary γ.

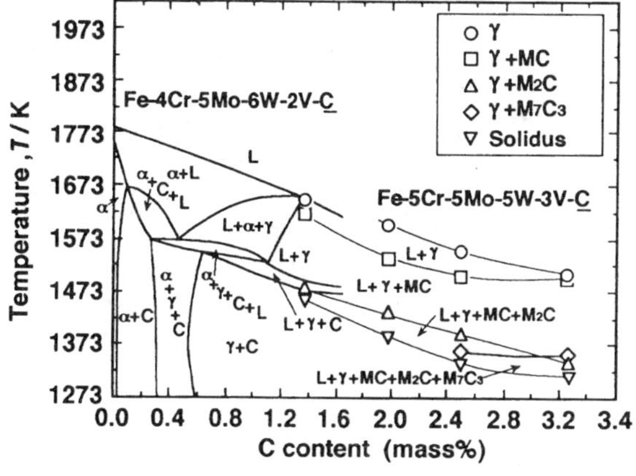

Fig.8 Phase diagram for Fe-5%Cr-
5%Mo-5%W-3~4%V-C
alloy system.

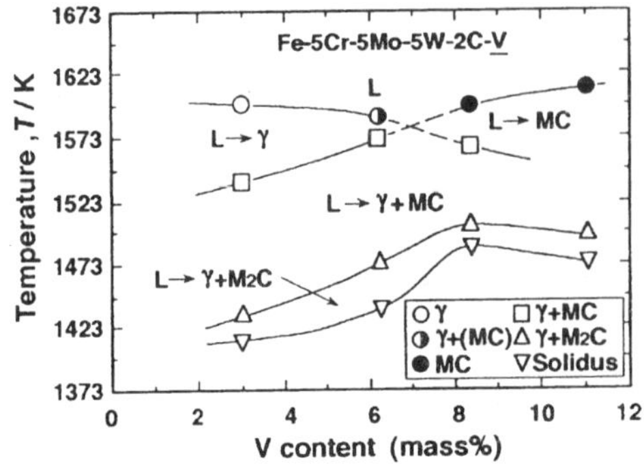

Fig.9 Influence of V content on crystallization temperatures of primary
and eutectic phases.

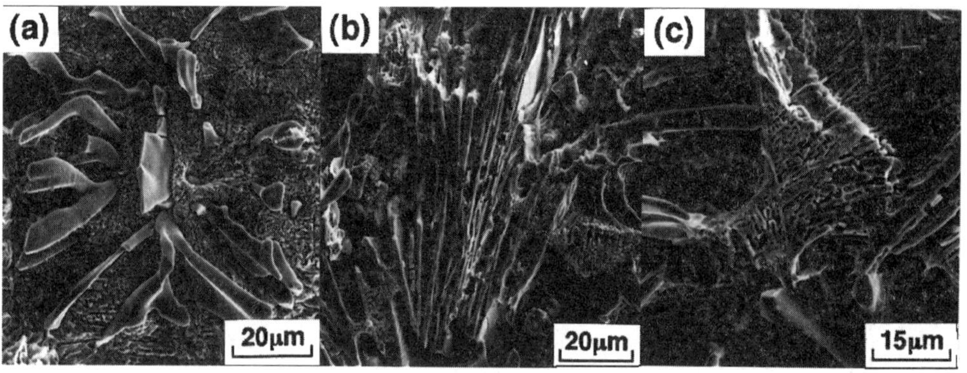

Fig.10 Morphologies of eutectic MC (a), M2C (b) and M7C3 (c).

also shown in Fig.9. The crystallization temperature of primary γ dropped and that of primary MC rose with increase in V content of alloy. As for hypo-eutectic alloys, the eutectic start temperatures of L $\rightarrow$ γ + M2C and L $\rightarrow$ γ + M7C3 rose as V content of alloy became larger. Eutectic MC was distributed more discontinuously than eutectic M2C and M7C3, as shown in Fig.10.

Influence of segregation on later eutectic reaction

The change in residual liquid during solidification and it's influence on the following crystallization phenomena were investigated on the specimen No.10 with 2.48%C-5.3%V. It was apparent from the cooling curve and changes in structure of quenched specimens A, B, C, D (Fig.11) how the solidification proceeded in this alloy. The area fractions of primary γ , (γ + MC), (γ + M2C) and (γ + M7C3) eutectics were measured as 0.25, 0.52, 0.17 and 0.06, respectively. The fraction of each phase in the eutectic are also shown in Fig.12. The smaller fraction of MC in (γ + MC) eutectic indicated that this carbide distributed finer than other eutectic carbides.

EPMA analysis on quenched specimens A, B, C, D revealed that composition of liquid changed during solidification as given in Fig.13. The partition coefficients of each alloying element

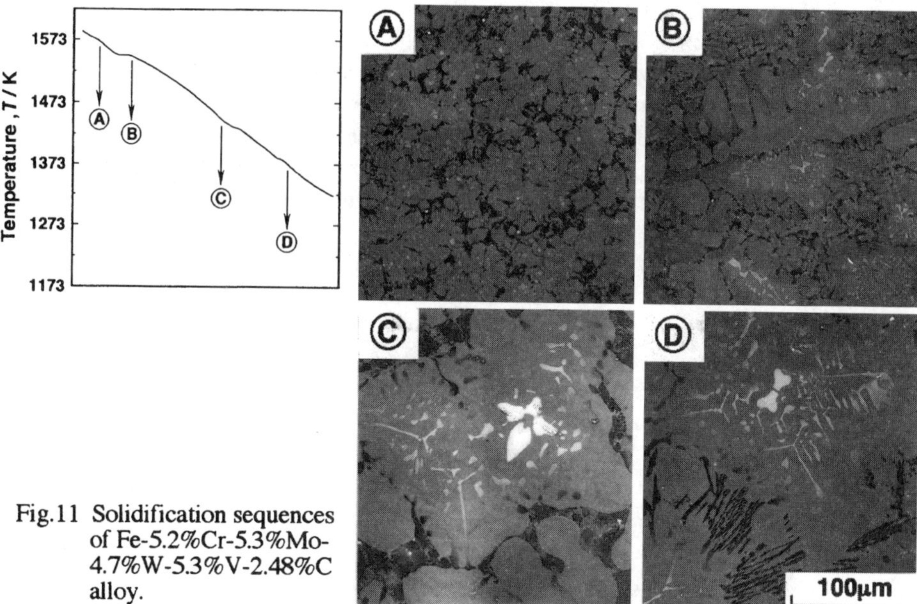

Fig.11 Solidification sequences of Fe-5.2%Cr-5.3%Mo-4.7%W-5.3%V-2.48%C alloy.

γ+MC Eutectic	γ 77.8		MC 22.2
γ+M₂C Eutectic	γ 59.4	M₂C 40.6	
γ+M₇C₃ Eutectic	γ 37.8	M₇C₃ 62.2	

Fig.12 The volumetric amounts of γ and carbide in eutectic.

Table 3 Partition coefficients of alloying elements to eutectic phases.

γ+MC eutectic

	k_V	k_{Cr}	k_{Mo}	k_W
γ	0.26	0.52	0.22	0.32
MC	11.03	0.67	1.70	2.68
γ+MC	2.65	0.56	0.55	0.84

γ+M₂C eutectic

	k_V	k_{Cr}	k_{Mo}	k_W
γ	0.17	0.31	0.23	0.43
M₂C	3.66	0.83	3.52	5.84
γ+M₂C	1.59	0.52	1.57	2.63

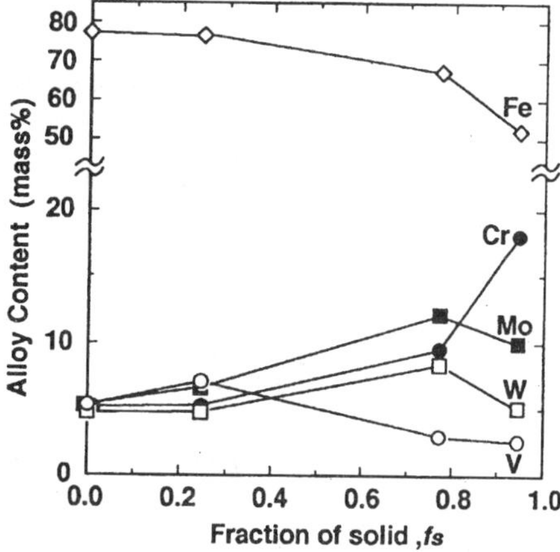

Fig.13 The changes in alloying elements in residual liquid during solidification of Fe-5.2%Cr-5.3%Mo-4.7%W-5.3%V-2.48%C alloy.

to eutectic phases were determined as the ratios of alloy contents of eutectic phases to those of residual liquid. The former was measured on the outer portion of eutectic cells which would crystallize just before quenching. From the obtained coefficients in Table 3, the solidification of this alloy should proceed as follows; Cr, Mo, W,and V were enriched to residual liquid during the crystallization of primary γ and (γ + MC) eutectic started crystallizing at fraction solid of 0.25. Then the content of V decreased and those of Mo and W increased in residual liquid and (γ + M2C) eutectic started crystallizing at fraction of solid 0.77 where the W equivalent (W + 2Mo) reached 32.5%. With the procedure of (γ + M2C) eutectic freezing, Mo and W contents decreased and Cr content increased to 20% in liquid. Therefore (γ + M7C3) could crystallize at the last stage of solidification.

CONCLUSION

The phase diagram on liquidus surfaces was made on Fe-5%Cr-5%Mo-5%W-0~11%V-1.4~4.4%C alloy system. The diagram was fairly similar to those of Fe-V-C and Fe-5%Cr-V-C alloy systems. To clarify the solidification process of this alloy system, the changes in residual liquid and its influence on following solidification phenomena were analyzed on a typical alloy by using the partition coefficients of alloying elements to each phase. These results could contribute the determination of the chemical composition for obtaining proper structures.

REFERENCES

[1] Kunio Goto, Yukio Matsuda, Kohichi Sakamoto and Yoshihito Sugimoto, ISIJ **32**, 1184 (1992).

[2] Yoshikazu Sano, Toshiyuki Hattori and Michio Haga, ISIJ **32**,1194 (1992).

[3] Junichi Kihara, Tetsu-To-Hagane (Journal of the Iron and Steel Institute of Japan) **80**, 386 (1994).

[4] Akira Sawamoto, Keisaku Ogi and Kimio Matsuda, Imono (Journal of Japan Foundrymen's Society) **54**, 726 (1982).

[5] Akira Sawamoto, Keisaku Ogi and Kimio Matsuda, A F S Transactions **94**, 403 (1986).

[6] Yukinori Ono, Noriko Murai and Keisaku Ogi, ISIJ **32**, 1150 (1992).

[7] Yukinori Ono, Tsutomu Takechi, Noriko Murai and Keisaku Ogi, J. of Japan Inst. Metals **56**, 802 (1992).

[8] Yukinori Ono, Tsutomu Takechi and Keisaku Ogi, J. of Japan Inst. Metals **57**, 432 (1993).

Advanced Materials Research Vols. 4-5 (1997) pp. 369-376
© *1997 Scitec Publications, Switzerland*

Unidirectional Solidification of Cast Iron:
Morphological Changes of Graphite Due to In-Situ Modification

A.N. Roviglione[1] and H. Biloni[2]

[1] Departamento de Física, Facultad de Ingeniería, Universidad de Buenos Aires (FIUBA-UBA)
Paseo Colón 850, (1063) Buenos Aires, República Argentina

[2] Laboratorio de Entrenamiento Multidisciplinario para la Investigación Tecnológica, Comisión de
Investigaciones Científicas de la Provincia de Buenos Aires (LEMIT-CIC), Calle 52 entre 121 y
122, (1900) La Plata, Provincia de Buenos Aires, República Argentina

Keywords: Unidirectional Solidification, Solidification of Flake and Compact Graphite, Structure of
Graphite

ABSTRACT

Ni alloyed cast iron was directionally grown with a Bridgman modified technique. Keeping growth
conditions constant, chemical composition changes were made, in front of the Solid-Liquid (S-L)
interfaces, by "in-situ" addition of Ce commercial alloy. Compact graphite cast iron continuously
evolves from flake graphite cast iron used as seed. S-L interfaces were frozen for flake and compact
morphologies. Optical microscopy and SEM with deep etching were used to study the microstructures.
An interpretation of the free flake eutectic grain growth is made. On the other hand, compact graphite
is composed by a large amount of randomly oriented fine faceted crystals and a progressive
"compactation" of these graphite crystals inside grooves of uncoupled γ-G-L interface seems to be the
mechanism of compact graphite formation.

INTRODUCTION

Basic knowledge about the mechanisms controlling the morphological features of nf-f eutectics is very
important due to the technological importance of systems such as Al-Si and iron-graphite (Fe-G). In the
Fe-G system, the morphology of the graphite is usually flake like. Adding Mg or Ce-Ca alloys to the
melt, the graphite morphology changes into other two main forms: nodular and compact or vermicular.
The modification mechanisms have been studied by many researchers and the different theories have
been extensively reviewed [1-3].
Many experimental works were carried out on flake cast iron, solidified under free and directional
conditions, [4-6]. Double and Hellawell [7], by making MET studies, determined the parallelism of the
graphite (00.2) planes with the surface of flakes randomly extracted from the matrix of Ni-C eutectic.
In a recent work by Roviglione and Hermida, [8], X-ray diffraction and a new metallographic technique
[9] were used on directionally solidified bulk samples of Ni alloyed cast iron. These authors determined
the crystallographic relationships between austenite (γ) and graphite (G), relative to the imposed growth
direction (GD), in strongly oriented flake and compact cast iron. Their result can be summarized as
follows: i) flake G and γ show a strong crystallographic relationship: $<11.0>_G$ // $<10.0>_G$//$<110>_\gamma$//GD;
ii) compact graphite shows a random crystallographic orientation, in spite of its highly oriented
microstructure relative to the GD.
Based on their own results and previous evidence by Double and Hellawell [7] those authors proposed
a model to account for the observed curvature and branching of the flakes during directional
solidification, *preserving its crystallographic orientation*. It consists of a large density of prismatic steps

of only a few cells' size leaving the $(00.2)_G$ planes in contact with austenite planes belonging to the $<100>_\gamma$ crystallographic zone.

In the present work that is a continuation of the previous publication [8] optical microscopy and SEM were used giving experimental information related to: 1) features of the frozen S/L interface in flake eutectics, 2) flake to compact graphite transition, 3) microstructure of the compact graphite, 4) nature of the frozen S-L interface of compact graphite.

EXPERIMENTAL

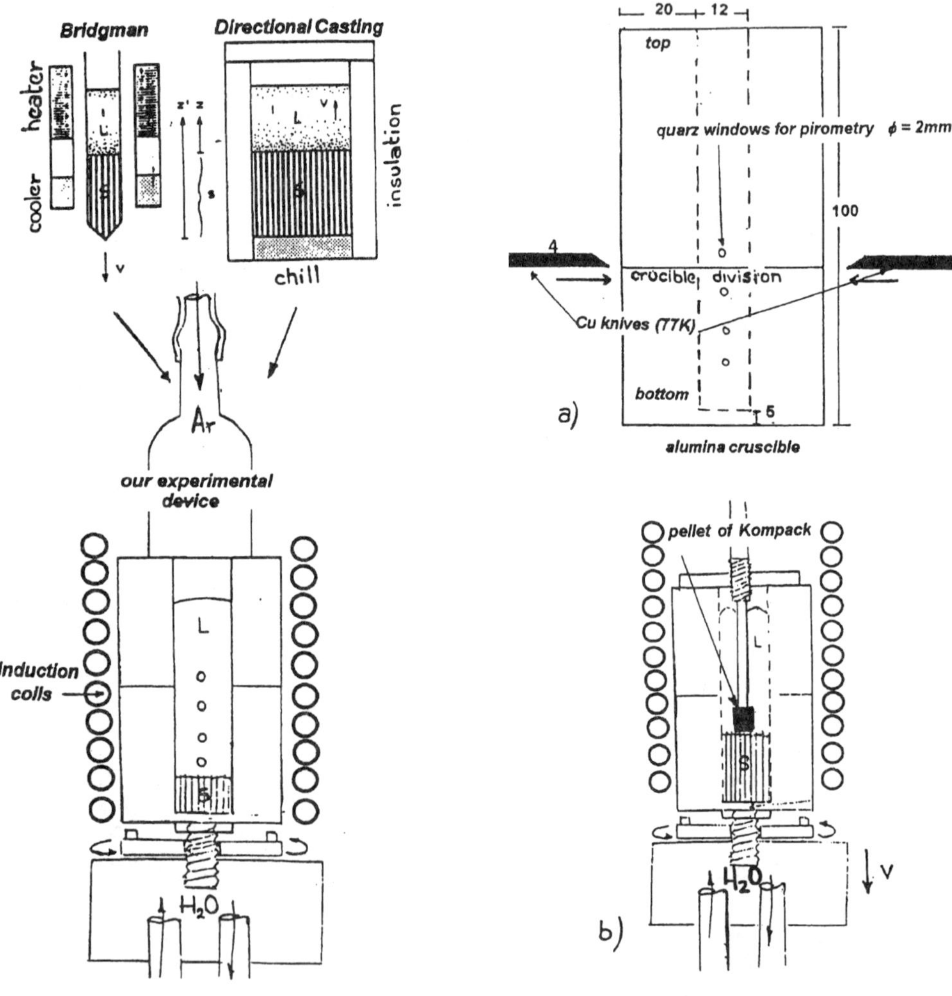

Fig.1. Sketck showing the device for restricted growth.

Fig.2. a)Details of the alumina crucible and Cu knives use to freeze the S-I interface. b) Details of the "in-situ"addition of the modifier in front of S-L liquid interface. The seed present an oriented flake eutectic structure.

Ingots of 70mm height and 13mm diameter were grown with a modified Bridgman technique, schematically shown in Fig.1. The composition of the ingots in wt% was:
C = 3.48%; Ni = 5.8%; Si = 2,51%; P = 0,024%; S < 0.03% and Fe balance. The Ni addition was used in order to eliminate the eutectoid transformation and to retain the γ phase. Fig.2a) shows schematically the alumina crucible used (100mm long; 20mm wall thickness and 5mm bottom thickness) and the device used to freeze the S-L interface during the solidification process. The thermal gradients imposed were measured by pyrometry with the aid of fine diameter quartz bars positioned in the crucible as "optical" windows.

The variables during the experiments were the liquid thermal gradient in front of the S-L interface (between 120 and 140°C/cm) and the growth speed (between 1 and 1.2 μ/sec). The alloy Ce-Ca-Si-Fe named "Kompack" by Foseco was used as modifier. Fig.2b) shows schematically the procedure of "in-situ" modification through small cylinders of Kompack alloy. The addition is made in such a way as to leave the remaining melt with 1.4% concentration of modifier.

The S-L frozen interface was obtained by cutting the liquid vein 2-3mm ahead of the S-L front with Cu knives cooled by liquid N_2 and then introduced into the crucible division, Fig.2b). In order to preserve strong connection between X ray crystallographic studies and metallographic characterization of microstructure half a longitudinal section of ingots for each one has been employed. After optical observation, and previously to SEM, deep etching was performed with 20wt % NO_3H in ethanol.

RESULTS AND DISCUSSION
1. Flake Graphite

Samples that were grown without modification show a flake graphite type A structure oriented along the growth direction, Figs. 3a) and 3b). Light micrography of the interface frozen during growth is shown in Fig.4a). As can be seen, the growth of the eutectic phases seems to be coupled and led by graphite. Details of the flake top belonging to the interface are shown in the SEM micrography in Fig.

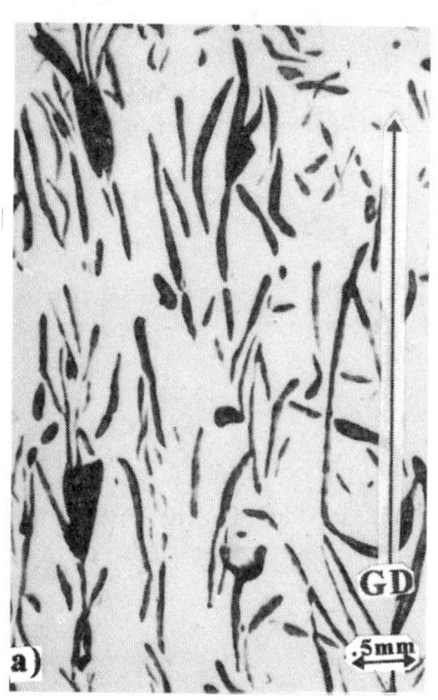

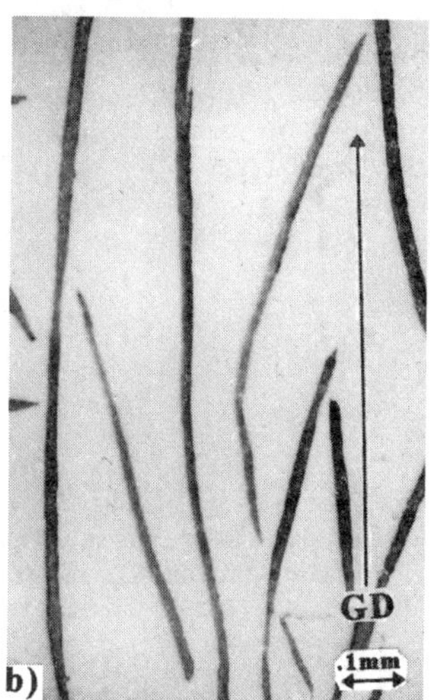

Fig. 3. a) Flake oriented eutectic type A. Optical microscopy (OM). b) Same structure at higher magnification.

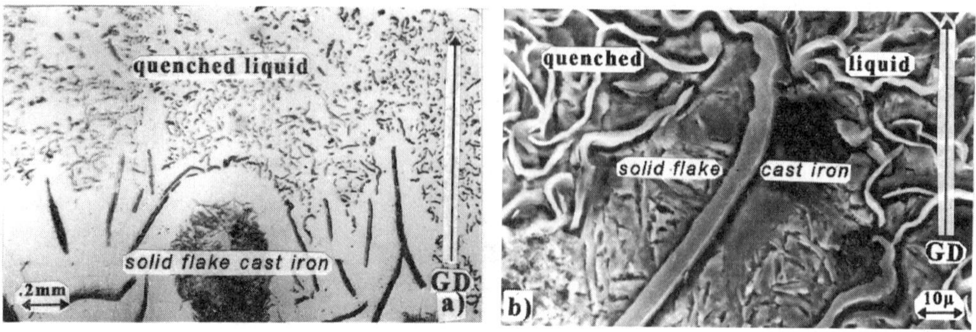

Fig.4. a) Frozen interface of flake type A eutectic. b) Details by SEM of the top of the flakes at the frozen S/L interface.

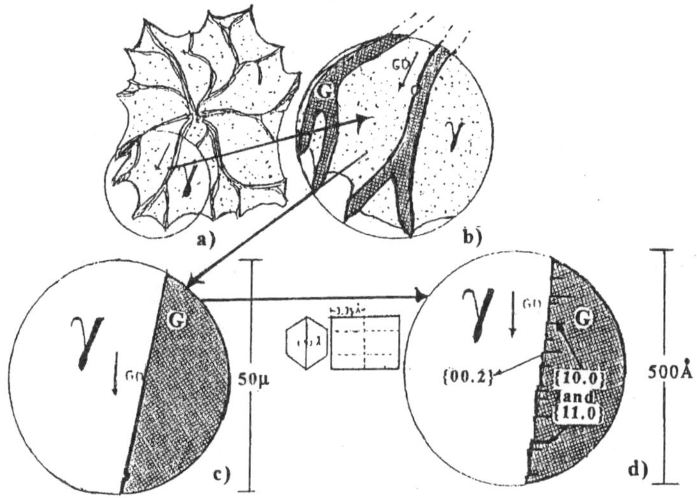

Fig.5. a) Current model of the flake eutectic grain: the basal planes bend following the flake curvature. When branching occurs the diverging flakes have different G.D. b) c) and d) coupled growth of G and γ phases, as an extrapolation from unidirectional solidification results [8] considering the X-ray experimental evidence.

4b); it is worth to notice the wavy character of the flakes growing ahead of the frozen interface. These features are similar to the results obtained by other authors during free and unidirectional solidification followed by a quench, see for example, Fig. 21, 22 and 27, ref. [4] and Figs. (5-7), ref. [10]. This suggests

it seems

[8]. This extrapolation is shown schematically in Fig.5 where the coupled growth of the γ and G phases is presented. The branching and curvatures of the flakes are accounted for by prismatic steps of only a few cell sizes. that leaves the (00.2)ᴳ planes in contact with γ planes of the <100> crystallographic zone.

According to this model all the observed changes in direction, including those provoqued by branching are only apparent if we look at them from a crystallographic point of view. These experimental evidences [8] suggest that the mechanism operating during bending and branching of flake cast iron is still open to discussion.

2. Flake-compact transition

The morphological modification was obtained by chemical composition changes in front of the S-L interface, maintaining the growth speed and thermal gradient constant. Flake graphite samples were used as seeds for compact graphite obtention through out the "in situ" modification. Figs. 6 a)-c) show the continuous change from flake to compact graphite. As can be seen when using SEM technique flakes do not stop growing when entering in the treated zone and the graphite seems to evolve without interruption from flake to pure compact graphite. Previous work by Liu *et al.,* see Fig.9, ref. [11], shows the same evidences using optical micrography. In our case the deep etching used with SEM reveals that the compact graphite seems to be composed of many faceted crystals.

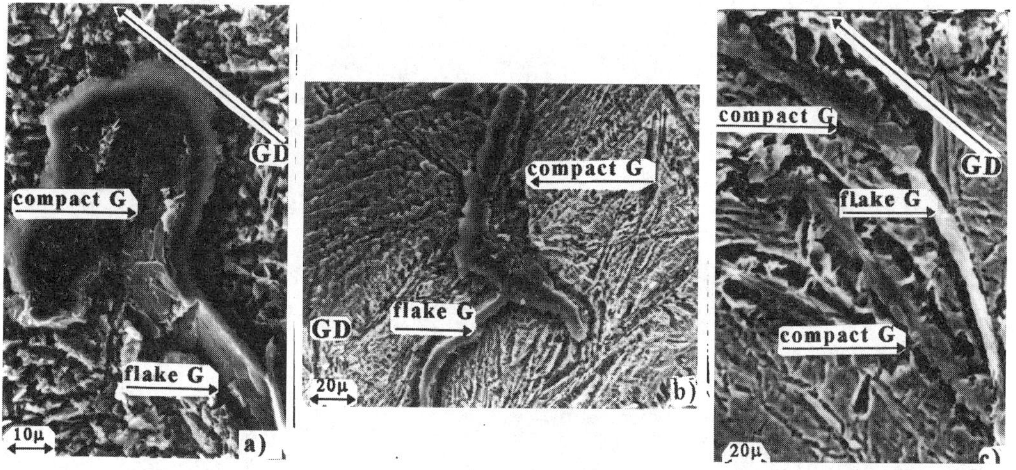

Fig.6. a), b) and c) different examples of a continuous change from flake to compact type graphite. In all the cases both structures are indicated.

3. Compact graphite microstructure

It was observed that some wormlike graphite phase stop growing , Fig.6 a) and b), but others keep growing with a strong alignment with the growth direction resulting in a microstructure as observed in Fig. 7a). In all cases the compact graphite is highly faceted as can be seen in Figs. 6 a)-c).
On the other hand, during vermicular unidirectional growth G is oriented in the solidification direction but some transitions occur as is shown in Fig.7a) and b). After the transition the size and spacing among wormlike graphite phase change. The nature of the structure at the transition zone marked in Fig.7a), Fig.8a), and b) is considered relevant. Figs. 9 a) and b) show, with higher magnification the nature of the flakes emerging from the wormlike graphite phase. They consist of planar faceted interconnected crystals having similar characteristics as the "foliated dendrites" described by Saratovkin [12] and quoted by Flemings [13], Fig.9c).

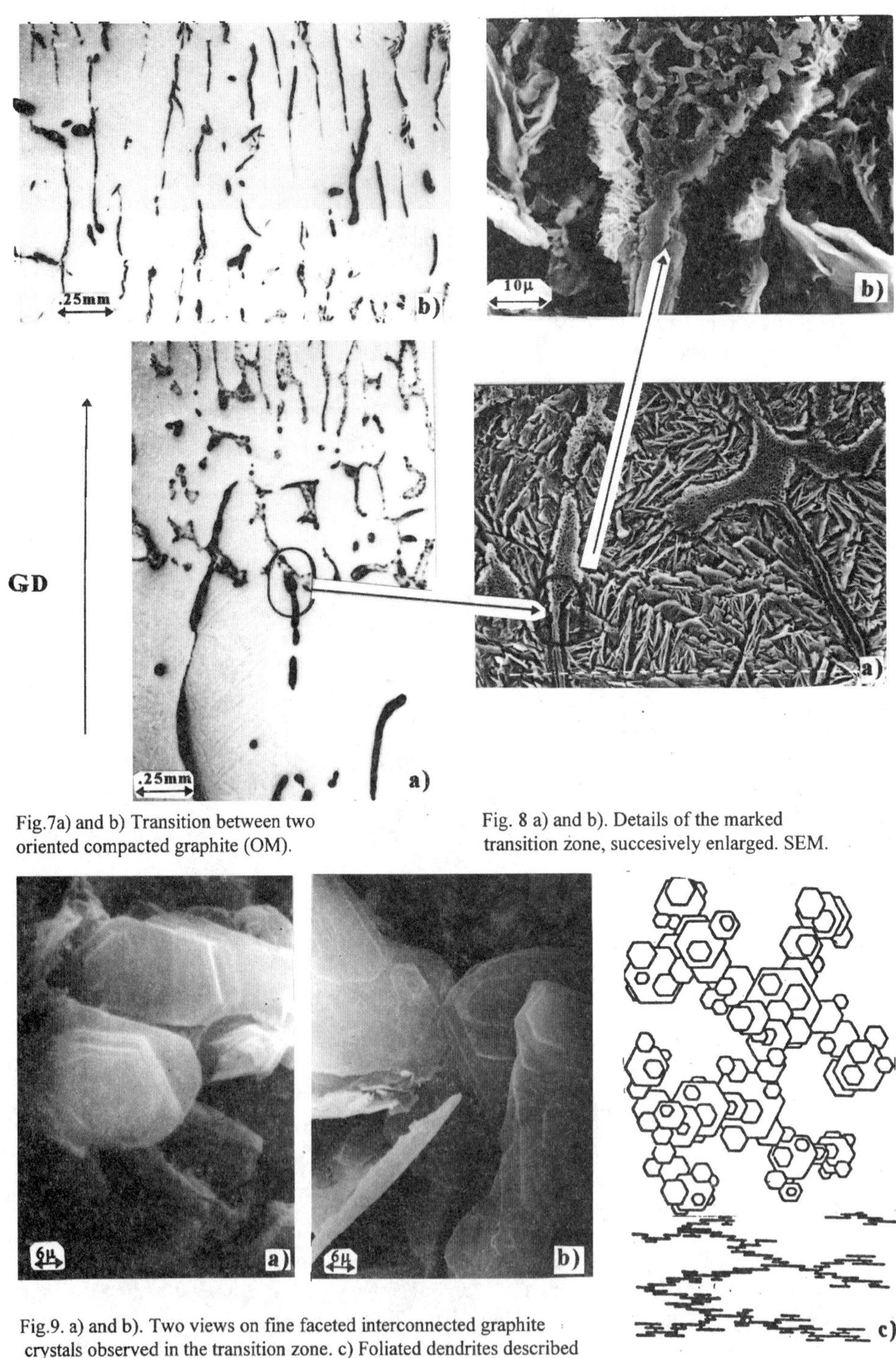

Fig.7a) and b) Transition between two oriented compacted graphite (OM).

Fig. 8 a) and b). Details of the marked transition zone, succesively enlarged. SEM.

Fig.9. a) and b). Two views on fine faceted interconnected graphite crystals observed in the transition zone. c) Foliated dendrites described by Saratovkin [12] and quoted by Flemings [13].

4. Frozen interfaces and growth mechanism of the compact graphite

Finally, a detail of the frozen S-L interface, obtained by the interruption of compact graphite directional solidification is shown in Figs. 10 a) and b), corresponding to optical and SEM micrographs, respectively. In the quenched "liquid zone" foliated dendrites type were found again, with the same characteristics as shown in Fig. 9 a) and b).

Fig.10 shows an austenite grain boundary groove and a γ morphological instability. In this case, the triple contact γ-G-L characteristic of the flake graphite growth morphology has been lost, probably due to chemical effects on surface tension of these phases when "in-situ" modification is performed. As a consequence a progressive *"compactation"* of free growing foliated dendrites type inside the γ-L groove would be the mechanism of the formation of compact graphite. This is supported by the results of Roviglione and Hermida [8]. These authors demonstrate the random oriented polycrystalline nature of the wormlike graphite phase.

Fig.11 shows partially compacted graphite inside different grooves of the γ phase. X-ray analysis [8] shows the random crystallographic orientation of these crystals compacted by the solidification of the γ walls of the groove.

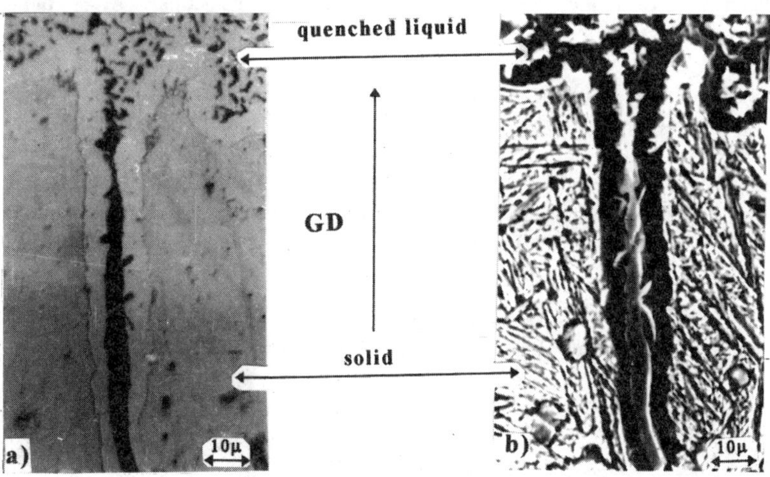

Fig.10. Frozen S-L interface during the growth of oriented compact graphite. a) OM b) SEM. The instability of the γ phase is noticiable giving as a result a groove in the grain boundary region

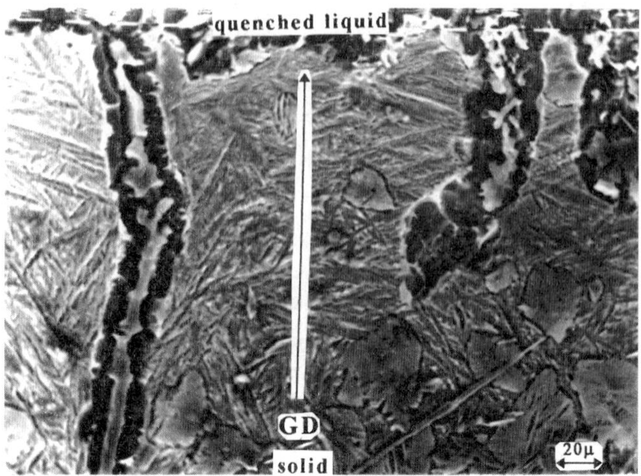

Fig.11. Foliated crystals partially compacted into γ grooves in a region of the frozen S-L interface

CONCLUSIONS

1.In free flake eutectic grains the fitness of curvature and branching zones of graphite may be accounted for by a high density of prismatic steps that leave the $(00.2)_G$ planes in contact with γ planes belonging to the <110> crystallographic zone as has been proposed for unidirectional solidification [8].

2.Compact graphite evolves without interruption from flake graphite.

3.During oriented growth morphological changes of compact graphite have been observed.

4.Compact graphite is composed by a large amount of fine and faceted crystals randomly oriented.

5.The basic element of the compact graphite seems to be the compacted foliated dendrites type graphite. In the present work the latter can be observed in the transition between different compact morphologies and in the frozen S-L interfaces where the *"compactation"* process is partial.

ACKNOWLEDGEMENTS

To Prof. J.D.Hermida for helpfull discussions. To the Argentine National Research Council (CONICET, PID 121) and Buenos Aires State Research Council (CIC) for partial support.

REFERENCES

[1] I.Minkoff, in "The Physical Metallurgy of Cast Iron", John Wiley & Sons; (1983)

[2] R.Elliott, in "Cast Iron Technology", Butterwords, (1988).

[3] D.M.Stefanescu, in "Metals Handbook", 9th.Ed., vol.**15**, "Casting", 168 (1988).

[4]K.D.Lakeland and L.M.Hogan, in "Recent Research on Cast Iron", Proceedings Seminar American Society for Metals, Detroit, Gordon & Breach Publ., 417 (1964).

[5] Ch.Yoshida *et al.*, Transactions ISSJ, **25**, 40, (1985).

[6] T.Calberg and H.Fredriksson, Proc. of International Conference in Solidification, Sheffield, The Metals Society, 115 (1977).

[7] D.D. Double and A.Hellawell, Acta Metall., **17**,1017 (1969).

[8] A.N.Roviglione, J.D.Hermida, Materials Characterization, **32**, 127 (1994).

[9] A.N.Roviglione, Materials Characterization, **31**, 209 (1993).

[10] H.Fredriksson and S.E.Watterfall, The Metallurgy of Cast Iron, eds. B.Lux, I.Minkoff, F.Mollard, Georgy Publ. Co., Switzerland, 278 (1975).

[11] Y.X.Li, B.C.Liu and C.R.Loper, AFS Transations, 90-112, **98**, 483 (1990).

[12] D.D.Sarovkin, in "Dendrite Solidification", Consultants Bureau Trans., New York, (1959).

[13] M.C.Flemings, in "Solidification Processing", McGraw Hill, N.Y., (1974).

Contribution to the Effects of Magnesium-Treatment, Inoculation and Section Thickness on the Spherulite Formation in Spheroidal Graphite Cast Iron

R. Doepp[1], B. Prinz[2], K.J. Reifferscheid[3], E. Schuermann[1] and T. Schulze[2]

[1] Institut für Eisenhüttenkunde und Giessereiwesen der Technischen Universität Clausthal, D-38678 Clausthal-Zellerfeld, Germany

[2] Zentrallaboratorium der Metallgesellschaft AG, D-60015 Frankfurt/Main, Germany

[3] SKW Giesserei-Technik GmbH, D-65760 Eschborn, Germany

Keywords: Spheroidal Graphite Cast Iron, Thermal Analysis, Inoculation, Fading Time, Nodule Number

ABSTRACT

Uniformly produced SG iron melts (CE = 4,3) have been cast into test plate specimens with 3 to 30 mm section thickness and into wedge test pieces with and without mould inoculation. Examination of the tested material by thermal analysis and structural investigation: solidification kinetics and inoculation effect (influence of type and quantity of inoculant); dissolution behaviour of mould inoculant blocks; time dependent inoculating effect; combination of treatment and inoculation by Mg_2Si. Possibilities for optimization of spherulite number and grey solidification on the one hand and fracture toughness on the other hand.

1. INTRODUCTION

Production of spheroidal graphite cast iron within the last 40 years increased up to about 30 % of the total production of iron castings in Germany (1993 899.000 t of 2.939.000 t = 30,6 %) [1]. Including austempered ductile cast iron (ADI) mechanical properties of nodular cast iron now range in the field of steel castings and steel [2]. Concerning casting and feeding advantages and heat transfer properties spheroidal graphite cast iron is similar to flake graphite cast iron. Its position between steel and grey cast iron generally corresponds to the position of malleable cast iron. Mainly whiteheart malleable cast iron has excellent mould filling properties.

Typical production stages are

- magnesium treatment for spherulization and
- inoculation for graphitization.

Furthermore cooling rate influences solidification behaviour, structure and properties.

2. OBJECTIVE

Aim of the present study was to examine

- type of the Mg-treatment,
- type and amount of inmould inoculant,
- influence of section thickness in plates and wedges,
- influence of pouring time (wedge specimens).

3. METHODS

Type and amount of charge materials, melt conduct and chemical composition (CE = 4,3) should be as constant as possible. Details see __Table 1__ and __Table 2__.

Serie I. Wanddickenproben – Kapitel 4.1 und 4.4
Serie II. Gießkeilproben und Wanddickenproben – Kapitel 4.2 und 4.3

Gattierung:	29000 g Roheisen (Sorel)
	1900 g Elektrolyteisen
	30 g (Serie I) bzw. 90 g (Serie II) Elektrolytmangan
	440 g FeSi75
	20 g Aufkohlungsmittel (Desulco, geringer Gehalt an Stickstoff und flüchtigen Bestandteilen)
Erschmelzung:	MF-Induktionsofen-Schmelzanlage mit Ton-Graphittiegel Schmelzegewicht ca. 32 kg
Magnesium-behandlung:	Tauchbehandlung mit 200 g (0,6 %) Vorlegierung bei Serie I Vorlegierung VL 55(M) – Tafel 2 bei Serie II Vorlegierung VL 55(O) oder VL 55(M)
Impf-behandlung:	bei Serie I: Formimpfkörper (Germalloy). 22 g – Tafel 2 oder Pfannenimpfung mit Granulat VP 216 bei Serie II: Formimpfung mit Germalloy

Erzielte chemische Zusammensetzung (in %)

C	3.5–3.6	Ce	< 0.020
Si	2.2–2.3		bei VL 55(O)
Mn	0.1 (I) bzw. 0.3 (II)		0.005–0.006
P	0.02		bei VL 55(M)
S	0.005	Ti	0.03–0.04
Mg	0.04–0.05	Sb	< 0.0004
Cr	0.05	Sn	< 0.0008
Ni	0.05	Pb	< 0.0020
Mo	0.02	Bi	< 0.0007
Cu	0.04	As	< 0.0010

Table 1. Charge material, melting, melt treatment applied to the casting of wall thickness test specimens and of wedge test pieces in the pouring car test

	Fe	Si	Al	Ca	Mg	SE	CeMM	Ti	Mn	Sr
VL 55(O)	5,7	56,8	0,35	3,6	31,2	0,03	–	–	–	–
VL 55(M)	5,7	56,8	0,35	3,6	31,2	0,03	1	–	–	–
Germalloy	22	72	4	0,6	–	–	–	0,05	–	–
Legierung A	21	73	1,5	0,5	–	–	–	0,06	4	–
Legierung B	30,8	66,7	0,16	0,04	–	–	–	n.b.	–	0,85
Legierung C	23	74,3	0,43	0,12	–	–	–	–	–	–

Table 2. Chemical composition (in weight %) of additions for modification and of mould inoculants used for wedge test series

Main investigation procedures are thermal analysis [3-21] and microstructural examination [22].

Two new experimental equipments were used [23,24,25]:

- the double stationary plate specimen with section thicknesses between 3 and 30 mm (__Fig. 1__) and

Figure 1. Double wall thickness test sample with two parallel treatment and runner systems

- the double movable wedge specimen (__Fig. 2__).

They allow to determine

- temperature-time cooling curves including minimum (T_{EU}) and maximum (T_{ER}) values of eutectic transformation (__Fig. 3__),
- nodule number per mm^2,
- grey structure respectively chill tendency,
- fading time (cooling time up to T_{EU}).

4. RESULTS

Minimum (T_{EU}) and maximum (T_{ER}) values of eutectic transformation increase with increasing section thickness (__Fig. 4__). Plotting these values against reciprocal values of section thickness (__Fig. 5__) and extrapolating to 1/section thickness = 0 (zero) one receives the theoretical eutectic temperature of the stable equilibrium T_E corresponding to the actual conditions, mainly following nodule number (in Fig. 5 two groups).

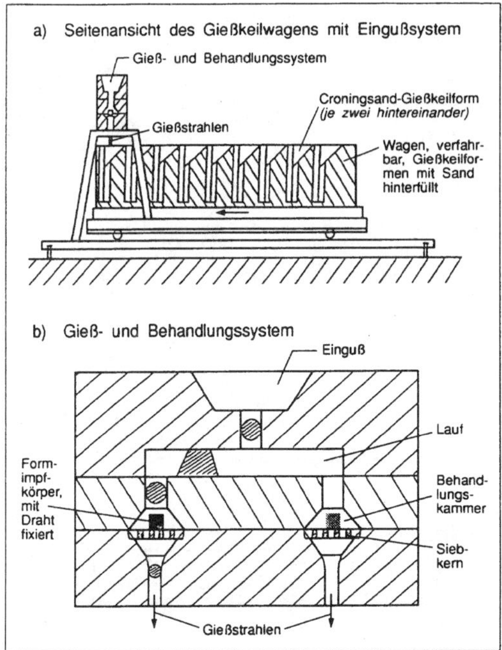

Figure 2. Experimental arrangement (pouring car test) for the determination of the dissolution behaviour of mould inoculants

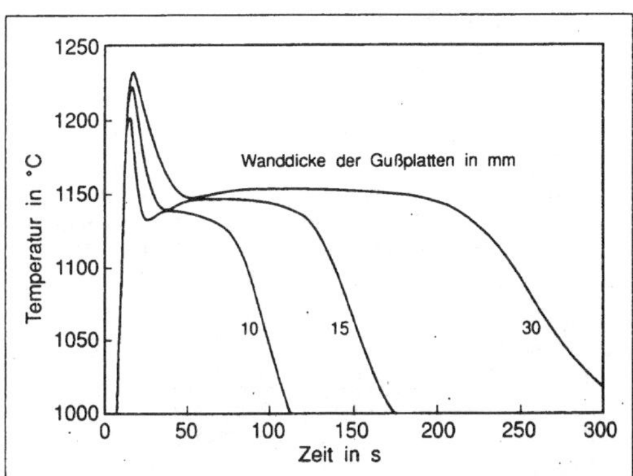

Figure 4. Cooling curves of test plates with different thickness taken from the same melt (after treatment and inoculation)

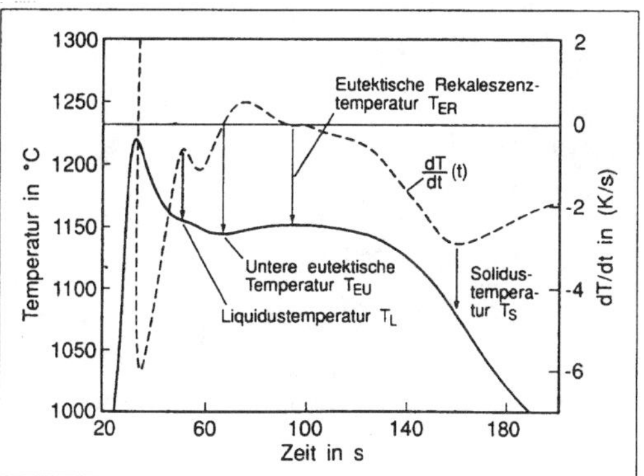

Figure 3. Typical cooling curve of a SG iron test plate (15 mm) and its first derivative dT/dt

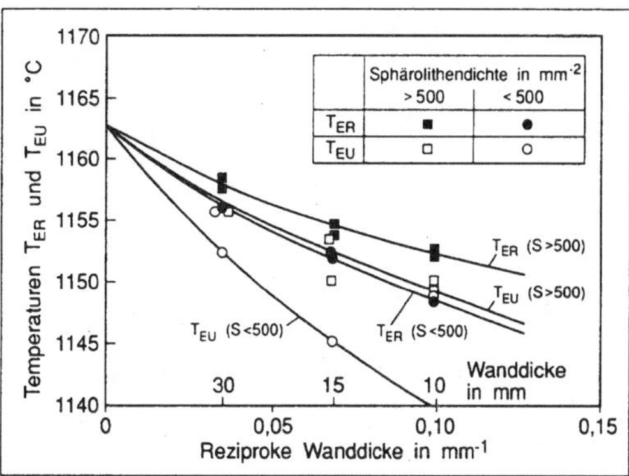

Figure 5. Maximum (T_{ER}) and minimum (T_{EU}) eutectic temperature correlated with spherulite density and section thickness

The difference between the extrapolated equilibrium values T_E and the measured minimum values T_{EU} are the maximum undercooling values ΔT_{EU} (**Fig. 6**). This difference increases with increasing cooling rate and decreasing nodule number. Nodule number corresponds to the nucleation state of the melt.

Increasing amount of the same graphitizing inoculant increases T_E (**Fig. 7**). This is due to the higher amount of silicon. Manganese and cerium decrease T_E.

Mould inoculation can be evaluated by measuring chill tendency in wedge specimens. Examples are shown in **Fig. 8**. Best results are coming from inoculant C (compare Table 2) because chill tendency remains low for the longest time. In this case for instance practical pouring time should not exceed about 8 s.

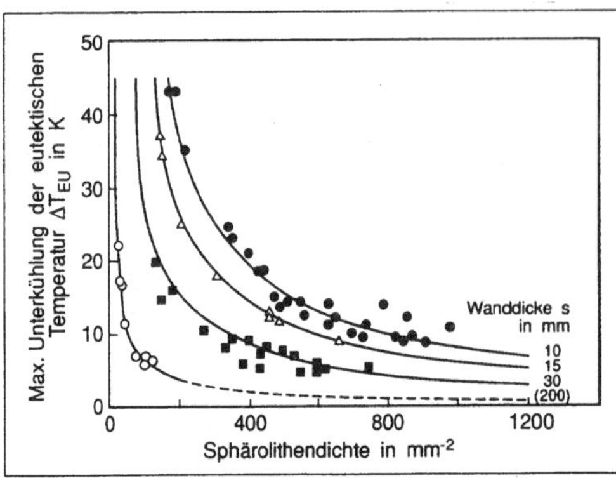

Figure 6. **Maximum eutectic undercooling** ΔT_{EU} **correlated with spherulite density and section thickness**

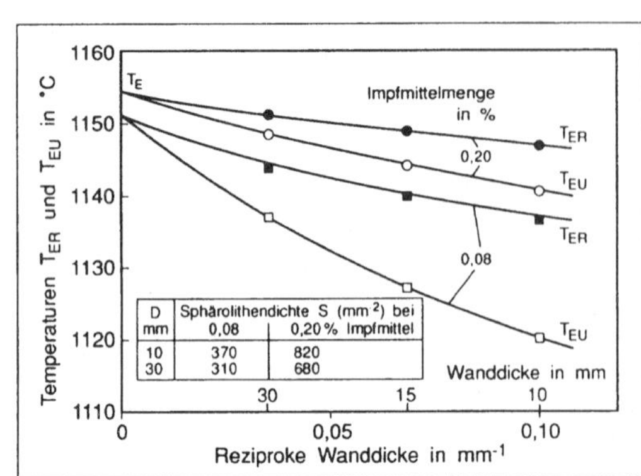

Figure 7. **Correlation between the maximum (T_{ER}) and minimum (T_{EU}) eutectic temperature and the section thickness and quantity of inoculant as measure of the nucleation state of the melt**

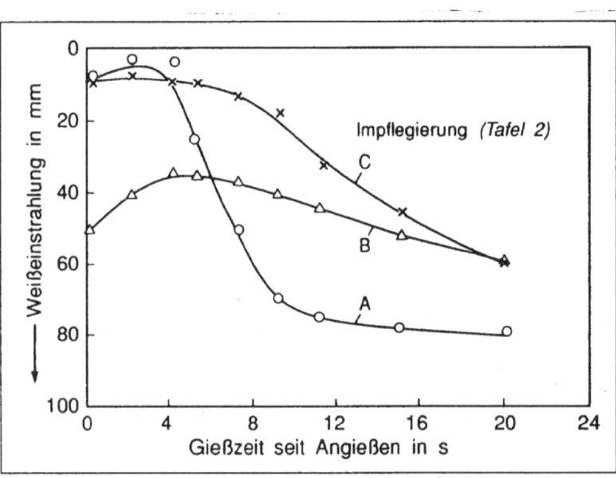

Figure 8. **Chill on wedge test pieces as measure of the time depending effect of mould inoculants**

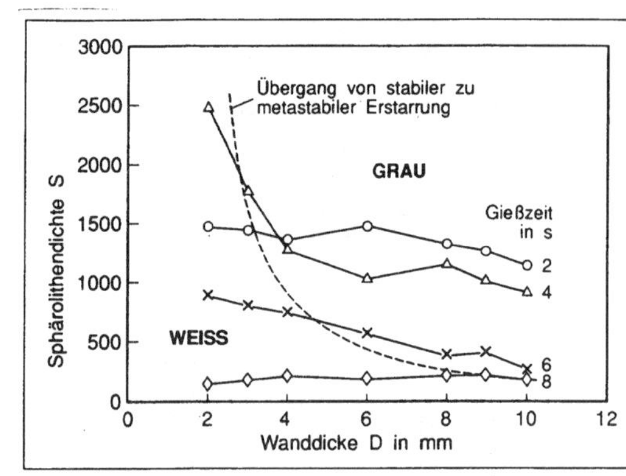

Figure 9. **Spherulite density (per mm^2) in the structure of wedge test pieces in correlation with the section thickness and the pouring time (= pouring time after test start)**

Fig. 9 shows that grey structure in a distinct section size needs certain maximum values of pouring time and minimum values of nodule number per square millimeter. The maximum time represents maximum time of mould filling. Nodule number decreases significantly with increasing pouring time and slightly with increasing section thickness.

The influence of amount of the same type of mould inoculant and the influence of section size are shown in **Fig. 10**.

In **Fig. 11** section size is replaced by cooling time up to T_{EU}. Mould inoculation is compared to a certain type of ladle inoculation. The difference may be less in case of other ladle inoculation techniques.

Fading time increases with increasing section size of test plates (**Fig. 12**). Nodule number normally decreases.

Initial nodule number S_0 increases with increasing amount of inoculant (Fig.13).
Actual nodule number S depends on S_0, fading time t_{EU} and fading exponent a
following the equation

$$S = S_0 \cdot (t_{EU})^a$$

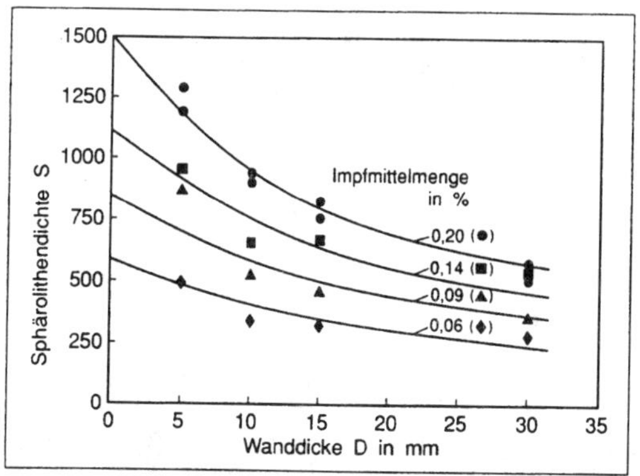

Figure 10. The spherulite density increases with a decreasing section thickness and an increasing
addition of inoculants

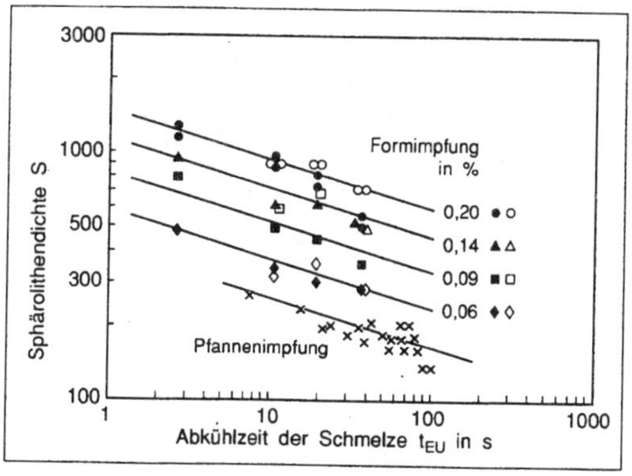

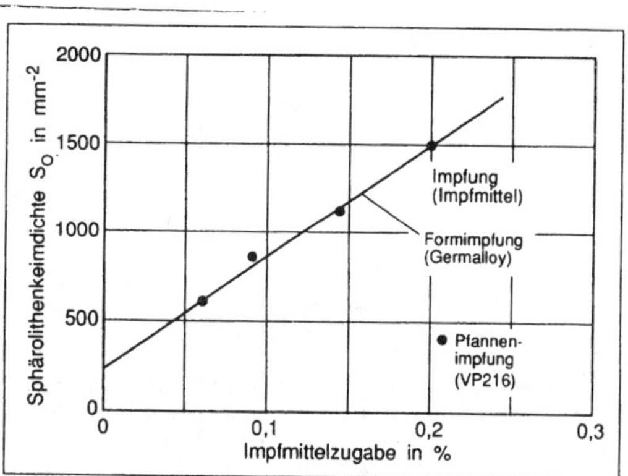

Figure 11. Variation of spherulite density S during the
cooling time of the melt t_{EU} (time until T_{EU}
is reached) with different inoculation treatments

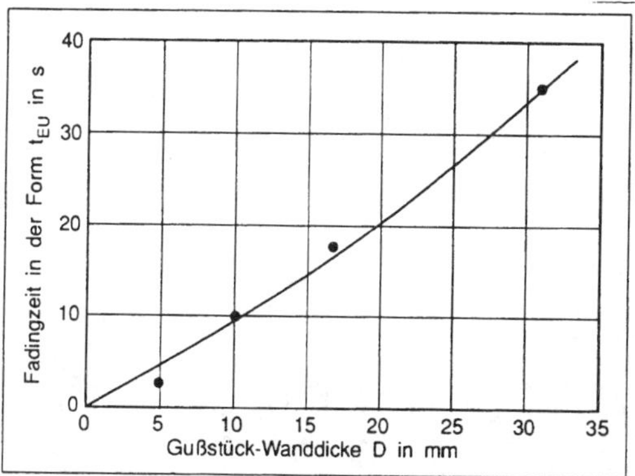

Figure 12. Correlation between fading time t_{EU} (time
until T_{EU} is reached) and the section thickness of the test plates

Figure 13. The initial density of spherulite nuclei S_0
(values from Figure 11) increases with increasing additions of inoculant; the nucleation
state with mould inoculation is better here
than with ladle inoculation

Mg_2Si as a combined material for treatment and inoculation increases nodule
number (Fig. 14a) and degree of completion in nodule shape (Fig. 14b).

5. CONCLUSIONS

Thermal analysis and microstructural examination are excellent ways to determine
details of solidification kinetics of spheroidal graphite cast iron. Stable
equilibrium eutectic temperature as the upper limit of the eutectic solidification range could be determined quantitatively. Together with provided measurements of metastable eutectic temperatures an actual information gap may be
closed in next years.

Furthermore, chill tendency in wedges as function of inoculation and pouring time could be determined. Fading time t_{EU} up to T_{EU} increases with section size and corresponds to decreasing nodule number.

Mg_2Si is an excellent but expensive combined material for Mg-treatment and inoculation.

Regarding nodule number in a certain section a compromise between sufficient graphitization and toughness will be useful [26-29].

Experiments should be completed by shrinkage and dilatation investigations [30-35].

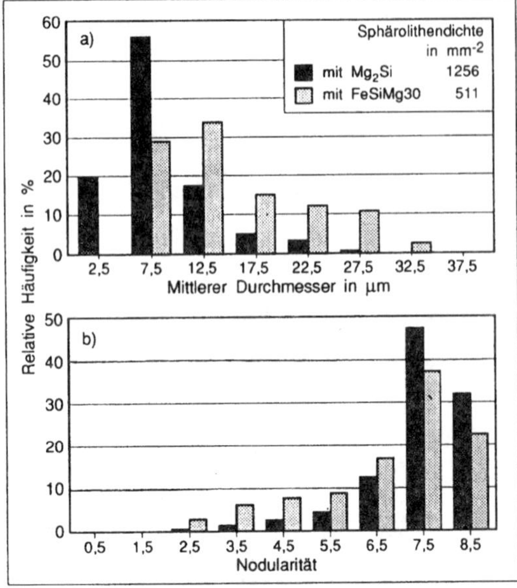

Figure 14. Relative frequency of a) the spherulite size, b) nodularity (10 = ideal spheroidal shape) after treating the melt with Mg_2Si or FeSiMg30 master alloy; mould inoculation with 0,2 % Germalloy (30 mm section thickness test pieces)

REFERENCES

[1]   Deutscher Giessereiverband (DGV), Düsseldorf (editor), Fakten und Daten Giesserei-Industrie 1993

[2]   Verein Deutscher Giessereifachleute (VDG), Deutscher Giessereiverband (DGV), Gesamtverband Deutscher Metallgiessereien (GDM), Düsseldorf (editors), Giesserei-Kalender 1993, 228

[3]   Piwowarsky, E.; Patterson, W.: Die Herstellung von Gußeisen mit Kugelgraphit. Vortrag 14. Aachener Gießerei-Kolloquium 9.-11.2.1950. - Giesserei, techn.-wiss. Beihefte 1950, Nr. 3, 91-103, s. bes. 99 - In: Piwowarsky, E.: Hochwertiges Gußeisen (Grauguß). 2. Aufl., Berlin-Göttingen-Heidelberg 1961, 244-245.

[4]   Scheil, E.: Giesserei, techn.-wiss. Beih., Nr. 24, April 1959, 1313-1338

[5]   Haensel, G.; Mitsche, R.; Dichtl, H.J.: Neue Untersuchungen über die Herstellung und Überwachung der Qualität von Gußeisen mit Kugelgraphit. 40. Internationaler Giesserei-Kongress Moskau, UdSSR, 1973. Vortrag Nr. 95

[6]   Titze, E.; Dichtl, H.J.: Giesserei 61 (1974), Nr. 20, 619-622

[7]   Knothe, W.: In: Praxis der thermischen Analyse zur Qualitätsüberwachung von Eisenschmelzen. Lehrgang Verein Deutscher Giessereifachleute, Düsseldorf, und Institut für Eisenhüttenkunde und Giessereiwesen der Technischen Universität Clausthal am 23.5.1986 und 8.5.1987 in Clausthal, Teil II, Bild 18

[8]   Esin, B.: Beitrag zur thermischen Analyse von Gußeisen mit Kugelgraphit.
      Diplomarbeit Institut für Eisenhüttenkunde und Giessereiwesen (R. Döpp)
      der Technischen Universität Clausthal, durchgeführt im Zentrallaboratorium
      (B. Prinz, K.J. Reifferscheid) der Metallgesellschaft AG, Frankfurt/Main,
      1986. In: Döpp, R.: Grundlagen der thermischen Analyse von Metallschmel-
      zen, besonders Kupfer und Aluminium. Schriftenreihe der GDMB, Clausthal-
      Zellerfeld. Heft 55 (1989), 19-60

[9]   Rödter, H.: Qualitätsverbesserung und Produktivitätserhöhung durch Lenkung
      der Erstarrungsmorphologie bei Gußeisen mit Kugelgraphit. Vortrag Große
      Giessereitechnische Tagung des Vereins Deutscher Giessereifachleute
      20.-22.5.1992 in Düsseldorf; s. Giesserei 79 (1992), Nr. 16, 686-687

[10] Bäckerud, L.; Nilsson, K.; Steen, H.: Study of nucleation and growth of
      graphite in magnesium-treated cast iron by means of thermal analysis.
      In: The Metallurgy of Cast Iron. Proc. Second International Symposium on
      the Metallurgy of Cast Iron, Geneva, Switzerland, 1974. St. Saphorin,
      Switzerland 1975, 625-637

[11] Bäckerud, L.: Die Bäckerud-Technologie macht Gußeisen zum Werkstoff der
      Zukunft. Vorträge Seminar der SinterCast GmbH "Gußeisen - der Werkstoff
      der Zukunft". Bad Homburg v.d.H., 22.1.1992;
      s. Herfurth, K.: Giesserei 79 (1992), Nr. 16, 691-692

[12] Döpp, R.: Diskussionsbeitrag zu [11] unter Bezug auf Bilder 3a, 3b und 1

[13] Radjef, Y.; Döpp, R.: Einige metallurgische Einflüsse bei Gußeisen mit
      Kugelgraphit - Einfluß der Gattierung und der Behandlungstemperatur auf
      Abkühlungsverlauf, Gefüge, Lunkerverhalten und Härte.
      Vortrag 13. Clausthaler Giesserei-Kolloquium 4./5.10.1990;
      s. Giesserei 78 (1991), Nr. 2, 67-68

[14] Sivula, J.; Rantala, T.; Louvo, A.: Possibilities of controlling the
      quality of SG-irons by computerized thermal analysis. Vortrag Inter-
      nationale Konferenz "Simulation, Designing and Control of Foundry Pro-
      cesses", Krakau, Polen, 9.-13.6.1986. Preprints 240-252

[15] Marincek, B.: Giesserei-Praxis 1982, Nr. 13/14, 217-229;
      Nr. 15/16, 235-246

[16] Monroe, R.; Bates, C.E.: Transactions American Foundrymen's Society 82
      (1974), 619-622. Giesserei-Praxis 1983, Nr. 4, 53-58

[17] Otte, W.: Giesserei-Erfahrungsaustausch 30 (1986), Nr. 2, 54-62

[18] Döpp, R.: Beitrag zur thermischen Analyse von Eisenlegierungen. Mittei-
      lungsblatt der TU Clausthal 1987, Nr. 64, 5-10

[19] Döpp, R.: Thermal analysis in ferrous foundries. In: Proceedings Workshop
      53rd International Foundry Congress, Prague, Czechoslovakia, 1986.
      Dordrecht-Boston-Lancaster 1987, 39-52

[20] Menk, W.; Speidel, M.O.; Döpp, R.: Giessereiforschung 44 (1992),
      Nr. 3, 95-105

[21] Döpp, R.; Blankenagel, D.; Lindemann, K.: Beitrag zur thermischen Analyse
      von Gußlegierungen. Giesserei-Praxis 1994, Nr. 7, 141-148

[22] Berghof-Hasselbächer, E.; Prinz, B.; Ribbens, A.; Schulze, T.: Praktische
      Metallographie, demnächst

[23] Reifferscheid, K.J.: Beiträge zur Metallurgie des Gußeisens mit Kugel-
      graphit. Dissertation Technische Universität Clausthal 1991

[24] Döpp, R.; Prinz, B.; Reifferscheid, K.J.; Schürmann, E.; Schulze, T.:
      60. Giesserei-Weltkongress, Den Haag, Niederlande, 1993, Vortrag Nr. 43

[25] Döpp, R.; Prinz, B.; Reifferscheid, K.J.; Schürmann, E.; Schulze, T.:
      Giessereiforschung 46 (1994), Nr. 1, 13-22

[26] Motz, J.M.; Berger, D.; Cohrt, G.; Godehardt, E.K.; Hüttebräucker, K.;
      Kuhn, G.; Reuter, H.; Schock, D.; Shakeshaft, W.; Wolters, D.:
      Giessereiforschung 32 (1980), Nr. 3, 97-111; Konstruieren + Giessen 6
      (1981), Nr. 1, 10-21

[27] Wolfensberger, S.; Uggowitzer, P.; Speidel, M.O.: Giessereiforschung 39
      (1987), Nr. 1, 17-24; Nr. 2, 71-80

[28] Pusch, G.; Liesenberg, O.; Stehmann, S.; Rehmer, B.: Giessereiforschung 42
      (1990), Nr. 2, 64-73

[29] Pusch, G.: Konstruieren + Giessen 17 (1992), Nr. 3, 29-35; Nr. 4, 4-12; 18
      (1993), Nr. 1, 4-11

[30] Hummer, R.: Giesserei-Praxis 1985, Nr. 17/18, 241-254

[31] Hummer, R.: Cast Metals 1 (1988), Nr. 2, 62-68

[32] Nandori, G.; Dul, J.: Untersuchungen der Längen-Temperatur-Änderungen der
      grauen und weißen Gußeisenlegierung während der Kristallisation. 40. In-
      ternationaler Giesserei-Kongress, Moskau, UdSSR, 1973, Vortrag Nr. 12

[33] Vigh, L.; Dul, J.; Szabo, Z.: Computerized analysis of the measurable
      processing variables influencing the dimensional accuracy of castings
      during the cristallization of cast iron. 57th World Foundry Congress,
      Osaka, Japan, 1990, report 20

[34] Yang, Y.; Alhainen, J.: Microstructure estimation of SG iron with a
      dilatometer. Institute of British Foundrymen, Proc. CASTCON '91,
      Harrogate, UK, June 6-7, 1991

[35] Yang, Y.; Alhainen, J.: Functional comparison of dilatation and thermal
      analysis in cast iron solidification studies. In: Advanced casting and
      solidification technology 1994. COST 504 Conference September 12-13, 1994,
      Espoo, Finland. European Cooperation in the Field of Scientific and Tech-
      nical Research. Editors H. Kleemola, B. Pukl. European Commission. Techni-
      cal Research Centre of Finland. Proceedings 1994, 281-293

Advanced Materials Research Vols. 4-5 (1997) pp. 385-390
© 1997 Scitec Publications, Switzerland

The Isothermal Transformation of Ductile Cast Iron

M.C. Leijten, H. Nieswaag and L. Katgerman

Delft University of Technology, Laboratory of Metallurgy,
Rotterdamseweg 137, NL-2628 AL Delft, The Netherlands

Keywords: Austempered Ductile Iron, Phase Transformation, Kinetics

Abstract Dilatation experiments were performed on unalloyed and alloyed cast irons. The resulting isothermal transformation diagram shows a remarkable difference between upper and lower bainite. Especially the results from upper bainite are very complex and two 'noses' in the IT-diagram representing two different morphologies of upper bainite can be seen. This results from the stepwise transition from upper to lower bainite-morphology as the austempering temperature decreases. The kinetics of the austempering reaction have also been determined.

Introduction

For the analysis of transformation kinetics a physical property (e.g. hardness, specific volume/length) of the material subject to investigation can be traced as a function of time and temperature. In the past a number of methods have been proposed to evaluate kinetic reaction parameters from such data. For example, the Johnson-Mehl-Avrami (JMA) theory has been developed to describe the progress of solid-state reactions during isothermal transformation, although the limitations of this theory are generally recognized [1]. This paper describes the kinetic analysis of the solid-state transformation of austenite to bainitic ferrite in ductile cast iron. In view of recent instrumental developments allowing high accuracy in data accumulation and processing, dilatometry can be very usefully applied for the analysis of kinetic reaction parameters.

Experimental Procedure

As-cast material

The chemical composition of the ductile irons studied are listed in Table I. The materials were cast into different samples. Image analysis techniques were used to measure the volume fraction of ferrite (f_α) and the nodule count in the as-cast structure.

Table I: Chemical composition of ductile irons

material	%C	%Si	%Mn	%Ni	%Cu	% Mo	nodule count (/mm²)	f_α
alloy no.0	3.62	2.86	0.033	-	-	-	100	0.87
alloy no.1	3.60	2.52	0.14	0.95	0.72	0.30	250-350	0.16
alloy no.2	3.62	2.46	0.14	1.49	0.76	0.34	250-350	0.16

Dilatometric study

The transformation has been registered with two different dilatometers. The first one is a Theta-dilatometer using specimens 3mm in diam. and 10mm high, which were machined of alloy no.0 and of alloy no.1. In the Theta-dilatometer the specimen is heated by induction in vacuum for austenitization. The quench to the transformation temperature T_b occurs by the inlet of helium in the chamber of the dilatometer. Secondly a Gleeble-dilatometer was used in which the specimen is heated by resistance heating using an electrical current through the specimen. The specimen used were 10 mm in diam. and 900 mm length machined from alloy no.1 and alloy no.2. Since the cooling in the Gleeble-dilatometer is slow (± 25 s from austenitization to austempering temperature) only alloyed cast irons with an estimated incubation time of more than 25 s can be used, otherwise transformation will start during cooling down. The displacement transducer of the dilatometer was calibrated by measuring the expansion coefficient of pure Ni-rods. Three austenitization temperatures T_a were chosen, namely 820, 860 and 900°C, with a soaking time of 1 hour, allowing a comparison to practical application. Isothermal transformation measurements were performed in the temperature range from 200-440°C.

<u>Metallography</u>

To obtain a good contrast between ferrite and austenite the specimens were etched in a Beraha etching which consist of a mixture containing 1000 ml H_2O dest. and 200 ml HCl conc., 24 g ammoniumbifluorid to which is added 18 g kaliumdisulfit [2]. Image analysis techniques (CBA-technique) were used to measure the volume fraction of retained austenite in the heat treated condition.

Theoretical Analysis

The initial matrix structure of most cast irons, prior to heat treatment, is either ferrite or pearlite, or a mixture of both phases. During heating above the metastable transformation temperature, the pearlite will initially transform to austenite. This occurs first in the intercellular regions of the matrix. Transformation of ferrite occurs more slowly and relies upon diffusion of carbon from graphite as to decrease the transformation temperature, so the nucleation of austenite at the ferrite/graphite interface is affected. Because the amount of carbon that should be present in the austenite after austenitization [$\%C_\gamma = (T_a/420)-0.17.wt\%Si-0.95$ [3]] is in most cases greater than the amount of carbon present in the as-cast matrix (ferrite and pearlite), graphite has to dissolve in the matrix and the mass fraction of it will decrease.

During austempering ferrite plates will grow into the austenite grains, after their nucleation on suitable sites at the grain boundaries and at the interface between austenite and graphite. The morphology of the ferrite depends on the transformation temperature. In the upper-bainite region (appr. > 350°C) the ferrite is lath-like white, in the lower-bainite region the ferrite is finer and more plate-like and has a martensite-like appearance. Since the carbon solubility in ferrite is low the excess of carbon present in the ferrite is pushed into the austenite. Consequently, the carbon content in the austenite increases and because silicon retards the precipitation of carbides from the austenite the carbon content in the austenite increases to a higher level resulting in an amount of non-transformed austenite at room temperature. This is the so-called retained austenite. In the lower-bainite region the excess of carbon present in the ferrite is only partially pushed to the austenite, due to the low diffusion coefficient. Carbides are formed within the ferrite even in cast iron; the amount of non-transformed austenite is therefore less than in the upper-bainite region. The here described processes can be distinguished as stage I of the transformation:

$$\gamma_o \rightarrow \alpha + [\gamma] \qquad\qquad (T_b > 350°C) \qquad\qquad\qquad 1a$$
$$\gamma_o \rightarrow (\alpha + carbides) + [\gamma] \qquad (T_b < 350°C) \qquad\qquad\qquad 1b$$

The symbols γ_o and $[\gamma]$ respectively represent the austenite after the austenitization and the carbon-enriched non-transformed austenite. At long austempering times it is not possible anymore to retard the precipitation of carbides from the austenite completely. This stage II in the process can be written as:

$$[\gamma] \rightarrow \alpha + carbides \qquad\qquad (450°C > T_b > 250°C) \qquad\qquad\qquad 2$$

For the analysis of solid-state transformation kinetics a physical property (e.g. specific volume/length) of the material subject to investigation can be traced as a function of time and temperature. The Johnson-Mehl-Avrami (JMA) equation for heterogenoues reactions is used [1,4]:

$$f = 1 - \exp\{-(kt)^n\} \qquad\qquad\qquad 3$$

with f the degree of transformation, n the order of the reaction or the morphology-dependent constant, k a function of nucleation rate and growth rate according to the Arrhenius equation:

$$k = k_0\exp(-E/RT) \qquad\qquad\qquad 4$$

with E the effective activation energy; t, k_0, R and T denote the transformation time, the pre-exponential factor, the gas constant and the absolute temperature, respectively. A plot of ln(k) versus 1/T yields values for the pre-exponential factor k_0 and the activation energy E. A modified JMA-equation was used to calculate the transformed fraction of ferrite during the stage I transformation [6]:

$$f = V_\alpha/\theta = 1 - \exp\{-(kt)^n\} \qquad\qquad 5$$

with V_α the volume fraction of transformed ferrite during stage I and θ the maximum volume fraction of ferrite transformed at the end of stage I that can be achieved at a particular isothermal transformation temperature. θ is derived from the retained austenite volume fraction.

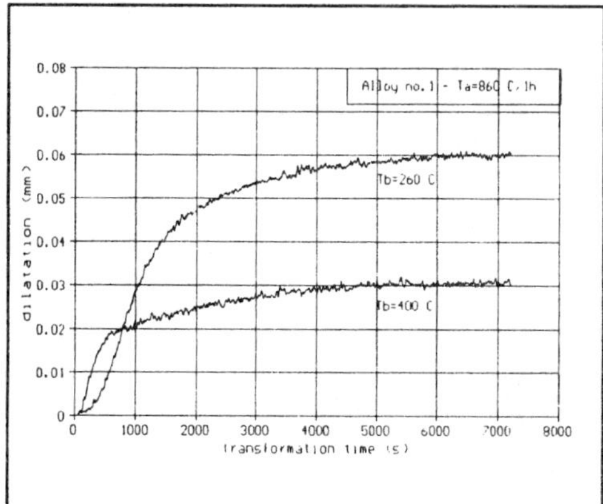

Fig.1. Dilatation during stage I of the isothermal transformation (Gleeble-dilatometer).

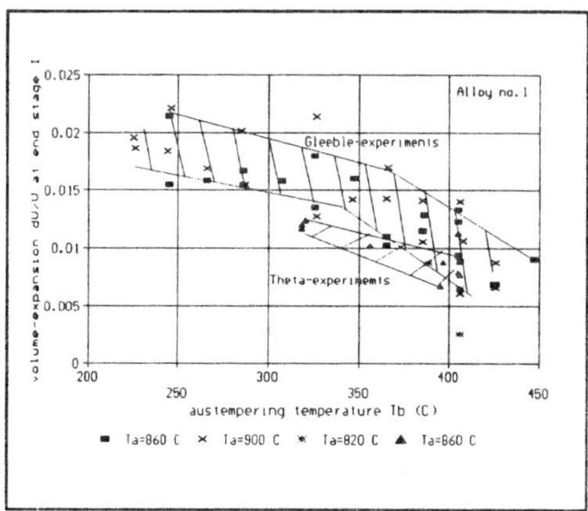

Fig.2. Volume-effect of stage I transformation (Gleeble- and Theta-experiments).

Results and discussion

<u>Dilatation experiments</u>

Fig.1 shows two typical dilatometer curves for alloy no.1 for transformation temperatures 400 and 260°C, thus in the upper- and lower-bainite region respectively. The dilatation is the summation of the volume-effects of the different subreactions during austempering [5]:

  a) $\gamma_o \rightarrow \alpha + [\gamma]$ $\Delta V > 0$
  b) increase of carbon in $[\gamma]$ $\Delta V > 0$
  c) formation of carbides $\Delta V > 0$ $\}\ \Delta V < 0$
  d) decrease of carbon in $[\gamma]$ $\Delta V < 0$

Subreaction c and d are only of importance for stage II transformation. There is however a great scatter in the maximum volume expansion from specimen to specimen (Fig.2) due to the inhomogeneity of cast iron (segregation of alloying elements, nodule count, amount of ferrite). Although the results of the Gleeble-experiments show a greater scatter than the results of the Theta-experiments, the Gleeble-dilatometer is prefered because of better data accumulation and processing. So a better analysis of the dilatation-experiment can take place.

From dilatation curves it is possible to trace the start and the end of stage I transformation, respectively f=0.01 and f=0.99. Fig.3 shows the experimentally obtained isothermal transformation (IT) diagrams of two different alloys. The unalloyed cast iron, alloy no.0, has a minimum incubation time of 15 s. There is also a clear distinction between upper and lower bainite (transition temperature T_b=350°C). Because of alloying with 0.95%Ni, 0.72%Cu and 0.30%Mo (alloy no.1) the beginning of the transformation is shifted to longer times. The shape of this curve does not indicate a clear difference between upper and lower bainite. There appears to be an extra 'nose' between upper and lower bainite (360 < T_b < 320°C) [7]. The bainitic ferrite goes from lath- to plate-like structure (upper and lower bainite resp.) as the length/thickness ratio of the bainitic ferrite will increase. Also the amount of retained austenite decreases, the white phase in Fig.5a and c. There is an additional morphology observed between lower and upper bainite (Fig.5b). It looks like lower bainite but with a lower length/thickness ratio than usually appears in the lower bainite region. The influence of the austenitization temperature on the IT-diagram is shown in Fig.4, the beginning of the transformation shifts to the right and the martensite temperature is lowered, because the carbon content of γ_0 is higher.

The amount of retained austenite at the end of stage I is measured using image analyses techniques (Fig.6). Although there is also a great scatter in the results, at lower austempering temperatures there will be less retained austenite because of the lower diffusion coefficient of carbon, so more carbon will remain inside the transformed ferrite. From Fig.6 it is clear that the maximum temperature where lower bainite will appear is 300°C. As in upper bainite practically no carbon is dissolved into the ferrite contrary to lower bainite, Fig.6 makes a transition region evident between upper and lower bainite, which is illustrated in Fig.5b.

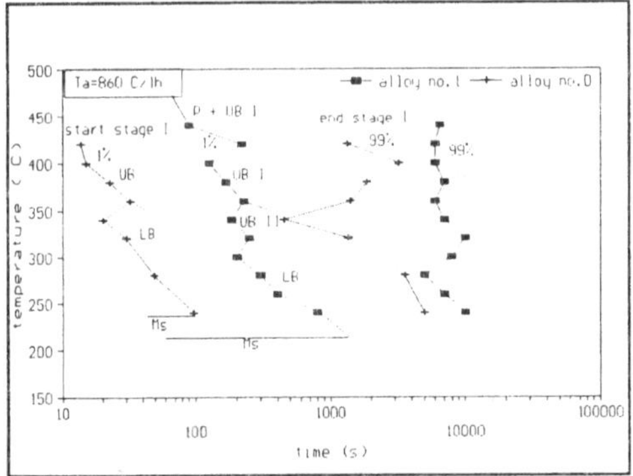

Fig.3.   IT diagrams of unalloyed (alloy no.0) and alloyed (alloy no.1) cast iron.

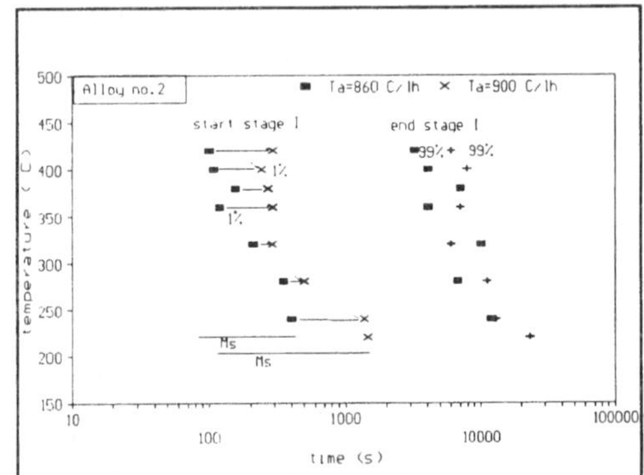

Fig.4.   Influence of austenitization temperature on IT diagram of alloy no.2.

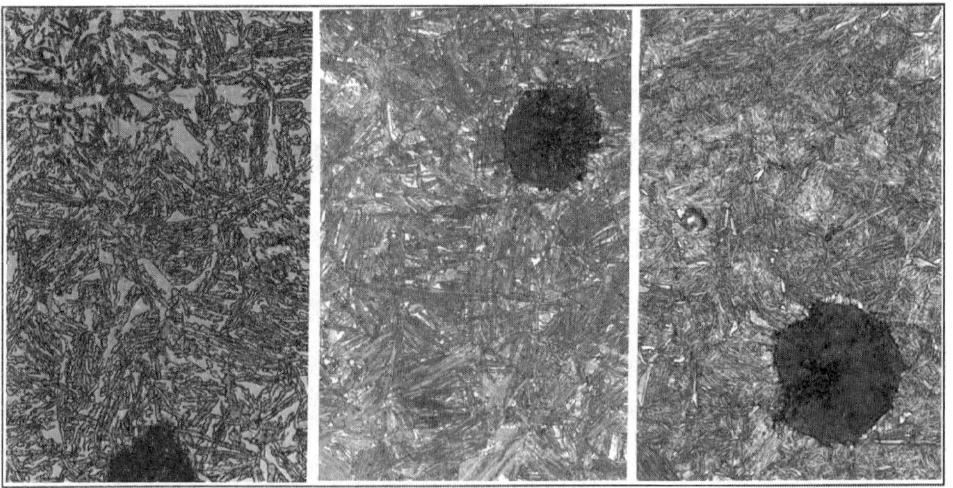

Fig.5.   Resulting structures of alloy no.1 at isothermal transformation temperature a) T_b=400°C, b) T_b=320°C, and c) T_b=240°C (T_a=860°C/1h / Beraha-etching / 500x).

<u>Kinetic analysis</u>

The resulting dilatation curves where also used to make a JMA-analysis with Eq.5. For this purpose the programme RRGRAPH is used which fits the JMA-equation to the measured length change. An example is shown in Fig.7. The resulting k (Eq.4) and JMA-exponent n are given as function of the austempering temperature in the Figs.8 and 9. The JMA-exponent n varies from 0.5 to 1.2. At 0.5 there is diffusion controlled growth and at higher values the growth will be interface reaction controlled [1], corresponding to the morphology change from upper to lower bainite. For unalloyed cast iron, alloy no.0, the JMA-exponent n is in the upper bainite region between 0.5 and 0.8. In the lower bainite region n is between 1 and 1.2 (transition temperature 350°C). For alloyed cast iron, alloy no.1 and no.2, the JMA-exponent n increases in the upper bainite region from 0.5 to 0.8 as the temperature decreases from 420 to 360°C. Then n decreases again to 0.5 at 320°C. Lowering the temperature further gives an increase for n, from 0.5 to 1.2. This indicates an additional morphology between (pure) upper and lower bainite.

The k-values will be higher (Fig.9) if the ductile cast irons are containing less alloy elements and held at higher temperatures. This is explained by the nucleation and growth of ferrite in austenite which is more activated in the conditions of high isothermal holding temperature and/or in ductile irons alloyed with less alloying elements. The rate value k remains almost the same for unalloyed and alloyed cast irons below 300°C. With the result of k, a plot of ln(k) versus 1/T yields values for the pre-exponential factor k_0 and the activation energy E. The accuracy of the activation energy E (Table II) is debatable since it is determined in a wide temperature-range with only a few measurements. Also the great scatter in the measurements of the Gleeble-dilatometer is of influence.

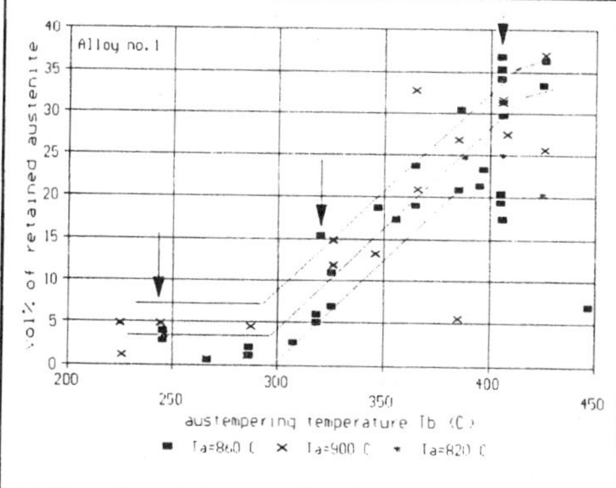

Fig.6. Volume fraction of retained austenite as function of austempering temperature T_b (alloy no.1).

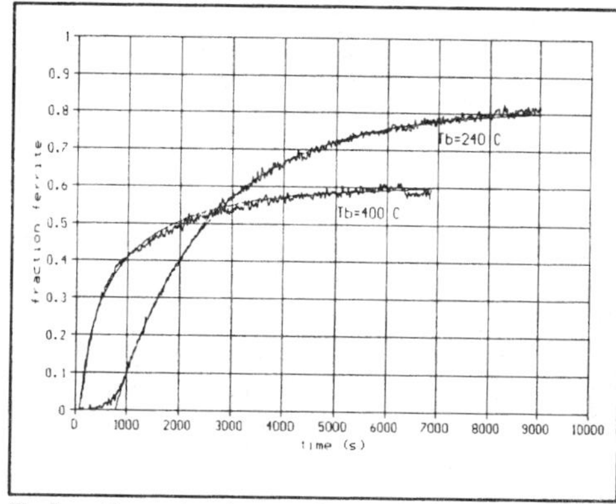

Fig.7. Comparison between the experiment and the RRGRAPH-fit (alloy no.1 - T_a=860°C/1h).

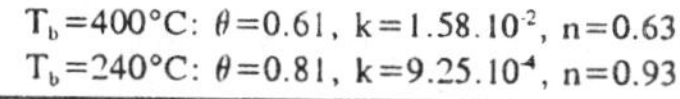

T_b=400°C: θ=0.61, k=1.58.10^{-2}, n=0.63
T_b=240°C: θ=0.81, k=9.25.10^{-4}, n=0.93

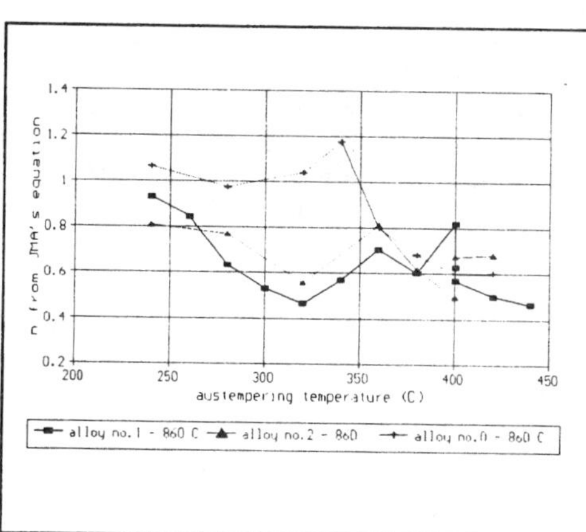

Fig.8: Results of JMA-analysis. The exponent n from the RRGRAPH-fit.

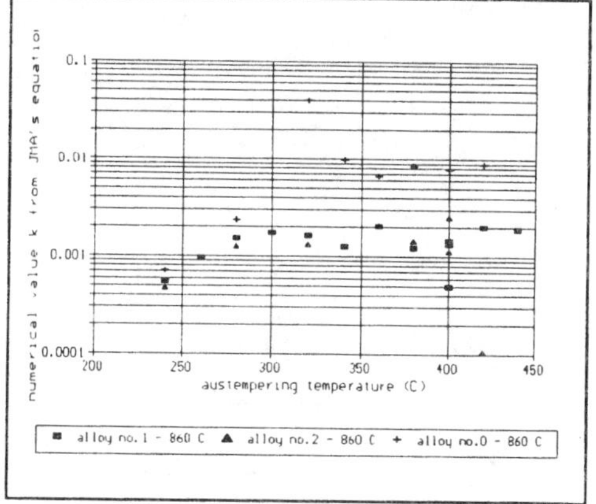

Fig.9. Results of JMA-analysis. The resulting numerical k-value from RRGRAPH-fit.

Table II: Results of JMA-analysis.

T_a/composition		T-range (°C)	E (kJ/mol)	n
860°C/alloy no.0	upper bainite	360-420	14	0.5-0.8
	lower bainite	240-340	90	1-1.2
860°C/alloy no.1	upper bainite I	380-440	35	0.5-0.8
	upper bainite II	320-360	17	0.5-0.7
	lower bainite	240-320	53	0.5-0.9
900°C/alloy no.1	upper bainite	360-420	15	0.5-0.7
	lower bainite	220-340	36	0.6-1.2
860°C/alloy no.2	upper bainite	380-420	?	0.5-0.8
	lower bainite	240-320	33	0.6-0.8
900°C/alloy no.2	upper bainite	360-420	20	0.6-0.7
	lower bainite	220-320	49	0.7-1.1

Conclusions

The kinitics of the isothermal decomposition of austenite in the matrix of ductile cast iron is affected by the chemical composition of the alloy, by austenitization conditions and by the transformation temperature. From the present investigation the following conclusions can be formulated:

1. Although the Gleeble-experiments give a great scatter in results, the data processing and analysis is very easy.
2. An additional morphology of bainitic ferrite is observed between the upper and lower bainite structure in alloyed ductile cast irons. This is verified by the experimentally determined IT-diagrams as well as the kinitic JMA-analysis.
3. The maximum temperature at which lower bainite appears is 300°C for alloyed ductile cast irons.
4. For a good JMA-analysis and an exact calculation of the activation energy E more experiments are required.
5. Because of the inhomogeneity of cast iron, it is difficult to predict the influence of alloying elements on the phase transformation.

Acknowledgements

This research was sponsered by the Dutch Ministry of Economic Affairs under contract number PBTS NM 89063; The following participants co-operated in this project : Globe/Globon, NEVE, DAF Trucks, TU Delft. The authors would like to thank Mr. Maas from UT Twente for giving the possibility to use the Theta-dilatometer.

References

[1] J.W.Christian: The Theory of Transformation in Metals and Alloys, 2nd ed., part I, Pergamon Press, Oxford, 1975, pp.525-548.
[2] K.-H. Bowe et al, Konstruieren + Gießen, 12 (1987), nr.4, 14-19
[3] R.C.Voigt, C.R.Loper, J. Heat Treating, 3, 1984, no.4, pp.291-309.
[4] Lui Cheng et al, Metall. Trans. A., 19A, 1988, pp.2415-2426.
[5] H.Nieswaag, J.W.Nijhof, Mat.Res.Soc.Symp.Proc. vol.34, 1985, pp.411-422
[6] C.H.Chang et al, Trans. Jap. Foundrymen's Soc. vol 12, Oct 1993, pp.107-114.
[7] S.Parent-Simonin et al, 2nd Int. Conf. on ADI, 17-19 March 1986, USA, pp.195-205.

Advanced Materials Research Vols. 4-5 (1997) pp. 391-398
© 1997 Scitec Publications, Switzerland

Control of Eutectic Carbide Precipitation and UAV Formation in 1% Mn Ductile Iron

A.M. Nili[1], E. Niyama[2] and T. Ohide[2]

[1] Dept. of Metallurgy and Materials Engineering, Tehran University, Tehran, Iran

[2] Dept. of Materials Science and Engineering, Tohoku University, Sendai 980-77, Japan

Keywords: Segregation, Eutectic Carbide, Austenite, Successive Austempering

ABSTRACT

A certain level of hardenability is required in austempering process of ductile iron to prevent pearlite formation. Manganese, an inexpensive and potent hardenability promoter, may be added to iron for that purpose. However, Mn, on the other hand, is known to segregate in intercellular regions and is harmful for mechanical properties of ADI(austempered ductile iron) because of two main reasons:

1-Precipitation of eutectic carbide in intercellular region in ductile iron[1-3], and

2- Formation of a heterogeneous microstructure during austempering process which leads to formation of untransformed austenite volume(UAV) in the intercellular region. UAV, that forms a network in the intercellular region, transforms to martensite upon cooling to room temperature[4,5].

Precipitation of brittle carbide plus formation of UAV in the intercellular region have a deteriorating effect on the mechanical properties of ADI. As a consequence of these negative contribution of Mn on the mechanical properties of ADI, some researchers suggested that Mn content of ductile iron should be kept around 0.3-0.35% [6,7], which does not provide sufficient hardenability, in particular, for thick-section castings.

The main objectives of this work are: 1) an attempt to control precipitation of eutectic carbide and formation of UAV in 1% Mn (Note that due to potent hardenability of Mn, 1% Mn provides sufficient hardenability for a wide range of castings) ductile iron by means of experiments and modeling, and 2) using some results of the first part, to improve mechanical properties of ADI by applying successive austempering process to keep the amount of retained austenite as high as possible and to minimize UAV.

THE MATHEMATICAL MODELING

Eutectic carbide formation

The chemical composition and structure of eutectic carbide in 1% Mn ductile iron were intensively discussed by Nili Ahmadabadi et al. [8] . They showed that this carbide forms due to segregation of Mn in intercellular region and its minimum Mn content is about 5%. They also suggested that if the Mn content of the last part of liquid to solidify is less than 5%, the casting is presumably carbide free. Using this suggestion, one bears in mind that by controlling Mn segregation carbide precipitation can be controlled.

Ohnaka[9] proposed an equation for microsegregation which can be shown as follows:

$$C_L = C_0(1-\psi f_s)^{(k-1)/\psi} \qquad (1)$$

$$\psi = 1 - \beta k/(1+\beta) \qquad (2)$$

It can be shown that for plate, columnar dendrites and spherical cells, $\beta=2\alpha$, 4α and 6α, respectively, where

$$\alpha=4D_s t_f/\lambda^2 \tag{3}$$

C_L: Solute content of liquid

k: Equilibrium distribution coefficient

λ: Dendrite arm spacing,

 t_f: Local solidification time,

 D_s: Solute diffusivity in the solid

By assuming a spherical model for cellular growth of ductile iron and geometrical calculation, it can be shown that[10]

$$N=1.15n^2 \tag{4}$$

where N is nodule count/mm^2 and n is linear density or number of graphite(cell)/mm. n could be calculated from λ (micrometer);

$$n=1000/\lambda \tag{5}$$

Now by using Eqs 1-5, knowing nominal Mn content, solidification time, $C_L \leq 5\%$ for a eutectic carbide free casting as explained earlier, and other required data a relation can be derived to control carbide precipitation.

UAV formation

Results of some researches show that UAV forms in the intercellular region where Mn segregates and the minimum value of UAV remains unchanged even after long austempering time when Mn segregation is severe[11]. So far it was assumed that severe Mn segregation in the intercellular region delays bainitic reaction in that region only. However, stability of UAV after a long austempering time suggests that formation of UAV should be discussed by considering mechanism of bainitic reaction. In this work, based on the results of Nieswaag and coworkers investigation [12], it was assumed that the *incomplete reaction phenomenon* which was first introduced by Bhadeshia and Edmond[13] can be a probable mechanism for bainitic reaction in ADI. Bhadeshia and Edmond[13] defined a T_0 temperature based on a proposal by Zener[14] that stress free austenite and ferrite of the same composition(with respect to both the interstitial and the substitutional alloying elements) are in metastable equilibrium. Thus any displacive transformation involving a full supersaturation of carbon can occur only below the appropriate T_0 temperature. Therefore, UAV in ADI should be formed in intercellular regions where carbon content is more than what is predicted by T_0 line. The T_0 curve can be calculated according to the model developed by [13,15,16]. Bhadeshia also introduced other term, T'_0, which is the modified version of T_0 in which the strain energy of bainitic transformation, 400J/mol[17], is taken into account.

In order to predict T_0, T'_0, and carbon content of austenite after austenitisation in ductile iron, local concentration of Mn, Si and C should be calculated. Local concentration of substitutional alloying elements can be calculated by using Eqs. 1-3, and concentration of C as a function of intercellular distance is deduced from the empirical equation developed by Wada et al.[18]. Their proposed equation for Fe-Mn-Si-C alloy is:

$$\log a_C/z_C = 2300/T - 0.92 + [(3860+10000y_{Si})/T]y_C - (2200/T)y_{Mn} + (2.1+3200/T)y_{Si} \tag{7}$$

where a_C=activity of C, y_C= the atom ratio$[n_C/(n_{Fe}+n_{Mn}+n_{Si})]$, y_{Mn} and y_{Si} are the atom fraction of manganese and Si in the C free alloy respectively, and $z_C = y_C/(1-y_C)$.

They utilize graphite as standard state, hence in ductile iron which graphites acts as a source of carbon, activity of C in austenite should be equal to one when austenite is saturated with graphite and it can be achieved if austenitisation time is long enough.

EXPERIMENTAL PROCEDURE

Eutectic carbide

High Mn ductile cast iron with composition given in Table 1 were poured in green sand molds.

Table1-Chemical composition of castings(wt%)

	C	Mn	Si	Cr	Mn	S
L-100	3.61	0.96	2.82	0.02	0.056	0.013
S-100	3.4	0.98	2.81	--	0.04	0.009

Table 2-Graphite structure, solidifucation time and eutectic carbide of different castings

	Graphite (vol%)	Nodule count/mm^2	Graphite ave.size (μm)	Nodularity (%)	Solidification time (sec)	Carbide (vol%)
UL*-100	10.7	55.3	20.1	73.7	292	0.08
LL**-100	12.4	125	14.1	79	167.5	0.027
S-100	12.7	356	9.0	84.5	80	very few

*-UL-100: Upper part of L-100 casting
**LL-100:Lower part of L-100 casting

Two different castings were made from differently treated melts. Larger Y block casting(L-100) was produced by a sandwich spheroidization treatment process, while the other(S-100) was produced by the inmold process, which is known to give higher graphite nodule count. For L-100 casting, base irons were treated by 5.5% MgFeSi alloy for spheroidization followed by post inoculation with 75%FeSi. S-100 casting was inoculated with a spheroidizing alloy containing 6.51%Mg, 45.1%Si, 0.2%Ca, 1.04%RE and 0.65%Bi.

The solidification time of L-100 specimen was measured by thermocouples which were placed in the vicinity of the riser(UL-100) and bottom part of casting(LL-100). In the case of S-100 specimen the thermocouple was placed in the centre of the casting.

Specimens were cut from different parts of the casting and austenitized at 900°C for 90 min and then quenched in ice water to eliminate eutectoid carbides. An image analyser was used to measure graphite morphology, nodule count and carbide percent in each casting. In the case of L-100 casting, measurements were conducted on two parts due to its size: upper part, in the vicinity of riser and bottom part. These two parts are expected to have different microstructure due to the difference in solidification time. The measurement of graphite structure was carried out at magnification of 100 on six different points in each specimen. The measurement of carbide was carried out at magnification of 400 on 20 different points in each specimen. The results of image analyser and solidification time of each casting are shown in Table 2

UAV measurement

Prior to untransformed austenite volume (UAV) measurements the specimens were austenitised for 90 min in an atmosphere-controlled furnace at 900°C, austempered for various times in salt bath at 375°C for high austempering temperature (HAT) process, or at 315°C for low austempering temperature (LAT) process, followed by quenching in ice water. Volume fraction measurements of the UAV were made on each specimen of 20x20x10 mm cut from Y block by using the image analyzer. In this method magnification of x50, x100 and x200 were selected and from each magnification two images from different points of each specimen were taken. The six images from each specimen were averaged to minimize the measurement error due to effect of etching and geometry variation of UAV in different specimens.

Charpy and tensile test

The heat treatment of impact and tensile test specimens for HAT and LAT processes was the same as the heat treatment of UAV specimens. In the case of successive austempering, high-low austempering temperature(HLAT) process , specimens were first austempered at 375°C for 30 to 600 min, and then austempered at 315°C for 10 to 10000 min, followed by quenching in ice water. In this study, only the casting of lower performance, i.e. L-100, was used to emphasize the effects of successive austempering.

Impact tests were carried out at room temperature in duplicate for each heat treatment condition with unnotched charpy test specimens of dimensions 10x10x55 mm.

Tensile test specimens were tested in duplicate for each heat treatment condition. The tests were carried out with a 5 tons tensile testing machine. Yield and tensile strength and elongation were determined.

EXPERIMENTAL RESULTS AND DISCUSSION

Eutectic carbide

Figure 1 shows the SEM micrograph of an eutectic carbide taken from UL-100 specimen. This figure clearly shows a big microvoid which was formed inside the carbide. Mechanical properties of casting become much worse due to formation of such microvoids inside the eutectic carbide.

To derive an equation to control carbide precipitation and based on the previous assumptions, the following figures are put into Eqs. (1-5):

$C_L \leq 5\%$, $f_{s} \cong 1$, $C_0 = 1\%$, $k_{Mn} = 0.7[19,20]$, $D_S = 1.53*10^{-10} cm^2 [21]$

Then the critical value of ψ, λ and N to avoid carbide is determined. The general form of the relation can be written as follow:

$$N \geq N_c = \delta \ t_f^{-1} \qquad (8)$$

where δ is about 21600 count.sec/mm^2.

Eq. (8) means that in any casting which satisfies this equation, Mn eutectic carbide does not precipitate. When t_f and nominal Mn content are kept constant, N could be affected by inoculation and other graphitizers. If these factors are chosen such that nodule count, N, becomes equal to or larger than the critical value calculated from Eq. (8), then casting is likely to be free of carbide, otherwise stable eutectic carbide may be formed.

In order to predict amount of precipitated carbide, two steps must be followed;

1- By substituting solidification time in Eq. (8), N_c is calculated, which then is compared with measured nodule count. If the measured value is more than the calculated value, carbide will not precipitate, otherwise there is a possibility of carbide formation.

2- In the case of carbide precipitation, by substituting the measured nodule count, the solidification time, C_L=5% and other data in Eq. (3), (2) and (1), the amount of liquid with Mn content more than 5% is obtained. Since the Mn content of this liquid is more than 5%, carbides will form.

In the case of UL-100 casting, substituting its solidification time in relation (5), N_c value is obtained;

$$N_c=74.1$$

Comparison of N_c with the measured nodule count of UL-100, namely 55.3, confirms that this part of the casting should have carbide.

By substituting measured N, t_f, and the other data in Eqs.(3), (2) and (1), it is found;

$$f_s=99.88\%$$

It means that when Mn content of liquid reaches 5%, 99.88% of the casting has been solidified. In other words, if we suppose no density changes during carbide precipitation, 0.12% of the casting may transform to carbide which is in good agreement with the result of Table 2 .

The same procedure was conducted for the other castings and the results are shown in Figure 2. They show good agreement between predicted and measured carbide volume fraction. Minor discrepancy can be due to measurement error and probable density changes during carbide precipitation.

Eq. (8) shows that with any solidification time it is possible, in principle, to prevent carbide precipitation. This matter is shown in Figure 2. If t_f is kept constant, the amount of carbide content can be changed by variation of nodule count. If the casting nodule count is equal to or more than the critical nodule count value, then carbides will not precipitate, otherwise carbides may be formed. Since solidification time is given in most practical cases, carbide precipitation can be prevented by control of nodule count. It was shown that by using better inoculation process nodule count can be increased and, as a consequence, the cell size is reduced[22,23]. Experiences such as this suggest that carbide precipitation can be controlled by inoculation for a specified solidification time and Mn content.

Table3-Comparison of the experimental and the calculated value of the minimum value of UAV when T_0 or T'_0 curve are the limiting values of ferritic bainite.

Austempering temperature (°C)	UAV % (measured)	UAV % (calculated)	
		T_0	T'_0
375	4.9	5.5	70
315	2.2	1.2	20

Critical nodule count vs. solidification time for different Mn content are shown in Figure 3. This figure shows that an increase in nominal Mn content increases the critical nodule count value. In other words, in the cases where higher Mn content is necessary, more efficient nucleation process is necessary to suppress carbide precipitation.

Prediction of UAV

Figure 4 shows the measured value of UAV vs austempering time for HAT and LAT specimens. This figure shows that UAV in the each specimens, after a long austempering time, approaches to a minimum value after a sharp decrease. In addition, Figure 4 indicates that the minimum value of UAV in the LAT process is less than the HAT process.

Figure 5 shows the results of calculated T_0, T'_0 curves and also carbon content of austenite after austenitisation.

Table 3 compares the experimental results with the calculated values when T_0 or T'_0 curve are the limiting values of bainitic ferrite. To calculate the minimum value of UAV from Figure 5, it was assumed that the regions where C_0 curve lies over T_0 or T'_0 curves are equivalent to the untransformed regions. Note that the measured UAV values are the area of the regions where bainitic ferrite has not been detected.

If one assumes that bainitic reaction is going to be stopped when carbon content of retained austenite reaches the T'_0 line, bainitic reaction in this study should be ceased in an early stage of austempering reaction and UAV in the HAT specimens should be very high, more than 70%. This, however, is not the cases shown by the experimental results given in Table 3. On the other hand, the good agreement between the measured minimum value of UAV and the calculation, assuming the T_0-curve to be the limiting value of bainitic ferrite, strongly suggests that bainitic transformation is stopped when carbon content of retained austenite reaches to T_0-curve.

The results of Table 3 not only endorses the displacive mechanism for bainitic reaction in ADI but also gives a very good criterion for selection of austempering and austenitising temperature to optimize the mechanical properties of ADI when Mn segregation exists. This aim can be achieved through application of the successive austempering to keep the amount of retained austenite as high as possible by using high temperature austempering, and then reducing the UAV by subsequent low temperature austempering.

<u>Austempering process</u>
Figure 6 shows the impact energy of three different austempering processes. HLAT-50 refers to the successive austempering specimens were austempered for 50 min. at 375°C followed by subsequent austempering at 315°C for different periods. This specimen shows the optimum impact energy amongst all successively austempered specimens, therefore, it was selected to compare with specimens austempered by conventional process. Dramatic improvement of impact energy in HLAT specimen is due to lower UAV compared with HAT specimens and higher retained austenite compare with LAT specimens.

Figure 7 shows the ultimate tensile strength vs. elongation for various ASTM and JIS grades, low Mn and 1% Mn ADI austempered by conventional and successive austempering. This figure along with figure 6 shows that control of UAV plus formation of high retained austenite by applying successive austempering can increase mechanical properties of high Mn ADI even more than conventional ADI.

CONCLUSIONS

1-The following relation between critical nodule count(N_C) and solidification time(t_f), for preventing carbide precipitation, was derived:

$$N \geq N_C = \delta\, t_f^{-1}$$

where δ is a function of nominal Mn content. The experimental data show good agreement with the predicted values.

2-The good agreement between the calculated and the measured minimum UAV value not only sheds light to the mechanism of bainitic reaction in ADI to some extent, but also bears an important implication to improve structural control of ADI.

REFERENCES

1- E. Dorazil, T. Podrabsky and J. Svejcar: AFS Trans., 98(1990), 765-774.

2-G. P. Faubert, D. J. Moore and K. B. Rundman: AFS Trans., 99(1991), 831-843.

3-M- Nili Ahmadabadi, T. Ohide and E. Niyama: Trans. of The Japan Foundrymen's Society,12(1992), 40-47.

4-D.J. Moore ,T.N. Rouns and K.H. Rundman: AFS Trans., 94(1986), 255-264.

5-M. Gange: AFS Trans., 93(1985), 801.

6- H. L. Morgan: British Foundryman, 80(1987), 2, 98-108.

7-Y. Ueda and M. Takita: Proceeding, 2nd Int. Conf. on Austempered Ductile Iron, Ann Arbor, Mi, USA(1986), March 17-19, 141-147.

8-M. Nili Ahmadabadi, E. Niyama, M. Tanino, T. Abe and T. Ohide: Metall. Trans., 25A(1994), 912-918.

9-I. Ohnaka: Trans. ISIJ, Vol. 26(1986). 1045-1051.

10-M. Nili Ahmadabadi, E. Niyama and T. Ohide: AFS Trans,(1994),

11-T. N. Rouns and K.B.Rundman: AFS Trans., 95(1987), 851.

12-H. Nieswaag and J. W. Nihof: Mat. Res. Soc. Symp., Vol. 34(1985), 412-422.

13-H.K.D.H. Bhadeshia and D.V. Edmonds: Acta Metall., Vol. 28(1980), 1265.

14- C. Zener: Trans. AIME, 167(1946), 50.

15-H. I. Aaronson and H.A. Dominan and G.M. Pound: Trans. AIME, 236(1966), 753.

16-G.J Shiflet, J. R. Bradley and H. I. Aaronson: Metall. Trans, 9A(1978), 999.

17-H.K.D.H. Bhadeshia: Acta Metall., 29(1981), 1171.

18-T. Wada, H. Wada, J.F. Elliott, and J. Chipman: Metall. Trans., 3(1972), 1657.

19-R. Boeri and F. Weinberg: AFS Trans., 97(1989), 179-184.

20-A. Kagawa and T. Okamoto: Mat. Res. Soc. Symp 34(1985), 201-210.

21- C. Wells and R. F. Mehl: Trans. of AIME 145(1941), 315-339.

22-H. Horie, T. Kowata and A. Chida: Cast Metals, Vol.1, no.2(1988).

23-H. Nakae, H. Koizumi, K. Takai, and K. Okauchi; Trans.of the Japan Foundrymen's Society. 12(1992),34-47.

24- M. Gagne: AFS Trans., 95(1987), 523.

25-D. J. Moore, T. N. Rouns and K. B. Rundman: AFS Trans., 93(1985), 705.

Figure1: SEM micrograph taken from carbides located in the intercellular region. The block carbide in the upper right part of photo is titanium carbide.

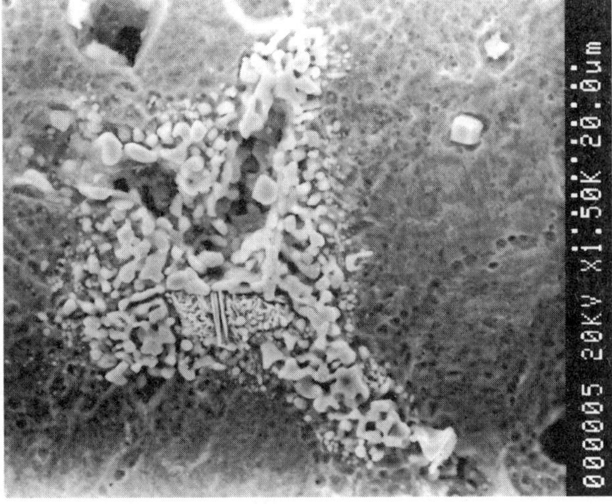

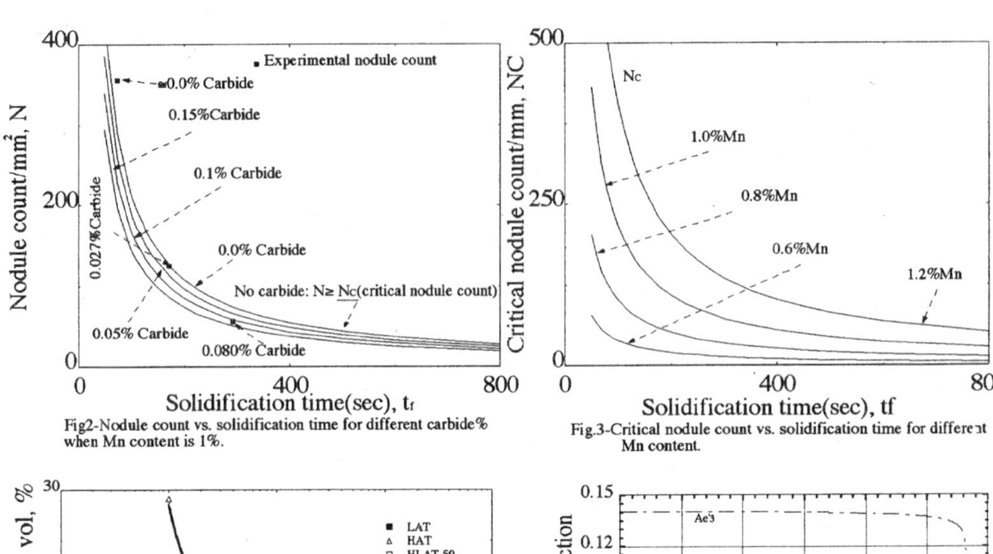

Fig2-Nodule count vs. solidification time for different carbide% when Mn content is 1%.

Fig.3-Critical nodule count vs. solidification time for different Mn content.

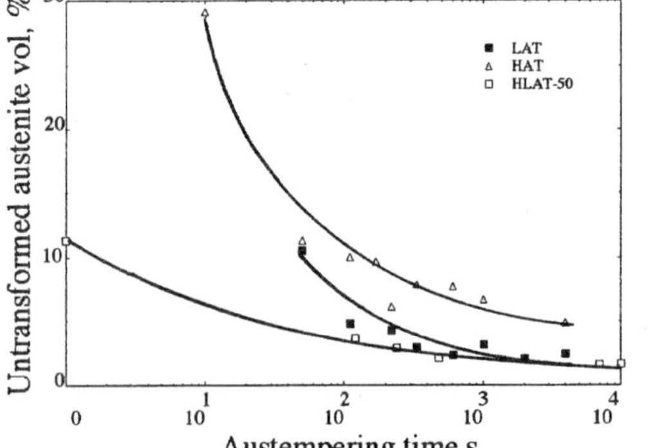

Figure4- UAV vs. austempering time of conventional and successive austempering

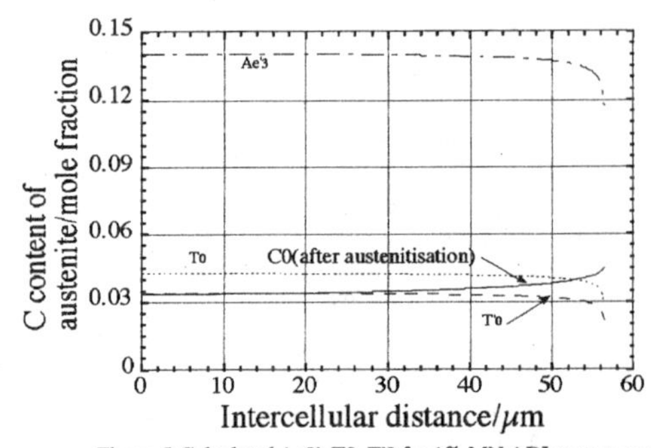

Figure 5-Calculated Ae3', T0, T'0 for 1% MN ADI austempered at 375°C and matrix carbon content after austenitization at 900° as a function of intercellular distance.

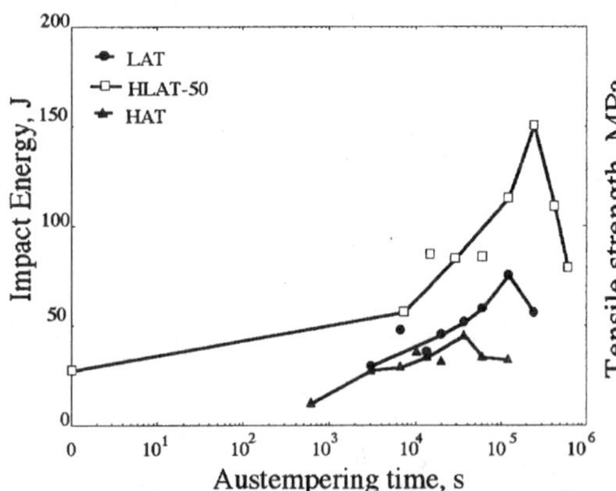

Figure6-Impact energy vs. austempering time for three typical specimens. "Austempering time refers to time at 375°C for HAT and time at 375°C for LAT and HLAT".

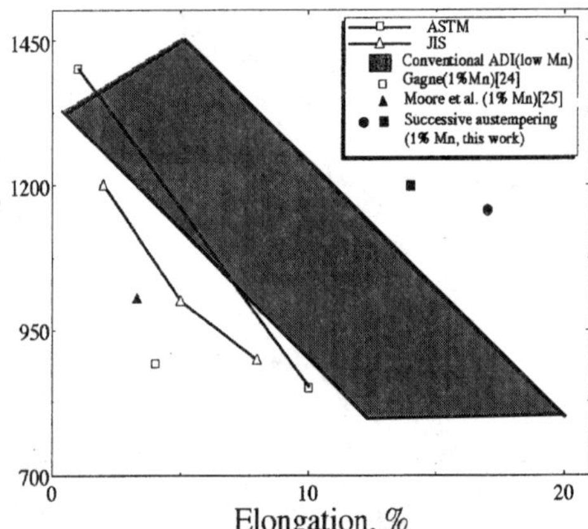

Figure7- Ultimate tensile strength vs.elongation for various ASTM and JIS grades, low Mn and1% Mn ADI austempered by conventional and successive austempereing. Black mark refers to HLAT-30 and black circular mark refers to HLAT-50 specimen.

Advanced Materials Research Vols. 4-5 (1997) pp. 399-406
© *1997 Scitec Publications, Switzerland*

A Stepped Austempering Heat Treatment for a
Mn Alloyed Ductile Cast Iron

H. Bayati and R. Elliott

Manchester Materials Science Centre, University of Manchester,
Grosvenor St., Manchester M1 7HS, UK

Keywords: Austempering, Stepped Heat Treatment, Mn Alloyed Ductile Cast Iron, Mechanic
Properties

ABSTRACT

Austempering kinetic measurements and microstructure observations show tha
failure to satisfy the higher ductility grades of the ADI standard in a single
heat treatment in an alloyed iron of composition
3.52%C, 2.64%Si, 0.25%Mo, 0.25%Cu, 0.67%Mn, 0.007%P, 0.13%S, 0.04%Mg
is due to the long delay in the completion of the stage I reaction in the
intercellular region caused by the segregation of Mn and Mo to these areas.
This delay means that the stage II reaction commences in the eutectic cell
before the stage I reaction is completed in the intercellular region and the
processing window is closed. A stepped heat treatment process is described in
which the stage I reaction in the intercellular region is completed before
the stage II reaction occurs in the eutectic cell. The improvement in
mechanical properties achieved by the stepped heat treatment is described.

INTRODUCTION

Austempering induces increased strength and ductility in ductile cast iron
As applications increase, the need to austemper thicker section castings will
arise. Thicker sections pose special problems because of the need to ensure
sufficient through hardenability or austemperability. The presence of pearlite
in the austempered structure reduces strength and ductility[1]. Alloying is th
main method of improving austemperability and Mo and Mn are the most potent
additions. However, the amount of Mo added is limited to about 0.3% because of
intercellular carbide formation and associated porosity which reduces mechanica
properties. Mn is troublesome because it delays the transformation kinetics
and, because of segregation to intercellular regions during solidification,
delays them to different extents in the eutectic cell and intercellular regions
The problem, the complexity of the stage I reaction, was identified by Schissle
and Chobaut [2]. This paper examines the problem quantitatively and suggests
one solution in the form of a stepped heat treatment.

EXPERIMENTAL RESULTS AND DISCUSSION

Fig. 1 and 2 were used by Schissler and Chobaut to explain the problem of
austempering a Mn alloyed iron. Fig. 1 illustrates schematically the
segregation of C,Si and Mn that occurs during solidification and remains after
austenitising. Fig. 2 shows how Mn segregation delays the transformation in
the intercellular boundary or region III. With the heat treatment path 1, the
stage I reaction is completed in the eutectic cell but does not start in the
intercellular boundary. The untransformed austenite in the intercellular
boundary is either thermally or mechanically unstable and transforms to
martensite causing a reduction in mechanical properties. It is suggested that

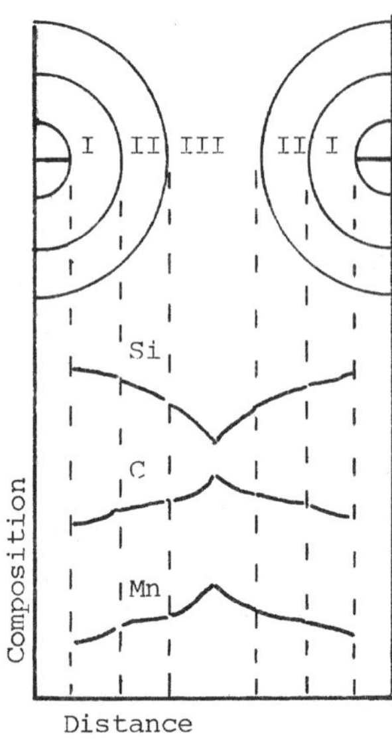

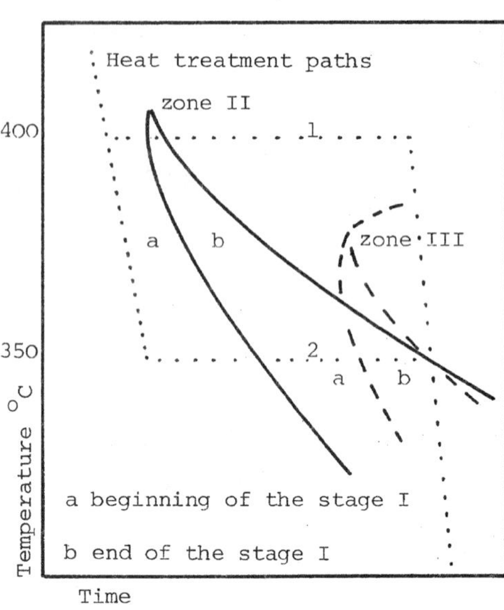

Figure 1

Figure 2

Schematic illustration of the microsegregation of alloying elements (Si,C,Mn) near to the nodule (zone I), inside the cell (zone II) and at the eutectic cell boundary (zone III).

Influence of Mn on the TTT curve of zones II and III;visualisation of the shifting initiated by Mn segregation in zone III.

the problem may be overcome by using a lower austempering temperature corresponding to the temperature at which the stage I reactions in the eutectic cell and intercellular boundary finish at the same time. Heat treatment path 2 in Fig 2 would produce this result. An austempering temperature of 350°C or above was suggested for the 1%Mn iron studied.

Austempering kinetics

We have studied this problem in an iron alloyed with 0.67%Mn, 0.25%Cu and 0.25%Mo. Detailed measurements of the austempering kinetics of the stage I and stage II reactions are reported elsewhere [3-6]. These results are summarised in Fig 3 in a slightly different form than in Fig 2. The time, t_1, represents the completion of the stage I reaction in the intercellular area. This is measured using point counting as the austempering time at which the untransformed austenite volume has fallen to 3%. The martensite is considered not to be continuous and harmful to mechanical properties below this level. In common with Fig 2, Fig 3 shows a long delay in the completion of the stage I reaction in the intercellular region at temperatures above 360°C. Similar measurements [7] for the eutectic cell show that the stage I reaction is completed in the eutectic cell at much earlier times. However, in contrast to Fig 2, it was found that the stage I reaction was always completed in the eutectic cell before the intercellular region.

The second concern in austempering is to avoid any stage II reaction producing a carbide precipitation initiating embrittlement. This is not considered in Fig 2. This is measured as the time, t_2, at which the volume of

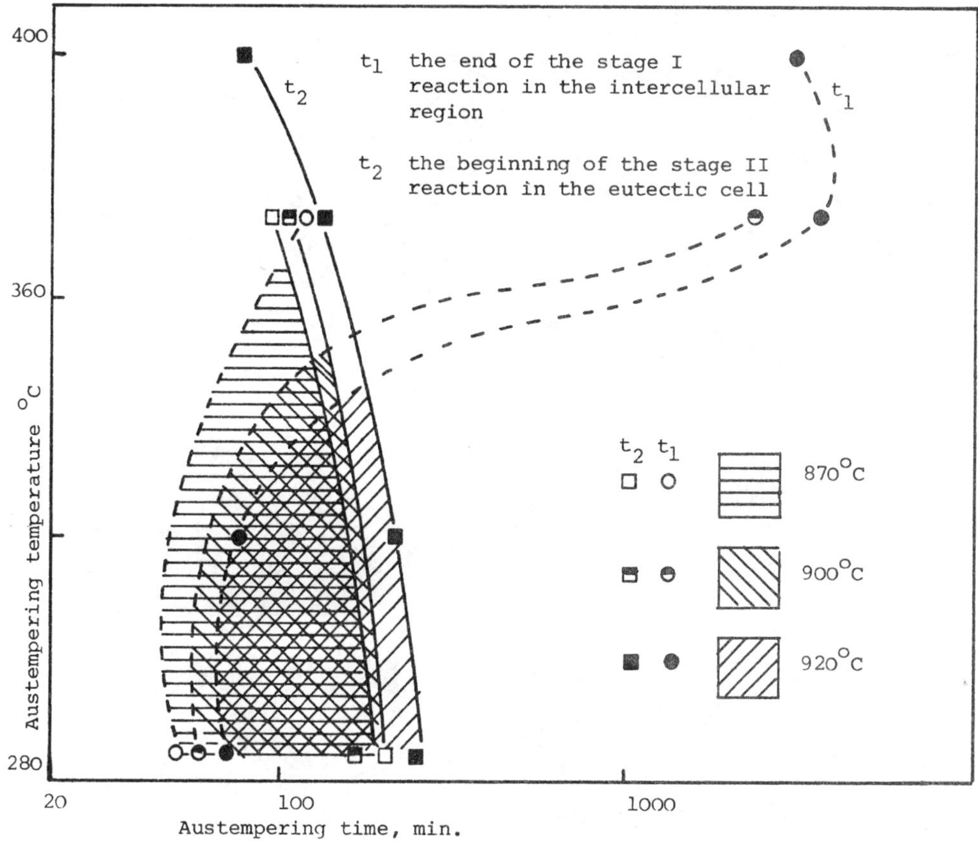

Figure 3

The variation of times, t_1 and t_2, and the width of the processing window with austempering temperature for different austenitising temperatures.

retained austenite in the structure begins to fall using x-ray measurements [6]. It is shown in Fig 3 as a function of austempering temperature and austenitising temperature. The desired austempered structure of retained high C austenite and bainitic ferrite or ausferrite is only obtained at a particular austempering temperature when $t_2 > t_1$. The time interval, t_2-t_1 is the heat treatment processing window and the differently shaded areas in Fig 3 show how reducing the austenitising temperature moves the processing window to shorter austempering times and increases the temperature at which the processing window closes.

Austempered microstructure development during a single heat treatment

Fig 4 shows the microstructure development with austempering time at 400°C after austenitising at 920°C for 120 minutes. In agreement with the stage I and II reaction times in Fig 3 the micrographs show ausferrite to form close to the nodules, then in the eutectic cell leaving large areas of eutectic cell boundary with untransformed austenite. After austempering for 4320 minutes to complete stage I in these areas, stage II is well advanced in the eutectic cell.

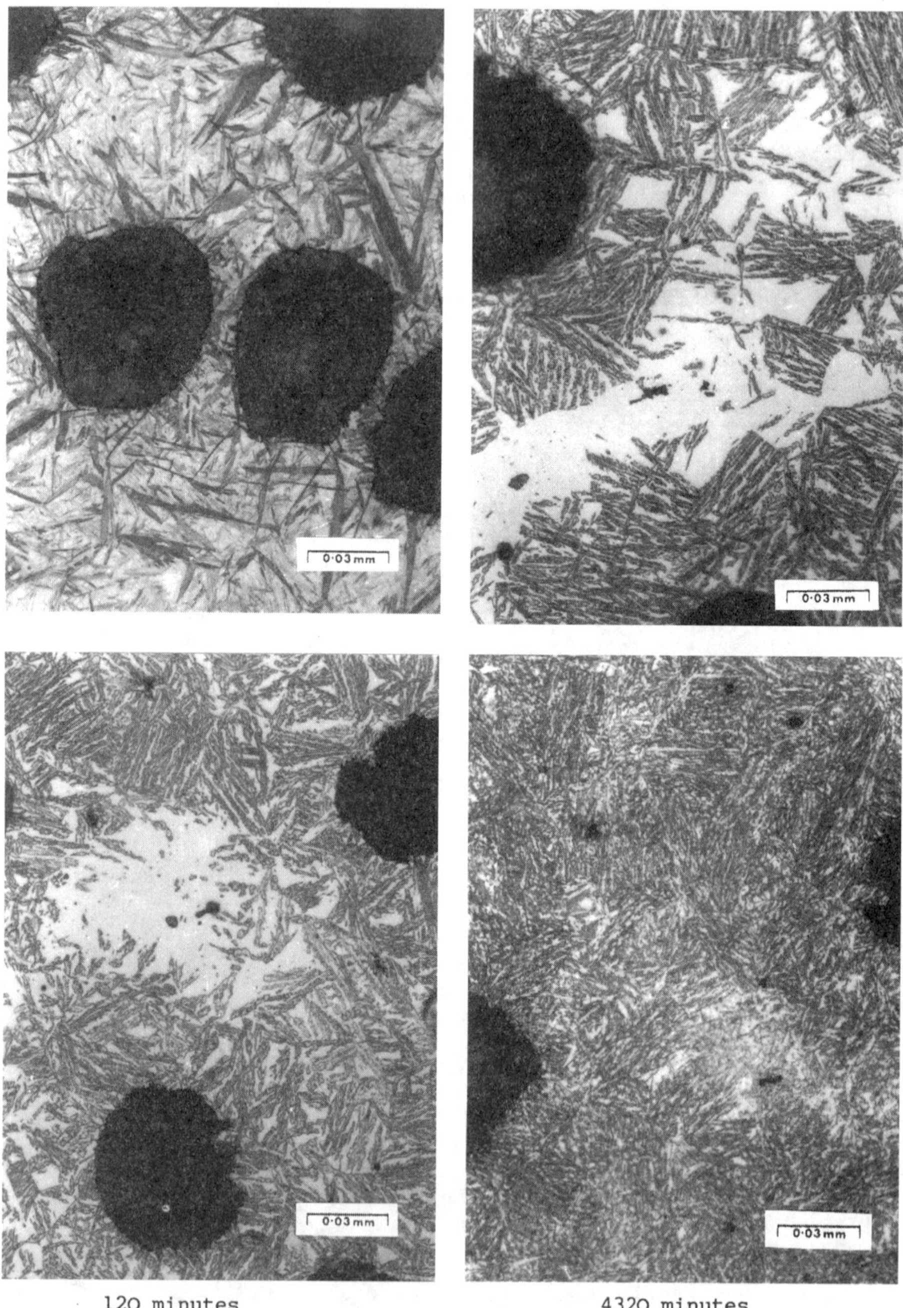

120 minutes   4320 minutes

Figure 4.   The change in microstructure with increasing austempering
time during a single austempering treatment at 400°C.
Austenitising temperature 920°C, austenitising time 120
minutes.

Mechanical properties of the austempered iron

The ADI standard ASTM A897M:1990 defines various grades of austempered iron in terms of UTS, proof strength, elongation and impact energy as shown in Fig 5. This standard is used to measure the success of austempering. Fig 3 shows that the processing window is open and wide at lower austempering temperatures. No difficulty was experienced in satisfying the high strength grades of the standard with a single heat treatment. We are only concerned with the high ductility grades. Fig 5 shows that austempering at $400^{\circ}C$ fails to satisfy the standard at all austempering times as would be expected when the processing window is closed. Following the suggestion that reducing the austempering temperature should improve mechanical properties measurements were made after austempering at $375^{\circ}C$. Fig 5 shows that although the ductility improved the standard was not satisfied. The properties can be improved further by reducing the austenitising temperature to $870^{\circ}C$ as shown in Fig 5. However, the impact energy still fails to satisfy the standard. It should be noted also that decreasing the austenitising and austempering temperatures moves the properties towards the higher strength grades. Also reducing the austenitising temperature reduces the iron austemperability.

A stepped heat treatment process

The difficulty in obtaining high ductility in a single heat treatment is the overlap of the stage I reaction in the intercellular region with the stage II reaction in the eutectic cell. This may be overcome with a stepped heat treatment. The principle is to austemper at a high temperature until the stage I is completed in the eutectic cell and then to step down to a lower austempering temperature. This will delay the stage II reaction in the eutectic cell and increase the driving force for the stage I reaction in the intercellular area. This is illustrated in Fig 6 for a 1st step austempering temperature of $400^{\circ}C$ and time of 120 minutes followed by a 2nd step austempering temperature of $285^{\circ}C$. It can be seen that with increasing time at $285^{\circ}C$ the intercellular regions are gradually filled with stage I lower bainite and even after 4320 minutes there is little evidence of the stage II reaction in the eutectic cell.

The change in mechanical properties during the stepped heat treatment is shown in Fig 7. A significant improvement in mechanical properties is evident particularly the impact energy. There is also an increase in the strength due to the formation of lower bainite. It will be noted that the optimum elongation and impact energy occur after different 2nd step austempering times. The former corresponds to a maximum retained austenite content and the latter to a maximum austenite C content. The selection of 1st and 2nd step austempering temperatures and times will allow control over the relative improvements in elongation and impact energy. This is being investigated at present.

REFERENCES

1. A.S. Hamid Ali, K.I. Uzlov, N. Darwish and R. Elliott, Mater. Sci. Technol, 10, 35 (1994).
2. J.M. Schissler and J.P. Chobaut, Conf. Proc., Cast Iron, MRS (1990).
3. H. Bayati and R. Elliott, Mater. Sci. Technol, 11, 284 (1995).
4. H. Bayati and R. Elliott, Mater. Sci. Technol, 11, 118 (1995).
5. H. Bayati and R. Elliott, Mater. Sci. Technol, in press.
6. H. Bayati and R. Elliott, Mater. Sci. Technol, in press.
7. M. Bahmani and R. Elliott, Mater. Sci. Technol, 10, 1068 (1994).

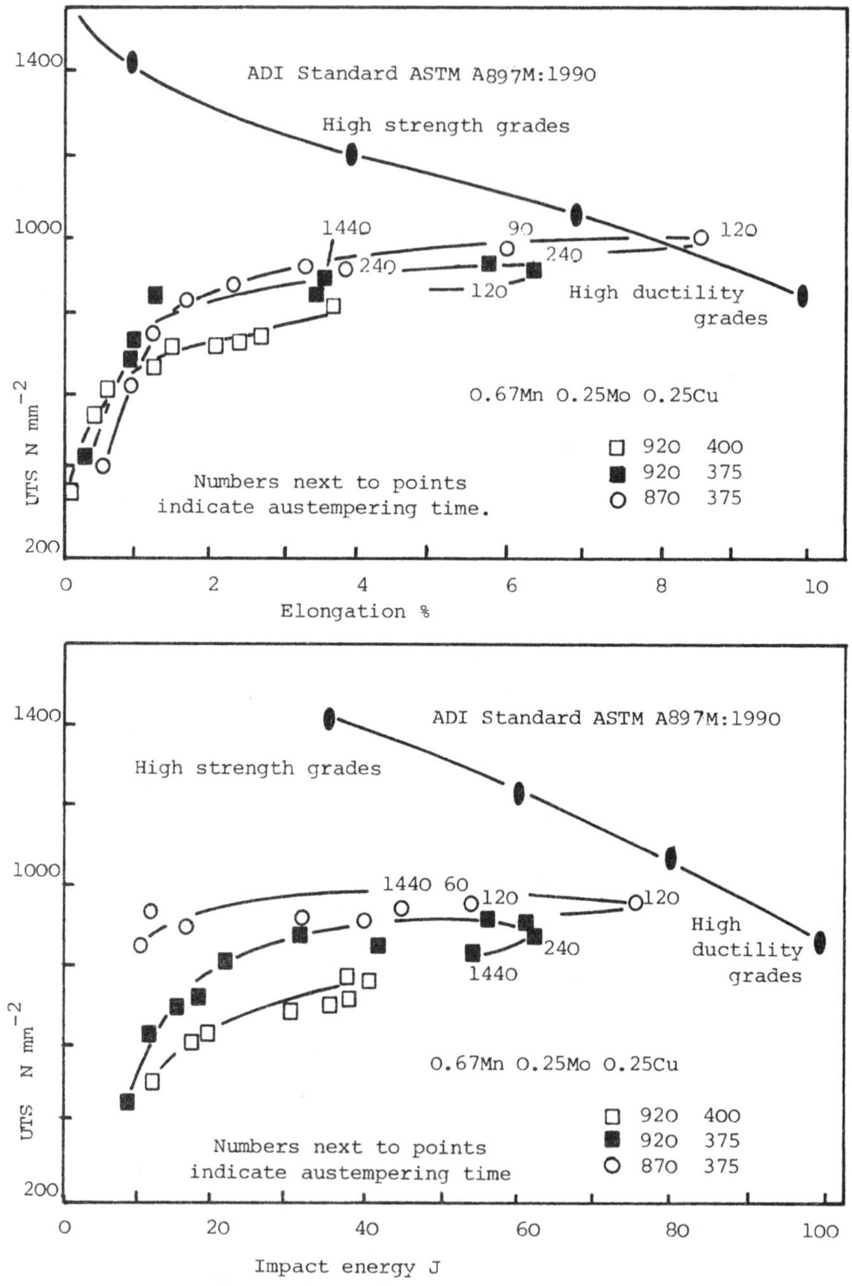

Figure 5.   The mechanical properties for different single step
             austempering heat treatments.

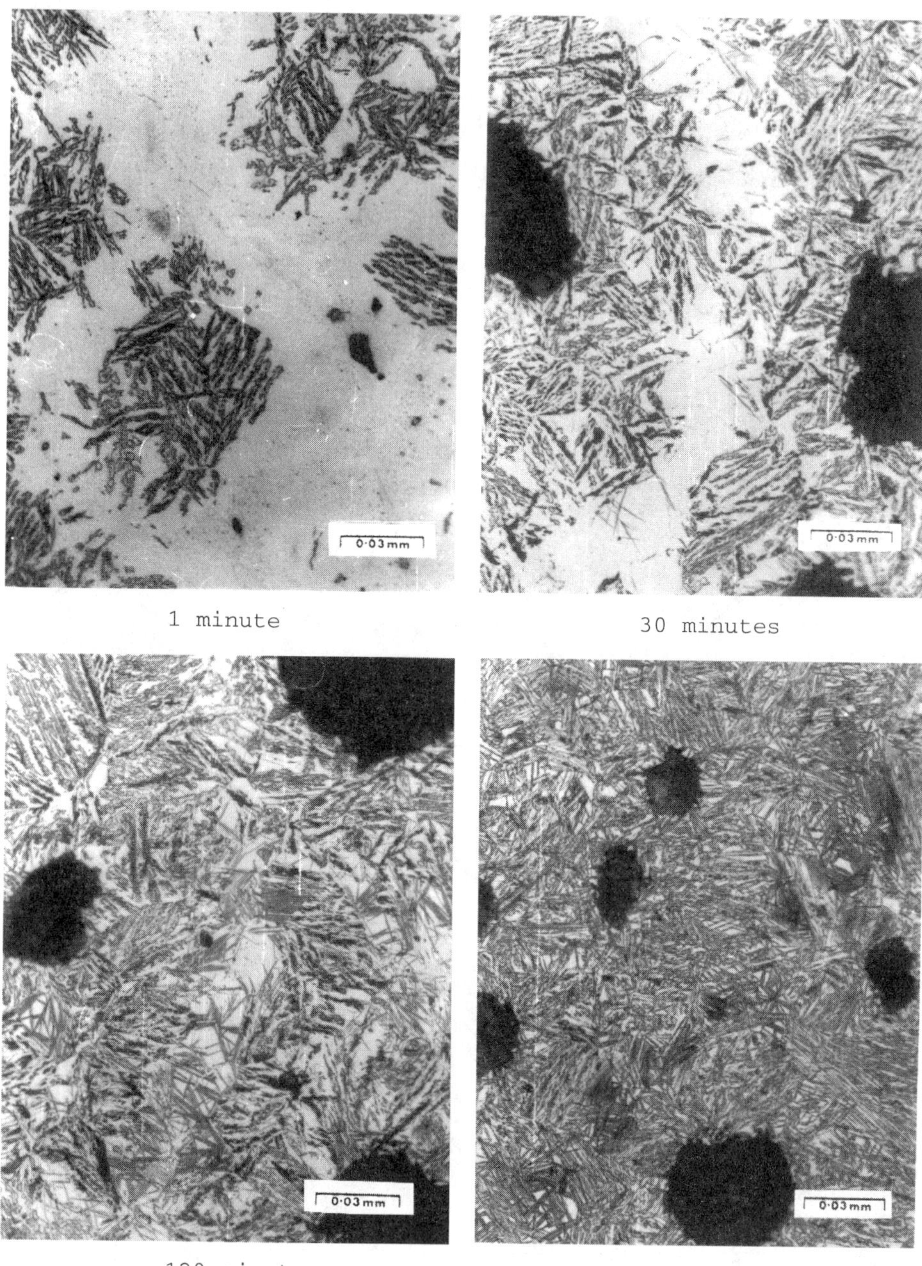

Figure 6.　The change in microstructure with increasing austempering time at the second step at 285°C of a stepped heat treatment. Austenitising temperature 920°C, austenitising time 120 minutes. First step austempering temperature 400°C, time 120 minutes. Second step temperature 285°C.

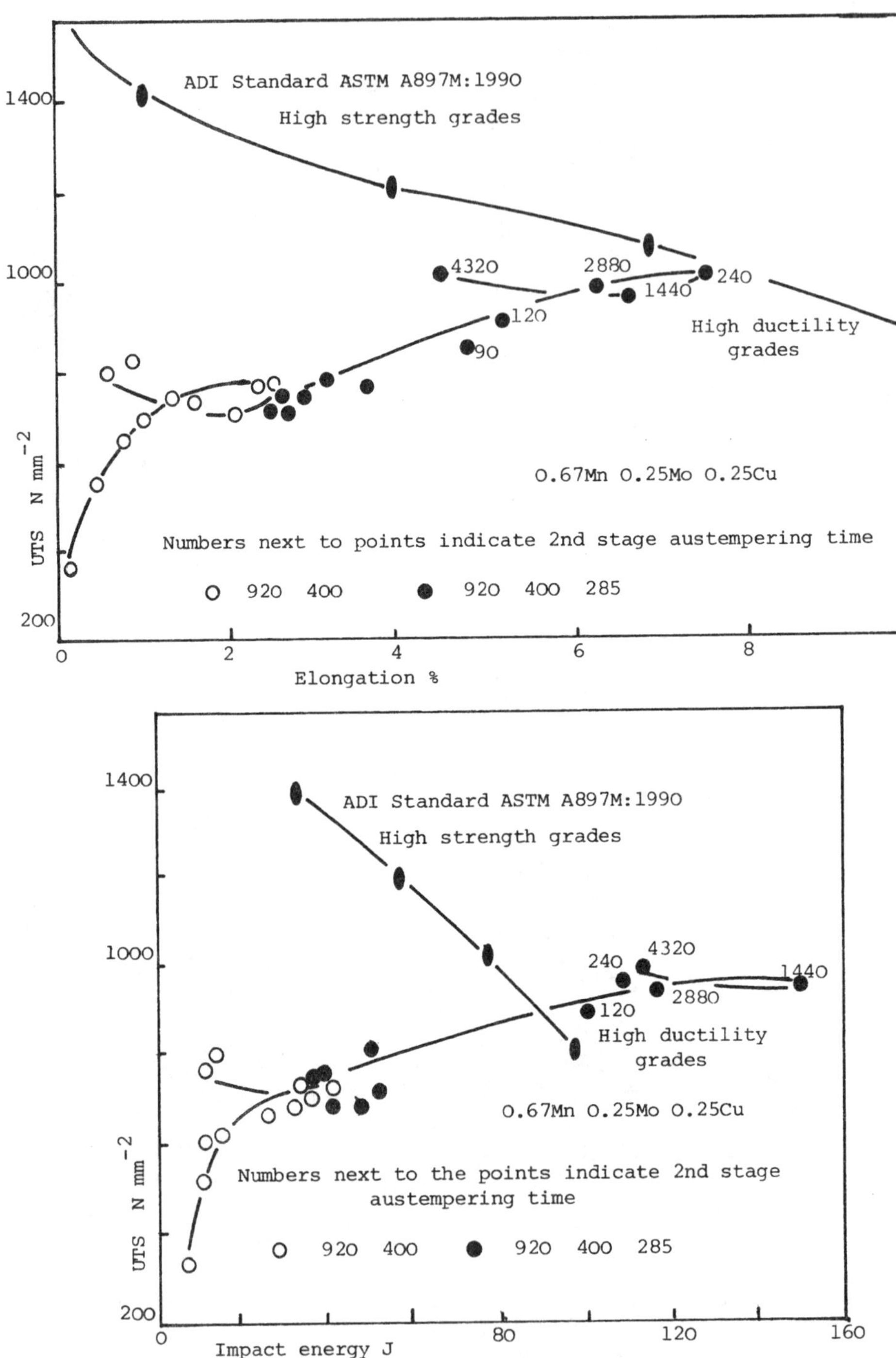

Figure 7.   The improvement in mechanical properties with a stepped heat treatment (400°C-285°C) compared with a single treatment at 400°C.

Advanced Materials Research Vols. 4-5 (1997) pp. 407-414
© 1997 Scitec Publications, Switzerland

Structural Evolution of Austempered Ductile Iron (ADI) during Tempering

Y.C. Liu[1,2], J.M. Schissler[1,2,3], J.P. Chobaut[1] and H. Vetters[4]

[1] CRITT METALL 2T, Parc de Saurupt, F-54042 Nancy, France

[2] LSG2M CNRS URA 159, Ecole des Mines de Nancy, Nancy, France

[3] Université de Nancy I, F-54500 Vandoeuvre, France

[4] Stiftung Institut für Werkstofftechnik (I.W.T.), Bremen 33, Germany

Keywords: ADI, Tempering, Transmission Electron Microscopy, Iron Carbide

ABSTRACT

The evolution on austempered ductile iron (ADI) structure during anisothermal tempering (300°C/hr) has been studied by thermodilatometry, X-ray diffraction and electron microscopy. The results permit the determination of the transformations occuring in different temperature ranges during tempering.

The results indicate that after tempering at temperatures below 450°C, the structures keep always their ausferritic characteristics. After tempering made from 460 to 550°C, the carbon enriched austenite is decomposed into ferrite and silico-carbides, and the oversaturated ferrite is transformed into equilibrium ferrite and carbides. The decrease of the austenite carbon content, associated to the carbide precipitation, leads to a secondary martensite formation obtained during quenching after tempering. Hardening and brittleness of ADI after tempering are caused by the presence of this secondary martensite. Several types of carbides have been identified in this tempering stage, orthorhombic silico-carbide (G) and monoclinic Fe-C-Si-Mn carbide. The morphology of the carbides is due to the different precipitation site (α or γ).

INTRODUCTION

In recent years with the achieved combination of high strength and high toughness, austempered ductile iron has already found many applications, such as, in gears, crankshafts etc.[1,2] The engineering use of these materials can be expected to continue and to expand. The characteristics of the ADI microstructure is the distribution of an acicular bainitic ferrite in a high carbon austenitic matrix. The continuous high carbon face centered cubic austenite phase plays the most important role for the high ductility of ADI.

In this case, and at room temperature, it is known that austenite is not a thermodynamically stable phase. Serious transformations of this phase can occur under certain conditions, eg: 1) The high carbon enriched austenite can transform to ferrite plus carbides during the second stage of the bainitic reaction,. This second stage takes place during long periods of austempering or during tempering of ADI at high temperature[3,4]. 2) The high carbon enriched austenite may transform to martensite under externally applied loads such as tensile stress, alternating fatigue stress and frictional stress [5,6]. Under these conditions, the strain-induced martensite is produced by plastic deformation. As found in earlier investigations [7] a reduction of the carbon enriched austenite level decreases the tensile ductility.

Although the ADI castings have been used more and more at elevated temperatures, little is known about the structural evolution during tempering in these complex and heterogeous materials. The aim of this study was the study of the ADI structure evolution during an anisothermal tempering.

EXPERIMENTAL METHODS

Materials

Cast iron production was carried out using a high frequency furnace. The chief raw material was pig iron, with an extremely low S content to which small quantities of steel scrap, Fe-Si (75%Si) alloy and electrolytic copper, nickel or manganese were added. The spheroidizing treatment has been carried out by using a commercial spheroidizing agent (Fe-Si-Mg 10% alloy). The molten metal has been poured into a 50mm Y-shape green sand mould.

In this study a Mn-Ni-Cu alloyed ductile iron has been used, and its chemical composition was 3.40% C, 2.35% Si, 0.55% Mn, 0.92% Ni, 0.49% Cu, 0.004% S, 0.063% Mg. In its as-cast state this alloy presented a mixed ferritic perlitic matrix and had a nodule count of 200 nodules/mm^2.

Heat treatment

Test specimens 10mm×30mm×35mm taken in an Y-block (10mm from its bottom) were machined and were austempered for various times and temperatures in order to produce a variety of austempered microstructures. The following heat treatment cycles has been used:
-- austenitisation at 930°C for 0.5 hours;
-- austempering in a salt bath at 380°C for various times;
-- water quenching at the end of austempering heat treatment.

Thermo-dilatometric study

Transfomations during tempering at a rate of 300°C/hrs under vacuum were registered using a differential thermo-dilatometer type "Adamel D.I. 24" (their specimens: 5mm×6mm ×20mm). After attaining the chosen temperature the specimens have been submitted at an air quenching operation.

Hardness and magnetic measurement

The Vickers hardness measurements were performed on each specimen before and after tempering (30 kg load). The magnetic intensity was measured by using a commercially available magnetic balance, known as the "Magne-guage". The Magne-guage measurement is closely related to the retained austenite content.

Microstructural analysis

General characterization of the microstructure has been made by optical microscopy and a JEOL 200CX transmission electron microscopy (TEM) operated at 120kV.

The structural evolution has also been identified by X-ray diffractometry and followed by using a Differential Thermal Analysis (DTA) apparatus.

RESULTS AND DISCUSSION

The evolution of upper bainitic structures during tempering (heating rate V=300°C/h) is shown in Fig. 1. The microstructures before tempering (fig.1a) present a classic feathery appearance of ferrite in a carbon enriched austenite matrix. Tempering at temperatures lower than 450°C doesn't start up a change of this characteristic structure. The carbon enriched austenite transforms into ferrite and silicocarbides after tempering at 450°C and higher temperatures (Fig.1b,c). In Fig. 1c a little amount of secondary martensite has been formed during the air quenching operation after tempering can be detected. The amount of this type of martensite depends on the tempering temperature. Figure 1d shows that secondary martensite is present in large amount in the specimens tempered at 490°C and 510°C. Tempering at 550°C produces a complete tranformation of the "austenite-ferrite" structure to an equilibrium ferrite and iron carbides structure (Fig.1e,f). The acicular structural characteristic, which was initially present in ADI, disappears. It is important to note that this final structure is not identical to the bainitic structure obtained after the second stage of the bainitic reaction (4) because the applied transformation temperature is higher.

The structural changes of ADI during tempering was also followed by differential thermodilatometric analysis which are represented on Fig. 2. The principal characteristic of the carbon enriched austenite is an expansion up to 450°C. For the early austempering time (5min), a very small contraction is also observed from 100 to 200°C and this is related to a decomposition of martensite. This reaction is classical and martensite α' is decomposed into ε (epsilon carbide) and pseudo cubic martensite containing a lower carbon content. The austenite begins to transform into ferrite and silicocarbides when the temprature attains 450°C. This transformation induces a contraction which is associated to an exothermic peak (b) in the DTA curve (Fig. 3). The decomposition temperature of the austenite becomes more and more higher and a new and additional expansion appears after austempered for 90min. The longer is the austempering time, the higher is the start temperature of the austenite decomposition range. Figure 3 shows very well that the exothermic peak (b) induced by the decomposition of austenite is shifted to higher temperatures when the austempering time at 380°C increases. This means that the carbon enrichment in austenite can delay the decomposition of this phase.

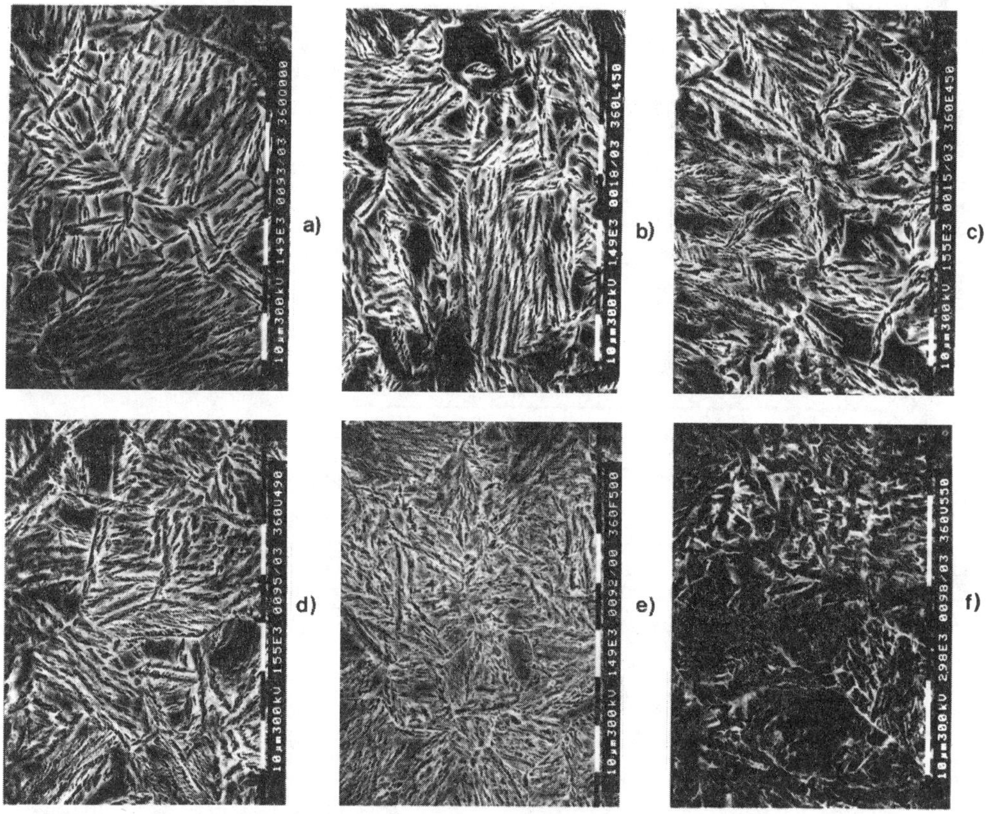

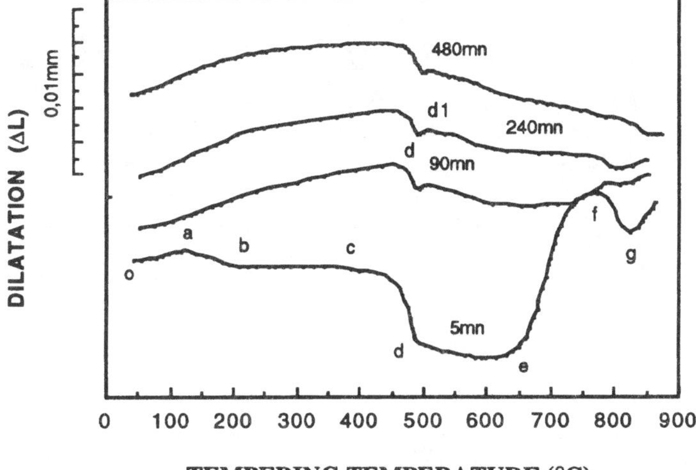

Fig. 1 *Influence of the tempering temperature upon the ADI microstructure*
(austenized at 930°C for 0.5 hrs. and austempered at 380°C for 6 hrs.)
(a) No tempering　　　　(b) Tempering at 450°C　　(c) Tempering at 480°C
(d) Tempering at 490°C　(e) Tempering at 515°C　　(f) Tempering at 550°C

TEMPERING TEMPERATURE (°C)

Fig. 2 *Thermodilatometric analysis of ADI after various preliminary bainitic heat treatments*

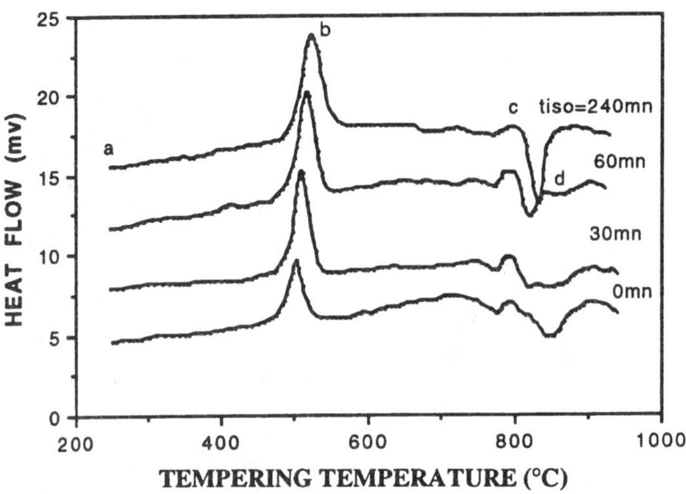

Fig. 3 DTA analysis of ADI after various preliminary bainitic heat treatments

Compared to a martensitic structure decomposition, the transition of unstable carbides to cementite during tempering begins at higher temprature for an ADI structure (480°C for α' structure and 580°C or more for ADI). This result may be related to different types of carbidse precipited during the decomposition of austenite. For an ex-martensitic structure, the carbide transition follows a process $\varepsilon{-}\chi{-}\theta$. It is known that the Hägg carbide "χ" is metastable and will be transformed to cementite at about 480°C during continuous heating conditions [8]. In the case of an austenite-ferrite structure, the silicocarbide precipitates are more stable than the χ carbide. The transition of silicocarbide to cementite occurs at higher temperatures.

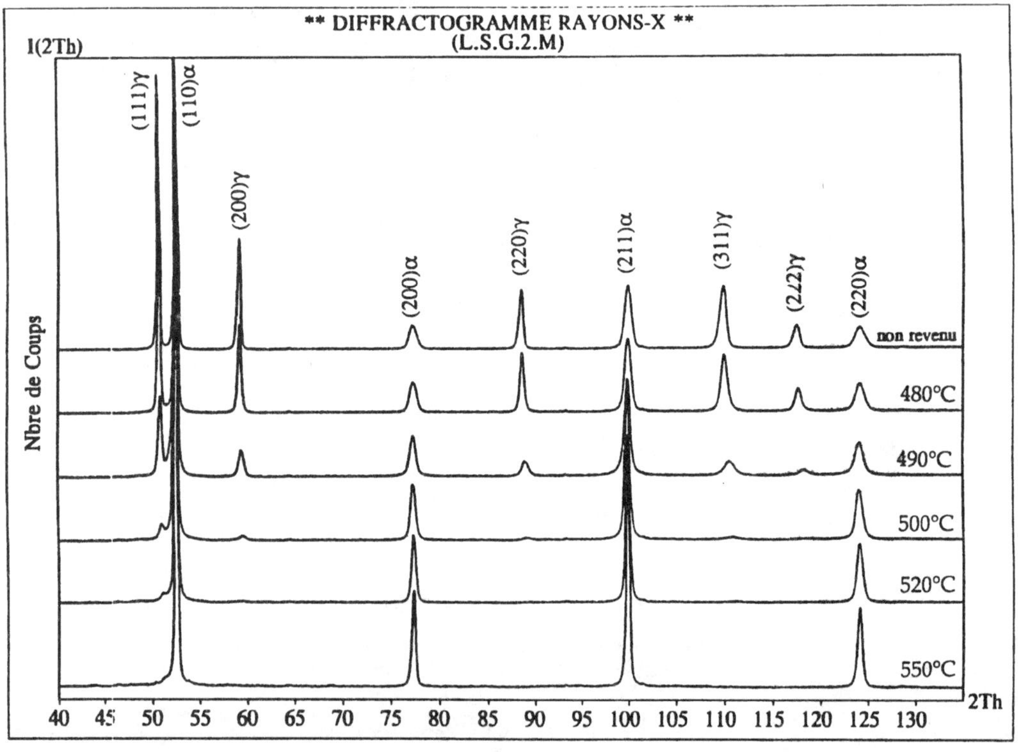

Fig. 4 Superimposed diffractometer traces from a number of ADI specimens tempered at different temperatures

X-rays diffraction, Vickers hardness and magnetic measurement has been used to determine macroscopically the evolution of the microstructure and the properties during the tempering process. This macroscopic structural information can be observed in Fig. 4, where several diffraction patterns of ADI, tempered at temperatures up to 550°C, have been shown. The austenite peaks for $(111)\gamma$, $(200)\gamma$, $(220)\gamma$, $(311)\gamma$ and $(222)\gamma$ begin to decrease from about 450°C. This change of the austenite content is very distinct after tempering at temperatures from 490 to 515°C. After tempering at 550°C, the X rays diffraction patterns shows a ferritic structure which means that a fully decomposition of austenite is obtained. However, it is necessary to indicate that the kinetics of the austenite decomposition depends on the heating rate during tempering. For example, it has been reported that 1% residual austenite was still found in the matrix after ADI tempering at 600°C [9].

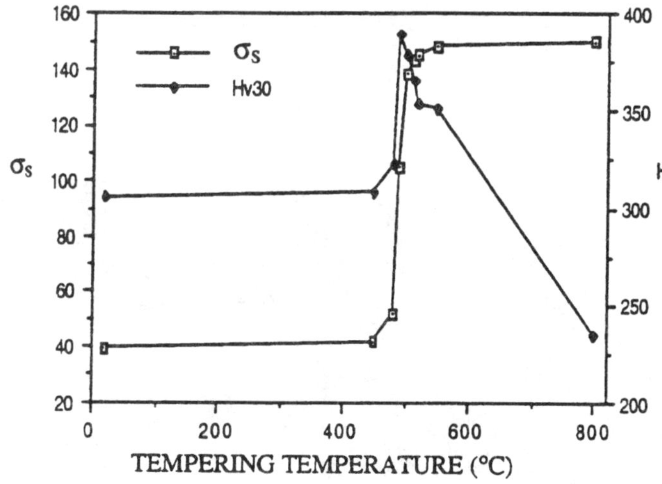

Fig. 5 Influence of tempering temperatures on hardness and magnetic intensity of ADI (austenized at 930°C for 0.5 hrs. and austempered at 380°C for 6 hrs.)

Fig. 5 represents the evolution of hardness (Hv30) and magnetic intensity (σ_s) after tempering at different temperatures in ADI. Both Hv and σ_s begin to increase at about 450°C and reach maximum values at 515°C and 600°C respectively.

This phenomena, which is referred to as secondary hardening, is probably associated to a carbide precipitation occuring during the tempering process or a secondary martensite formation during the subsequent quenching after tempering. In fact, the secondary hardening has been much studied in as-quenched martensite alloyed Cr, Mo, V, W, and Ti steels during tempering [10]. This hardening process is considered as a type of age-hardening reaction, in which a relatively coarse cementite dispersion is replaced by a new and much finer alloy carbide dispersion.

In ADI condition, the chemical composition and microstructure make the transformation characteristics different. It is noted in Fig. 5 that the maximum hardness corresponds to a magnetic intensity $\sigma_s=100$ and the full decomposition of austenite ($\sigma_s>145$) accompanies a fall of the hardness. This means that the secondary martensite formation is the most important factor of the secondary hardening. Early works have given more precision about the relation between the secondary martensite formation and the alloying elements addition [11]. It has been demonstrated that the Mn addition in ADI can lead to secondary martensite formation after tempering.

This martensitic transformation can also be observed by transmission electron microscopy (Fig.6). Martensite needles with the so-called "zig-zag" morphology form in austenite between ferrite plates. In Fig. 6 b, one martensite plate cuts across a austenite grain. This martensite, formed by a sudden shear process in austenite lattices is always at the origin of microcracks and therefore at the origin of loss of matter [12].

Transmission electron micrographs (Fig. 7) show changes in the matrix structure that develop during tempering of ADI at various tempratures. The microstructures always keep their ausferritic characteristics after tempering at a temperature under 450°C (Fig. 7a). The ferrite presents a plate form with a pointed end that extends into austenite. The presence of carbide precipitates has not been detected by electron diffraction neither in austenite nor in ferrite during this tempering stage.

diffraction the presence of carbide precipitates neither in austenite nor in ferrite during this tempering stage.

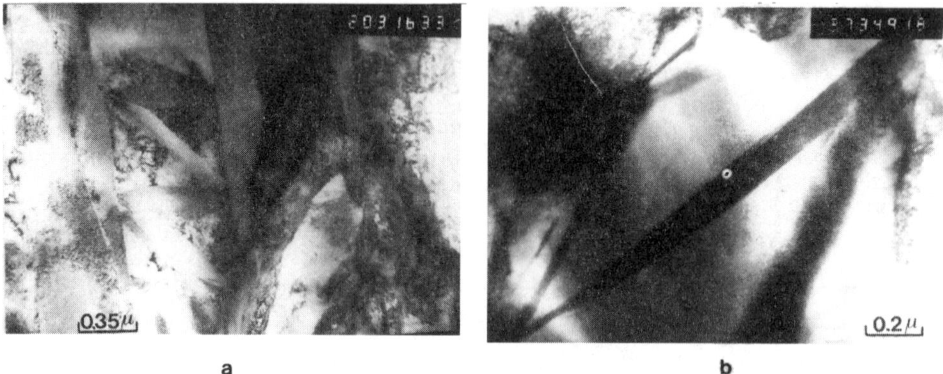

a   b

Fig. 6 TEM images showing the secondary martensite in austenite matrix
    (a) Tempering at 490°C (b) Tempering at 480°C

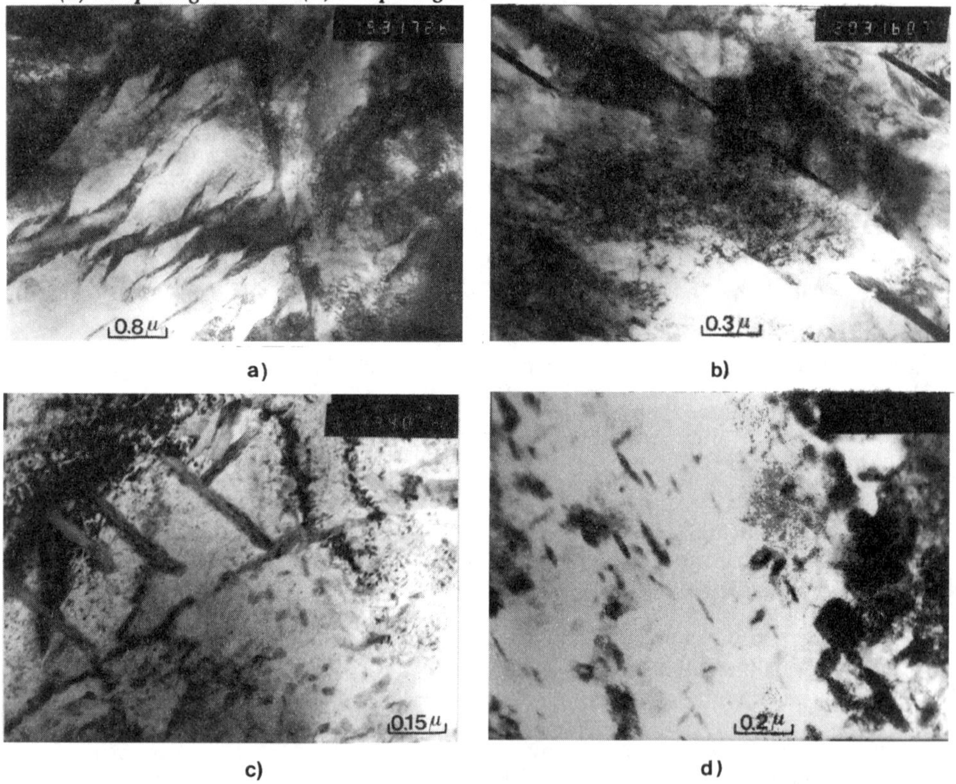

a)   b)

c)   d)

*Fig. 7 TEM images showing the influence of tempering temperature on **ADI** microstructure*
    (austempered at 380°C for 6 hrs.)
    (a) Tempering at 450°C   (b) Tempering at 490°C   (c) Tempering at 512°C
    (d) Tempering at 600°C

Fig. 7b shows a bright field image from the specimen tempered at 490°C. The carbides precipitate at α/γ interface then grow in austenite. This carbide nucleation at α/γ interface may be due to the fact that the carbon diffuse from supersaturated ferrite to this zone during the heating. Sometimes, the carbide nucleation in austenite is also found.

Reasons for the instability of ADI microstructure at elevated temperatures include the supersaturation of carbon in the face-centered cubic austenite and in the body-centered ferrite, the strain energy associated with the fine dislocation or twin structure of the austenite and the interfacial energy associated with the α/γ boudaries. The supersaturation of carbon provides the driving force for a carbide formation.

Fig. 7c shows the morphology of precipitated carbides, which present a well-defined rod morphology in austenite and an irregular form at a/g boudaries.

Electron diffraction analysis revealed that the nature of carbide precipitates is not uniform and are neither Hägg (χ) carbide nor cementite for tempering at 460 to 550°C. All efforts to extract carbides for X-ray investigation were resultless, probably because of the plate shaped morphology. By using the crystalline specimen tilting method, two types of carbide structures have been indexed. One corresponds to a Fe-C-Si orthorhombic carbide, which was initially determined by SCHISSLER from a upper bainitic Fe-C-Si alloy[4]. Another corresponds to a new Fe-C-Si-Mn monoclinic carbide[13]. Details of the microscopy analysis and other micrographs will be published elsewhere. Coexistence of several carbides may be due to chemical composition heterogeneity in ADI matrix and different decomposition phases (γ or α).

From about 560°C, cementite first appears in the microstructure with particles up to 12 nm long and 4 nm in diameter (Fig. 7d). During tempering, the most likely sites for the nucleation of the cementite are the silico-carbide interfaces with the matrix, and as the Fe_3C particles grow, the silicon-carbides gradually disappear. With an increasing in tempering temperature, the cementite particles undergo a coarsening process, which is encouraged by the resulting decrease in surface energy.

CONCLUSIONS

1. The microstructure of ADI is unstable under certain conditions. With a continous heating at 300°C/hr, the austenite phase begins to decompose at 450°C. This decomposition leads to a qualitative change of mechanical properties, especially the ductility and toughness which decrease dramatically.

2. A secondary hardening phenomena has been observed during tempering at about 500°C. This phenomena is principally relatated to the secondary martensite formation which occurs during the subsequent quenching after tempering.

3. The decomposition of ADI microstructure begins by the transformation of carbon enriched austenite into ferrite and silico-carbides, followed by the decomposition of supersaturated ferrite into equilibrium ferrite and carbides. Several types of carbide were identified in tempering between 460 and 550°C, eg: orthorhombic silico-carbide and monoclinic Fe-C-Si-Mn carbide. These carbides will transform into cementite at tempering temperatures higher than 550°C.

REFRENCES

[1]   E. Dorazil, High Strength ADI, England, (1991).
[2]   Y.C.LIU, J.M. SCHISSLER and J.P. CHOBAUT, Traitement Thermique, June-July (1994) 49-55.
[3]   Y.C.LIU, J.M. SCHISSLER and A. MUNTEANU, Revue de Métallurgie, 5 (1994) 815-826
[4]   J.M. SCHISSLER, Thèse d'état, Nancy (1972), France.
[5]   P. MAYR, H. VETTERS and J.WALLA, Proceedings of 2nd World Conference on ADI, Ann Arbor, March 1986 p171-178.
[6]   R. BÖSCHEN, H. VETTERS, P. MAYR and P.C. RYDER, Praktische metallographie, 25, 11 (1988) 524-531.
[7]   J.M. SCHISSLER, Y.C. LIU, J.P. CHOBAUT and H. VETTERS, 9th Inter. Congress on Heat treatment and Surface Engineering, 26-28 Sept. 1994, Nice, France.
[8]   A. Simon, Thèse d'état, Nancy (1972), France.
[9]   B. Q. WEI, World Conference on ADI, 1991 Chicago, USA.
[10] G. KRAUSS, Principles of Heat Treatment of Steel, ASM (1980).
[11] J.M. SCHISSLER, J.P. CHOBAUT and P. BRENOT, AFS Trans. 1988 475-488.
[12] R.C. VOIGT, H. DHANE and L. ELDOKY, 2nd Int. Conf. on ADI, Ann Arbo, MI (1988).
[13] Y.C. LIU, Doctoral Thesis of Nancy University, Nancy, March 1994, France

Advanced Materials Research Vols. 4-5 (1997) pp. 415-420
© *1997 Scitec Publications, Switzerland*

Austempering Kinetics in Cu-Mo Alloyed Ductile Iron: A Dilatometric Study

M.M. Cisneros-Guerrero[1], R.E. Campos-Cambranis[2], M.J. Castro-Román[3] and M.J. Pérez-López[4]

[1] Instituto Tecnológico de Saltillo,

Blvd. V. Carranza 2400, A.P. 600, 25280 Saltillo, Coahuila, México

[2] Universidad de San Luis Potosí,
Av. Dr. Manuel Nava No. 2, 78210 San Luis Potosí, S.L.P., México

[3] Formerly Instituto Tecnológico de Saltillo,
now at CINVESTAV Saltillo, Carr. Saltillo-Monterrey km 13, A.P. 663, Saltillo, Coahuila, México

[4] Instituto Tecnológico de Zacatecas,
Dom. conocido "La Escondida", A.P. 245, 98000 Zacatecas, Zac., México

Keywords: Ductile Iron, Austempering, Dilatometry, Mechanical Properties, Microstructural Evolution

ABSTRACT

The austenite transformation during austempering was studied in a ductile iron containing 0.28% Cu and 0.28% Mo. Progress of the austenite decomposition was monitored in-situ by dilatometry and from quenched specimens. In both cases, austenitization at 870 °C for 120 minutes was followed by isothermal reaction at either 370 °C or 315 °C. As-quenched specimens were obtained for austempering times in the range of 3-420 minutes. The outcome of this work suggests that the dilatometric curves enable the determination of an acceptable processing window for the investigated ductile iron. Also, the improvements in tensile strength, ductility and impact energy can be correlated with the dilatometric curves.

INTRODUCTION

Austempered ductile irons, (ADI), are potentially important in engineering applications due to their unique combination of high strength, good ductility, and toughness. However, the heat treatment parameters used in making ADI's with improved properties are strongly influenced by the alloy additions. Hence, it is critical to obtain a processing window for each ADI alloy. In the present work, the microstructural evolution of a 0.28%Cu-0.28%Mo ductile iron during austempering was investigated. Dilatometric experiments were carried out and the dilatometric curves were associated with the microstructural evolution. Also, tensile and impact strength measurements coupled with X-ray diffraction measurements of austenite fractions were correlated with the dilatometric results.

EXPERIMENTAL PROCEDURE

The base metal was produced from nodular cast iron and steel scraps in a coreless frequency induction furnace. The nodulizing treatment was carried out in a tundish covered ladle using 1.8 % of Fe-Si-5%Mg alloy. The melt was inoculated using 0.53% ferrosilicon (75%Si) during the transfer to the ladle and poured into Y shaped block silicate and CO_2 bonded molds. Tensile and impact specimens were machined from the bottom section of the Y blocks according to ASTM 897M-90 and ASTM 327M-91 standards, respectively. Also, 2 mm diameter dilatometric specimens of 12 mm in length were cut from these sections.

The specimens were austenitized at 870 °C for 2 hours, quenched and then held isothermally in a salt bath at either 370 °C or 315 °C for various times ranging from 3 to 420 minutes. This was followed by a water quench. A similar thermal cycle was given to the dilatometric specimens using a quench dilatometer Adamel Lhomargy LK-02. A Rigaku D diffractometer was employed for X ray diffraction evaluations on sample wafers using Cu Kα radiation. The austenite and ferrite fractions were estimated by direct comparison [1] using $(220)\gamma$ and $(211)\alpha$ peaks. Finally, the resultant tensile and impact properties of the as-heat treated specimens were experimentally determined.

RESULTS AND DISCUSSION

<u>As cast specimens</u>

The chemical composition of the as-cast alloy was: 3.56% C, 2.76% Si, 0.29% Mn, 0.28% Cu, 0.28% Mo, 0.042% Mg, 0.008% S and 0.017% P. The structure presented a nodularity higher than 95 %, with a nodule count of 160 nodules/mm², and a matrix of about 29.1% ferrite and 62.4% pearlite. The mechanical properties were: UTS: 697 MPa, Yield Strength: 392 MPa, %Elongation: 5.8 and Impact Energy: 26 Joules.

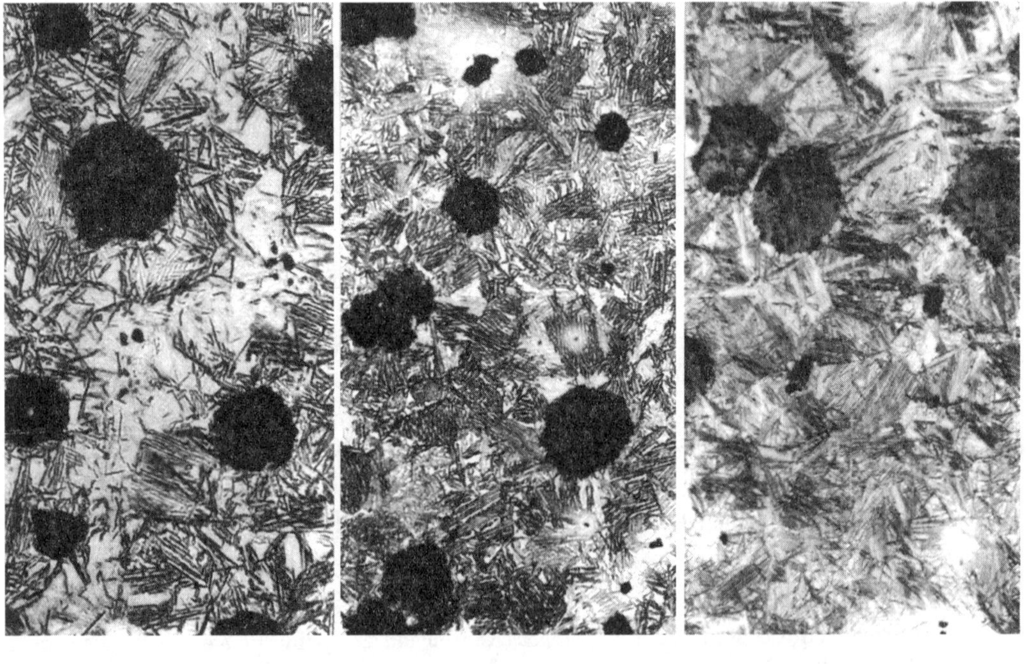

Figure 1   Optical micrographs of 0.28%Cu and 0.28%Mo alloys austenitized at 870 °C, austempered at: a) 370 °C (30 min), b) 370 °C (90 min) and c) 315 °C (90 min).

<u>Microstructural evolution</u>

Photomicrographs of the Cu-Mo alloy austempered at either 370 °C or 315 °C are given in Figure 1. The microstructure consists of ausferrite, stage I transformation product, and martensite formed during water quenching of the untransformed austenite; the austenite which does not contains enough carbon to become stable at room temperature transformes to martensite. The ausferrite structure is made up of carbon-supersatured ferrite platelets (α), mixed with high carbon austenite, (γ), according to the reaction: $\gamma = (\alpha)+(\gamma)$. The ausferrite is found next to or near graphite nodules and continue to grow in the intercellular regions, which are the last to transform. Hence, the fraction of martensite continually diminishes as the austempering time increases as illustrated in Table 1.

This table also shows that the stage I kinetics is faster at 315 °C than at 370 °C. This result emphasizes the importance of temperature on the kinetic stage I. Moreover, tempering at 315 °C led to a relatively fine ausferrite structure. This can be explained from the higher degrees of austenite undercooling coupled with a reduction in carbon diffusivity, both of which strongly favour the nucleation of ferrite needles.

Time (min.)	10	15	30	45	60	90	120	180	300	420
% Martensite (315 °C)	48.5	12.6	3.1	2.4	*	*	*	*	*	*
% Martensite (370 °C)	81.5	55.1	14.5	10.4	6.3	4.6	2.8	1.6	*	*

Table I　Martensite volume percent at different austempering times. * No martensite detection

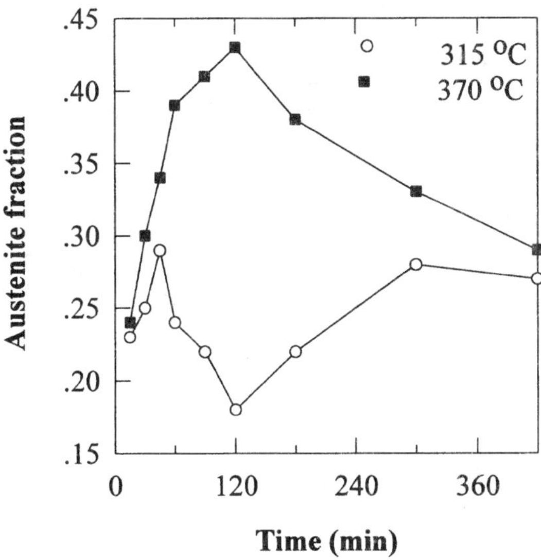

Figure 2.　Effect of austempering time on the austenite fraction
estimated by diffraction analysis.

<u>Diffraction Analysis</u>

The estimated volume fraction of high carbon retained austenite, X_γ, was deduced from diffraction patterns. The parameter X_γ first increased at the beginning of the treatment at both austempering temperatures, as illustrated in Figure 2. This increment is associated with the development of stage I. At 370 °C, X_γ exhibited a maximum at 120 min., followed by a drop which can be related to stage II progress. In the case of specimens treated at 315 °C, the X_γ maximum was found 75 min. earlier than in the 370 °C treatment; the time taken by the decomposition of high carbon austenite diminished as the austempering temperature decreased. In the last tempering period (at 315 °C), X_γ increased again. An explanation of the exhibited behavior can be given by the observations of carbide interference peaks in the (220) austenite peak used to estimate X_γ. Reflections of the important MoC and Mo_3C_2 carbides coincide with the (220) austenite reflection, however, the existence or absence of these carbides needs to be corroborated by transmission electron microscopy.

<u>Mechanical Properties</u>

The effect of temperature and duration of austempering on the tensile properties is shown in Figure 3. The relatively bad properties at the beginning of the treatment are attributed to the high martensite content. A good combination of properties is obtained after tempering for 120 minutes at 370 °C. In this case the resultant mechanical properties are: UTS = 1026 MPa, Yield = 754 MPa, %Elongation = 7.9% and Impact Energy = 101 Joules. Also, this austempering treatment is associated with a relatively small martensite content (2.8%), and a large X_γ value of 0.43. Further austempering does not result in better properties (i.e. the elongation and the impact energy tend to

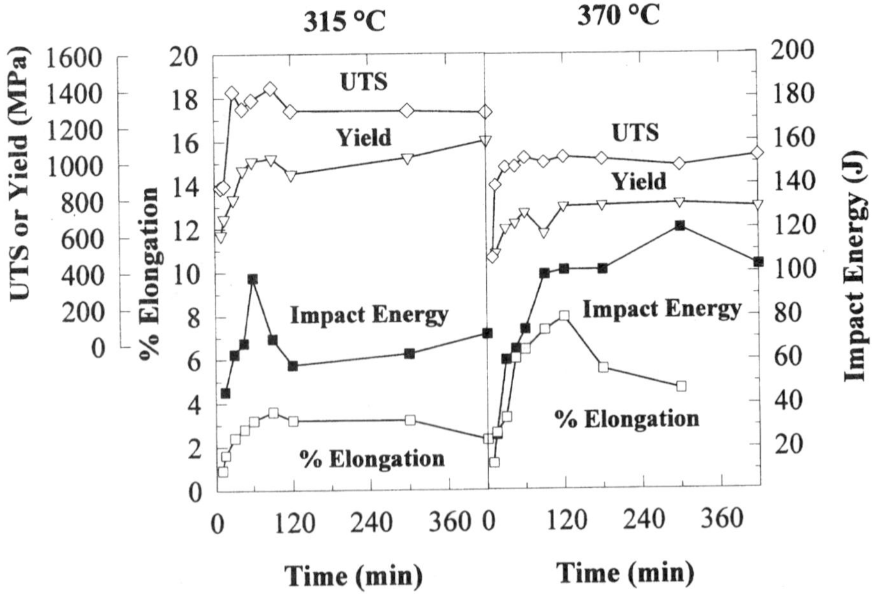

Figure 3. Mechanical properties versus austempering time at either 370°C or 315 °C treatments.

decrease as a consequence of carbide precipitation reactions (stage II)). In the 315 °C treatment a good combination of properties is obtained by tempering for 60 to 90 minutes. The UTS and Yield values are superior to those obtained at 370 °C, but this is at the expense of the elongation and the impact energy which are inferior in the 315 °C treatment.

<u>Dilatometric study</u>

The effect of the austempering temperature on the Cu-Mo cast dilatometric curves is illustrated in Figure 4. After the first minutes of isothermal transformation at either 315 °C or 370 °C the curves exhibit an important expansion that is associated to stage I development. This in accordance with previous reports on alloyed and unalloyed ductile irons [3-7].

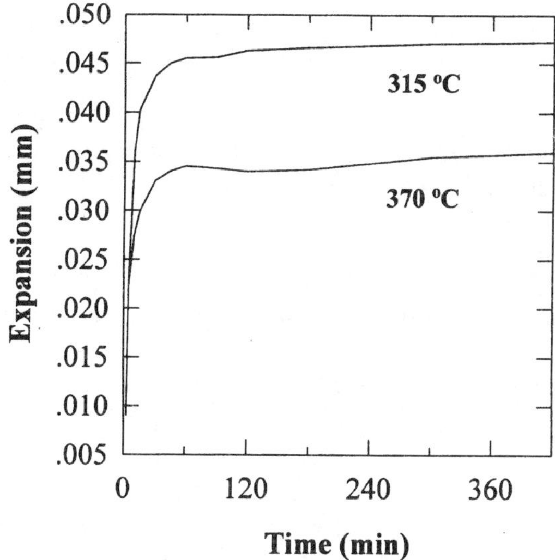

Figure 4   Dilatometric   curves   obtained   during   the austempering of alloy studied here.

The end of the stage I is marked by a plateau in the curve. Moreover, the second expansion after the plateau is related to carbide precipitation reactions. In the studied Cu-Mo alloys the carbides appear earlier at lower temperatures according with previous reports in the case of Cu-Ni alloys [8,9]. In both temperature treatments, the second expansion marks the begining of ductility losses.

CONCLUSIONS

Austempering treatment of the 0.28% Cu and 0.28% Mo ductile irons improves mechanical properties such as tensile strength, ductility and toughness. In the case of isothermal treatments at 315 °C, the evolution of austenite fraction during the stage II reaction is not clear due to possible interference from carbide peaks with the austenite diffraction peak used to estimate X_γ. The process window was determined from dilatometric curves for the ADI studied at 370 °C and 315 °C austempering treatment temperatures. Moreover, in both treatments the austempering time used to obtain the optimum combination in tensile strength, ductility, and impact energy can be inferred from dilatometric curves. The decrease in ductility observed in the isothermal treatments given coincides with the second expansion recorded in the dilatometric curve.

ACKNOWLEDGMENTS

The authors would like to thank CONACyT for its financial support of this work. We are also grateful to H. López for critical review of the manuscript.

REFERENCES

[1] B. D. Cullity, "Elements of X-ray Diffraction", Addison-Wesley Publishing Co., (1978).

[2] T. N. Rouns, K. B. Rundman, D. M. Moore, AFS Trans., **92**, 815 (1984).

[3] B. V. Kovacs, J. R.Keough, Conf. Proc. On Advances in High Integrity Castings, ASM International, 91 (1988).

[4] H. Nieswaag, J. W.Nijhof, "The Physical Metallurgy of Cast Iron", MRS Symp. Proc., **34**, Elsevier Sci. Publ. Co., 411 (1984).

[5] Y. Ueda, M. Takita ,Conf. Proc. on Austempered Ductile Iron, ASME-AMAX, 141 (1986).

[6] R. E. Campos-Cambranis, "Estudio Cinético y Microestructural del Proceso de Austemperizado de Hierros Nodulares", Master Sc.Tesis, Instituto Tecnológico de Saltillo, México (1993).

[7] Y. C Liu, J. M. Schissler, A. Munteanu, La Revue de Métallurgie-CIT/Science et Génie des Matériaux, 815 (Mai 1994).

[8] M. Grech, AFS Trans., **98**, 345 (1990).

[9] M. Grech, P. Bowen, J. M. Young, Proc. 1991 World Conference on ADI, AFS Inc., **2**, 338 (1991).

Advanced Materials Research Vols. 4-5 (1997) pp. 421-426
© *1997 Scitec Publications, Switzerland*

On the Characterization of Nodule Size Distributions of Commercial Ductile Iron Alloys

S.-Y. Huang and F.J. Bradley

Department of Materials Science and Engineering, University of Wisconsin-Madison;
1509 University Avenue, Madison, WI, USA

Keywords: Ductile Iron, Nodule Spatial Distributions, Stereological Parameters, Section Diameter Method, Section Area Method, Stochastic Simulation

ABSTRACT

Stochastic simulation using input unimodal and bimodal lognormal spatial distributions is applied to assess several methods for determining spatial particle distributions and estimating stereological parameters. The conventional Schwartz-Saltykov section diameter and Saltykov section area methods and modified versions of these two methods are considered. The modification essentially involves determining the class interval sizing on the basis of a constant minimum planar section diameter rather than the maximum observed planar section diameter. Stochastic simulation results indicate that section area methods are preferable to section diameter methods in the analysis of multimodal lognormal particle distributions such as those observed in ductile iron microstructures. This is primarily because of the existence of a mode in the distribution at the small size range that is not effectively resolved using the relatively coarse uniform class interval sizes employed in the section diameter methods. Of the two section area methods, the modified version is recommended as it possesses an additional advantage of a fixed class interval scheme which facilitates direct comparison of microstructural characterization results for a wide variety of specimens.

INTRODUCTION

Accurate determination of spatial nodule size distributions of ductile iron alloys is important with regard to understanding and interpreting the influence of processing variables on solidification behavior, to developing solidification and solid state transformation models which realistically account for the nucleation characteristics and time-dependent evolution of microstructure, and to correlating mechanical properties and microstructure on the basis of fundamental stereological parameters. In this paper, conventional and modified versions of the section diameter and section area methods for determining spatial nodule size distributions are assessed.

SECTION DIAMETER AND SECTION AREA METHODS

In previous work published in the foundry literature, the Schwartz-Saltykov section diameter method has been used most often to determine spatial nodule size distributions [1-3]. An alternative method developed by Saltykov is the section area method. In each case an estimate of the spatial distribution of spherical particles in a polydispersed system is obtained from the planar distribution of sections measured on a plane of polish. A fundamental assumption of both methods is that the maximum observed section diameter (d_{max}) is equal to the maximum sphere diameter (D_{max}) in the spatial distribution. The methods differ in that the section diameter method utilizes linearly scaled class intervals while the section area method uses geometrically scaled class intervals.

In the Schwartz-Saltykov section diameter method [4], the measured section diameters are subdivided into k equally-spaced discrete size groups or class intervals of width Δ, where $\Delta = D_{max}/k$. Expressed in matrix form, the fundamental equation of the section diameter method is

$$\{(N_A)_i\} = [a_{ij}]\{(N_V)_j\} \tag{1}$$

where

$$[a_{ij}] = \Delta \sum_{i=1}^{k} \sum_{j=1}^{k} \left[\sqrt{j^2 - (i-1)^2} - \sqrt{j^2 - i^2} \right] \tag{2}$$

and $(N_A)_i$ is the number of planar sections per unit area in class interval i, $(N_V)_j$ is the number of nodules per unit volume in class interval j, and $[a_{ij}]$ is the upper triangular transformation matrix.

Details of the Saltykov section area method are given in reference [5]. A logarithmic particle diameter scale is used for the class intervals with the scale factor ξ being given by $\xi = 10^{-0.1} = 0.7943$. Saltykov presented a formula for determining the size distribution of spherical particles where the number of class intervals k is twelve or less. Vander Voort [6] extended Saltykov's original analysis to develop the following general equation for determining the number of particles per unit volume for the case of 15 or fewer classes:

$$\begin{aligned}
(N_V)_j = \frac{1}{D_j}(&1.645N_{A,i} - 0.4542N_{A,i-1} - 0.1173N_{A,i-2} - 0.0423N_{A,i-3} \\
&-0.01561N_{A,i-4} - 0.0083N_{A,i-5} - 0.0036N_{A,i-6} - 0.0019N_{A,i-7} \\
&-0.00090N_{A,i-8} - 0.00044N_{A,i-9} - 0.00036N_{A,i-10} - 0.00010N_{A,i-11} \\
&-0.00003N_{A,i-12} - 0.00003N_{A,i-13} - 0.00001N_{A,i-14}
\end{aligned} \tag{3}$$

where D_j is the maximum particle diameter of class j, and the $N_{A,i}$, etc. represent the measured number of sections per unit area associated with the various class intervals. In applying equation (3) to the determination of $(N_V)_j$ for a given class, the calculation is continued until the index of $N_{A,i}$ reduces to zero.

One limitation in basing class interval sizing on D_{max} is that the class intervals are then unique for each specimen and thus direct comparisons of calculated spatial nodule diameter frequency distributions are difficult unless an interpolation algorithm is used to convert the result to a standard class interval sizing. Another consideration in determining nodule spatial distributions and in interpreting and comparing results for different microstructural specimens is the value of minimum planar section diameter d_{min} below which planar section diameters are not measured or incorporated in the spatial distribution calculation. For example, Skaland and Grong [3] used a filter of $d_{min}=7.0$ μm in order to avoid counting planar sections associated with carbides and inclusions. In practice, measurement resolution limits will require the specification of some value of $d_{min}>0$ below which section diameters are not measured. As the lower limit of the smallest class interval (which contains a proportionately large number of planar sections in the case of ductile iron microstructures) is always less than d_{min}, a truncated measure of the number of planar sections in the smallest class interval always results. The extent of the truncation is dependent on the measure value of D_{max} and the minimum filter setting d_{min} chosen for a particular analysis.

In an effort to develop a standard procedure which minimizes the truncation error and yields standard class interval sizings for every analysis the section diameter and section area methods were modified as follows. In both cases class interval sizing is based on d_{min}, rather than d_{max}. For the modified section area method, the lower and upper limits of each class interval $[D_j, D_{j-1}]$ are given by $D_{j-1}=d_{min}(\xi^{1-j})$ and $D_j=d_{min}(\xi^{j})$. The number of class intervals is determined by incrementing j until the upper bound of the highest class interval is greater than the observed d_{max}. In comparison, for the conventional section area method, k is automatically determined by the method and the class interval limits are $D_{j-1}=d_{max}(\xi^{k-j+1})$ and $D_j=d_{max}(\xi^{k-j})$. For the modified section diameter method, class interval spacing is set to $\Delta=5.0$ μm (the value of d_{min} normally employed in our nodule size characterization studies), and the number of class intervals is incremented until the upper bound of the highest class interval is greater than d_{max}.

Each of the methods described above yields an estimate of the number of particles per unit volume in each class interval, $(N_V)_j$. The total number of particles per unit volume N_V obtained by summing over all class intervals is used to determine the relative frequency f_j of particles of diameter D_j in class interval j. Once the relative frequency distribution is obtained, the moments of the spatial distribution can be estimated using

$$\mu_m(D) = \int f(D)D^m dD \approx \sum f_j D_j^m \tag{4}$$

where $\mu_m(D)$ is the *mth* moment about the origin of the particle diameter distribution. With knowledge of N_V and the moments, all of the basic stereological parameters for the spatial distribution can be estimated [4,7-8]. For example, mean particle diameter, surface area, and volume are computed using

$$\overline{D} = \mu_1(D), \qquad \overline{S} = \pi\mu_2(D), \qquad and \qquad \overline{V} = \frac{\pi\mu_3(D)}{6} \tag{5}$$

with total surface area per unit volume and volume fraction of particles being given by

$$S_V = N_V \overline{S} \qquad and \qquad V_V = N_V \overline{V}. \tag{6}$$

STOCHASTIC SIMULATION

A stochastic simulation model has been developed to assess the various methods for estimating spatial distributions. The algorithm consists of two main sections, one involving the generation and testing of specified input spatial distributions of spherical particles and the other involving random sectioning of the simulated particles and analysis of the resulting planar distribution of section diameters. Performance of the estimation methods is evaluated by comparing the resultant distributions with the known spatial distribution. Input to the algorithm includes type of distribution, parameters of distribution, permissible range of diameters $[D_{min}, D_{max}]$, and number of particles per unit volume (N_V) or volume fraction of particles (V_V). To limit the number of particles to be generated to approximately 10,000 and produce simulated spatial distributions with particle densities similar to that observed in typical ductile iron microstructures, a simulation space of matrix volume 1 mm^3 is considered.

As in the case of many particulate systems [9], spatial nodule size distributions are lognormal in nature. The polar method developed by Marsaglia and Bray [10] and described by Law and Kelton [11] is used to generate input log-normal, $LN(\mu,\sigma^2)$, distributions of spherical particles. In stage two of the simulation, a specified number of randomly oriented section planes are generated and the diameters of the resulting intercepted planar sections are determined and stored for subsequent analysis using the methods outlined above. Due to space limitations, details of the algorithm for the stochastic simulation will be presented in a forthcoming paper.

DISCUSSION OF RESULTS

Simulation results are presented for the case of a unimodal input lognormal distribution $LN_1(2.938,0.1833)$ and a bimodal lognormal input distribution which is generated as a composite of two unimodal distributions — $LN_2(1.471,0.2776)$ and $LN_3(3.162,0.1133)$ — where the probability of occurrence of each is 0.50. The input distributions were truncated such that generated sphere diameters fell in the range 5.0-85.0 microns. Pertinent data for the simulations are summarized in Table 1. Input data was selected to give values of volume fraction of spheres and planar section density similar to values of volume fraction graphite and nodule count associated with typical thermal analysis ductile iron microstructures.

Table 1
Data for Unimodal and Bimodal Lognormal Stochastic Simulations

Simulation Data	Unimodal	Bimodal
No. of spheres generated	10894	14974
Volume fraction of spheres	0.0869	0.088
No. of section planes	48	48
No. of intercepted planar sections	10352	10705
Planar section density (mm^{-2})	216	223
Minimum planar section diameter (mm)	5	5
Maximum planar section diameter (mm)	80.4	80.0
k : SD(con), SA(con), SA(mod)	13	13
k : SD(mod)	17	16

Tables 2 and 3 contain estimated first, second, and third moments and basic stereological parameters obtained using the conventional (con) and modified (mod) section area (SA) and section diameter (SD) methods for the unimodal simulation. In comparing the estimates of the moments with the known values, it is apparent that the section area methods yield the better estimates with little difference between estimates obtained using the conventional and modified versions. The modified section diameter method gives a slightly better result for moments than does the conventional algorithm. All of the methods yield higher estimates of the moments.

Table 2
Estimated and Known Spatial Moments for Unimodal Simulation

Moment	SD (con)	SD (mod)	SA (con)	SA (mod)	Known
μ_1 (μm)	22.28	21.58	20.95	20.99	20.68
μ_2 (μm^2)	581.7	553.4	537.5	539.0	513.1
μ_3 (μm^3)	17864	16825	16732	16776	15240

Table 3
Estimated and Known Stereological Parameters for Unimodal Simulation

Parameter	SD (con)	SD (mod)	SA (con)	SA (mod)	Known
N_V (mm^{-3})	9922	8352	10688	10662	10894
$\overline{D}$ (μm)	22.3	21.6	20.9	21.0	20.7
$\overline{S}$ (μm)	1827	1739	1689	1693	1612
$\overline{V}$ (μm)	9353	8809	8761	8784	7980
S_V (mm^{-1})	18.1	14.5	18.0	18.1	17.6
V_V	0.0928	0.0736	0.0936	0.0937	0.0869

With regard to estimation of N_V, the two section area methods yield excellent results. The conventional section diameter method under-estimates N_V somewhat, but still gives a reasonable value for the total number of particles per unit volume. The results for mean particle diameter, surface, and volume directly mirror the moment results as indicated by Eqn. (5). The low estimate of S_V and V_V obtained in the case of the modified section diameter method is directly related to the extremely low estimate of N_V. This is attributed to the high number of class intervals (k=17) which occurred as a result of setting the class interval spacing at Δ=5.0 μm. According to Vander Voort [6], 8-12 class intervals generally provide the optimal result for the section diameter method, with 7 as minimum and 15 as maximum.

Tables 4 and 5 give corresponding results for the bimodal simulation. For this case, the section area methods over-estimate the moments and under-estimate N_V, but yield significantly better estimates of these parameters in comparison to the section diameter methods. This is attributed to the relatively coarser class interval sizes in the case of the section diameter methods at the smaller size range where a significant proportion of the measured planar sections are observed in the case of the input bimodal distribution. With regard to the section area methods, both appear to estimate the stereological parameters equally well. The results suggest that one possible approach to improving the moment estimates (and thereby the estimates of the stereological parameters) is to incorporate a calibration factor in the Eqn. (4) to appropriately adjust the moment estimates.

Table 4
Estimated and Known Spatial Moments for Bimodal Simulation

Moment	SD (con)	SD (mod)	SA (con)	SA (mod)	Known
μ_1 (μm)	21.28	20.11	17.50	17.46	16.30
μ_2 (μm^2)	559.5	509.3	433.1	431.7	381.8
μ_3 (μm^3)	17377	15536	13563	13517	11225

Table 5
Estimated and Known Stereological Parameters for Bimodal Simulation

Parameter	SD (con)	SD (mod)	SA (con)	SA (mod)	Known
N_V (mm^{-3})	10481	9455	13661	13705	14974
$\overline{D}$ (μm)	21.3	20.1	17.5	17.5	16.3
$\overline{S}$ (μm)	1758	1600	1361	1356	1198
$\overline{V}$ (μm)	9011	8135	7102	7078	5878
S_V (mm^{-1})	18.4	16.6	18.6	18.6	17.9
V_V	0.0954	0.0860	0.0970	0.0973	0.088

Table 6 gives estimates of the stereological parameters of a typical thermal analysis ferritic ductile iron microstructure. Planar section data were obtained using image analysis from 48 randomly selected

Table 6
Estimated Stereological Parameters for Typical Ductile Iron Microstructure

Parameter	SD (con)	SD (mod)	SA (con)	SA (mod)
N_V (mm^{-3})	11783	9398	13986	15463
$\overline{D}$ (μm)	19.2	19.4	17.0	16.0
$\overline{S}$ (μm)	1525	1497	1286	1178
$\overline{V}$ (μm)	7981	7637	6728	6150
S_V (mm^{-1})	18.0	14.1	18.0	18.2
V_V	0.0940	0.0718	0.0941	0.0951

fields about a 1 cm. radius surrounding the thermocouple tip. In comparing the results in Table 6 with those of the bimodal simulation in Table 5, comparable trends are seen to exist in the estimates of the stereological parameters obtained using the four methods. On the basis of these results, section area methods are recommended for the determination of spatial nodule size distributions. It is noted that lognormal probability plots indicate that ductile iron spatial nodule size distributions consist of truncated mixtures of lognormal distributions, with the components of the mixture presumably being related to distinct graphite nodule nucleation and growth characteristics associated with the various stages of primary and eutectic solidification.

SUMMARY

Stochastic simulation has been used to assess the accuracy of section diameter and section area methods and modifications of these methods for determination of spatial particle size distributions and the estimation of basic stereological parameters. Results show that the appropriate method for a particular analysis is strongly dependent on the nature of the underlying spatial distribution. Vander Voort [6] has previously noted that if the underlying spatial distribution is skewed better results may be obtained using the logarithmically-scaled class intervals employed in the section area methods. Results in this study demonstrate quantitatively that this is the case. However, it is noted in the case of the unimodal input lognormal simulation that the discrepancy between the conventional section diameter and section area methods is not as great as might be expected.

In the case of the bimodal input lognormal simulation, which more closely approximates the type of multimodal lognormal spatial distributions observed in ductile iron microstructures, both section area methods yield significantly better estimates of the moments and N_v than the section diameter methods. Of the two section area methods, the modified version in which class interval spacing is determined on the basis of minimum resolution planar section diameter is recommended. This approach has the additional advantage of facilitating direct comparison of spatial relative frequency distributions. Results to date indicate that estimates of the moments of the partial distribution can be improved by incorporating a calibration factor in the moment equation. Research in improving methods for estimating spatial nodule size distributions is ongoing.

ACKNOWLEDGMENTS

Support for this research provided by the Cast Iron Research Consortium at UW-Madison and NSF grant DDM-9102682 is gratefully acknowledged.

REFERENCES

[1]   E.E. Underwood and J.T. Berry, AFS Transactions, Vol. 89, 755-766 (1981)

[2]   T. Noguchi and K. Nagaoka, AFS Transactions, Vol. 93, 115-122 (1985)

[3]   T. Skaland and O. Grong, AFS Transactions, Vol. 99, 153-157 (1991)

[4]   E.E. Underwood, Quantitative Stereology, Addison-Wesley Publishing Co, 119-123 (1970)

[5]   S.A. Saltykov, Proc. of 2nd Intl. Congress for Stereology, Chicago, 163-167 (1967)

[6]   G.F. Vander Voort, Metallography Principles and Practice, McGraw-Hill, 465-471 (1984)

[7]   J.E. Hilliard, Trans. AIME **242,** 1373-1380 (1968)

[8]   R.T. DeHoff, Trans AIME **233** 25-29 (1965)

[9]   H. Exner, International Metallurgical Reviews, Vol. 17, 25-42 (1972)

[10]   G. Marsaglia and T.A. Bray, SIAM Rev., **6:** 260-264 (1964)

[11]   M.G. Kendall and A. Stuart, Advanced Theory of Statistics Vol. 1 Charles Griffith and Co., London 75-77 (1963)

Advanced Materials Research Vols. 4-5 (1997) pp. 427-432
© *1997 Scitec Publications, Switzerland*

Characterizing the Form of Graphite in Cast Irons Using an Image Analyser

J. Fargues, M. Hecht and M. Stucky

Centre Technique des Industries de la Fonderie,
44 avenue de la division Leclerc, F-92310 Sèvres, France

Keywords: Image Analysis, Shape Criterion, S.G. Cast Iron, Graphite Classification, Mechanical Properties

Following a preliminary study to assess several shape criteria, a method of characterizing graphite that is especially suitable for spheroidal graphite cast iron has been developed.
This method is based on the application of a very rigorous metallographic polishing procedure and on the use of two shape criteria chosen to be complementary.
The sensitivity of the various criteria to elongation of the particles and to irregularities of their contours was the basis for this choice.
Six classes inspired by those of the ISO 945 standard are proposed for the automatic characterization of the shape of the graphite.
After using an original "membership" test, the analysis programme estimates the percentage of the particles in each of the classes.
As an application, the programme was used to study the influence of graphite morphology on the tensile properties of forty specimens of various commercial SG cast irons.

INTRODUCTION

The mechanical and functional characteristics of cast irons are governed, other things beeing equal (matrix, chemical composition, eutectic cells or particles sizes, ...), by the graphite particles morphology.
The use of the graphite form type charts proposed by the standards has a subjective character which is often pointed out. On the contrary, image analysis offers the possibility of an objective and repeatable graphite characterization.
Such a characterization appears to be particularly interesting in the case of cast irons with modified graphite, like SG and CVG cast irons.
Our study aims at establishing a graphite form analysis programme for cast irons.

I - NORMALIZATION

Two almost similar documents are commonly used for characterizing graphite: the French NF A 32 100 document (July 1967) [1] and the ISO 945 international standard (1975) [2]. The ASTM 247 American testing method is quoted in some specifications [3].
All these documents propose different plates (images) which allow the characterization of graphite according to form, distribution and size.
The relation between the different kinds of cast irons and the graphite form type charts proposed by NF A 32 100 and ISO 945 documents are implied. Forms VI and V are mainly corresponding to SG cast irons, forms V and IV to malleable cast irons, form III to CVG cast irons.
One of the conditions imposed by the French standard NF A 32 201 (June 1986) for the supplying of SG iron castings concerns the microstructure, which must contain a minimum of 85% of graphite of V or VI type and no graphite of I or II type at all (according to NF A 32 100 document).

II - MICROSTRUCTURE ASSESSMENT OF GRAPHITE SHAPE IN THE CASE OF SG CAST IRONS

Whenever there is no particular agreement between the partners, the graphite assessment is carried out visually with a microscope.

The operator polishes correctly the sample, chooses a suitable magnification and evaluates the overall of graphite particles corresponding to graphite of VI and V types.

A systematic counting of unsuitable particles can be necessary if, for example, the fraction of that graphite seems to be close to 15%.

The method has an obviously subjective character, although it appears fairly accurate, if practised by an experienced operator.

The precision of the determination depends greatly of the time spent (number of counted particles, fraction and distribution of irregular graphite, ...) and it is to be evaluated by a statistical analysis.

III - POSSIBILITIES OF THE IMAGE ANALYSIS

The latest generation of image analysers which are often coupled with powerful computers permit to treat very efficiently the characterization of metallographical structures.

Generally, the image captured by microscope + camera needs to be submitted to transformations allowing the relevant measurement. These treatments contribute to improve the contrast between the objects and the background, but also to suppress the causes of parasitic informations.

All the required operations can be rapidly carried out. For example, the treatment of 300 particles takes 7 to 8 minutes with our graphite analysis programme.

However, the sample polishing which must be very carefully done will remain a necessity.

IV - SELECTION OF THE FORM CRITERIA

Preserving the principle of a bidimensionnal approach of the problem, the use of several form criteria is possible. They are undimensional numbers calculated from various parameters. Six form criteria were studied (see table 1)

DESIG- NATION	FORMULATION	MEASURED PARAMETERS
PAM [4]	$\dfrac{3 \times Rg^2}{S}$	Geodesic radius Area
PSA	$\dfrac{P^2}{4 \times \pi \times S}$	Perimeter Area
RA	$\dfrac{Fmax}{Fmin}$	Maximal Feret Diameter Minimal Feret Diameter
1/ICA	$\dfrac{Sconvex}{S}$	Convex area Area
INSC	$\dfrac{S}{Sinscr.circle}$	Area Area of the largest inscribed circle
JAP	$\dfrac{Scirc.circle}{S}$	Area of the circumscribed circle Area

TABLE 1   Summary of the criteria studied in this work

Each criterion, when used for the characterization of a particle with a circular contour, has a value close to unity.

On account of their probable influence on the mechanical properties of cast irons, two aspects of the form were particularly studied: the elongation of the particles and the aspect of their contour (irregular or indented).
In order to establish a comparison between the sensitivities of the first five criteria, a characterization of the fourteen geometrical particles shown in figure 1 has been performed. This characterization has been realised in two different ways: - i) calculation from the geometry of the particles - ii) measurement carried out by the image analyser.

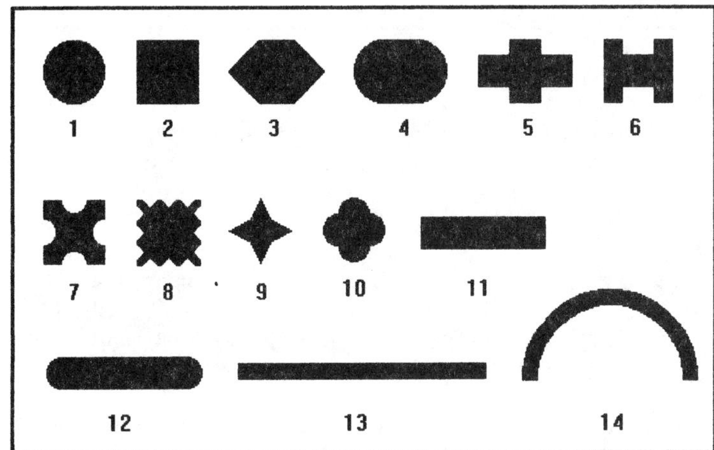

FIGURE 1 : Theoretical pictures used for testing shape factors.

Except for geodesic radius for which a programme has been developped, all parameters used for the calculation of the criteria were measured by using the specific algorithms of the Quantimet 570 image analyser of LEICA.
The results obtained show that PAM and RA criteria are more sensitive to the elongation of the particles than to the aspect of their contour.
On the contrary, PSA criterion takes into account with a good sensitivity the anomalies of the contour. INSC criterion is sensitive to the different studied aspects and 1/ICA presents some limitation in its behaviour.
Concerning the RA criterion, the values determined with the image analyser are very close to the theoretical values. Indeed, the differences which were noted are about 1 to 5%. Furthermore, these values are not very sensitive to the size of the particles.
On the contrary, the values determined for PAM and PSA criteria show a sensitivity to size, especially for the particles with small dimensions. It appeared that the selection of a well adapted magnification allows the determination of satisfactory values for these two criteria, values that differ by 5 to 10% at most from theoretical values.
Study of geometrical particles showed that the geodesic radius can be advantageously replaced by half the maximum Feret diameter (Fmax/2) for the determination of the PAM criterion. Thus, an appreciable saving of time is obtained.
Measures of 1/ICA criterion carried out on image analyser confirm a total lack of sensitivity to elongation of particles if these ones do not present a contour with concavities. So, this criterion was not considered for ulterior investigations.
Table 2 gives for each criterion the sensitivities that were determined in this way.
The various tests carried out have shown that none of the studied criteria can on its own take into account satisfactorily the two aspects of the form that are important in order to perform a correct classification of particles.
   This difficulty, in relation with an imperfect behaviour of the criteria, may be overcome by using two criteria with complementary sensitivities.

CRITERIA	PARTICULARITY OF THE FORM		
	ELONGATION	INDENTATION	INCURVATION**
PAM (Rg)*	XX	X	-
PAM (Fmax/2)*	XX	X	X
PSA	X	XX	-
1/ICA	-	X	XX
RA	XX	-	X
INSC	XX	X ou XX	-
* PAM (Rg): PAM criterion calculated with geodesic radius * PAM (Fmax/2): PAM criterion calculated with Fmax/2 **Concerning particles with elongated form XX: High sensitivity X : Moderate sensitivity - : Low sensitivity			

TABLE 2   : Comparison between the sensitiveness of the various shape factors studied.

V - STUDY OF THE GRAPHITE FORM TYPE CHARTS OF ISO 945 STANDARD

The values of the different shape criteria were determined for each particle belonging to the six images.

It appears clearly that almost all the particles having the same shape criterion value, although belonging to different images (classes), can be classified unambiguously when characterized with two shape criteria having complementary sensitivities.

A careful statistical study pointed out the shape criteria which permit the best identification of the six form classes proposed by the standard [5].

Because the use of only one criterion presents different disadvantages, the method which consists in characterizing each particle by two criteria was considered.

The choice of PAM and PSA criteria was based on the results given by the studies concerning the geometrical particles and the graphite form type charts.

VI GRAPHITE DESIGNATION BY IMAGE ANALYSIS

In order to guarantee a correct evaluation of graphite form and comparative results (from an apparatus and/or laboratory to an other one), it would be necessary to define, control and specify well the working conditions and procedures, that is to say [5]:
- preparation of the samples (respect of the contour and of the total graphite fraction),
- microscope adjustment,
- magnification adjustment,
- image treatment (avoid altering the image and segmentation of the objects),
- definition of minimal and maximal sizes of the analysed particles.

Whatever the chosen adjustment of those factors may be, the evaluation of the graphite form by image analysis appears to be a precious method of controlling a fabrication or of characterizing it. After agreement between the different partners, it could also be used as a reception control or for a standardized document.

VII - METHOD SET UP BY C.T.I.F.

A graphite form analysis programme was realized, operating with two shape criteria (the PAM and the PSA one) and classifying the cast iron graphite particles in six classes inspired by ISO 945 standard.

The classification uncertainties which result from overlapping of the class ranges can be removed by using two different methods. One consists in classifying the litigious particles in the most unfavourable class. The other one is an original "membership" test.

Fully automatic, the programme permits the analysis of all samples up to about 50x 200 mm. The fields can be joined or not. The results are given per field and per zone with statistical data.
As the output cards are compatible with an Excel spreadsheet, a complementary treatment of the results is possible.

VIII - EXAMPLE OF APPLICATION TO SG CAST IRONS.

Forty tensile specimens taken from normalized test blocks and issued from different foundries were studied. The cast irons belong to main normalized grades (400-15 to 700-2). Each specimen was subjected to following characterizations:

- tensile properties: Rm, A %,
  - chemical composition: Si, Mn, Cu,P,
  - microstructural features:
    - pearlite fraction, p,
    - graphite fraction, Ag,
    - count (%) of different types of graphite (I to VI),
    - mean area of graphite particles, ag,
    - mean area of different types of graphite (I to VI).

In order to simplify the description of the morphological graphite structure of each specimen and to make easier the search of correlations, we proceeded to the following groupings:
- graphite of types I + II
- graphite of types III + IV
- graphite of types V + VI

A) Influence of graphite morphology on mechanical properties.

We tried at first to establish, by multi-linear regression study, the relations between Rm or A% and the following parameters: pearlite fraction, count (%) of graphite of I + II or III + IV or V + VI types, count of particles (NN/mm^2), silicon content (% pond.).

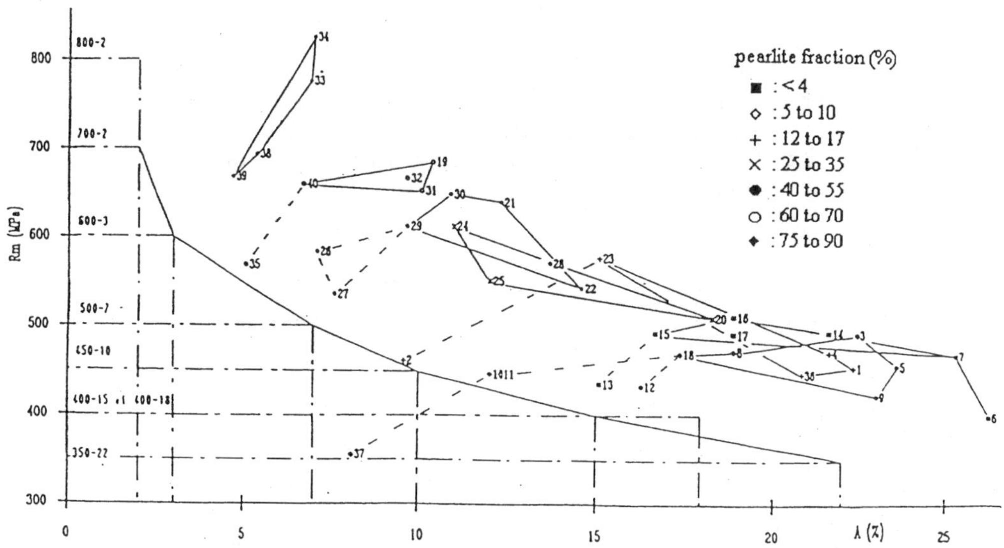

FIGURE **2** : Mechanical properties of the different cast irons in relation with the pearlite % and the classification of SG cast irons according to NF A 32 201 standard.

The study shows that Rm and A % depend strongly and mainly on pearlite fraction. It confirms also the influence of graphite morphology on these two properties. The count of particles and the silicon content are only related to Rm.

However the standard deviations and the differences between calculated and measured properties are generally too large to permit a practical prediction. The best correlations are those concerning Rm.

B) Classification of SG cast irons according to NF A 32 201 standard and morphology of graphite.

It is known that the classification of SG cast irons is based on minimal guaranteed values of tensile strength and tensile elongation. The Rm - A% diagram (Fig.2) on which these minimal values have been placed permits to determine the number of grades,which each cast iron is belonging to. By setting in increasing order, on a same graphite, the count (%) of graphite of VI or V + VI or I + II types and so on, corresponding to each cast iron (specimen), we achieve an evaluation of the concordance between the normalized criteria and the morphology of graphite (see for example figure 3). The following results appear:

· the quasi totality of cast irons which do not belong to one single grade contains less than 65% of graphite of VI type, less than 85% of graphite of V + VI types or more than 15% of graphite of III + IV types.

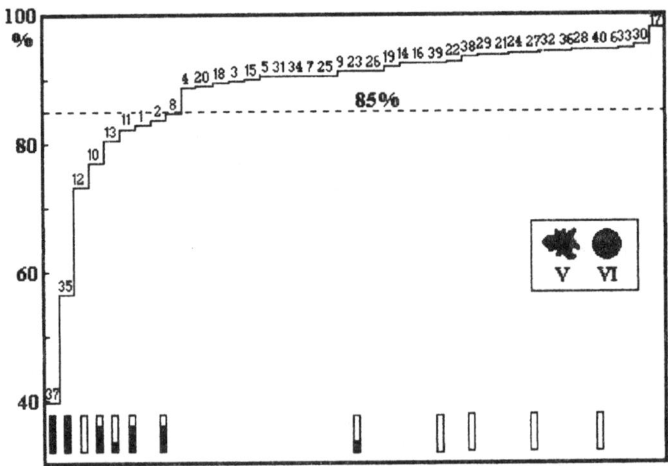

FIGURE 3 : Percentages of V + VI types graphite corresponding to the different cast irons, set in increasing order. Situation of the cast irons belonging to only one grade(▌), to none (▌) or situated at the limit of one grade (in it: ▌ ; out of it: ▌)

· the presence of 1 to 2% of graphite of I + II types is not necessarily incompatible with good mechanical properties (Rm, A%) according to one or more normalized grade(s).

· the ferritic grades of SG cast irons seem to tolerate a higher count (%) of defectuous types of graphite than the pearlitic grades.

References

1. Fascicule français de documentation NF A 32 100 (juillet 1967)
2. ISO 945 international standard (1975)
3. ASTM 247 american testing method
4. C. Lantuejoul, G. Toniolo : journal de microscopie et de spectroscopie électronique, vol 12, 75, (Feb. 1987)
5. J. Fargues, M.Stucky, Fonderie-Fondeur d'aujourd'hui, 129, 13, (Nov. 1993)

Advanced Materials Research Vols. 4-5 (1997) pp. 433-438
© *1997 Scitec Publications, Switzerland*

Functional Improvement of Dilatation Analysis on Thermal Analysis for S.G. Iron Quality Estimation

Y. Yang and J. Alhainen

VTT Manufacturing Technology, P.O. Box 1703, FIN-02044 VTT, Espoo, Finland

Keywords: Dilatation Analysis, Quality Control, S.G. Iron

ABSTRACT

A new type of dilatometer with computer-aided derivative analysis is developed to reveal the graphite precipitation details of cast iron in terms of the precipitation time, intensity, and growth style, which is currently impossible for thermal analysis. Relation of the dilatometer parameters to some quality factors of SG iron, like the graphite morphology, nodule count, matrix composition, and the shrinkage tendency, is discussed.

INTRODUCTION

Graphite precipitation plays the most important role in the structure forming of cast irons and is decisive to their mechanical properties. However, this key solidification process can not be normally observed on a thermal analysis cooling curve due to the small latent heat evolved by graphite precipitation [1,2]. The primary liquidus arrest (TGT) appears only when the carbon equivalent is over 4.60 % [3]. Such fact is obviously an obstacle for further application of thermal analysis in either theoretical solidification studies or practical quality control of cast irons.

Solidification of all graphitic cast irons is accompanied by a voluminal expansion. Different devices have been developed to measure the expansion forces [4,5] and mould wall dilatation [4,6-8]. These devices have proved useful in the studies on solidification characteristics of cast irons. During the last few years, the authors tried to improve the accuracy of the dilatation measurement and exploit its industrial application. Consequently, A new type of dilatometer with derivative dilatation analysis is developed [9,10]. With this method, the present work aims to clarify some details of the graphite precipitation, showing a major advantage of the dilatation analysis over thermal analysis.

VERIFICATION OF THE GRAPHITIC EXPANSION

The developed dilatometer is shown in Fig. 1. And Fig. 2 shows a typical dilatation and its derivative curve measured from an SG iron with a composition of 3.66% C, 2.1% Si, 0.27% Mn, 0.018% P, 0.007% S and 0.037% Mg. In order to verify the graphitic expansion, a cast steel melt with 0.06% C is also measured as a comparison, with the results shown in Fig. 3. The steel should have no graphite precipitation in the solidification process. The dilatation at the beginning of measurement in Fig. 3 could only be a result of the thermal expansion of the moulding materials. This thermal expansion is found proportional to the pouring temperature. In the case of cast iron, besides a similar thermal expansion represented by the initiating pulse of the derivative curve, there is also a derivative peak which obviously stands for the graphitic expansion during the solidification and it is nominated as the main derivative peak. Definitions of the parameters used in the paper are also listed in Fig. 2.

The accuracy of the dilatometer measurement is examined by computer simulation with MAGMA Soft. The results show that the error of the key dilatometer parameter MP value caused by temperature gradient in the dilatometer casting is only about one second, quite within the allowance. Numerous repeating tests also give satisfactory assessment to the measurement reliability.

SENSITIVITY TO THE GRAPHITE PRECIPITATION

The superiority of the dilatation analysis method in revealing the graphite precipitation is further proved by the following experiment. A slightly hyper eutectic (CE = 4.42%) SG iron melt was prepared with a composition of: C, 3.79%; Si, 1.86%; Mn, 0.30%; P, 0.026%; S, 0.010%; Mg, 0.050%; and Cu, 0.97%. The melt was cast into a 300 x 300 x 300 mm cube in a water-glass-sand mould with dilatometers and a thermal couple, as shown in Fig. 4, and the measured dilatation or contraction of the casting surface as well as the cooling curve are shown in Fig. 5. The cooling curve, just like what is reported in literature [1-3], does not show any sign of primary graphite precipitation when the carbon equivalent and thus the latent heat evolved by the graphite precipitation is not high enough. However, the contraction curve has a clear arrest before the eutectic reaction, revealing the primary graphite precipitation. Metallography examination has confirmed an overall existence of primary graphite nodules in the cube casting.

Similar experiments have been done also with hypo-eutectic and eutectic composition, but no such primary phase arrest has been found in those contraction curves.

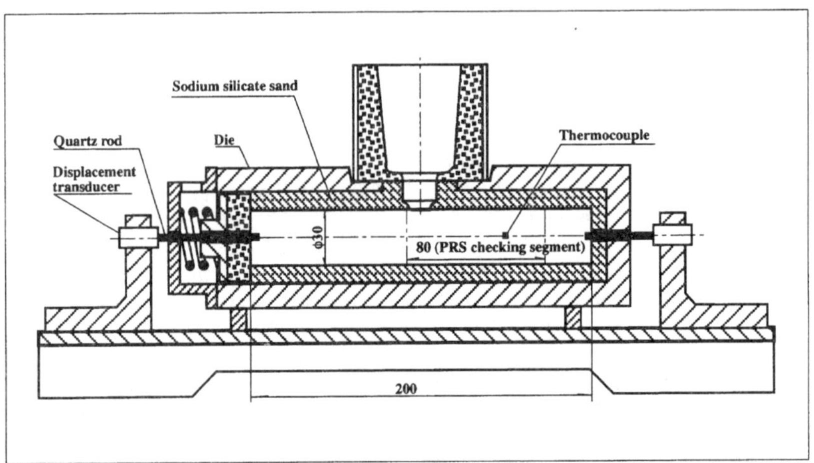

Fig. 1. Structure of the dilatometer.

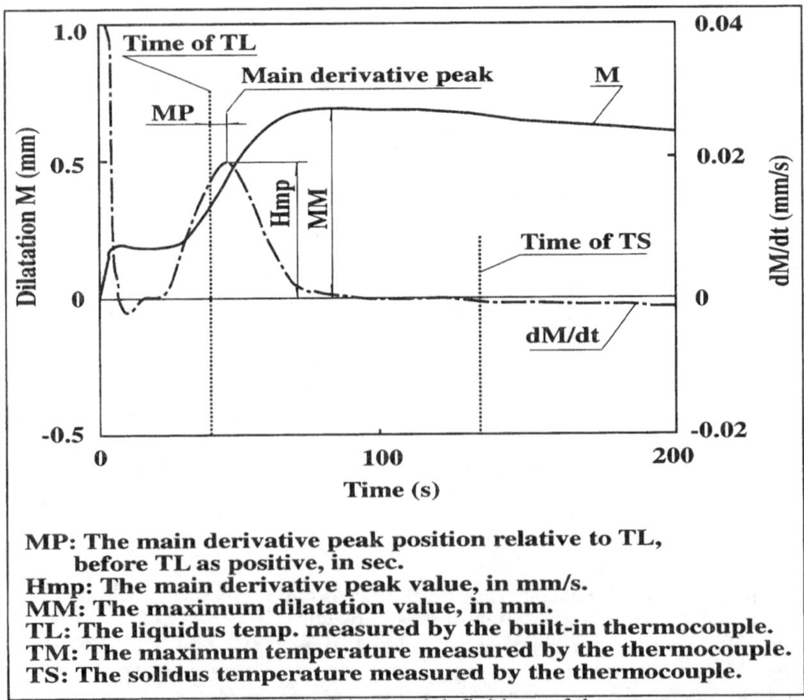

MP: The main derivative peak position relative to TL, before TL as positive, in sec.
Hmp: The main derivative peak value, in mm/s.
MM: The maximum dilatation value, in mm.
TL: The liquidus temp. measured by the built-in thermocouple.
TM: The maximum temperature measured by the thermocouple.
TS: The solidus temperature measured by the thermocouple.

Fig. 2. Dilatation curves of an SG iron and definitions of the parameters.

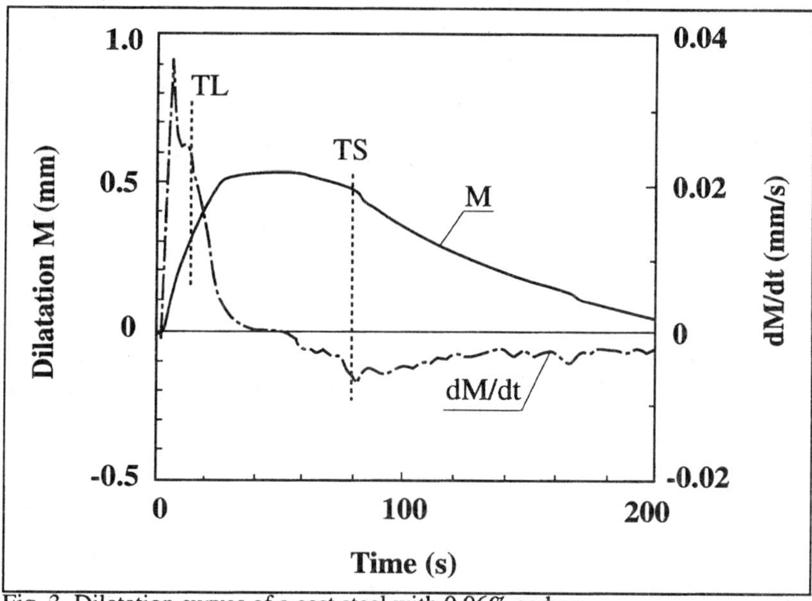

Fig. 3. Dilatation curves of a cast steel with 0.06% carbon.

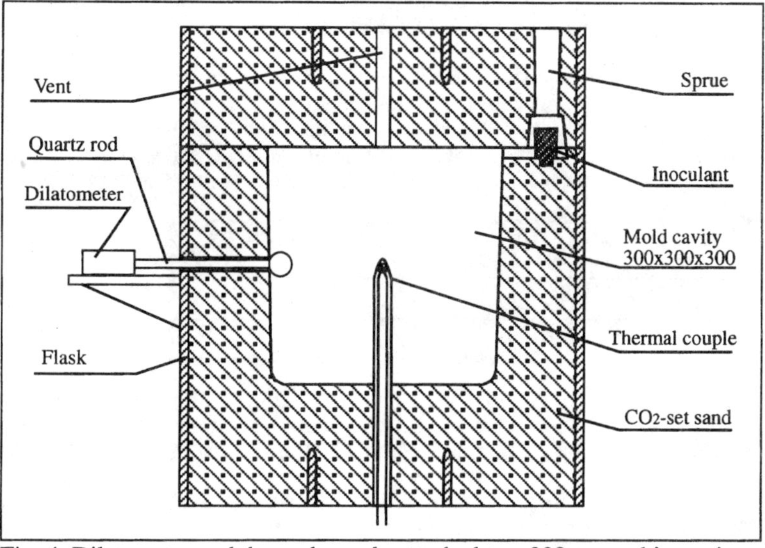

Fig. 4. Dilatometer and thermal couple attached to a 300 mm cubic casting.

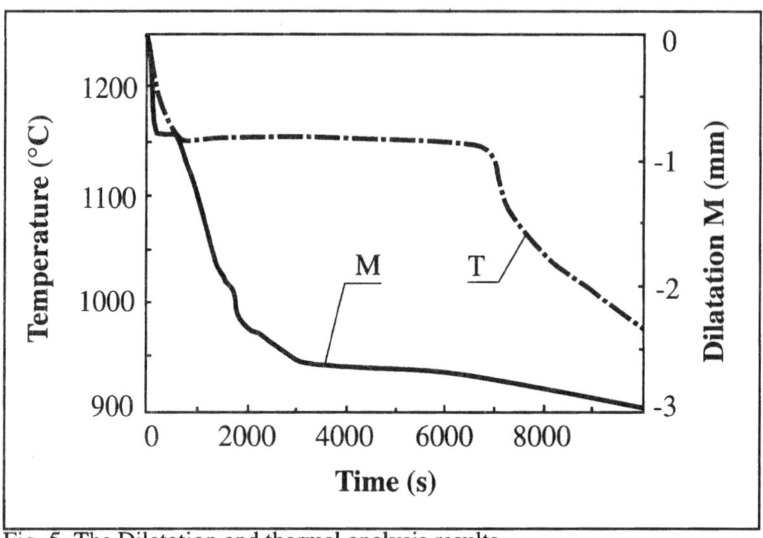

Fig. 5. The Dilatation and thermal analysis results.

PROSPECTS FOR INDUSTRIAL APPLICATION

<u>Evaluation of inoculation intensity or graphitization tendency</u>

The derivative dilatation curve clearly indicates at what time and at what rate the graphite nucleates and grows. The MP value shows the bulk graphite precipitation time and Hmp indicates the growth rate. A stronger inoculation results in higher MP and Hmp values. Traditionally, eutectic temperature TEU is used to estimate the inoculation intensity. However, correlation between the eutectic temperature and the inoculation intensity is often confusing. As an example, two SG iron melts, numbered P5 and P7, were tested with the dilatometer at a foundry automatic production line. These two melts were from the same base melt with the same melt treatment process except some variation in the post-inoculation amount. The test results are listed in the following table, showing that the MP values correspond well with the nodule count and ferrite amount, while TEU values not. A series of test data relating the MP values to the nodule count and free carbide content in the matrix are shown in Fig. 6. (Note: The metallographic examination was carried out at 7.5 mm from the circumferential surface, 40 mm from the end of the dilatometer casting shown by Fig. 1.)

Sample No.	MP (s)	TEU (°C)	Nodule Count (/mm^2)	Ferrite (%)
P5	8.8	1145.8	220	60
P7	-6.4	1148.3	140	40

The dilatometer also measures a dilatation during the eutectoid reaction of the iron although the magnitude is smaller than that during the eutectic reaction. Good correlation exists between the eutectoid expansion amount and the ferrite content in SG iron, and the test results from two Finnish foundries are also shown in Fig. 6.

A good deal of experimental evidence has shown that the main derivative peak can either precede or succeed the time of liquidus temperature even with a hypo-eutectic iron melt. The MP value can be up to plus 10 seconds, implying that the primary phase in hypo-eutectic SG iron can be graphite. Such result supports some earlier research findings: Graphite spheroids are present at temperatures of 1320-1350 °C [11]; experiments recording volume changes in SG iron balls confirm that the expansion starts well above the liquidus temperature, graphite may be present not only in liquid hyper eutectic, but also in hypo eutectic cast irons up to a temperature of 1450 - 1500 °C[12].

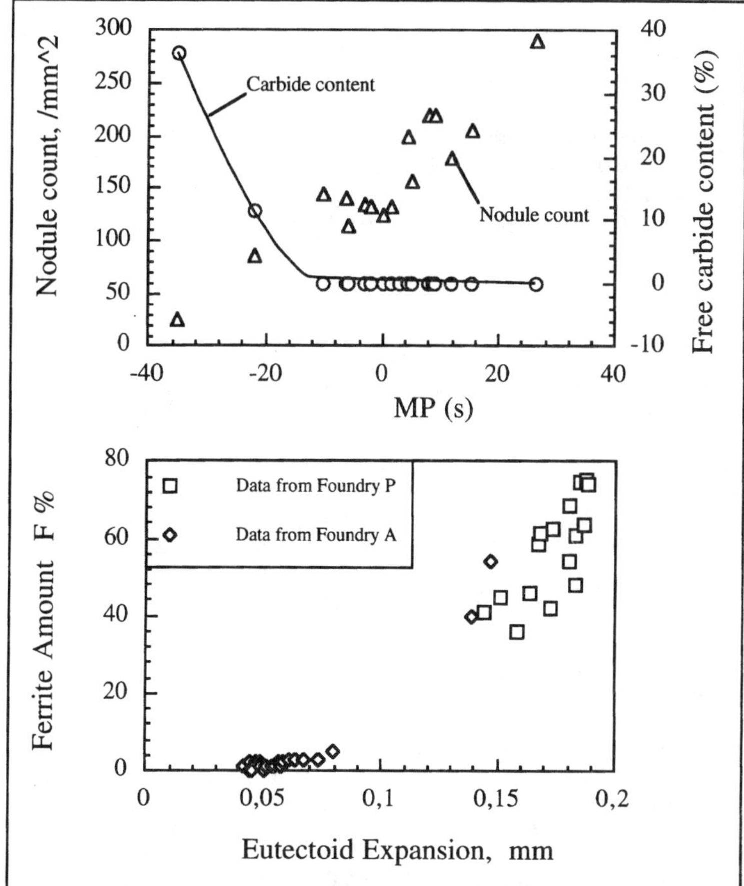

Fig. 6. Relationship between the graphitization tendency and the dilatometer parameters.

<u>Revealing the Graphite Growth Style</u>

It has been reported that SG iron exerts higher graphitic expansion pressure than grey iron against the mould wall[13]. The eutectic grey iron is found to expand less than half of that of SG iron of nearly the same composition[14]. This is also proved by our experiments. Repeated test results by the dilatometer have shown that grey iron has about half the MM value (the maximum dilatation, ref. Fig. 2) of the SG iron though the carbon equivalent of the two kinds of iron is about the same. This may create a fairly simple way to estimate the graphite morphology of the irons.

Differences in the dilatation amount between grey and SG irons have earlier aroused argument. As a fact, eutectic grey iron has nearly the same amount of graphite as SG iron. If the expansion is due to the graphite, it is difficult to understand the great difference in the dilatation magnitude normally measured. To look into this problem, a special experiment was done. A grey iron melt and an SG iron melt of about the same carbon equivalent were separately poured into the dilatometers shown by Fig. 1. The feeder necks of the dilatometers were reduced in size so that they froze quickly to prevent the expansion releasing from the feeder. In this case, the maximum dilatation value, MM, of the grey iron increased up to the same level as that of SG iron. Obviously, the smaller expansion behaviour of grey iron in normal practice should be caused by a relatively earlier expansion in the solidification process and/or the fluidity of the melt remaining higher than that of the SG iron (SG iron is known as mushy solidification style) so that the graphitic expansion pressure could be released through the feeder neck or ingate.

<u>Quantitative relations to SG iron quality</u>

The dilatometer parameters show very close correlation with some key quality factors of SG iron. Multiple regression is used to make quantitative expressions for the relations and some of the results are listed below [9]:

Ferrite percentage (F%) in the matrix of SG iron,

$$F\% = 57.06 + 0.6801 \cdot MP + 0.00737 \cdot MP^2 + 0.02268 \cdot TM - 0.05611 \cdot TL$$

The nodule count (NC, per mm^2) of SG iron,

$$NC = -96.85 + 4.423 \cdot MP + 0.03698 \cdot MP^2 - 0.1587 \cdot TM + 0.3790 \cdot TL$$

The hardness (HB) of SG iron,

$$HB = 226.3 - 2.252 \cdot MP + 0.0123 \cdot MP^2 + 0.1796 \cdot TM - 0.1494 \cdot TL$$

The melt tendency towards porosity (PRS, specified in [9]),

$$PRS = -1859 - 7.365 \cdot MP + 0.8927 \cdot TM + 0.7456 \cdot TL - 64.01 \cdot Hmp$$

The above regression equations also qualitatively indicate the effects of some main processing factors on the quality of SG iron. Taking the PRS equation as an example, it shows that the porosity decreases with increasing MP and Hmp values, which means intensifying inoculation. TM is directly related to the pouring temperature, and a higher pouring temperature causes higher porosity. High liquidus temperature TL means not only high solidification contraction of the melt accompanying the formation of austenite, but also less fluidity of the feeding melt and more difficulties for the feeding melt in going through the densely grown austenite dendrites. Lower carbon equivalent and higher magnesium contents in the iron increase the liquidus temperature[15], thus increasing the porosity.

SUMMARY AND CONCLUSIONS

The main advantage of the dilatation analysis over thermal analysis is that it can directly reveal graphite precipitation details during the solidification of cast irons like the bulk nucleation and growth time, the graphite growth rate and style. Depending on the inoculation intensity or the graphitization tendency, the bulk graphite precipitation time of the iron melts, even in cases of hypo-eutectic composition, can be either precede or succeed the austenite liquidus temperature.

The dilatometer easily distinguishes grey iron from SG iron because the former normally exhibits about half as high expansion as the latter. This appearance should be a result of better melt fluidity and/or relatively earlier graphitic expansion in the solidification of grey iron. The absolute expansion value for the both is about the same if the carbon equivalent is similar.

The graphitization tendency measured by the dilatometer is closely related to many quality factors of SG iron. In a couple of minutes, the dilatometer can estimate the melt tendency towards porosity, the ferrite or pearlite percentage in the matrix, the nodule count and the hardness, and may offer a promising method for instant quality control of SG iron in foundries.

REFERENCES

1   D. M. Stefanescu, C. R. Loper, Jr., R. C. Voigt and I. C. Chen, AFS Trans. 90, (1982), 333.
2   J. Van Eeghan, et al., AFS Int. Cast Metals J., 2 (1977), 2, 57.
3   C. R. Loper, Jr., R. W. Heine, and M. D. Chaudhari, The Metallurgy of Cast Iron, Geneva, Switzerland, May 29-31, 1974, 639.
4   G. Nandori, et al., 45th International Foundry Congress. Budapest, 1978, CIATF.
5   G. Nandori, et al., 55th International Foundry Congress. Moscow, 1988, CIATF.
6   R. Hummer, Giesserei-Praxis 1985, 17-18, 241.
7   S. Engler, et al., Cast Metals Research Journal 1973, 20.
8   B. P. Winter, et al., AFS Trans. 92 (1984), 551.
9   Y. Yang and J. Alhainen, 60th World Foundry Congress. Hague, 1993, CIATF.
10 Y. Yang and J. Alhainen, Proceedings of CASTCON 91, Harrogate, June 6-7, 1991, IBF.
11 C. R. Loper, Jr., R. W. Heine: AFS Trans. (1963), 135.
12 S. I. Karsay, Ductile Iron, The State of the Art 1980. QIT-Fer et Titane Inc.
13 S. I. Karsay, 45th International Foundry Congress. Budapest, 1978, CIATF.
14 C. E. Bates, et al., AFS Trans. 85 (1977), 289.
15 Y. Yang and J. Alhainen, Cast Metals, 5 (1992), 2, 73.

Advanced Materials Research Vols. 4-5 (1997) pp. 439-444

Using Thermal Analysis to Predict the Microstructure of Cast Iron

P. Zhu and R.W. Smith

Department of Materials and Metallurgical Engineering, Queen's University,
Kingston, Ontario K7L 3N6, Canada

Keywords: Microstructure Prediction, Thermal Analysis

ABSTRACT

The thermal analysis of test samples of iron melts has always been the traditional method of predicting the likely microstructure to be found in a grey iron casting produced from the melt. This study was concerned with examining the reliability of existing thermal analysis-based microstructural predictive schemes for nodular graphite cast iron, with a view to determining the effects of minor amounts of impurities upon the microstructure, based on its thermal analysis results. Computer-Aided Thermal Analysis (CATA) has been used to carry out the interpretation process.

The metallurgical process variables studied for their effects on solidification using the CATA technique included nodulizer type and melting additions of Ti, Sb, Pb, Bi and rare-earths (RE). It was found that the thermal analysis results are not only microstructure- (graphite morphology) sensitive, but also nodulizer- and trace element-sensitive. Based on these findings, a predictive scheme of graphite morphology for cast iron melts treated with different nodulizers has been proposed.

1. INTRODUCTION

Many attempts have been made to correlate the data obtained from the cooling curve with the various forms of graphite in order to develop a method of controlling the liquid treatment of nodular graphite cast iron [1-6]. However, some of the results are in conflict, e.g. [1] and [2,3]. In similar vein, much work has been done in an attempt to unravel the general action of Ce and other rare earth (R.E.) elements in cast iron. Morrogh claims it first desulphurises the melt and then inhibits carbide decomposition [7], they are then supposed to neutralise any traces of subversive elements in the melt [8,9].

Bi, Pb, Sb and Ti, common trace elements, are all considered to be "subversive" in the production of nodular cast iron. Much research has been carried out in terms of their effects on graphite morphology [6-12]. However, insufficient data are available to specify how they affect the cooling curve of nodular cast iron.

From the brief review above, it is apparent that there are many inconsistencies regarding the conclusions reached with respect to graphite shape prediction from cooling curve data. Therefore, it was decided that some clarification was needed in order to develop an effective graphite morphology predicting technique.

2. EXPERIMENTAL PROCEDURE

Temperature information was obtained from a K-type thermocouple horizontally positioned in a test cup (manufactured by ElectroNite). The speed of collecting data (controlled by LabTech NoteBook™ Software) is 4/2 data per second. The collection time was 4 minutes, which allowed any sample to cool from the casting temperature to one below the eutectic temperature.

The composition of the pig iron used was 3.90% C, 1.38% Si, 0.01% Mn, 0.020% P and 0.007% S. The composition of the NiMg nodulizer was 15.4% Mg, 0.02% C, 0.05% Co, 0.26% Fe, 0.23% Si and the balance of Ni. Four types of Mg-Ferrosilicon (MgFeSi) with different ratios of Mg and RE were used as nodulizers, namely:

MR-55 (RE/Mg = 1.0):   4-5% Mg,   40-43% Si and the balance of Fe;
MR-84 (RE/Mg = 0.4):   8.5-9% Mg, 3.5-4.0%RE,   41-45% Si and the balance of Fe;
MR-92 (RE/Mg = 0.1):   8.5-9%Mg, 0.81-1.0% RE,   41-45% Si and the balance of Fe;
7.5RE (RE/Mg = 0.075): 8% Mg,   0.6-0.7% Ce,   2.5% Si and the balance of Fe;

A 65% FeSi was used as both a silicon composition adjuster and a postinoculant. This actually contained 67.4% Si, 0.92% Al and 0.14% Ca. All charge materials were melted down using a clay graphite crucible installed in a 10 kg induction furnace. Any required additions were placed at the bottom of a preheated ladle before pouring. The nodulization process was performed just after the base iron was tapped into the ladle, using the "deep plunge" method. The treatment temperature varied in the range of 1350 - 1400°C. After the nodulization reaction had subsided, the 65% FeSi was added into the ladle as postinoculant. The melt was stirred to ensure the complete solution and reaction of inoculants.

The molten sample for thermal analysis was taken from the preheated clay-graphite ladle and poured into the sample cup (38 x 38mm top, 32 x 32mm bottom, 42mm height). The accuracy of the temperature measurement was ± 1.0°C. The weight of the casting ranged from 360 to 400 g. Subsequently, metallographic analysis, was carried out. In this study, following Ref. [13], an average graphite shape factor (AGSF) [14] was introduced to provide an objective evaluation of the graphite morphology, thereby eliminating operator subjectivity when carrying out visual nodularity determinations.

3. EXPERIMENTAL RESULTS

3.1 Graphite Morphology vs. Residual Magnesium (Using a NiMg Nodulizer)

In practice, insufficient residual Mg is one of the reasons for nodulizer-treated melt not to display a full nodularity; two factors: an insufficient addition of nodulizer and/or its excessive consumption caused by too much sulphur being present in the melt, are the most common causes. They were investigated in the following manner. Firstly a gradual decrease of NiMg addition was made whenthe sulphur content was fixed; secondly a gradual increase of sulphur level was produced (by the addition of FeS) while the NiMg addition was kept at a fixed amount sufficient to cause the iron to solidify as nodular iron.

Fig. 1 shows the relationship between the residual Mg and AGSF, in which it is clearly observed that when the residual Mg in the cast iron is above 0.04% wt, the graphite structure will be in the nodular form (AGSF > 0.7). When the residual Mg decreases into the range of 0.04% to 0.025% wt, then the graphite structure will correspondingly change from irregular to the compacted form. With less than 0.025% Mg residual in the cast iron, flake graphite will be the main graphite form present. These observations reinforce the well-accepted conclusion that a certain amount of residual Mg is required in order to get quality nodular irons.

According to previous studies [1,2], as nodular graphite deteriorates from a fully spheroidal to vermicular/compacted form, the difference (ΔT) between the eutectic recalescence temperature (TER) and the eutectic undercooling temperature (TEU) will get larger. However, our results indicate that there is no direct relationship between ΔT and

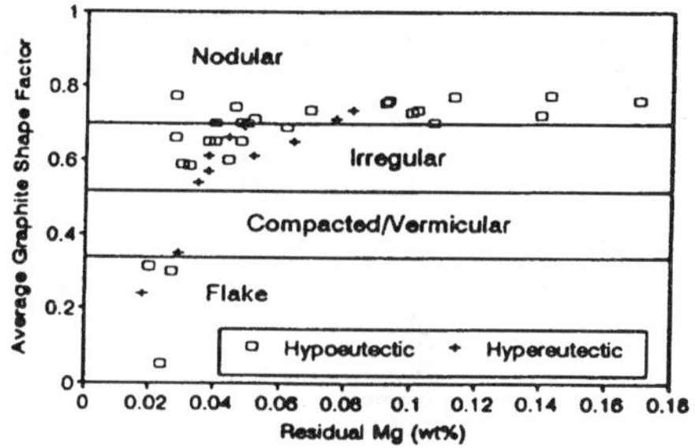

Fig. 1: Results between residual Mg and Average Graphite Shape Factor (AGSF).

AGSF for either of hypoeutectic and eutectic nodular cast iron treated with a NiMg nodulizer master alloy. The TEU for a NiMg-treated melt is usually above 1155°C and TER about 1158-1160°C.

Thus it may be concluded that ΔT cannot be used effectively to define the nodularity for a NiMg-treated cast iron. This implies that the graphite morphology might be only one of the more important variables influencing the shape of the cooling curve, a fact not appreciated by earlier workers [1-3] who were looking for significant casual features.

With respect to the increase in FeS addition, the residual Mg gradually decreases, and consequently the graphite morphology changes from nodular to compacted and finally to flake. However, while the graphite morphology is degraded, we could detect no apparent response in their corresponding thermal analysis results, i.e. no direct relationship was found between AGSF and ΔT (TER - TEU) as a function of S content. This result can be better interpreted as that the thermal analysis method cannot be used to detect the graphite degradation caused

by too much sulphur present in the NiMg-treated melt if no other measure is taken.

3.2 The Effects of Trace Elements on Thermal Analysis (using a NiMg Nodulizer)

As shown in Fig. 2(a), when additions of increasingly large amounts of Pb and Bi are made, the graphite shape deteriorates correspondingly from nodular to compacted and finally to flake form. The carbon equivalent values for the alloys shown in Fig. 2(a) are roughly eutectic or slightly hypoeutectic. It should be pointed out that the residual Mg value in all of these samples would have caused solidification as fully nodular graphite iron had no addition of these "subversive" elements been made. However, in contrast to the earlier work reported in Section 1.2, additions of Sb and Ti appear to have little effect on the AGSF.

It is interesting for us to find in 2(b) that a much higher recalescence (>10°C) is created with the Bi addition as compared with other trace additions. A close examination of cooling curves indicate that Bi decreases both TEU (~ 1132°C) and TER (>1140°C), but decreases TEU much more significantly that TER, which allows a huge recalescence (ΔT = TER-TEU) to appear. It is also seen in Fig. 2(b) that an Sb or Pb addition to an eutectic/hypoeutectic melt also causes a ΔT (4-5°C) to occur, but it is smaller than that caused with Bi addition. Ti is seen to produce little recalescence.

Fig. 3(a) shows the relationship between the NiMg additions and the recalescence with 0.01% wt Bi added to the thermal analysis cup, brought about for both hypo and that

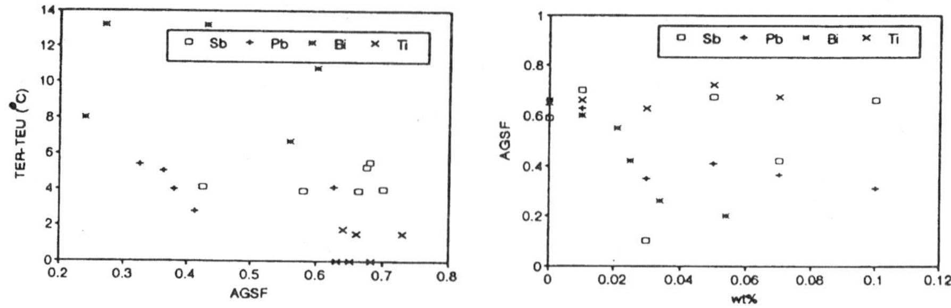

**Figs. 2(a) The additions of Ti, Sb, Pb and Bi vs. AGSF.
and 2(b) AGSF vs. ΔT (TER - TEU) with Ti, Sb, Pb and Bi additions.**

hypereutectic nodular cast iron. The more interesting effect of adding Bi to the melt is when nodulization is not successful i.e. not enough residual Mg is present in the melt, the recalescence falls. This indicates that the ΔT criterion may be used as a nodularity predicting parameter for NiMg-treated melts with the help of a Bi-added sample cup. The higher the ΔT of the cooling curve with the Bi addition in the testing cup, the better will be the nodularity of the original NiMg-treated melt before the Bi addition. A ΔT of 9°C. or higher with the Bi-added testing cup indicates the original NiMg-treated melt will solidify as fully nodular graphite iron.

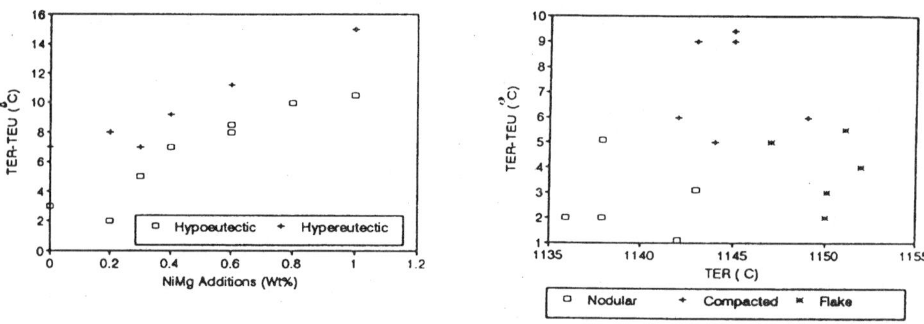

Fig 3(a)   The relationship between the NiMg additions and the recalescence with a Bi-added cup (0.01% Bi) .

Fig. 3(b)   Correlation between the TER and ΔT (TER-TEU) for cast irons with different structures, treated with rare earth-containing nodulizers.

3.3 <u>The Effect of RE/Mg Nodulizers on Thermal Analysis</u>

It has been shown that, without a Bi addition to a NiMg-treated melt in the testing cup, there are some difficulties with respect to the prediction of the graphite morphology using the thermal analysis method. It should be kept in mind that the above results may not necessarily be true for cast iron treated with other nodulizer systems [1,2].

Fig. 3(b) shows the relationship between the eutectic recalescence temperature (TER) and ΔT (TER-TEU) for cast irons treated with rare earth-containing nodulizers (7.5RE,MR-55, MR-84 and MR-92). It can be seen that the ΔT for vermicular/compacted graphite iron (>6°C) is usually higher than that of nodular and flake graphite iron (<6°C), but the TER for vermicular/compacted graphite iron is statistically located between those of nodular and flake graphite irons. The above results agree with Bäckerud's [1], but disagree with Stefanescu's [2] conclusion.

From the results obtained for melts treated with rare earth-containing nodulizers, it can be concluded that a ΔT above 6°C indicates compact graphite growth. Therefore nodular graphite growth can be predicted by considering both ΔT and TER. It should be pointed out that these conclusions, especially for TER, are based upon the experimental conditions used in this study.

Had the data of Fig. 3(b) been plotted as TEU vs. TER, it would have been seen that the TEU and TER for vermicular/compacted graphite iron are statistically located between those of the nodular and flake varieties for rare earth-containing nodulizers.

3.4 The Effect of Bi on RE-Containing Nodulizers

No recalescence was obtained by adding Bi to the melt treated with RE-containing nodulizers, presumably due to it having been neutralised by the rare earth elements present in these melts.

4. DISCUSSION

The undercooling temperature (TEU) of a cooling curve reflects the initial eutectic growth temperature. In the case of cast iron, this temperature can be used to determine the form of graphite, i.e. carbide or nodular, vermicular or flake if the postinoculation conditions are kept consistent. On the other hand, TER indicates the bulk eutectic growth temperature, and it is higher than the corresponding nucleation temperatures. This is schematically demonstrated in Fig. 4. From this, it is understood that if the initial carbon-rich phase growth and the bulk growth forms are the same, only a small recalescence (ΔT) may be observed. A large ΔT will occur only if a chilling tendency (low TEU) is accompanied with graphite growth mode change during the eutectic phase transformation. The combination of chilling tendency and carbon growth change may be achieved by three methods: i) putting enough RE (high in Ce -- a strong carbide stabilizer) into the melt; ii) by holding a nodulizer-treated melt to let it fade in both nodulization and postinoculation processes and iii) adding Bi into the NiMg-treated melt.

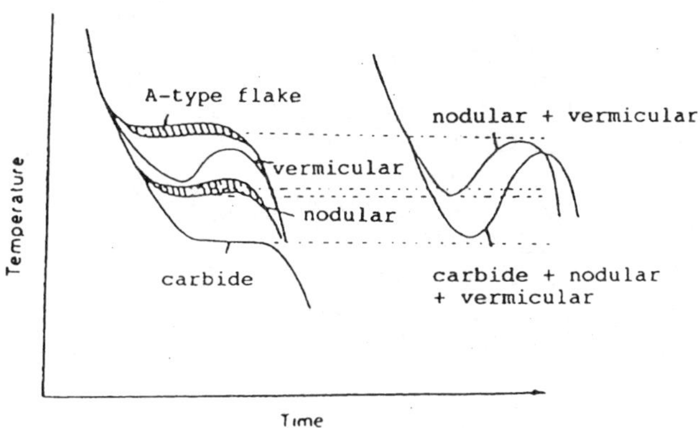

Fig. 4: The schematic relationship for the different nucleation and growth temperature in cast iron.

5. CONCLUSIONS

Based on the results of this study, the following conclusions may be reached.

i) The thermal analysis results are more nodulizer sensitive than graphite morphology sensitive and the most influential elements in nodulizers are the rare earth (RE) compositions. In applying thermal analysis to predict thegraphite shape/morphology in nodular cast iron, it is necessary to take note of the kind of nodulizers being used.

ii) Bi acts as both a carbide stabilizer and a nodular graphite degrader during solidification. The carbide stabilization effect makes TEU lower, and the nodular graphite degradation effect allows a higher TER. These two effects causes a large ΔT in a well-nodulized melt.

iii) For those cases when using a NiMg nodulizer, the graphite morphology for either hypoeutectic or hypereutectic nodular cast iron can be predicted by the eutectic solidification recalescence (ΔT=TER-TEU) only after a small addition (0.01%wt) of Bi is put into the melt. When the $\Delta T > 10°C$, goodnodularity will be obtained. Based upon this observation, a sampling cup with a small mass of Bi is being developed for NiMg-treated cast iron.

iv) For RE-containing nodulizers, if ΔT is larger than 6°C, then (a) a TER at 1150°C and above indicates graphite growth; (b) a TER of 1140-1145°C indicates vermicular graphite formation. Nodular graphite microstructure can be predicted from a small ΔT combined with TEU/TER below 1140°C.

v) Vermicular/compacted graphite growth is not necessarily associated with a large recalescence (ΔT). A large ΔT for vermicular growth appears only for a melt associated with a large undercooling tendency and nodulization degradation during solidification. This combination can be achieved by holding an otherwise nodular graphite iron during solidification, adding RE alloy to a melt, and/or adding a strong carbide stabilizer and graphite deteriorator (Bi) into the melt.

REFERENCES

[1] L. Bäckerod, K. Nilsson and N. Steen, The Metallurgy of Cast Iron, St. Saphorin, Switzerland, Georgi Publishing Company, 625-637, (1975).

[2] D.M. Stefanescu, C.P. Loper Jr. et al. Trans. AFS, 90, 333-348 (1982).

[3] I-G. Chen and D.M. Stefanescu, Trans. AFS, 92, 947-964, (1984).

[4] R. Monroe and C.E. Bates, Trans. AFS, 909, 307-311, (1982).

[5] C.R. Loper, Jr., R.W. Heine and M.D. Chaudhari, The Metallurgy of Cast Iron, St. Saphorin, Switzerland, Georgi Publishing Company 640-657, (1975).

[6] P.K. Basutkar and C.R. Loper Jr., Trans. AFS 79, 176-185, (1979).

[7] H. Morrogh and W.J. Williams, JISI, v. 170, 306, (1948).

[8] H. Morrogh, Trans AFS, 439-452, 1952.

[9] J.C. Sawyer and J.F. Wallace, Trans. AFS, 76, 386, (1968).

[10] D.G. Schelling, Trans. AFS, 74, 700-708, (1966).

[11] R. H. Döpp, The Metallurgy of Cast Iron, St. Saphorin, Switzerland, Georgi Publishing Company, 640-657, (1975).

[12] C.E. Bates and J.F. Wallace, Trans. AFS, 76, 815, (1967).

[13] T.L. Capeletti and J.R. Hornaday, Trans. AFS, 82, 59-64, (1974).

[14] Panping Zhu, (Ph.D. Thesis, Queen's University at Kingston, 1992).

[15] P. Zhu and R.W. Smith, Cast Metals, 5(3), 158-162, (1992).

Advanced Materials Research Vols. 4-5 (1997) pp. 445-450
© 1997 Scitec Publications, Switzerland

Application of Fourier's Thermal Analysis to the Determination of Kinetics Solidification of Cast Iron

E. Fraś, W. Kapturkiewicz, A.A. Burbielko and E. Guzik

University of Mining and Metallurgy, Reymonta 23, PL-30-059 Cracow, Poland

Keywords: Solidification Kinetics, Thermal Analysis, Simulation

ABSTRACT

A multi-point thermal analysis, called the Fourier method, was tested and compared with the widely used one-point analysis, otherwise called the Newtonian method.

From the experimental measurements carried out with the use of two-point temperature measurement, from the additional verification which consisted in freezing of the solidifying samples, and from the computer simulation, it follows that the Newtonian method is applicable in determining of a total effect of, e.g., heat of solidification, but whenever the measurement of a local kinetics of solidification is to be done, the Fourier method proves to be much more useful.

INTRODUCTION

Thermal analysis is a very useful tool in investigating the thermal events that take place when a given material is cooled (or heated). In the case of liquid solid transformations, the whole solidification process can be analyzed and recorded from cooling curves. The information thus generated can be used in quantitative or qualitative sense to discern the solidification phenomena [1]. Among the thermal analysis technique currently used are:
- Cooling curves and derived cooling curves (cooling rate curves)
- Differential Thermal Analysis (DTA)
- Differential Scanning Calorimetry (DSC)
- Newtonian Thermal Analysis (NTA)
- Fourier's Thermal Analysis (FTA)

The short characterization of the above methods is done in [2,3] and [4].

The main difference between the one-point Newtonian method NTA and multi-point Fourier's method FTA is in procedure of preparation the baseline (zero curve Z). In the Newtonian method the Z_N is fitted by extrapolation a curve type $T' = -A\exp(-Bt)$ between the onset and the end of the cooling curve. The A and B constants are determined from data on the cooling rate curve at the beginning and the end of solidification [2,3]. It does not into account the temperature gradient existing in the casting. This assumption can give rise to a source of error in interpreting the solidification kinetics, as was shown in [4]. In the FTA method the preparation of zero curve Z_F is based on calculation of the value of temperature laplacian. The fundamentals of the FTA method were given by W. Longa [5,6].

FOURIER THERMAL ANALYSIS

Assuming that the heat transfer by heat conduction is dominant in metal-mold systems, the temperature of the metal after mold pouring can be determined from the Fourier equation, which includes a heat source:

$$\frac{\partial T}{\partial t} = a\nabla^2 T + \frac{q_s}{c_v} \tag{1}$$

This equation can be described in terms of the heat generated during solidification q_s:

$$q_s = c_v T' - ac_v \nabla^2 T \tag{2}$$

where a is the thermal diffusivity;
c_v is volumetric specific heat.
Equation (2) can be rewritten as:

$$q_s = c_v (T' - Z_F) \tag{3}$$

where Z_F is the above mentioned zero curve, baseline or gradient curve (the name proposed by Longa [5]) and it is given by:

$$Z_F = a\nabla^2 T \tag{4}$$

The kinetics of the solidification process can be established from:

$$f_s(t) = \frac{1}{L} \int_{t_b}^{t} q_s(t) dt \tag{5}$$

where f_s is the fraction of solid phase;
L is volumetric latent heat;
t_b is the time at the onset of solidification.
According to the above expressions, when the thermal diffusivity and the temperature field $T(x,t)$ are known inside the metal, the magnitudes of the various heat rate contribution can be established. Experimentally, a minimum data of three temperature points is necessary for the determination of the laplacian. Nevertheless, in symmetric temperature fields, these data points are reduced to only two. Thus, by introducing two thermocouples in the casting, it is possible to find $\nabla^2 T$.

The laplacian for cylindrical casting with two point measure T_1 and T_2 in distance (radius) R_1 and R_2 from axis of the cylinder can be calculated from [4]:

$$\nabla^2 T = \frac{4(T_2 - T_1)}{\left(R_2^2 - R_1^2\right)} \tag{6}$$

Determination of heat rates generated are influenced by a and c_v. In gray cast iron, it is found that the thermal diffusivities and specific heats vary from 0.0379 to 0.0701 cm^2/s, and 5.278 to 5.859 J/(cm^{3}K), respectively.

When there is no heat generated during the solidification of a given cast metal (i.e., during melt cooling from the pouring temperature to the liquidus temperature, or after solidification), $q_s=0$ and the thermal diffusivity can be determined from Eq. 1 as:

$$a = \frac{T'}{\nabla^2 T} \tag{7}$$

Thus, when T' and $\nabla^2 T$ are experimentally known before and after solidification, the thermal diffusivities of the liquid and solid can be found. If both thermal diffusivities have similar values, the kinetics of the solidification process can be established by:
- determining the zero curve Z_F from Eq. 4;
- determining the rate of heat generation during solidification (Eq. 3);
- determining the solidification kinetics by means of Eqs. 4 and 5.
When the thermal diffusivities of liquid and solid are different, it can be assumed that the thermal diffusivity is a function of the volume fraction of solid phase. In this case the iterative procedure, described in [4] can be used for determination f_s. The same assumption can be applied with reference to the specific heat.

TESTING OF THE METHOD

The FTA method was tested through a two-point temperature measurement in the test cylinder made of cast iron characterized by the following composition, % weight:

C	Si	P
3.8	1.58	0.08

Using the test device, showed in Fig. 1, the cast test cylinders of dimensions dia 40x200 mm, solidifying in sand molds, were pushed out at the successive stages of solidification into a container with water. Through metallographic analysis, the freezing of the solidifying metal in water enabled determination of the volume fraction of the phase solidified at a given time instant. Due to this an experimental relationship V=f(t) was obtained.

In a simultaneously poured cylindrical specimen with installed thermocouples the cooling curves were measured in two points distant by $x = 0$ and 12 mm from the casting center, to make computations according to the method discussed here.

Applying own computer program COMCAST®, a simulation of the solidification process was done along with additional computation, related to the FTA method. The parameters used in the experiment and adopted for computations were compiled in Table 1.

Table 1

Thermophysical and Initial Parameters

Name and units	Value
Thermal conductivity, W/(cm K):	
liquid cast iron	0.18
solid cast iron	0.37
mold	$0.097 - 1.24 \cdot 10^{-5}\,T + 1.20 \cdot 10^{-8}\,T^2$
Specific heat, J/(g K):	
liquid cast iron	0.837
solid cast iron	0.754
mold	1.30
Latent heat, J/cm^3	
austenite	1905
graphite eutectic	1833
Linear growth coefficient, cm/(K^{2}s)	
austenite	$7.0 \cdot 10^{-5}$
graphite eutectic	$2.0 \cdot 10^{-6}$
Nucleation coefficient, 1/(cm^{3}K^2)	
austenite	$0.5 \cdot 10^{3}$
graphite eutectic	3.5
Initial temperature, °C	
cast iron	1270
mold	20

RESULTS

In Fig. 2 the cooling curves from experiment and simulation are shown. The results reveal a similar course of the curves. The derivatives of the curves shown in Fig. 2 are shown in Fig. 3 and are compared with the "zero curves" calculated according to Eq.4 - Fourier method and Newtonian method. The course of the zero curves is quite different. Fig. 4 reflects the different course of the zero curves from Fig. 3. According to the Newtonian method (both experiment and simulation) there is a distinct maximum on the curve at the onset of solidification; then the kinetics curves drop down.

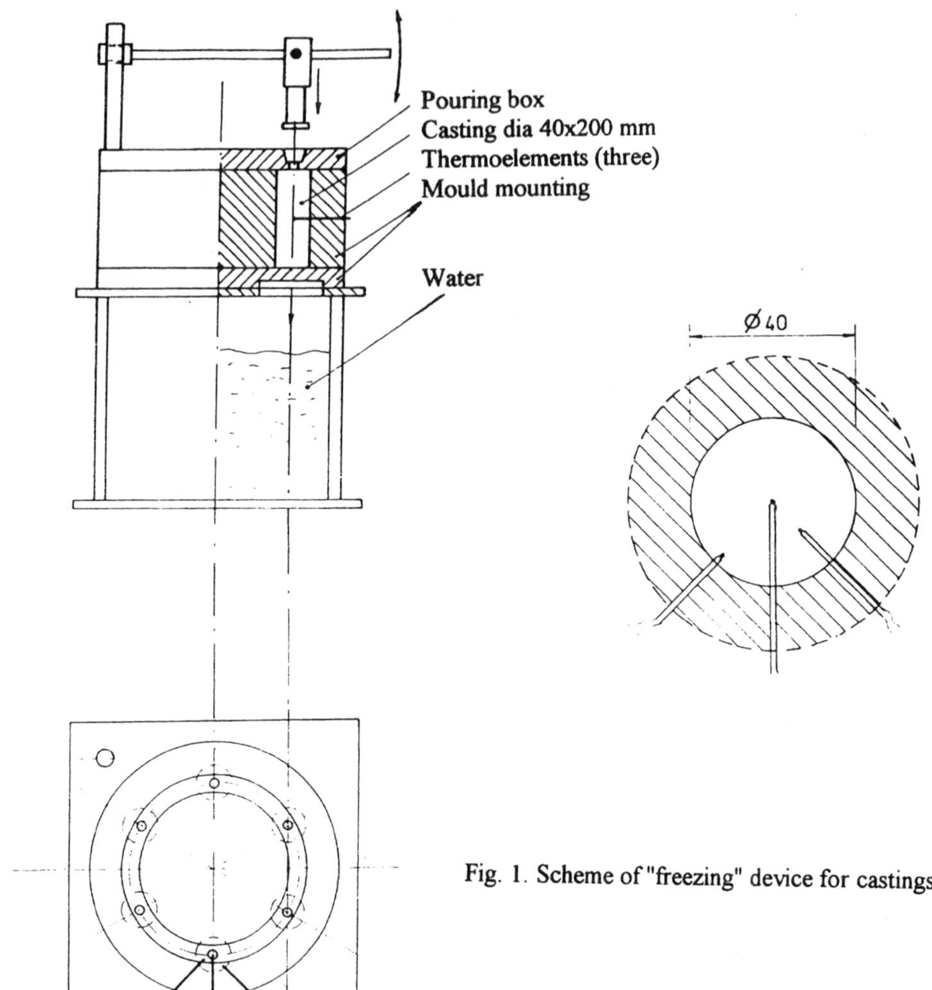

Fig. 1. Scheme of "freezing" device for castings

In Fourier method the initial maximum is much smaller than the next maximum, observed to occur at the end of solidification. What follows next from these differences is the different course of kinetics in the solidified volume, which can be seen in both Fig. 4 and Fig. 5. In Fig 5, additionally, the points from a metallographic analysis of the frozen specimen were marked.

The reason of differences in both methods consist in that fact that the Newtonian method refers in its calculations to the casting taken as whole, while Fourier method refers to this region only in which the temperatures are measured to calculate the laplacian. In Newtonian method we get information about the kinetics averaged over the entire casting; Fourier method enables an assessment of the local kinetics.

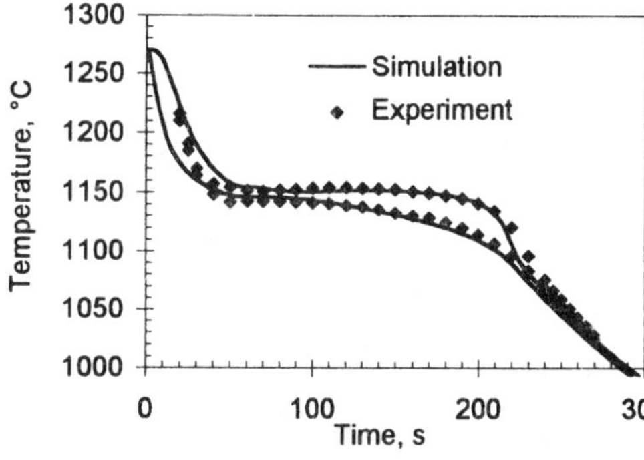

Fig. 2. Experimental and modelling cooling courves for two points of a casting $\varnothing$ 40 x 200. Points: 0 and 12 mm from center of casting

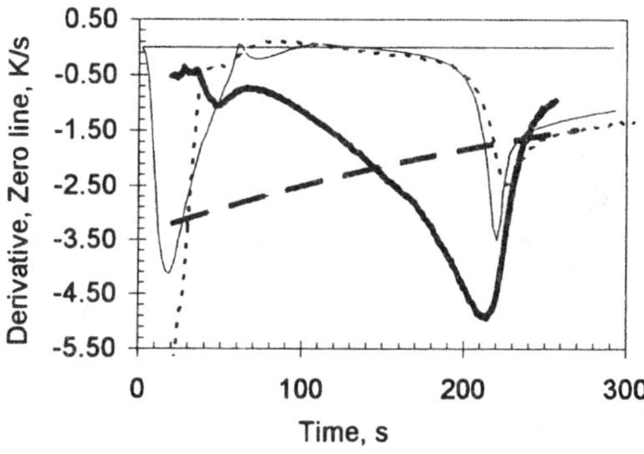

Fig. 3. Temperature derivatives for casting center and zero lines according to different calculation methods

Derivatives ——— Simulation
· · · · · · Experiment
Zero lines ━━━ Fourier's Method
━ ━ Newton's Method

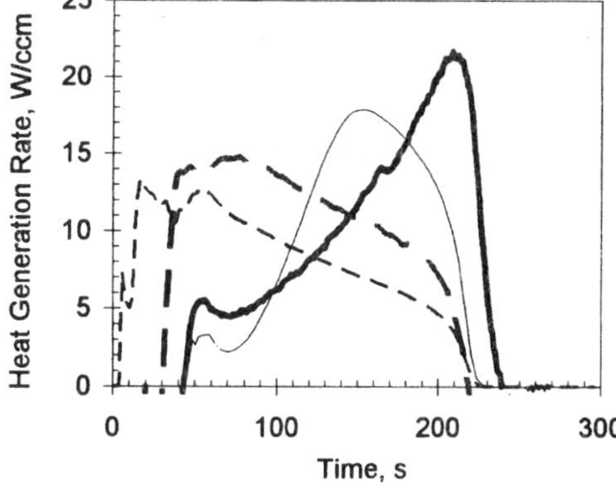

Fig. 4. Heat generation rate

Experiment ━━━ Fourier's method
━ ━ Newton's method
Simulation ——— Local volume
– – – Total volume

Due to the simulation program it was possible to compute both local kinetics within the assumed measuring points) and the total kinetics (referred to the whole casting), which is a confirmation of the explanation of source of divergencies in both methods. It has to be emphasized that the solidification heat calculated from Eq. 3 and Fig. 4 gives similar results: according to the Newtonian method L=2090 J/cm^3, according to Fourier method L=2042 J/cm^3. From this analysis it can be concluded that both methods give correct results; the difference consist in the fact that the Newtonian method they are referred to the casting taken as a whole, while in Fourier method they cover only the area measured by thermocouples.

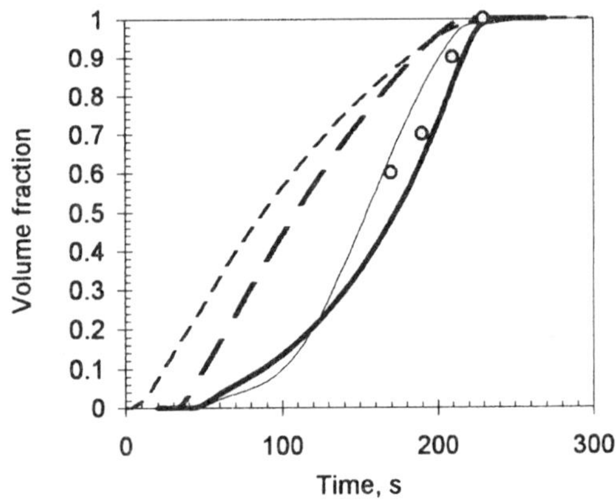

Fig. 5. Kinetics of solidification

Experiment ——— Fourier's method
— — Newton's method
o Metallography
Simulation ——— Local volume
– – – Total volume

SUMMARY

From the experimental measurements carried out with the use of two-point temperature measurement, from the additional verification which consisted in freezing of the solidifying samples, and from the computer simulation, it follows that the Newtonian method is applicable in determining of a total effect of, e.g., heat of solidification, but whenever the measurement of a local kinetics of solidification is to be done, the Fourier method proves to be much more useful.

REFERENCES

1. T. Russel: The Interpretation of Thermal Curves and some Application to Ferrous Alloy" J.I.S.I. 139, 147 (1939).
2. I.G. Chen, D.M. Stefanescu: AFS Trans. 92, 947 (1984).
3. K.G. Upadhaya et al.: AFS Trans. 87, 61 (1989).
4. E. Fras et al.: AFS Trans. 131, 505 (1993).
5. W. Longa: Archives of Metallurgy 28, 281 (1982).
6. W. Longa et al.: "53 World Foundry Congress" Prague, 14 (1986).

STRUCTURE AND PROCESS MODELLING

Microstructure formation

Modelling related to processes

Advanced Materials Research Vols. 4-5 (1997) pp. 453-460
© *1997 Scitec Publications, Switzerland*

3D Stochastic Modelling of Nodular Cast Iron Solidification

Ch. Charbon and M. Rappaz

Laboratoire de Métallurgie Physique , EPFL, MX-G, CH-1015 Lausanne, Switzerland

Keywords: Nodular Cast Iron, Solidification, Modelling, Stereology

ABSTRACT

A three-dimensional (3D) stochastic model previously developed for the solidification of eutectic alloys [1] has been extended to the case of nodular cast iron. The main output of such a stochastic model is to provide a realistic picture of the 3D grain structure, thus giving access to stereological parameters of interest and to grain size distributions. In the present contribution, the emphasis is placed upon the relationship between the area density of the graphite nodules in metallographic sections and the volumetric density of the grains. The influence of the cooling conditions and of the radius-dependent growth rate on the solidified grain structure is also shown.

INTRODUCTION

It is widely accepted that the solidification of nodular cast iron primarily occurs via the coupled growth of a nodule of graphite and of a surrounding austenite shell [2-6]. The growth rate of a grain is thus limited by the rate of diffusion of carbon from the liquid to the graphite nodule through the austenite shell.

Microscopic models of the solidification of nodular cast iron have been developed over the past with the aim of predicting the cooling curve, the evolution of the volume fraction of solid and average microstructural parameters (e.g. mean grain size). Under the assumption of a uniform temperature, these deterministic models couple a nucleation law (i.e. the dependence of the grain density on the undercooling), a growth law and a grain impingement model.

The growth of the grain being dictated by the diffusion of carbon through the austenite shell, the growth rate of the austenite shell, v_γ, is given by [2-6] :

$$v_\gamma = A \cdot \frac{\Delta T}{r_\gamma} \qquad (1)$$

where A is a constant equal to 2.87×10^{-13} [$m^2 K^{-1} s^{-1}$], ΔT is the undercooling and r_γ is the radius of the grain. The growth rate being a function of the radius r_γ, it may differ from grain to grain if they have not nucleated at the same time. Most deterministic models usually neglect this dependence and use an average grain radius for the computation of an average growth rate of the grain population.

Stochastic models have been developed recently for the modelling of the solidification of eutectic alloys [1, 7-8]. Although the grain growth is made according to a deterministic law, the location of the nucleated grains is chosen randomly within the volume of the specimen and the evolution of each grain during growth is tracked. Therefore, such models do not need any impingement model and furthermore can easily account for radius-dependent growth rate. Moreover, they produce realistic 3D microstructures, thus allowing to compute during and after solidification the grain size distributions, the distributions measured in 2D cross-sections and any stereological parameter of interest. The present paper describes a stochastic model for the solidification of nodular cast iron.

MODEL

The present model is a simple extension of a 3D stochastic model previously developed for the solidification of eutectic alloys [1]. A cubic volume of liquid cast iron of uniform temperature is divided into a large number, N_{tot}, of cubic elementary cells, of size Δl. An index is associated with each cell indicating whether the cell belongs to the liquid, to the austenite or to graphite nodules. In a time-stepping scheme, the temperature history and the evolution of the volume fraction of solid of the system are solved, assuming that the heat-flux leaving the system as well as the thermal properties of the metal are known.

The nucleation is assumed to be a function, f, of the undercooling only, and thus, at each time step, the density of grains, n, is computed, according to the relationship :

$$n(\Delta T(t)) = \int_{0}^{\Delta T(t)} f(\Delta T')\cdot d(\Delta T') \tag{2}$$

The location of the new grains is randomly chosen among the cells which are still liquid. As soon as a grain forms, it begins to grow as a sphere with the kinetics, $v_{\gamma,i}$, given by Eq. 1 (i is the index of the grain). In order to avoid the divergence occurring at $r_{\gamma,i} = 0$, the initial value of the grain radius is set to $r_{\gamma,i} = 1\mu m$. The radius of each grain is updated at each time step, δt, using the relationship :

$$r_{\gamma,i}(t+\delta t) = r_{\gamma,i}(t) + v_{\gamma,i}(\Delta T(t))\cdot\delta t \tag{3}$$

All the liquid cells which are inside the spherical envelope of a grain are captured by this grain. Thus, the solid fraction at a given time is given by the total number of cells which have been solidified (either by growth or by nucleation).

The radius of the graphite nodule within each grain is given by a simple mass balance of carbon which shows that $V_{gr,i}/V_{g,i} \approx 13.8$ [2-6], where $V_{gr,i}$ and $V_{g,i}$ are the volumes of the grain and of the graphite nodule, respectively. The first volume is simply given by the total number of cells belonging to this grain multiplied by the volume of a cell. Assuming that the graphite nodule remains spherical even during impingement, its radius is thus simply given by :

$$r^t_{\gamma,i} = \left(\frac{3}{4\pi}\cdot\frac{V^t_{gr,i}}{13.8}\right)^{\frac{1}{3}} \tag{4}$$

All the cells of a grain which are inside the spherical envelope of the graphite nodule are changed to graphite. It is further assumed that the centre of the nodule coincides with the centre of the grain during the whole solidification process.

At each time step, the knowledge of the state indices of all the cells allows one to compute the solid fraction as well as the stereological parameters of the microstructure in the volume and in 2D sections. The surface distributions of the grain structures are obtained by averaging the results over 10 parallel and equidistant section planes. Please note that, in the following section, the volumes, V, (or surfaces, S) of the grains will be normalised by the average volume, V_{av} (surface, S_{av}) at the corresponding time.

RESULTS AND DISCUSSION

Four different cases have been computed according to the numerical values of the parameters listed in Table 1. Cases A and B correspond to a constant heat extraction rate, $\dot{H}$, whereas a constant cooling rate, $\dot{T}$, is applied for cases C and D. The growth rate of each grain is a function of its radius for cases A and C (Eq. 1), whereas it is computed with the mean grain radius for cases B and D.

Since the cooling curves and the evolutions of the volume fraction of solid predicted with the stochastic model are very close to the results obtained with the deterministic approach, they will not be displayed here. The results and the discussion will be focused on the predicted microstructures and size distributions, two outputs which are specific to the stochastic model.

As an example, Figure 1 shows the evolution in a section plane of the simulated grain structure for four values of the volume fraction of solid, f_s (case A).

Volume distributions

Comparison A-B

The use of a constant heat extraction rate, rather than a fixed cooling rate, for these two cases induces a recalescence, at about 2% of solid fraction, which stops the nucleation. Since the growth rate is inversely proportional to the grain radius in case A, the small grains which have nucleated just before the recalescence grow faster and tend to catch up with the larger ones nucleated earlier. This effect is clearly seen on the volume distributions calculated at $f_s = 0.1$ (Fig. 2). For case A, this distribution is very narrow whereas in case B, the corresponding distribution is wider, since all the grains grow at the same rate. At that time of solidification, the distributions reflect the combined effects of various nucleation times and growth rates. As solidification proceeds, the impingement of the grains tends to broaden the "initial" distribution of case A and to narrow that corresponding to case B. At the end of solidification, the two distributions are almost identical. This shows that, in the case of a quasi instantaneous nucleation, the final volume distribution of the grains mainly reflects the effect of impingement rather than the influence of their growth rate.

	A	**B**	**C**	**D**
Specimen size, l [m]	5×10^{-4}		3×10^{-4}	
number of cells N_{tot}	8×10^6			
Cooling conditions	$\dot{H} = -3 \times 10^6$ [Jm^{-3}s^{-1}]		$\dot{T} = -1.6$ [K/s]	
ρc_p [Jm^{-3}K^{-1}]	4.65×10^6			
L [Jm^{-3}]	1.4×10^9		0	
growth rate	$v_i = A\Delta T / r_{\gamma,i}$	$v_i = A\Delta T / \bar{r}_\gamma$	$v_i = A\Delta T / r_{\gamma,i}$	$v_i = A\Delta T / \bar{r}_\gamma$
nucleation [m^{-3}K^{-1}]	$f(\Delta T) = dn/d(\Delta T) = 5 \times 10^{11}$			
number of grains 3D	664	685	675	659
number of grains 2D	2193	2203	2163	1889

Table 1 *Values of the different parameters used in simulations A to D and number of grains obtained.*

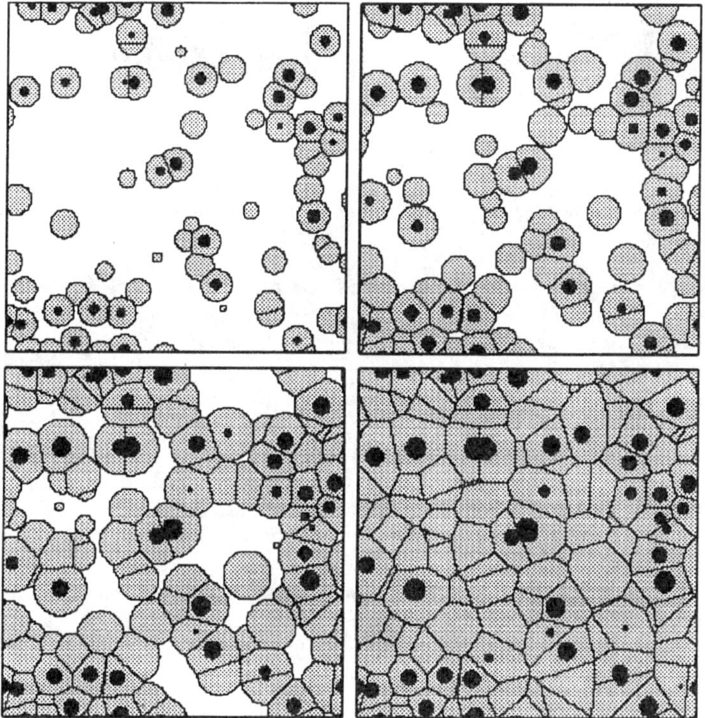

Figure 1 : *Grain structures calculated for case A at $f_s = 0.25$, 0.5, 0.75 and 1.*

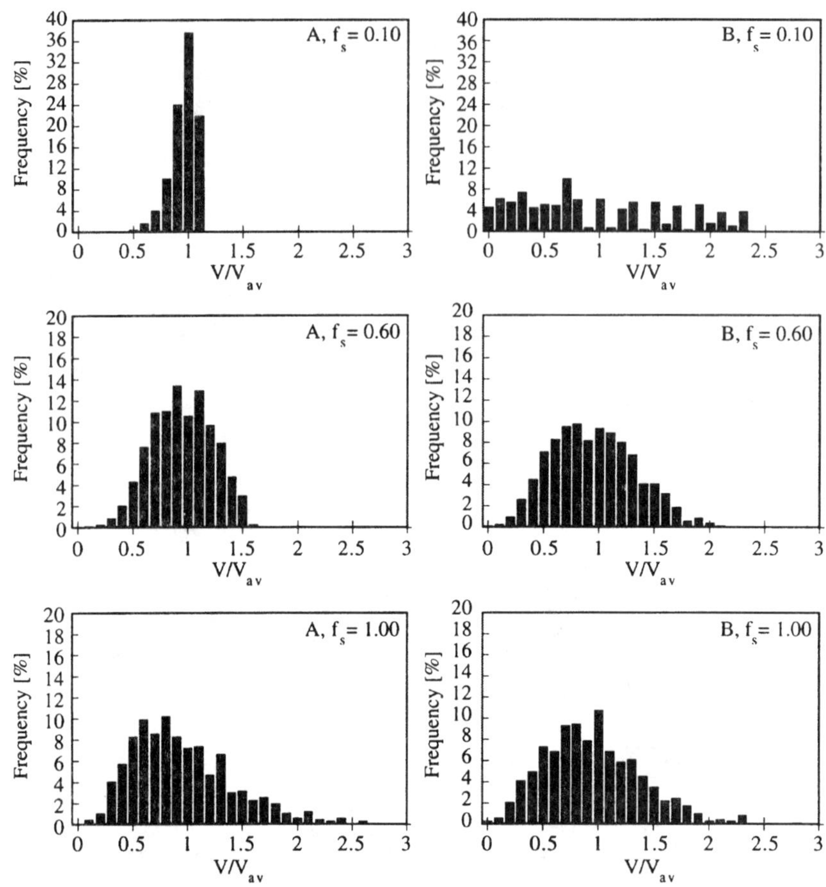

Figure 2 *: Volume distributions for cases A and B, calculated at f_s = 0.1, 0.6 and 1.*

Comparison C-D

In both cases, the cooling rate is constant and according to the nucleation law, new grains constantly appear in the remaining liquid. (It should be noted that, even though the nucleation rate is fixed, the effective number of new nuclei is proportional to the number of remaining liquid cells and thus decreases with $(1 - f_s)$).

At low volume fraction of solid, the volume distribution for case C is nearly uniform over a certain range, all the classes being occupied by an almost equal number of grains (see Fig. 3). This is due to the fact that new grains constantly appear with a radius $r_{\gamma,i} = 1\mu m$, thus "feeding" the smallest size class. The population of grains of course drifts in volume as a result of the growth process. At the end of solidification, the distribution is similar to that of case A. However, the classes of small sizes are more populated for case C as a result of the nucleation of grains in the remaining pockets of liquid near the end of the solidification.

For case D, the volume distribution of the grains is much more spread than for case C since the smaller grains cannot catch up with the larger ones (identical growth rate for all the grains). The increased values of the distribution at smaller grain volumes can be explained as follows. Without grain impingement, a constant nucleation rate would give a uniform distribution in *radius* since the velocity is the same for all the grains. In such a case, the *volume* distribution would decrease as $V^{-2/3}$.

The volume distributions at the end of the solidification for cases C and D are very different. This shows that, in the case of a continuous nucleation, the final volume distribution is more influenced by the growth law than by impingement.

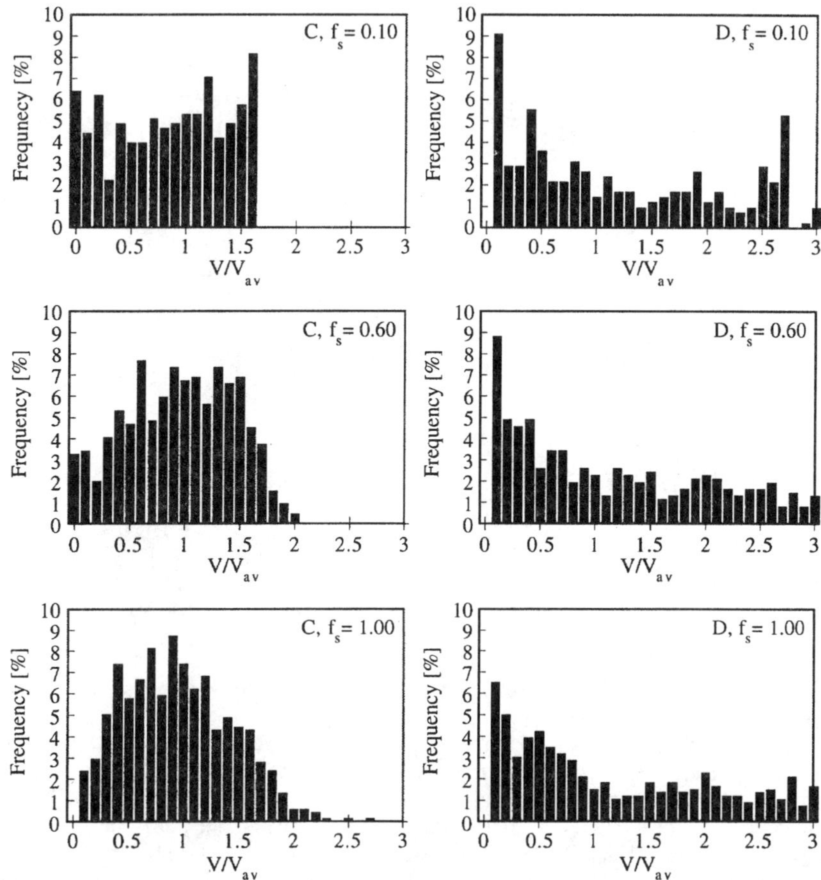

Figure 3 : *Volume distributions for cases C and D, calculated at f_s = 0.1, 0.6 and 1.*

Surface distributions

Experimentally, the volume distribution of the grains can only be obtained by means of very costly and painful methods such as serial sectioning. Furthermore, this method is usually restricted to a very small number of grains. For those reasons, the surface distributions of the grains or nodules are usually determined in 2D cross-sections. The present stochastic model gives the opportunity to compute such surface distributions and to compare them with volume distributions. Please note that in the following, the surface distributions are related to the grain sizes (and not the nodule sizes) as "measured" in ten sections of the computed 3D microstructures.

Comparison A-B

At 50% volume fraction of solid (Fig. 4), the two distributions can hardly be differentiated. For case B, one can notice that the classes $S/S_{av} > 2$ are slightly more populated. The effect of the section through the 3D grain structure is to smear the differences seen in the volume distributions. As a result, the surface distributions are almost identical at the end of solidification.

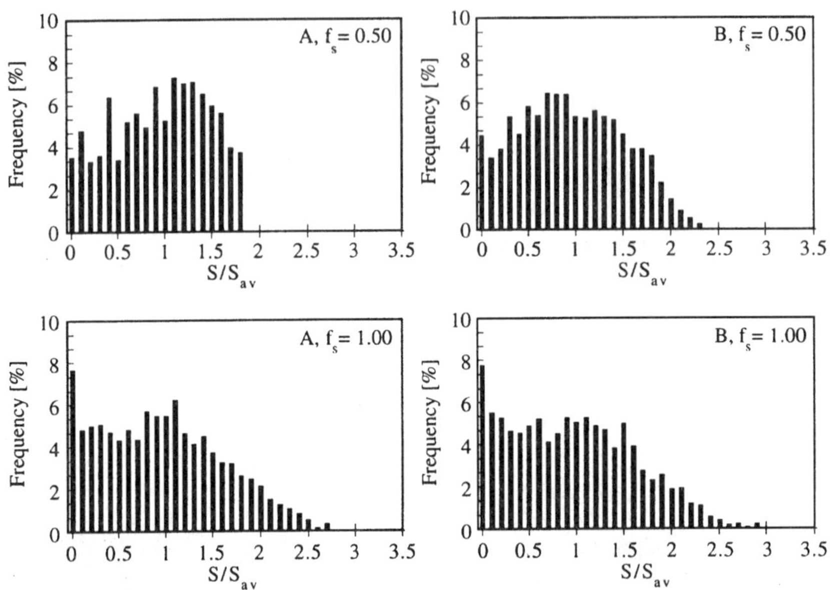

Figure 4 : Surface distributions for cases A and B, calculated at f_s = 0.5 and 1.

Comparison C-D

The large differences exhibited by the volume distributions of these two cases (Fig. 3) almost disappear when the statistics of the grains are made for 2D cross-sections (Fig. 5). At $f_s = 0.5$, the surface distributions have trapezoidal and triangular shapes for cases C and D, respectively. After complete solidification, the differences are minor compared to those in Fig. 3.

This clearly demonstrates that the sectioning of a 3D grain structure dramatically smears out any differences seen in the grain size distribution. This seriously puts into question the use of 2D distributions to access 3D distributions (e.g., Saltykov method [4, 10, 11]), since the tiny differences seen in the calculated surface distributions (Fig. 5) are certainly smaller than the experimental uncertainties.

Relationship between volumetric grain density and surface nodule count

In experimental 2D cross-sections, the density of the graphite nodules is much easier to determine than the grain density. It is then interesting to obtain a stereological relationship between the 2D nodule density, N_A^*, and the volumetric grain density, N_V. Two possible methods can be used for that purpose. The first method makes the hypothesis that the population of grains can be described by an average grain for which stereological rules are then used. Using such a method, Owadano [9] derived the relationship :

$$N_V = \sqrt{\frac{\pi}{6f_c}} \left(\alpha N_A^* \right)^{\frac{3}{2}}$$

(5)

where f_C is the volumetric fraction of carbon and α is a coefficient which ranges in the interval [1,1.4]. Richoz [2] obtained a similar relationship :

$$N_V = 2.69 \cdot \left(N_A^* \right)^{\frac{3}{2}} \qquad (6)$$

which is identical to the relationship of Owadano when $\alpha = 1$ and $f_C = (13.8)^{-1}$.

The second method is based upon the Saltykov [10] technique: from the measured 2D nodule size distribution, the 3D nodule size distribution is first computed starting from the larger nodule class seen in 2D and then progressively deconvoluting the smaller classes. The relationship between N_V and N_A^* can then be deduced by solidifying several specimens of the same alloy at various cooling rates in order to vary the final grain densities. This method was used by Saltykov [10], Noguchi [11] and Castro [4] who found the relationships plotted in fig. 6.

It may be seen in Fig. 6 that the curve which fits best the results obtained with the stochastic model is the simple stereological model based on an average grain. This shows that the influence of the size-distribution on the relationship between the 2D nodule count and the 3D grain density is weak. This implies that the use of the Saltykov's method is not necessary in this case.

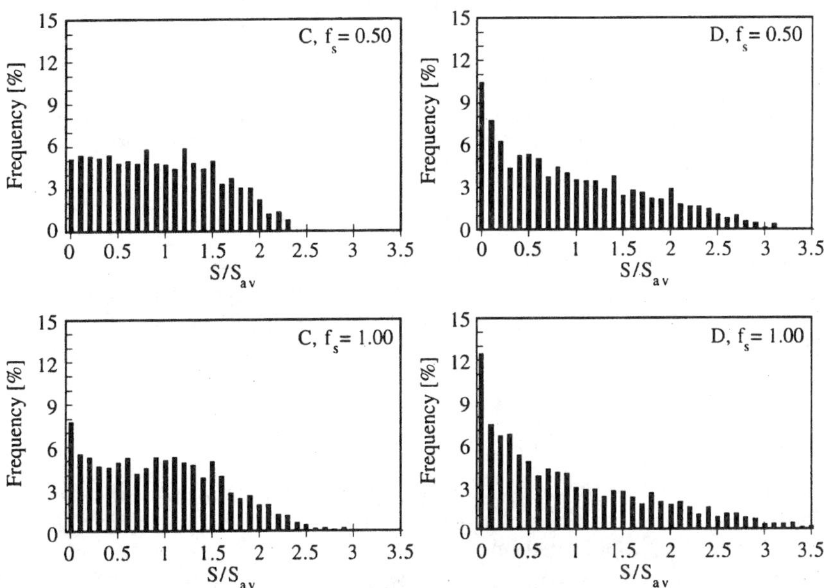

***Figure 5** : Surface distributions for cases C and D, calculated at $f_s = 0.5$ and 1.*

CONCLUSION

The stochastic model presented here gives valuable information on the grain size distributions in three dimensions. When the nucleation is rather instantaneous (cases A and B), it has been shown that the final grain size distribution mainly reflects the influence of the impingement of the grains, regardless of the growth kinetics law. However, for a nearly constant nucleation rate (cases C and D), the influence of the growth rate is clearly seen. When the growth rate is specific to each grain, the smaller grains tend to catch up with the larger ones and the final distribution is very close to that obtained with an instantaneous nucleation law. However, when the same growth rate is used for all the grains, the volume distribution is much wider.

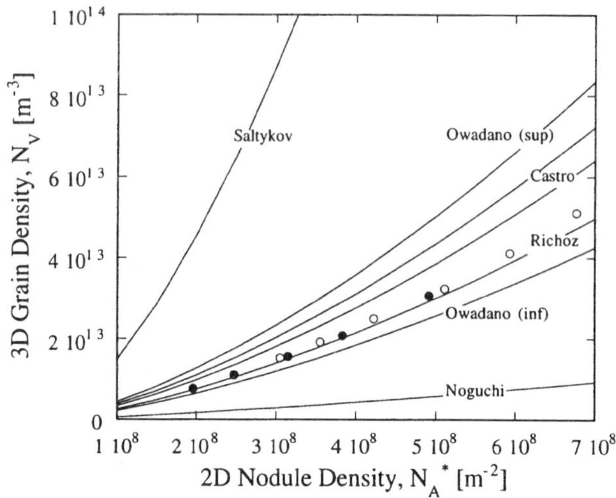

Figure 6 *3D grain density as a function of the 2D nodule count. The black dots and circles are the results obtained with the stochastic model and correspond to cases A and C, respectively. The curves correspond to different models. Owadano (inf) and Owadano (sup) are obtained with ($f_c = 1$ and $\alpha = 1$) and ($f_c = 0.07$ and $\alpha = 1.4$), respectively.*

It has also been shown that the differences seen in volumetric distributions are smeared ou when the statistics are made in 2D cross-sections. This finding implies that it is certainly very difficu to access the parameters of the nucleation law only from the final grain size distribution measured i metallographic cross sections.

Finally, the influence of the nodule size distribution on the relationship between the 2D nodul count and the 3D grain density has been shown to be weak.

REFERENCES

[1]   Ch. Charbon and M. Rappaz, Modelling Simul. Mater. Sci. Eng. **1**, 455 (1993).

[2]   M. Rappaz, J.-D. Richoz and Ph. Thévoz, in *Euromat 89*, Eds. H. E. Exner an V. Schumacher (DGM 1990), p. 135.

[3]   S. E. Wetterfall, H. Fredriksson and M. Hillert, J. Iron Steel Inst. **210**, 323 (1972).

[4]   R. Castro, PhD Thesis, Ecole des Mines, Nancy (1992)

[5]   D. M. Stefanescu and C. Kanetkar, in *State of the Art of Computer Simulation of Casting an Solidification Processes*, Ed. H. Fredriksson, (les éditions de physique, 1986), p. 255.

[6]   K. C. Su, I. Ohnaka, I. Yamauchi an d T. Fukusako, in *The Physical Metallurgy of Cast Iror* Eds. H. Fredriksson and M. Hillert (MRS Proc., North Holland, 1985, Vol. 34), p.181.

[7]   M. Rappaz, Ch. Charbon and R. Sasikumar, Acta Met. **42**, 2365 (1994).

[8]   Ch. Charbon, A. Jacot and M. Rappaz, Acta Met. **42**, 3953 (1994).

[9]   T. Owadano, Trans. J. Japan Foundrymen's Soc. **45**, 193 (1973).

[10] S. A. Saltykov, Metallurg. **14**, 10 (1939).

[11] T. Noguchi and K. Nagaoka, Trans. Amer. Foundrymen's Soc. **93**, 115 (1985).

Advanced Materials Research Vols. 4-5 (1997) pp. 461-468
© *1997 Scitec Publications, Switzerland*

Graphite Growth in S.G. Cast Irons: Simulation vs. Experiment

Y. Zhang, S.V. Subramanian and G.R. Purdy

Department of Materials Science and Engineering, McMaster University,
Hamilton, Ontario L8S 4L7, Canada

Keywords: Spheroidal Graphite Cast Iron, Growth of Graphite, Capillarity, Interfacial Kinetics, Comparison of Model Results with Experiment

ABSTRACT:

We have endeavoured first to develop a numerical model that includes a realistic heterogeneous nucleation sequence, and takes into account the effects of capillarity and of interfacial kinetics on the growth of spheroidal graphite in cast iron, then to compare the computed spheroid size distributions with those observed in experimental heats subject to differing and well-characterized thermal histories. An empirical continuous nucleation model was required to simulate the effects of heterogeneous nucleation of graphite. The results of the growth modelling were compared with the size distributions of graphite spheroids taken from finned castings of known thermal history. Certain of the parameters in the growth model were varied systematically in order to determine the sensitivity of the final results to each. The interfacial mobility of the graphite/liquid interface, for example, has a prominent effect on the growth of graphite, particularly at the earliest stages. This quantity is estimated to be of order 10^{-4} m/at%^{2}.sec., for both graphite/liquid and graphite/austenite interfaces. The model also supports the concept that the formation of austenite envelopes around the spheroids has a strong effect on graphite growth.

INTRODUCTION:

For nearly a half-century, since the publication of the first detailed studies of spheroidal graphite cast irons by Morrogh and Williams [1], metallurgists have been intrigued, and sometimes frustrated by their inability to understand quantitatively and in depth this simplest of growth forms. We do not wish to suggest that no progress has been made; only that there are many unanswered or partially answered questions surrounding this topic, and that a fully quantitative and predictive model for the development of the final microstructures has not yet been made available to the materials community.

From studies of graphite growth from the vapour phase, we find that many growth forms are dictated by the anisotropy of attachment of carbon atoms to the graphite crystal, and that the basal planes of graphite have a rather surprising ability to bend during growth [2]. This results in scrolls, fibres and other unlikely forms. In the case of graphite growth from the melt, it again appears that these characteristics are manifested in the development of graphite flakes. Spheroids, however, are expected to grow in a relatively isotropic manner: In an ideal spheroid, the basal plane normal is radially directed, and the basal plane is everywhere presented to the melt. The spherical geometry has perhaps prompted the development of diffusional growth models which assume that various transformation interfaces (graphite-melt, graphite-austenite, austenite-melt) are in local equilibrium [3,4]. Indeed, it is one of the objectives of the current work to extend these analyses, and to define the growth regimes within which they are valid.

Based on microstructural observation, it is now widely accepted that graphite spheroids are heterogeneously nucleated [5,6], and that austenite often envelopes the spheroids, especially at later stages of growth [3]. Electron spectroscopy has confirmed the role of nonmetallic inclusions (e.g. oxy-sulphide particles) in the nucleation of the spheroids [6], and high resolution transmission electron microscopy has been used to demonstrate that the spheroids very likely grow by a lateral process, in which carbon atoms are added primarily at the growth ledges [5].

This latter observation prompts the question: If the number of possible attachment sites for carbon atoms on the spheroid surfaces is severely restricted, what is an appropriate value of the interfacial kinetic coefficient to associate with the growth interface? (There is already in the literature, strong indirect evidence for a significant kinetic inhibition in the growth of flake graphite [7].) A related question: In regimes dominated by interfacial attachment, how is the driving force for growth partitioned between the interfacial reaction and the volume diffusion process? In the following, we will address these questions.

EXPERIMENTAL:

A top-poured finned casting (Figure 1) was used in order to vary the cooling rate and the undercooling for a given chemistry, and therefore to obtain a database relating heterogeneous nucleation rates to the imposed conditions of solidification. A chromel-alumel thermocouple was placed in the centre of each of the fins.

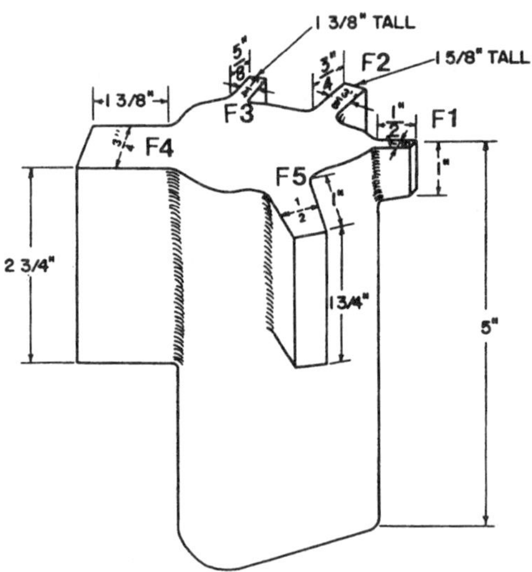

Figure 1: The pattern used for the experimental castings

The material studied began as a low sulphur base iron supplied by Sorrel Metal, melted in an MgO-lined crucible, and alloyed when molten with 75% grade ferrosilicon. The base chemistry was then adjusted to near-eutectic composition. Nodularisation was effected by plunging a Ni-15%Mg alloy, and post-innoculation by adding Renod grade ferrosilicon to the stream. Table 1 gives the final analysis of the casting.

TABLE 1; Chemical Analysis of the Casting (wt%)

C	Mn	P	S	Si	Cu	Ni	Cr	Mo	Mg
3.018	0.0299	0.0785	0.0029	2.895	0.0204	0.6755	0.0300	0.0045	0.0060
σ	0.0003	0.0138	0.0012	0.0438	0.0006	0.0110	0.0015	0.0011	0.0005

The thermocouple signals were acquired and stored using an industrial IBM PC I/O card model TC15/RTD15, and a portable PC. Microstructures of the resulting castings were analyzed quantitatively using a LECO image analyzer with L2001 software; the smallest graphite spheroid detectable was 1.5 μm in radius.

MODELLING:

i) Physico-Chemical Basis:

Our goal has been to model the size distribution in the casting as a function of thermal history; the basis of the model and the assumptions implicit in it will first be described qualitatively. The computed behaviour will be discussed, and comparisons will be drawn between the numerical results and the observed microstructures.

First, the model assumes that a broad spectrum of potential nucleation sites is avaliable for spheroidal graphite formation, and that additional sites are activated as the temperature is lowered, in the manner first proposed by Oldfield [8], and developed recently by Castro et al [9]. It is assumed that the number of new nuclei per unit volume of mushy zone, dN_v, at a specific temperature range (T, T-dT) is proportional to the residual volume of the liquid, f_L, and some function of the eutectic undercooling, ΔT_{eut}. Oldfield's model does not include the term f_L

$$dN_v = A \times f_L \times (\Delta T_{eut})^n \tag{1}$$

A further assumption concerning free growth of the spheroids from the liquid phase is that the eutectic reaction is strongly divorced; austenite dendrites form first, and reject carbon to the liquid in accordance with the liquid/austenite phase equilibria. (A local equilibrium is assumed for this process.) If no graphite spheroids begin to grow immediately below the eutectic temperature, austenite solidification will continue to follow the metastable austenite/liquid and liquid/austenite boundaries as cooling proceeds; i.e. the liquid is continually enriched in carbon, the highest carbon concentrations located in the immediate vicinity of the growing austenite. This leads to the development of undercooling, and to an increasing thermodynamic driving force for graphite growth.

The way in which this driving force is assumed to act is shown schematically in Figure 2, a b and c. Figure 2(a) illustrates the basic concepts of the model for free growth of the graphite from the liquid, first, for a local equilibrium boundary condition which ignores capillarity, then for a departure from local equilibrium due to an interfacial kinetic inhibition, still ignoring capillarity. In the latter case, the thermodynamic driving force for growth, which is derived from the total undercooling of the graphite ΔT, is partitioned between the diffusion field (δ) and the interfacial reaction (ρ) in the ratio δ/ρ. (In Figure 2(a), this ratio is projected (in approximation) onto the G axis at $C = 1$, for purposes of illustration.) The net effect is to detract from the concentration difference available to drive the diffusion process, and to slow the growth of the spheroid. In this situation, the growth velocity must at all times satisfy the diffusion and the interfacial conditions:

$$v_i = \frac{dr}{dt} = D_L^c \times \frac{V_m^{gr}}{V_m^L} \times \frac{C^{L/\gamma} - C_i}{C^{gr} - C_i} \times \frac{1}{r} \tag{2}$$

and

$$v_i = \frac{dr}{dt} = K_c \times (C_i - C^{L/gr})^m \tag{3}$$

where the C's are mole fractions, D's are diffusion coefficients, V's are molar volumes, r the radius of the spheroid, K_c the interfacial kinetic coefficient for graphite growth normal to the basal plane, and the exponent m determined by the mechanism of supply of growth ledges. (Here m will be taken = 2, implying dislocation spiral sources.) The matching of the two velocities is accomplished by the instantaneous choice of C_i, the average carbon concentration in the liquid at the interface.

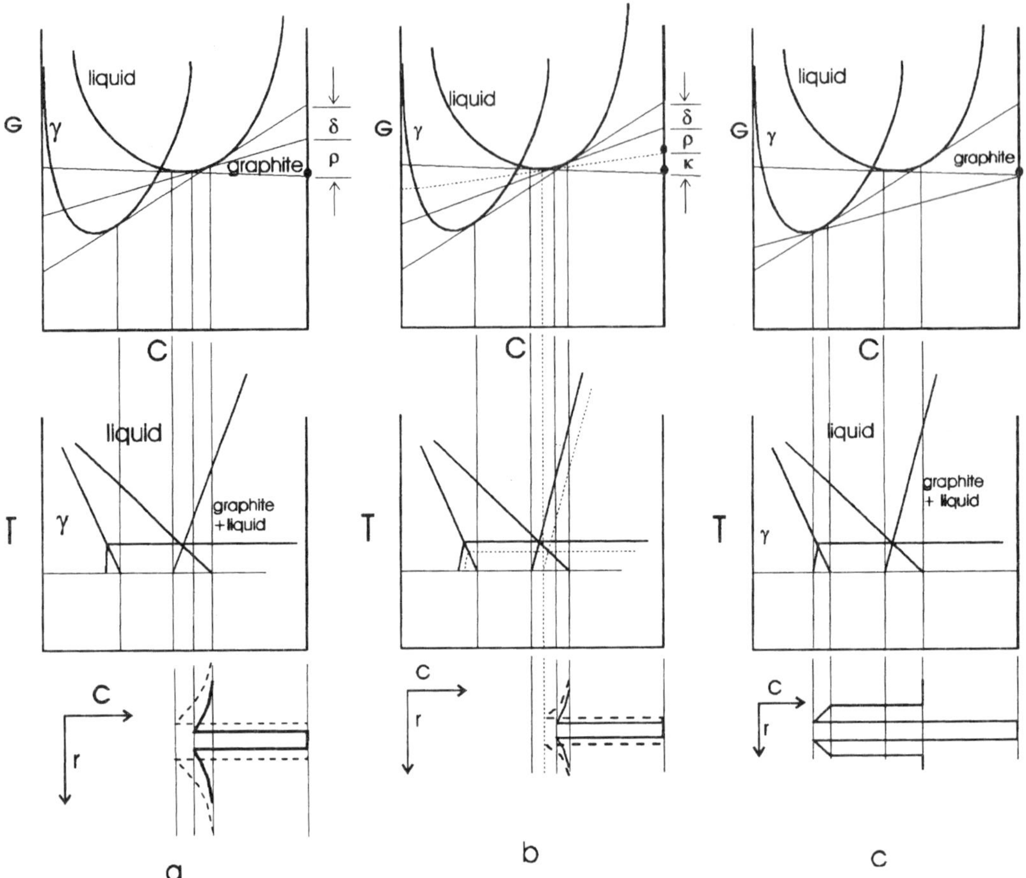

Figure 2: Free energy vs. composition, phase diagram, and concentration profiles through growing spheroids: 2(a); comparison of the local equilibrium (broken line) and mixed diffusion-reaction interfacial conditions, neglecting capillarity, the total driving force is projected to the graphite axis, to illustrate the partitioning between competing dissipative processes. 2(b) is similar, but includes capillarity. 2(c) represents thelocal equilibrium condition in the presence of an austenite envelope.

A further diminution of growth rate results when capillarity is introduced as shown in Figure 2(b). Here, the total potential driving force is first reduced by a capillary term, κ, and the equilibrium graphite liquidus is shifted by an amount related to the curvature of the spheroid (2/r) and the specific interfacial free energy of the graphite/liquid interface. The equilibrium eutectic temperature is also affected. Then the remaining driving force is partitioned as before into diffusion (δ) and reaction (ρ) components. The form of the Gibbs-Thompson equation appropriate to this situation gives the difference in carbon concentrations in liquid in equilibrium with small spheroids of radius r and large spheroids (r = ∞) as

$$C_r = C_\infty^{L/gr} \times (1 + \frac{2\sigma V_m^{gr}}{rRT}) \qquad (4)$$

where σ is the interfacial free energy of unit area of graphite/liquid interface.

Once the graphite spheroid is enveloped by austenite, the boundary conditions for growth are changed as shown schematically in Figure 2(c). Here, for simplicity, a local equilibrium boundary condition is assumed; the same interfacial reaction and capillary conditions can be added to this construction as well, but the resulting graphical construction is not informative. Under local equilibrium conditions, the rate of growth of the spheroid within the austenite shell is given in approximation by:

$$\frac{dr}{dt} = D_\gamma^C \times \frac{V_m^{gr}}{V_m^\gamma} \times \frac{C^{\gamma/L} - C^{\gamma/gr}}{C^{gr} - C^{\gamma/gr}} \times \frac{1}{r^2(1/r - 1/s)} \tag{5}$$

where the relevant diffusion coefficient is now that for carbon in austenite, and r and s are the radii of the spheroid and the shell respectively. As before, the capillary and interfacial reaction terms can be added, replacing $C^{\gamma/gr}$ by C_i in the above equation, and introducing a second equation (analogous to eq.3) for the relation between the interface velocity and the interface reaction.

ii) Numerical Modelling:

Numerical routines were developed to permit the investigation of the growth of the graphite spheroids; the routines included the effects of interfacial kinetics and capillarity on the relevant interfacial concentrations. It was assumed that all interfaces but the graphite basal plane were in local equilibrium. The interface kinetic coefficent for graphite growth was varied as part of a search for the best agreement between experimental and computed results.

In the numerical work, the following thermodynamic and kinetic coefficients were assumed: The diffusion coefficients of carbon in liquid iron and in austenite were taken as 2.5 x 10 $^{-9}$ and 1.0 x 10 $^{-10}$ m^2/sec respectively at 1153 C [10,11] (a correction for temperature variation was applied in those cases where undercooling was significant); the molar volumes of the liquid, austenite and graphite phases were taken as 7.0, 7.0, and 5.5 (x 10 $^{-6}$ m^3); and the specific interfacial free energy of the graphite/liquid and graphite/austenite interfaces as 1.5 J/m^2.

In order to model spheroid growth, it was necessary to make an assumption about the initial radius of the spherulites; exploration of the sensitivity of the final size to this quantity, using growth laws appropriate to free growth from the liquid, and to growth from an austenite envelope, indicated that the computed size of the spheres after five seconds growth was almost completely insensitive (within 1%) to choices of initial radius in the range 5 - 100 nm. An initial radius of 100 nm was chosen for subsequent calculations.

The validity of the local equilibrium boundary condition was also investigated using the numerical routines. It was found, for all reasonable choices of thermodynamic and kinetic parameters, that the neglect of the interface reaction invariably gave final graphite radii much larger (e.g., by and order of magnitude) than those observed. The interface reaction is therefore included in all of the numerical results presented below.

RESULTS and DISCUSSION:
i) Experimental:

A set of cooling curves is shown in figure 3, along with an optical micrograph of the material of fin #2. The solidification times were estimated from the lengths of the plateaus for fins #'s 2, 4 and 5 as 105, 210 and 125 sec. respectively. The microstructures from these fins yielded histograms, representing numbers of spheroids per square cm. vs. spheroid radius; these will be compared with the results of numerical calculations in the next section.

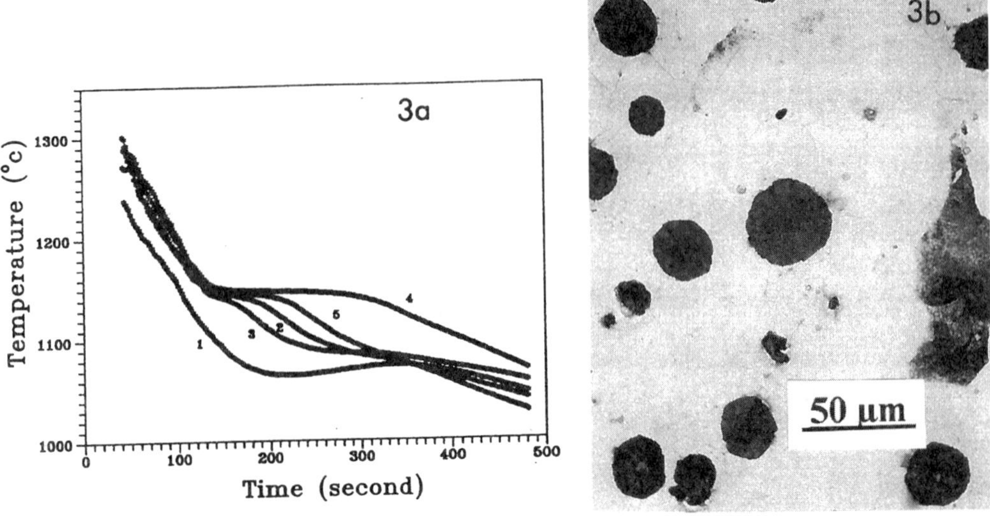

Figure 3(a): Cooling curves for the various sections of the test casting; fins #'s 2 (10 mm), 4 (20 mm) and 5 (13 mm) were chosen for further examination. The microstructure of fin #2 is shown in figure 3(b).

ii) Numerical:

The heterogeneous nucleation model employed is necessarily rather empirical; nevertheless, the form first suggested by Oldfield [8]seems to represent the observed behaviour rather well. The optimal values of the coefficient A and of the exponent n in equation 1 are found to be 90 and 2.5.

The effect of variations in the interface kinetic coefficient K_c were investigated using a simplified free growth model in which the capillary effect was neglected. If the interfacial mobility is taken as approximately 10^{-2} m/at%^{2}.sec, the reaction is essentially diffusion controlled for all radii above 0.1 μm, and no agreement is found between observed and computed size distributions; a choice of approximatedly 10^{-4} m/at%^{2}. sec. is required for general consistency. This was found to be the case for the more sophisticated growth models (those including capillarity, and growth in the presence of an austenite envelope) as well. We infer that growth is dominated by the low mobility of the graphite interface at radii of about 1.5 μm or less; for larger radii, diffusion plays an increasingly greated role.

If we compare the relative calculated rates of growth with and without the austenite shell, it is found that the rate with the shell is very roughly four times lower than the rate of free growth, all other factors being equal. Considering the nucleation or acquisition of the austenite envelope after a period of free growth from the melt, the ratio of the two radii, s/r, is (obviously) small when the envelope first forms. It then increases to a computed limiting value of 2.4, provided only that the time for growth is sufficiently long.

With these findings and assumptions, the comparison of the calculated and experimental size distributions for fins #'s 2, 5, and 4 are shown in Figure 4. One further assumption is made here: It is found that the agreement can be optimized by permitting the maximum size to which the spheroids grow before being enveloped by austenite to be 14 μm when the temperature is 1148 C and above, zero when the temperature is 1138 C and less, and to vary according to:

$$\bar{r} = 14 - 700 \times \frac{1148 - T}{5} \qquad (6)$$

between these limits. This assumption is consistent with the observation [3] that graphite spheres can grow freely to sizes of approximately 10 μm at higher temperatures. We note also that previous computer simulation studies have assumed that the formation of the austenite envelope accompanies that nucleation of graphite under all conditions [7,12].

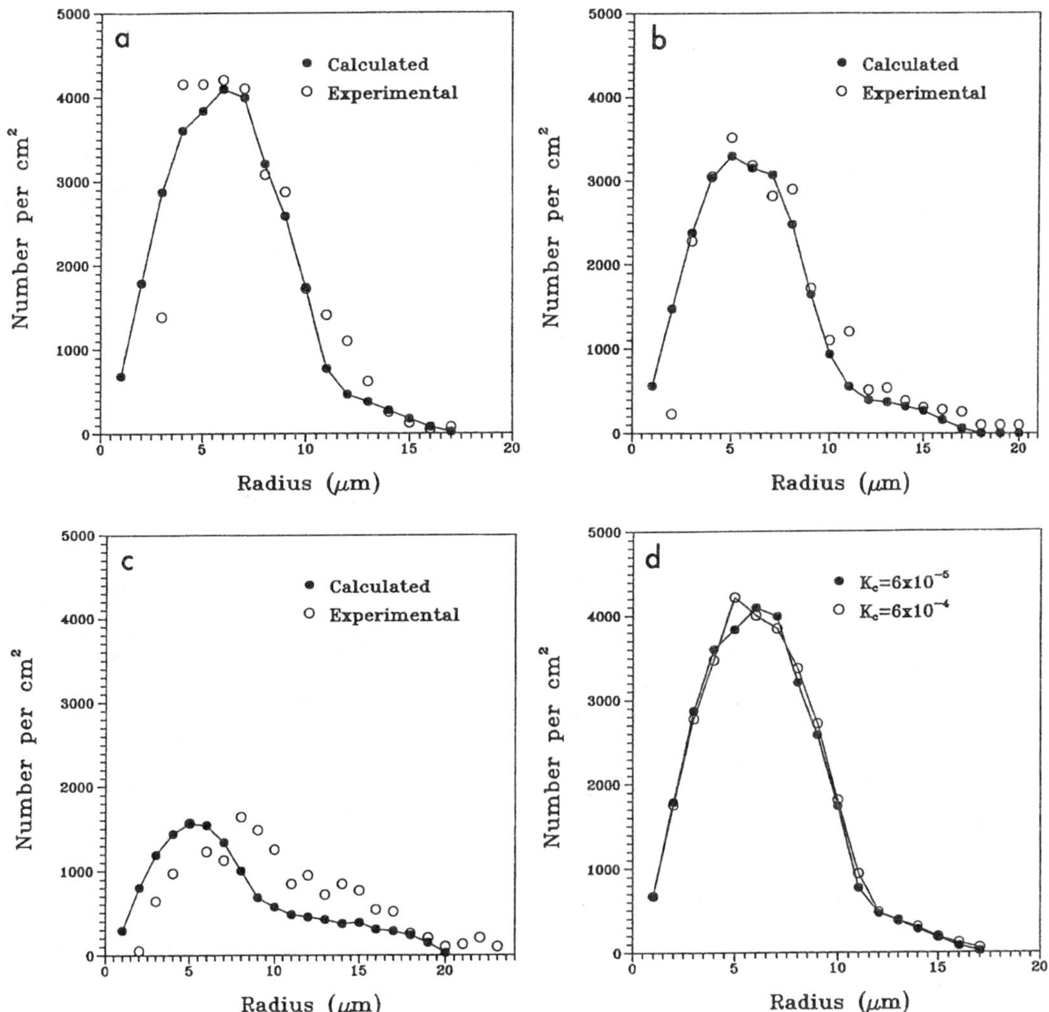

Figure 4 (a, b, c, d): Comparisons of experimental data and computed results for fins #2, #5, and #4; figure 4(d) is computed for fin #2 for two different K_c values.

CONCLUSIONS:

Based on comparisons of calculated and experimental graphite spheroid size distributions, we find:

- A continuous nucleation model of the type proposed by Oldfield [8] appears to best describe the present results. The number of new nuclei per unit volume is related to the undercooling by: $dN = A \, (\Delta T)^n$ with $A = 90$ and $n = 2.5$.

- Initial growth, either from liquid or from austenite, is very rapid; the calculated graphite radii after 5 seconds are quite insensitive to reasonable choice of post-nucleation initial radius. For radii less than about 1.5 µm, growth appears to be dominated by interfacial attachment kinetics; for larger radii, growth is essentially controlled by volume diffusion.

- The computed growth rate is very sensitive to the presence (or absence) of an austenite envelope. Once the austenite shell forms, the growth rate of the graphite slows, and the growth of the austenite accelerates rapidly as the ratio of the two radii approaches a limiting value of 2.4.

- The formation of the austenite shell seems to depend not only on the graphite spheroid radius, but also on the degree of undercooling. The largest graphite sphere grown directly from the melt and without an austenite shell is estimated to be < 14 µm in radius. At lower growth temperatures however, the austenite shell is thought to envelop the graphite immediately upon nucleation.

ACKNOWLEDGEMENTS:

The authors acknowledge with thanks the support of the Natural Sciences and Engineering Research Council of Canada, and valuable discussions with Dr. J. R. Dryden, of the University of Western Ontario.

REFERENCES:

1. H. Morrogh and W. J. Williams, J. I. S. I., London, **155**, 321 (1947).
2. I. Minkoff, in "The Physical Metallurgy of Cast Iron, Wiley Interscience, John Wiley and Sons Ltd. New York, (1983).
3. S. E. Wetterfall, H. Fredriksson and M. Hillert, J. I. S. I., (May 1972), ASM Source Book on Ductile Iron, p 162.
4. E. Scheil and D. Schöbel, Arch. Eisen., **4**, (1955)
5. G. R. Purdy and M. Audier, in "The Physical Metallurgy of Cast Iron", Editors H. Fredriksson and M. Hillert, MRS symposium proceedings **34**, North Holland, p13 (1985)
6. S. V. Subramanian and G. R. Purdy, F. Weinberg Symposium on Solidification Processing, Editors: J. Lait and I. V. Samarasekar, Pergamon , (1990), p 289.
7. H. Fredriksson and S. E. Wetterfall, in "The Metallurgy of Cast Iron" Eds. B. Lux, F. Mollard and I. Minkoff, Georgi Publications, Switzerland, (1975), p 377.
8. W. Oldfield, Trans. ASM, **59**, 949, (1966).
9. M. Castro, P. Alexander, J. Lacaze and G. Lesoult, in "The Physical Metallurgy of Cast Iron", Proceedings of 4th International Symposium, Eds. G. Ohira, T. Kusakawa and E. Niyama, MRS, (1989), p 433.
10. M. Hillert and N. Lange, J. I. S. I. **203**, 273, (1965).
11. C. Wells, W. Batz and R. F. Mehl, J. Metals, **2**, 553, (1950).
12. E. Scheil and L Hutter, Arch. Eisen., **4**, 24, (1953).

Advanced Materials Research Vols. 4-5 (1997) pp. 469-478
© *1997 Scitec Publications, Switzerland*

Modeling of the Stable-to-Metastable Structural Transition in Cast Iron

L. Nastac* and D.M. Stefanescu

Solidification Laboratory, Dept. of Metallurgical and Materials Engineering, University of Alabama, Tuscaloosa, AL 35487, USA

Keywords: Gray Iron, Chill, Intergranular Carbide, Gray / White Transition, Mottled, Microsegregation, Modeling of Structural Transition

ABSTRACT

A frequent reason for rejecting gray iron castings is the occurrence of chill, in particular in the thin sections and at corners. The origin of this common defect is the inadvertent gray / white (G/W) transition, resulting from poor control of process or material variables. In this paper it was proposed to develop a model for the G/W transition that includes the effects of nucleation and grain growth for both eutectics, together with the influence of microsegregation. The model predicts formation of chill and of intergranular carbides.

To model the G/W structural transition, the nucleation and growth of both the stable and metastable eutectics, the growth competition between the same eutectics, and the change in equilibrium temperatures and solubility limits because of the microsegregation of various elements occurring during solidification, were described.

A spherical geometry was assumed for both the white and gray eutectic grains. A previously developed micro-segregation model was modified to describe solute redistribution of a third element in the two competing eutectic phases. It was used to calculate the change in stable and metastable equilibrium temperatures during solidification.

The model was used to evaluate the influence of chemical composition and process variables such as cooling rate and inoculation on the G/W transition. The model was incorporated in ProCAST ^{TM} and validation was done by comparing theoretical predictions with experimental values for a shaped casting. Using a sensitivity analysis of the main variables, such as nucleation, growth, and partition coefficients, liquidus slopes, and heat transfer coefficient, it was found that the silicon concentration and the nucleation and growth of the gray eutectic have the main positive effects on the G/W structural transition.

INTRODUCTION

One of the frequent reasons for rejecting gray iron castings is the occurrence of chill, in particular in the thin sections and at corners. The origin of this common defect is the inadvertent gray / white (G/W) transition, resulting from poor control of process or material variables. This is a structural transition from the stable austenite-graphite eutectic to the metastable austenite-cementite eutectic (ledeburite). The resulting microstructure may be entirely white, or a combination between white and gray phases, called mottled structure. The basic variables affecting the G/W transition in cast iron are the cooling rate of casting, the nucleation potential, and the chemical composition of the melt. Typically, the G/W transition is favored by increased cooling rate, low nucleation potential, and low carbon equivalent and/or higher amounts of carbide promoting elements. In some cases, because of segregation during solidification, inverse chill may occur. These are regions of white structure formed in the last regions to solidify, which should in principle be gray because of their lower cooling rates.

In order to model the G/W structural transition, the following phenomena should be described through appropriate physical and mathematical models:

---

* Presently with Caterpillar Inc., Peoria, IL.

- the nucleation and growth of both the stable and metastable eutectics;
- the growth competition between the same eutectics;
- the change in equilibrium temperatures and solubility limits because of the microsegregation of various elements occurring during solidification.

2. Background

The eutectic structure of white cast iron is a rather complex one. According to Hillert and Steinhauser [1] growth of the austenite-iron carbide eutectic (ledeburite) begins with the development of a cementite plate on which an austenite dendrite nucleates and grows. This destabilizes the Fe_3C, which then growth through the austenite. As a result, two types of eutectic structure develop: a lamellar eutectic with Fe_3C as a leading phase in the edgewise direction, and rod eutectic in the side wise direction. Cooling rate significantly influences the morphology of the $\gamma + Fe_3C$ eutectic. Later, Hillert [2] proposed the following relationship between growth velocity, V, and interface undercooling, ΔT, for the cooperative eutectic growth of ledeburite:

$$V = \mu_w \Delta T^2 \qquad\qquad (1)$$

where $\mu_w = 2.5 \cdot 10^{-5}$ $m \cdot s^{-1} \cdot K^{-2}$ is the growth coefficient for the white eutectic.

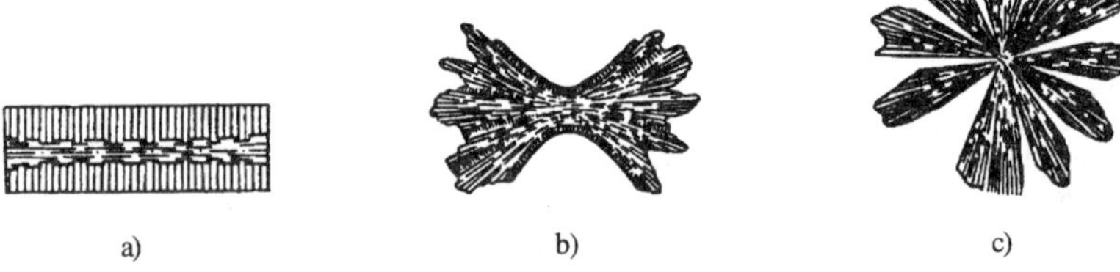

a)   b)   c)

Fig. 1 Schematic illustration of the change in the morphology of the $\gamma + Fe_3C$ eutectic grain at increasing cooling rate after [3].

Bunin *et al.* [3] suggested that, as the cooling rate increases, the lamellar part of the original parallelepipedic eutectic grain becomes larger at the expense of the rod eutectic (Fig. 1a), the grain starts bending inward (Fig. 2b), and eventually a spherolitic eutectic grain results (Fig. 1c).

Snagovski [4] suggested that for hypoeutectic medium alloy cast iron containing some carbide promoting elements such as Cr, V, Ti or Mo, below the amount required for the formation of special type carbides (MC, M_7C_3, or M_6C), the solidification structure results from a peritectic-eutectic reaction.

Directional solidification experiments by Jones and Kurz [5] and by Magnin and Kurz [6] confirmed that ledeburite behaves like a regular eutectic under certain conditions. They also evaluated the critical growth velocities for the G/W transition and the growth constants for pure Fe-C, and low alloy Fe-C-X systems.

The G/W structural transition has been described through numerical models on the basis of growth competition between the gray and white eutectics (*e.g.*, Stefanescu and Kanetkar [7], Fredriksson *et al.* [8,25]). The growth velocities of the two eutectics were calculated as a function of undercooling, and the two equilibrium eutectic temperatures were calculated as a function of the initial melt composition (compositional change due to microsegregation was ignored). It was further assumed that there is no "nucleation barrier" in the system, that is that each type of eutectic will nucleate as the respective eutectic temperature is reached. Under these assumptions the G/W transition occurs when the melt undercools below the metastable eutectic temperature, since the growth velocity of white grains is faster than that of gray grains.

Another model for G/W transition proposed by Upadhya *et al.* [9] was based on the concept of the existence of a critical cooling rate at which this transition can occur. The critical cooling rate is a function of chemical composition and nucleation potential and must be experimentally evaluated. This model predicts chill formation in cast iron, but can not describe mottled structure and intergranular carbide formation.

The purpose of the present work is to develop a model for the G/W transition that includes the effects of nucleation and grain growth for both eutectics, and the influence of microsegregation. The new model should predict formation of chill as well as of intergranular carbides.

3. A Model for the Growth of the White Eutectic

Based on the literature review it was concluded that the white eutectic may solidify with parallelepipedic (Fig. 1a) or spherical eutectic grains (Fig. 2c), within which the morphology can be lamellar, or rod, or both. Accordingly, the analysis should include all these eutectic grain shapes and morphologies.

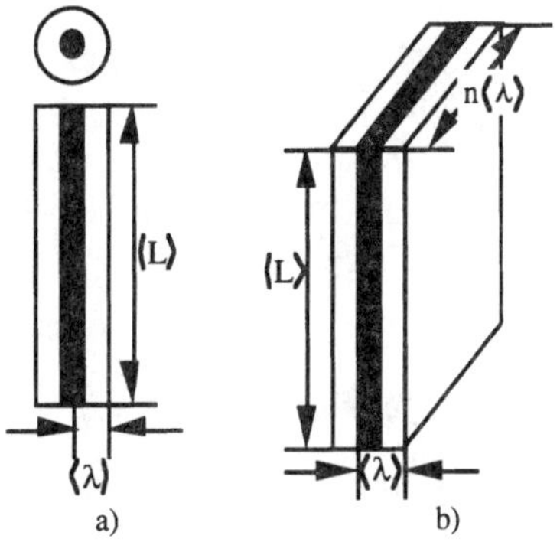

3.1 Models for lamellar and rod white eutectics.

For the rod eutectic a cylindrical geometry was assumed (Fig. 3a). For the lamellar eutectic a

Fig. 2 Assumed geometry of white eutectic basic units growing with different morphologies: a) rod eutectic basic unit (cylindrical geometry); b) lamellar eutectic basic unit(parallelepipedic geometry).

parallelepipedic geometry was assumed, with the edgewise direction as the main growth direction, and the sidewise direction as the secondary growth direction (Fig. 2b).

Assuming that the rod eutectic basic unit grows in both radial and longitudinal direction the change in the average volume, $\langle v \rangle$, with time is:

$$\frac{\partial \langle v \rangle}{\partial t} = 2\pi \langle \lambda \rangle \langle L \rangle \frac{\partial \langle \lambda \rangle}{\partial t} + \pi \langle \lambda \rangle^2 \frac{\partial \langle L \rangle}{\partial t} \qquad \text{where} \qquad \frac{\partial \langle L \rangle}{\partial t} = V \qquad (2)$$

where $\langle \lambda \rangle$ is the average radius (spacing), and $\langle L \rangle$ is the average length. Following the Jackson and Hunt theory [10], modified by Magnin and Kurz [11], the correlations between the average spacing, the interface undercooling, and the interface velocity, after some manioulation, are as follows:

$$\frac{\partial \langle \lambda \rangle}{\partial t} = \frac{\partial \langle \lambda \rangle}{\partial T} \dot{T} = \frac{2\,a\,(\varphi^2 + 1)}{\Delta T^2} \dot{T} = \langle \lambda \rangle \frac{\dot{T}}{\Delta T} \qquad (3)$$

where, a, b, φ are constants for a given system, and $\dot{T}$ is the cooling rate. Multiplying Eq. (2) by the stereological volumetric density (number of basic units per m^3), N_V, the evolution of solid fraction of white eutectic can be written as:

$$\frac{\partial f_S}{\partial t} = f_S \left[2\,\frac{\dot{T}}{\Delta T} + \frac{\mu_w \Delta T^2}{L} \right] \qquad \text{with} \qquad f_S = \pi \langle \lambda \rangle^2 \langle L \rangle N_V \qquad (4)$$

where f_S is solid fraction of white eutectic.

For lamellar white eutectic, assuming that the width of the basic unit is $n\langle \lambda \rangle$, where n is a constant (see Fig. 3b), Eq. (4) can still be used. However, the fraction of solid will be:

$$f_S = \langle \lambda \rangle^2 \langle L \rangle N_V \qquad (5)$$

Calculations with this model require experimental evaluation of N_V as a function of cooling rate, as well as of a, b and φ. For pure Fe-Fe$_3$C systems it was found that $\varphi = 2.0$ [5]. The diffusion coefficient of carbon in the melt at 1148 °C was taken to be 2×10^{-9} m^2 s^{-1} [5]. The liquidus slope of the cementite phase was calculated from ref. [13] to be 170 (K/wt%). The other physical constants were rigorously taken from [14]. Then, the growth parameters for both rod and lamellar eutectics are calculated as:

for rod growth $\quad\quad 2\,a\,(\varphi^2 + 1) = 4.9\ 10^{-6} \quad\quad \mu_w = 2.3\ 10^{-6}$ (6)

for lamellar growth $\quad 2\,a\,(\varphi^2 + 1) = 6.8\ 10^{-6} \quad\quad \mu_w = 2.5\ 10^{-6}$ (7)

The question is know which of the two models must be used. Experimental evidence demonstrates that both the rod and lamellar eutectics may simultaneously occur in white irons. However, according to calculations by Jackson and Hunt [10], who used the minimum undercooling criterion to described the lamellar to rod transition, systems in which the ratio between the fractions of the two phases of the eutectic is one, the morphology should be lamellar. Thus, since for the pure $Fe\text{-}Fe_3C$ white eutectic the fraction of austenite in the eutectic is 51.4 %, only lamellar eutectic should be expected. Nevertheless, for multicomponent $Fe\text{-}Fe_3C$ alloys microsegregation will affect the ratio between the two phases of the eutectic, as well as the growth conditions, resulting in a transition from lamellar to rod morphology.

3.2 Model for spherical grains.

To circumvent the difficulties associated with the stereological measurement, and with the lamellar to rod transition, a kinetic model that assumes spherical grain growth can be derived. It can be shown through a simple geometrical conversion from rod or lamellar morphology to spherical grains, that the growth coefficient, μ_w, remains the same, although the length scale changes from that of eutectic basic unit to that of a grain. Thus, the same growth coefficients as described by Eqs. (6) and (7) can be used. An average value of 2.4×10^{-6} m s^{-1} K^{-2} was taken for the calculation of growth of white eutectic. Note that this value is 3 times higher than that obtained by Krivanek and Mobley [15] through fitting calculations for spherical geometry to experimental data. Their values ranged from 6 to 8×10^{-7} m s^{-1} K^{-2}. Also, the calculated growth coefficient is one order of magnitude lower than that found by Hillert [2], as shown in Eq. (1). With these in mind, the evolution of fraction of solid for the white eutectic can simply be calculated as:

$$\frac{\partial f_S}{\partial t} = 4\,\pi\,R^2\,N_w\,\mu_w\,\Delta T^2\,(1 - f_S)$$ (8)

where N_w is the grain density (grains/m^3) of the white eutectic and R is the actual grain radius. The same equation can be applied for the gray eutectic, with different nucleation and growth parameters. In this work the growth coefficient of gray iron was taken to be 3×10^{-8} m s^{-1} K^{-2} [16].

4. Nucleation of stable and metastable eutectics

The model assumes instantaneous nucleation (site saturation) for both eutectics at the nucleation temperature [17]. The grain density data used in the present model are as follows:

for the gray eutectic: N_g (m^{-2}) $= 1.0 \times 10^5 + 3.36 \times 10^4\ \dot{T}$ $\quad$ (ref. [8]) $\quad$ (9a)

for the white eutectic: N_w (m^{-2}) $= 5.0 \times 10^5 + 1 \times 10^5\ \dot{T}$ $\quad$ (assumed) $\quad$ (9b)

5. Microsegregation of the third element and its influence on the G/W transition

As previously discussed, the final structure results from the competition between the gray and white eutectic. This is directly influenced by the two equilibrium temperatures, since each phase can only start nucleating and growing under its equilibrium temperature. The equilibrium temperatures, T_{stable} and $T_{metastable}$, for the gray and white eutectic, respectively are calculated as follows [18]:

$$T_{stable}\,(°C) = 1154 + 4\,\langle Si_L\rangle^L - 2\,\langle Mn_L\rangle^L - 30\,\langle P_L\rangle^L$$ (10)

$$T_{metastable}\,(°C) = 1148 - 15\,\langle Si_L\rangle^L + 3\,\langle Mn_L\rangle^L - 37\,\langle P_L\rangle^L$$ (11)

where $\langle x_L \rangle^L$ represents the intrinsic volume average concentration of the liquid phase.

Consider a macro-volume element within the solidifying metal. Within this macro-element the temperature is assumed uniform, and is obtained from the solution of the thermal field. The macro-element is further subdivided in a number of spherical volume micro-elements. Within each of these elements both gray and white eutectic can competitively grow into the liquid. The problem to solve is to calculate the volume average concentration of the third element in the liquid during solidification. The liquid concentration controls both the stable and metastable temperatures bellow which growth of gray and, respectively white eutectic can occur. The final liquid fraction will have a volume average concentration that depends not only on the diffusion coefficients in the solid phases, but also on the growth velocities (or solid fractions) of the gray and/or white grains.

The solute concentrations in the solid- gray, C_S^g, and solid-white, C_S^w, phases must each satisfy Fick's second law in spherical coordinates:

$$\frac{\partial C_S^g}{\partial t} = \frac{1}{r^2} \frac{\partial}{\partial r} \left(r^2 D_S^g \frac{\partial C_S^g}{\partial r} \right) \text{ and } \qquad \frac{\partial C_S^w}{\partial t} = \frac{1}{r^2} \frac{\partial}{\partial r} \left(r^2 D_S^w \frac{\partial C_S^w}{\partial r} \right) \tag{12}$$

where, D_S^w and D_S^g are the diffusion coefficients in the gray and white phases, respectively. The concept of this model is similar to that previously developed by the authors for one solid phase ([19]. However, because of the complications of tracking two different liquids, it was assumed that complete mixing in liquid occurs. Two liquid-solid interfaces must be considered: liquid-white eutectic and liquid-gray eutectic. The interface compositions of these interfaces are:

$$C_g^* = k_g^L \langle C_L \rangle^L \qquad\qquad C_w^* = k_w^L \langle C_L \rangle^L \tag{13}$$

where $\langle C_L \rangle^L$ is the intrinsic volume average liquid concentration, k_g^L and k_w^L are the partition ratios between the liquid and gray and white, respectively, and the superscript $*$ denotes values at the interface. Then, mass conservation is used to couple the concentration fields in the solid and liquid phases:

$$C_0 = \int_0^{f_g} C_S^g(f_g') \, df_g' + \int_0^{f_w} C_S^w(f_w') \, df_w' + f_L \langle C_L \rangle^L \tag{14}$$

where C_0 is the initial concentration of the alloy, $f_w + f_g = 1 - f_L$, and f_g, f_w, and f_L are the fraction of gray, white, and liquid phases, respectively. The intrinsic volume average liquid concentration is:

$$\langle C_L \rangle^L = C_0 \frac{1 - 3 \left(f_g I_S^g k_g^L + f_w I_S^w k_w^L \right)}{1 - f_g \left(1 - k_g^L + 3 k_g^L I_S^g \right) - f_w \left(1 - k_w^L + 3 k_w^L I_S^w \right)} \tag{15}$$

where and I_S^g and I_S^w have similar expressions and are expressed as follows:

$$I_S^k = \frac{2 f_k}{\pi^2} \sum_{n=1}^{\infty} \frac{1}{n^2} \exp\left[-\left(\frac{n\pi}{R_k^*} \right)^2 D_S^k t \right] \tag{16}$$

where t is the time, R_k^* is the interface position, and the subscript k denotes either g or w subscripts.

6. Calculation procedure

The solidification kinetics model that includes the nucleation and growth competition between the gray and white eutectics was coupled to a macro-transport code through the latent heat method [20].

Using heat balance at the interface and overall balance within the control volume element, and assuming quasi steady state approximation and spherical geometry, the interface temperature, T^*, can be calculated as:

$$T^* = T_b + \frac{\rho\,L\,V}{2\,K_L}\left[\,2 + n^3 - 3\,n\,\right] \qquad \text{with} \qquad n = \frac{R^*}{R_f} \tag{17}$$

where T_b is the bulk temperature, V is the growth velocity (for gray or white eutectic), K_L is the liquid thermal conductivity, L is the latent heat, ρ is the density, R^* is the interface radius, and R_f is the final radius of the grain.

7. Model analysis

In this work, spherical grain geometry was assumed for both gray and white eutectics. The subroutines for gray and white eutectic solidification, as well as the proposed microsegregation model have been incorporated in ProCAST [TM].

A theoretical analysis of the model predictions was conducted on a cylindrical step casting poured into a pepset sand mold. The exact geometry of this casting, as well as pertinent macro-transport data for gray iron, have been given in a previous paper [20]). Data for macro-transport relative to white iron were taken from ref. [15]. As stated in the introduction, the G/W transition is controlled primarily by the chemical composition of the melt, the nucleation potential, and the heat removal rate of the sample.

7.1 Influence of the silicon content

Silicon is the main element that controls the transition in plain cast iron. Therefore, for the theoretical analysis.of the model, silicon was used as a variable. The range was from 0.5 to 3.0 %. The other elements of the chemical composition were maintained constant. The main data used in computation are given in Table 1.

Tab. 1 Main data used in the modeling of the G/W transition.

Data	units	value	Refs.
Latent heat of gray eutectic	J / Kg	2.6×10^5	[15]
Latent heat of white eutectic	J / Kg	1.5×10^5	[15]
K_L	W / m/ K	27.5	[15]
Mn	wt. %	0.50	-
S	wt. %	0.025	-
P	wt. %	0.05	-
C	wt. %	3.60	-
Si	wt. %	0.5 - 3.0	-

The effect of silicon can be illustrated through the graphs in Fig. 3. From the cooling curves it may be implied that at 1% Si the iron solidifies completely white, since the solidification part of the cooling curve lies completely under the metastable eutectic temperature. However, from the graphs describing the evolution of the fraction of white and gray eutectic (Fig. 4) it is apparent that this is not true. Indeed, a mottled structure is predicted with about 40% gray eutectic. The reason for this apparent discrepancy is that once nucleated, the gray eutectic can in principle grow under T_{met}. The final structure will result from the nucleation and growth competition between the gray and white eutectic. Unfortunately, because of the paucity of data related to the nucleation and growth of the white eutectic, an accurate prediction is difficult.

When the Si content was raised to 2%, some intergranular carbide form at the end of solidification. This is evident from the fact that the cooling curve lies under T_{met}, for a limited time before the end of solidification (Fig. 3b). Clear recalescence is observed on the cooling curve, as opposed to the 1% Si case. Overall the silicon effect is, as expected, to increase the amount of gray phase (Fig. 4b). There is no chill, but the total amount of white phase is still 10%. Note that the two equilibrium temperatures diverge in time for the 1% Si sample, but converge for the 3% Si sample. They also converge for 2% Si. This effect is a result of the evolution of the average silicon, manganese and phosphorus concentrations. Indeed, as seen in Fig. 5, for 1% Si the silicon concentration increases continuously with time (fraction of solid). For 2% and 3% Si a reverse trend is observed. This change in the redistribution of silicon results from opposite partitioning in the two eutectics, and the nucleation and growth competition between them. For the higher silicon content of 3% it is still possible to obtain intergranular carbides, as seen from the cooling curve in Fig. 3b, and from the evolution of fraction of solid in Fig. 4b.

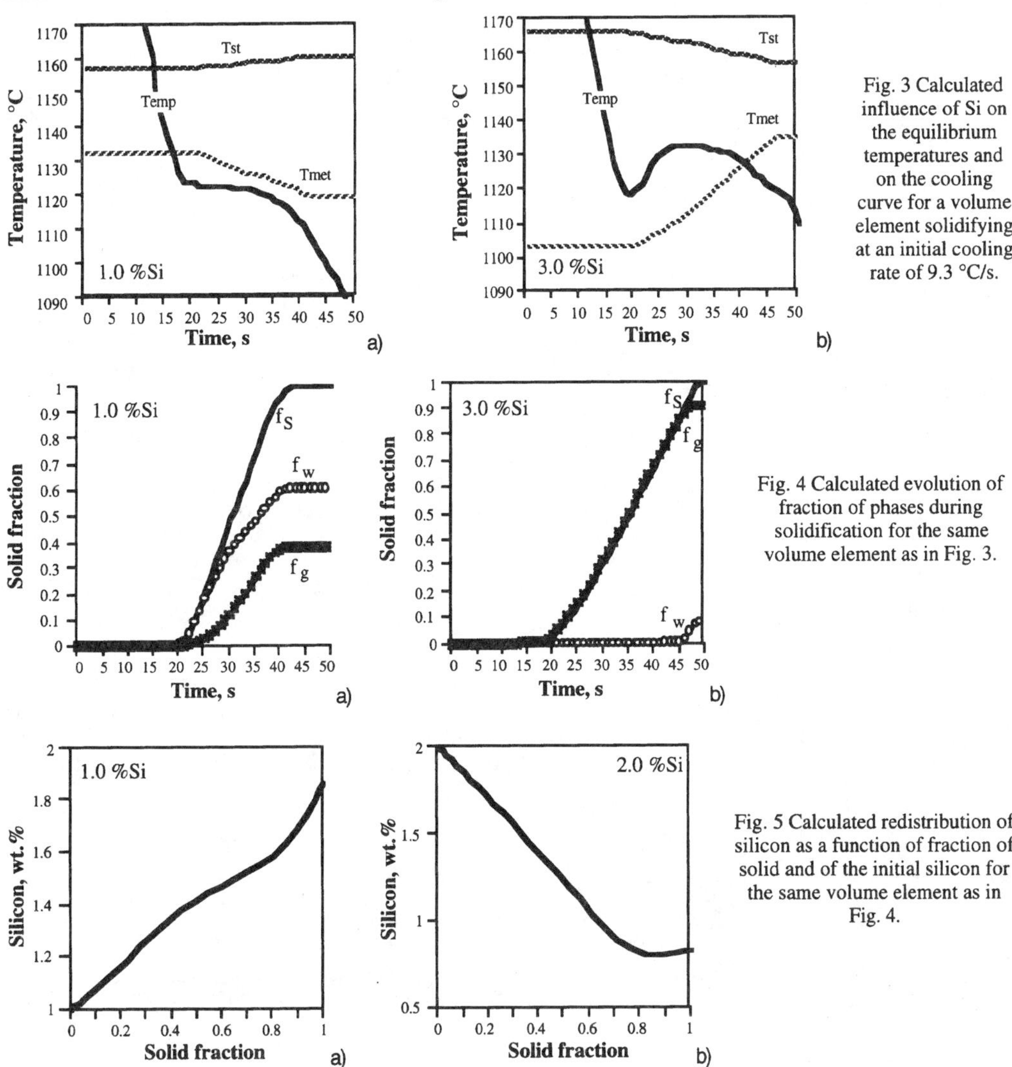

Fig. 3 Calculated influence of Si on the equilibrium temperatures and on the cooling curve for a volume element solidifying at an initial cooling rate of 9.3 °C/s.

Fig. 4 Calculated evolution of fraction of phases during solidification for the same volume element as in Fig. 3.

Fig. 5 Calculated redistribution of silicon as a function of fraction of solid and of the initial silicon for the same volume element as in Fig. 4.

7.2 Combined influence of silicon and cooling rate

Fig. 6 illustrates the combined influence of silicon and of the cooling rate on the structure of a cast iron having 3.6% C, 0.5% Mn, 0.05% P, 0.025% S. Two structural transitions must be defined: a gray-to-white transition resulting in a completely white structure, and a white-to-gray transition, resulting in a completely gray structure. In between these two, a mottled zone exists. The three structural zones on Fig. 7 can also be defined in terms of the type of carbide formation mechanism. A fully white structure is the result

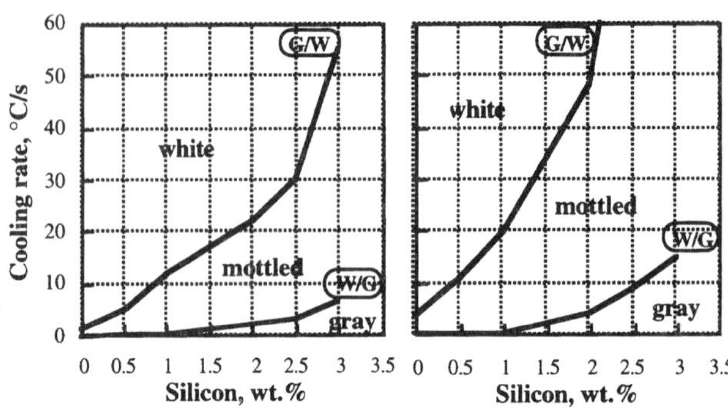

Fig. 6 The influence of silicon and initial cooling rate on structural transitions. a) uninoculated (N_g from Eq. 11); b) inoculated ($3 \times N_g$).

of chill formation. A mottled structure results when both chill and intergranular carbides are formed. It is seen that the size of the mottled zone increases with silicon and inoculation. It will also be affected by other elements such as chromium, aluminum, etc., and by inoculation. As expected, inoculation increases the gray region and decreases the white on (the curves are pushed upward on Fig 7b). The mottled region is also increased.

7.3 Sensitivity analysis

For a better understanding of the controlling variables of the G/W transition a sensitivity analysis was performed. It was found that selecting different literature values of the slope of the eutectic valley for silicon influences the fraction of white eutectic. On the contrary, choosing to use the bulk rather than the interface undercooling in calculation does not affect the outcome.

It was concluded that the major variables are the nucleation potential and growth coefficient of the gray phase. Thus, the fact that the solidification kinetics parameters of the white phase are not very well documented, may not be an insurmountable handicap in the modeling of the G/W transition.

8. Model validation

To validate the model, results of calculations were compared with experimental results obtained on a step casting whose geometry and structure are given in Fig. 7. The casting was poured in a pepset resin bonded sand mold, from an uninoculated iron, having the following chemical composition: 3.47% C, 1.78% Si, 0.182% Mn, 0.05% P, 0.02% S. Complete experimental details can be found in [9].

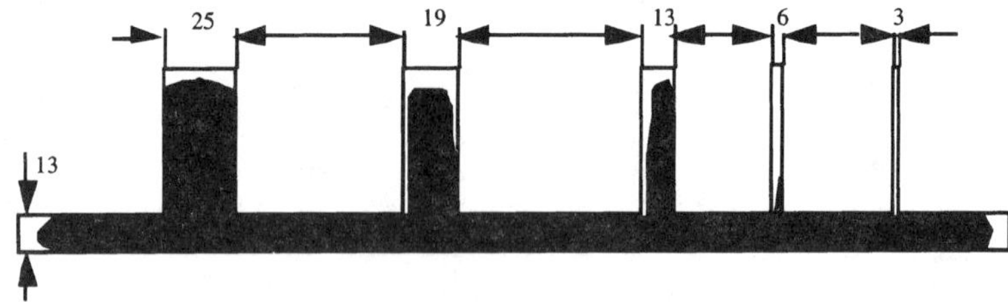

Fig. 7 Geometry and map of experimental G/W transition on the step casting (dimensions in mm).

Some typical examples of calculated structures, illustrating the G/W transition are given in Fig. 8. Fig. 8a shows the results of calculations using the same composition and nucleation law as for the experimental casting. In addition, the following data were used in calculations: $\mu_g = 3 \times 10^{-8}$ m s^{-1} K^{-2}, $\mu_w = 8 \times 10^{-7}$ m s^{-1} K^{-2}, $h = 581 + 2.05\,T + 2.19 \times 10^{-5}\,T^2 - 7.9 \times 10^{-5}\,T^3$, W m^{-2} K^{-1}, where T (in degree K) is the metal - mold interface temperature. It can be seen that the calculated map is reasonably close to the experimental one. An increase in the silicon content from 1.78% to 2.5% (Fig. 8b) resulted in some decrease of the white regions, but as expected, inoculation was the most efficient way to decrease chill (Fig. 8c). This computer experiment is an example of the procedure than could be followed on the foundry floor to decrease or eliminate chill formation in specific castings.

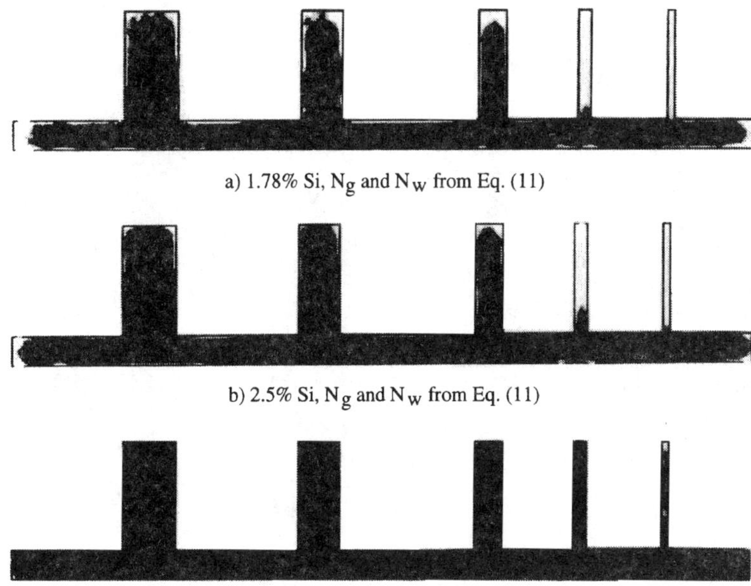

a) 1.78% Si, N$_g$ and N$_w$ from Eq. (11)

b) 2.5% Si, N$_g$ and N$_w$ from Eq. (11)

c) 2.5% Si, 3xN$_g$, N$_w$ from Eq. (11)

Fig. 8 Map of calculated G/W transition.

9. Conclusions

A physical and mathematical model that describes the gray to white transition in cast iron was developed. It takes into account the nucleation and growth of both the white and gray eutectics, as well as the effects of microsegregation of the third element. The model was incorporated in the commercial software ProCAST ^{TM}. Using a sensitivity analysis of the main process variables, it was found out that the silicon concentration of the melt and the nucleation and growth of the gray eutectic have the main positive effects on this structural transition.

The model was used to evaluate the influence of cooling rate, silicon content, and nucleation potential on the G/W and W/G structural transitions. It was found that both silicon and inoculation increase the mottled and gray zones.

Successful validation of the model was performed against an experimental casting. However, it must be noted that because of the incertitude associated with some of the physical parameters, it is necessary to calibrate the model if useful predictions are desired.

References

1   M. Hillert and H. Steinhauser, Jernkontorets Ann., *144* (1960) 520
2   M. Hillert, in *Recent Research on Cast Iron*, H. D. Merchant editor, Gordon and Breach, New York, 1969 p. 101

3   K. P. Bunin, I. N. Malinotchka and I. N. Taran, *Osnovi Metallographyia Tchuguna* (Fundamentals of the Metallography of Cast Iron), Metallurghyia, Moscow, 1969

4   L. M. Snagovski, in *Microstructural Design by Solidification Processing*, E. J. Lavernia and M. Gungor editors, TMS, Warrendale, Pa., 1992, p. 231-239

5   H. Jones and W. Kurz, Z. Metallkde., *11* (1981) 792 - 97

6   P. Magnin and W. Kurz, Metall. Trans. A, *19A* (1988) 1955-63 and 1965-71

7   D. M. Stefanescu and C. S. Kanetkar, AFS Trans., *95* (1987) 139-44

8   H. Fredriksson *et al.*, in *Physical Metallurgy of Cast Iron III*, H. Fredriksson and M. Hillert editors, MRS, 1984, p. 273 - 84

9   G. Upadhya, D. K. Banerjee, D. M. Stefanescu and J. L. Hill, AFS Trans., *98* (1990) 699 - 706

10   K. A. Jackson and J. D. Hunt, Trans. of AIME, *236* (1966) 1129-42

11   P. Magnin and W. Kurz, Acta Metall., *35*, 5 (1987) 1119-1128

12   E. Underwood, *Metals Handbook*, Vol. 9: Metallography and Microstructures, 9th edition, ASM Metals Park, Ohio, 1985, p. 123 - 34

13   F. Neumann, in *Recent Research on Cast Iron*, H. D. Merchant editor, Gordon and Breach, New York, 1968, p. 659-705

14   P. Magnin and R. Trivedi, Acta Metall., *39*, 4 (1991) 453 -67

15   R. A. Krivanek and C. E. Mobley, AFS Trans., *92* (1984), 311-17

16   Tian H. and D. M. Stefanescu, *Modeling of Casting, Welding and Advanced Solidification Processes VI*, Edited by T. S. Piwonka et al., The Minearls, Metals and Materiasl Soc., Warrendale, Pa, 1993, p. 639-46

17   D. M.Stefanescu, G. Upadhya, and D. Bandyopadhyay, Metall. Trans., *21A* (1990) 997-1005

18   A. Kagawa and T. Okamoto, *The Physical Metallurgy of Cast Iron*, H. Fredriksson and M. Hillert eds., Materials Research Society Symposia Proceedings, *34* (1985) 201-210

19   L. Nastac and D. M. Stefanescu, Metall. Trans., *24A* (1993) 2107-18

20   L. Nastac and D. M. Stefanescu, *Micro/Macro Scale Phenomena in Solidification*, HTD-Vol. 218/AMD-Vol. 139, ASME (1992) 27-34

21   R. W. Heine, AFS Trans., *94* (1986), 391-402

22   A. Kagawa and T. Okamoto, Metal Science, (1980) 519-24

23   D. M. Stefanescu, *Metals Handbook*, Vol. 15: Castings, ASM Metals Park, Ohio, 1988, p. 61-70, 168-81

24   C. Y. Wang and C. Beckermann, in *Micro/Macro Scale Phenomena in Solidification*, HTD-Vol. 218/AMD-Vol. 139, ASME, 1992, p. 43-56

25   H. Fredriksson, J. T. Thorgrimsson and I. L. Svensson, *in State of the Art of Computer Simulation of Casting and Solidification Processes*, H. Fredriksson ed., Les Edition de Physique, Les Ulis, France, 1986, p. 267-275

Advanced Materials Research Vols. 4-5 (1997) pp. 479-484
© *1997 Scitec Publications, Switzerland*

Modelling and Simulation of Ferrite Growth in Nodular Cast Iron

M. Wessén[1,2] and I.L. Svensson[1,2]

[1] Former affiliation: Dept. of Materials Processing, Royal Institute of Technology,
S-100 44 Stockholm, Sweden

[2] Div. of Component Technology, Jönköping University,
Box 1026, S-551 11 Jönköping, Sweden

Keywords: Ferrite, Interface Control, Kinetics, Modelling, Nodular Cast Iron, Simulation

ABSTRACT

In nodular cast iron it is well known that a major factor determining the hardness is the relative amount of ferrite and pearlite in the structure. Ferrite forms around the graphite nodules and growth proceeds until pearlite nucleates and consumes the remaining austenite. In order to simulate the hardness it is therefore necessary to have accurate models for the ferrite growth. Other investigators have proposed that the growth is completely governed by carbon diffusion through the ferrite shell.
In the present work, it is shown that the ferrite growth in nodular cast iron can be divided into three different stages where the growth initially is governed by carbon diffusion in the austenite until the graphite nodule is entirely enveloped by a ferrite shell. In the second stage it is proposed that the growth is interface controlled at the ferrite/graphite interface corresponding to the incorporation of the carbon atoms on the graphite nodule. During the later stages of the transformation the diffusion distance has increased considerably and therefore the diffusion of carbon through the ferrite shell will determine the growth rate.

INTRODUCTION

In nodular cast iron the structure after completed solidification consists of graphite nodules dispersed in an austenitic matrix which is non-homogeneous with respect to alloying elements due to micro segregation during solidification. It is of major importance to know how different alloying elements are distributed in the matrix since it has a very large influence on the solid state transformations. Elements like Si, Cu and Ni are concentrated in the vicinity of the graphite (negative segregation), while the highest levels of Mn, Cr and Mo are found in the last solidified areas (positive segregation), i.e. half-way between two nodules [1,2]. Two of the most common alloying elements in nodular cast irons are Si and Mn. Silicon is known to drastically increase both the stable and the metastable eutectoid temperatures while manganese has an opposite effect [3]. Bearing in mind the segregation pattern of Si and Mn, we find that there will be a gradient in eutectoid temperatures between the nodules with the highest values at the graphite/austenite interface. From this consideration it is natural to conclude that ferrite nucleates at the graphite/austenite interface.
When the metastable eutectoid temperature is reached there is also a possibility for pearlite to nucleate and grow in competition with the ferrite. The growth of pearlite depends only on short-range diffusion of carbon corresponding to the distance of the lamellar spacing. Therefore its growth is considerably faster than that of ferrite. Since pearlite is a metastable structure, it is also possible that the formed pearlite might decompose into the stable phases ferrite and graphite during subsequent cooling [4].
Consequently, there will only be a relatively small increase in ferrite content after the moment when pearlite has nucleated. In order to predict the ferrite/pearlite ratio in a casting it is therefore necessary to have a good understanding of what mechanisms are governing the ferrite growth, and to present models which take all the relevant mechanisms into account.
A limited number of papers in the literature have been treating this topic [5-10]. All presented models for the ferrite growth in nodular cast irons are based on the assumption that the growth rate is controlled by the rate of carbon diffusion through the ferrite shell. In one of the most recent

publications [9], it is suggested that another mechanism operates at the same time as the one mentioned above. Assuming that local equilibrium prevails at the ferrite/austenite interface, the carbon content in austenite in equilibrium with ferrite, $C_c^{\gamma/\alpha}$, is at a temperature below the stable eutectoid temperature higher than the bulk carbon content, C_0^{γ}. Therefore, it exists a driving force for carbon diffusion into the austenite, thereby increasing the growth rate. However, the author further concludes that the contribution of diffusion in austenite to the overall transformation rate is expected to be significant only very early in the transformation. This is due to the lower diffusivity of carbon in austenite than in ferrite, and to the fact that the enrichment of carbon in the austenite successively will lead to a decreasing carbon gradient.

EXPERIMENTAL PROCEDURE

Plates with thicknesses of 4, 8, 15, 30 and 50 mm have been cast in furan-bonded quartz sand. The melt was taken from a 3000 kg ladle which had been treated with FeSiMg by the sandwich method. The melt was then inoculated in the stream with Inoculin. Thermocouples (Pt-10% Rh) were situated at a distance of 1/4 of the plate-thickness from the centreline in each plate. The composition was: 3.73% C, 2.6 % Si, 0.23 % Mn, 0.02 % P, 0.007 % S, 0.044 % Mg, 0.041 % Cu. Samples were taken from the areas of the thermocouples for metallographical examination. In each sample the nodule count and the ferrite and pearlite fractions were measured.

RESULTS

In order to derive the kinetics of the ferrite growth from the cooling curves, a new method based on thermal analysis has been developed. The basic principles are the following: The heat flow away from the plate into the mould is calculated at each time from the derivative of the cooling curve before the start of the ferrite growth. By fitting these data to an exponential expression, the cooling rate of a fictive reference sample can be calculated during the transformation interval which then can be used to extract the transformation rate. By assuming that ferrite grows radially away from the graphite and that the impingement can be described by the Avrami/Johnson/Mehl equation, it is now possible to calculate the growth rate of the ferrite shell from the transformation rate and the nodule count.

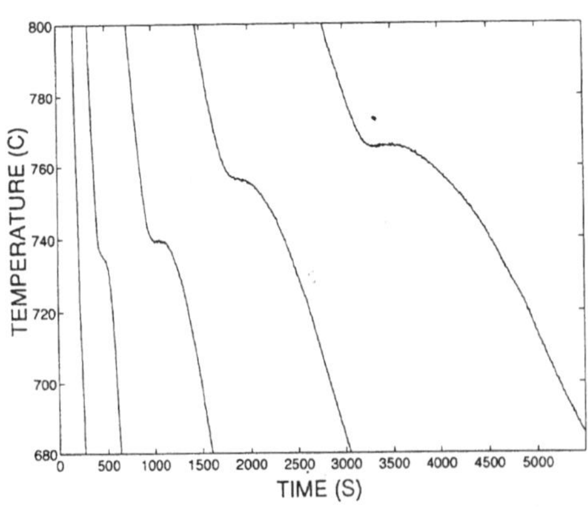

Figure 1. Cooling curves in plate castings

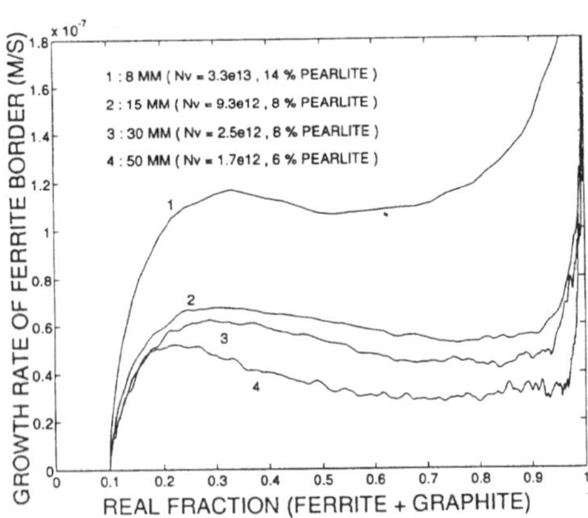

Figure 2. Growth rate of ferrite shell

In Fig. 1 the cooling curves at temperatures around the solid state transformations for the different plates are shown. The cooling curves from the four thickest plates were analysed by the above given procedure. The temperatures for start of the reaction, T_s, at maximum transformation rate, T_m, and at the end of the reaction, T_e, were evaluated, see Table 1. It is interesting to note that the reaction starts at approximately the same temperature (785 $\pm$ 5 C) for all plates but that T_1 decreases with decreasing plate thickness. The total time for the transformation was also measured.

PLATE	4 MM	8 MM	15 MM	30 MM	50 MM
T_s (C)		780	785	791	785
T_m (C)		736	740	757	766
T_e (C)		715	705	715	725
t(trans.) (s)		230	630	1100	1750
f (ferrite)	0.44	0.74	0.82	0.77	0.81
f (pearlite)	0.50	0.12	0.08	0.08	0.06
N_V (m^{-3})	$7.8 \cdot 10^{13}$	$3.3 \cdot 10^{13}$	$9.3 \cdot 10^{12}$	$2.5 \cdot 10^{12}$	$1.7 \cdot 10^{12}$

Table 1. Experimental data of plate castings

In Fig 2 the calculated growth rate of the ferrite shell is plotted versus the total fraction of graphite and ferrite. Principally it is possible to distinguish three different parts in the curves. Initially, the growth rate seems to increase rapidly to a maximum value. The first part represents a totally different growth morphology than when the ferrite shell grows radially away from the graphite, which has been assumed in previous works. Instead, ferrite nucleates successively on the graphite surface and starts to grow in the tangential plane until the nodules are totally surrounded by a ferrite shell. In the second part, the growth rate decreases during the transformation, the amount depending on the nodule count (section thickness). From now on, the ferrite is assumed to grow in the radial direction. In the third part the growth rate increases very rapidly to an infinite value. This has no physical significance for the ferrite growth, and is simply explained by the fact that the Avrami/Johnson/Mehl correction for grain impingement poorly describes the reality at high fractions. In addition, pearlite forms, and due to its high growth rate the heat evolution is higher than it would have been if only ferrite had formed.

FERRITE GROWTH MODELS

When modelling ferrite formation in nodular cast iron it will be necessary to divide the growth into three different stages. The *first stage* consists of nucleation of a number of small ferrite grains on the graphite nodule and a subsequent growth in the tangent plane which proceeds until the whole nodule is surrounded by a ferrite shell. In the *second stage* the growth is assumed to be controlled by an interface reaction at the ferrite/graphite interface. When the ferrite shell has reached a certain thickness, the diffusion of carbon through the ferrite will be the rate controlling mechanism; this is the *third stage*.

Due to the lower solubility of carbon in ferrite than in austenite, carbon has to diffuse away when ferrite forms. In Fig 3 an isopleth of a schematic Fe-C-Si phase diagram is shown. At a certain supercooling below the upper stable eutectoid temperature, T_U^α, it is seen that four different carbon solubility's has to be distinguished:

$C_c^{\alpha/gr}$, solubility in ferrite in equilibrium with graphite

$C_c^{\alpha/\gamma}$, solubility in ferrite in equilibrium with austenite

$C_c^{\gamma/\alpha}$, solubility in austenite in equilibrium with ferrite

C_0^γ, carbon content in the bulk austenite.

We find that there is a driving force for carbon diffusion through the ferrite $(C_c{}^{\alpha/\gamma} - C_c{}^{\alpha/gr})$ as well as into the bulk austenite $(C_c{}^{\gamma/\alpha} - C_0{}^\gamma)$. However, when the temperature is above the lower stable eutectiod temperature, T_L^α, there is no driving force for diffusion through the ferrite, as pointed out by Lacaze [11], which makes it reasonable to assume that ferrite in the *first stage* grows by bulk diffusion of carbon into the austenite. In Fig. 4 a schematic sketch shows the situation when one ferrite grain has nucleated on the graphite.

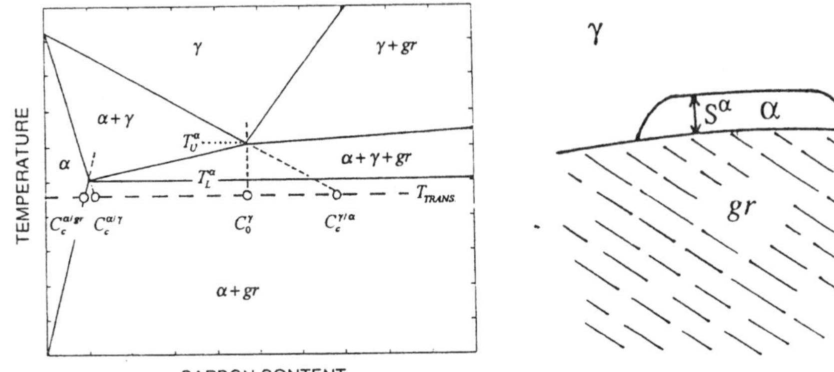

Figure 3. Schematic isopleth of the Fe-C-Si phase diagram

Figure 4. Schematic sketch of the first stage of the ferrite growth

The diffusion rate of carbon into the austenite is given by Fick's 1st law:

$$\frac{dm}{dt} = -D_c^\gamma \cdot A \cdot \left[\frac{dC_c^\gamma}{dy}\right]_{y=0} \cdot \frac{\rho^\gamma}{M_c} \qquad \left[\frac{moles\ c}{second}\right] \qquad \text{Eq. 1}$$

where y is a co-ordinate in the plane of the graphite nodule.
A carbon mass balance at the austenite/ferrite interface gives us the amount of carbon which has to diffuse away:

$$\frac{dm}{dt} = A \cdot \left[\frac{C_c^{\gamma/\alpha} \cdot \rho^\gamma}{M_c} - \frac{C_c^{\alpha/\gamma} \cdot \rho^\alpha}{M_c}\right] \cdot \frac{dy}{dt} \qquad \left[\frac{moles\ c}{second}\right] \qquad \text{Eq. 2}$$

The carbon gradient in the austenite can be estimated with Zener's equation

$$\left[\frac{dC_c^\gamma}{dy}\right]_{y=0} = \frac{C_c^{\gamma/\alpha} - C_0^\gamma}{S^\alpha} \qquad \text{Eq. 3}$$

where S^α is the thickness of the primary ferrite.
Equation 1 to 3 now yield the growth rate of the primary ferrite ($\rho^\alpha = \rho^\gamma$)

$$\frac{dy}{dt} = \frac{D_c^\gamma\left(C_c^{\gamma/\alpha} - C_0^\gamma\right)}{S^\alpha\left(C_c^{\gamma/\alpha} - C_c^{\alpha/\gamma}\right)} \qquad \text{Eq. 4}$$

When the graphite nodule is totally surrounded by a ferrite shell, further growth has to proceed by diffusion of carbon through the ferrite. The upper limit for the growth velocity of the ferrite shell is given by this diffusion rate. However, at an early stage, the thickness of the ferrite shell, which is a major restrictive factor, is small and therefore the calculated growth rate, assuming diffusion control, becomes considerably higher than what is obtained in the experiments. Consequently, the growth of the ferrite during this *second stage* is here assumed to be controlled by an interface reaction at the ferrite/graphite interface. The consequence of this assumption will lead to a carbon concentration profile where the gradient within the ferrite is small and where a discontinuity in carbon content will be found over the interface. Hereby, the carbon content at the interface is only somewhat smaller than $C_c^{\alpha/\gamma}$, and the amount of carbon which can be incorporated on the graphite can now be calculated as:

$$\frac{dn}{dt}=\frac{4\pi r_g^2 \rho^\alpha \left(C_c^{\alpha/\gamma}-C_c^{\alpha/gr}\right)}{M_c}\cdot\mu \qquad\qquad \left[\frac{moles}{\sec ond}\right] \qquad\qquad \text{Eq. 5}$$

where μ, having the dimension of velocity, is an unknown parameter controlling the interface reaction.

At the ferrite/austenite interface, carbon is released when the interface progresses with the velocity dl^α / dt; the amount being given by:

$$\frac{dn}{dt}=\frac{4\pi R_\alpha^2 \rho^\gamma \left(C_c^{\gamma/\alpha}-C_c^{\alpha/\gamma}\right)}{M_c}\cdot\frac{dl^\alpha}{dt} \qquad\qquad \left[\frac{moles}{\sec ond}\right] \qquad\qquad \text{Eq. 6}$$

where R_α is equal to $r_g+\ell^\alpha$.

However, the impingement of the ferrite spheres will lead to a situation where the amount of released carbon decreases during the transformation. Again, assuming that the Avrami/Johnson/Mehl equation can describe the impingement we obtain:

$$\frac{dn}{dt}=\frac{4\pi R_\alpha^2 \rho^\gamma \left(C_c^{\gamma/\alpha}-C_c^{\alpha/\gamma}\right)}{M_c}\cdot\exp\left(-\frac{4\pi R_\alpha^3 N_v}{3}\right)\cdot\frac{dl^\alpha}{dt} \qquad \left[\frac{moles}{\sec ond}\right] \qquad \text{Eq. 7}$$

From Eq. 5 to 7 we thus obtain the growth rate of the ferrite shell.

$$\frac{dl^\alpha}{dt}=\frac{\rho^\alpha \left(C_c^{\alpha/\gamma}-C_c^{\alpha/gr}\right)}{\rho^\gamma \left(C_c^{\gamma/\alpha}-C_c^{\alpha/\gamma}\right)}\cdot\left(\frac{r_g}{R_\alpha}\right)^2\cdot\exp\left(\frac{4\pi R_\alpha^3 N_v}{3}\right)\cdot\mu\approx k_f\cdot\Delta T\cdot\left(\frac{r_g}{R_\alpha}\right)^2\cdot\exp\left(\frac{4\pi R_\alpha^3 N_v}{3}\right)\cdot\mu \qquad \text{Eq. 8}$$

where $k_f\approx 1.02\cdot 10^{-5}\cdot(\%Si)^2 - 3.63\cdot 10^{-5}\cdot\%Si + 2.19\cdot 10^{-4}$, evaluated with a commercial softvare [12], and ΔT is the stable eutectoid supercooling.

Below, the following value of the μ-factor have been used: $\mu=1.5375\cdot 10^{-10}\cdot r_g^{-1.0383}$, which is the best fit for the data from the four thickest plates in the plate castings. The variation of with the graphite radius is not yet fully understood.

By the use of the above given equation, the growth rate at later stages of the transformation will increase drastically as a consequence of the impingement correction. This has no physical relevance since at the same time the increased thickness of the ferrite shell will cause the transformation to become diffusion controlled.

From now on (*third stage*) the growth cannot proceed any faster than what is obtained by treating the transformation as totally controlled by carbon diffusion through the ferrite shell. A carbon mass balance at the ferrite/austenite interface, combined with Fick's first law assuming a quasi-steady-state carbon profile within the ferrite, will yield the growth rate [10]:

$$\frac{dl^\alpha}{dt}=\frac{\rho^\alpha \left(C_c^{\alpha/\gamma}-C_c^{\alpha/gr}\right)}{\rho^\gamma \left(C_c^{\gamma/\alpha}-C_c^{\alpha/\gamma}\right)}\cdot\frac{r_g D_c^\alpha}{1^\alpha \left(r_g+1^\alpha\right)} \qquad\qquad \text{Eq. 9}$$

In Fig. 5 the calculated growth rate of the ferrite shell in the 15 mm plate is shown for the case of interface controlled growth (Eq. 8) and for diffusion controlled growth (Eq. 9). It is seen that the interface reaction will control the growth up to a total fraction of about 0.8 and that diffusion will be the rate controlling mechanism after this moment.

SIMULATION

The above presented models have been implemented as a subroutine into the commercial simulation program MAGMASOFT in order to perform the micro-macro simulation. This is most advantageous since it makes it possible to simulate the mould filling and thereby obtain a realistic initial temperature distribution in the liquid and the mould before the solidification starts. Of course, this will also have a major effect in the temperature interval around the solid state transformations.

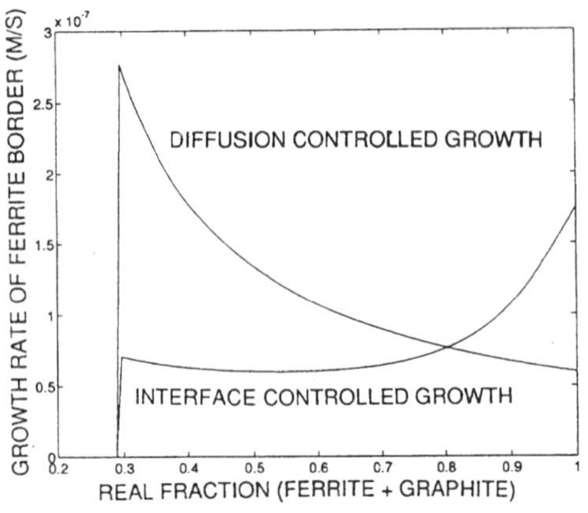

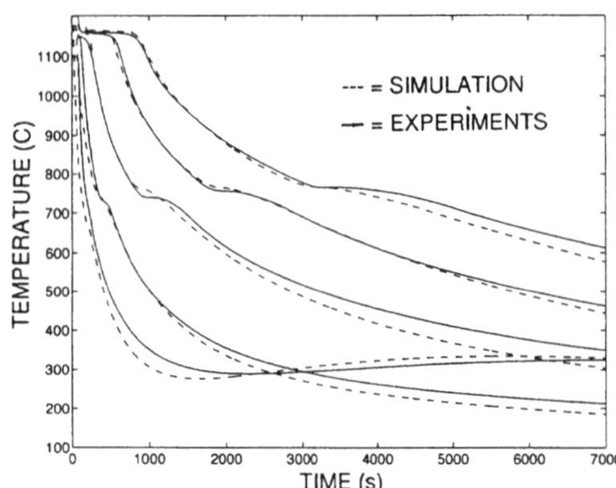

Figure 5. Calculated growth rate of the ferrite shell: interface & diffusion control

Figure 6. Simulated and experimental cooling curves in plate castings

As a demonstration of the capability of simulation, the plate castings in this investigation have been simulated including mould filling, solidification and solid state transformations. The simulated and experimental cooling curves are shown in Fig. 6. In the simulation of the solid state transformations it was assumed that four primary ferrite nuclei were formed on each nodule when the temperature reached 790 C. The agreement is seen to be satisfactory for the 4, 8, 30 and 50 mm plates. However in the 15 mm plate it seems that the primary ferrite growth is completed somewhat earlier than what was the case in the experiment. The explanation to this could be that the nucleation of ferrite in this plate occurred during a more extended temperature interval than in the others. The reason for the increasing temperature in the thinnest plate after about 1500 seconds is that the heat from the thicker plates at this moment reaches thi s plate.

REFERENCES

1. J.M. Schissler, D. Le Dily, J.P. Chobaut, 54th International Foundry Congress, New Delhi 1987.
2. R. Boeri, F. Weinberg, Cast Metals, vol. 6, p 153-157 (1993).
3. J.E. Rehder, AFS Trans.73, p 473-487 (1965).
4. M. Wessén, I.L. Svensson, Modeling of Casting, Welding & Advanced Solidification Processes VI,TMS Warrendale, p 71-78 (1993).
5. D.M. Stefanescu, C. Kanetkar, Proceedings, The Metallurgical Society in Toronto, Canada, oct. 1985, p. 171-188.
6. D.M. Stefanescu, C. Kanetkar, Proceedings of 54th International Foundry Congress, CIATF, New Dehli, India (1987), paper 19.
7. D.M. Stefanescu, G. Upadhya, D. Bandyopadhyay, I.G. Chen, JOM, feb. 1989, p 22-25.
8. D. Venugopalan, Proceedings, Physical Metallurgy of Cast Iron IV, MRS (1990).
9. D. Venugopalan, Met. Trans. 21A, p 913-918 (1990).
10. E. Lundbäck, Thesis, Royal Institute of Technology, Stockholm (1991).
11. J. Lacaze, G. Lesoult, Proceedings, COST 504, Espoo, Finland, sept. 12-13 (1994).
12. B. Sundman, B. Jansson, J-O. Andersson, Calphad, vol. 9, p 153-190 (1985).

Advanced Materials Research Vols. 4-5 (1997) pp. 485-490
© *1997 Scitec Publications, Switzerland*

Columnar Dendrite Growth in Cast Iron

F. Mampaey and Z.A. Xu

WTCM Foundry Centre, Technologiepark 9, B-9052 Zwijnaarde, Belgium

Keywords: Dendrite Growth, Modelling

ABSTRACT

A numerical model has been developed which allows to simulate the austenite dendrite growth in hypoeutectic cast iron. This model is based on dendrite tip growth for a fraction solid below 0.335 and dendrite arm thickness growth by diffusion at higher fraction solid. The model results in cooling curves which are in agreement with foundry experience. Moreover the simulated liquidus temperature is in accordance with experimental data obtained by analyzing cooling curves recorded in commercial cups. Simulated fractions of austenite dendrites agree with experimental ones which have been obtained by quenching experiments.

INTRODUCTION

Simulation of the solidification of austenite dendrites in hypoeutectic compositions has received considerably less attention than the solidification of the eutectic phases. On the other hand, quite some research has been carried out in the field of the thermal analysis in order to obtain the relation between the austenite liquidus temperature and the carbon equivalent of cast iron. Several formulas have been published which give a linear relation between the carbon equivalent and the liquidus temperature [1,2]. However, considerable differences, up to 39 K, may be calculated for the same carbon equivalent [2]. Oxidizing melting conditions increase the liquidus temperature with up to 10 K [2]. Menk et al [3] have established the influence of several elements (Te, S, P, Mn, Mg) on the liquidus temperature.

The solidification type of the primary austenite dendrites may change too. Patterson and Döpp [4] introduced a 6 type classification series with the 'exogenous' and 'endogenous' solidification type as limits. Exogenous dendrites grow from the mould to thermal centre of the casting and will be denoted here as columnar dendrites. Endogenous dendrites are smaller and randomly distributed in the metal interior. The endogenous type of dendrites is favoured by lower temperatures (melting, holding and pouring temperature) and longer holding times in the furnace and will produce lower liquidus temperatures [4]. The morphology of the austenite dendrites may be related to the mechanical properties of the iron at room temperature [5]. Exogenous dendrites give anisotropic tensile strength and elongation in blackheart malleable cast iron, with the highest values parallel to the primary dendrite axes.

Examination of the literature allows some conclusions. When charge materials and melting practice are reproducible, the carbon equivalent may be predicted quite accurately from the liquidus temperature measurement. However, the liquidus temperature is considerably lower than the equilibrium liquidus

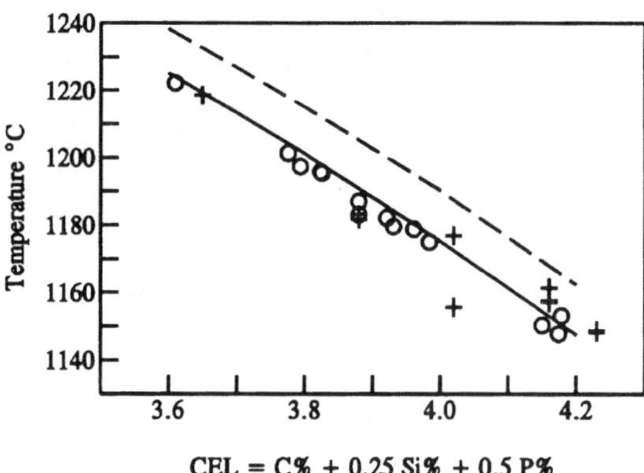

Fig. 1. Relation between the austenite liquidus temperature and the carbon equivalent. From the Fe–C diagram (dashed line), as simulated by the present model (solid line) and by experiment for lamellar graphite cast iron (+) and nodular cast iron (o).

temperature taken from the Fe–C diagram which is illustrated in Fig. 1. The experimental data listed in Fig. 1 will be discussed later. The value of 0.25 % Si in the expression of CEL is accepted to indicate the effect of silicon on the austenite liquidus temperature [2]. Finally, examination of the cooling curves reveals a liquidus arrest when according to the phase diagram only a change in the slope should be recorded.

In the present research a simulation model for the austenite dendrite solidification has been developed and its results have been compared with experimental evidence.

EXPERIMENTS

Metallographic specimens have been obtained by quenching experiments used to study the solidification morphology of cast iron [6]. In these experiments, three cylindrical bars measuring 38 mm in diameter and 120 mm in height are vertically cast without any gating system. During solidification, the specimens are quenched in water, which is highly agitated by the injection of pressurized air. Simultaneously, temperature at mid–length is recorded. A supplementary specimen is allowed to cool to room temperature without quenching in order to record the cooling curve integrally. As a result of the quenching the remaining melt in each bar solidifies as a fine structured white cast iron. This structure can easily be distinguished from preliminary solidified material: austenite dendrites, the lamellar graphite eutectic or the spheroidal graphite eutectic. The presence of gray solidification (normal cooling rate) or white solidification (quenching) makes cast iron an excellent material to 'freeze' the solid – liquid mixture during solidification.

Different amounts of austenite dendrites have been visualized by quenching lamellar graphite cast iron with variable hypoeutectic composition. More details are given in a recent publication on grain impingement in cast iron [7]. Both ordinary silicon cast iron and aluminum cast iron were quenched. Metallographic examination reveals that for the present casting conditions, the dendrites are columnar; they grow from the casting – mould interface towards the centre of the casting. The primary trunks length is often equal to half of the casting radius; some trunks have been found which reach till the casting centre. Equiaxed dendrite structures do not appear in the central part of the casting. The secondary dendrite arm spacing has been measured on primary trunks which are cut by the metallographic surface close to their trunk axis. The thickness of the dendrite arms has been measured on all trunks visible in the metallographic plane. Secondary dendrite arm spacing and dendrite arm thickness increase during solidification (Fig. 2), a well known phenomenon [8–10]. Secondary dendrite arm spacing and dendrite arm thickness are nearly equal during solidification. Metallographic examination gives the impression that dendrite tip growth is possible till approx 0.35 fraction solid. Up to 0.34 fraction solid, secondary dendrite arm spacing and dendrite arm thickness increase slowly; the austenite volume increase is mainly due to dendrite tip growth. At higher fractions of solid the austenite dendrite volume increases by thickening of the dendrite arms although some liquid remains between the dendrite arms.

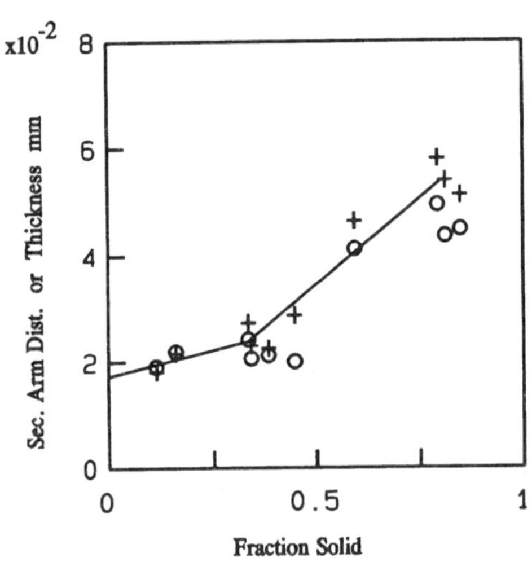

Fig. 2. Secondary arm distance (o) and dendrite arm thickness (+) as a function of the austenite fraction solid.

SIMULATION MODEL

The experimental data of the dendrite free surface may be related to a simple model [7]. In this model, dendrite arms are represented by cylinders which are arranged in a simple cubic arrangement.

Impingement of the cylinders starts from a fraction solid equal to 0.785 (= π/4). The cylinder diameter is set equal to the dendrite arm thickness (solid line in Fig. 2) and the total length of all cylinders l_d in a volume V_e is derived from the fraction solid ($f_s < 0.78$)

$$l_d = f_s V_e / (\pi R_d^2) \tag{1}$$

with R_d the radius of a cylinder. One obtains for the free surface S_d per unit volume ($f_s < 0.78$)

$$S_d = 2 \pi R_d l_d = 2 f_s V_e / R_d \tag{2}$$

For fractions solid above 0.785, l_d is kept constant at its maximal value and the free surface is calculated by elementary geometry.

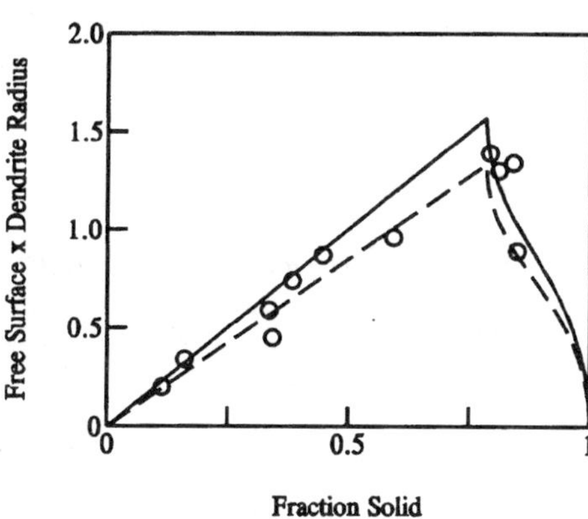

Fig. 3. Free surface × dendrite radius product during solidification of hypoeutectic cast iron; by experiment (o) and by the dendrite model (solid line).

As shown by Eq. 2, any change of the dendrite arm thickness influences the free surface area considerably. Since the dendrite arm thickness changes continuously during solidification (Fig. 2) it is very difficult to directly compare the experimental data with the theoretical model for the free surface. This problem may be eliminated by comparing per unit volume, the free surface multiplied by the dendrite arm thickness, which gives a dimensionless number equal to 2 times the fraction solid. In Fig. 3, experimental data are compared with the theoretical model (solid line). Examination of this figure reveals that on the average, the experimentally obtained free surface area is equal to 0.85 of the theoretical one (dashed line). This may be attributed to the fact that secondary and ternary dendrite arms can impinge within a primary trunk and that the primary trunks themselves mutually interfere in a real casting. From 0.785 fraction solid the dendrite arms impinge and the free surface decreases.

Dendrite tip growth is based on the formulas for a the parabolic tip shape, as given in detail by Kurz and Fisher [11]. The product of the dendrite tip radius R and the dendrite tip velocity V is calculated from

$$\Delta T = m C_o [1 - 1 / (1 - p I(P_c))] + I(P_t) \Delta h_f / c \tag{3}$$

$$P_c = V R / (2 D) \tag{4}$$

$$P_t = V R / (2 \alpha) \tag{5}$$

with P_c the solute and P_t the thermal Péclet number, D the diffusion coefficient in the liquid, α the thermal diffusivity, m the liquidus slope, C_o the initial alloy concentration, p the complementary distribution coefficient, Δh_f the latent heat of fusion per volume, c the volumetric specific heat, I the Ivantsov function and ΔT the supercooling.

The dendrite tip radius R is given by

$$R = \frac{-4 \pi^2 \Gamma}{\dfrac{P_c m C_o p}{1 - p I(P_c)} - \dfrac{P_t \Delta h_f}{c}} \tag{6}$$

with Γ the Gibbs – Thomson coefficient. Equations 3–6 allow to determine the dendrite tip growth velocity V. The new values of V allow to update the length of the primary trunk in each element. This information allows to determine when the dendrite growth in neighbour elements can start. In the present model, initial nucleation of dendrites only occurs at the mould wall. The primary dendrite trunk spacing is taken from the experiments (3.6×10^5 /m^2). The dendrite tip velocity V of secondary tips which start to grow from the primary trunk is taken as 40 % of the axial growth rate [8]. The number of the sidebranches is equal to four times the primary trunk length divided by the secondary arm distance, as given by the experimental measurements in Fig. 2.

Dendrite tip growth is stopped at a fraction solid of 0.335. At higher fraction solid the dendrite fraction increases by thickness growth of the dendrite arms. This has been taken into account by developing a simple model which is based on diffusion of carbon in the remaining liquid and associating the dendrite arms with cylinders as explained before. Eq. 7 gives the mass balance of the carbon atoms in an element during a time step Δt. The left hand side of this equation gives the amount of carbon which is rejected by the solid; the right hand side represents the transport of carbon in the liquid by diffusion (Fick's first law).

$$V_e \ \Delta f_s \ (C_l - C_s) \ = \ D \ \frac{C_l - C_i}{\Delta R} \ S \ \Delta t \qquad (7)$$

In the above equation, V_e is the element volume, Δf_s the increase of the austenite fraction solid, C_l and C_s the carbon concentration at the liquidus and the solidus plane resp., C_i is the bulk concentration of carbon in the liquidus in the element, D de diffusion coefficient, ΔR is the distance from the dendrite surface to the interdendritic liquid and S the dendrite arm surface. $C_l - C_i$ is related to the element supercooling ΔT and the slope of the austenite liquidus

$$C_l - C_i = \Delta T / m \qquad (8)$$

The dendrite arm surface S may be expressed in terms of the dendrite arm radius R_d and the austenite fraction solid f_s (Eq. 2). Combining Eq. 7, 8, 2 gives

$$\Delta f_s \ = \ \frac{2 \ f_s \ D}{\Delta R \ R_d \ m \ (C_l - C_s)} \ \Delta t \ \Delta T \ = n \ \Delta t \ \Delta T \qquad (9)$$

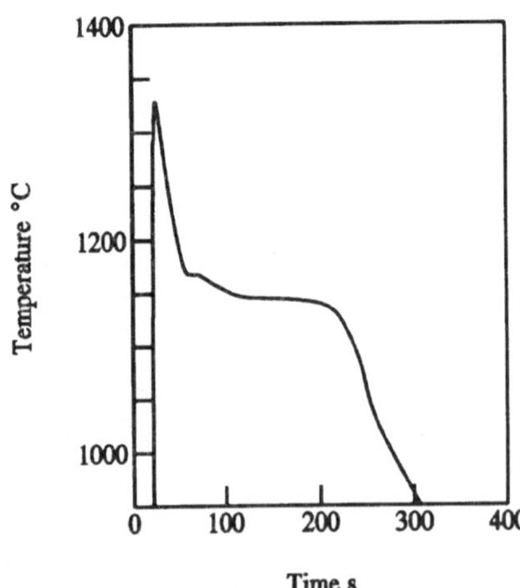

Fig. 4. Simulated cooling curve of hypoeutectic spheroidal graphite cast iron.

The value of n may be calculated by substituting appropriate numerical values in Eq. 9. Experimental measurement of quenched specimens at a fraction solid of 0.8 allows to estimate ΔR at 4×10^{-6} m, R_d is taken 4×10^{-5} m (Fig. 2). For f_s equal to 0.6 one obtains for n 0.36 at 3.50 % C and 6.41 at 0.6 % C. This means that the undercooling ΔT during dendrite arm thickness growth will be very low (< 0.1 K) and the solidification is very close to equilibrium conditions. This is confirmed by the numerical calculations. Segregation of carbon and silicon is taken into account by the Clyne and Kurz model [12].

RESULTS

The present model allows to simulate cooling curves of hypoeutectic cast iron compositions (Fig. 4). The geometry model of Fig. 4 corresponds to a commercial

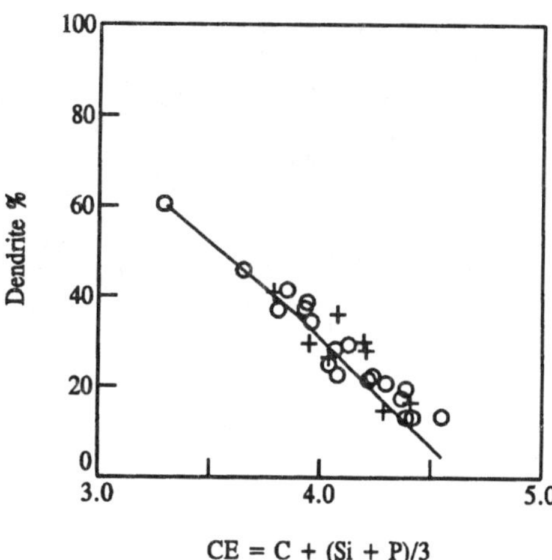

Fig. 5. Dendrite fraction as a function of the carbon equivalent as simulated by the present model (solid line). Experimental measurements are obtained by quenching lamellar graphite cast iron samples, uninoculated (o) and inoculated (+).

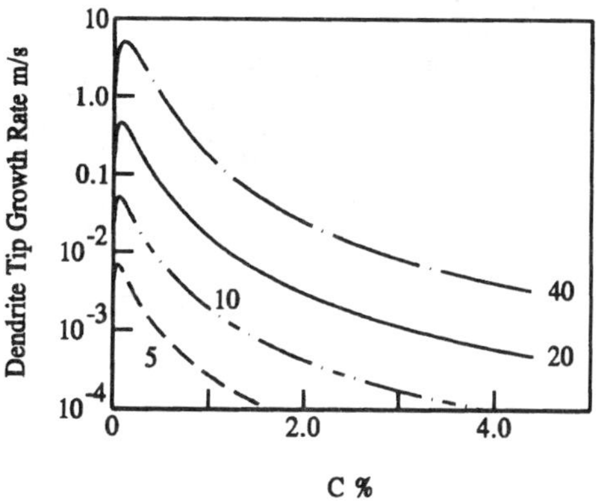

Fig. 6. Dendrite tip growth rate as a function of the carbon content in Fe–C for undercoolings of 5, 10, 20 and 40 K.

available thermal analysis sand cup [13], including the thermocouple located in the centre of the cup.The simulated curve is in good accordance with experimental experience showing a clearly visible liquidus arrest. From the simulated cooling curve, the liquidus temperature may be determined for a given carbon equivalent. The results which are plotted in Fig. 1 are in good agreement with the experimental measurements obtained in commercial cups [13]. Simulations are based on the same geometry model, including the thermocouple. The liquidus temperature as determined with thermal analysis cups is about 14 K lower than the equilibrium liquidus temperature from the Fe–C diagram. By combining the dendrite model with a white eutectic solidification model, the simulated fraction of austenite dendrites may be compared with experimental data obtained by quenching lamellar graphite castings during solidification (Fig. 5). Since in this figure, CE is used to estimate the amount of the eutectic, the value of 0.33 % Si in the expression of CE is applied. Experimental measurements of the austenite fraction have been done just ahead of the eutectic cells. In this figure, no difference is noticed in the experimental dendrite fraction between inoculated and non–inoculated lamellar graphite cast iron. Indeed in both cases the remaining liquid completely solidifies white during quenching. The simulated dendrite fraction is in accordance with the experimental data.

The model for the dendrite tip growth [11] predicts a growth rate which changes as a function of the carbon content (Fig. 6). For the carbon content range of cast iron, dendrite tip growth becomes substantial for undercoolings in the order of 20 K. However, the growth velocity varies non linear as a function of the carbon content. Moreover, the austenite liquidus temperature in the Fe–C diagram changes non linearly with the carbon content too. Both facts allow to conclude that the relation between the carbon equivalent and the experimentally measured liquidus temperature will be non linear too. This might partly explain why substantial differences appear in the published formulas. The application of a particular formula should be limited to a restricted carbon equivalent range and not extrapolated outside the limits for which the formula was originally obtained. Of course, the metallurgical reasons reviewed in the introduction, will also influence the relation between the carbon equivalent and the liquidus temperature.

At fractions solid above 0.335, solidification is modelled by thickness growth of the dendrite arms. The calculations predict a diameter increase of the dendrite arms of 4.05×10^{-2} mm for an increase of the fraction solid of 1 when applied to the experimental conditions of Fig. 2. The experiments give in the same conditions a value of 6.44×10^{-2} mm (Fig. 2, fraction solid > 0.335). This thickness increase is the result of growth as well as of a coarsening effect [9]. In the hypothesis that at fractions solid lower than 0.335 dendrite arm thickness increase is only due to coarsening; the coarsening effect may be estimated at 1.93×10^{-2} mm if the fraction solid would increase by 1 in the present experimental conditions. As a result, the experimental thickness growth in the absence of any coarsening would be $(6.44 - 1.94) \times 10^{-2}$ mm $= 4.50 \times 10^{-2}$ mm. This value is relatively close to the calculated one.

CONCLUSIONS

The model developed for simulating the austenite dendrite growth produces results which are in good agreement with experimental data. Simulated cooling curves show a liquid arrest temperature which is about 14 K below the equilibrium temperature of the liquidus line in the Fe–C diagram. This is in accordance with the present experiments and the formulas published in literature. This behaviour is due to the undercooling which is necessary to create substantial dendrite tip growth. Dendrite tip growth is the main growth mode below a fraction solid of 0.335. At higher fractions solid dendrite arms increase in thickness. The present model takes this into account by a simple diffusion driven growth model which takes place close to the equilibrium temperature. The simulated dendrite fraction as a function of the carbon equivalent of the iron, is in good agreement with the experimental one obtained by quenching experiments.

REFERENCES

[1] R. Döpp, D. Blankennagel, K. Lindemann. Giessereiforschung **33**, 119 (1981).
[2] R. W. Heine, Trans. AFS **85**, 537 (1977).
[3] W. Menk, M.O. Speidel, R. Döpp, Giessereiforschung **44**, 66 (1992).
[4] W. Patterson, R. Döpp, Giesserei **51**, 414 (1964).
[5] W. Patterson, H. Geilenberg, Giessereiforschung **20**, 101 (1968).
[6] F. Mampaey, La Fonderie Belge **53**, I 3, III 4 (1983).
[7] F. Mampaey, Z.A. Xu, accepted for publication in Cast Metals.
[8] S.–C. Huang, M.E. Glicksman, Acta Metall. **29**, 717 (1981).
[9] L.M. Colucci–Mizenko, M.E. Glicksman, R.N. Smith, JOM **46**, Jan. 51 (1994).
[10] M.C. Flemings, T.Z. Kattamis, B.P. Bardes, Trans. AFS **99**, 501 (1991).
[11] W. Kurz, D.J. Fisher, Fundamentals of Solidification, Trans Tech Publ. Aedermannsdorf – Switserland (1986).
[12] T.W. Clyne, W. Kurz, Metall. Trans. **12A**, 965 (1981).
[13] Electro–Nite Quick Cup QC4010.

Advanced Materials Research Vols. 4-5 (1997) pp. 491-498
© *1997 Scitec Publications, Switzerland*

Microsegregation in Spheroidal Graphite Iron: A Numerical Model and its Validation

S. Chang and D.M. Stefanescu

Solidification Laboratory, Dept. of Metallurgical and Materials Engineering, University of Alabama
Tuscaloosa, AL 35487, USA

Keywords: Microsegregation, Spheroidal Graphite Iron, Equiaxed Dendrite, Equiaxed Eutectics

ABSTRACT

A numerical model for evaluation of microsegregation during solidification has been developed. Its purpose is to assess solute redistribution occurring during the growth of plate, columnar or equiaxed dendrites, and of eutectics, when considering diffusion in both solid and liquid, and variable physical parameters (temperature, concentration, and interface velocity). Results calculated by the proposed model have been compared with calculations based on some existing models, *e.g.*, the lever rule, Scheil equation, Brody-Flemings model, etc. Validation of the proposed model was performed for the case of solidification of spheroidal graphite cast iron under variable growth rate, partition, and diffusivity.

INTRODUCTION

It has long been recognized that microsegregation in metallic alloy systems, resulting from the partitioning of solute between liquid and solid, is detrimental to mechanical properties, particularly to ductility, because of inhomogeneous microstructures. Indeed, second phases (*e.g.*, eutectics, carbides, sulfides, etc.), form in the interdendritic, intercellular, or grain boundary regions during solidification. The occurrence of microsegregation is inevitable in normal solidification of casting alloys. This defect must be minimized.

Considerable amount of effort has been made to develop analytical or numerical models for prediction of microsegregation [1-15]. The major assumptions used in the existing microsegregation models are summarized in Table 1.

Table 1   Major assumptions in existing microsegregation models.

Model	Geometry	Solid diffusion	Liquid diffusion	Partition coefficient	Growth	Coarsening
Analytical Models						
Lever rule	all	complete	complete	constant	any type	no
Scheil [1]	all	no	complete	constant	any type	no
Brody-Flemings (BF) [2]	all	estimated	complete	constant	any type	no
Clyne-Kurz (CK) [3]	all	estimated	complete	constant	any type	no
Ohnaka [8]	linear columnar	quadratic	complete	constant	linear parabolic	no
Sarreal-Abbaschian [9]	all	any	complete	constant	any type	no
Kobayashi [10]	columnar	any	complete	constant	linear	no
Nastac-Stefanescu [14]	all	any	any	variable	variable	no
Numerical Models						
Kirkwood [5]	linear	estimated	complete	constant	linear	yes
Voller-Sundarraj [13]	plate	any	any	variable	linear	yes
proposed	all	any	any	variable	variable	no

In these models, with the exception of the NS model, only plate and columnar dendrites were tackled. No attempts were made to solve numerically solute distribution for a spherical grain, such as eutectic grain in cast iron. Most of the models were established on the basis of the profile-defined approach, since the diffusion equation for moving boundary cannot be solved analytically without some simplifying assumptions. Hence, the aim of this research work is to develop a microsegregation model which is capable of solving problems that include: (a) the growth of plate, columnar, or equiaxed dendrites, and of eutectics, (b) diffusion of solute in the solid and in the liquid, while conserving mass balance, and (c) variable partition coefficient as a function of composition, temperature, and growth velocity.

THE MICROSEGREGATION MODEL

Diffusion Path Length

The diffusion path length in the plate geometry is shortest normally and can be regarded as the distance between the stalk of dendrite arm to the midway of two neighboring arms (Fig. 1a). In the columnar dendrite, the diffusion path length is obtained from the center of columnar dendrite to the equivalent radius, R_f (Fig. 1b).

It can be calculated from:

$$R_f = \left[\pi N_g(x,y)\right]^{-1/2} \qquad (1)$$

where $N_g(x,y)$ is the number of columnar dendrite arms per unit area in the cross-section perpendicular to the growth direction. A similar approach can be applied for spherical growth (Fig. 1c). The diffusion path length can be written as:

$$R_f = \left[4\pi/3 \cdot N_g(x,y,z)\right]^{-1/3} \qquad (2)$$

where $N_g(x,y,z)$ is the number of grains per unit volume.

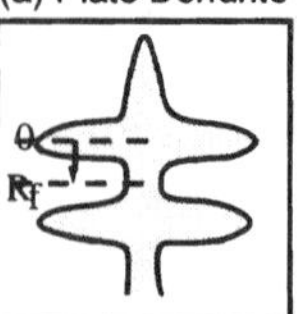
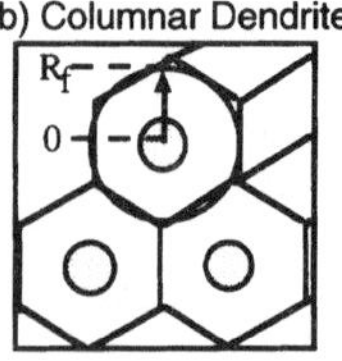
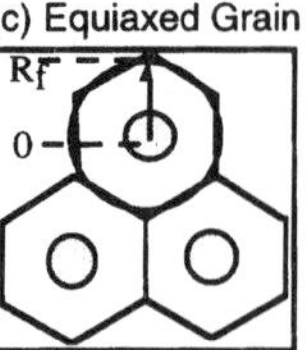

Fig. 1 Representation of different growth geometry: (a) plate dendrite, (b) columnar dendrite, and (c) equiaxed grain.

<u>Governing Equations and Initial, Interface, and Boundary Conditions</u>

The governing equations and boundary conditions are described on the basis of the following assumptions:

(1) Solute transport in both phases is by diffusion alone. Thus, the governing diffusion equations in the solid and in the liquid can be expressed as:

$$\frac{\partial C_S}{\partial t} = \frac{1}{r^m}\frac{\partial}{\partial r}\left(r^m D_S \frac{\partial C_S}{\partial r}\right) \quad \text{and}$$

$$\frac{\partial C_L}{\partial t} = \frac{1}{r^m}\frac{\partial}{\partial r}\left(r^m D_L \frac{\partial C_L}{\partial r}\right) \qquad (3)$$

where D_L is the diffusion coefficient in the liquid, t is time, and the superscript m represents an index number for geometry (for example, 0 for plate, 1 for cylinder, and 2 for sphere).

(2) The relationship between solute concentrations in the liquid, C_L^*, and in the solid, C_S^*, can be expressed as:

$$C_S^* = k_{ef} C_L^* \qquad (4)$$

where the effective partition coefficient, k_{ef}, depends on temperature, concentration, and interface velocity.

(3) Densities of solid and liquid are equal. The shrinkage-induced liquid flow at the microscopic level is neglected. The fraction of solid, f_S, can be calculated as

$$f_S = \left(r^*/R_f\right)^{m+1} \qquad (5)$$

where r^* is the interface position.

(4) There is no mass flow in and out of the volume element. Thus, the boundary conditions in this closed system can be written as:

$$\frac{\partial C_S}{\partial r} = 0, \quad \text{at } r=0 \quad \text{and} \quad \frac{\partial C_L}{\partial r} = 0, \quad \text{at } r=R_f \qquad (6)$$

The interface condition can be derived through the flux balance at the interface, that is:

$$\left(C_L^* - C_S^*\right)\frac{dr}{dt} = D_S\left(\frac{\partial C_S}{\partial r}\right)^* - D_L\left(\frac{\partial C_L}{\partial r}\right)^* \qquad (7)$$

where dr/dt is the interface velocity, V, and the * sign denotes interface position. From the fourth assumption, initial concentration is constant for each volume element, and can be written as:

$$C_L = C_o, \quad \text{for } 0 \le r \le R_f \text{ and } t=0 \qquad (8)$$

where C_o is the initial concentration.

<u>The Effective Partition Coefficient</u>

The equilibrium partition coefficient is obtained at very low solidification velocities. When the velocity

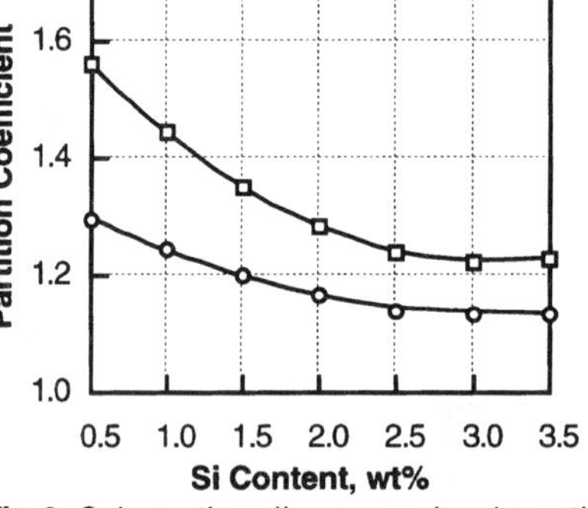

Fig. 2 Schematic diagram showing the variation of partition coefficient with the silicon content of an Fe-C-Si eutectic alloy, and with the solidification velocity. The equilibrium partition coefficient (ko) is taken from ref. [15] and the effective partition coefficient (kef) is calculated with Eq. (9).

increases to a certain value, it can be expected that there exists substantial departure from equilibrium. Burton *et al.* [16] proved that when the liquid/solid interface is planar, the effective partition coefficient can be expressed as a function of growth velocity:

$$k_{ef} = k_0 \left/ \left[k_0 + (1-k_0)\exp\left(-\frac{V\delta}{D}\right) \right] \right.$$ (9)

Here, δ is the thickness of the boundary layer and D is the diffusion coefficient in the liquid at the interface. Fig. 2 shows the effect of growth velocity ($V=5\times10^{-6}$ m/s) on the partition coefficient as a function of the silicon content in the stable Fe-C eutectic system. The values of D and δ used are 1.9×10^{-9} m^2/s and 1.729×10^{-4} m, respectively [15]. As seen, the effective partition coefficient is moving from the equilibrium partition coefficient toward unity. Thus, less segregation will occur. However, this equation is not valid for rapid solidification.

Another model for the calculation of velocity-dependent effective partition coefficient was formulated by Aziz [17], involving the theory of solute trapping. The dependence of the effective partition coefficient on growth velocity for continuous growth can be written as:

$$k_{ef} = \left[k_0 + \left(\frac{V\lambda}{D^*}\right) \right] \left/ \left[1 + \left(\frac{V\lambda}{D^*}\right) \right] \right.$$ (10)

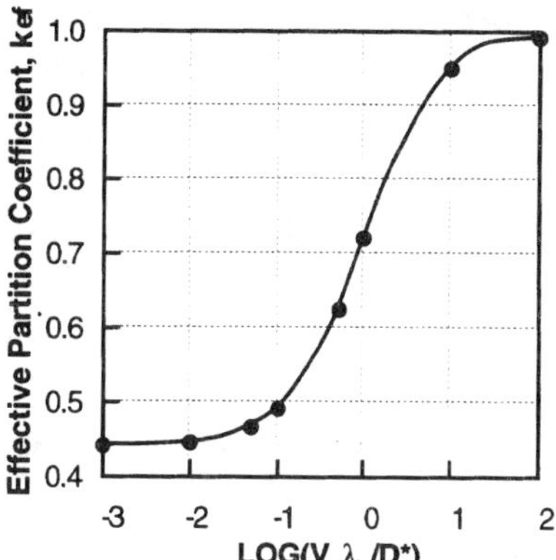

Fig. 3 Curve (ko=0.44) showing the growth velocity dependent of effective partition coefficient, kef.

where λ is a length scale related to the interatomic distance, normally ranging from 0.5 to 5 nm, and D* represents the interface diffusion coefficient. It can be seen that the effective partition coefficient is k_0 when V=0, and 1 when V»D*/λ. Fig. 3 shows the variation of k_{ef} with growth velocity for λ=5 nm and D*=1x10^{-9} m^2/s. It is concluded that the effective partition coefficient remains the same with the equilibrium partition coefficient for growth velocity up to 0.2 cm/s. From this analysis, it is concluded that Burton's model should be used for casting solidification modeling, where moderate cooling rates are involved.

Numerical Analysis

The diffusion equations in finite difference form is derived by the control volume method [18] for two different regions, that is in the liquid (or solid) region and in the interface region, as shown in Fig. 4. Assuming no second phase formed during grain growth, the concentration in volume element i (in the liquid or solid region) can be expressed as:

In the solid:
$$C_i^{t+\Delta t} = \frac{1}{V_i^{t+\Delta t}} \left[V_i^t C_i^t + A_{i+1/2}D_S\Delta t\left(\frac{C_{i+1}^t - C_i^t}{dr}\right) - A_{i-1/2}D_S\Delta t\left(\frac{C_i^t - C_{i-1}^t}{dr}\right) \right]$$ (11)

In the liquid:
$$C_i^{t+\Delta t} = \frac{1}{V_i^{t+\Delta t}} \left[V_i^t C_i^t + A_{i+1/2}D_L\Delta t\left(\frac{C_{i+1}^t - C_i^t}{dr}\right) - A_{i-1/2}D_L\Delta t\left(\frac{C_i^t - C_{i-1}^t}{dr}\right) \right]$$ (12)

where $A_{i+1/2}$ is the east surface area of element i and V_i is the volume of element i.

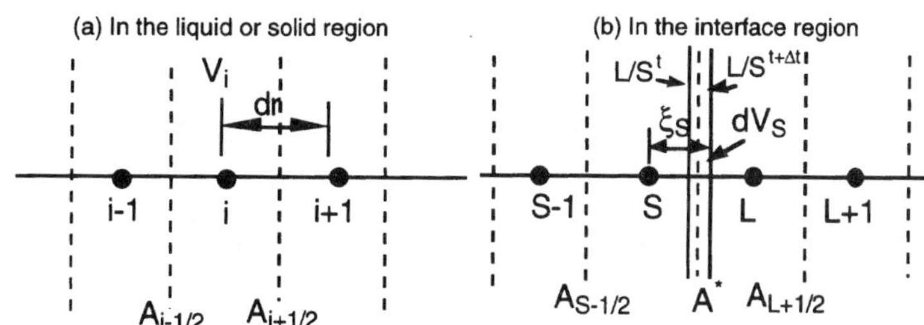

Fig. 4 Geometry discretization of the numerical model for solute redistribution:(a) in the liquid or solid region and (b) in the interface region.

At the liquid/solid interface region, larger amounts of solute will flow into the liquid, building up solute in front of the moving interface. The remainder of solute will diffuse back into the solid. The balance must obey the interface condition. The concentrations in volume element S and L can be calculated as follows:

<u>In the S element:</u>

$$C_S^{t+\Delta t} = \frac{1}{V_S^{t+\Delta t}}\left[V_S^t C_S^t + k_{ef}C_L^{*\,t}dV_S + A^*D_S\Delta t\left(\frac{C_S^{*\,t}-C_S^t}{\xi_S}\right) - A_{S-1/2}D_S\Delta t\left(\frac{C_S^t - C_{S-1}^t}{dr}\right)\right] \qquad (13)$$

<u>In the L element:</u>

$$C_L^{t+\Delta t} = \frac{1}{V_L^{t+\Delta t}}\left[V_L^t C_L^t - k_{ef}C_L^{*\,t}dV_S - A^*D_S\Delta t\left(\frac{C_S^{*\,t}-C_S^t}{\xi_S}\right) - A_{L+1/2}D_L\Delta t\left(\frac{C_{L+1}^t - C_L^t}{dr}\right)\right] \qquad (14)$$

where ξ_S is the distance between the solid/liquid interface and node S.

Discussion and Validation

The particular cases studied are described in Table 2 through the diffusion coefficients, diffusion length path, partition coefficient, and local solidification time. The first case involved plate dendrites growing at constant velocity, with low diffusivity in the solid (1×10^{-14} m^2/s) and short diffusion path length (64μm). Scheil assumptions, no solid diffusion and complete liquid diffusion, were observed (Fig. 5a). The calculated results were almost identical with Scheil model predictions (Fig. 5b).

Table 2: Data used in the calculation of concentration profiles for Figs 5 to 11.

Case No.	Fig. No.	C_o (wt%)	D_L (m^2/s)	D_S (m^2/s)	R_f (mm)	tf (s)	k
1[20]	5	5.25	1×10^{-9}	1×10^{-14}	64	40	0.48
2[19]	6&9	1.3	9.0×10^{-10}	2.0×10^{-13}	1000	200	0.7
3	7	2.5	1×10^{-9}	2×10^{-10}-2×10^{-12}	200	50	0.7
4[20]	8	5.25	1×10^{-9}	1×10^{-12}	64	40	0.48
5[22]	10	2.45	Eq. 16	Eq. 16	75	30	Eq. 15
6[22]	11	1.36	3×10^{-9}	5×10^{-12}	75	30	Eq. 17

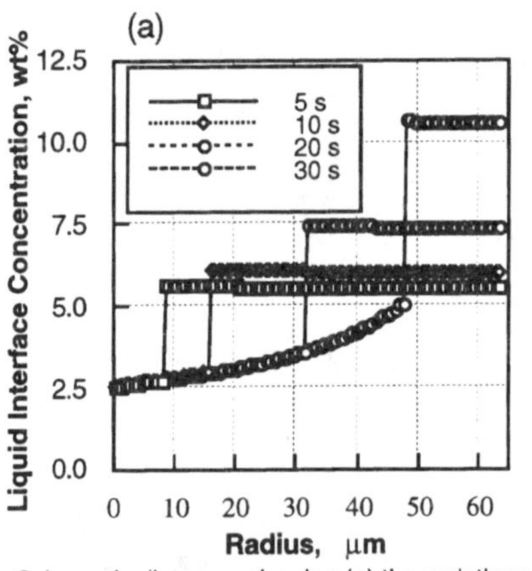

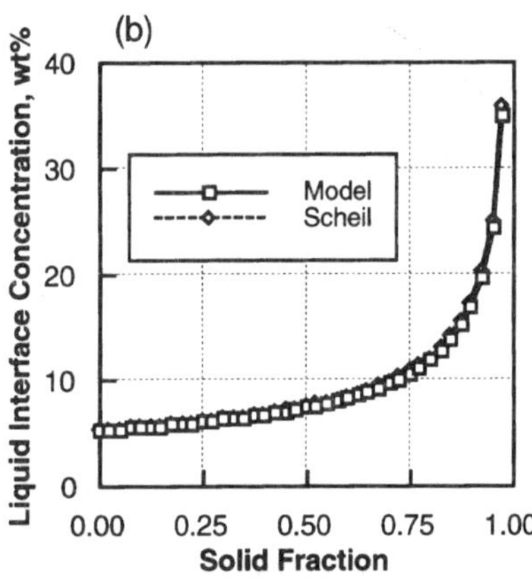

Fig. 5   Schematic diagrams showing (a) the variations of the concentration profiles with a moving interface and (b) the comparison of liquid interface concentration obtained from the proposed model and Scheil equation for plate geometry (case 1 in Table 2).

However, for cast iron, the diffusion path length, calculated with Eq. (2), can be relatively large (e.g., 0.1 mm) [19]. A long diffusion path is hardly compatible with the assumption of complete diffusion in liquid. Hence, a spherical eutectic grain of flake graphite cast iron with the diffusion path length of 0.1 mm was selected as case 2. The calculated results (Fig. 6a) showed incomplete liquid diffusion. Considerable differences between the solutions calculated with the proposed model and the Scheil model were observed (Fig. 6b).

The influence of solid state diffusion may not be significant if the solid diffusion coefficient is rather small. However, for some interstitial solutes (*e.g.*, carbon), the diffusion coefficient is rather high so that diffusion in the solid cannot be ignored. Thus, calculations were performed for various solid diffusivities (case 3), and the results were shown in Fig. 7a. The solute profile in the liquid at the interface is approaching that calculated from the lever rule as the solid diffusion coefficient increases. By contrast, it is close to that from the Scheil equation, when a small solid diffusion coefficient is used. The calculated liquid concentration profiles obtained from the proposed model, Scheil equation, lever rule, BF and CK models were compared in Fig. 7b. Note that the profiles calculated by the BF and CK models are superimposed for the case of $D_S=2\times10^{-11}$ m^2/s, and lower than that given by the proposed model. This is a consequence of the additional assumptions included in the BF and CK models. This can be attributed to the assumption of complete diffusion in the liquid.

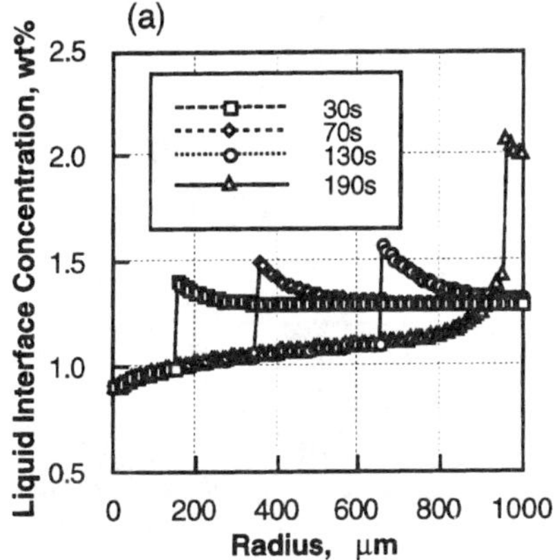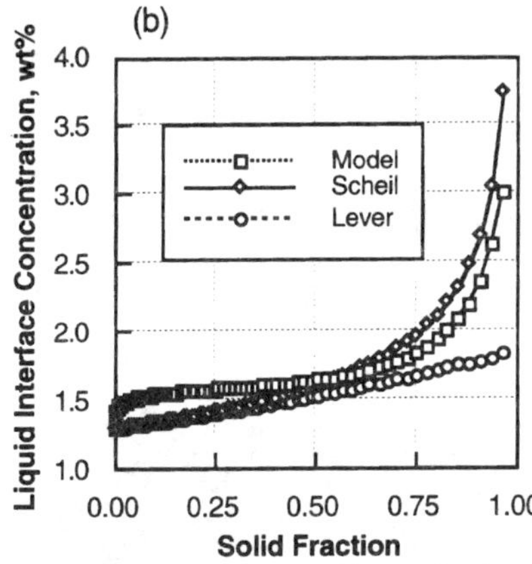

Fig. 6   Schematic diagrams showing (a) the variations of the concentration profiles with a moving interface and (b) the comparison of liquid interface concentration obtained from the proposed model, Scheil equation, and lever rule for spherical eutectic grain (case 2 in Table 2).

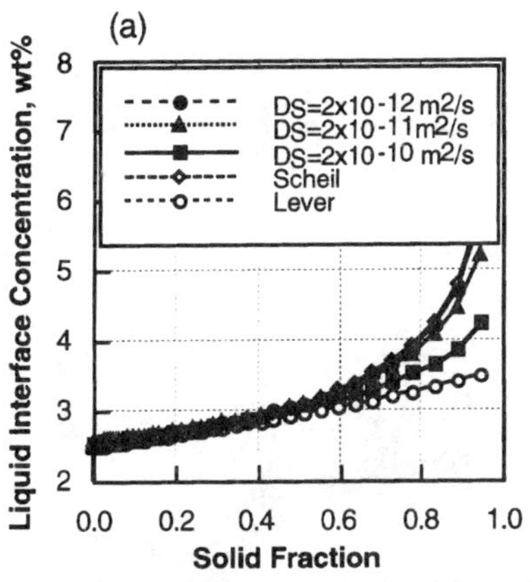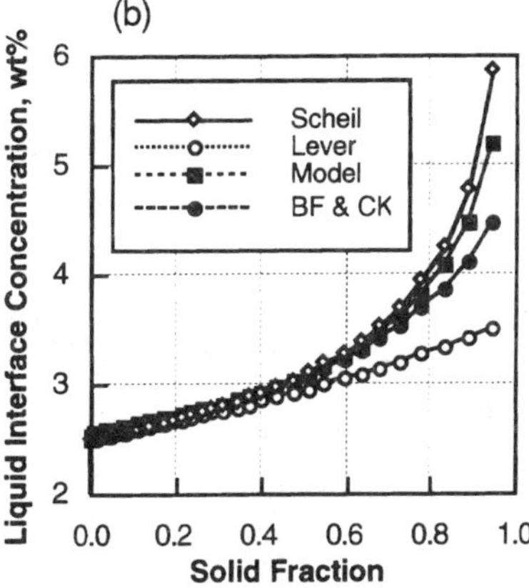

Fig. 7   Schematic diagrams showing (a) the effect of solid diffusion coefficient on liquid interface concentration and (b) the comparison of the liquid interface concentration calculated from the proposed model with those from the BF and CK models (case 3 in Table 2).

The proposed numerical model was also compared with the analytical NS model [14] for plate, cylindrical, and spherical geometry (case 4). First, a solidification velocity much smaller than the diffusion velocity, $V < D_L/R_f$, was selected This satisfies the assumption made when solving the equations in the NS model. Fairly good agreement was found, as shown in Fig. 8, but slight deviations can be clearly seen in all three cases, and in particular for the plate geometry.

When a solidification velocity larger than the diffusion velocity, $V > D_L/R_f$, was selected (case 2), great discrepancy was found between the numerical and analytical solutions (Fig. 9). The analytical solution lies very close to the results calculated with the Scheil equation. However, it can be seen from Fig. 6a that complete liquid diffusion cannot be established until the later stages of solidification. Higher liquid concentration at the interface is thus expected in the early solidification.

To prove the validity of the proposed model for the case of variable physical parameters, an evaluation was made for the case of spheroidal graphite cast iron solidifying with spherical geometry. The proposed model was implemented into the existing Heat Transfer-Solidification Kinetics model [21]. The interface velocity and temperature, was calculated on the basis of solidification kinetics of the alloy of interest. Thus, diffusion and partition coefficients, as functions of temperature and velocity, respectively, can be calculated and then used in the numerical model.

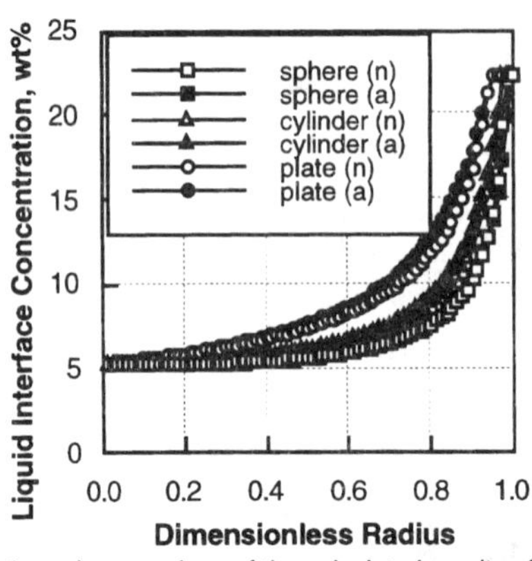

Fig. 8   A comparison of the calculated results of liquid interface concentration from the proposed numerical model and the analytical model by Nastac and Stefanescu for plate, cylindrical and spherical geometry (case 4 in Table 2).

Fig. 9   A comparison of the calculated results of liquid interface concentration from the proposed numerical model and the analytical model by Nastac and Stefanescu for eutectics with long diffusion path (case 2 in Table 2).

First, an analysis of silicon was conducted (case 5). The general data was obtained from the work of Boeri and Weinberg [22]. The diffusion path length was approximately 75 µm and the local solidification time was around 30 seconds. The equilibrium partition coefficient depending on the concentration at the liquid/solid interface for stable eutectic transformation has been reported by Kagawa and Okamoto [15] to be:

$$k_o = 1.70 - 0.31(\%Si) + 0.05(\%Si)^2 \tag{15}$$

where %Si is the weight percent of silicon at the interface. This partition coefficient is larger than unity. Accordingly, solute depletion occurs during solidification. Temperature-dependent diffusion coefficients were also used. The values can be expressed as:

$$D_L = 5.14 \times 10^{-8} \exp\left(\frac{-9100}{RT}\right) \qquad \text{and} \qquad D_S = 8.0 \times 10^{-4} \exp\left(\frac{-59500}{RT}\right) \tag{16}$$

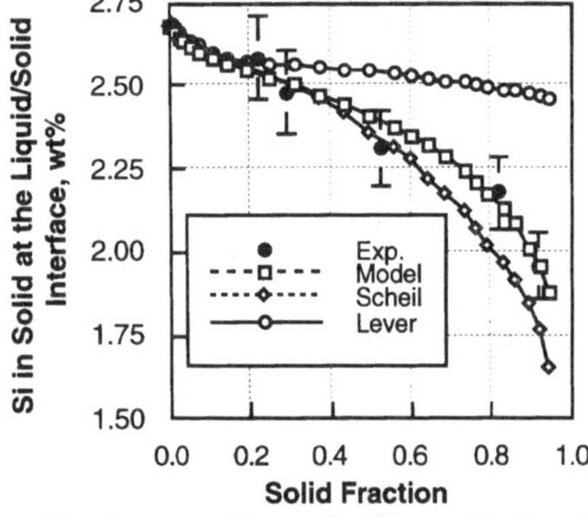

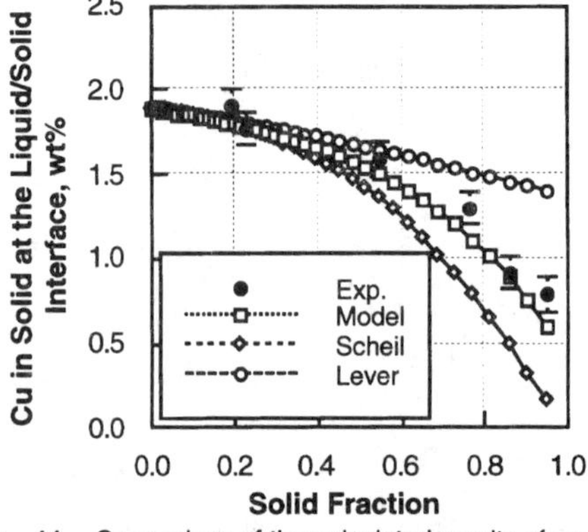

Fig. 10   A comparison of the solid interface concentrations calculated from the present model, Scheil equation, and lever rule and the experimental data for silicon segregation in spheroidal graphite cast iron (case 5 in Table 2).

Fig. 11   Comparison of the calculated results of copper solid interface concentrations from the present model, Scheil equation, and lever rule and the experimental data for Cu segregation in spheroidal graphite cast iron (case 6 in Table 2).

where R is the gas constant and T is the temperature in Kelvin. When incorporating these equations into the various models, solutions for silicon concentration in solid at the interface are obtained, as shown in Fig. 10. It can be seen that the results are in fairly good agreement with the experimental data, but are slightly higher than the solution of the Scheil equation.

An analysis of copper distribution was also performed (case 6). Due to the lack of boundary layer thickness data, the effective partition coefficient was taken from the literature as [22,23]:

$$k_0 = 2.08 - 0.51(\%Cu) \tag{17}$$

where %Cu is the weight percent of copper at the interface. Fig. 11 shows the comparison of the calculated results for copper concentration in the solid at the interface from the proposed model, the Scheil equation, and the lever rule. Satisfactory agreement is seen between the experimental data and the calculated solution from the present model. It is to be noted that constant diffusion data was used because of the lack of accurate temperature-dependent data. This is not unreasonable since the variation of temperature during eutectic solidification is rather small.

CONCLUDING REMARKS

A numerical model has been developed to assess microsegregation occurring during the growth of plate, columnar, or equiaxed dendrites, and of eutectics. Diffusion of solute both in the solid and in the liquid was taken into consideration, together with the interface flux balance. A closed system was assumed. From the theoretical analysis, along with the validations of the segregation of silicon and copper in spheroidal graphite cast iron, it is apparent that the application of the proposed model is manifold. It can deal with the microsegregation problem for different grain geometry, for different diffusion conditions, for partition coefficients greater or less than unity, and for variable physical parameters. The model was also implemented into an existing Heat Transfer-Solidification Kinetics model to allow the calculation of the growth velocity and solid fraction based on the nucleation and growth of alloys involved.

ACKNOWLEDGMENT

The authors would like to acknowledge the financial support from Concurrent Technologies Corporation under contract number 921100058, and the technical support of Dr. G. Uphadya.

REFERENCES

[1] E. Scheil, *Z. Metallk.* 34, 70 (1942).
[2] H. D. Brody and M. C. Flemings, *Trans. Met. Soc. AIME* 236, 615 (1966).
[3] T. W. Clyne and W. Kurz, *Metall. Trans. A* 12A, 965 (1981).
[4] M. C. Flemings, *Solidification Processing*, McGraw-Hill, New York (1974).
[5] D. H. Kirkwood, *Mat. Sci. and Eng.* 65, 101 (1984).
[6] A. J. W. Ogilvy and D. H. Kirkwood, *Appl. Sci. Res.* 44, 51 (1987).
[7] T. Matsumiya, H. Kajioka, S. Mizoguchi, Y. Ueshima, and H. Esaka, *Trans. Iron Steel Inst. Jpn.* 24, 873 (1984).
[8] I. Ohnaka, *Trans. Iron Steel Inst. Jpn.* 26, 1045 (1986).
[9] J. A. Sarreal and G. J. Abbaschian, *Metall. Trans. A* 17A, 2863 (1986).
[10] S. Kobayashi, *Trans. Iron Steel Inst. Jpn.* 28, 728 (1988).
[11] K. S. Yeum, V. Laxmanan, and D. R. Poirier, *Metall. Trans. A* 20A, 2847 (1989).
[12] T. P. Battle and R. D. Pehlke, *Metall. Trans. B* 21B, 357 (1990).
[13] V. R. Voller and S. Sundarraj, *Mat. Sci. and Techn.* 9, 474 (1993).
[14] L. Nastac and D. M. Stefanescu, *Metall. Trans. A* 24A, 2107 (1993).
[15] A. Kagawa and T. Okamoto, *Metal Science*, November, 519 (1980).
[16] J. A. Burton , R. C. Prim, and W. P. Slichter, *J. Chem. Phys.* 21, 1987 (1953).
[17] M. J. Aziz, *J. Appl. Phys.* 52, 1159 (1982).
[18] S. V. Patankar, *Numerical Heat Transfer and Fluid Flow*, Hemisphere, Washington (1980).
[19] G. Upadhya, D. K. Banerjee, D. M. Stefanescu, and J. L. Hill, *Trans. AFS* 90, 699 (1990).
[20] R. G. Thompson, D. E. Mayo, and B. Radhakrishnan, *Metall. Trans. A* 22A, 557 (1991).
[21] S. Chang, D. Shangguan, and D. M. Stefanescu, *Metall. Trans. A* 23A, 1333 (1992).
[22] R. Boeri and F. Weinberg, *Trans. AFS* 89, 179 (1989).
[23] L. Nastac and D. M. Stefanescu, *Trans. AFS* 93, 933 (1993).

Advanced Materials Research Vols. 4-5 (1997) pp. 499-504
© *1997 Scitec Publications, Switzerland*

Computer Modeling of Primary Structure Formation in Ductile Iron

E. Fraś, W. Kapturkiewicz and A.A. Burbielko

University of Mining and Metallurgy, Reymonta 23, PL-30-059 Cracow, Poland

Keywords: S.G. Cast Iron, Eutectic Solidification, Simulation

ABSTRACT

The micro-macro model of solidification of ductile iron was prepared. The main assumption of the model was that the growth of the graphite and austenite in the grain was controlled by transient diffusion of carbon through the austenite shell and by undercooling at the graphite-austenite interfaces. The computer program was elaborated and the results were compared with the experiment, referring to the temperature field, grain density and diameter of spheroidal graphite in the casting.

INTRODUCTION

Basing on microscopic examination of ductile freezing at different stages of eutectic transformation [1-4], one can reproduce the sequence of formation of eutectic grain. After undercooling of cast iron in respect to the equilibrium point of liquidus for graphite, in the liquid the nucleation of spheroidal graphite begins. During the growth of a spheroid of graphite, the carbon content in solidification front at the side of the liquid is decreasing, and therefore the undercooling for austenite nucleation assumes in this place the highest value. Austenite quickly nucleates on the graphite surface, forming a shell which separates graphite from the liquid [4]. It has been proved that the envelope of austenite starts appearing when the radius of graphite spheroid reaches the dimensions of from 7.5 to 15 μm [3]. The eutectic grain forms in this way is characterized by the radius R_g of the growing spheroid of graphite and by the radius $R\gamma$ of growing shell of austenite, and $R_\gamma/R_g=2.4$ [3].

The growth of eutectic grain is mainly controlled by the diffusion of carbon from the liquid to graphite through the austenite shell. For calculation of radii of graphite and austenite a solution resulting from stationary diffusion under conditions of pseudoequilibrium at the phase boundaries [3,5,6] is the most often used. The relationship obtained in this way can be combined with an equation of heat transfer which will give the curves of kinetics of solidification [7-11].

The problem of grain growth with respect to the non-stationary diffusion is a very general nature, and as such has been discussed in a few studies only [12-14]. It is restricted to determining the dimensions of the individual phases under stationary conditions of heat transfer.

The aim of this study was to give an algorithm of the kinetics of ductile iron solidification which would combine the process of solidification of the eutectic grains under non-stationary conditions of diffusion with a non-stationary heat transfer in the casting-mold system.

MODEL OF PROCESS

We assume that the casting made of an eutectic Fe-C alloy is solidifying in a casting mold. After reducing the temperature below the eutectic equilibrium point the spheroids of graphite are precipitated from the molten metal. The graphite is covered by a shell of austenite with the radius constant in respect to that of graphite. The rate of nucleation and rate of growth depend on the undercooling below the eutectic temperature [15]. The undercooling is growing till its maximum is reached and at this point the nucleation is interrupted [16]. At this first stage of the process a non-diffusional model of the grain growth, typical of the above mentioned works is assumed. We assume that from this point of maximum undercooling the diffusion growth begins. The spherical diffusion of carbon occurs between the sphere of graphite and the liquid through the austenite shell within the closed spherical volume, called Elementary Diffusion Field (EDF). The radius of the EDF depends on the density of grain, different for each point of casting. The casting is divided into the typical difference elements and each of the elements has own EDF.

SET OF EQUATIONS OF THE PROCESS

Temperature field in casting-mold system

$$\frac{\partial T}{\partial \tau} = \frac{1}{x^m} \frac{\partial}{\partial x}\left(x^m a \frac{\partial T}{\partial x}\right) + \frac{q_s}{c_v} \tag{1}$$

where m = 0, 1, 2 for planar, cylindrical and spherical geometry, respectively;

a - thermal diffusivity (for metal or for mold);   x - coordinate in casting - mold system;

q_s - heat generation rate of solidification; c_v - specific heat.

The spherical field of concentrations for austenite (γ phase) and liquid, respectively:

$$\frac{\partial C_\gamma}{\partial \tau} = D_\gamma\left(\frac{\partial^2 C_\gamma}{\partial r^2} + \frac{2}{r}\frac{\partial C_\gamma}{\partial r}\right) \tag{2}$$

$$\frac{\partial C_L}{\partial \tau} = D_L\left(\frac{\partial^2 C_L}{\partial r^2} + \frac{2}{r}\frac{\partial C_L}{\partial r}\right) \tag{3}$$

where r - coordinate in Elementary Diffusion Field.

D_γ, D_L - interdiffusion coefficient of carbon in austenite and liquid Fe-C.

Each grain grows in its own separated EDF which radius r_0 is calculated as:

$$r_0 = 1/\sqrt[3]{N} \tag{4}$$

where N is the grain density.

Mass balance for graphite - austenite interface (Figs. 1 and 2)

$$u_{\gamma/gr}\left(C_{gr} - C_{\gamma/gr}\right) = -D_\gamma \left.\frac{\partial C_\gamma}{\partial r}\right|_{r_{gr}^+} \tag{5}$$

where u_{gr} - interface γ/gr velocity; C_{gr} - concentration of carbon in graphite (100 %).

Mass balance for austenite-liquid interface (Figs. 1 and 2)

$$u_{L/\gamma}\left(C_{L/\gamma} - C_2\right) = D_\gamma \left.\frac{\partial C_\gamma}{\partial r}\right|_{r_\gamma^-} - D_L \left.\frac{\partial C_L}{\partial r}\right|_{r_\gamma^+} \tag{6}$$

Condition for border of EDF (r = r_0) :

$$\left.\frac{\partial C_L}{\partial r}\right|_{r_0} = 0 \tag{7}$$

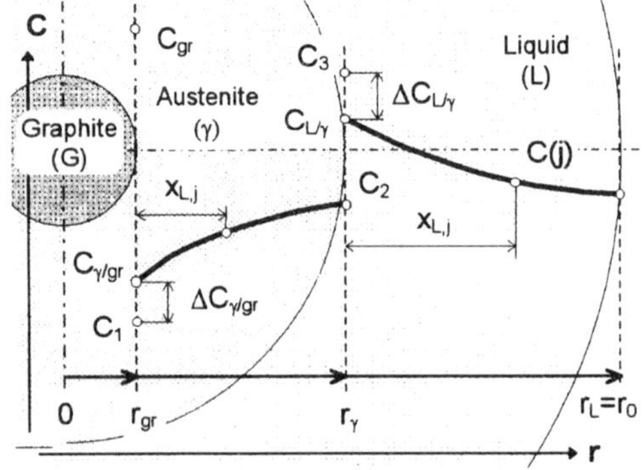

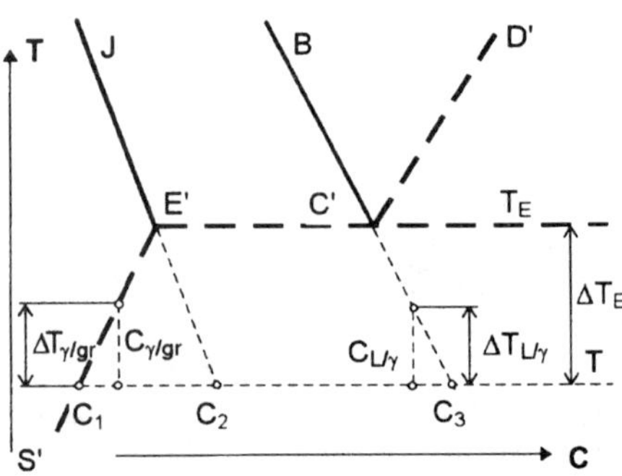

Fig. 1. Scheme of field of concentration: r_{gr} - radius of graphite; (r_γ-r_{gr}) - austenite shell; (r_L- r_γ) - liquid shell; $x_{\gamma,j}$, $x_{L,j}$ - distance of point "j" in austenite and liquid.

Fig. 2. Fragment of Fe-C equilibrium phase diagram.

Non-equilibrium concentrations $C_{\gamma/gr}$ and $C_{\gamma/L}$ depend on the undercooling $\Delta T_{\gamma/gr}$ and $\Delta T_{L/\gamma}$, respectively. From Fig. 2:

$$\Delta T_{\gamma/gr} = m_{S'E'}\left(C_{\gamma/gr} - C_1\right) \tag{8}$$

$$u_{\gamma/gr} = \mu_{gr}\left(\Delta T_{\gamma/gr}\right)^{n_{gr}} \tag{9}$$

$$u_{\gamma/gr} = \mu_{\gamma/gr}\left(m_{S'E'}\left(C_{\gamma/gr} - C_1\right)\right)^{n_{gr}} \tag{10}$$

Identically for austenite-liquid interface: $u_{\gamma/L} = \mu_{\gamma/L}\left(m_{BC'}\left(C_{L/\gamma} - C_3\right)\right)^{n_\gamma}$

where μ_{gr}, $\mu_{\gamma/gr}$, - kinetic coefficient of growth for graphite at the graphite-austenite interface and for austenite, respectively;

$m_{S'E'}$, $m_{BC'}$ - slope of lines S'E' and BC' in equilibrium diagram;

C_1, C_3 - function of actual temperature T (Fig. 2).

The influence of moving interfaces for rate of change of concentration at internal point of growing grain may be represented, after Murray-Landis [13] by substantial derivative :

$$\frac{dC_j}{d\tau} = \frac{\partial C_j}{\partial r_j} \cdot \frac{dr_j}{d\tau} + \frac{\partial C_j}{\partial \tau} \tag{11}$$

where $\partial C/\partial \tau$ is the contribution of Eqs. (2) and (3).

The rate of travel $dr_j/d\tau$ of an internal point "j", whose location is always a constant percentage of the instantaneous phase thickness, is functionally related to the interface velocity.

For austenite shell and for the liquid shell the $dr_j/d\tau$ is [13] (Fig. 1), respectively:

$$\frac{dr_{\gamma.j}}{d\tau} = u_{gr/\gamma}\left(1 - \frac{x_{\gamma.j}}{X_\gamma}\right) + u_{\gamma/L}\left(\frac{x_{\gamma.j}}{X_\gamma}\right) \tag{12}$$

$$\frac{\partial r_{L.j}}{\partial \tau} = u_{\gamma/L}\left(1 - \frac{x_{L.j}}{X_L}\right) \tag{13}$$

Combining Eqs. (11) and (12,13) one arrives at the relationship for the time rate of change of the γ phase concentration at the internal node, j :

$$\frac{dC_{\gamma.j}}{d\tau} = \frac{\partial C_{\gamma.j}}{\partial r_j}\left(u_{gr/\gamma}\left(1 - \frac{x_{\gamma.j}}{X_\gamma}\right) + u_{\gamma/L}\frac{x_{L.j}}{X_\gamma}\right) + \frac{\partial C_{\gamma.j}}{\partial \tau} \tag{14}$$

and for liquid

$$\frac{dC_{L.j}}{d\tau} = \frac{\partial C_{L.j}}{\partial r_j}u_{\gamma/L}\left(1 - \frac{x_{L.j}}{X_L}\right). \tag{15}$$

The heat generation rate q_s is calculated from :

$$q_s = L\frac{dV}{d\tau} \tag{16}$$

where L - solidification heat; V - solidification volume.

The solidified volume is calculated from Kolmogorov 's [17] equation :

$$V = 1 - \exp(1 - \Omega) \tag{17}$$

where Ω is a geometrical volume :

$$\Omega(\tau) = \frac{4}{3}\pi\int_0^\tau \alpha(t')\left(\int_{t'}^\tau u \cdot dt\right)^3 dt' \tag{18}$$

where u is the velocity of solid - liquid front: $u = u_{L/\gamma}$;

$$\alpha \text{ - rate of nucleation:} \quad \alpha(\tau) = \frac{dN}{d\tau}\,; \qquad\qquad N \text{ - density of grains.}$$

The density of grains is calculated as a function of undercooling to the eutectic equilibrium temperature T_E [15]

$$N = \Psi(\Delta T_E)^n \tag{19}$$

where ψ - nucleation coefficient; $\Delta T_E = T_E - T$; T - temperature of metal.

Variable Ω may be treated as a sum of geometrical volume of spherical grains neglecting of possibility of its interpenetration.

The set of equations (1) - (19) is solved numerically for each time step $\Delta\tau$ after transformation them according to the finite difference method.

It is possibly to use the implicit scheme, especially according to the calculation the q_s function but then would be necessary the longer computing time (for using e.g. iteration method). The difference form of the above diffusion equations is showed in Appendix.

RESULTS

The calculations were made for dia 40 mm cylinder casting of a eutectic Fe-C alloy (4.26%C) assuming that the spheroidal grains were formed. The experiment was made with ductile iron (C=3.36%, Si=2.84%, Mn=0.2%, P=0.05%, S=0.01%, Mg=0.054%) on casting of dia 40x200 mm, solidifying in the sand mold.

Fig. 3 shows the course of cooling curves in casing and the kinetics of growth of the phases in the casting grains. Fig. 4 shows concentration field in a selected grain during the successive stages of the grain growth. A comparison of grain density obtained in computer simulation and in experiment is shown in Fig. 5. The results of simulation are rough estimetes only due to roughly calculated values of a number of the process parameters, especially of nucleation coefficients, growth coefficients and due to the fact that the diffusion of the other alloying elements like Si, P etc. has been neglected.

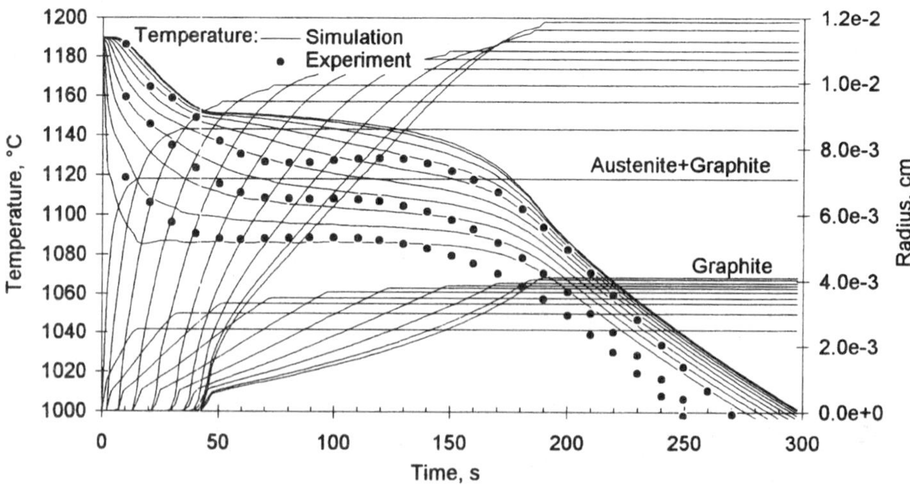

Fig. 3. Simulated and experimental curves of temperature and kinetics of grain growth in 10 points of casting.

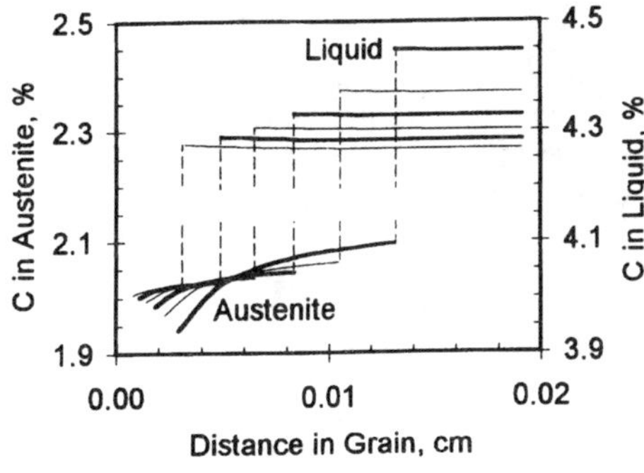

Fig. 4. Carbon concentration in the growing grain (simulation).

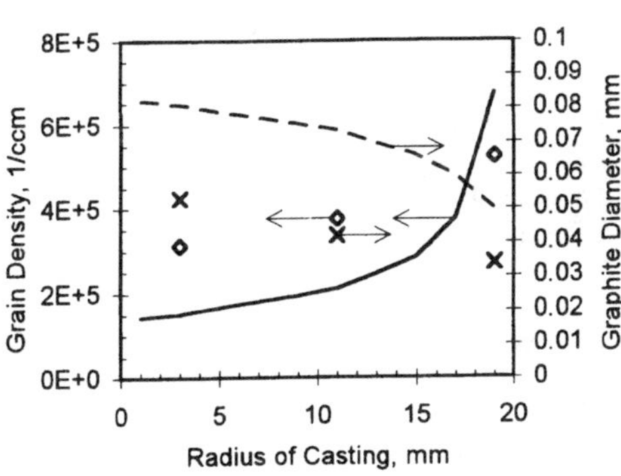

Fig. 5. Grain density and graphite diameter with casting radius. Solid and dashed lines - simulations; points - experiment.

CONCLUSIONS

The results of computations obtained from a simulation program prove the correct operation of elaborated micro-macro model. The distinct quantitative differences in results obtained from simulation and from the experiment point out to the fact that it is advisable to analyze in a more precise way the parameters adopted for calculations and that the effect of complex diffusion of additional elements in cast iron, like Si and P, should be taken into consideration.

REFERENCES

1. B. Lux, F. Mollard, J. Minkoff: "The Metallurgy of Cast Iron" Georgi Publ. Co., 371 (1975).
2. J. Schobel: "Recent Research on Cast Iron Gordon & Breach Publ., 303 (1968).
3. S. Wetterfal, H. Fredriksson, M. Hillert: J. Iron Steel Inst., 393 (May 1972).
4. Zhou Jiyang et al.: "Physical Metallurgy of Cast Iron" MRS, 141 (1990).
5. C. Birchenal, H. Head: J. of Metals, 1004 (1956).
6. E. Fras: J. of Cryst. Growth **73**, 460 (1985).
7. K. Su et al. : "The Physical Metallurgy of Cast Iron" - MRS Symp. Proc. **34** Elsevier Publ. Co., 181 (1985).
8. H. Fredriksson and I. Svenson: idem, 273 (1985).
9. E. Fras: idem, 191 (1985).
10. S. Chang et al.: Metall. Trans. **22A**, 915 (1991).
11. D.M. Stefanescu et al.: "Physical Metallurgy of Cast Iron" MRS, 15 (1990).
12. R. Sekerka et al.: "Lectures on the Theory of Phase Transformation" American Inst. of Mining, Metalurgical and Petroleum Eng. Inc., 117 (1975).
13. R.A. Tanzili, R.W. Heckel: Trans. of the Metall. Soc. of AIME, **242**, 2313 (1968).
14. R.D. Lanam, R.W. Heckel: Metall. Trans. **2**, 2255 (1971).
15. W. Oldfield: Trans. of ASM, **59**, 945 (1966).
16. E. Fras et al. : AFS Trans. **100**, 583 (1992).
17. A.N. Kolmogorov. Izvestia AN SSSR. Seria matematic., **3**, 355 (1937).

APPENDIX

For transformation the equation (1) to the difference form we can use one of the typically methods explicit or implicit. Rewriting the above equations of diffusion growth in finite-difference form leads to the following expressions, where k is the index for concentration after the last time step and $k+1$ is the index for the next time step: austenite (γ phase) for index i (which is omited in notation of equations) of thermal element of casting (Fig. A.1)

$$\frac{C_j^{k+1} - C_j^k}{\Delta\tau} = \frac{C_{j+1}^k - C_j^k}{\Delta x_\gamma}\left(u_{gr/\gamma}\left(1 - \frac{j-1}{n_1 - 1}\right) + u_{\gamma/L}\left(\frac{j-1}{n_1 - 1}\right)\right) +$$

$$+ D_\gamma\left(C_{j-1}^k - 2C_j^k + C_{j+1}^k\right)\Big/\left(\Delta r_\gamma\right)^2 + \left(C_{j+1}^k - C_{j-1}^k\right)\Big/\left(r_j \Delta r_\gamma\right) \qquad (A1)$$

where j - index of diffusion element in EDF (Fig.   );

liquid:

$$\frac{C_j^{k+1} - C_j^k}{\Delta\tau} = \frac{C_{j+1}^k - C_j^k}{\Delta x_L}\left(u_{\gamma/L}\left(1 - \frac{j - n_1 - 1}{n_2 - n_1 - 1}\right)\right) +$$

$$+ D_L\left(C_{j-1}^k - 2C_j^k + C_{j+1}^k\right)\Big/\left(\Delta r_L\right)^2 + \left(C_{j+1}^k - C_{j-1}^k\right)\Big/\left(r_j \Delta r_L\right) \qquad (A2)$$

Mass balance for graphite-austenite interface:

$$u_{\gamma/gr}\left(C_{(0)}^k - C_{(1)}^k\right) = D_\gamma\left(C_{(2)}^k - C_{(1)}^k\right)\Big/\Delta x_\gamma \qquad (A3)$$

where $C_{(0)}^k = C_{gr}$; $C_{(1)}^k = C_{\gamma/gr}$

and for austenite-liquid interface:

$$u_{\gamma/L}\left(C_{n_1+1}^k - C_{n_1}^k\right) = D_\gamma\left(C_{n_1}^k - C_{n_1-1}^k\right)\Big/\Delta x_\gamma - D_L\left(C_{n_1+2}^k - C_{n_1+1}^k\right)\Big/\Delta x_L \qquad (A4)$$

where $C_{n_1+1}^k = C_{\gamma/L}$ $C_{n_1}^k = C_2$ $\left(\text{Fig.2}\right)$;

$$C_2 = f(T_i)$$

where T_i - actual temperature of i element;

$C_{\gamma/gr}$ and $C_{\gamma/L}$ — nonequilibrium concentration of carbon at austenite-graphite and austenite-liquid interface, respectively.

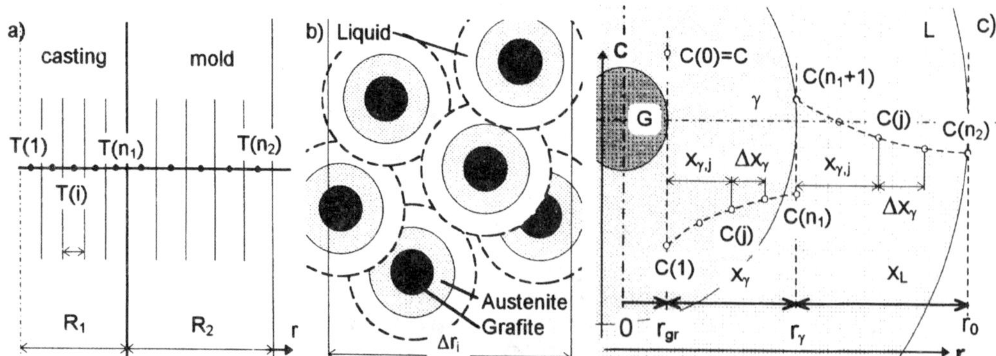

Fig. A1. Scheme of geometry of calculated problems: a - scheme of casting - mold system; b - one casting difference element; c - scheme of the grid in the one EPD

Advanced Materials Research Vols. 4-5 (1997) pp. 505-512
© *1997 Scitec Publications, Switzerland*

Metastable Solidification of Eutectic Grey Cast Irons

H. Fredriksson

Royal Institute of Technology, Dept. of Materials Processing, S-100 44 Stockholm, Sweden

Keywords: Gray Cast Irons, Lattice Defects, Free Energy, Latent Heat

ABSTRACT

Lattice defects and vacancies are formed in austenite and graphite during solidification. The free energy in each phase is increased with the fraction of lattice defects formed. The melting points of the phases are changed.

Temperature time curves and heat flow models are used to evaluate the latent heat during solidification of eutectic cast iron alloys. The changes in the free energy of the solid with changes in latent heat are discussed.

INTRODUCTION

Thermal analysis of cast iron has been performed over many years in order to determine the heat of fusion and analyse the solidification process. In many different studies experimental cooling curves are compared with numerical models in order to simulate the solidification process and the structures formed. Growth equations are used in the models in order to describe structures. All the models are based primarily on a heatbalance, where the heat of fusion is assumed to be constant (1-4).

In two recent papers (5, 6) the influence of the latent heat by the cooling rate (especially high rates) was discussed. It was experimentally verified that the latent heat varied with the cooling rate. It was proposed that this is an effect of the formation of lattice defects in the solid phase during solidification. The decrease in the heat of fusion also decreases the melting point of the alloy. This paper discuss the changes in latent heat for eutectic cast iron alloys as a function of cooling rate.

THERMAL ANALYSIS

Cooling curves are normally used to analyse solidification processes. The heat extracted from the sample or mould either cools the liquid or allows it to solidify. The cooling curves give information about the heat extracted from the liquid. The heat transfer from a unit volume of liquid alloy before the start of solidification can be expressed as:

$$\frac{dQ}{dt} = \rho \cdot C_p \cdot \frac{dT_l}{dt} \tag{1}$$

where ρ is the density of the liquid, C_p is the heat capacity and dT_l/dt is the cooling rate of the liquid and is evaluated from the cooling curve.

Once the temperature of the liquid is below the liquidus temperature, crystals will nucleate and grow. The heat transfer from a unit volume of a liquid will now include another term due to the latent heat released during solidification, in addition to the heat released due to temperature decrease. The following equation can thus be obtained:

$$\frac{dQ}{dt} = \rho \cdot C_p \frac{dT_2}{dt} + \frac{d(F,\rho,\Delta H)}{dt} \qquad (2)$$

where dT_2/dt is the cooling rate of the alloy during primary solidification, F is the volume fraction of solid in the unit volume, as well as in the system and ΔH is the latent heat.

Cooling curves presented in a number of papers and experimental results from four different investigations (1,2,7,8) are used in our analysis. Four different experimental methods have been used. However, in all experiments the extracted heat has been evaluated from the slope of the cooling curves prior to solidification by combining equation 1 and 2. The latent heat can be evaluated by defining the end of the solidification process. It is normally assumed that the latent heat is known, and the fraction of solid and/or the kinetics during solidification can be evaluated. In our case we have assumed that the latent heat depend on the cooling rate. The results (figure 1) show that the latent heat decreases with increasing cooling rate.

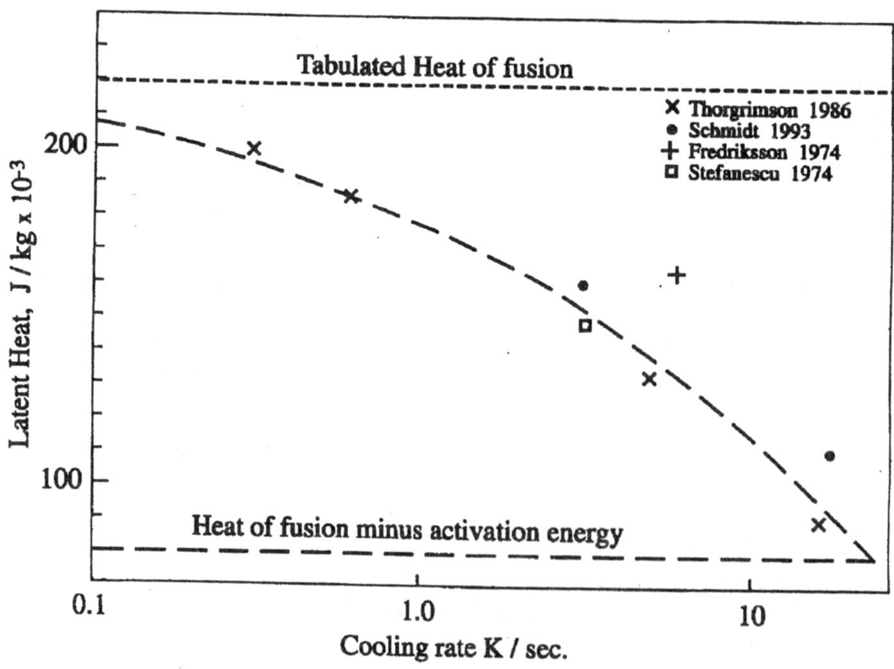

1.   Latent heat as a function of cooling rate.

The decrease in the latent heat with cooling rate, shown in figure 1 results from a higher free energy in the solid during solidification than under equilibrium conditions. The increase in free energy is probably related to the formation of lattice defects such as vacancies. The enthalpy stored in the solid as a function of lattice defects can be described by the following heat balance:

$$\Delta H_{l.d.} = \Delta H^o - \Delta H_{Meas.} \tag{3}$$

$$\Delta H_{l.d.} \quad = \quad \text{Enthalpy of lattice defects}$$
$$\Delta H^o \quad = \quad \text{Heat of fusion}$$
$$\Delta H_{Meas.} \quad = \quad \text{Measured latent heat}$$

For a eutectic reaction, the fraction of crystal defects can either be found in graphite or in austenite. It is difficult to estimate which of the two phases will be mandatory. The increase in free energy in the solid will change the phase diagram fo Fe-C, which will be discussed in the next section.

<u>The Fe-C phase diagram</u>

The Fe-C system has been investigated theoretically and experimentally for many years. A phase diagram is normally derived by analysing the free energy of the phases. At equilibrium the free energies are equal, which makes it possible to construct a phase diagram using the effect on the heat of fusion as discussed earlier. It it possible to conclude that the free energy of the solid phase increases with the fraction of vacancies or lattice defects. An increase in the free energy of one of the phases results in a change in the melting point. These changes can be calculated in the following way.

The free energy in the liquid and in the solid are equal at melting point.

$$G_A^L = G_A^S \tag{4}$$

where G_A^L = free energy in the liquid A

G_A^S = free energy in the solid A

The free energy of the liquid will not change but the enthalpy in the solid will, so we get:

$$G_A^L = {}^oG_A^L = G_A^S = {}^oG_A^S + \Delta H_{l.d.} \tag{5}$$

where ${}^oG_A^L$ = standard free energy of liquid A

${}^oG_A^S$ = standard free energy of solid A

since

$${}^oG_A^L - {}^oG_A^S = \frac{\Delta H^o\left(T_m - T_m^{St}\right)}{T_m} \tag{6}$$

Where

T_m = melting point of element A

T_m^{St} = melting point of element A with lattice defects included.

A combination of equations (5) and (6) gives:

$$T_m - T_m^{l.d} = \frac{T_m}{\Delta H^o} \Delta H^{l.d.} \tag{7}$$

Equation 7 is valid for a pure element but can also be used to analyse the decrease in eutectic temperature in a system with a eutectic reaction. The undercooling as a function of latent heat can then be caluclated. The results for eutectic cast iron alloys are presented in figure 2, and can be compared with the estimated undercoolings in the experimental results shown in figure 3. The measured undercoolings plotted in figure 3 and show a large variation. This is mainly an effect of a variation in the number of crystals formed during solidification. However, there is a clear tendency that the undercooling increases with increasing cooling rate, and that the measured undercooling is much smaller than the calculated one.

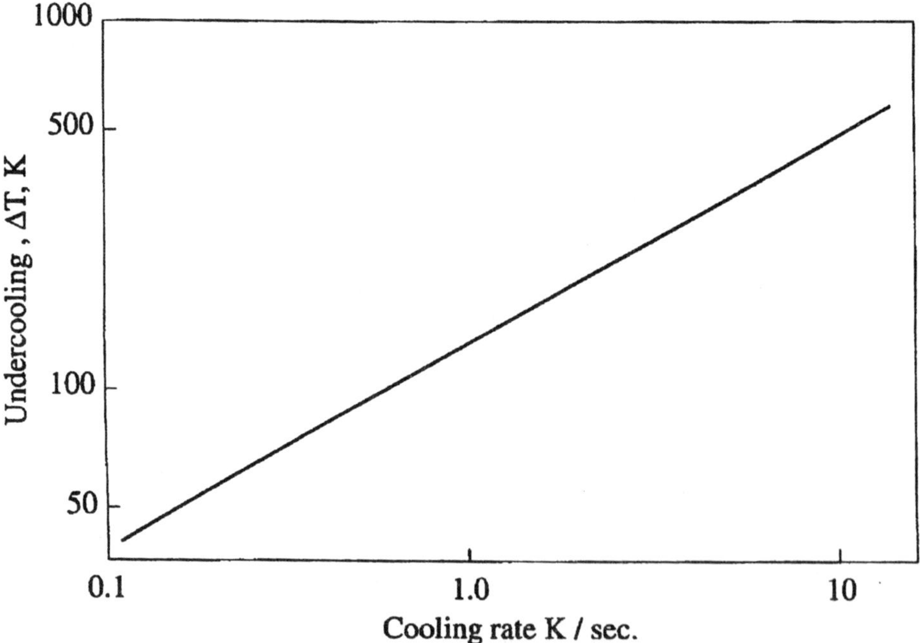

2.   Calculated latent heat as a function of undercooling for eutectic cast iron alloys.

The deviation between measured and calculated values might be explained if the assumption that the lattice defects does not contribute with an entropy term to the free energy is wrong. If an entropy term is added to equation 7 the changes in the melting point will be:

$$\left(T_m - T_m^{l.d.}\right) = \frac{T_m}{\Delta H^o}\left(\Delta H^{l.d.} - T\Delta S^{l.d.}\right) \qquad (8)$$

The entropy of stacking faults was estimated from the measured values of undercooling and the changes in the heat of fusion. The results are shown in figure 4. The entropy changes can be a result of a disordering of atoms. It is interesting to note that the entropy increases with cooling rate as shown in the figure. A reasonable explanation would be that the disordering of the atoms decreases with increasing cooling rate as a result of a reduced mobility or a shorter time for rearrangement of the atoms.

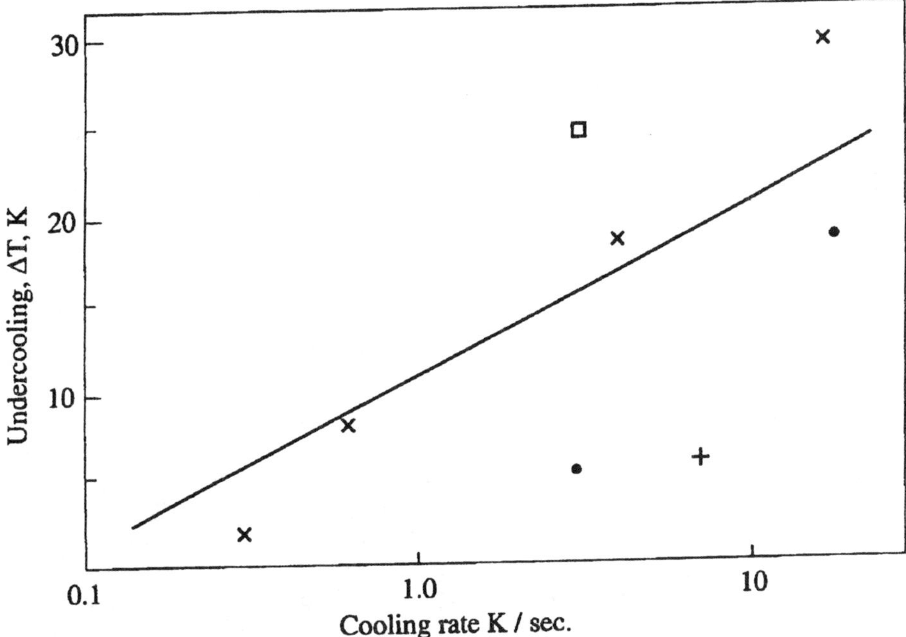

3. Measured undercooling as function of cooling rate.

<u>Interface processes</u>

Undercoolings in relation to the growth rate has been intensively studied for many years. Fredriksson et al (1) analysed the experimental data available in 1974 theoretically. They found that the data could not be described by a pure diffusion model and introduced an interface kinetic constant, μ. It was described by a relation of the type

$$v = \mu\left(X^{gr/L} - X^{\gamma/L}\right) \qquad (9)$$

where v = growth rate
    $X^{gr/L}$ = mole fraction of carbon at the interface graphite/liquid
    $X^{\gamma/L}$ = mole fraction of carbon at the interface austenite/liquid

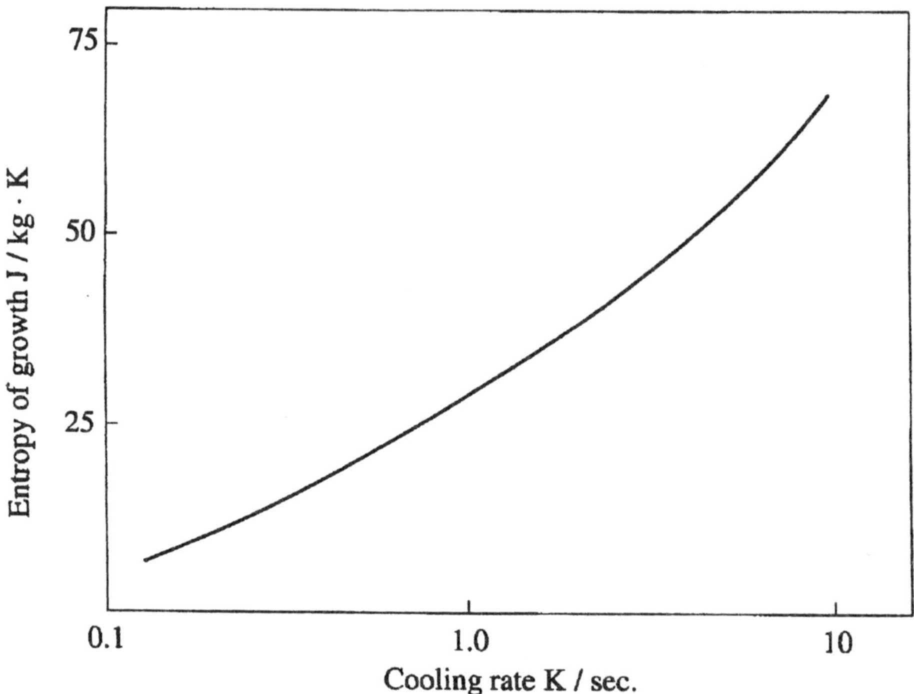

4.   The entropy of lattice defects as a function of cooling rates.

The experimental results showed that the constant varied between 10^{-2} to 10^{-3} m/second in order to describe the interface reaction. This value is so small that it is unlikely that it describes the ordering diffusion process when atoms are transferred from the liquid unordered state to the solid ordered one. From the absolute reaction rate theory it is reasonable to assume that the kinetic coefficient is described by the following relation:

$$\mu = \cdot \frac{D}{\delta} \exp \frac{\Delta S}{R} \exp - \frac{\Delta H}{RT} \tag{10}$$

where D   = diffusion coefficient over the interface
       δ = interface thickness
       ΔS = growth entropy
       ΔH = activation enthalpy

The constants ΔS and ΔH are difficult to estimate, but should be related to the lattice faults formed. In our case we have taken the maximum value of the entropy of lattice faults found in figure 4 and an average value of $5 \cdot 10^{-3}$ for μ calculated ΔH. This ΔH value was found to be 140 000 J/kg. The value is now defined as the activation energy for formation of lattice defects and can be compared with $\Delta H_{l.d.}$ in equation (8). The activation energy for formation of lattice defects minus the heat of fusion is indicated by a line in figure 1. It falls below the values for latent heat given from thermal analysis experiments but not very far below.

It is presumed that the latent heat in question has evolved at the solidification front (during solidification). This is not always the case, and the reason will be discussed in the next section. Before we discuss this effect we will point out that the $\Delta H_{l.d.}$ mentioned before will never go below this value. The reason for this is that when the solid reaches the energy state corresponding to the line in figure 1, the growth process will be purely diffusion controlled.

<u>Homogenisation</u>

Vacancies or other types of defects will be condensed during the growth process or dislocations will be formed. The condensation of vacancies will probably occur at the interfaces and at the grain boundaries. In our case the interfaces or the boundaries between austenite and graphite. The concentration of condensed vacancies can be described by the same laws as are used for homogenisation. The condensed fraction is exponentially relation and is dependent on the diffusion distance λ.

λ is related to the lamellar distance between graphite and austenite, which is related to the growth rate and given by the theory for growth rate of eutectic cells. The growth rate is related to the heat of extraction, and the following heat balance will be valid.

$$\frac{dQ}{dt} = v\rho \cdot c_p \frac{dT}{dt} + \rho(\Delta H)\frac{df}{dt} + f \cdot \Delta H_v \cdot \frac{dx}{dt} \qquad (11)$$

The total heat of fusion measured from a cooling curve is the heat evolved at the solidification front plus the heat evolved during the condensation of vacancies or reordering of the lattice defects (the third term in equation 11). Very little is known about the formation of stacking faults and its effect on the thermodynamics of the alloys. More experimental information is therefore necessary to get a good picture of this process. The homogenisation process is probably the explanation for the fact that the tabulated values are the sum of the heat evolved at the front and the heat evolved from the reorganisation of crystal defects.

<u>Concluding remarks</u>

The latent heat is found to depend on the cooling rate. Changes in latent heat with the cooling rate are used to explain the formation of crystal defects. The relationship between growth rate and undercooling found in solidification experiments is explained by the measured latent heat. A new concept for explaining of crystal morphology is indicated. This will be developed further in future work.

REFERENCES

1. H. Fredriksson, S.E. Wetterfall "The Metallurgy of Cast Iron", Georgi Publ. Co, 277 (1975).
2. D. Stefanescu, S. Trufinescu, Zeitskrift für metallkunde 65 (1974), 610-615.
3. H. Fredriksson et al., E-MRS Proc. 14, Edit. de Physique, 267 (1986).
4. M. Rappaz, Int. Mater. Rev. 34 (1989), 93-123.
5. N. E.lMahallawy et al., Mater. Scie. and Eng. A179/A180 (1994), 587-591.
6. M. Haddas Sabzevar, to be publ. in Met. Trans.
7. Jon-Thor Thorgrimson, Thesis, KTH, Stockholm 1986.
8. P. Schmidt, Thesis, KTH, Stockholm 1994.

Advanced Materials Research Vols. 4-5 (1997) pp. 513-520
© *1997 Scitec Publications, Switzerland*

Morphological Stability of Graphite Growth from Cast Iron Melt

S. Long and Y. Niu

Chongqing University, China

Keywords: Graphite Growth, Morphological Stability, Cast Iron

ABSTRACT

An approach is made in the present work to kinetically understand the morphological development of graphite and its relationship with the growth environment and the interfacial properties as it grows from the melt of cast iron. Analysis results show that graphite will gradually lose its morphological stability as it grows in a supercooled, or supersaturated cast iron melt, but the maximum dimension, within which graphite remains its original spherical configuration, is dominated by its interfacial properties and crystal structure of the graphite.

INTRODUCTION

The morphology of the graphite phase has been one of the primary concerns of the production of cast iron engineering structures due to its dominant effect on the mechanical performance of iron castings. In the past decades extensive research activities on morphological control of the graphite have been carried out and many techniques, such as, melt treatment of spheroidisation and inoculation, solidification control, have been developed to obtain desirable graphite morphologies. Consequently, different mechanisms about the formation of the graphite with various configurations have been suggested based on the morphological crystallographic, observations of graphite crystals, and physico-chemical phenomena and graphite-melt interfacial property investigations[1~26].

Different types of graphite have an original spherical configuration as it nucleates from the melt, which indicates that the final morphologies of the graphite phase in various cast irons are developed gradually during its subsequent growth[27~32]. It is, therefore, reasonable and essential to kinetically understand the morphological development of the graphite and the effects of melt treatment and the growth environment on graphite morphological development.

As a part of an approach to kinetically understand the relationship between its morphological development and growth environment, morphological stability of the graphite growth from the melt has been theoretically analysed under the frame work of morphological stability established[33,34], and the effects of the variations of interfacial crystallography and interfacial thermodynamic property of graphite caused by melt treatment have been discussed.

MORPHOLOGICAL STABILITY OF A SPHERICAL GRAPHITE CRYSTAL

The graphite growth in a cast iron melt is a phase transformation regulated by the diffusion of the carbon onto, and the potential heat transfer out of its growth front. Assuming that local thermodynamic equilibrium at graphite-melt interface is maintained and all interfacial properties are independent of interfacial crystallography the effect of thermal and constitutional fields and interfacial properties on the morphological stability of graphite phase can be mathematical analysed.

An originally spherical nodule of graphite loses its spherical configuration as transport phenomena give rise to thermodynamically stable protuberances on the graphite-melt interface that will subsequently develop into dendrites. Therefore, from the view point of morphological kinetics, the morphological stability of the original spherical graphite can be measured by the growth tendency of an interfacial protuberances, called perturbation thereafter.

A spherical interface slightly distorted by a spherical harmonic perturbation Y_{lm} with an infinitesimal altitude δ, as shown in Fig. 1, can be linearly expressed as

$$r = \rho(\theta, \varphi) = R + \delta Y_{lm}(\theta, \varphi) \qquad (1)$$

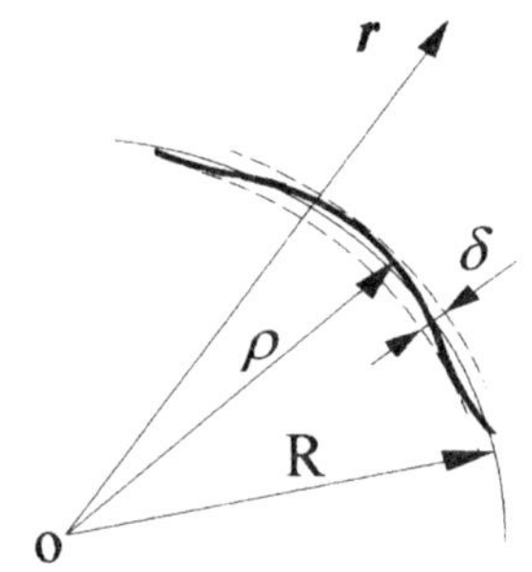

Fig. 1 Schematic of a perturbation on a spherical interface

where, ρ, δ and R are time dependent, l is the number of the perturbations with same power needed to fully cover the geometric surface of the sphere in question. Only a single perturbation is discussed here because the behaviour of an arbitrary distortion is the linear superposition of its all harmonic components. In all the components the effects of the perturbations with powers higher than first, i.e., $m>1$ ($m=1, 2, 3, ...$), may be neglected due to $\delta \to 0$ and their thermodynamic instability as indicated by the analysis in following sections.

1. Growth Tendency of a Perturbation in Constitutional Field

Following Mullins[35], the carbon concentration field surrounding the distorted graphite expressed by Eq. 1 is, if the average carbon content of the melt is c_∞ and the equilibrium interfacial concentration at the point in question is $c_s = c_o + c_o \Gamma_D K$

$$c(r, \theta, \varphi) = \frac{(c_o - c_\infty)R + 2c_o\Gamma_D}{r} + \frac{\left[(c_o - c_\infty)R^l + c_o\Gamma_D R^{l-1}l(l+1)\right]\delta Y_l}{r^{l+1}} + c_\infty \qquad (2)$$

where, $\Gamma_D = \gamma\Omega/kT$, and c_O is the equilibrium concentration of carbon of the melt at a flat graphite-melt interface; K, mean curvature of surface at the point in question; γ, interfacial tense; T, absolute temperature; k, Boltzmann's gas constant; and Ω, the volume increment of graphite for per mole carbon.

The growth velocity, v, of a graphite in a volume constitutional field is

$$v = \frac{dR}{dt} + \frac{d\delta}{dt}Y_{lm} = \frac{D}{C - c_i}\left(\frac{\partial c}{\partial n}\right)_{r=\rho}$$

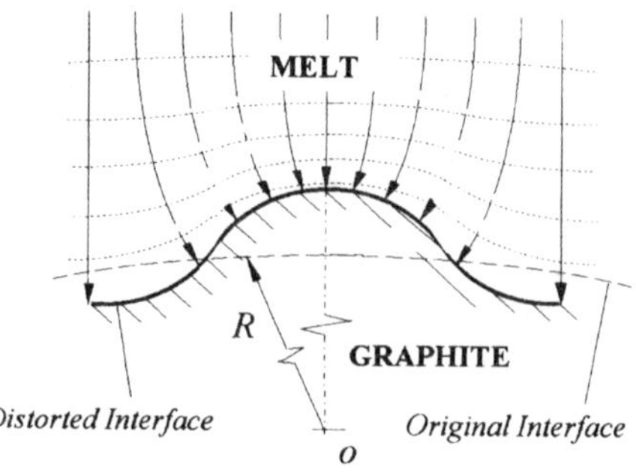

where, D, volume diffusion coefficient of carbon in the melt; C, the carbon concentration on the graphite; c_i, equilibrium concentration of carbon in the interfacial melt; $\partial c/\partial n|_{r=\rho}$ the interfacial concentration gradient in the melt. Substituting the differential of Eq. 1 and 2 into above equation gives the growth tendency of the perturbation in the first order approximation as follows[35]

Fig. 2 The constitutional field in front of the perturbation

$$\dot{\delta}_l = \frac{D(l-1)}{(C - c_R)R}\left\{\frac{c_\infty - c_R}{R} - \frac{c_o\gamma\Omega}{R^2 kT_i}\left[(l+1)(l+2) + 2\right]\right\}\delta_l \qquad (3)$$

where, $c_R = c_o(1 + 2\Gamma_D / R)$ is the equilibrium carbon concentration of the melt at the interface of melt and the original spherical graphite. Eq. 3 shows that the growth tendency of a perturbation, δ_l, is determined by two opposing factors: one is the interfacial constitutional gradient that encourages the growth of perturbation, the other is the Gibbs-Thomson effect, which tends to dissolve away the perturbation under the driving of the requirement of minimising the interfacial energy, or, to stabilise the original spherical morphology, and is proportional to l^3. This indicates that the morphological stabilising effectiveness of the interfacial energy is strongly dependent on the curvature variation of the distorted interface, as well as the area increment of the interface. The Gibbs-Thomson effect being proportional to R^{-3} is related to the reduction of the fraction of the interfacial energy in the growth system as the graphite volume increases, which indicates the original spherical morphology tends to be unstable as the graphite grows.

2. Growth Tendency of a Perturbation in Temperature Field

After the nucleation of a graphite crystal from a supercooled melt, the resultant release of latent heat induces a spherical thermal field surrounding the growing graphite. The temperature field can be described as follows if the equilibrium temperature on the distorted graphite-melt interface can be expressed as $T_i = T_m - T_m \Gamma_T K$ and the temperature of the melt far from the interface is T_∞

$$T(r,\theta,\varphi) = \frac{(T_m - T_\infty)R - 2T_m\Gamma_T}{r} + \frac{\{(T_m - T_\infty)R^l - T_m\Gamma_T R^{l+1}l(l+1)\}\delta Y_l}{r^{l+1}} + T_\infty \qquad (4)$$

where, $\Gamma_T = \gamma/L_V$; T_m is the equilibrium temperature for a flat graphite-melt interface; T_i, equilibrium temperature at the point of the interface in question with a mean curvature K; L_V, the volume latent heat of graphite crystallisation[35].

The growth rate of a crystal in a supercooled melt is $v = (k_m / L_V)(\partial T / \partial n)_i$, where, k_m is the thermal conduction coefficient of the melt, and $(\partial T/\partial n)_i$, the normal interfacial temperature gradient in the melt. Therefore, the growth rate of a perturbation, δ_l, can be approximately expressed as follows[35] if the thermal flow inside the graphite is negligible

$$\dot{\delta}_l = \frac{k_m(l-1)}{RL_V}\left\{\frac{T_o - T_\infty}{R} - \frac{T_i\gamma}{R^2 L_v}[(l+1)(l+2)+2]\right\}\delta_l \qquad (5)$$

Eq. 5 indicates that due to the geometrical similarity of the temperature field to the volume diffusion field their effects on the morphological stability of the graphite are the same.

Eq. 3 and Eq. 5 both indicate the increase of melt supercooling will reduce the morphological stability of the graphite

3. Effect of Interfacial Diffusion

The same Gibbs-Thomson effect also drives the interfacial flow in the distorted interface if interfacial diffusion is permitted, as shown in Fig. 3. Assuming that all interfacial properties are

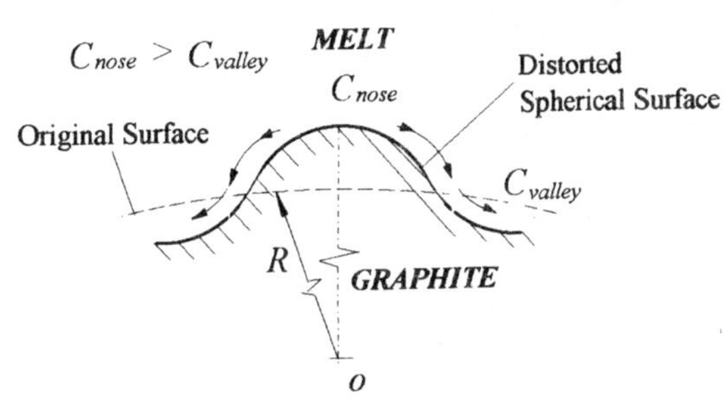

Fig. 3   Schematic of interfacial diffusion driven by difference of curvature on the distorted sphere

independent of crystallographic orientation of the graphite crystal surface and the Gibbs-Thomson equation gives all the interfacial atoms the correct chemical potential, the normal growth velocity outwards at any point of an arbitrary surface as a result of interfacial diffusion is $\partial n / \partial t = \left(D_i \gamma \nu \Omega^2 / kT_i \right) \nabla_i^2 K$ [36], where, D_i is the isotropic interfacial self-diffusion coefficient; ν, number of diffusing atoms per unit surface area; $\nabla_i^2 K$, the interfacial Laplacian at the point in question with a mean curvature K. So, the growth tendency of the perturbation can be expressed by the first order approximation as[37].

$$\dot{\delta}_l = -\frac{D_i \gamma \nu \Omega^2 l\left(l^2 - 1 \right)\left(l + 2 \right)}{kT_i R^4} \delta_l ; \qquad \dot{R} = 0 \qquad (6)$$

where, $\dot{R} = 0$ indicates no contribution of the interfacial diffusion to the average growth rate of the spheroid.

The negative growth velocity of the perturbation shown in Eq. 6 indicates that as interfacial diffusion exists, the Gibbs-Thomson effect, or curvature, always drives the surface diffusion flow from the noses of perturbations downward, as shown in Fig. 3, it dampens the perturbation and increases the stability of the original spherical graphite.

The effect of l on the morphological stabilising ability of the interfacial diffusion is directly related to the difference of curvatures on the distorted interface. The higher the l, the bigger the curvature difference and, thereby, the diffusion driving force. However, the morphological stabilising effect of the interfacial diffusion more dramatically decreases as the crystal grows because it is proportional to R^{-4} instead of R^{-3} in the cases of thermal and constitutional fields.

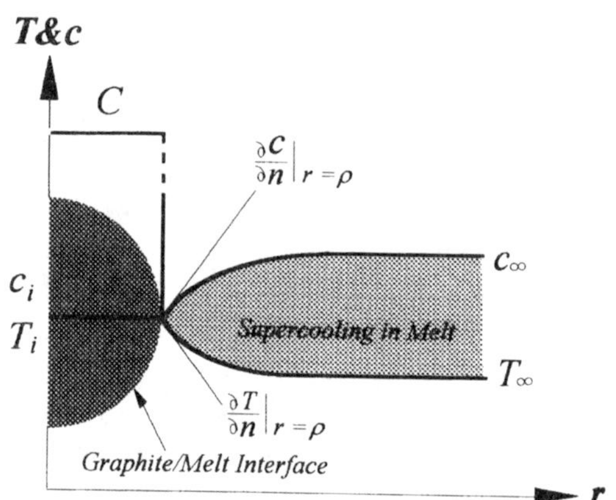

Fig. 4 The sketch of growth driving force distributions in front of interface during solidification

4. Morphological Stability in Solidification

In the practical solidification of a cast iron melt, constitutional and thermal supercooling co-exist, as shown in Fig. 4, and the growth rate of the perturbation on the graphite spheroid, according Eq. 3, Eq. 5 and Eq. 6, is

$$\dot{\delta}_l = \frac{(l-1)}{R^2}\left\{ \left[\frac{k_m}{L_v}\left(T_o - T_\infty \right) + \frac{D\left(c_\infty - c_R \right)}{\left(C - c_R \right)} \right] - \left[\left(\frac{k_m T_i \gamma}{RL_v^2} + \frac{c_o D \gamma \Omega}{\left(C - c_R \right) kT_i R} \right)\left(l+1 \right)\left(l+2 \right) + \frac{D_i \gamma \nu \Omega^2 l}{kT_i R^2}\left(l+1 \right)\left(l+2 \right) \right] \right\}\delta_l, \quad (7)$$

According to Eq. 7, $\dot{\delta}_l = 0$ defines a critical radius R_c, its relationship with solidification condition and the interfacial energy is given as follows.

$$\frac{k_m}{L_v}\left(T_o - T_\infty \right) + \frac{D\left(c_\infty - c_R \right)}{\left(C - c_R \right)} = \left(\frac{k_m T_i \gamma}{R_c L_v^2} + \frac{c_o D \gamma \Omega}{\left(C - c_R \right) kT_i R_c} \right)\left(l+1 \right)\left(l+2 \right) + \frac{D_i \gamma \nu \Omega^2 l}{kT_i R_c^2}\left(l+1 \right)\left(l+2 \right) \quad (8)$$

Obviously, The physical meaning of the R_c is the maximum radius of the graphite, within which the perturbation δ_l is kinetically unstable and the graphite can hold its standard spherical configuration. As $R > R_C$, the stable perturbations can quickly grow and develop into dendrites under the favourable growth environment and the graphite spheroid will degenerate permanently. So, R_C is a measure of the morphological stability of the graphite.

$$\text{If} \quad G = \frac{k_m}{L_v}(T_o - T_\infty) + \frac{D(c_\infty - c_R)}{(C - c_R)}, \quad C = \left(\frac{k_m T_i}{L_v^2} + \frac{c_o D\Omega}{(C - c_R)kT_i}\right) \quad \text{and} \quad S = \frac{D_i v\Omega^2}{kT_i} \quad \text{are}$$

introduced, the above equation becomes

$$R_c = \frac{1}{G}C\gamma(l+1)(l+2)\left[1 + \left(1 + \frac{4GSl}{C^2\gamma(l+1)(l+2)}\right)^{1/2}\right] \tag{9-1}$$

For simplification, we assume that as $l >> 1$ the term $\dfrac{4GSl}{C^2\gamma(l+1)(l+2)} << 1$, the above

equation becomes

$$R_c = \frac{2}{G}C\gamma(l+1)(l+2) \tag{9-2}$$

EFFECTS OF MELT TREATMENT ON THE MORPHOLOGICAL STABILITY

A graphite crystal with a hexagonal structure is anisotropic. It has been clear to date that the spheroidisation treatment of the melt not only increases the interfacial energy between graphite and the melt, but also the interfacial energy between the melt and basal plane, γ_{bm}, and the interfacial energy between the melt and prism plane, γ_{pm}, have been reversed from $\gamma_{pm} > \gamma_{bm}$ in the untreated melt to $\gamma_{pm} < \gamma_{bm}$ in the treated melt. It is, therefore, necessary to take into account the effects of these variations on the morphological stability of the spherical graphite.

1. Effect of Average Interfacial Energy Variation

Eq. 9 clearly indicates that an increase of the interfacial energy will proportionally increase R_C, i.e., the morphological stability of the graphite.

As the melt has experienced the spheroidisation the average graphite-melt interfacial energy will be significantly increased from 82~950 ergs/cm^2 in untreated sulphur-bearing irons up to 1400~1620 ergs/cm^2 according the degree of spheroidisation treatment in question[2, 13]. Correspondingly, R_C, i.e., the morphological stability of the graphite, increases about 1.5~20 times. If a minimum critical radius is defined as $R_c(\gamma=200\text{ergs/cm}^2, l=2)$ the relationship between R_C, γ and l can be schematically illustrated in Fig. 5 via $n = R_c(\gamma, l)R_c^{-1}(\gamma = 200, l = 2)$.

2. Effect of Crystal Structure Variation of the Graphite

Due to the interfacial energy reversal from $\gamma_{pm} > \gamma_{bm}$ in the sulphur-bearing melt to $\gamma_{pm} < \gamma_{bm}$ in the spheroidised melt, the interfacial crystallography of the graphite has been changed from growth in the direction normal to prism planes to growth in the direction normal to basal planes, as shown in Fig. 6.

It is a widely accepted fact that the graphite growth from the spheroidised melt consists of many considerably even developed single crystal segments separated by crystal boundaries, as shown in Fig. 6. However, the random deposition nature makes the crystallites in the graphite growth from sulphur-bearing melt less evenly developed during its growth.

Under thermal and constitutional disturbances during the growth of the graphite, the response of the interface to the perturbation is thermodynamically preferential. That is, only the perturbations of $\delta(l)$ that causes minimum energy rise in the growth system can probably exist on the growth front with thermodynamic stability, otherwise, the perturbation will be dissolved soon after its

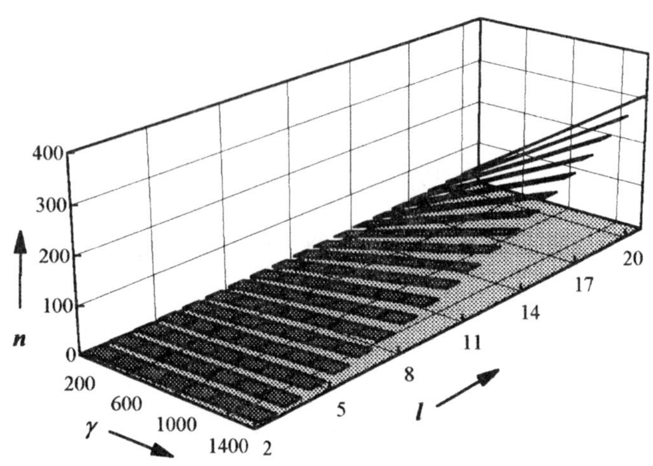

Fig. 5 Relationship between R_c, γ and l.

occurrence. So, to determine the possible range of l of the perturbations, to which the graphite with different crystal structures preferentially respond, it is essential to understand the effect of the crystal structure on the morphological stability of the graphite.

The morphological stability theory has already indicated that the preferential $\delta(l)$ of the perturbation tends to be close to 3 downward for a single crystal spheroid because it causes lowest interfacial energy increase[8~11]. The higher the value of l, the greater is the increment of interface energy caused, and, therefore, less opportunity to become thermodynamically stable on the interface.

However, for spherical graphite consisting of number of crystallites, if l is much lower than the number of the crystallites exposed on the interface, a fraction of the perturbation will geometrically cross the grain boundaries and dislocations must be introduced in the perturbation to accommodate the lattice orientation incompatibility of the neighbouring crystallites. This will significantly raise the energy level of the perturbation and make it thermodynamically unstable.

From the requirement of minimising the energy increase, therefore, the most

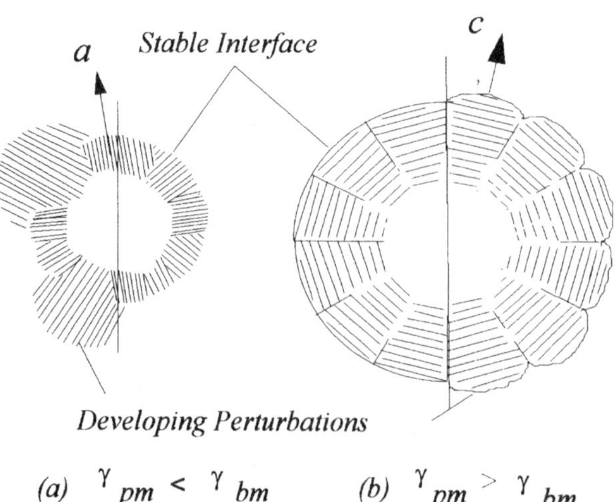

Fig. 6 The interfacial crystallography and crystal structure of the graphite spheroids before and after instabilisation in untreated (a) and spheroidised melt (b)

probable growth mode of the perturbation on a multi-crystalline spherical graphite is one of direct growth from the growth front of the crystallites comprising the spherical graphite, as shown in Fig. 6. So, for a multi-crystal graphite spheroid, the most likely value of l should be approximately the number of the crystallites exposed on the interface.

Morphological observation of the pro-eutectic graphite shows that the number of the sub-crystals in the graphite growth from the spheroidised melt is usually more than 12, much larger than that of the graphite in sulphur-bearing melt which is usually less than 6. Furthermore, if graphite in sulphur-bearing melt nucleates on a large inclusion, the lowest value of l might be 2, which results in the commonly observed floating flake graphite in proeutectic pig iron.

As a summary of the above discussions, the relationship between the morphological stability of the spherical graphite growth from the melt experiencing different degrees of the spheroidisation treatment, the graphite-melt interfacial energy γ and the possible l has been given in Fig. 7, which clearly shows the influence of the spheroidisation on the morphological stability.

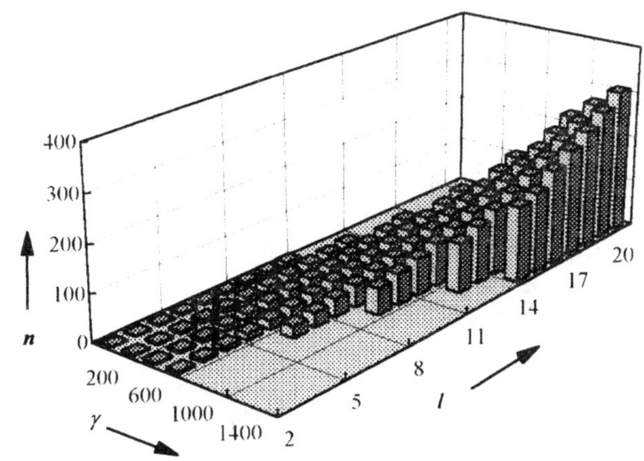

Fig. 7 Relationship between n, γ, and possible l.

THE MORPHOLOGY OF THE DENDRITE

In principle, the growth of the dendrites, developing from the unstabilised original spherical graphite, is still governed by the theory of the morphological stability discussed above. So, before the radius of curvature of the growth fronts of the dendrites grow up to the critical radius defined by Eq. 9, the interface can only grow to a larger radius, while conserving a configuration of a segment of a sphere.

However, as the austenite precipitated in the graphite, the thermal and constitutional fields, as well as the geometrical space available for graphite growth have been changed, which make the analysis of morphology development complicated and unpredictable in some cases. The more significant effect of the appearance of the austenite on the graphite morphology is that due to the partition nature of the spheroidising elements, such as, Mg, R_E, Ca, Sr, etc., and anti-spheroidising elements, such as, S, O, etc., concentration of them will be altered and eventually change the spheroidisation degree of the melt in front of the graphite growth front, and the interfacial crystallography of the graphite. This is suggested as the reason why so many different graphite shapes have been observed in various cast iron castings.

CONCLUSIONS

The morphological stability analysis of a spherical graphite growth during solidification shows that graphite is bound to lose its original spherical configuration as it grows in a supercooled melt. The supercooling decreases the morphological instability of the graphite, but the graphite-melt interfacial energy and the interfacial diffusion tend to enhance the stability.

The increases of interfacial energy and the variation of the crystal structure caused by spheroidisation treatment of the melt will significantly enhance the morphological stability, which cause significant morphological difference between proeutectic flake graphite growth from sulphur-bearing melt and the nodular graphite growth from spheroidised melt.

REFERENCES

1. P. K. Basutkar, C. S. Park, R. E. Miller, and C. R. Loper, Jr., **AFS Trans.**, Vol. 81 (1973), p180.
2. N. A. Sidorov, **Russian Casting Production**, June, 1964.
3. R. K. Buhr, **Modern Casting,** Vol. 54 (1968), No.6, P497.
4. J. F. Ellis, and C. K. Donoho, **AFS Trans.**, Vol.66 (1958), p203.
5. Matsujiro Hamasumi, **J. Cast Metals Research**, Vol. 1 (1965), No. 3, p9.
6. B. Lux, et al, **The Metallurgy of Cast Iron**, 1974, p495.
7. S. V. Subramanian, D. A. R. Kay, and G. R. Purdy, **AFS Trans.**, Vol.90 (1982), p589.
8. S. V. Subramanian, et al, **Iron and Steel**, March 1980, p18.
9. P. C. Liu, C. R. Loper, Jr., T. Kimura, and E. N. Pan, **AFS Trans.**, Vol.89 (1981), p65.
10. P. C. Liu, C. R. Loper, Jr., T. Kimura, and H. K. Park, **AFS Trans.**, Vol.88 (1980), p97.
11. G. S. Cole, **AFS Trans.**, Vol. 80,(1972), p335.
12. P. K. Basutkar, C. R. loper, Jr., and C. L. Babu, **AFS Trans.**, Vol.78 (1970), p429.
13. R. H. Mcsain, C. E. Bates, and W. D. Scott, **AFS Trans.**, Vol.82 (1974), p85.
14. G. F. Ruff, and J. F. Wallace, **AFS Trans.**, Vol.85 (1977), p167.
15. B. Lux, and M. Grages, **Praktische Metallographie**, Vol. 3, 1968.
16. B. Lux, W. Bollman, and M. Grages, **Praktische Metallographie**, Vol. 6, 1969.
17. H. Itofuji, Y. Kawano, *et al*, **AFS Trans.**, Vol.91 (1981), p831.
18. S. Yamamoto, Y. Kawano, *et al*, **AFS Trans.**, Vol.83 (1975), p217.
19. K. M. Htum, and R. W. Heine, **J. Cast Metals Research**, Vol. 5 (1969), June, p51.
20~23. B. Chang, S. Yamamoto, *et al*, **J. Inst. of Met. of Japan**, Vol. 41 (1977), No. 5; No. 6; No.10.
24. C. R. Loper Jr., and R. W. Heine, **Trans. AMS**, Vol. 56 (1961), p583.
25. C. R. Loper Jr., and R. W. Heine, **Trans. AMS**, Vol. 69 (1963), p135.
26. I. E. Bolotov, V. I. Syreischikova, and S. G. Guterman, **Urals Scientific Research Institute of Ferrous Metals**, May 1956, p144.
27. G. X. Sun, and C. R. Loper Jr., **AFS Trans.**, Vol.83 (1975), p841.
28. I. Minkoff, and B. Lux, **The Metallurgy of Cast Iron**, 1974, p473.
29. C. R. Loper Jr., and R. W. Heine, **AFS Trans.**, Vol.69 (1961), p583.
30. Wachenko, and Rudoy, **AFS Trans.**, Vol.70 (1962), p855.
31. X. S. Cheng, **Ductile Iron**, No.1 (1979), p1.
32. S. Long, and Y. Niu, '***Effects of Strontium on Morphology of Graphite in Cast Iron***', a presentation on '**National Cast Iron Conference**', Fuzhou, China, 1987.
33. H. Biloni: '***Solid-liquid Interfacial Morphology***' in **Physical Metallurgy** ed. by R. W. Cahn, and P. Haasen, Elsevier Science publishers B. V., 1983, p500.
34. R. F. Sereka **Crystal Growth: An Introduction**, edited by P. Hartman, p 403.
35. W. W. Mullins, and R. F. Sekerka, **J. Appl. Phys.**, Vol. 34 (1963), p323.
36. W. W. Mullins, **J. Appl. Phys.**, Vol. 28 (1957), p333.
37. F. A. Nichols and W. W. Mullins, **Trans. AIME**, Vol.233 (1965) p1843.

Advanced Materials Research Vols. 4-5 (1997) pp. 521-528
© 1997 Scitec Publications, Switzerland

Modelling the Filtration of Cast Iron by Lattice Gas

R. Bremond[1], D. Jeulin[1], M. Abouaf[2] and C. His[2]

[1] CMM, École Nationale Supérieure des Mines de Paris,
35 rue St. Honoré, F-77305 Fontainebleau, France

[2] Saint Gobain Fine Ceramic Research Center, Z.I. 1, F-27025 Evreux, France

Keywords: Ceramic Filters, Cast Iron, Lattice Gas, Clogging

Abstract. A model of the filtration process with a lattice gas is proposed. A lattice gas is a cellular automaton which is able to simulate hydrodynamical flows. It is completed by adding aggregation and disaggregation rules, which appear to be the key factor of the clogging process. A reference simulation corresponding to the particular physical conditions of the problem was defined. The simulated results were compared to experimental data obtained by observation of the polished sections of the filters, and by recording the flux of cast iron during the filtration.

1 INTRODUCTION

Filtration of suspensions is a physical process which appears in many industrial fields, like for example the elimination of druss in castings, water purification, or food filtration. The purpose of this study is to model the filtration of iron, in order to have a tool for the design of new filters. In order to obtain metal alloys of a good quality, one needs to master the elimination of inclusions contained in the alloy. This is one of the purposes of filtration. The filter efficiently eliminates large suspensions, of order of several millimeters, but the case of very small particles, of a few microns, is not well understood yet. The company Saint-Gobain, in its subsidiary Hamilton Porcelains, produces ceramic filters devoted to molten metal filtration. These are made of several parallel channels, with a diameter close to one millimeter (Fig. 1). One particular property of these filters is a wall effect. The diameter of the trapped suspensions (of order several microns) is much smaller than the size of the channels. The aim of this work is to simulate the filtration of cast iron. The study also gives a simulation tool which helps the design of new ceramic filters. From the model of filtration, it is now possible to change the physical and geometric parameters in order to test new configurations.

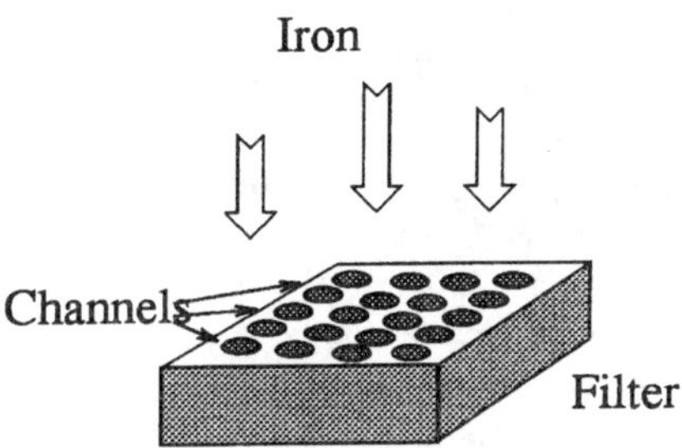

Figure 1: Geometry of the studied filters.

2 EXPERIMENTAL DATA

To study the clogging of ceramic filters during the filtration of molten ductile iron, experimental data and SEM[1] micrographs of filters were obtained from monitoring castings. Experimental castings of iron were performed through filters at the CRPAM[2]. These experiments were made of several interrupted castings, the amount of iron crossing the filters being too low to clog them completely. The experiments make it possible to see, on the polished sections of the filters, how the aggregates evolve during the clogging process. Castings performed until the filter is clogged completely allow to compare simulated clusters with real clusters.

Filters. The diameter of the channels is between 1 and 3 millimeters and their depth ranges from 9.5 to 24 millimeters. For ductile iron, the most common size is 2 millimeters in diameter and 15 millimeter in depth. This gives a ratio $\frac{diameter}{depth} = 0.133$. The ceramic used in these experiments is a mix of mullite and cristoballite. These filters are used under a pressure of the order of 50 cm. of iron (around one third atmospheric pressure).

Castings. Experimental castings use a clean laboratory iron, with spheroidal graphite, made of C = 3,65%, Si = 2,2%, Mg = 0,035%, S = 0,005%. The casting temperature is around 1380°C. This iron was chosen because it allows one to observe the clogging resulting from micro-suspensions, keeping the process away from large suspensions clogging effects. We used a ductile iron comming from a deslagged laddle in order to eliminate larger suspensions (laddle slag, dross).

Chemical analysis. After the casting, the filter was removed from the gating by cutting. Chemical analysis of the clogging material was performed through electron microprobe analysis (energy dispersive analysis). The chemical analysis shows that the clusters are made of an impurity network, among which the largest part consists in oxides, sulfides and alumino-silicates of Mg and Ca.

Hydrodynamics. Experimental castings have a homogeneous flow rate during the first 10 seconds leading to a cumulative filtered mass of 30 kg. The iron viscosity is 5 centipoises and its density is 7000 kg.m^{-3}. With an average flow velocity of 1 m.s^{-1}, this gives a Reynolds number equal to $R_e = 1400$ during this first step of the casting. Ductile iron was poured through the filters, and both the pour rate and the amount of metal that passed before clogging was measured for comparisons.

Images. The clogged filters are cut and polished after the casting ends with a clogging. The polished sections are then observed in a SEM which shows the geometrical structure of the clusters causing the clogging. Figure 2 gives two examples of such images. Dark rectangles denote the ceramic, white areas correspond to the iron, grey regions are clusters and black spots are particles of spheroidal graphite. The black areas close to the ceramic are holes due to the iron shrinkage during cooling.

3 SIMULATIONS USING LATTICE GAS MODELS

The lattice gas model is a recent method to simulate hydrodynamics [5]. It is a discrete version that is free of the inherent defects in classical numerical methods (boundary conditions, round-off errors, etc.) and is easily implemented on general purpose computers. Each vertex of an hexagonal graph is connected to its six nearest neighbours. It can contain between 0 and 6 particles with a mass and a velocity of one (at most one particle per velocity, corresponding to the six velocities of the grid). At every time step the location of particles is changed according to the following local operations: a propagation of particles to the nearest neighbour in the direction of their velocity and a collision redistributing the velocities.

---

[1] Scanning Electron Microscope
[2] Centre de Recherche de Pont-à-Mousson.

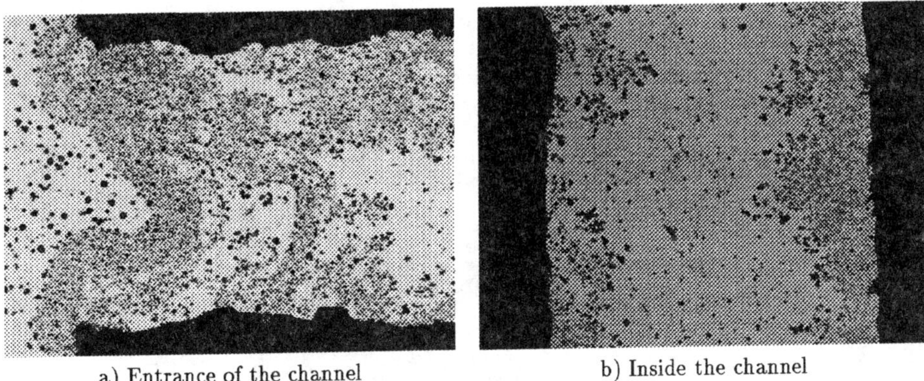

a) Entrance of the channel   b) Inside the channel

Figure 2: SEM images of one channel of the filter.

The *collisions* are local operations that preserve the total mass and momentum (and consequently the kinetics energy) of each node. In [3], it was shown that this model, called FHP, follows the Navier-Stokes equations on a macroscopic scale. Therefore the simulation of complex flows can be made using lattice gases. The rules used in this study are the FHP IV rules given in [4]. The *physical boundary conditions* are easily handled. Each vertex of the graph can be referenced as a pore or as an obstacle. In the latter case, the collision is replaced by bounce-back conditions. This is equivalent to a no-slip condition on the boundary. To simulate a mixture of fluid and suspensions we use *marks*. They are redistributed at random after the collisions. The behavior of the two types of particles differs only during the aggregation process. The following rule of *aggregation* is introduced. The field contains pores and obstacles (including clusters). When a particle becomes the nearest neighbour of an obstacle, it is a candidate to aggregate, with the probability of aggregation, p^+ (it is bounced back with the probability $1 - p^+$). In order to simulate a *disintegration*, particles in the aggregate can leave the aggregate for the fluid with a probability p^-.

A lattice gas, like any other model in physics, has a domain of validity. The main limit in our case is the value of the Reynolds number. The best lattice gas with respect to this (i.e. the less viscous) cannot exceed $R_e = 50$ with the system size we use, which is much lower than the actual case (around 1400). We assume that the process is not qualitatively different. This is based on the idea that both values correspond to laminar flows. A software was written for this purpose, and simulations were performed on a SUN Sparc 1 workstation. For a 150×400 image, each iteration of the simulation needs 4 seconds. With the current implementation, a simulation of 10000 time steps is performed in 11 hours.

4 STANDARD SIMULATION

From the simulation model, the study of filtration can be divided into several parts. First, reference conditions are defined, corresponding to a simulation called the "standard" simulation. This simulation is characterized in this section in terms of physical measurements and by the morphology of the clusters. The "standard" simulation was determined in a heuristic way to closely match experimental data. From this simulation a parametrical study was performed in order to characterize the role of each parameter[2].

4.1 Reference system

We were interested in simulating what happens in the entrance of the channels because the clogging seems to occur upstream. Due to limited computer power, our simulations are two-dimensionnal:

150×400 image, standing for the entrance of a single channel (Fig. 3). The system is periodic vertically, and aperiodic horizontally. A single channel is simulated, but by vertically periodizing the system we take into account the neighbour channels. The channel itself has 114 pixels width and begins at the first quarter of the image ($x = 100$). These sizes corresponds to a distance between two consecutives pixels of the order of 15 μm, and so we can only simulate flows in which the largest suspensions do not exceed this size, which is physically correct for the studied process, where the largest suspensions do not exceed 10 μm.

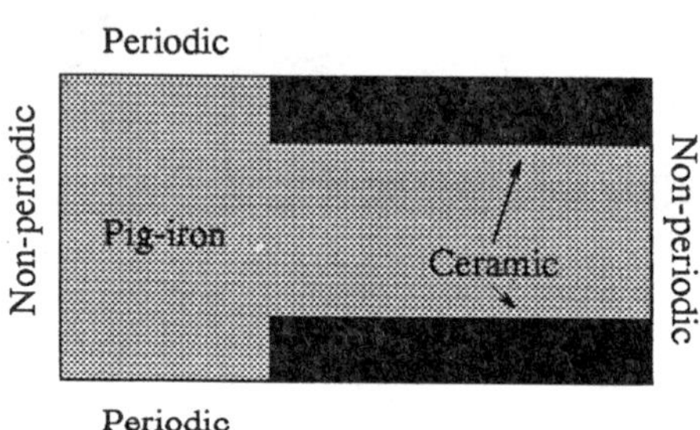

Figure 3: Physical system.

Compared to the real physical phenomena, our simulation stands on several simplifications, among which the main ones are:

- The simulation is two-dimensional.

- Chemical or thermal effects are not taken into account.

- The Reynolds number is lower than the real one (but we nevertheless simulate a laminar flow, which corresponds to the observed hydrodynamics).

- The size distribution of the suspensions is not taken into account.

- The clusters which go back to the flow have no internal cohesion force.

The boundary condition upstream and downstream are set constant all along the simulations, in order to fix the pressure gradient and the suspension density in the iron. The initial conditions are:

- Upstream suspension density : $\rho_S^{up} = 0.1$

- Upstream iron density : $\rho_F^{up} = 2.9$

- Downstream iron density : $\rho_F^{down} = 2.7$

We should point out here that the density of suspensions is mainly conventionnal, and is not the true concentration of impurities. In fact in the model a "suspension" is made of a given concentration of impurities surrounded by the fluid. For example, if we assume that each "suspension" contains a 4 μm particle, the ρ_S suspension density of 0.1 corresponds to an actual particle density of 0.002. Setting the upstream and downstream density constant allows us to control the pressure gradient, which corresponds to an experimental constraint. Initial conditions consist in an empty channel. During a first step of the simulation, which lasts 5000 time units, iron is introduced upstream and downstream until the hydrodynamical equilibrium is reached. Suspensions are introduced in the second step of the simulation. Aggregation and disaggregation probabilities are as follows:

p_{ss}^{+}	p_{sc}^{+}	p_{ss}^{-}	p_{sc}^{-}
0.015	1.000	0.005	0.000

Here p_{sc}^{+} and p_{sc}^{-} denote the aggregation and disaggregation probabilities for suspensions sticking to the ceramics and p_{ss}^{+} and p_{ss}^{-} for suspensions sticking to other suspensions. Note that the suspensions which stick to the ceramic are trapped forever ($p_{sc}^{+} = 1$). The value of all these parameters was chosen in an heuristic way, in order to reproduce as well as possible the experimental conditions.

4.2 Images

Images of the evolution of the channel during the clogging simulation process (Fig. 4) shows the ceramic walls, the clusters and the moving suspensions. These images correspond to the observed physical behaviour, ending with a clogging at the entrance of the channel (compare with Fig. 2). We observe the limited mobility of the clusters, which grow with time without dis-agreggating large clusters.

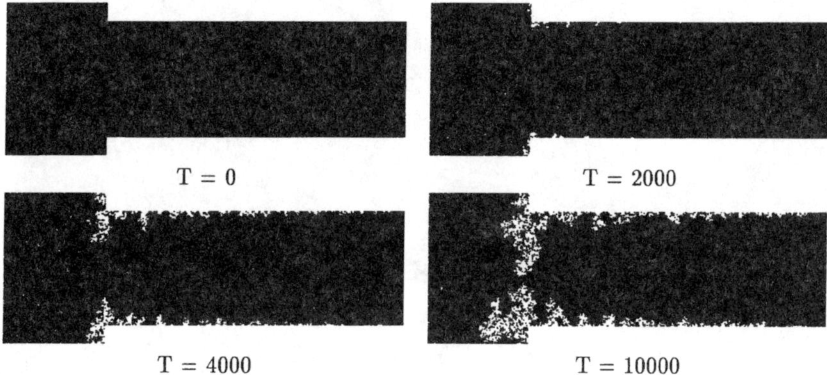

Figure 4: Standard simulation.

4.3 Filter efficiency

To design new filters geometries, it is necessary to be able to estimate quantitatively the properties of the filters. From an industrial point of view, two aspects are important: the flux and the efficiency.

	p_{ss}^{+}	p_{sc}^{+}	p_{ss}^{-}	p_{sc}^{-}	E
α	0.000	0.000	1.000	1.000	0.000
β (Standard)	0.015	1.000	0.005	0.000	0.742
γ	1.000	0.100	0.001	0.000	0.875

Table 1: Aggregation parameters and filter efficiency.

One way to estimate the efficiency of the filters is the ratio between the number of suspension trapped into the clusters and the total number of suspensions: those in the clusters (A) plus those which passed through the filter (D_s). This estimator is given by

$$E = \frac{A}{A + D_s}, \tag{1}$$

where A is the final area of the clusters. Table 1 gives the value of these estimators with the previous simulations, called β and γ (the simulation α, without aggregation, leads to $E=0$).

With simulation γ, the filtered iron is two times cleaner than with simulation β (standard simulation). Indeed, from 0.742 to 0.875 the iron is two times cleaner. Unfortunately, no experimental results are presently available for comparisons. At least it will be possible, as soon as experimental data will be available, to adjust the aggregation parameters and thus to match with the real process.

5 COMPARISON BETWEEN SIMULATION AND EXPERIMENTS

A comparison with experiments is made, in order to test the validity of the model.

Qualitative comparisons. The main characteristics of the physical process are found in the simulations: the clogging of the filters comes from an accumulation of tiny inclusions and it happens on the upstream face with a growing cluster which ends up in clogging the filters channel.

5.1 Flow rate

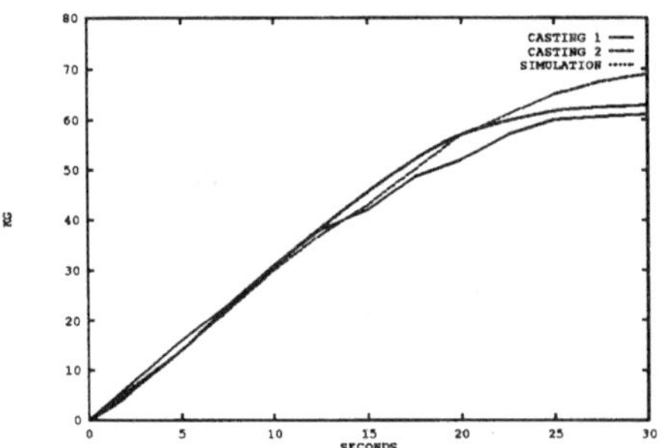

Figure 5: Comparison between simulation and the physical process (cast iron flow rate).

A comparison of the actual and simulated iron flow rate (Fig. 5) is based on an estimation of the physical value of the units used throughout the simulations. The distance between two points was about 15 μm. The linear part of the flow rate graph allows us to compute the mass and time value: the time unit is of the order of $3 \ 10^{-3}$ s. while the mass unit is around $1.5 \ 10^{-6}$ kg. The simulated graph occurs to be located between the measured graphs, which means a good agreement.

5.2 Cluster geometry

Cluster images. Some of the measurements performed previously on the simulated clusters can also be done on the SEM images. But for this it is necessary to isolate the clusters in these images, in order to get a binary image which could be compared with the binary image of the simulated clusters. For this purpose, classical tools from mathematical morphology [6][1] were used, which allows to identify the relevant clusters in each of the SEM image [2]. The following images (Fig. 6) present an example of segmented clusters. Some measurements (porosity, longitudinal profile) are made from the binarized images.

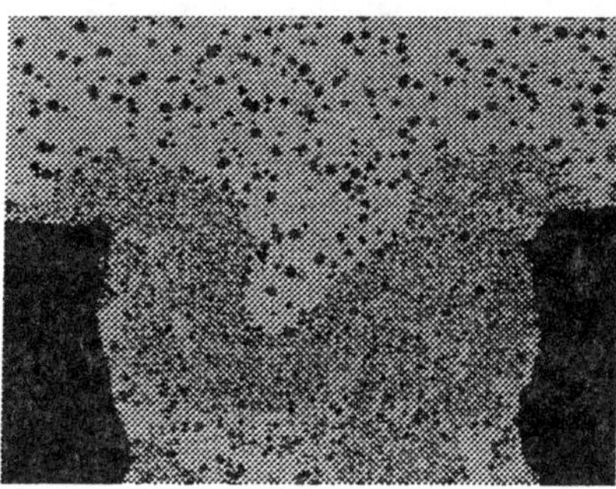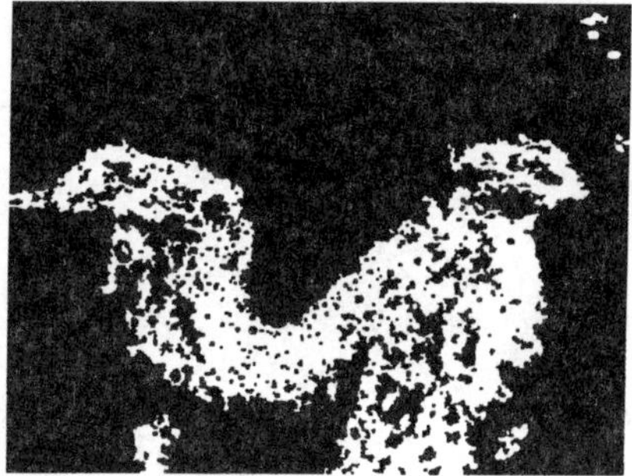

Figure 6: Cluster segmentation from SEM images.

Porosity.   Porosity measurements of these segmented clusters are compared to those of simulated clusters. An average over 5 segmented images is compared to an average over 10 simulations using the "standard" conditions. The results are close enough to conclude to a quantitative agreement between simulation and experiment with respect to porosity.

SEM images	Simulations
0.14	0.16

Table 2: Porosity measurements.

Distribution.   The area profiles of the clusters along the channel was also compared: the simulated data are more regular (Fig. 7), because the number of SEM images used here was small. The important remarks to be done are firstly that there is a good agreement between the two graphs and secondly that simulated clusters have their main part (the higher peak on the graph) slightly before the entrance of the channel, while on SEM images this main part stands slowly after. This is due to one of the limitations of our model, which is the rigidity of the simulated clusters. The examination of SEM images shows that real clusters have a certain plasticity, which allow them to move slightly under the strong pressure they hold.

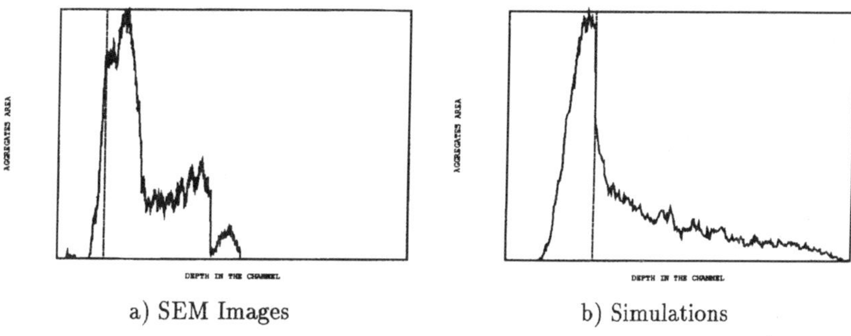

a) SEM Images   b) Simulations

Figure 7: Comparison of the cluster's profiles.

6   CONCLUSION

Using computer simulations implementing a lattice gas model, a satisfactory model of the complex processes involved during the filtration of iron in a cellular filter was obtained. This model can be used as a tool in order to design new filter geometries. Exploratory simulations can help to make a first choice before starting experiments at a pilot scale. With the kind of filter geometry used in this study, the two-dimensionnal simulation seems not to be a strong limitation of the model. However similar studies for more complex filters, such as granular beds, would require the use of three dimensionnal simulations to reproduce more accurately the percolation effects involved in the clogging process. Appropriate three-dimensionnal lattice gas models are available [7], and could easily integrate the presence of suspensions particles as well as the aggregation rule introduced in the present study; this would involve a higher computationnal cost. Other developments can be made, concerning the size (and shape) distribution of suspensions, as well as the reproduction of the filtration of other materials.

*This work has been partially supported by the French Ministry of Research and Space (contract n°
90A511)*

References

[1] S. Beucher. *Segmentation d'images et morphologie mathématique.* PhD thesis, Ecole des Mines de Paris, (1990).

[2] R. Brémond. *Modélisation de la filtration par gaz sur réseau.* PhD thesis, Ecole Nationale Supérieure des Mines de Paris, (1993).

[3] U. Frish, D. d'Humières, B. Hasslacher, P. Lallemand, Y Pomeau, and J.P. Rivet. Lattice gas hydrodynamics in 2 and 3 dimensions. *Complex System 1*, pages 649–707, (1987).

[4] U. Frish, B. Hasslacher, and Y. Pomeau. Lattice-gas automata for the navier-stokes equation. *Phys. Rev. Lett. 56*, pages 1505–1508, (1986).

[5] J. Hardy, Y. Pomeau, and O. de Pazzis. Molecular dynamics of a classical lattice gas : transport properties and time correlation functions. *Phys. Rev. A, 13*, pages 1949–1961, (1976).

[6] J. Serra. *Image Analysis and Mathematical Morphology.* Academic Press, London, (1982).

[7] J. A. Somers and P. C. Rem. Workshop on numerical methods for the simulation of multi-phase and complex flow. *Lecture notes in Physics, Ed. T.M.M. Verheggen, Springer-Verlag,* (1992).

Advanced Materials Research Vols. 4-5 (1997) pp. 529-534
© *1997 Scitec Publications, Switzerland*

Prediction of Microstructures in Freezing of Cast Irons

A. Kagawa[1] and Y. Ohta[2]

[1] Department of Materials Science and Engineering, Nagasaki University,
1-14 Bunkyo-machi, Nagasaki 852, Japan

[2] Technology Center of Nagasaki, Ikeda 2-1303-8, Ohmura 856, Japan

Keywords: Computer Simulation, Microstructure, Chill, Gray and Ductile Irons

ABSTRACT

The prediction method for the formation of chilled structure in freezing of gray and ductile cast irons was investigated by computer simulation. The validity of the simulation was examined in comparison with the results of chill tests using a wedge-shaped mold. It has been found that the dependence of nodule count on supercooling should be taken into account to give a reasonable distribution of chill fraction and to explain the variation of observed nodule count with cooling rate. Predicted distribution of chill fraction is significantly influenced by the introduction of a silicon concentration dependence of eutectic temperature. The simulation predicts critical cooling rates and critical nodule counts for the formation of chill in both the irons.

INTRODUCTION

The formation of chill is one of the problems encountered in the production of thin-sectioned cast irons. Hitherto, several works have been carried out on the factors of cooling rate[1-4] and alloy composition[5, 6]. It is demonstrated that there are critical cooling rates and critical nodule counts for the formation of chill and the influence of alloying elements is expressed by the index of their graphitizing ability. The kinetics of transitions from gray(G) to white(W) and from W to G has been discussed in terms of the effect of alloying elements on the stable and metastable eutectic temperatures[7-11]. Recently, some applications of computer simulation have been reported for the prediction of chill formation in relation to eutectic cell count[12] and cooling rate[13].

To predict the formation of chill, the quantitative evaluation of nucleation rates and growth rates for both the eutectics should be systematically taken into account. Since these equations are given as a function of supercooling, the results of simulation depend on the accuracy of evaluated supercooling and hence the eutectic temperatures need to be given as a function of alloy composition which varies during solidification[14].

In the present work, to establish the prediction method for the formation of a chilled structure in freezing gray and ductile cast irons, the validity of theoretical equations for nucleation and growth as well as evaluation method of the eutectic temperatures have been examined by comparing the results of simulation with the experimental ones obtained by a chill test.

EXPERIMENTAL

The gray iron (FC) melt and ductile cast iron (FCD) melt subjected to a spheroidizing treatment and subsequent inoculation treatment with ferrosilicon were cast at 1653K and 1623K, respectively, in air into a wedge-shape mold illustrated in Fig.1. The variation of temperature during solidification was measured at locations on the center line of the casting (The locations of three thermocouples for FC and four thermocouples for FCD are shown in Fig.1). The wedge-shape FC casting (3.72%C, 1.70%Si, 0.43%Mn, 0.03%P and 0.004%S) and FCD casting (3.62%C, 2.42%Si, 0.42%Mn, 0.03%P, 0.006%S and 0.045%Mg) were cut longitudinally and horizontally near the thermocouples. Eutectic cell size and cell count for FC, graphite nodule size and nodule count for FCD and the fraction of chill for both the irons were measured on the horizontal sections.

COMPUTER SIMULATION

The present simulation was carried out on the left half of a longitudinal section of the wedge-shaped casting (Section A in Fig.1), with meshed elements and boundary conditions shown in Fig.2.

In the simulation program, the following nucleation rate equation given for heterogeneous nucleation[15] was employed.

$$I = 5 \times 10^{27} N_0 \, r\{1600/(1.45\Delta T + 600)\} \times \exp\{-46080/(1.45\Delta T + 600)\} \quad \cdots \quad (1)$$

, where I is nucleation rate, N_0 is nucleation rate constant, r is diameter of SiO_2 particles as a heterogeneous nucleus, and ΔT is supercooling.

The growth rates for ledeburite and for flake graphite eutectic are expressed in the following equations.

$$V_W = K_W \, (\Delta T)^2 \quad \cdots \quad (2)$$
$$V_G = K_G \, (\Delta T)^2 \quad \cdots \quad (3)$$

, where K_W and K_G are the growth rate constants with the values of 3×10^{-5} $(m/s/K^2)$ and 4×10^{-8} $(m/s/K^2)$, respectively, given in literature[12,16] employed in the present work.

The growth rate of nodular graphite is given, on the basis of the carbon diffusion model through the surrounding austenite shell[17] by:

$$V_G = D_C^{\gamma}\{R_g \, (C^{\gamma/L} - C^{\gamma/g})/R_G(R_G - R_g \,)(C^{L/\gamma} - C^{\gamma/L})\} \quad \cdots \quad (4)$$

, where $V_G(=dR_G/dt)$ is the growth rate of the austenite shell, D_C^{γ} is diffusion coefficient of carbon in austenite, R_g and R_G are radii of nodular graphite and austenite shell, respectively, and $C^{\alpha/\beta}$ ($C^{\gamma/L}$, $C^{\gamma/g}$ and $C^{L/\gamma}$) is volumetric concentration of carbon in α phase equilibrating with β phase.

In the present work, the following relationships and values were used.

$$R_G = 2.4 \, R_g \, [17],$$
$$D_C^{\gamma} = 9 \times 10^{-11} (m^2/s) \, [18],$$
$$C^{\gamma/L} - C^{\gamma/g} = 3.44 \times 10^{-4} \Delta T \, [19],$$
and
$$C^{L/\gamma} - C^{\gamma/L} = 0.084 \, [19]$$

Equation 4 then becomes:

$$V_G = 0.26 \, \Delta T/R_G \quad \cdots \quad (5)$$

The equations 1-3 and 5 are given as a function of supercooling ΔT and the supercooling should vary during eutectic solidification since the equilibrium eutectic temperatures for the stable and metastable systems would change as a result of solute enrichment. For the alloy composition of cast irons in the present work, the effects of minor elements such as manganese, phosphorus and sulfur may be neglected[6]. Hence, the dependence of eutectic temperatures on silicon concentration is considered in the calculation.

The changes in the stable and metastable eutectic temperatures by an addition of silicon, ΔT_E^S and ΔT_E^M, respectively, are given as a function of silicon concentration in the liquid[20], and the silicon concentration in the liquid(L) is approximately given by the Scheil equation[21] as follows.

$$\Delta T_E^S \doteqdot 5 \, (\%Si)^L \quad \cdots \quad (6)$$
$$\Delta T_E^M \doteqdot -10(\%Si)^L \quad \cdots \quad (7)$$

$$(\%Si)^L = (Si)_0(1-f_s)^{(k_{Si}-1)} \quad \cdots \quad (8)$$

, where $(Si)_0$ is silicon content of the alloy, f_s is solid fraction and k_{Si} is the equilibrium partition coefficient of silicon between eutectic solid and liquid.

For the evaluation of fraction solid in Eq.8, the Johnson-Mehl equation is generally applied for the growth of nodular graphite. Here, we examined two cases : i.e. Johnson-Mehl equation for both the growth of graphite eutectic and ledeburite(Eq.9)[12], and the case where fraction solid is given by a sum of volumetric solid fractions(VSF) of both eutectics(Eq.10).

$$f_s = 1 - \exp\{-4\pi(N_GR_G^3+N_WR_W^3)/3\} \quad \cdots \quad (9)$$
$$f_s = 4\pi(N_GR_G^3+N_WR_W^3)/3 \quad \cdots \quad (10)$$

, where N_G and N_W are the stable and metastable eutectic cell counts (per m^3), and R_G and R_W are the radii of the eutectic cells(m), respectively.

Thermophysical data employed in the calculation are summarized in Table 1.

RESULTS AND DISCUSSION

Macro and microstructures of the castings are shown in Figs.3-5 for FC and FCD irons. On the macroscopic scale in Fig.4a, the thickness of chill in the FCD casting was about 10 mm and the portions with the thickness less than 20 mm showed a chilled white structure, while in the FC iron casting(Fig.3a) having a similar composition, a thinner chill depth was observed. As seen in Fig.5, the microstructure near the top thermocouple(No.4) for the FCD iron contained a white structure, showing that a mottled structure was obtained in a wide range of cooling rate. In Tables 2 and 3, the observed maximum size of eutectic cell (D_c), cell count (C_A, C_V) for FC, the maximum size of graphite nodule (D_g), nodule count (N_A,N_V) for FCD and chill fraction (f_W) for both the irons are tabulated for the locations of thermocouples, where C_A (N_A) is cell (nodule) count per unit area, corrected by a gray fraction, i.e., multiplied by $1/(1-f_W)$, and C_V (N_V) is cell (nodule) count per unit volume, converted by the following equations[22].

$$C_V = 0.72\ C_A^{1.5} \quad \cdots \quad (11)$$
$$N_V = 2.35\ N_A^{1.5} \quad \cdots \quad (12)$$

In the simulation, the nucleation rate constants in Eq.1 for graphite eutectic and ledeburite were selected to be $N_0^G = 1.2 \times 10^{14}$ and $N_0^W = 8 \times 10^{14}$ on the basis of the observed nodule count in Tables 2 and 3 and the relationship between nodule count and cooling rate observed for non-inoculated gray cast iron[23]. In Fig.6, calculated chill fractions for the elements corresponding to the locations of thermocouples are given for both the irons. In the case of FCD iron, the application of the Johnson-Mehl equation tends to give relatively higher chill fractions in comparison with the observed ones. This may be attributable to the exponential term in Eq.9 where an increasing rate of fraction solid is reduced exponentially. Using the VSF equation, the effect of silicon concentration dependence of eutectic temperature(T_E) was examined by comparing with the case when the eutectic temperature remained constant during solidification. When applying the constant T_E condition, lower chill fractions were obtained. In cooperation with the silicon concentration dependence of T_E, the VSF equation gives excellent agreement with the observed chill fraction. As seen in Fig.6a for the FC iron, a similar good agreement was obtained when employing the VSF equation. In Figs.3b and 4b, the calculated distributions of chill fraction for both the irons are illustrated by tiling and a good accordance is seen with the macroscopic observation in Figs.3a and 4a.

Employing the VSF equation(Eq.10), the dependence of chill fraction on nodule count was calculated for different nucleation rate constants of graphite eutectic and the results for FC and FCD irons are illustrated in Fig.7. The curves indicate that, for a certain cooling rate corresponding to the location of thermocouples, there is a critical nodule count where the formation of chill is completely suppressed. Figure 8 shows the relationship between the critical cell or nodule count, C_v^* or N_v^*, and cooling rate for both the irons. This figure indicates that the requirement for obtaining a sound ductile iron casting is more severe than in gray iron casting. The critical cooling rate for the production of a ductile iron casting completely free of chill is less than 1 K/s (average cooling rate during the solidification interval) for the range of nodule count, less than $10^{14}/m^3$, obtained by an ordinary inoculation, while for FC iron it is around 10K/s in the range of cell count $\sim 10^{11}/m^3$.

CONCLUSION

The chill fraction, the size and number of eutectic cells (or graphite nodule) observed in gray and ductile irons cast in a wedge-shaped mold were compared with the results of simulation. It was indicated that the introduction of the Johnson-Mehl equation as an expression of fraction solid tended to give a higher chill fraction. Provided that the rates of nucleation for graphite eutectic and ledeburite were given as a function of supercooling and the dependence of the eutectic temperatures on alloy composition was taken into acount, predicted chill fraction and its distribution in the casting reflected well the observed one. A relationship between chill fraction and critical eutectic cell count or graphite nodule count was deduced by simulation. The simulation predicted that a critical cooling rate for the production of chill-free ductile cast iron would be less than 1 K/s which was about 10 times smaller than that of gray iron.

REFERENCES

[1] P.Magnin and W.Kurz, Metall.Trans A,19A,1955(1988).
[2] P.Magnin and W.Kurz, idem 19A,1965(1988).
[3] H.Horie, T.Miyate, M.Saito and T.Kowata, Imono 56,91(1984).
[4] T.Kitsudo, K.Fujita and Y.Kawano, idem 60,13(1988).
[5] E.Piwowarsky:"Hochwertiges Gusseisen 2.Auflage" Springer Publ.,p769(1951).
[6] A.Kagawa and T.Okamoto, Imono 57,113(1985).
[7] W.Oldfield, BCIRA J.9,506(1961).
[8] W.Oldfield, idem 10,17(1962).
[9] J.Charbonnier and J.C.Margerie,Fonderie No.259,333(1967).
[10] M.Hillert and V.V.Subba Rao:in "The Solidification of Metals" I.S.I.Publ.
     n°110,p204(1968).
[11] B.Lux and W.Kurz, idem p193(1968).
[12] H.Fredriksson and I.L.Svensson, in "The Physical Metallurgy of Cast Iron"
     -MRS Symposia Proceedings.Vol34 Elsevier Publ.Co.,p273(1985).
[13] A.A.Blanco in "Modeling of Casting and Solidification Processes 1991"
     Yonsei Univ.Press,p295(1991).
[14] A.Kagawa and T.Okamoto, J.Mater.Sci.22,643(1987).
[15] K.C.Su,I.Ohnaka,I.Yamauchi and T.Fukusako:MRS Symp.Proc.34,see[11],
     181(1985).
[16] K.Kishitake, T.Owadano and K.Miyamoto, Imono 53,295(1981).
[17] T.Owadano, K.Yamada and K.Torigoe, Trans.Jpn Inst.Met.18,871(1977).
[18] Nihon-Kinzoku-gakkai ed.:"Kinzoku Data Book" Maruzen Publ.,p26(1984).
[19] T.B.Massalski:"Binary Alloy Phase Diagrams" ASM Publ.,p562(1986).
[20] A.Kagawa and T.Okamoto, Imono 57,113(1985).
[21] E.Scheil, Z.Metallkde 34,70(1942).
[22] T.Ohwadano, Imono 45,193(1973).
[23] Y.Kondo, M.Isotani and Y.Ueda, idem 45,962(1973).

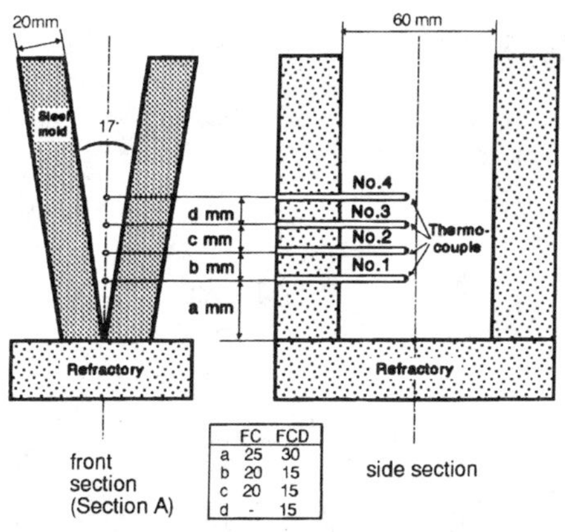

	FC	FCD
a	25	30
b	20	15
c	20	15
d	-	15

Fig. 1 Vertical sections of the wedge-shaped mold

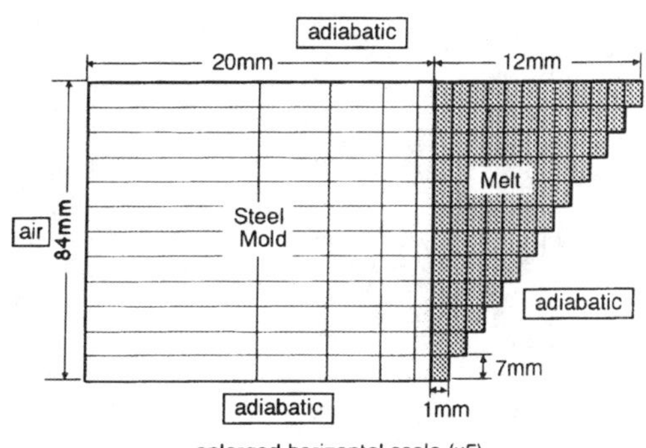

Fig. 2 Meshed elements on the vertical section of the wedge-shaped casting.

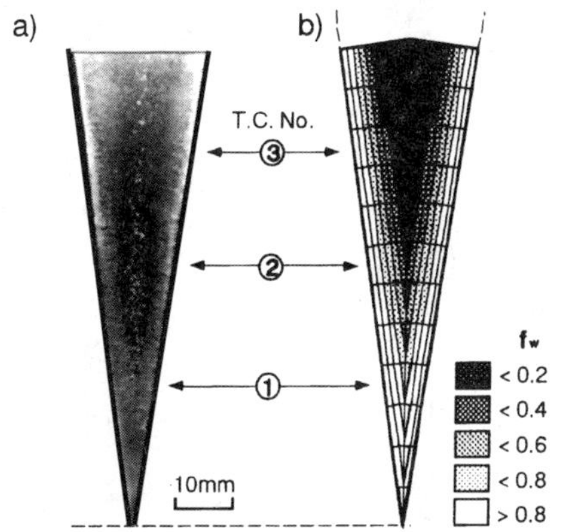

Fig.3 a) Appearance of the vertical section of a wedge-shaped FC iron casting and b) the predicted distribution of chill fractions.

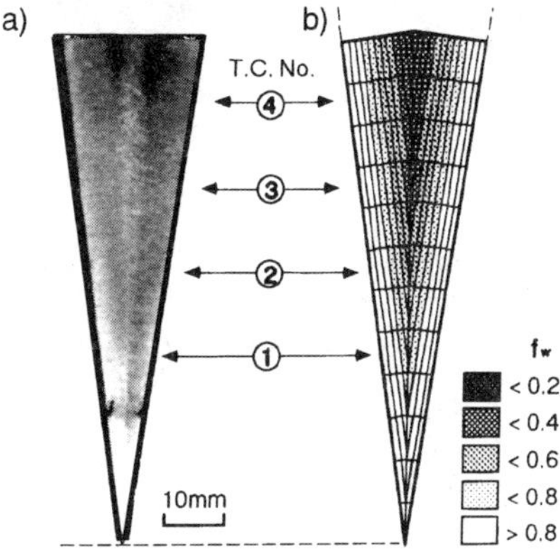

Fig.4 a) Appearance of the vertical section of a wedge-shaped FCD iron casting and b) the predicted distribution of chill fractions.

Table 1 Thermophysical properties of FC, FCD irons and steel mold

	FC Iron , FCD	Steel mold
density (kg/m^3)	6900	7500
specific heat (J/(kg·K))	754	670
thermal conductivity (W/(m·K))	8.38	50.3
stable eutectic temperature (K)	1426	-
metastable eutectic temperature (K)	1420	-
latent heat(kJ/kg)	210	-
partition coefficient of silicon k_{si}	1.2 - 1.4 (stable) 0.8 (metastable)	-

Table 2 Cell size (D_C), cell count (C_A,C_V) and chill fraction (f_w) observed near the thermocouples

T.C. No	D_C (10^{-6}m)	C_A (10^6/m^2)	C_V obs. (10^9/m^3)	C_V cal. (10^9/m^3)	f_w
1	200	27.8	106	101	0.68
2	370	14.6	40	49	0.10
3	480	8.5	18	21	0.04

Table 3 Nodule size (D_g), nodule count (N_A,N_V) and chill fraction (f_w) observed near the thermocouples

T.C. No	D_g (10^{-6}m)	N_A (10^8/m^2)	N_V obs. (10^{12}/m^3)	N_V (VSF) (10^{12}/m^3)	N_V (JM)	f_w
1	12.0	12.3	101	94	97	0.65
2	13.3	9.9	73	73	79	0.42
3	16.0	8.1	54	57	61	0.31
4	21.7	6.5	39	45	49	0.24

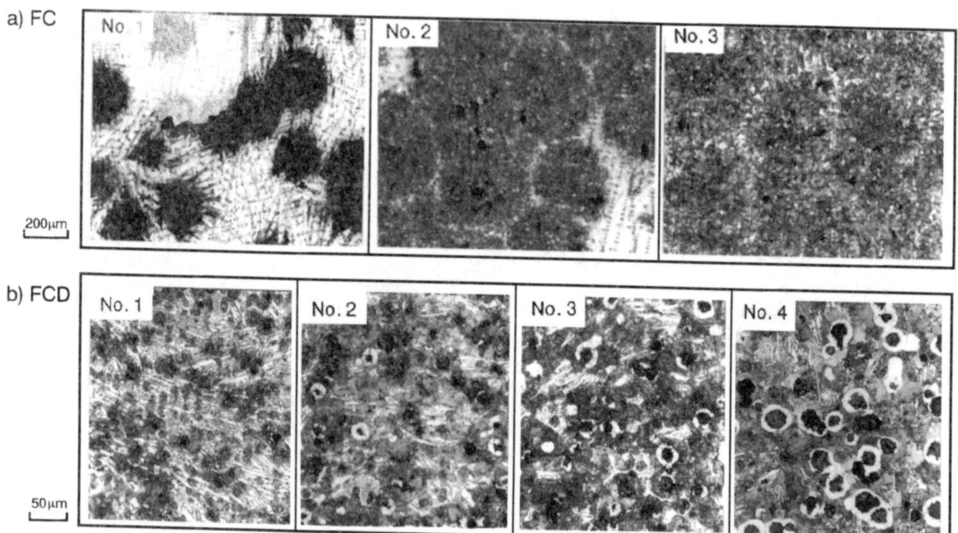

Fig.5 Microstructures near the thermocouples for FC iron (a) and FCD iron (b).

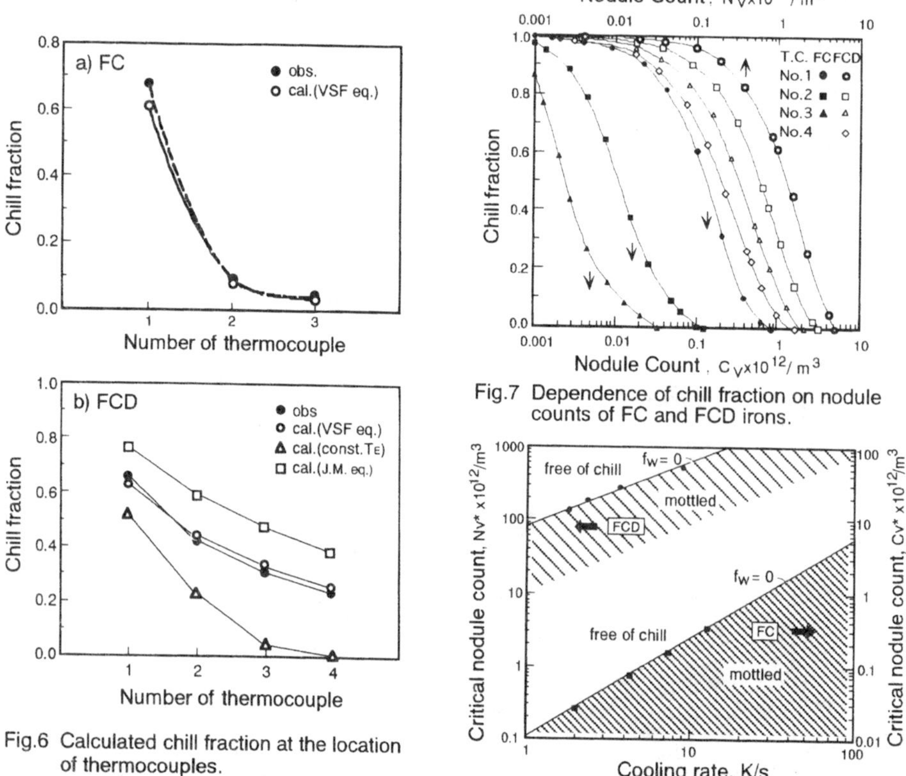

Fig.6 Calculated chill fraction at the location of thermocouples.
(a) FC iron and b) FCD iron)

Fig.7 Dependence of chill fraction on nodule counts of FC and FCD irons.

Fig.8 Critical nodule count as a function of cooling rate for FC and FCD irons.

Advanced Materials Research Vols. 4-5 (1997) pp. 535-542
© *1997 Scitec Publications, Switzerland*

Integration of Casting Process and Microstructure Modelling in an Industrial Nodular Iron Casting by Computer Simulation

I.L. Svensson[1,2] and M. Wessén[1,2]

[1] Former affiliation: Dept. of Materials Processing, Royal Institute of Technology,
S-100 44 Stockholm, Sweden

[2] Div. of Component Technology, Jönköping University,
Box 1026, S-551 11 Jönköping, Sweden

Keywords: Casting Simulation, Process Simulation, Nodular Cast Iron, Mechanical Properties, Microstructure Modelling

ABSTRACT

The paper aims to demonstrate the capability of simulation today and to study the complex phenomena in the casting process which affect the microstructure formed at solidification and solid state transformations. Temperature measurements were performed in a wheel hub in an industrial production line. The results were used in the validation of the simulation. The simulation will include mould filling, different orientation of the casting in the mould, with and without feeder, influence of inoculation treatment and fading, microstructure development, and the related mechanical properties. The microstructure development will include primary precipitation of austenite, eutectic grey and white solidification, segregation of elements and change of equilibrium temperature for both the stable and metastable systems as well as ferrite and pearlite growth. The hardness of each phase depends on the chemical composition. Other properties as e.g. ultimate tensile strength, yield strength and ultimate deformation may also be calculated.

INTRODUCTION

Ductile iron is used more and more in most countries of the world due to its excellent mechanical properties and castability. These alloys can be given a wide range of properties by changing the heat treatment, alloy composition, inoculation treatment or cooling conditions. To optimize the properties by casting experiments in a production line is very time consuming and costly. However, with the use of computer simulation programs the optimizartion will be possible in a shorter time and thus saving a lot of money.

To model the properties, both the solidification and the solid state transfomations have to be considered. These phase transformations are to a large extent affected by the silicon content. The silicon addition has several impacts on the nodular cast iron material, one of them being the solution hardening effect [1] and another the increased austenite transformation rate during the ferritic reaction [2]. The ferrite formation depends on carbon diffusion from the austenite through the ferrite to the graphite nodule. Due to segregation of alloying elements during solidification, an inhomogeneous composition between the nodules is obtained. This influences the equilibrium temperatures and driving force for the solid state transformations. Silicon has an inverse segregation in Fe-C-Si alloys, which means that the concentration will be higher in the first precipitated austenite than in the last solidified areas. Carbon diffuses much faster, but calculations have shown that small deviations from equilibrium can be obtained in the austenite at lower temperatures [3]. The growth and the final content of ferrite will be dependent on nodule count, segregation profile and time for precipitation before the formation of pearlite

MODELLING OF MOULD FILLING AND HEAT FLOW

In order to simulate the filling sequence, the temperature distribution in the metal and the mould after mould filling and subsequent cooling of the metal and the mould, conservation equations for continuity, momentum and energy in the form of Governing Differential Equations (GDEs) must be solved [4]. The heat flow is calculated by a Finite Difference Method in the casting simulation program MAGMAiron [5]. Appropriate relations describing initial and boundary conditions are added for both mould filling and heat flow calculations.

MODELLING OF STRUCTURE FORMATION AND HARDNESS

For the simulation of the structure formation in the casting, models for nucleation and growth of all relevant phases are implemented as a subroutine, which is linked to the macroscopic simulation of filling and heat flow.

Nucleation of graphite nodules

The number of graphite nuclei in the melt is calculated by Eq.1 [6], where the fading of the inoculant is taken into account by an exponential expression. The constants were evaluated from plate casting experiments, which had been inoculated in the stream [7].

$$N_v = k \cdot \exp\left(-t/t^*\right) \cdot \Delta T^n = 1.0 \cdot 10^{13} \cdot \exp\left(-t/40\right) \cdot \Delta T^{1.25} \qquad \left[m^{-3}\right] \qquad (1)$$

where t is the elapsed time from the moment when the grapite liquidus is passed and ΔT is the eutectic supercooling. Equation 1 is a version of Oldfield's original model [8], adding the fading of nuclei. Several relations have been presented by different authors, e.g. Lacaze [9].

Primary austenite and eutectic solidification

The primary austenite precipitation is calculated according to the phase diagram. This precipitation influences the undercooling and thereby also the number of nucleated graphite nodules in the melt.

To describe the eutectic solidification of nodular cast iron the model for the binary Fe-C system derived by Wetterfall et al. [10] has been modified to take into account the effect of silicon. The model is based on quasi stationary carbon diffusion through the austenite shell surrounding the graphite nodule. The growth rate, dR_g/dt, of the graphite nodule is calculated from:

$$\frac{dR_g}{dt} = \frac{C_n}{R_g} \cdot \left(T_E(\%Si) - T\right) \qquad [m/s] \qquad (2)$$

The value used for C_n is $2.87 \ 10^{-13}$ (m^2/s/K) [11]. The eutectic temperature, T_E, depends on the silicon content in the liquid, which is calculated by the Sheil segregation model. The partitation

coefficient for silicon was taken from ref. 12; $k_{Si}^{\gamma/l} = 1.09$.

Austenite decomposition into ferrite

To calculate the decomposition of austenite into ferrite and graphite, the growth have been divided into three different stages [13,14].

First stage: At low supercoolings below the upper stable eutectoid temperature, growth proceedes by carbon diffusion into the austenite. The ferrite grows in the plane of the graphite surface until the graphite is totally surrounded by a thin ferrite shell (~6 μm).

Second stage: The rate controlling mechanism during the second stage is assumed to be related to an interface reaction at the graphite surface, corresponding to the incorporation of carbon atoms. It was found that this mechanism controlled ferrite growth up to a transformed austenite fraction of ~0.8.

Third stage: During ferrite growth, the diffusion distances for carbon through the ferrite shell increases continuously. As a consequence, the diffusion rate of carbon will control ferrite growth at later stages of the transformation. There is no sharp transition between the second and third stage. Instead, the increased diffusion distances will continuously cause the transformation to become more diffusion controlled.

Calculation of hardness (HB) from the volume fraction of phases

In previous sections, models for simulation of the microstructure development have been given. The obtained structure, i.e. the volume fraction of each phase, can then be used to estimate the mechanical properties of the casting. For cast iron it is known that the static mechanical properties are affected mainly by the volume fraction of ferrite and pearlite and to a small extent by the number and size of nodules [15]. The hardness-relations used in this work are based on the simple models given in ref. 16, which gives the upper and lower limits of the properties.

From experiments [1], the following equations have been derived, describing the relation between the fraction ferrite, f^α, pearlite fraction, f^p, and Brinell hardness, HB.

$$HB(f^\alpha, f^p, \%Si) = HB^\alpha(\%Si) \cdot f^\alpha + HB^p(\%Si) \cdot f^p \tag{3}$$

$$HB^\alpha(\%Si) = 60 + 34 \cdot \%Si \tag{4}$$

$$HB^p(\%Si) = 167 + 31 \cdot \%Si \tag{5}$$

These equations are valid for a silicon content in the interval 1.7 – 4.9 % Si and 0.25% Mn.

RESULTS OF SIMULATIONS AND COMPARISON WITH EXPERIMENTS

Casting conditions, metallurgical treatment and chemical composition

A rear wheel hub for trucks has been cast (Arvika Gjuteri AB) with high silicon content, which gives a fully ferritic structure. The experiment was carried out in the production line and one mould was prepared with thermocouples. Six hubs are moulded in the same flask and different arrangements were made to investigate the need of feeders and the influence of the existance of a feeder on the microstructure formation. Due to the difficulties to make temperature measurements in a foundry production line, thermocouples were only placed in one wheel hub. The temperature in the pressure pouring furnace was 1400 °C, and the temperature was assumed to be somewhat lower in the pouring basin.

Casting	% C	% Si	%Mn	% Cu	% Mg	% Cr	% S
1	3.26	4.00	0.21	0.029	0.030	0.03	0.022

Table 1   Chemical composition of the wheel hub alloy.

The melt was inoculated in the metal stream entering the mould and an inoculant body was inserted at the bottom of the downsprue. A green sand mould was used and the core was made of cold-box bonded quartz sand. The green sand contained 7.5 % bentonite, 3.5 % carbon, 3.2 % water and a typical sand density was 1500 kg/m^3. In figures 1a and 1b, the position of the thermocouples are shown.

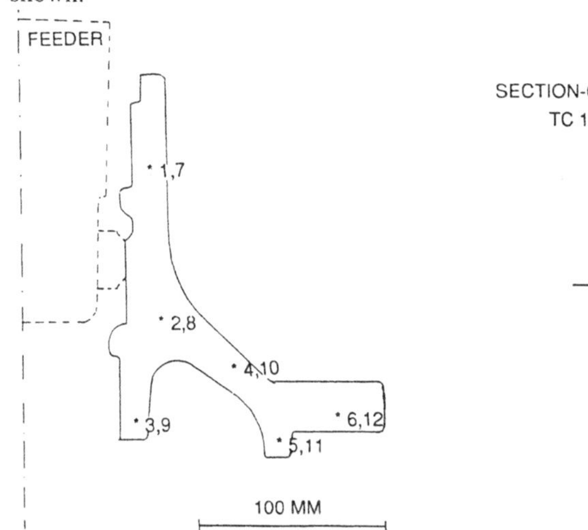

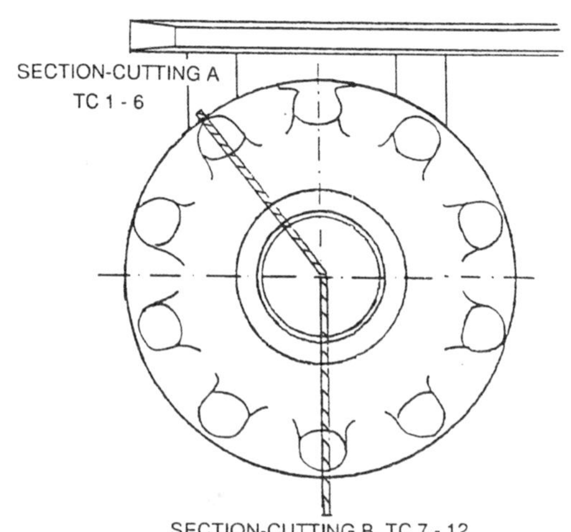

Figure 1a. Cross section of hub with feeder. The numbers show location points used in the comparison with simulation.

Figure 1b. Section A and B where the solidification times are compared.

Cooling curves of solidification and solid state transformation

The thermocouple type was Pt-10%Rh and the thermocouples were inserted in the mould cavity through holes which were drilled in the cope. The thermophysical properties of the mould and the metal used in the simulation were values provided by MAGMA and were not specially adapted to the experimental condition. Figures 2a and 2b shows the cooling curves measured in point 4 and 12 compared with simulation. When comparing the cooling curves from the simulation with the experiments the agreement is seen to be satisfactory. In spite of the complex heat flow in this 3D geometry, the temperature in the casting can be predicted within about °60 C after a time of almost 2 hours in the mould. The solidification times are fairly well calculated. It is noted that the 3D heat flow has caused the heat evolution from the eutectoid transformation to be hardly discernible in the cooling curve from point 4 while point 12 exhibits normal heat evolution. This effect is fairly well discribed in the simulation.

Simulated solidification time as influenced by casting condition

Four different cases have been simulated: upright, with and without feeder, and upside down, with and without feeder. The pouring temperature was 1360 C and the pouring time was 20 s for the casting with feeder and 17 s without feeder.

In figures 3a and 3b the influence of a feeder on the simulated solidification times have been illustrated for the two ways of positioning the casting in the mould. The solidification time in the upright position is not changed to any large extent by the feeder. However, the presence of a feeder will increase the transformation time for the solid state transformations. In this case, it did not make any difference in the ferrite/pearlite ratio due to the fully ferrritic microstructure in all cases. For ferritic/perlitic nodular cast irons this effect will be very important. The reason is that the growth time for ferrite will become larger and the resulting material close to the feeder will become softer and present a lower strength. In the upside down position, the effect of a feeder on the solidification time is more pronounced as seen in figure 3b.

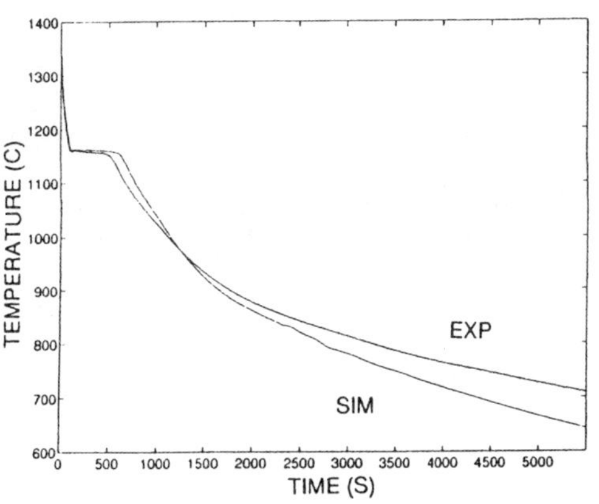

Figure 2a. Measured and simulated cooling curve from point 4. Geometry with feeder.

Figure 2b. Measured and simulated cooling curve from point 12. Geometry with feeder.

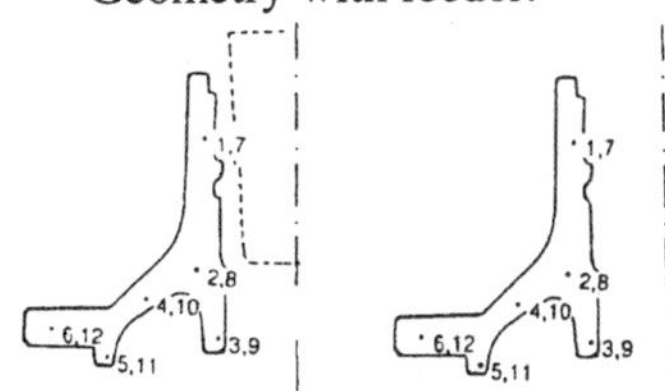

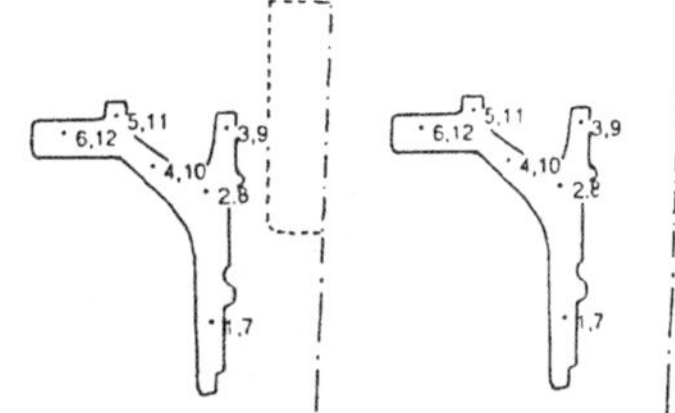

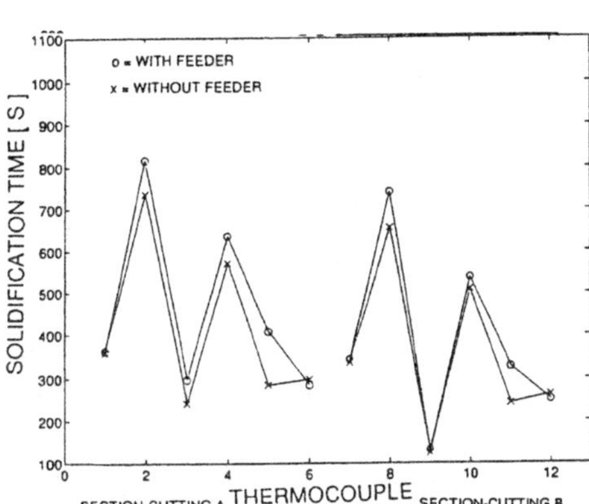

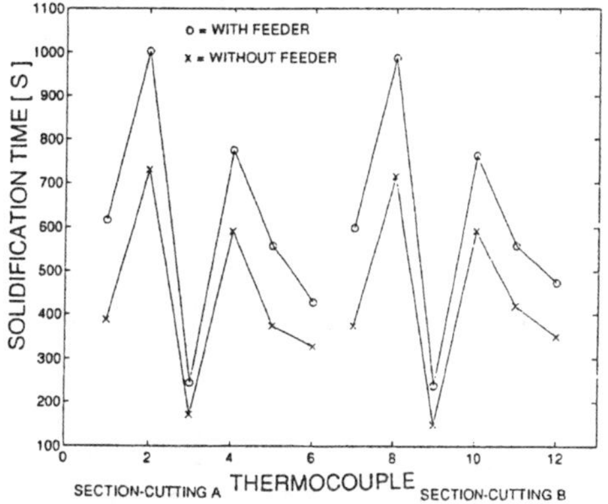

Figure 3a. Simulated solidfication time, Upright position.

Figure 3b. Simulated solidfication time, Upside down position.

Microstructure analysis and nodule count

The microstructure of the hub was fully ferritic which was to expected due to the high silicon content. The nodule counts were measured in 6 positions in each section cutting (A and B) and in figure 4a and 4b the results are shown from section cutting B in a hub with and without feeder respectively. In the same figures the nodule counts obtained in the simulation are given within brackets.

It was found that the presence of a feeder had a very small effect on the nodule count except for position no. 1 (152 and 248 mm^{-2}). In order to understand this effect it is necessary to consider what happens during the mould filling. Until the moment when the liquid reaches the feederneck there will be no difference between the two cases. However, when the liquid starts to flow into the feeder, the liquid velocity above the feederneck will decease significantly. As a result of the low velocity of the rising metal, the mould will at this point be heated more than in the case of no feeder, and naturally the cooling rate becomes reduced. It is well known that this will cause a decrease in the nodule count. In the most upper point it is seen that the nodule counts are very alike (340 and 344 mm^{-2}). This position was not reached by the liquid until the feeder was filled and the cooling rate will not be affected by the presence of a feeder. A possible way to avoid the decreased nodule count in this position when introducing a feeder would have been to decrase the size of the feederneck. Under those conditions, the velocity of the rising liquid in the casting would have been more alike in the two cases and the feeder should be the last part of the casting to fill.

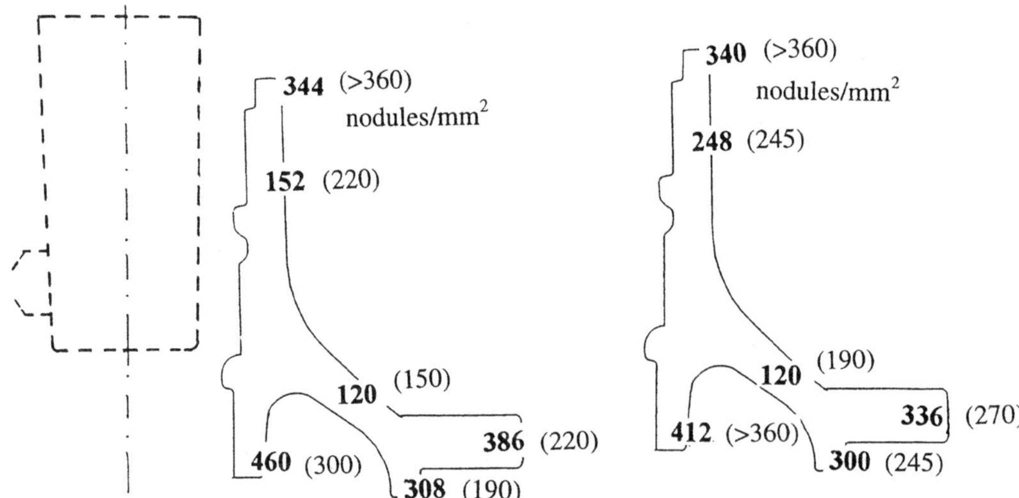

Figure 4a. Section B of hub with feeder showing simulated and measured nodule counts (mm^{-2}). Simulated values given within brackets.

Figure 4b Section B of hub without feeder showing simulated and measured nodule counts (mm^{-2}). Simulated values given within brackets.

Simulation of hardness and mechanical properties

The structure was fully ferritic in all experimental castings. The measured hardness at different positions was 200 ± 5 (HB), the ultimate tensile strength was 550 N/mm^2, the yeild strength was 450 N/mm^2 and the elongation at failure was 12 %. The simulation resulted in a fully ferritic structure which, with the silicon content of 4 %, gave a hardness of 197 (HB).

DISCUSSION

When simulating microstructure formation in a complex 3D geometry, a necessary pre-condition is that the macroscopic heat flow calculation results in a realistic temperature evolution in the casting. The validation of simulation with temperature measurements in industrial castings have not been reported very much in the literature. Nevertheless, this comparison must be done when extending the simulation from a laboratory-scale casting to industrial casting conditions which actually is what simulation will be used for in the foundry industry.

It was found that the temperature in the casting could be fairly well predicted in the simulation and that the soldification times were in good agreement with the experiments. However, the nodule count distribution obtained in the simulation diverged considerably from the experiments. The reason is probably that the nucleation law used in this work has been evaluated from plate casting experiments which only had been inoculated in the pouring stream, whereas the present hub casting was inoculated both in the pouring stream and using an inoculant body at the bottom of the ingate sprue. The action of an inoculant body cannot at present be included in the nucleation law, and further work is needed within this field.

CONCLUSIONS

Today it is possible to simulate mould filling and heat transfer in a complex casting with coupling to the microstructure formation. Mechanical properties in nodular cast iron can then be calculated from the predicted microstructure. It is necessary to take into account the cooling of the melt and the heating of the mould during mould filling since this influences the cooling rate at solidification and thereby also the nucleation of graphite nodules in the melt. It was observed that the orientation of the hub in the mould generally had a larger influence on the solidification time than had the presence of an internal feeder. This is an effect of a change in mould filling sequence when the hub is turned upside down.

The nucleation law used in this work was not sufficient to predict the variation in nodule count in the casting with an acceptable accuracy. The reason is probably that an inoculant body was used together with stream inoculation and that the two groups of inoculants have different characteristics, as e.g. fading time. Further work is needed to improve the nucleation laws so that different inoculation methods can be taken into account in a reasonable way.

Even if there is some discrepancy between measured and simulated cooling curves and micro-structures, the results are good enough for many industrial applications, and alternative ways to make experiments are in most cases not possible.

ACKNOWLEDGEMENTS

The authors wish to thank Volvo Technical Development and Arvika Gjuteri AB for their kindness to get access to samples and drawings, and to MAGMA Giesseritechnologie GmbH for proving the simulation code.

This work has been carried out within the framework of the COST 504 and SEGLÄTT programmes, which has been supported by Nutek and the Nordic Industrial Fund, which are gratefully acknowledged.

REFERENCES

1   I.L. Svensson, 57th WFC, paper 2, Osaka, Japan (1990).
2   M. Wessén, Modelling of Casting, Welding and Advanced Solidification Processes VII, TMS, p.671-78 (1995).
3   E. Lundbäck, Lic. Thesis, Royal Inst. of Technology, Stockholm, Sweden (1989).
4   M. Lipinski, W. Schaefer, S. Andersen, Proc. Modelling of Casting, Welding and Advanced Solidification Processes V, TMS, p. 771-76 (1991).
5   MAGMASOFT User Manual, Release 2.1, MAGMA Giessereitechnologie GmbH, Alsdorf, Germany (1991).
6   E. Lundbäck, Thesis, Royal Inst. of Technology, Stockholm, Sweden (1991).
7   M. Wessén, Unpublished work, Royal Inst. of Technology, Stockholm, Sweden (1994).
8   W. Oldfield, Trans ASM, 59, p 945 (1966).
9   J. Lacaze, M. Castro, G. Lesoult, Proc. Euromat'89, DGM, vol. 1, p. 147-52 (1989).
10 S.E. Wetterfall, H. Fredriksson, M. Hillert, J. Iron and Steel Inst., vol. 210, p. 323-33 (1972).
11 F. Mampaey, Proc. Modelling of Casting, Welding and Advanced Solidification Processes V, TMS, p. 403-10 (1991).
12 R. Boeri, F. Weinberg, Cast Metals, vol. 6, p. 153-57 (1993).
13 M. Wessén, I. L. Svensson, This volume.
14 M. Wessén, I. L. Svensson, Accepted for publication in Met. Trans. A (1995).
15 I. Ohnaka, MRS-symposium, State of the art of computer simulation of casting and solidification processes, Strasbourg, France (1986).
16 G. Grimvall, Thermophysical properties of materials, North-Holland, Amsterdam (1986).

Advanced Materials Research Vols. 4-5 (1997) pp. 543-550
© 1997 Scitec Publications, Switzerland

A New Shrinkage Cavity Prediction Method for Solidification Simulation of S.G. Iron Casting

B. Liu, J. Li and R. Liu

Tsinghua University, Beijing 100084, China

Keywords: Shrinkage Cavity Prediction, S.G. Iron Casting, Numerical Simulation, Solidification Simulation

ABSTRACT

A series of experiments with inverse T-type sample castings were carried out to study comprehensively the effect of metallurgical and technological parameters including carbon equivalent, modulus, inoculation intensity as well as mold rigidity on solidification behavior of S.G. iron.

A new three dimensional quantitative method so called dynamic expansion contraction accumulation method (DECAM) was established based on experimental results and mathematical modeling. The three dimensional temperature fields are computed by using finite difference method (FDM) with rational selection of thermal properties and boundary conditions. The volume variation of each single element of S.G. casting during each time step of the solidification process is calculated and accumulated considering the following stages of solidification process, namely: 1. liquidus contraction, 2. contraction or expansion due to the precipitation of primary austenite or graphite of S.G. iron, 3. contraction or expansion due to the eutectic transformation, 4. volume changes due to the wall movement of mold depending on the mold rigidity. Finally, the location, size and volume of the shrinkage cavity is computed quantitatively and colorfully displayed by using the advanced post-processing module.

All the simulation results match the experimental results of sample castings quite well, and the new method has been put into application in a number of foundries with some very promising results.

INTRODUCTION

The simulation of solidification process of shaped castings has been widely used in foundry industry to assure the soundness of castings, to optimize the casting design, to improve the casting yield and hence to save the manufacturing cost of castings. One of the most important purposes of solidification simulation is to predict and then to avoid shrinkage cavity and porosity if they will occur in shaped castings. Many works have been done for the purpose of predicting shrinkage cavity and porosity of castings and many prediction criteria or methods have been proposed as well[1-3]. However, for practical application most of them are used qualitatively instead of quantitatively and especially for steel casting.

Since S.G. iron castings become more and more important especially for automobile and metallurgy industries, study of the shrinkage behavior of S.G. iron and prediction of shrinkage cavity of S.G. iron casting are of important significance and become one of the hottest topics for solidification simulation[4-8].

Therefore, the object of the present work is not only to approach the influences of metallurgical and technological factors comprehensively on the shrinkage behavior and the formation of the shrinkage cavity and porosity of S.G. iron castings but also to develop a new prediction method especially for shrinkage cavity prediction of S.G. iron casting.

EXPERIMENTAL STUDY OF SHRINKAGE BEHAVIOR OF S.G. IRON CASTING

Since 1950s, many foundrymen have been studying the shrinkage behavior of S.G. iron castings. Since the mid-1980s, foundrymen have been concentrating the study of the formation mechanism of the shrinkage cavity and porosity of S.G. iron castings. It is concluded by previous studies that shrinkage behavior and formation of the shrinkage cavity and porosity during freezing of S.G. iron castings can be affected by many metallurgical and technological factors. However, only a few investigators have comprehensively studied the effects of these factors on the shrinkage behavior and the formation of the shrinkage cavity and porosity of S.G. iron castings.

Procedure

The shrinkage behavior and the formation of the shrinkage cavity and porosity in S.G. iron castings are closely associated with carbon equivalent, inoculation, casting modulus and mold rigidity of S.G. iron castings. Experiments designed based on the orthogonality principle with four factors, namely: carbon equivalent, modulus, inoculation intesity and mold rigidity and three levels were carried out in order to study comprehensively the effects of the above mentioned factors on the shrinkage behavior and the formation of the shrinkage cavity and porosity in S.G. iron castings with both dry and green sand mold[9].

Inverse T-shaped castings with different carbon equivalents such as CE= 3.92, 4.65 and 4.95%, and different modulus such as M=1.22, 1.94 and 2.71cm were designed for experiments. Besides, all the other composition is controlled as follows: 0.15-0.35% Mn, <0.05% P, <0.03% S, 0.005-0.015% RE, and 0.040-0.050% Mg. Both dry sand mold and green sand molds with different hardness from 45-50, 60-65 and 75-80 were prepared for experiments.

The pouring temperatures as well as the temperatures of the different positions of the castings were measured by using thermal couples and recorded by using a computerized data logger.

Then, the density and the volumes of shrinkage cavities and porosities of all the sample castings were also measured and calculated precisely based on buoyancy principle.

Results

Rubbing pictures of actual shrinkage cavity and macro-porosity of all the eighteen sample castings after dissection were made accordingly. All the results of the shrinkage cavity ratio of those sample castings are listed in Table 1. In the Table, V_1/V represents the shrinkage tendency of the primary shrinkage cavity, V_2/V the shrinkage tendency of the secondary shrinkage cavity, while shrinkage cavity ratio is equal to the total shrinkage cavity tendency, i.e., $(V_1 + V_2)/V$.

According to the range analysis of the experimental results, the effect of some main factors such as carbon equivalent and modulus on the shrinkage cavity ratio are shown in Fig. 1 to 2 respectively.

Table 1 Shrinkage Cavity Ratio of S.G. Iron Sample Castings

Sample	C.E. (%)	Modulus (cm)	Mold	Specific density (g/cm^3)	Total shrinkage cavity ratio (%)	V_1/V (%)	V_2/V (%)
213	3.92	1.22	green	7.0076	1.400	0	0.686
212	3.92	1.94	green	7.1659	1.832	0.087	1.072
211	3.92	2.71	green	7.1388	3.173	2.893	0.246
223	3.92	1.22	dry	7.1823	0.306	0	0.119
222	3.92	1.94	dry	7.1714	1.178	0.144	0.616
221	3.92	2.71	dry	7.1315	1.586	1.588	0
313	4.65	1.22	green	7.0308	1.137	0.447	0.690
312	4.65	1.94	green	7.0083	2.543	2.360	0
311	4.65	2.71	green	7.0131	1.699	1.652	0
323	4.65	1.22	dry	7.0298	0.074	0	0.074
322	4.65	1.94	dry	7.0130	0.137	0.137	0
321	4.65	2.71	dry	7.0374	0	0	0
513	4.94	1.22	green	6.9951	1.108	0.759	0.044
512	4.94	1.94	green	6.9933	1.908	1.908	0
511	4.94	2.71	green	6.9927	1.909	1.909	0
523	4.94	1.22	dry	7.0250	0.088	0	0
522	4.94	1.94	dry	6.9860	0.404	0.404	0
521	4.94	2.71	dry	6.9775	0.399	0.399	0

Besides, under the dry sand mold condition, the order of the effects of those factors is: carbon equivalent, modulus and inoculation, while for green sand mold the order is: modulus mold rigidity, carbon equivalent and inoculation.

Finally, two regression equations considering the comprehensive effect of those metallurgical and technological factors on shrinkage cavity ratio within the pouring temperature from 1280 to 1320 °C has been derived.

for green sand mold:

$$Y (\%) = 0.00641Tp - 0.556CE_1 + 10.645I - 4.659I^2 + 2.712M - 0.5156 M^2 - 0.163H + 0.0012H^2 - 7.857 \qquad [1]$$

for dry sand mold:

$$Y (\%) = 0.00037Tp - 16.069CE_1 + 1.904CE_1^2 + 7.770I - 3.620I^2 + 1.891M - 0.386 M^2 - 27.406 \qquad [2]$$

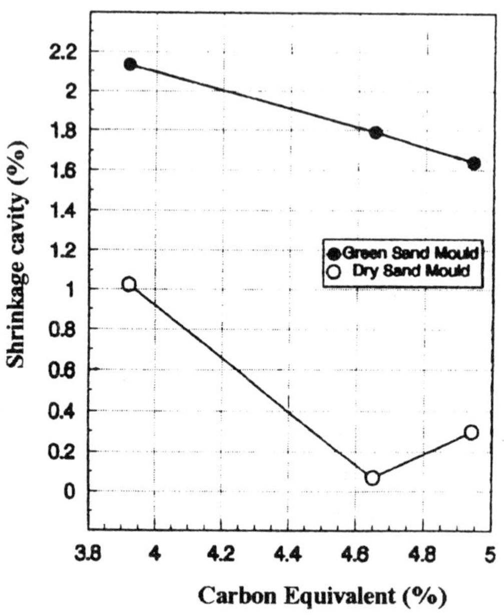

Fig. 1 Effect of CE on Shrinkage Cavity Tendency by Range Analysis

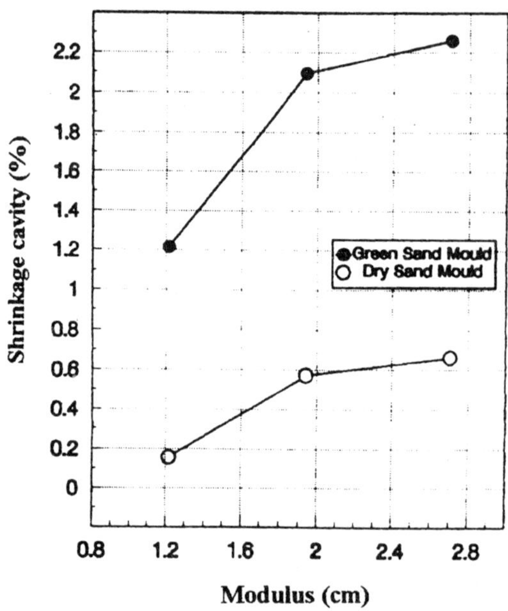

Fig. 2 Effect of Modulus on Shrinkage Cavity Tendency by Range Analysis

where: $CE_1 = C + 1/7\,Si\ (\%)$

SHRINKAGE CAVITY PREDICTION OF S.G. IRON CASTING

A new prediction method so-called Dynamic Expansion and Contraction Accumulation Method (DECAM) for quantitative shrinkage cavity prediction of S.G. iron casting was then studied, developed and put into practical application. The three dimensional temperature fields are computed by using finite difference method with rational selection of thermal properties and boundary conditions.

Assumptions for Physical and Mathematical Model

The new method to predict the shrinkage cavity of S.G. iron casting was established under the following assumptions:

1) Liquid S.G. iron is solidified under good metallurgical condition, i.e., good nodularity and inoculation with no free carbide (<3.0%) in matrix.
2) Liquid iron can flow if the solid fraction ratio is less than critical solid fraction ratio f_{sc}, while liquid iron can expand if the solid fraction ratio is less than critical expansion solid fraction ratio f_{ec}.
3) During solidification process mold wall will move due the expansion of graphitization, and this phenomena is taken into consideration for the new model.
4) The volume of shrinkage cavity of S.G. iron casting will be the sum accumulating liquidus contraction, contraction or expansion of primary phase, contraction or expansion of eutectic graphite and austenite, and the volume change due to the mold wall movement.

Mathematical Model

Based on the above experiments as well as some assumptions, a new method for predicting the shrinkage cavity of S.G. iron casting so-called Dynamic Expansion Contraction Accumulation Method (DECAM) was proposed in the paper.

The total volume variation between time step Δt can be calculated according to the following equations:

$$\Delta V = \Sigma \Delta V_{iSL} + \Sigma \Delta V_{iGP} + \Sigma \Delta V_{iGl} + \Sigma \Delta V_{iAl} + \Delta V_{nE} \qquad [3]$$

where, ΔV_{iSL} - liquidus contraction of element i,

ΔV_{iGP} - expansion of primary graphite of element i,

ΔV_{iGl} - expansion of graphite during eutectic solidification of element i,

ΔV_{iAl} - contraction of eutectic austenite of element i,

ΔV_{nE} - mold wall movement.

Among them,

$$\Delta V_{iSL} = \alpha_{sl}\,(\,T_i^{\,t} - T_i^{\,t+\Delta t}\,)\,V_i \qquad [4]$$

where: α_{sl} - liquid volume contraction coefficient,
       V_i - volume of element i.

$$\Delta V_{iGP} = \alpha_{GP}\{(\,C_X - C_E\,)\,/\,(\,100 - C_E\,)\}\Delta f_{iGP}V_i \qquad [5]$$
where: α_{GP} - volume expansion coefficient of primary graphite,
       Δf_{iGP} - increment of primary graphite of element i.

$$\Delta V_{iGl} = \alpha_{Gl}\,\{(\,C_E - C_\gamma\,)\,/\,(\,100 - C_\gamma\,) \times (\,100 - C_X\,)\,/\,(\,100 - C_E\,)\}$$
$$\times\{1 - \Sigma\,(\,V_{iSL}\,/\,V_i\,)\}\,\Delta f_{iGl}\,V_i \qquad [6]$$
where: Δf_{iGl} - increment of eutectic graphite of element i.

$$\Delta V_{iAl} = \alpha_{Al}\,\{(\,100 - C_E\,)\,/\,(\,100 - C_\gamma\,) \times (\,100 - C_X\,)\,/\,(100 - C_E\,)\}$$
$$\times\,\{1 - \Sigma\,(\,V_{iSL}\,/\,V\,)\}\,\Delta f_{iAl}\,V_i \qquad [7]$$
where: Δf_{iAl} - increment of eutectis austenite of element i.

$$\Delta V_{nE} = V_{nE}{}^{t+\Delta t} - V_{nE}{}^{t} \qquad [8]$$

Then, a software program was developed to realize the computation of the new prediction method and added to a commercial solidification simulation system.

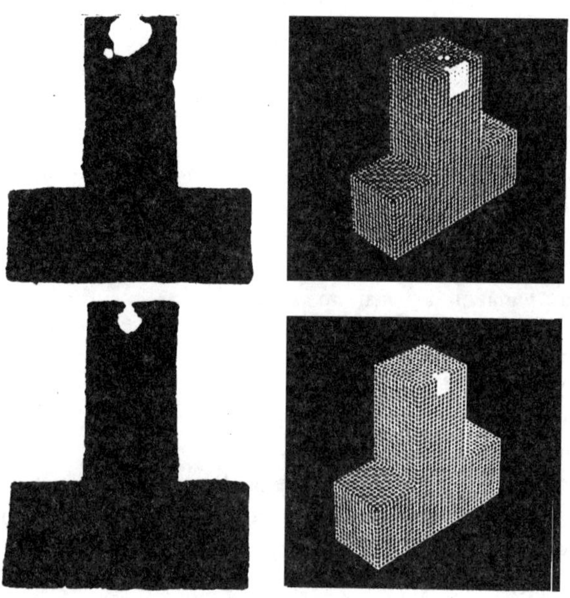

**Fig. 3 Experimental and Simulated Results of Prediction of Shrinkage Cavity of
S.G. Iron Sample Castings   a) sample No. 211   b) sample No. 221**

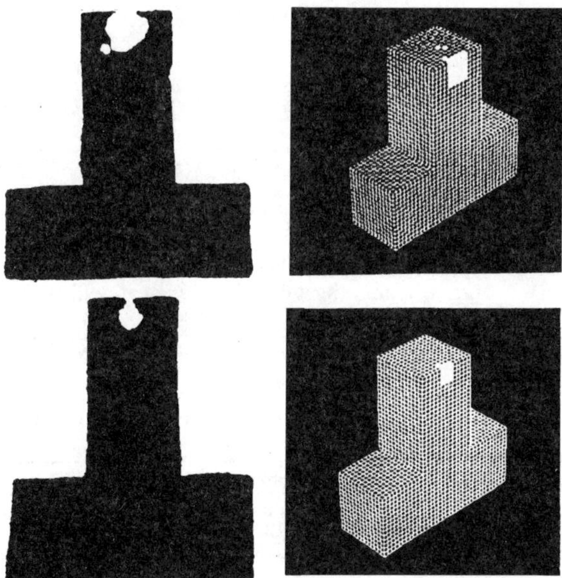

**Fig. 4 Experimental and Simulated Results of Prediction of Shrinkage Cavity of
S.G. Iron Sample Castings   a) sample No. 511   b) sample No. 521**

Verification of New Prediction Method

Shrinkage cavities of all sample castings were calculated by using the new method and then compared with that of experimental results. In addition, the visualized display of parts of the results in comparison with the real sectioned sample castings is shown in Fig. 3 and 4. It is indicated from those Figures that the quantitative predication of the volume and the location of shrinkage cavity match those of the actual sectioned sample castings quite well.

Practical Application

The new shrinkage cavity prediction method for S.G. iron casting was added to a three dimensional solidification simulation system for shaped casting so-called FTSolver System[10]. This new methods together with the solidification simulation system has been put into application in a number of steel and S.G. iron castings foundry plants with some very good results as listed in Table 2.

CONCLUSIONS

1. The effects of metallurgical and technological factors on the formation of shrinkage cavity of S.G. iron casting have been studied comprehensively for the purposes of further understanding

the formation of shrinkage cavity and porosity of S.G. iron and providing experimental basis to predict shrinkage defects of S.G. iron casting by solidification simulation.

Table 2 Case Study for Shrinkage Cavity Prediction of S.G. Iron Casting

No.	Plant	Casting	Feature	Purpose	benefit
1	turbine plant	upper piston	heavy section	new product	successful at once
2	turbine plant	turbine blade holding ring	heavy section	evaluate old plan	optimized
3	S.G. iron foundry	die plate	heavy section	optimization	shorten lead time
4	auto-works	small crankshaft	small size	evaluation	optimized
5	auto-works	hub	small size	evaluation	
6	auto-works	crankshaft	medium size	evaluation	optimized

2. A new Dynamic Expansion Contraction Accumulation Method (DECAM), considering not only the expansion of graphite precipitation or contraction of liquid and austenite but also the volume variation due to the mold wall movement, was developed and proposed to predict the shrinkage cavity of S.G. iron casting.

3. The results of non-destructive test and practical dissection to castings show that the prediction methods of shrinkage cavity for S.G. iron castings are in good agreement with the result of experiments and practical castings.

Acknowledgment: The paper is partly supported by the Chinese National Natural Science Foundation under the Key Project No. 59235102. `

REFERENCES

1. E. Niyama, Annual Review Mater. Sci., (1990), 20, 101-105.
2. M. kikuchi et al, Trans. of JFS, (1990) 9, 63-66.
3. Y. Nagasaka et al, AFS Trans., (1989), Vol. 97, 553-564.
4. S. I. Karsay, Ductile Iron III: Gating and Risering, QIT-Fer et Titane. Inc., (1981).
5. J. F. Wallace et al, AFS Trans., 92 (1984), 765.
6. K. C. Su, Itsuo Ohnaka et al, Imono (Japanese) 58 (1986), 702.
7. M. S. Chandrasekhara Rao et al, AFS Trans., 96 (1988), 551.
8. S. Takamori and E. Niyama, Imono (Japanese) 64 (1992), 338.
9. J. R. Li, B. C. Liu, C. R. Loper, Jr. et al, AFS Trans., (1994), 94-086.
10. Jing Tao, Mei Qibo, Liu Baicheng, 60th World Foundry Congress, The Netherland, (1993).

Advanced Materials Research Vols. 4-5 (1997) pp. 551-556
© *1997 Scitec Publications, Switzerland*

Residual Stress Computation and Analysis of Machine Tool Bed Casting

B. Liu[1], R. Zhu[1], S. Xiong[1], Y. Gao[2] and Y. Zhang[2]

[1] Tsinghua University, Beijing, China

[2] Beijing Machine Tool Research Institute, Beijing, China

Keywords: Machine Tool Bed Casting, Solidification Simulation, Residual Stress, Temperature

ABSTRACT

Residual stress computation is very important in manufacture of precision iron casting, because it can help foundry engineers to optimize the casting process, to reduce the residual stress and to improve dimensional stability of castings. In this paper, a FDM / FEM combined software system was developed for practical application in foundry industry.

In this system, temperature field was calculated by using FDM, while stress computations were carried out by using FEM according to the temperature load transferred from FD model.

The temperature and stress computation of a lathe bed iron casting were carried out both under primary and modified cooling conditions, and results of computation are in agreement with experimental results.

INTRODUCTION

Residual stress, which occurs during uneven cooling and solidification, can cause deformation of iron casting. Those iron castings used as beds or slideways of precision machine tools require strict dimensional stability, and therefore it is very important to find ways to improve its stability. Stress computation can be a powerful tool for foundry engineers to have visual comprehension about stress development in casting during solidification and cooling, and thus they can evaluate or improve the structure and cooling condition efficiently for the purpose of reducing residual stress.

For practical application, a lathe bed iron casting was cast and measured in foundry plant, and then residual stress was computed by using an integrated Finite Deference Method (FDM) / Finite Element Method (FEM) combined software system. The experimental and computing results are in agreement. Furthermore, stress computations were carried out under modified cooling condition, results shows that residual stress could be reduced and be forced to distributed more evenly.

ANALYSIS METHOD AND SOFTWARE SYSTEM

Both FDM and FEM can be used to compute the temperature field of casting. However, up to now, most of commercial software packages, such as SOLDIA, SOLSTAR, FTSolver etc., are programmed by using FDM. FDM has become the dominant method widely used in temperature field computation, fluid flow computation, shrinkage cavity prediction as well as microstructure prediction.[1]

On the other hand, FEM as a classical numerical method for stress computation has been developed unprecedentedly. It is predicted that FEM will still be the dominant method used to compute stress distribution in many engineering problems including foundry engineering.

In order to carry out both the temperature and stress computation of shaped casting economically, an integrated FDM / FEM software system was developed. In this system, the temperature field is computed by using FDM, while the stress field is computed by using FEM. Both FD model and FE model can be generated according to the same geometric model positioned in the mold, the two models were positioned and matched in Cartesian coordinate system , therefore, temperature load for stress computation can be transferred from FD model to FE model correspondingly and automatically. The flowchart of this system is shown in Fig. 1.

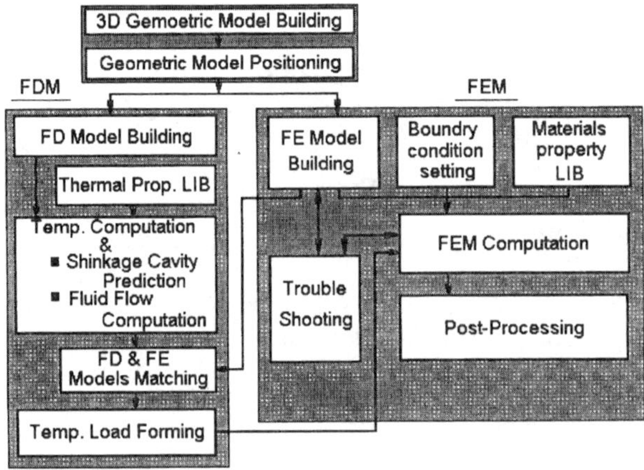

Fig. 1 Flowchart of the software system

COMPUTATION OF TEMPERATURE AND RESIDUAL STRESS

In stress computation, boundary condition was simplified and constitutive relationship was approximated. It is assumed that: 1) moulding sand has no resistance to casting during cooling because of good collapsibility of resin sand at high temperature, 2) material is homogeneous and continuous and obeys Thermo-elastic-plastic constitutive relationship and Von Mises yield function.

A MAZAK lathe bed iron casting was computed for the purpose of optimizing the cooling condition as well as to improve the dimensional stability, as shown in Fig. 2. Information of stress computation is shown in Tab. 1. Fig. 3 shows the accuracy of the temperature computation. The difference of cooling speeds at different location in the casting is shown in Fig. 4.

Tab. 1

Casting process	FD model	FE model	Computer
Furan resin sand mold	178,266 elements	1,512 nodes, 800 8-node elements	PC 486 DX / 66

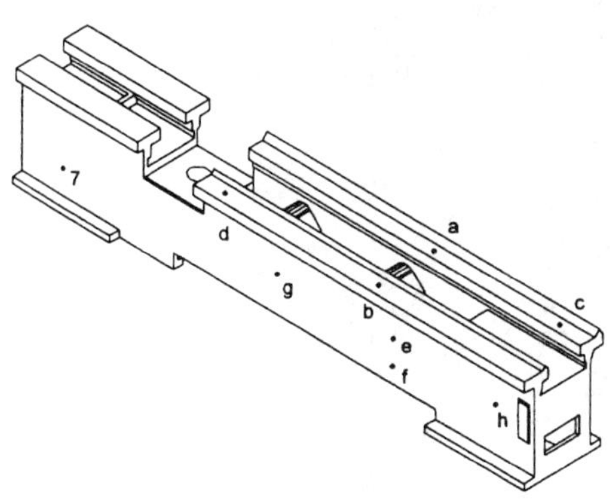

Fig. 2   Geometric model of the lathe
bed iron casting

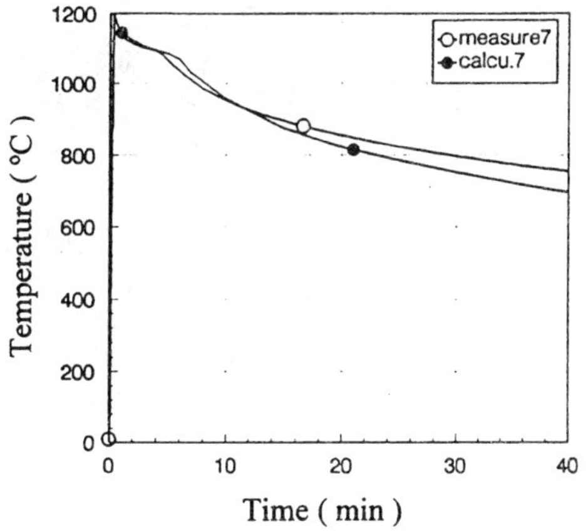

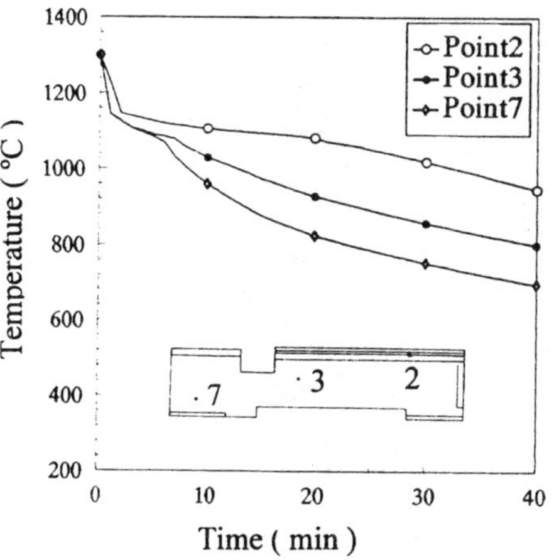

Fig. 4 Cooling curves of the casting under
primary cooling condition

Fig. 3 Accuracy of temperature computation

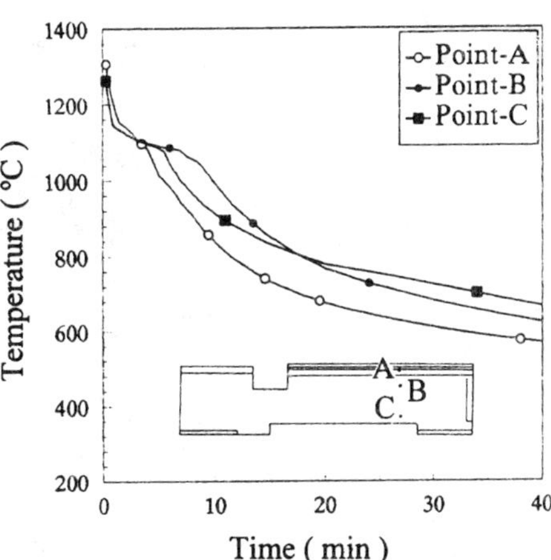

Fig. 5 Cooling curves of the casting under
modified cooling condition

Since the slideway is thick and the side wall is thin, the side wall will cool faster than the slideway. In the early stage, the lower part of the side wall shrinks ahead of the slideway, and therefore the side wall is in tension and the slideway is in compression, resulting in the occurrence of plastic yield in the casting. When the slideway begins to shrink faster later, the slideway is in tension, and the side wall is in compression. This state remains until the casting cools to ambient temperature to form residual stress, as shown in Fig. 6 and Fig. 7. Animation of computation shows that the tension area moves from lower part of the side wall to slideway with time going on.

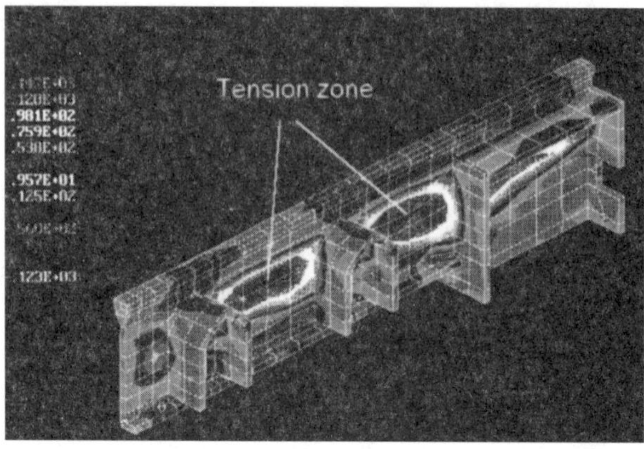

Fig. 6 Stress distribution ($\sigma_{y\text{-}y}$, alone the slideway, at 40 min, primary cooling condition)

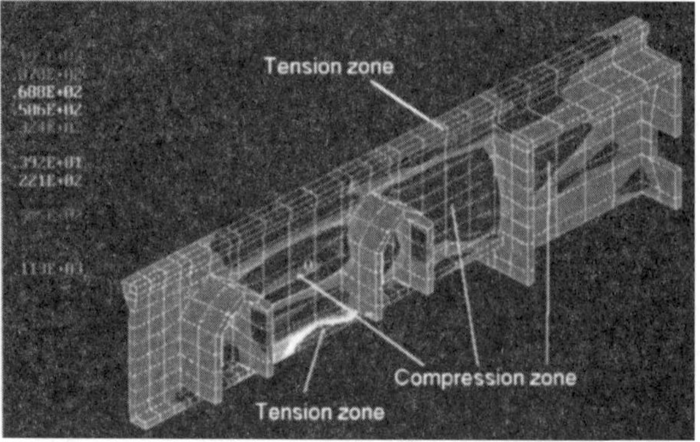

Fig. 7 Residual stress distribution ($\sigma_{y\text{-}y}$, alone the slideway, primary cooling condition)

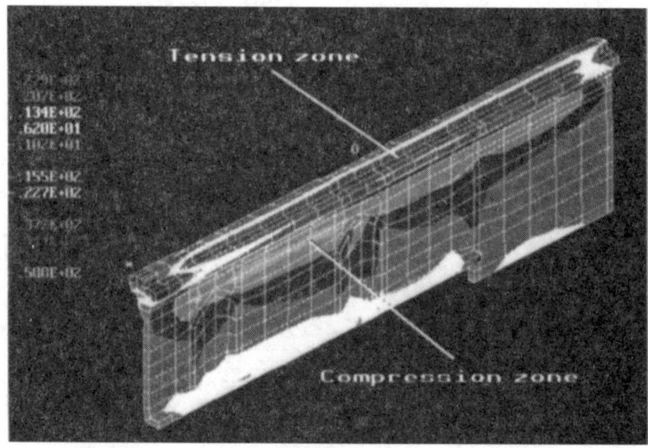

Fig. 8 Stress distribution ($\sigma_{y\text{-}y}$, alone the slideway, at 10 min, modified cooling condition)

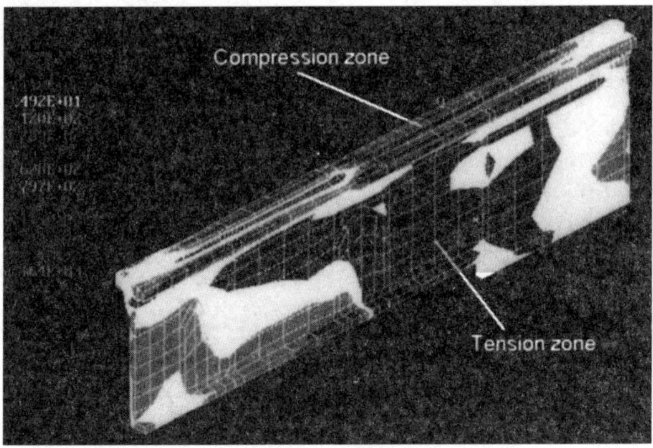

Fig. 9 Residual stress distribution ($\sigma_{y\text{-}y}$, alone the slideway, modified cooling condition)

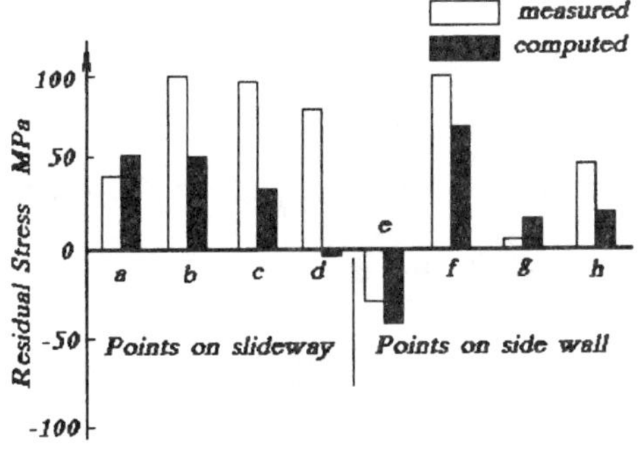

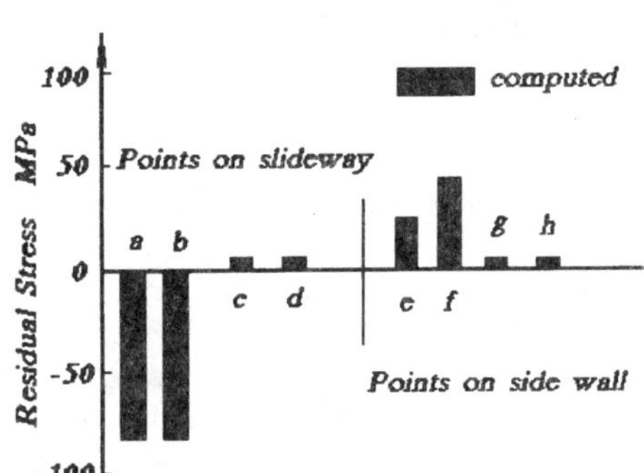

Fig. 10 Comparison of computed and measured   Fig. 11 Computed results under modified
      results, primary cooling condition   cooling condition

To measure the residual stress, 19 strain gauges were attached to the casting. Residual stress was measured by drilling holes and letting stress to be released. Comparison of y-y component (along slideway)computed by the software and residual stress measured at different locations on the casting is shown in Fig. 10.

To reduce residual stresses, the cooling condition should be modified. The cooling could become even by adding chill under the slideway, and then residual stresses would be lower. Under modified cooling condition , further stress computation of the casting were carried out.

In the temperature computation, the FD model of the casting, initial conditions and thermal parameters of sand and casting are the same as those in the former computation except a FD model of a chill. Fig. 5 shows the temperature curves under modified cooling conditions.

Stress computations are almost the same as in the former computation both in FE model and in boundary conditions except for the new temperature load computed under modified cooling conditions. results are shown in Fig. 8 , Fig. 9 and Fig. 11.

Here, Δ TD is assumed to stands for temperature differences between different location. When chill is used, large ΔTD occurs in small range of slideway, The chill can prevent large Δ TD to occur in a large range of the whole structure as we can see in the former computation, therefore, the residual stress is smaller and more even than in former computation under primary cooling conditions.

Cast iron has higher strength in compression than in tension, and the strain is smaller in compression than in tension under constant stress. If we attempted to create compression residual stresses in iron casting by modifying the cooling conditions and structure, the dimension stability would be improved. The results of the computation show this possibility to a certain extent.

Because of the complexity of the casting and the interaction between mould materials and casting during solidifying and cooling, it is rather difficult to set constraints. To compute stress development in complex shape casting accurately and efficiently, further research work must be carried out in boundary conditions and constitutive relationship. J. W. Wises and YANG Bingjian have done valuable work in this area.[2]

CONCLUSIONS

1. A combined FDM and FEM numerical simulation system for stress computation of casting was developed.
2. Boundary conditions and constitutive relationship should be simplified so as to reach a compromise of efficiency and accuracy in practical use.
3. Residual stress could be reduced and be forced to distribute evenly by modifying coolin conditions and dimension stability could be improved.

ACKNOWLEDGMENTS

Thanks to Wang Fenghua, Liang Dongmei, Liu Jianjun of Beijing Machine Tool Research Institute and Qian Qiuxian, Qiao Jindong of Jinan First Machine Tool Plant for their support of this work. This project is partly financed by NNSF under the key project No.59235102.

REFERENCE

1. LI Jiarong: Study on Formation and Prediction of Shrinkage Cavity and Porosity of S.G. iro Castings, Doctoral Thesis of Tsinghua University, China, 1994.
2. J.W.Wises and J.A.Dantzig.: Modeling Stress Development during the Solidification of Gray Iro Casting, Metallurgical Transactions A, Vol. 21A , 489-497(1990).

Advanced Materials Research Vols. 4-5 (1997) pp. 557-562
© 1997 Scitec Publications, Switzerland

Constitutive Equations to Simulate the Casting of S.G. Cast Iron Parts

R. Billardon[2], N. Hamata[2], D. Marquis[2] and D. Visconte[1]

[1] Renault, Direction de la Recherche, Sevice 0968,
860 quai Stalingrad, F-92109 Boulogne Billancourt, France

[2] Laboratoire de Mécanique et Technologie, E.N.S. de Cachan, C.N.R.S., Université Paris 6,
61, Avenue du Président Wilson, F-94235 Cachan Cedex, France

Keywords: Spheroidal Graphite Cast Iron, Constitutive Equations, Phase Transformation, Transformation Viscoplasticity

Abstract This paper is devoted to the modelling of the thermo−mechanical behaviour of nodular graphite ferritic cast iron from solidus to room temperature. The ultimate aim of this study is to implement these constitutive equations in a finite element code designed to simulate casting in order to optimize this process and in particular to predict the defects and residual stresses created in the casts. The constitutive equations and the procedure adopted for their identification are presented in details.

1. INTRODUCTION

During the last decade, computer simulation has become an integral part of foundry practice for the design of new castings at a better cost and in reduced time. The study presented in this paper constitutes part of a project the aim of which is to develop and validate an integrated computer software in order to simulate the casting process of spheroïdal graphite cast iron components. In this software the finite element technique is used to model the whole process including mould filling, solidification and cooling.

In the case of nodular graphite ferritic cast iron, the typical cooling rates observed during the casting process lie between 0.01 and 0.2 $°Cs^{-1}$. For such low cooling rates, the material can be considered as: first, at high temperature, a γ−phase (i.e. an austenitic matrix with graphite nodules); second, at low temperature, an α−phase (i.e. a ferritic matrix with graphite nodules); and third, during the transformation, an evolutive mixture of these two phases.

2. MODELLING OF THE BEHAVIOUR OF EACH PHASE

The thermo−mechanical behaviour of each phase is described by an anisothermal elasto−viscoplastic model [1] based on classical continuum thermodynamics concepts [2]. *As a first approximation*, it is assumed that *the evolutive carbon fraction* of the austenitic and ferritic matrixes is modelled through *the dependence of this model upon temperature*. The thermo−mechanical state of each phase is described by the following set of independent state variables:

$$\mathbb{V} \equiv (\varepsilon, T, \varepsilon^{p}, \mathbf{V}) \qquad \text{with} \qquad \mathbf{V} \equiv (p, \alpha) \qquad (1)$$

where ε, ε^p denote the total and plastic strain tensors respectively, T the temperature, $(\varepsilon^p, \mathbf{V})$ the set of the internal variables, p the accumulated plastic strain and α a tensorial variable.

The reversible behaviour is described by the Helmholtz specific free energy $\Psi = \Psi(\varepsilon, T, \varepsilon^p, \mathbf{V})$, so that the thermodynamic forces $A \equiv (\sigma, s, -\sigma, \mathbf{A})$ are defined from the following *state laws:*

$$\sigma = \rho \frac{\partial \Psi}{\partial \varepsilon} = -\rho \frac{\partial \Psi}{\partial \varepsilon^p} \, , \, s = -\rho \frac{\partial \Psi}{\partial T} \, , \, \mathbf{A} \equiv (R, X) = \rho \frac{\partial \Psi}{\partial \mathbf{V}} \tag{2}$$

where ρ denotes the mass density, σ the stress tensor, s the entropy, R the isotropic hardening, and X the kinematic hardening.

The elastic domain within which no irreversibility is possible is defined by a yield function such that:

$$f(A \, ; \, \mathbb{V}) \le 0 \tag{3}$$

The irreversible behaviour is described by a dissipation potential

$$\Phi = \Phi(A \, ; \, \mathbb{V}) = \Phi(\sigma, \mathbf{A}, \overrightarrow{\mathrm{Grad}}T \, / \, T \, ; \, \mathbb{V}) \tag{4}$$

from which the following *evolution and Fourier's laws* are derived:

$$\dot{\varepsilon}^p = -\frac{\partial \Phi}{\partial \sigma} \, ,$$

$$\dot{\mathbf{V}} = -\frac{\partial \Phi}{\partial \mathbf{A}} \, , \tag{5}$$

$$\vec{q} = -\frac{\partial \Phi}{\partial [\overrightarrow{\mathrm{Grad}}T \, / \, T \,]} \, ,$$

where $\vec{q}$ denotes the heat flux.

3. COMPLETE CONSTITUTIVE EQUATIONS

General formalism

In the model proposed in this report, the description of *the behaviour of a two–phase material* is based on *the theory of mixtures.* A simplified approach consists in taking *the state variables* ε, T and ε^p, *as variables common to both phases.* To model the phase transformation (from γ to α), *an additional scalar internal variable* x ($x \in [0,1]$) is introduced. It represents *the evolutive volumetric fraction of the α–phase* and the associated variable Z is the phase transformation "energy release rate". Therefore, the set of the thermodynamical variables becomes:

$$\mathbb{V} \equiv (\varepsilon, T, \varepsilon^p, x \, , \, \mathbf{V}) \text{ with } \mathbf{V} \equiv (p, \alpha) \, ; \, A \equiv (\sigma, s, -\sigma, Z, \mathbf{A}) \text{ with } \mathbf{A} \equiv (R, X) \tag{6}$$

Further assumptions must be made to define the *couplings* [3] between the transformation mechanism on the one hand, and the mechanisms involved in the anisothermal elasto–viscoplastic behaviour of each phase on the other hand. These assumptions lead to particular forms of the potentials $\Psi(V)$ and $\Phi(A\ ;\ V)$, and consequently, of the evolution and Fourier's laws ($\dot{\varepsilon}^P, \dot{V}, \dot{x}, \vec{q}$).

The following *state couplings* are assumed:

$$\rho\Psi(\varepsilon, T, \varepsilon^P, x, p, \alpha) = [(1 - x)\, \rho_\gamma\Psi_\gamma + x\, \rho_\alpha\Psi_\alpha] + \rho\Psi_{vtr}(\varepsilon, T, \varepsilon^P, x) + \rho\Psi_{Ltr}(T, x) \quad (7)$$

where $(\rho_\gamma\Psi_\gamma)$ and $(\rho_\alpha\Psi_\alpha)$ correspond to the γ– and α– phases respectively. The first term corresponds to a linear mixing law, and the last two terms, to the volumetric variation and to the latent heat emission associated to the phase transformation respectively. It is assumed that the last two terms take the following linear forms:

$$\rho\,\Psi_{vtr}(\varepsilon, T, \varepsilon^P, x) = x\,\rho_\alpha\,\Psi_{vtr}^*(\varepsilon, T, \varepsilon^P) \quad (8)$$

$$\rho\,\Psi_{Ltr}(T, x) = (1 - x)\,L(T)$$

where $L(T)$ denotes the latent heat associated to the phase transformation.

The following usual *state uncouplings* are assumed:

$$\rho_i\,\Psi_i = \rho_i\,\Psi_i(\varepsilon, T, \varepsilon^P, p, \alpha) = \rho_i\,\Psi_i^{ce}(\varepsilon, T, \varepsilon^P) + \rho_i\,\Psi_i^{ih}(T, p) + \rho_i\,\Psi_i^{ykh}(T, \alpha) \quad (9)$$

where $i = \alpha, \gamma$. The first term correspond to classical thermoelasticity, whereas the last two terms correspond to isotropic and kinematic hardenings respectively.

The following *dissipation uncouplings* are assumed.

First, it is assumed that viscoplasticity, thermics and phase transformation are independent so that:

$$\Phi = \Phi_{vp}(\sigma, X, R, V) + \Phi_{th}(\vec{Grad}\,T\,/\,T, V) + \Phi_{tr}(Z, V) \quad (10)$$

where the first term refers to viscoplasticity, the second term to thermics, and the third term to phase transformation.

Second, following the assumption that the viscoplastic strain ε^P is split into a classical viscoplastic strain ε^{cp} and a transformation viscoplastic strain ε^{trp}, the first term takes the following particular form:

$$\Phi_{vp} = \Phi_{cvp}(\sigma, X, R\ ;\ V) + \Phi_{tr\text{-}vp}(\sigma\ ;\ V) \quad (11)$$

The following *evolution couplings*, modelled through the introduction of state variables as parameters in the dissipation potential, are assumed.

First, it is assumed that the influence of the phase transformation on the viscoplastic flow is modelled by a linear mixing law:

$$\Phi_{cvp} = \Phi_{cvp}(\sigma, X, R \;; T, x) = (1 - x)\,\Phi_{\gamma cvp}(\sigma, X, R \;; T) + x\,\Phi_{\alpha cvp}(\sigma, X, R \;; T) \qquad (12)$$

where $\Phi_{\gamma cvp}$ and $\Phi_{\alpha cvp}$ correspond to the $\gamma-$ and $\alpha-$ phases respectively.

Second, it is assumed that the influence of the phase transformation on the evolution of the transformation viscoplastic strain is modelled by:

$$\Phi_{tr\text{-}vp} = \Phi_{tr\text{-}vp}(\sigma \;; T, x) = F(\sigma)\,H(T, x) \qquad (13)$$

where F and H are functions to be identified from experiments.

Third, it is assumed that the influence of the phase transformation on the coefficients of the Fourier's law is modelled by a linear mixing law:

$$\Phi_{th} = \Phi_{th}(\overrightarrow{Grad}T \,/\, T, T, x) = (1 - x)\,\Phi_{\gamma th}(\overrightarrow{Grad}T \,/\, T, T) + x\,\Phi_{\alpha th}(\overrightarrow{Grad}T \,/\, T, T) \qquad (14)$$

Fourth, it is assumed that the influence of the thermomechanics on the kinetics of the phase transformation is a priori modelled by:

$$\Phi_{tr} = \Phi_{tr}(Z \;; \varepsilon^{p}, T, x) = <-Z>\,G(\varepsilon^{p}, T, x) \qquad (15)$$

where G is a function to be identified from experiments.

<u>Particularization</u>

Finally, the proposed model corresponds to the following partition of the strain tensor:

$$\varepsilon = \varepsilon^{e} + \varepsilon^{p} \quad \text{with} \quad \varepsilon^{e} = \varepsilon^{ce} + \varepsilon^{vtr} \quad \text{and} \quad \varepsilon^{p} = \varepsilon^{cp} + \varepsilon^{trp} \qquad (16)$$

where ε^{ce}, ε^{vtr}, ε^{cp} and ε^{trp} denote the classical thermoelastic, the transformation volumetric, the classical viscoplastic and the transformation viscoplastic strain tensors respectively.

The thermo–elastic law is written as:

$$\begin{aligned}
\sigma = &\left\{ (1 - x)\,\frac{v_{\gamma}\,E_{\gamma}(T)}{(1 + v_{\gamma})(1 - 2\,v_{\gamma})} + x\,\frac{v_{\alpha}\,E_{\alpha}(T)}{(1 + v_{\alpha})(1 - 2\,v_{\alpha})} \right\}\,\mathrm{tr}\,[\varepsilon^{e}]\,\mathbb{1} \\[2mm]
&+ \left\{ (1 - x)\,\frac{E_{\gamma}(T)}{(1 + v_{\gamma})} + x\,\frac{E_{\alpha}(T)}{(1 + v_{\alpha})} \right\}\,\varepsilon^{e} \\[2mm]
&- \left\{ (1 - x)\frac{E_{\gamma}(T)}{(1 - 2\,v_{\gamma})}\,\alpha_{\gamma}(T) + x\,\frac{E_{\alpha}(T)}{(1 - 2\,v_{\alpha})}\,\alpha_{\alpha}(T) \right\}(T - T_{0})\mathbb{1} - \left\{ x\,\frac{E_{\alpha}(T)}{(1 - 2\,v_{\alpha})}\,\delta \right\}\mathbb{1}
\end{aligned} \qquad (17)$$

where E_{i}, v_{i} and α_{i} with $i = \alpha, \gamma$ denote the Young's moduli, Poisson's ratios and dilatation coefficients of each phase, respectively. The volumetric variation when the material transforms from phase γ to phase α is equal to $(3\,\delta)$ when measured at room temperature T_{0} at null stress state.

The yield function for each phase is written as:

$$f_i = J_2(\sigma - X) - R - \sigma_{yi}(T) \quad \text{with} \quad i = \alpha, \gamma \tag{18}$$

where σ_{yi} denotes the yield stress and s the stress deviator so that:

$$J_2(\sigma - X) = \sqrt{\frac{3}{2}\ (s - X) : (s - X)} \tag{19}$$

The total viscoplastic strain rate is given by:

$$\dot{\varepsilon}^p = \frac{3}{2}\left\{\frac{(s - X)}{J_2(\sigma - X)}\right\}\left[(1 - x)\mu_\gamma(T)<f_\gamma>^{n_\gamma} + x\,\mu_\alpha(T)<f_\alpha>^{n_\alpha(T)}\right] - \frac{3}{2}\left\{\frac{\sigma_{eq}}{\lambda}\right\}^N \frac{s}{\sigma_{eq}}\dot{x} \tag{20}$$

where $\mu_i = (K_i)^{-n_i}$, n_i, λ and N denote material dependent parameters and σ_{eq} the Von Mises' equivalent stress defined by $\sigma_{eq} = \sqrt{\frac{3}{2}\ s : s}$

The rate of the internal variable associated to isotropic hardening is given by:

$$\dot{p} = (1 - x)\,\mu_\gamma(T)\,<f_\gamma>^{n_\gamma} + x\,\mu_\alpha(T)\,<f_\alpha>^{n_\alpha(T)} \tag{21}$$

whereas the rate of the internal variable associated to kinematic hardening is given by:

$$\dot{\alpha} = \frac{3}{2}\left\{\frac{s - X}{J_2(\sigma - X)}\,\dot{p} - \left\{(1 - x)\frac{\mu_\gamma(T)}{g_\gamma(T)}<f_\gamma>^{n_\gamma(T)} + x\,\frac{\mu_\alpha(T)}{g_\alpha(T)}<f_\alpha>^{n_\alpha(T)}\right\}X\right\} \tag{22}$$

where g_i denote material dependent parameters.

A variable ζ is introduced to model the delay for the transformation to occur. Its rate is defined as:

$$\dot{\zeta} = \frac{<T_s - T>^2}{k_{c0}\,T} \tag{23}$$

where k_{c0} denotes a material dependent coefficient and T_s the transformation temperature at infinitively slow cooling rate. The phase transformation kinetics is given by:

$$\dot{x} = G(T, \dot{T}, x)\,H(1 - \zeta) \tag{24}$$

where H denotes the Heavyside function.

The evolutions of the isotropic and kinematic hardening variables are given by:

$$R = (1 - x)\,c_\gamma(T)\left\{1 - \exp(-\gamma_\gamma\,p)\right\} + x\,c_\alpha(T)\left\{1 - \exp(-\gamma_\alpha\,p)\right\} \tag{25}$$

and

$$\dot{X} = X \left\{ \dot{x} \left\{ \frac{g_\alpha(T)b_\alpha - g_\gamma(T)b_\gamma}{(1-x)g_\gamma(T)b_\gamma + xg_\alpha(T)b_\alpha} \right\} + \dot{T} \left\{ \frac{(1-x)g_\gamma'(T)b_\gamma + xg_\alpha'(T)b_\alpha}{(1-x)g_\gamma(T)b_\gamma + x g_\alpha(T)b_\alpha} \right\} \right\}$$

$$+ \frac{2}{3} \left\{ (1-x)g_\gamma(T)b_\gamma + x g_\alpha(T)b_\alpha \right\} \dot{\alpha} \tag{26}$$

where b_i, c_i and γ_i denote material dependent parameters.

4. EXPERIMENTAL IDENTIFICATION

The constitutive equations described in the previous sections of this report have been completely identified from room temperature to 1000 °C for a nodular graphite ferritic cast iron supplied by Renault. The strategy adopted for the experimental identification of this complex model is briefly described below.

The viscous behaviour at different temperatures of the α–phase is identified from isothermal *relaxation tests* subsequent to tensile tests at various strain rates. Since *the hardening of the γ-phase appears practically nill*, the viscous behaviour at different temperatures of this phase is identified from isothermal *creep tests* at different stress levels. The elasto–plastic behaviour of the α–phase is identified from isothermal *tension tests with elastic unloadings performed down to the yield stress in compression at different values of the plastic strain*. The elastic behaviour of the γ–phase is identified from isothermal *tension tests performed at relatively high stress rates*. The thermal expansion coefficients and the transformation volumetric variation are identified from *dilatometric tests performed at constant cooling rates without any mechanical load*. The phase transformation kinetics is identified from *electrical resistivity* and strain measurements made during dilatometric tests performed at constant cooling rates without any mechanical load. The transformation viscoplastic behaviour is identified from *dilatometric tests performed at constant cooling rates and at different constant mechanical loadings*.

Constitutive equations have also been developed for green sand, and further validations of the proposed model for cast iron are currently in progress by comparison between experimental results on casting experiments and the corresponding numerical simulations.

ACKNOWLEDGMENTS

The first three authors gratefully acknowledge the financial support of Renault through contract CNRS with the Laboratoire de Mécanique et Technologie, Cachan.

REFERENCES

[1] A. BENALLAL and A. BEN CHEIKH, Constitutive Equations for Anisothermal Elasto–Plasticity, in C.S. DESAI et al. (edts), Constitutive laws for Engineering Materials : Theory and Applications, Elsevier, pp. 667–674 (1987).

[2] P. GERMAIN, Q.S. NGUYEN and P. SUQUET, Continuum Thermodynamics, J. Appl. Mech, A.S.M.E., **50**, 1010–1020 (1983).

[3] J. LEMAITRE, and D. MARQUIS, Modeling Complex Behavior of Metals by the 'State-Kinetic Coupling Theory', J. Engng Mat. and Techn., **114**, 250–254 (1992).

AUTHOR INDEX

KEYWORD INDEX